THE MOLECULAR
BIOLOGY OF THE YEAST
SACCHAROMYCES

LIFE CYCLE
AND
INHERITANCE

THE MOLECULAR BIOLOGY OF THE YEAST SACCHAROMYCES

LIFE CYCLE AND INHERITANCE

Edited by

Jeffrey N. Strathern
Cold Spring Harbor Laboratory

Elizabeth W. Jones
Carnegie-Mellon University

James R. Broach
State University of New York at Stony Brook

Cold Spring Harbor Laboratory
1981

**COLD SPRING HARBOR
MONOGRAPH SERIES**

The Lactose Operon [1]
The Bacteriophage Lambda [2]
The Molecular Biology of Tumour Viruses [3]
Ribosomes [4]
RNA Phages [5]
RNA Polymerase [6]
The Operon [7]
The Single-Stranded DNA Phages [8]
Transfer RNA:
  Structure, Properties, and Recognition, 9A
  Biological Aspects, 9B
Molecular Biology of Tumor Viruses, Second Edition:
  ~~The Transformed Cell, 10A~~ (Publication canceled)
  DNA Tumor Viruses, 10B
  RNA Tumor Viruses, 10C
The Molecular Biology of the Yeast Saccharomyces:
  Life Cycle and Inheritance, 11A
  Metabolism and Gene Expression, 11B

THE MOLECULAR BIOLOGY OF THE YEAST SACCHAROMYCES
LIFE CYCLE AND INHERITANCE

© 1981 by Cold Spring Harbor Laboratory

Printed in the United States of America

Book design by Emily Harste

Library of Congress Cataloging in Publication Data

Main entry under title:

The Molecular biology of the yeast
  saccharomyces, life cycle and inheritance.

  (Cold Spring Harbor monograph series,
ISSN 0270-1847 ; 11A)
  Includes index.
  1. Saccharomyces.   2. Molecular biology.
  I. Strathern, Jeffrey N.   II. Jones, Elizabeth W.
  III. Broach, James R.
  IV. Series.
  QK623.S23M64   589.2'33   81-68895
  ISBN 0-87969-139-5   AACR2

All Cold Spring Harbor Laboratory publications are available through booksellers or
may be ordered directly from Cold Spring Harbor Laboratory, Box 100, Cold Spring
Harbor, New York 11724.

SAN 203-6185

Contents

Contents of companion volume
THE MOLECULAR BIOLOGY OF THE YEAST SACCHAROMYCES
METABOLISM AND GENE EXPRESSION

Preface

At the Molecular Biology of Yeast meeting at Cold Spring Harbor Laboratory in 1979, it was apparent that investigation of the biology of yeast was in a period of rapid and exciting growth. The meeting was alive with excitement over the fruits to accrue from the union of the developing technologies in molecular biology with the elegant genetic and phenomenological studies possible in yeast. Jim Watson perceived that it was an appropriate time to review the biology of yeast and attempt to define the direction in which the field would progress in the 1980s. He encouraged us to assemble this monograph.

Most of the volumes in the Cold Spring Harbor Monograph Series are organized as a combination of reviews and research papers. We felt, however, that the core of accumulated knowledge in yeast molecular biology was too large and the rate at which new information was being added was too rapid for the field to be covered adequately in a research paper format. Therefore, we decided to limit the monograph to reviews that would provide both a background in yeast molecular biology and an indication of the direction in which individual areas were proceeding. Our basic goal was about 24 reviews in about 750 pages, to be published in late 1980. The vigor of the field has been reflected in the rapid expansion of this monograph. The reviews have been divided into two volumes, which together will number nearly 1400 pages. In this volume, the cell cycle, the life cycle, non-Mendelian elements, and the mechanics of nuclear inheritance are discussed. The second volume will deal with metabolism and gene expression. These books are meant to serve both as general reviews and as references. Accordingly, the authors attempted to provide in their texts frameworks for the data from which guiding principles emerge. At the same time, the data had to be reviewed in sufficient detail so that exceptions to such generalities would be apparent, both to guard against pitfalls and to serve as impetus for further research. With that in mind, rather extensive tables are included in several of the chapters. The authors have our deep appreciation for their efforts in attaining these twin goals.

There is a recurring theme in these chapters that the tools now exist to understand these topics in molecular detail. One gets the impression that the authors of these reviews returned to their laboratory benches with enthusiasm and great expectations. It is our hope that these volumes leave the reader with the same feeling of inspiration.

We wish to thank Nancy Ford and Nadine Dumser for their roles in editing the manuscripts and making this publication a reality. Furthermore, we would like to thank our colleagues Jim Hicks, Amar Klar, and George Zubenko for their support during these long months of editing.

Jeffrey N. Strathern
Elizabeth W. Jones
James R. Broach

THE MOLECULAR
BIOLOGY OF THE YEAST
SACCHAROMYCES

LIFE CYCLE
AND
INHERITANCE

Development of Yeast as an Experimental Organism

Herschel Roman
Department of Genetics
University of Washington
Seattle, Washington, 98195

The editors of this volume have asked me to describe the landmark discoveries that have led to the prominence of *Saccharomyces cerevisiae*[1] as a eukaryotic organism of choice for a number of problems in molecular genetics. To place the beginnings of yeast genetics in proper perspective, it should be recalled that at the time that Ö. Winge was beginning his researches with yeast in the mid-thirties, the principal organisms then in use were *Drosophila*, corn, and *Neurospora*. The dazzling successes of the prokaryotic era were more than a decade away.

Yeast was already well known as an important tool in biochemical research. The interest in yeast stemmed at first from its role in alcohol production in the making of wine and beer. Early in this century, when it was found that alcoholic fermentation could be carried out by an extract from yeast, the fractionation of the extract led to the discovery and characterization of enzymes and coenzymes. The glycolytic pathway—the breakdown of glucose as a result of the fermentation process—was worked out in detail. The individual enzymes, their specific coenzymes, and the products formed at each step of the pathway were identified. As a rich source of the water-soluble B complex, yeast also was a principal contributor to early research in vitamins. Thus, many of the important concepts of biochemistry owe their origins to investigations in which yeast played a key role. The reemergence of yeast as an important organism in biochemical research is described in a review by de Robichon-Szulmajster and Surdin-Kerjan (1971).

The early period of yeast genetics was concerned mainly with establishing that yeast followed the rules of Mendelian inheritance. In a series of papers (see, e.g., Winge and Roberts 1952), Winge and his collaborators demonstrated the 2:2 segregation of certain enzymatic markers in four-spored asci from diploid cells. There was one important exception to orthodoxy. Lindegren and his colleagues reported departures from 2:2 segregation, i.e., 4:0, 3:1, 1:3, and 0:4, that occurred too frequently to be ignored. Lindegren (1949) interpreted these cases of irregular segregation as examples of "gene

---

[1]The strains of yeast now in use have been derived from many sources (see, e.g., Lindegren 1949). Claims to call them *Saccharomyces cerevisiae* come from their morphologies and the fact that they are interfertile with the species (for taxonomic considerations, see Lodder and Kreger-van Rij 1952).

conversion" after Winkler's (1930) theory of recombination bearing the same name and based on similar data cropping up at that time. Other explanations were put forward to explain these observations on more conventional grounds (see, e.g., Roman et al. 1951; Winge and Roberts 1954), until findings in *Neurospora* (Mitchell 1955) and in yeast (Roman 1956) provided unmistakable evidence of the essential correctness of the Lindegren observations.

Gene conversion is interpreted quite differently today than it was by Lindegren and Winkler, who thought of it as a mutational process. Now we think of it as being due to DNA heteroduplex formation and the subsequent repair of noncomplementary sites in the heteroduplex (Holliday 1964). Historically, it is a curious twist that gene conversion was originally proposed by Winkler to account for recombination in opposition to the then-emerging theory of recombination by the breakage and reunion of homologous chromosomes. Although Winkler was wrong in his interpretation of gene conversion and, therefore, in its application to recombination, it has turned out that gene conversion as we understand it today has provided the basis for a molecular theory of recombination that does in fact involve the breakage and reunion of chromosomes. The modern theory and its variations are discussed later in this volume.

The discovery of mating types in yeast (Lindegren and Lindegren 1943) was a technical advance that turned out to have an importance beyond its mere utility. Winge, in his experimental protocol, used spore-to-spore fusion to obtain the diplophase. Some couplings were successful, whereas others were not, for no apparent reason. By utilizing the vegetative cells derived from individual spores, the Lindegrens recognized that yeast had two mating types, which were designated **a** and α (Lindegren and Lindegren 1943). Cells of mating-type **a** mated only with cells of mating-type α. A new technique therefore became possible for crosses in which vegetative cells (not spores) of one mating type were mixed with vegetative cells of the other mating type. This meant that the parents of a cross could be retained for later testing and for subsequent crosses, if that proved to be desirable. The diploid zygotes that were formed in such a mixture could then be isolated with a micromanipulator or, as nutritional markers became available, they could be selected for on appropriate media. A refinement of the technique was the isolation of individual diploid cells from which clones were derived for sporulation and dissection. This improvement avoided the chief criticism of the mass-mating technique, namely, that spontaneous changes in either one of the two haploid parents could result in a heterogeneous mixture of zygotes and, therefore, tetrads obtained from the mixture did not come from a defined genetic source.

It immediately became evident that the strains of *Saccharomyces* used in genetic research belonged to two classes: the heterothallic class with its two mating types, **a** and α; and a homothallic class that was characterized by cell fusion among the cells derived from a single spore and by the absence of mating type, or so it seemed. Spore-to-spore matings between heterothallic and homothallic strains produced a diplophase that in turn produced asci with two spores of the homothallic type and two spores of the heterothallic type (Winge and Roberts 1949). Thus, a single pair of alleles, designated D/d for diploidization (now called the HO gene for homothallism), is responsible for the homothallism/heterothallism difference. Segregation of the alleles in tetraploids showed that homothallism is dominant (D. C. Hawthorne, pers. comm.; later confirmed by Hopper and Hall 1975). Oeser (1962) and Hawthorne (1963b) further found that homothallism masked the presence of the **a** and α genes, and that in fact, the mating-type alleles were present in homothallic strains. A spore of mating-type **a** on germination gives rise to cells of the same mating type until a mutation to the opposite mating type occurs early in the clone's lineage. There follows cell fusion between the cells of opposite mating type to restore the diplophase of genotype **a**/α and a loss of mating ability. Thus, the apparent failure to mate is due to the replacement of competent haploid cells with diploid cells. The same sequence of events occurs when an α spore germinates, but in this case, the mutation is to the **a** mating type with the subsequent restoration of the diplophase (Fig. 1).

A related observation, for which the stimulus arose from attempts to find alternative explanations for the irregular segregations found by Lindegren, was the discovery that polyploidy occurred in yeast (Lindegren and Lindegren 1951; Roman et al. 1951; Roman and Sands 1953). When a haploid culture of mating-type **a** was grown in nutrient broth, larger cells began to appear in the culture and, because of their superior growth rate, ultimately became dominant in the culture. The same was found to be the case with α cells. When the larger cells were isolated and analyzed, it was found that they were of two types: One type was heterozygous for mating type, was incapable of mating, and could be induced to sporulate; the other was capable of mating and was incapable of sporulation. The first type was not detectably different from an ordinary diploid and was undoubtedly due to mutation at the mating-type locus to the opposite mating type and subsequent fusion. Mating-type mutation was rare, occurring with a frequency of approximately 10^{-7} or less per cell generation. The second type, occurring with about equal frequency, was found to be diploid also, either **a**/**a** from **a** cells or α/α from α cells, and these were able to mate but not to sporulate. Segregations typical of **a**/**a**/α/α tetraploids were obtained from cells resulting from the mating of the putative **a**/**a** and α/α diploids. Similarly, triploids could be constructed by crossing the diploid homozygote with a haploid cell of opposite mating type. The triploid gave a high frequency of inviable spores in tetrads, as would be expected from studies of triploidy in higher organisms. Disomics could be generated in this way, and these became useful in chromosome mapping (Mortimer and Hawthorne 1973).

When diploidization leading to homozygosis for mating type was found, it

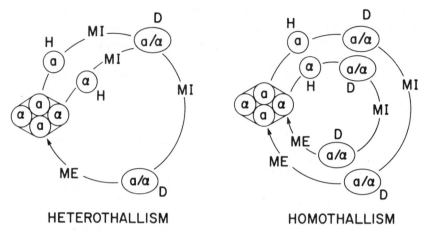

HETEROTHALLISM   HOMOTHALLISM

Figure 1 Schematic showing principal differences between heterothallism and homothallism. The individual spores are dissected from the ascus. In heterothallism, the haplophase (H) may exist more or less indefinitely, the cells multiplying by mitosis (MI). At any time when confronted by cells of opposite mating type, the diplophase (D) will be formed, and this in turn can be sustained indefinitely, again multiplying by mitosis, until conditions are such that the diploid cells undergo meiosis (ME) to produce the ascus and spores that are once again haploid. In homothallism, the cells of the haplophase are not stable. In the early divisions following spore germination, a genetic switch to the opposite mating type occurs in some of the cells; then mating occurs to produce the diplophase. The diplophase is stable and multiplies by mitosis. As in heterothallism, the diplophase can be interrupted at any time by inducing meiosis to restore the ascus.

became clear that diploidy was not a sufficient condition for sporulation, since the cells do not sporulate. The fact that they mate and **a**/α cells do not suggested that the genetic conditions for mating were antagonistic to those for sporulation. An additional key finding was that of Hawthorne (1963a), which suggested that the mating-type locus was separable into two parts, one responsible for **a** function and the other responsible for α function.

The complex control governed by the mating-type system was only vaguely hinted at in these early experiments. The genetic basis for a much more complicated pattern of controlling elements was worked out principally by Oshima and his collaborators (Harashima et al. 1974), and the interaction of these elements was postulated to account for changes of mating-type specificity. A further elaboration of the nature of these interactions led to a novel hypothesis of gene duplications, some distance removed from the mating-type locus and interacting with the latter to cause "switches" in mating type (Hicks et al. 1977). The "cassette" model of mating-type control and the pattern of that control are discussed by Herskowitz and Oshima (this volume).

In the development of an organism for use in genetic research, it is helpful to know the number of chromosomes that are present and the distribution of genes on these chromosomes. Chromosome mapping became an important activity in yeast, since the chromosomes were difficult to see, and an estimate of the number of linkage groups was the best and most certain indicator of the number of chromosomes. We need not dwell on the controversies over chromosome number in the early days of yeast genetics. As the number of linkage groups increased, so did the number of chromosomes reported. Through the use of light microscopy, Tamaki (1965) estimated the haploid number to be 18. Today, with some 300 genes mapped, it is fairly well established that there are 17 chromosomes in the haploid set (R. K. Mortimer and D. Schild, pers. comm.). Byers and Goetsch (1975) counted synaptonemal complexes in meiotic cells of diploids and tetraploids; their observations were in general agreement with this number. The subject of mapping is discussed by Mortimer and Schild (this volume).

Ordinarily, the geneticist is not pleased to find that the organism he or she is working with has a large number of chromosomes. In this case, however, the finding that yeast also had a low nuclear DNA content—about 10^{10} daltons (Bicknell and Douglas 1970)—was an important factor in bringing yeast to the attention of molecular biologists. The average yeast chromosome was thus substantially smaller than the chromosome of *Escherichia coli*. Thus, although yeast is a eukaryotic organism possessing traits associated with such organisms, e.g., meiosis and the alternation of haplophase and diplophase, its chromosomes are sufficiently small to make them readily available for studies of replication and structure. Helpful for this purpose and others was the finding that the tough cell wall could be enzymatically removed (Eddy and Williamson 1957). In a key paper by Petes and Fangman (1972), the feasibility of using yeast for molecular studies of chromosomal DNA was demonstrated.

The fact that yeast can grow under either aerobic or anaerobic conditions renders it particularly suitable for studies of mitochondrial inheritance. In 1949, Ephrussi published the discovery that acridines induced, with high frequency, a class of mutants that he called "petite colonie," or simply petite. They were thus named because cells that were petite utilized glucose (or other fermentable carbon sources) less efficiently than did normal cells and therefore gave rise to colonies that were distinctly smaller than normal colonies. It was concluded from tetrad analysis that the petite phenotype was due to a cytoplasmic mutation. The consequence of the mutation was the simultaneous loss of several elements in the electron transport system, including cytochromes *a* and *b*. The mutation was stable, i.e., it did not revert to normal even under severe selective pressure. For routine diagnostic testing, advantage was taken of the fact that petites could not utilize carbon sources such as glycerol or ethyl alcohol for growth. The early history of cytoplasmic inheritance is given by Ephrussi (1953).

It was suspected that the mitochondrion was the site of the petite mutation.

Morphological and biochemical studies revealed irregularities in the structure of the mitochondrion and in its DNA content. However, an investigation of the genetics of this organelle awaited a technical breakthrough: the finding of mutations resistant to various antibiotics known to inhibit protein synthesis in bacteria, such as chloramphenicol and erythromycin (Wilkie et al. 1967). These were found to reside in mitochondrial DNA and formed the basis for early studies of transmission and recombination in cytoplasmic inheritance (described in Wilkie 1970). Slonimski, who earlier made important contributions to the biochemistry of the petite phenotype, immediately saw the power of the new technique, and he and his collaborators became the leading contributors to what has developed into a new and exciting research area in which *Saccharomyces* has been the organism of choice (Coen et al. 1970). The present status of mitochondrial genetics is given by Dujon (this volume).

Yeast has also played a distinctive role in the study of the control of cell division. By using time-lapse photography as a means of detecting arrest of cell division, Hartwell and his collaborators were able to accumulate some 150 temperature-sensitive mutants capable of growth at 22°C but not at 36°C. The mutants represent some 35 genes that control blocks in various stages of the cell-division cycle (Hartwell et al. 1974). Advantage was taken of the fact that yeast reproduces by budding. In the haploid strain used by Hartwell's group, the onset of budding was closely correlated in time with the onset of DNA replication. The size of the bud was a reliable indicator of the stage in the cell-division cycle of mother and daughter cells. More recent results of the cell-division cycle studies are discussed by Pringle and Hartwell (this volume).

Another area of research in which yeast has played an important part is in the elucidation of the action of supersuppressors. These, also demonstrated in bacteria, are nonsense suppressors that are allele-specific and not locus-specific (Hawthorne and Mortimer 1963). Three general classes of supersuppressors have been found in yeast, corresponding to the UAA, UAG, and UGA nonsense codons. The findings were taken as an illustration of the universality of the genetic code, since the same nonsense or termination codons were found in *E. coli*. Ultimately, the supersuppressors, of which a large variety was obtained at different loci by mutation, were shown to be altered tRNA genes whose products were now able to read the corresponding nonsense codons during the translation process. This subject is discussed in more detail by Sherman (1982).

There is one other subject that should be mentioned for its importance in propelling yeast as a major experimental eukaryotic organism. It also serves to point up again the flexibility of yeast as a facultative aerobe. After the amino acid sequence of iso-1-cytochrome *c* was worked out by Narita and Titani (1969), Sherman and his collaborators (1970) took the opportunity to investigate the question of colinearity between DNA and its protein product in a eukaryotic organism. Later studies embraced a variety of problems, such as the dependence of recombination on gene distances of known nucleotide lengths and the control of protein synthesis, as well as other ancillary inquiries. Differences were indeed found, in relation to prokaryotic organisms, particularly as regards the mechanism of the control of gene action and the specificity of mutagens.

I have chosen topics that appear to be important in the early development of yeast as a potentially useful organism for investigations of certain types of fundamental problems. Not less important is the application of yeast in the investigation of problems in which other organisms also had a role to play, such as the regulation of enzyme biosynthesis, the structure and function of the cell surface, the control of transcription and translation, the mechanisms of macromolecular repair, the recent advances in recombinant DNA, and gene cloning.

Of special note in connection with gene cloning is the recent establishment of the technique for transformation in yeast (Hinnen et al. 1978). Oppenoorth (1960) was the first to report the achievement of transformation in this organism, but his paper was subject to criticism on grounds of experimental design, and efforts to repeat his results were unsuccessful. Hinnen et al. (1978) succeeded in transforming yeast by incorporating genetically marked yeast DNA into an *E. coli* plasmid, concentrating the DNA by passage through *E. coli*, and exposing yeast cells to the high concentration of specific DNA under conditions known to favor alterations in cell membranes. Subsequent to these observations, DNA fragments from yeast have been identified that are capable of autonomous, extrachromosomal replication when reintroduced into yeast by transformation. As a consequence, various yeast-transformation vectors have been constructed, and techniques have been developed that permit the isolation of a number of specific yeast genes by complementation of genetic lesions. Thus, through the wedding of yeast genetics and yeast transformation, one has the potential to isolate DNA corresponding to any yeast gene for which there exists a scorable phenotype for mutations in that gene. In addition, cloning strategies, such as cDNA cloning, probing with homologous sequences, etc., that are applicable to other organisms are also applicable to yeast. The techniques involved in cloning and transformation are discussed in detail by Botstein and Davis (1982).

Altogether, the subject matter of this volume shows quite clearly that *Saccharomyces* is for the most part a typical eukaryotic organism that offers the advantages of simplicity of organization and ease of handling for investigations involving classical techniques as well as techniques that are at the forefront of molecular biology.

REFERENCES

Bicknell, J.N. and H.C. Douglas. 1970. Nucleic acid homologies among species of *Saccharomyces. J. Bacteriol.* **101**: 505.

Botstein, D. and R.W. Davis. 1982. Principles and practice of recombinant DNA research with yeast. In *The molecular biology of the yeast* Saccharomyces. II. *Metabolism and gene expression* (ed. J. Strathern et al.), Cold Spring Harbor Laboratory, Cold Spring Harbor, New York. (In press.)

Byers, B. and L. Goetsch. 1975. Electron microscopic observations on the meiotic karyotype of diploid and tetraploid *Saccharomyces cerevisiae. Proc. Natl. Acad. Sci.* **72:** 5056.

Coen, D., J. Deutsch, P. Netter, E. Petrochilo, and P.O. Slonimski. 1970. Mitochondrial genetics. I. Methodology and phenomenology. *Symp. Soc. Exp. Biol.* **24:** 449.

de Robichon-Szulmajster, H. and Y. Surdin-Kerjan. 1971. Nucleic acid and protein synthesis in yeasts: Regulation of synthesis and activity. In *The yeasts* (ed. A.H. Rose and J.S. Harrison), vol. 2, p. 335. Academic Press, New York.

Eddy, A.A. and D.H. Williamson. 1957. A method for isolating protoplasts from yeast. *Nature* **179:** 1252.

Ephrussi, B. 1953. *Nucleo- cytoplasmic relations in microorganisms.* Clarendon Press, Oxford, England.

Harashima, S., W. Nogi, and W. Oshima. 1974. The genetic system controlling homothallism in *Saccharomyces* yeasts. *Genetics* **77:** 639.

Hartwell, L.H., J. Culotti, J.R. Pringle, and B.J. Reid. 1974. Genetic control of the cell division cycle in yeast. *Science* **183:** 46.

Hawthorne, D.C. 1963a. A deletion in yeast and its bearing on the structure of the mating type locus. *Genetics* **48:** 1727.

―――. 1963b. Directed mutation of the mating type alleles as an explanation of homothallism in yeast. *Proc. Int. Congr. Genet.* **1:** 34.

Hawthorne, D.C. and R.K. Mortimer. 1963. Super-suppressors in yeast. *Genetics* **48:** 617.

Hicks, J., J. Strathern, and I. Herskowitz. 1977. The cassette model of mating type interconversion. In *DNA insertion elements, plasmids, and episomes* (ed A. Bukhari et al.), p. 457. Cold Spring Harbor Laboratory, Cold Spring Harbor, New York.

Hinnen, A., J.B. Hicks, and G.R. Fink. 1978. Transformation of yeast. *Proc. Natl. Acad. Sci.* **75:** 1929.

Holliday, R. 1964. A mechanism for gene conversion in fungi. *Genet. Res.* **5:** 282.

Hopper, A.K. and B.D. Hall. 1975. Mutation of a heterothallic strain to homothallism. *Genetics* **80:** 77.

Lindegren, C.C. 1949. *The yeast cell, its genetics and cytology.* Educational Publishers, St. Louis.

Lindegren, C.C. and G. Lindegren. 1943. A new method for hybridizing yeast. *Proc Natl. Acad. Sci.* **29:** 306.

―――1951. Tetraploid *Saccharomyces. J. Gen. Microbiol.* **5:** 885.

Lodder, J. and N.J.W. Kreger-van Rij. 1952. *The yeasts, a taxonomic study.* North-Holland, Amsterdam.

Mitchell, M.B. 1955. Aberrant recombination of pyridoxine mutants of *Neurospora. Proc. Natl. Acad. Sci.* **41:** 215.

Mortimer, R.K. and D.C. Hawthorne. 1973. Genetic mapping in *Saccharomyces.* IV. Mapping of temperature-sensitive genes and use of disomic strains in localizing genes. *Genetics* **74:** 33.

Narita, K. and K. Titani. 1969. The complete amino acid sequence in baker's yeast cytochrome *c. J. Biochem.* **65:** 259.

Oeser, H. 1962. Genetische Untersuchungen über das Paarungstypverhalten bei *Saccharomyces* und die Maltose-Gene einiger untergäriger Bierhefen. *Arch. Mikrobiol.* **44:** 47.

Oppenoorth, W.F.F. 1960. *Modification of the hereditary character of yeast by ingestion of cell-free extracts.* Eur. Brewery Convention, p. 180. Elsevier, Amsterdam.

Petes, D.D. and W.L. Fangman. 1972. Sedimentation properties of yeast chromosomal DNA. *Proc. Natl. Acad. Sci.* **69:** 1188.

Roman, H. 1956. Studies of gene mutation in *Saccharomyces. Cold Spring Harbor Symp. Quant. Biol.* **21:** 175.

Roman, H., and S.M. Sands. 1953. Heterogeneity of clones of *Saccharomyces* derived from haploid ascospores. *Proc. Natl. Acad. Sci.* **39:** 171.

Roman, H., D.C. Hawthorne, and H.C. Douglas. 1951. Polyploidy in yeast and its bearing on the occurrence of irregular genetic ratios. *Proc. Natl. Acad. Sci.* **37:** 79.

Sherman, F. 1982. Suppression in the yeast *Saccharomyces cerevisiae*. In *The molecular biology of the yeast* Saccharomyces. II. *Metabolism and gene expression* (ed. J. Strathern et al.), Cold Spring Harbor Laboratory, Cold Spring Harbor, New York. (In press.)

Sherman, F., J.W. Stewart, J.H. Parker, G.J. Putterman, B.B.L. Agrawal, and E. Margoliash. 1970. The relationship of gene structure and protein structure of iso-1-cytochrome *c* from yeast. *Symp. Soc. Exp. Biol.* **24:** 85.

Tamaki, S. 1965. Chromosome behavior at meiosis in *Saccharomyces*. *J. Gen. Microbiol.* **41:** 93.

Wilkie, D. 1970. Analysis of mitochondrial drug resistance in *Saccharomyces cerevisiae*. *Symp. Soc. Exp. Biol.* **24:** 71.

Wilkie, D., G. Saunders, and A.W. Linnane. 1967. Inhibition of respiratory enzyme synthesis in yeast chloramphenicol: Relationship between chloramphenicol tolerance and resistance to other anti-bacterial antibiotics. *Genet. Res.* **10:** 199.

Winge, Ö, and C. Roberts. 1949. A gene for diploidization in yeasts. *C. R. Trav. Lab.* **24:** 341.

———1952. The relation between the polymeric genes for maltose, raffinose, and sucrose fermentation in yeasts. *C. R. Trav. Lab.* **25:** 141.

———1954. On tetrad analyses apparently inconsistent with Mendelian law. *Heredity* **8:** 295.

Winkler, H. 1930. *Die Konversion der Gene*. Verlag Gustav Fischer, Jena.

Genetic Mapping in
Saccharomyces cerevisiae

Robert K. Mortimer and David Schild
Department of Biophysics and Medical Physics and
Donner Laboratory, University of California
Berkeley, California 94720

The first genetic map of *Saccharomyces cerevisiae* was published by Lindegren (1949), and several revisions of the map have appeared since then (Hawthorne and Mortimer 1960, 1968; Lindegren et al. 1962; Mortimer and Hawthorne 1966, 1973, 1975). The latest version of the map is included in Appendix II. In this paper the various techniques used to locate genes on the genetic map of the yeast *S. cerevisiae* will be reviewed. Some of these techniques have been described in detail elsewhere (Mortimer and Hawthorne 1969, 1975; Wickner 1979) and will be discussed only partially here.

Both meiotic and mitotic approaches have been developed to map yeast genes. The meiotic approaches include tetrad analysis, random spore analysis, and trisomic analysis. The mitotic cell-cycle events that yield mapping information are mitotic crossing-over and mitotic chromosome loss.

MEIOTIC MAPPING TECHNIQUES

Tetrad Analysis

The life cycle of *S. cerevisiae* normally alternates between diplophase and haplophase. Both ploidies can exist as stable cultures. In heterothallic strains the haploid cultures are of two mating types, **a** and α. Mating of **a** and α cells results in **a**/α diploids that are unable to mate but can undergo meiosis. The four haploid products resulting from meiosis of a diploid cell are contained within the wall of the mother cell (the ascus). Digestion of the ascus wall and separation of the spores by micromanipulation yield four clones that represent the four haploid meiotic products (reviewed in Mortimer and Hawthorne 1969). Analysis of the segregation patterns of different heterozygous markers among the four spores constitutes tetrad analysis. Linkage between two genes or between a gene and its centromere is revealed by this kind of analysis. In addition, both chiasma and chromatid interference can be assayed. Chromosome aberrations can also be analyzed by examination of the patterns of spore lethality in a sample of tetrads (Perkins and Barry 1977).

Monofactorial Crosses

In a normal meiosis of a diploid cell, each heterozygous nuclear gene (A/a) is expected to segregate 2 A:2 a. This is the usual finding. However, gene conversion and other events result in a low frequency of deviation from this expected ratio. The gene-conversion events are usually of the 3 A:1 a and 1 A:3 a type, although several other aberrant ratios are seen in cases involving postmeiotic segregation. The average frequency of gene-conversion events in a sample of 221,751 segregations from 30 heterozygous sites was 4.54% (Fogel et al. 1979). Ratios of 4 A:0 a and 0 A:4 a are also encountered, and it has been shown that these aberrant segregations are usually due to a mitotic event in a division preceding meiosis.

Multifactorial Crosses

If a diploid is heterozygous at more than one site, each pairwise combination of markers, e.g., AB x ab, can be considered independently. If both A/a and B/b segregate in a 2:2 fashion, only three types of asci can arise. These ascus types are: parental ditype (PD), $AB\ AB\ ab\ ab$; nonparental ditype (NPD), Ab $Ab\ aB\ aB$; and tetratype (T), $AB\ Ab\ aB\ ab$.

The relative frequencies of PD, NPD, and T for a given pair of markers depend on the separation of the two markers if they are on the same chromosome, or on the distances between both markers and their respective centromeres if they are on different chromosomes.

Two Markers on the Same Chromosome. Consider two heterozygous sites, A/a and B/b, located on the same chromosome. If no crossovers occur in the interval between the markers, only PD asci will result. A single crossover results in a T ascus. Additional crossovers yield all three ascus types. If there is no chromatid interference, the fractions of PD, NPD, and T asci as a function of the number of crossovers, r, in the interval between the markers are given by the following equations (see Snow 1979a):

$$PD = NPD = \tfrac{1}{6} + \tfrac{1}{3} \cdot (-\tfrac{1}{2})^r, \ r > 1$$
$$T = \tfrac{2}{3} \cdot \mid 1 - (-\tfrac{1}{2})^r \mid \tag{1}$$

For example, double crossovers (r = 2) yield $\tfrac{1}{4}$ PD : $\tfrac{1}{4}$ NPD : $\tfrac{1}{2}$ T.

Calculation of Map Distances. The map distance x, in centiMorgans (cM), between the markers is defined as the average number of crossovers per chromatid times 100. Since only two of the four chromatids are involved in a single recombination, the equation is as follows:

$$x = 100 \cdot \tfrac{1}{2} \cdot \sum_{r=0}^{\infty} r \cdot p(r), \tag{2}$$

where $p(r)$ is the probability that r crossovers occur in the interval between the markers. In order to calculate map distances, equations are needed that describe x as a function of PD, NPD, and T. The first such equation was derived by Perkins (1949). This derivation assumes that $p(r)$ for $r > 2$ is zero. That is, only zero, one, or two crossovers can occur in the interval. The NPD class represents one fourth of the double crossovers. Also, the zero- and single-crossover class can be deduced from the PD and T frequencies corrected for the asci of these classes that arise from double exchanges. The resultant equation is:

$$x = \frac{100}{2} \left| \frac{T + 6NPD}{PD + NPD + T} \right| \tag{3}$$

This equation is valid within a few percent for map distances up to about 35–40 cM. For greater distances, use of this equation leads to an underestimation of the true map distance, because triple and higher-order exchanges become significant.

Snow (1979a,b) has derived a more general set of equations to determine map distances. This derivation is an extension, using maximum likelihood methods, of the derivation of Barratt et al. (1954). All crossover classes are considered, and chiasma interference is also taken into account. The equations are complex and must be solved by iteration using a computer. These equations have been applied to a large set of tetrad data from yeast. For each set of PD, NPD, and T values, estimates of x_p, x', and k were determined (Mortimer and Schild 1980). These variables are, respectively, the map distance derived from equation 3, the corrected map distance, and the chiasma interference value ($k = 0$ indicates complete interference; $k = 1$ indicates no interference). A plot of x_p versus x' is presented in Figure 1. It can be seen that for distances up to about 40 cM, equation 3 is reasonably accurate. At 50 cM, this equation underestimates the map distance by about 10%, and at greater distances the error becomes progressively larger. To obtain an estimate of corrected map distance, one needs simply to calculate x_p and then determine the corresponding x' value from Figure 1. Alternatively, the equations derived by Snow (1979a,b) can be used if one has access to a computer.

Centromere Linkage. If there has been no exchange between a heterozygous marker and its centromere, both alleles from one parent will go to one pole, and the other parent's alleles will go to the opposite pole at the first meiotic division. This is known as first division segregation (FDS). If an exchange has occurred in this interval, one allele from each parent will go to both poles, and

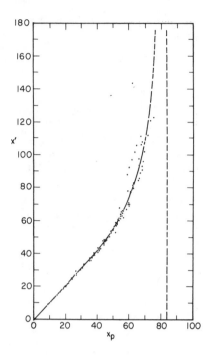

Figure 1   Map distances calculated according to the method of Snow (1979a,b) plotted against the map distances calculated for the same sets of data (Mortimer and Schild 1980), using the formula derived by Perkins (1949). The line is drawn as "best-fit." The scatter of points about this line is due to different interference values; the relationship between x' and x_p depends on k.

the two parental alleles will not segregate until the second meiotic division. This is called second division segregation (SDS). FDS can result from no crossovers or from multiple crossovers; SDS arises from single or multiple crossovers. SDS frequencies increase from zero at the centromere to 67% for genes far out on the chromosome arm. If there is chiasma interference, which is present in most regions in yeast (Snow 1979a), SDS frequencies will rise above 67% at approximately 40–60 cM from the centromere (super-recombination) and then fall to 67% at greater distances (Fig. 2; Barratt et al. 1954). A gene is considered to be centromere-linked if it shows an SDS frequency significantly less than two-thirds. Genes that exhibit super-recombination should also be considered in this class because they will show linkage to genes near their respective centromeres (Mortimer and Hawthorne 1973).

There are three approaches in *S. cerevisiae* to determine whether a gene is centromere-linked. Two of these methods are now rarely used and are primarily of theoretical interest. (1) Certain hybrids produce linear instead of oval or tetrahedral asci. Hawthorne (1955) found that centromere-linked genes segregated predominantly in an alternating array (*AaAa*) in such linear asci, whereas genes not linked to their centromeres also exhibited the other possible segregation arrays (*AAaa, AaaA, aAAa*). (2) Tetraploid hybrids that are in a duplex state (*A/A/a/a*) for a genetic marker will yield three types of

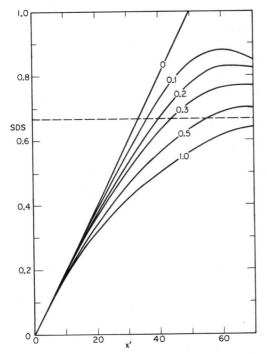

Figure 2 Plot of SDS frequencies vs. map distance (x′) for different interference values. The curves were calculated from the equations derived by Snow (1979a,b).

asci, i.e., *Aa, Aa, Aa, Aa; AA, Aa, Aa, aa* and *AA, AA, aa, aa*. The phenotypic segregations of *A:a* in these asci are 4:0, 3:1, and 2:2, respectively. Roman et al. (1955) showed that noncentromere-linked genes yielded these three segregation classes in the ratios 4/9: 4/9: 1/9, whereas for centromere-linked genes (SDS = 0) the ratios were 2/3:0:1/3; thus, the 3:1 asci signal exchange between the marker and its centromere. (3) Most hybrids are diploid and yield unordered asci, and so neither of the methods above is applicable. However, it is still possible to determine centromere linkage in such cases if more than one centromere-linked gene is segregating. For two heterozygous sites, *A/a* and *B/b*, each on different chromosomes and with SDS frequencies x and y, respectively, it can be shown that the frequency of tetratype asci for the two markers is:

$$f(T_{ab}) = x + y - \tfrac{3}{2} \cdot xy \tag{4}$$

If the SDS frequency for at least one centromere-linked gene is known, SDS values for all other segregating markers can be determined from equation 4. In practice, a gene such as *trp1, pet8*, or *met14* (all of which have SDS values of $< 1\%$) is included in a cross in order to determine centromere linkage of other markers. To a good approximation, the tetratype frequency of a gene against any of these centromere markers corresponds to the SDS frequency of that gene. The map distance of a gene from its centromere can then be determined from the SDS frequency. To a first approximation, this distance is half the SDS frequency. However, this approximation is good only for low SDS values. In Figure 2, a plot of SDS frequencies is presented as a function of map distance from the centromere for different chiasma interference values. The average interference value for several regions near centromeres is approximately 0.3 (Snow 1979a), so the curve corresponding to this value can be used to estimate man distances from SDS values.

If a gene is shown to be centromere-linked, it is then tested for linkage to markers near the centromeres of all of the established chromosomes. If it fails to show linkage to any of these centromere markers, this gene then defines a new chromosome. In *S. cerevisiae*, chromosomes I–XVI were identified in this manner.

Random Spore Analysis

Genetic analysis may also be carried out on a random sample of spores obtained from a hybrid. This method is useful if one is not set up for micromanipulation or if one wishes only to obtain a segregant from a cross.

Preparation of random spore samples involves dispersal of spore aggregates and separation of the spores from the unsporulated diploid cells. The first step involves treating the asci with Glusulase (Endo Laboratories) or Zymolyase (Kirin Breweries), followed by sonication to disrupt the spore tetrads (Magni 1963; Siddiqi 1971). Several methods may then be employed to separate the spores from the diploid cells.

Physical Separation

Emeis and Gutz (1958) showed that spores could be separated from vegetative cells by phase separation; spores separated into the paraffin phase of a paraffin-water mixture. Resnick et al. (1967) also showed that spores were electrophoretically distinct from vegetative cells and could be separated in a flow-electrophoresis apparatus.

Selective Inactivation of Diploid Cells

Some reports indicate that diploid cells are inactivated more rapidly than spores by heat, ethanol, or ether (Dawes and Hardie 1974; Zakharov and Inge-Vechtomov 1964).

Genetic Selection of Spores

Diploids may be constructed so that they are heterozygous for a recessive marker that confers resistance to a drug. By plating the sonicated mixture on a medium containing the drug, only those spores that get the marker will grow. Canavanine resistance is frequently used. Another approach is to include in heterozygous state, markers that affect colony color or morphology. Colonies from the random spore mixture that express the phenotype associated with the markers are then restreaked for testing.

Trisomic Analysis

This method of analysis is relatively efficient for assigning a gene to a particular chromosome. In singly or multiply trisomic strains, a gene in duplex condition on a trisomic chromsome ($A/A/a$) will segregate 4:0, 3:1, and 2:2, in contrast to the normal 2:2 segregation pattern seen for heterozygous sites on a diploid chromosome. If an unmapped gene segregates in trisomic fashion, it can be excluded from all chromosomes that segregate 2:2, and vice versa.

Known Chromosome Is Disomic

If a strain that is disomic for a particular chromosome is crossed to a haploid strain carrying an unmapped gene, 4:0, 3:1, and 2:2 segregations of the unmapped gene serve to place this gene on the chromosome present with an extra copy. Also, 2:2 segregations exclude the gene from this chromosome. This method is reliable only if the cross is constructed so that one or more markers on the trisomic chromosome is in duplex condition ($A/A/a$). This serves to monitor loss of the extra chromosome. Singly disomic strains also may become disomic for one or more other chromosomes. Unless the haploid strain carrying the unmapped gene carries enough additional markers to monitor such events, the use of this method may produce misleading results.

Chromosome Bearing the Unmapped Gene Is Disomic

If the chromosome on which the unmapped gene is present can be made disomic, it is relatively simple to identify this chromosome by crossing the disome to a strain or set of strains that carry markers on all of the chromosomes (Mortimer and Hawthorne 1973).

Methods for constructing disomic strains are usually based on recovering such strains as rare meiotic products of diploid strains. If the diploid is heteroallelic for complementing or noncomplementing alleles at a particular locus, disomes are detected as those spore clones that show either a complementation or a heteroallelic reversion response for the phenotype associated with that locus.

Multiply Disomic Strains

When sporulated, triploid strains yield spores with chromosome numbers distributed between those of haploids and diploids. Because of chromosomal imbalance, most (~85%) of these aneuploids fail to develop into visible colonies. The spores that do form colonies generally have chromosome numbers near those of haploids or diploids; near-haploidy is the most frequent class. The probability that different chromosomes are present in disomic condition is nonrandom, as are the patterns of multiple disomy (Parry and Cox 1971). The number of extra chromosomes in the near-haploid class range from 1 to 5 with a mean of 2. If a haploid that carries markers on many of the chromosomes plus an unmapped gene is crossed to a set of these multiply disomic strains, it is possible to deduce on which chromosome the unmapped gene is located. Those markers that are present on chromosomes that are disomic in the aneuploid parent will segregate 4:0, 3:1, and 2:2, whereas markers on other chromosomes will segregate 2:2. By examining the segregation pattern of the unmapped gene relative to those of the various chromosome markers, it is possible to exclude all chromosomes that segregate differently than the one on which the unmapped gene is located. As many as 4 or 5 chromosomes per cross can be excluded by this method. Several groups of linked genes were assigned to specific chromosomes through the use of this method (Mortimer and Hawthorne 1973), and recently Hilger and Mortimer (1980) showed, through the use of a set of multiply aneuploid strains, that *arg1* and *arg8* were on chromosome XV.

A variation on the method above that eliminates the necessity of coupling the unmapped gene with chromosome markers has been developed by Wickner (1979). This approach is known as the "super triploid method." The super triploid is constructed with at least one marker on each chromosome in duplex ($A/A/a$) condition. Random spores from this triploid are mated with haploid cells carrying the unmapped gene, and several crosses are selected and analyzed genetically. Each cross serves, on the average, to exclude 4.2 chromosomes as the location of the unmapped gene.

MITOTIC MAPPING TECHNIQUES

Mitotic Crossing-over

Diploid yeast cells that are heterozygous for a genetic marker, A/a, may yield A/A and a/a cells during mitotic division. The spontaneous frequency of such events is on the order of 10^{-4} to 10^{-5}/division. However, if the cells are exposed to ionizing radiation, UV radiation, or any of a large number of alkylating agents, the frequency of mitotic segregation can be increased to 10^{-1} to 10^{-2} (Nakai and Mortimer 1969). Certain mutations such as *chl1* (Liras et al. 1978) and *rad18* (Boram and Roman 1976), when homozygous, also increase the frequency of spontaneous mitotic crossing-over.

If a mitotic crossover event occurs between a gene, A/a, and its centromere, subsequent chromosome disjunction during mitosis may result in two daughter cells that are AA and aa. Descendants of these two cells will then form the two sides of a sectored colony. All genes distal to the crossover event are expected to become homozygous. Thus, concomitant sectoring of two heterozygous markers is evidence that they are on the same chromosome arm.

Mitotic gene conversion also occurs and can result in homozygosity for previously heterozygous sites. The mitotic gene-conversion event results in two descendant cells that are either A/a and A/A or A/a and a/a. Mitotic gene-conversion events may also result in homozygosity for linked markers.

Sectoring of genes that affect colony color or morphology is relatively easy to detect. Sectoring of nutritional markers can be detected by replicating colonies to various omission media (Fogel et al. 1979).

Several genes have been assigned to specific chromosome arms using induced mitotic crossing-over (Nakai and Mortimer 1969; Mortimer and Hawthorne 1973; Wickner 1979).

Chromosome Loss and Transfer

Chromosome Loss

Chromosome loss has been used successfully to assign genes to specific chromosomes in several organisms, including *Schizosaccharomyces pombe* (Flores da Cunha 1970; Kohli et al. 1977), and similar techniques have been developed in *S. cerevisiae*. The basic method used for *S. cerevisiae* is to construct a diploid strain in which each of the chromosomes to be tested is heterozygous for an easily scored marker and in which the recessive allele of the mutation to be mapped is coupled with the other recessive markers. If a recessive mutation (e.g., *cdc6, cdc14,* or *chl1*) is used to induce chromosome loss, the strain must also be homozygous for this mutation. The frequent expression of two recessive markers is an indication that these markers are on the same chromosome. Other mapping techniques, such as mitotic recombination and tetrad analysis, are then used to confirm the chromosomal assignment and to determine the location on that chromosome of the gene. The specific approaches to mapping genes by chromosome loss are discussed below.

chl1. The recessive chromosome loss mutation (*chl1*) causes diploids to lose spontaneously specific chromosomes (i.e., I, III, V, VII, VIII, IX, XII, XV, and XVI) during vegetative growth as well as causing a general tenfold increase in mitotic recombination (Liras et al. 1978; J. McCusker and J. Haber, pers. comm.). Chromosome loss in the *chl1* homozygous diploid can be used either to assign a gene to one of the chromosomes frequently lost or to eliminate these chromosomes as possible locations of the gene to be mapped. The *chl1*-induced chromosome-loss method was used to assign *gal4* to chromosome XVI, and this assignment was confirmed by mitotic recombination (J. Haber and J. McCusker, pers. comm.). The enhanced mitotic recombination in the *chl1* homozygous diploid can also be used in mapping genes (see Mitotic Crossing-over).

cdc6 and cdc14. When vegetative diploids homozygous for either of the temperature-sensitive cell-division-cycle mutations *cdc6* and *cdc14* are heat shocked (~ 6 hr at 36° C), they undergo marker loss (Kawasaki 1979). This has been shown to be due to chromosome loss in the *cdc14* homozygote, and evidence indicates that the *cdc6* homozygote is undergoing chromosome loss as well. Approximately half of the *cdc6* homozygous cells and between 1% and 10% of the *cdc14* homozygotes underwent marker loss following heat shock. Homozygotes for either *cdc6* or *cdc14* underwent marker loss on all chromosomes tested (16 in the case of *cdc6* and 15 in the case of *cdc14*) and, although single losses predominated, multiple losses were also observed. Concomitant marker loss in *cdc6* homozygotes was used to place *MAL3* on chromosome II, *SUC* on chromosome IV (Kawasaki 1979), and *rad2* on chromosome VII (D. Schild, unpubl.). Marker loss in *cdc14* homozygotes was used to place *cdc6* on chromosome X and to determine that an insertional translocation was involved in the transfer of part of chromosome III to chromosome I (Kawasaki 1979). Strains specially constructed for marker loss mapping with *cdc6* are available from the Yeast Genetic Stock Center (Donner Laboratory, University of California, Berkeley).

Methyl Benzimidazole-2-yl Carbamate. Chemically induced chromosome loss in *S. cerevisiae* has been reported using para-fluorophenylalanine (Emeis 1966; Strömnaes 1968), $CdCl_2$, $CoCl_2$, and acriflavine (Takahashi 1972). These chemicals also have other genetic effects, such as increased mitotic recombination and mutation, and have not been used for mapping. Methyl benzimidazole-2-yl carbamate (MBC, a derivative of benomyl) has been shown to cause chromosome loss in *S. cerevisiae* without an increase in mitotic recombination or mutation (J. Wood, pers. comm.). Approximately 50% of cells treated with MBC lose one or more chromosomes, and all of the 13 chromosomes tested were lost with near equal frequencies (2.5–6.9% per chromosome in surviving cells). Although no gene has yet been mapped using MBC, these experiments are currently being attempted (J. Wood, pers. comm.). One advantage of chemically induced chromosome loss is that it is not necessary to have, in homozygous state, a recessive mutation such as *chl1* or *cdc6*. The diploid strains to be treated with MBC need only have the recessive allele of the gene to be mapped in coupling with recessive chromosomal markers. If the frequency of loss of more than one chromosome is sufficiently high, as it appears to be for MBC, it may also be possible to map genes that are in repulsion to the standard chromosomal markers. If the unmapped gene and a particular chromosomal marker in repulsion to that gene are never expressed together, this is evidence that these two genes are located on the same chromosome.

Chromosome Transfer

A mapping method is currently being developed that uses chromosome exchange in heterokaryons involving one *kar1* (karyogamy-defective) parent (S. Dutcher, pers. comm.). When *KAR1* and *kar1* haploids of opposite mating types are mated, they produce heterokaryons that bud off cytoductants or heteroplasmons (haploids with a nucleus from either one of the parents but cytoplasm from both parents) (Conde and Fink 1976). When both parents have many chromosomes genetically marked, it has been found that in about 2% of the cytoductants, one or more chromosomes from one nucleus has been transferred to the other nucleus (S. Dutcher, pers. comm.). These have been termed exceptional cytoductants. Most of the exceptional cytoductants have replaced their original chromosome(s), but some are disomic for the transferred chromosome(s). The frequency of transfer for a particular chromosome shows an inverse correlation with the genetic size of that chromosome, i.e., the smallest chromosomes are transferred the most frequently. Genetic mapping using *karl* could be accomplished by mating a near-wild-type haploid containing the mutation to be mapped by a canavanine (or other drug)-resistant *karl* haploid, with each chromosome marked by a recessive marker. If examination of, for example, canavanine-resistant exceptional cytoductants reveals the frequent coordinate appearance of a dominant marker and the recessive marker being mapped (both from the nonselected nucleus), it would indicate that both genes are located on the same chromosome.

FINE-STRUCTURE MAPPING

Fine-structure maps of many genes of *S. cerevisiae* have been constructed. The approaches that have been used are to (1) determine the number of prototrophs produced in a heteroallelic diploid after treatment of mitotic cells with a recombinagenic agent and (2) determine the frequency of prototrophic spores among the meiotic products of such diploids. Manney and Mortimer (1964) demonstrated that X-ray-induced mitotic recombination could be used to order different alleles of a gene, and Manney (1964) constructed a map of the *TRP5* gene that was both internally consistent and consistent with intracistronic complementation results of different *trp5* alleles. Methylmethanesulfonate (MMS) (Snow and Korch 1970) and near-UV radiation (Lawrence and Christiansen 1974) have also been used to construct fine-structure maps by the induction of intracistronic recombination in mitotic cells. Esposito (1968) constructed a fine-structure map of the *ade8* gene both by the X-ray-induced heteroallelic reversion method and by the meiotic approach. The two maps were approximately colinear. However, Moore and Sherman (1975) demonstrated that recombination frequencies between closely spaced alleles in the *cyc1* gene were almost completely unrelated to the physical separation of the alleles, for all of five different fine-structure mapping methods. It is possible that marker effects predominate for closely spaced alleles, and this could result in inaccuracies in fine-structure maps. For example, most conversion events involving closely spaced allele pairs would involve both alleles and would not be expected to generate prototrophs unless the mismatches associated with one allele remain uncorrected. Another problem associated with fine-structure mapping is the dependence of prototroph frequency on the plating medium. The frequency of arginine prototrophs from a pair of *arg4* heteroalleles was found to vary as much as 20-fold depending on whether or not a small amount of arginine was added to the selective plating medium and whether or not the cells had been starved for arginine before treatment (Murthy et al. 1976). These results suggest that there can be a phenotypic lag in expression of prototrophy. Because of this, it is likely that leaky alleles would give anomalous results in fine-structure mapping.

Deletion mapping is the method of choice for fine-structure mapping. Fink and Styles (1974) successfully used deletion mapping with UV treatment to determine the sequence of alleles in the *HIS4* region. They found that, although most of the relative positions assigned by X-ray mapping were correct, some were not. Sherman et al. (1975) used deletion mapping in combination with X-ray treatment to form a fine-structure map of *CYC1*. Except for one allele, this map was colinear with the physical map. The advantage of deletion mapping is that the relative positions of alleles are determined by the presence or absence of recombinants, rather than by the relative frequency of recombinants.

The recently introduced methods of gene cloning and DNA sequencing very likely will replace the traditional approaches described above for determining the relative positions of alleles within a gene (Struhl and Davis 1977; Smith et al. 1979).

MAPPING STRATEGIES

Locating genes on the genetic map of yeast may involve only one or most of the approaches described above. Because each of these methods yields different and complementary information, there is likely an optimum strategy for locating a gene with the minimum of effort. Two possible strategies that may be used to map a gene in *Saccharomyces* are described below.

Method I

The first step is to cross the mutant to a strain carrying a centromere-linked marker such as *trp1*. If the unmapped gene shows less than two-thirds SDS (less than 2/3 tetratype asci relative to *trp1*), it is then crossed to a set of centromere-tester strains available from the Yeast Genetic Stock Center. The unmapped gene should show significant linkage to one of the markers near the centromere of the established chromosome. If it does not, it serves to establish a new chromosome (or the centromere of chromosome XVII).

If the unmapped gene is not centromere-linked, the next step is to determine on which chromosome it is located. This can be done by either

cerevisiae by trisomic analysis combined with interallelic complementation. *J. Bacteriol.* **141:** 270.

Kawasaki, G. 1979. "Karyotypic instability and carbon source effects in cell cycle mutants of *Saccharomyces cerevisiae.*" Ph.D. thesis, University of Washington, Seattle.

Kohli, J., H. Hottinger, P. Munz, A. Strauss, and P. Thuriaux. 1977. Genetic mapping in *Schizosaccharomyces pombe* by mitotic and meiotic analysis and induced haploidization. *Genetics* **87:** 471.

Lauer, G.D., T.M. Roberts, and L.C. Klotz. 1977. Determination of the nuclear DNA content of *Saccharomyces cerevisiae* and implications for the organization of DNA in yeast chromosomes. *J. Mol. Biol.* **114:** 507.

Lawrence, C.W. and R. Christiansen. 1974. Fine structure mapping in yeast with sunlamp radiation. *Genetics* **76:** 723.

Lindegren, C.C. 1949. *The yeast cell, its genetics and cytology.* Educational Publishers, St. Louis.

Lindegren, C.C., G. Lindegren, E. Shult, and Y.L. Hwang. 1962. Centromeres, sites of affinity and gene loci on the chromosomes of *Saccharomyces. Nature* **194:** 260.

Liras, P., J. McCusker, S. Mascioli, and J.E. Haber. 1978. Characterization of a mutation in yeast causing nonrandom chromosome loss during mitosis. *Genetics* **88:** 651.

Magni, G.E. 1963. The origin of spontaneous mutations during meiosis. *Proc. Natl. Acad. Sci.* **50:** 975.

Manney, T.R. 1964. Action of a supersuppressor in yeast in relation to allelic mapping and complementation. *Genetics* **50:** 109.

Manney, T.R. and R.K. Mortimer. 1964. Allelic mapping in yeast by x-ray-induced mitotic reversion. *Science* **143:** 581.

Moore, C.W. and F. Sherman. 1975. Role of DNA sequences in genetic recombination in the iso-1-cytochrome *c* gene of yeast. I. Discrepancies between physical distances and genetic distances determined by five mapping procedures. *Genetics* **79:** 397.

Mortimer, R.K. and D.C. Hawthorne. 1966. Genetic mapping in *Saccharomyces. Genetics* **53:** 165.

———. 1969. Yeast genetics. In *The yeast I* (ed. A.H. Rose and J.S. Harrison), p. 385. Academic Press, New York.

———. 1973. Genetic mapping in *Saccharomyces.* IV. Mapping of temperature-sensitive genes and use of disomic strains in localizing genes. *Genetics* **74:** 33.

———. 1975. Genetic mapping in yeast. *Methods Cell Biol.* **11:** 221.

Mortimer, R.K. and D. Schild. 1980. The genetic map of *Saccharomyces cerevisiae. Microbiol. Rev.* **44:** 519.

Murthy, M.S.S., B.S. Rao, and V.V. Deorukhakar. 1976. Dependence of the expression of the radiation-induced gene conversion to arginine independence in diploid yeast on the amino acid concentration: Effect on allelic mapping. *Mutat. Res.* **35:** 207.

Nakai, S. and R.K. Mortimer. 1969. Studies of the genetic mechanism of radiation-induced mitotic segregation in yeast. *Mol. Gen. Genet.* **103:** 329.

Nasmyth, K.A. and S.I. Reed. 1980. Isolation of genes by complementation in yeast: Molecular cloning of a cell cycle gene. *Proc. Natl. Acad. Sci.* **77:** 2119.

Parry, E.M. and B.S. Cox. 1971. The tolerance of aneuploidy in yeast. *Genet. Res.* **16:** 333.

Perkins, D.D. 1949. Biochemical mutants in the smut fungus *Ustilago maydis. Genetics* **34:** 607.

Perkins, D.D. and E.G. Barry. 1977. The cytogenetics of *Neurospora. Adv. Genet.* **19:** 134.

Resnick, M.A., R.D. Tippetts, and R.K. Mortimer. 1967. Separation of spores from diploid cells of yeast by stable-flow free-boundary electrophoresis. *Science* **158:** 803.

Roman, H., M.M. Phillips, and S.M. Sands. 1955. Studies of polyploid *Saccharomyces.* I. Tetraploid segregation. *Genetics* **40:** 546.

Sherman, F., M. Jackson, S.W. Liebman, A.M. Schweingruber, and J.W. Stewart. 1975. A deletion map of *cyc1* mutants and its correspondence to mutationally altered iso-1-cytochromes *c* of yeast. *Genetics* **81:** 51.

Siddiqi, B.A. 1971. Random-spore analysis in *Saccharomyces cerevisiae*. *Hereditas* **69**: 67.

Smith, M., D.W. Leung, S. Gillam, C.R. Astell, D.L. Montgomery, and B.D. Hall. 1979. Sequence of the gene for iso-1-cytochrome *c* in *Saccharomyces cerevisiae*. *Cell* **16**: 753.

Snow, R. 1979a. Maximum likelihood estimation of linkage and interference from tetrad data. *Genetics* **92**: 231.

———. 1979b. Comment concerning maximum likelihood estimation of linkage and interference from tetrad data. *Genetics* **93**: 285.

Snow, R. and C.T. Korch. 1970. Alkylation induced gene conversion in yeast: Use in fine structure mapping. *Mol. Gen. Genet.* **107**: 201.

Strathern, J.N., C.S. Newlon, I. Herskowitz, and J.B. Hicks. 1979. Isolation of a circular derivative of yeast chromosome III: Implications for the mechanism of mating type interconversion. *Cell* **18**: 309.

Stro≫mnaes, O. 1968. Genetic changes in *Saccharomyces cerevisiae* grown in media containing DL-para-fluoro-phenylalanine. *Hereditas* **59**: 197.

Struhl, K. and R.W. Davis. 1977. Production of a functional eukaryotic enzyme in *Escherichia coli*: Cloning and expression of the yeast structural gene for imidazole-glycerolphosphate dehydratase (*his3*). *Proc. Natl. Acad. Sci.* **74**: 5255.

Takahashi, T. 1972. Abnormal mitosis by some ρ-mutagens in *Saccharomyces cerevisiae*. *Bull. Brew. Sci.* **18**: 37.

Wickner, R. 1979. Mapping chromosomal genes of *Saccharomyces cerevisiae* using an improved genetic mapping method. *Genetics* **92**: 803.

Zakharov, I.A. and S.G. Inge-Vechtomov. 1964. Ascospore isolation of yeast for genetic analysis without a micromanipulator. *Issl. Genet.* **2**: 134.

Genome Structure and Replication

Walton L. Fangman
Department of Genetics, SK-50
University of Washington
Seattle, Washington 98195

Virginia A. Zakian
Hutchinson Cancer Research Center
Seattle, Washington 98104

INTRODUCTION

Saccharomyces cerevisiae is an ideal model system for studies on the structure and replication of the eukaryotic chromosome. Yeast cells are easy to grow and handle biochemically, and they are amenable to sophisticated genetic manipulations. The availability of mutants is especially critical for a detailed understanding of the replication process. Without them it is almost impossible to determine the in vivo role of a particular protein. Although the small size of yeast chromosomes makes cytological studies difficult, it provides the possibility of isolating intact the DNA corresponding to a given chromosome. Moreover, chromosomal substructures such as telomeres and centromeres are found in far higher concentrations in yeast DNA than in that of higher eukaryotes and can therefore be studied with greater ease. The small genome and low percentage of repetitive sequences has enabled the relatively rapid isolation and characterization of specific sequences by recombinant DNA technology. The ability to transform yeast with recombinant DNA plasmids is invaluable for investigating the functions of specific DNA sequences, including those involved in replication. Finally, yeast contains an endogenous multiple-copy extrachromosomal plasmid called 2-micron (μ) DNA, which provides a tractable model for some features of chromosomal DNA replication.

In this paper we have attempted to provide an overview of chromosomal organization and DNA replication. More detailed information on the following topics can be found in other papers in this volume: Mitochondrial Genetics and Function (Dujon); The Yeast Plasmid 2μ Circle (Broach); The *Saccharomyces cerevisiae* Cell Cycle (Pringle and Hartwell). Information on the structures of tRNA genes, genes coding for mRNAs, and ribosomal DNA can be found elsewhere (Warner 1982; also see Guthrie and Abelson 1982; Hall and Sentenac 1982).

STRUCTURE OF CHROMOSOMAL DNA

The *S. cerevisiae* haploid genome consists of approximately 14,000 kbp[1] ($\pm 20\%$) on the basis of renaturation kinetics and the DNA content of stationary-phase cells (Lauer et al. 1977). Approximately 95% of the sequences are single copy (Whitney and Hall 1974; Lauer et al. 1977). A base composition of 60% A + T gives the DNA a buoyant density in CsCl of 1.699 g/cm[3] (Moustacchi and Williamson 1966). The DNA contains approximately 1 5-methylcytosine residue/200 cytosines (Hattman et al. 1978), but it has not been reported whether these occur at specific sites in the genome.

Chromosomal DNA Size

Because of the small size and apparent lack of condensation of the 17 yeast chromosomes, routine karyotypic analysis has not been possible. However, complete chromosomes have been observed by thin-section analysis of synaptonemal complexes in meiotic cells (Byers and Goetsch 1975). Sedimentation, electron microscopy, and viscoelastic studies have revealed a broad distribution of chromosomal DNA sizes from 150 kbp to about 2500 kbp (Blamire et al. 1972; Petes and Fangman 1972; Petes et al. 1973, 1974; Cryer et al. 1974; Lauer et al. 1977; Forte and Fangman 1979). The smallest yeast chromosome is about the size of the T4 or vaccinia virus genome. The 16-fold range in chromosomal DNA size is reflected in the broad range of recombination distances and number of genes per chromosome (Mortimer

---

[1]In this article, data from the literature are converted to kilobase pairs (kbp) by the relationships 2×10^6 daltons = 1 μm = 3 kbp.

and Hawthorne 1975) and in the lengths of synaptonemal complexes (Byers and Goetsch 1975).

Since the average size of a yeast chromosomal DNA molecule (Petes and Fangman 1972; Petes et al. 1973, 1974) is close to the calculated average DNA content of a chromosome (about 800 kbp), most chromosomes must contain a single DNA molecule. RNA or protein linkers do not appear to be in the DNA (Blamire et al. 1972; Petes and Fangman 1972).

Centromeres and Telomeres

It follows from the deduction that a yeast chromosome contains a single DNA molecule that chromosomal DNA is continuous through the centromere. Direct evidence for this fact was obtained by isolating circular forms of chromosome III (Strathern et al. 1979). These chromosomes apparently arise by recombination between the *HML* locus on the left arm and the *MAT* locus that is near the centromere on the right arm of chromosome III. In strains with this chromosome, a 190-kbp supercoiled circular DNA molecule containing chromosome-III DNA sequences was detected. A DNA fragment containing the centromere of chromosome III has been cloned (Clarke and Carbon 1980). The centromere was identified by its promotion in yeast of 2:2 meiotic segregation of genes on the same plasmid. Hybridization failed to detect sequences elsewhere in the genome that are homologous to the chromosome-III centromere.

The termini, presumably the telomeres, of yeast chromosomal DNA molecules have cross-linked DNA strands. The cross-links, which were detected by the rapid renaturation of large (500 kbp) DNA molecules, are not destroyed by protease or RNase treatments (Forte and Fangman 1976). Further analyses showed that a significant fraction of chromosomal DNA molecules of all sizes contain cross-links and that there are about two cross-links per molecule (Forte and Fangman 1979). *Eco*RI fragments containing cross-links are a subset of the fragments from total yeast DNA, which shows that the cross-links are at specific sites in the genome. Finally, electron microscopy of partially denatured DNA fragments revealed that the cross-links are at the DNA termini (Fig. 1). These observations are consistent with the chromosome's terminus being a continuous thread of DNA without free 3' and 5' ends. Such a structure would allow replication around the terminal loop without the requirement for a special primer (Bateman 1975).

Chromatin

Yeast chromatin contains the four nucleosomal inner histones H2A, H2B, H3, and H4 (Wintersberger et al. 1973; Franco et al. 1974; Brandt and von Holt 1976; Moll and Wintersberger 1976; Thomas and Furber 1976; Nelson et al. 1977; Mardian and Isenberg 1978). H4 has an amino acid composition

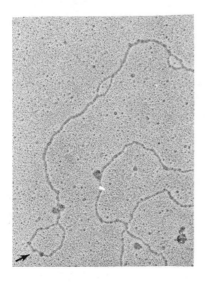

Figure 1 Single-strand denaturation loop revealing a cross-linked terminus of chromosomal DNA. (Reprinted, with permission, from Forte and Fangman 1979). Fragments of 150 kbp were partially denatured during preparation for electron microscopy. 14% of the DNA termini in the preparation exhibited single-strand loops (→), indicating that the cross-links are at or within 50 bp of the last nucleotide pair. This value was close to that expected (18%) if there are cross-links at the ends of all 17 yeast chromosomes.

close to that of calf and pea, but the other yeast histones have diverged considerably from those of other eukaryotes (Mardian and Isenberg 1978). However, calf-yeast and yeast-calf H2A-H2B interactions suggest that the interacting surfaces of these proteins are also highly conserved. The H3 histone of yeast is the most divergent from that of pea and calf. H1 histone has not been demonstrated clearly. Its absence may account for the lack of condensation of the chromosomes. Several proteins have been isolated from yeast chromatin that are analogous in electrophoretic mobility and amino acid composition to the high mobility group (HMG) proteins of higher organisms (Sommer 1978; Spiker et al. 1978; Weber and Isenberg 1980). The HMG proteins have been proposed to be structural components of chromatin (Goodwin et al. 1977; Javaherian et al. 1978) or to have a role in transcriptional activation (Weisbrod et al. 1980).

Yeast chromatin is organized in typical nucleosomal subunits (Lohr and Van Holde 1975; Lohr et al. 1977a,c; Nelson et al. 1977). Digestion of yeast chromatin with micrococcal nuclease indicates that the nucleosome core contains about 140 bp, similar to that of animal chromatin. However, the spacer is fairly uniform at about 20 bp, 20–40 bp smaller than the more variable spacer in most animal repeats. The smaller nucleosome repeat size (about 160 bp) in yeast is similar to that of two other fungi, *Neurospora crassa* (Noll 1976) and *Aspergillus nidulans* (Morris 1976), and to that of plant chromatin (Phillips and Gigot 1977). DNase I digestion results in single-stranded fragments with lengths that are multiples of 10 bp extending for at least 300 bp. Thus, regular spacing persists from one nucleosome core to the next, which suggests that core units occupy specific positions on the DNA with respect to adjacent ones (Lohr et al. 1977b); this phasing appears to be more extensive in yeast than in other organisms. Yeast chromatin also may be unique in that the set of fragments produced by DNase I exhibits an abrupt size discontinuity in the 10-bp multiples pattern (Lohr and Van Holde 1979). The size step has led to the deduction that adjacent nucleosomes, which are disk-shaped, are arranged side-by-side rather than stacked (Lohr and Van Holde 1979).

Compared with higher eukaryotes, a large fraction of the yeast genome ($\geq 40\%$ of the duplex DNA) is transcribed (Hereford and Rosbash 1977). Using cDNA made from total mRNA as a probe, transcribed DNA in yeast chromatin has been found to have the same sensitivity to DNase I digestion as DNA in total chromatin (Lohr and Hereford 1979). This result contrasts with that found in animal chromatins, where actively transcribed genes are preferentially digested (Weintraub and Groudine 1976; Lilley and Pardon 1979). Either nearly 100% of the yeast genome is actively transcribed or DNase I sensitivity monitors a state that is necessary but not sufficient for transcription (Lohr and Hereford 1979).

Rapidly sedimenting ($> 1000S$) forms of chromosomal DNA that presumably reflect, in part, the intracellular conformation of chromatin have been described previously (Piñon and Salts 1977). The effect of ethidium bromide on their sedimentation properties indicates that the DNA in them is rotationally constrained and negatively superhelical, but the nature of these complexes is not yet clear. Alterations in the sedimentation properties have been observed for complexes released from cells under a variety of physiological conditions (Piñon 1978, 1979a,b; Piñon and Pratt 1979).

REPEATED CHROMOSOMAL ELEMENTS

rRNA Genes

On the basis of saturation hybridization values (Schweizer et al. 1969; Rubin and Sulston 1973), the size of the rRNAs (Rubin 1973; Valenzuela et al. 1977; Philippsen et al. 1978) and the genome size, it can be estimated that the *S. cerevisiae* genome contains 120 ± 40 copies of the genes for 25S, 18S, 5.8S, and 5S ribosomal RNAs. These genes occur as directly repeated tandem units, with each unit containing a gene for each of the rRNAs in the order 18S, 5.8S, 25S, 5S (Bell et al. 1977; Philippsen et al. 1978; see Warner 1982). The linkage of 5S rRNA genes with the other rRNA genes is uncommon in eukaryotic organisms (Federoff 1979) but is not unique to yeast (Cockburn et al. 1976; Maizels 1976; Cihlar and Sypherd 1980). Transcription of the 5S rRNA occurs from the strand complementary to that transcribed into the 18S-5.8S-25S precursor (Arstad and Øyen 1975; Valenzuela et al. 1977; Kramer et al. 1978).

Because of their high GC content, DNA fragments composed of rRNA genes (rDNA) can be purified partially by isopycnic banding in cesium density gradients (Cramer et al. 1972). *Sma*I restriction endonuclease cleaves once per repeat to give units of about 8.4 kbp (Cramer et al. 1976). Electron microscopy analysis of reassociated strands failed to detect length or sequence heterogeneity (Cramer et al. 1976). Homogeneity of rDNA units has been confirmed by studies with cloned fragments (Petes et al. 1978; Philippsen et al. 1978); no intervening sequences have been found. The number of clusters of 8.4-kb repeat units is uncertain. rDNA separates from total chromosomal DNA in cesium gradients only when the DNA size is less than 120 kbp, which suggests that the rRNA genes are contained in about ten clusters separated by AT-rich DNA (Cramer et al. 1972). However, an extensive analysis of cloned rDNA fragments so far has revealed only two rDNA single-copy DNA junctions (Petes et al. 1980). Moreover, an electron microscopy study of chromosomal DNA R-looped with 18S rRNA detected very few DNA molecules (one or two per haploid genome) containing both rRNA genes and non-rDNA sequences (Kaback and Davidson 1980). R-looped molecules with up to 26 evenly spaced 18S rRNA genes were seen (Kaback et al. 1979).

The discovery of rDNA variants (Petes et al. 1978)—one with seven *Eco*RI sites per repeat and one with six—has allowed the genetic mapping of rDNA. A diploid strain, produced by mating haploids containing primarily one variant or the other, gave the 2:2 meiotic segregation pattern expected for a single Mendelian unit (Petes and Botstein 1977). This result suggests that the frequency of meiotic recombination within the rDNA is extremely low compared to the rest of the yeast genome. Linkage tests subsequently showed that the rDNA is on chromosome XII (Petes 1979; Szostak and Wu 1979).

The number of rRNA genes appears to be reduced in strains monosomic (2n-1) for chromosome I (Finkelstein et al. 1972; Kaback et al. 1973), which suggests that dosage of genes on chromosome I controls the rDNA level in a cell. These strains amplify their rDNA content back to the euploid level during prolonged culturing (Kaback et al. 1976; Kaback and Halvorson 1977, 1978). In addition, varying amounts of rDNA have been found among the four spores in an ascus (Øyen et al. 1978). Cells with altered levels of rDNA may arise by unequal crossing-over. Unequal sister chromatid recombination has been demonstrated during meiosis (Petes 1980) and during mitotic cell growth (Szostak and Wu 1980). In both cases, recombination was detected by using strains in which the gene *LEU2* was inserted into the rDNA by transformation. It was estimated (Szostak and Wu 1980) that the frequency of recombination is adequate to account for the maintenance of rDNA homogeneity. Other experiments (Brewer et al. 1980) show that all rDNA units are transmitted through meiosis to the haploid generation. Therefore, transmission of a subset of units, a possibility raised by the observation that a portion of the parental nucleolus is excluded from meiotic spores (Moens and Rapport 1971; Heywood and Magee 1976), does not contribute to the maintenance of rDNA homogeneity. A low frequency of circular extrachromosomal copies of rDNA has been observed (Meyerink et al. 1979; Clark-Walker and Azad 1980). The function of these extrachromosomal rDNA circles, if any, is unknown; they may arise by intrachromosomal recombination between rDNA repeats.

tRNA Genes

On the basis of saturation hybridization with total tRNA preparations (Schweizer et al. 1969) and the genome size, there are 240–300 genes for the tRNAs, an average of 12–15 genes per amino-acid-specific tRNA. Over 20 tRNA genes have been identified by their mutation to nonsense suppressors and have been mapped genetically (Hawthorne and Leupold 1974; Sherman 1982). At least one gene, that for $tRNA^{Ser}_{UCG}$, is present in single copy as shown by the occurrence of suppressor mutations that are haplolethal (Hawthorne and Leupold 1974; Brandriss et al. 1976; Etcheverry et al. 1979). Six of eight suppressor loci corresponding to genes for $tRNA^{Tyr}$ have been mapped to specific EcoRI fragments from total yeast DNA by the meiotic cosegregation of EcoRI fragment size variants and suppressor activities (Olson et al. 1979a). The genes for tRNAs do not appear to be clustered extensively (Beckmann et al. 1977; Olson et al. 1979b). However, a $tRNA^{Arg}$ gene and a $tRNA^{Asp}$ gene are separated by only 9 bp and transcribed from the same DNA strand (Abelson 1979). Intervening sequences, located next to the anticodon and ranging in size from 14–60 bp, have been found in some but not all tRNA genes (Goodman et al. 1977; Valenzuela et al. 1978; Abelson 1979; Knapp et al. 1979; Venegas et al. 1979).

Other Repeated Genes and Elements

The *S. cerevisiae* genome contains a small fraction of repeated DNA sequences compared with higher eukaryotes. Renaturation kinetic analyses with 500-bp fragments indicate that only about 5% of the DNA is repeated (Whitney and Hall 1974; Lauer et al. 1977). On the basis of an rDNA repeat size of 8.4 kbp, a minimum of about 6% of the genome corresponds to rRNA gene units. Therefore, the repeated sequences detected by renaturation kinetics are accounted for by rDNA. However, the tRNA genes ($< 0.1\%$ of the genome) are also repeated, and the abundant mRNAs (Hereford and Rosbash 1977; Holland et al. 1977) are likely to be coded for by moderately repeated genes.

The abundant mRNAs code for the major cellular proteins, including the glycolytic enzymes pyruvate kinase, enolase, aldolase, and glyceraldehyde-3-phosphate dehydrogenase (Holland et al. 1977). Three copies of the gene for glyceraldehyde dehydrogenase have been found (Holland and Holland 1980). Two copies of a gene for a ribosomal protein (Woolford et al. 1979) and two copies each of the genes for the H2A and H2B histones exist (Hereford et al. 1979). The histone genes are arranged in two sets, each made up of one H2A and one H2B gene, which are separated by 700 bp and divergently transcribed. The two sets are widely separated ($>$ 35 kbp) from each other and from the genes for histones H3 and H4. The low copy number and dispersion are quite different from what has been found for the histone genes of other eukaryotes (Kedes 1979). DNA sequencing studies reveal that the two H2B genes code for proteins that differ at four amino acid positions (Wallis et al. 1980). The products of both genes are found in chromatin, but cells are viable when either gene is replaced by a mutant copy (M. Grunstein et al., pers. comm.). None of these repeated yeast structural genes contains intervening sequences.

An additional 1–2% of the yeast genome consists of inverted-repeat sequences that renature with zero-order kinetics (Klein and Welch 1980). Electron microscopy of 23-kbp DNA fragments, denatured and renatured at low DNA concentrations, revealed one hairpin or lariat per 54 kbp of DNA, giving an estimate of 260 inverted repeats per genome. The average hairpin and lariat stem length was 300 bp. Hydroxyapatite chromotographic analysis of DNA sheared to 500 bp, denatured and renatured to low C_0t, suggests that there are $\leq$ 560 inverted repeats per genome. Both approaches indicate that the inverted repeats are somewhat clustered in the genome.

Two families of repeated elements of unknown function have been identified in clones of yeast DNA and studied in some detail (Cameron et al. 1979). ϵ is a sequence of approximately 5 kbp, present in about 35 copies (1.2% of the DNA). The copies show some sequence divergence, and most of them are dispersed in the genome. ϵ is bounded by a direct repeat of about 300 bp called δ, forming an element called Ty1 (Fig. 2). δ sequences are found at more than 100 sites dispersed throughout the genome. They are also found as tandem direct repeats, as tandem inverted repeats, and as inverted repeats flanking sequences other than ϵ (Cameron et al. 1979). Inverted repeat sequences of δ probably account for many of the 300-bp hairpins observed

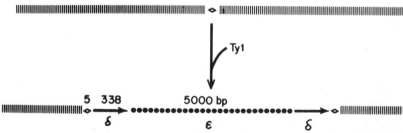

Figure 2 Structure and insertion of the transposable element Ty1. The size of ϵ is approximate. The 338-bp size of δ and the 5-bp duplication in the target DNA are based on the sequence data of Gafner and Philippsen (1980). Farabaugh and Fink (1980) observed, with a different Ty1 element, a 334-bp δ and a 5-bp duplication in the target DNA.

by electron microscopy of total DNA fragments (Klein and Welch 1980). The sequences of δ are also divergent, and tandem repeats are sometimes less homologous than distantly spaced ones. However, the two δ sequences that bound a given ε are identical (Gafner and Phillipsen 1980).

The structure of Ty1 (Fig. 2) is similar to that of bacterial transposons and to the dispersed repeated gene families in *Drosophila (copia, 412)* that code for abundant mRNA species (Finnegan et al. 1978). Ty1 also codes for an abundant mRNA species (Cameron et al. 1979). Moreover, like the *Drosophila* genes (Potter et al. 1979; Strobel et al. 1979), the Ty1 sequences undergo transposition. They are found at different sites in the genomes of different yeast strains, and their locations change during long-term culture (Cameron et al. 1979). Inactivation of the promotor of the *HIS4* gene by insertion of a Ty1 element has been reported (Farabaugh and Fink 1980). Analyses of the *sup4* region of chromosome X suggest that δ or Ty1 sequences can promote deletions. A region flanked by δ sequences is deleted at high frequency (Rothstein 1979), and a mutation, *cyc1-1*, has been shown to be a deletion of 12 kbp with one breakpoint at or near a Ty1 element (P. Shalit et al., in prep.). Comparison of DNA sequences near *sup4* and *HIS4* in strains with and without a Ty1 has demonstrated that, as with bacterial transposons, insertion of Ty1 results in a duplication of 5 bp in the target DNA (Farabaugh and Fink 1980; Gafner and Philippsen 1980).

DNA SYNTHESIS PHASE

S-phase Position and Duration

The position and duration of the S phase in the cell cycle (Fig. 3) can be estimated by autoradiography of pulse-labeled yeast cells (Williamson 1965). An inherent limitation of whole-cell autoradiography experiments is the possibility that incorporation reflects the kinetics of precursor uptake or pool equilibration in addition to the rate of DNA synthesis. A second approach

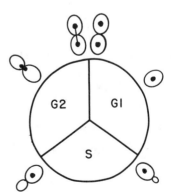

Figure 3 The G_1, G_2, and S phases of the cell cycle. The G_2-to-G_1 transition is defined as the time of complete separation of nuclei.

used by one group to estimate cell-cycle phases is flow microfluorimetry (Slater et al. 1977; Johnston et al. 1980). In this technique, the DNA content of thousands of cells is recorded; however, the resolution has so far been inadequate for application to haploid cells (Slater et al. 1977).

The duration of the S phase in cells growing with the highest growth rates ($\sim$ 1½–2 hr doubling times) has been determined by several laboratories. Three different diploid cells had S phases of 20–25% of the cell cycle (Williamson 1965; Slater et al. 1977; B. Brewer and W. Fangman, unpubl.). Cells of undetermined ploidy had an S phase of approximately 50% of the cell cycle (Barford and Hall 1976). Data for only one haploid strain have been reported, and they indicate the S phase is 45–50% of the cell cycle (Rivin and Fangman 1980a). The differences in S-phase lengths appear to be significant and indicate that a diploid amount of DNA can be replicated in half the time taken to replicate a haploid amount. The length of the G_1 phase varies from less than 10% (Williamson 1965; Barford and Hall 1976) to 30% or more (Slater et al. 1977; Rivin and Fangman 1980a) of the cell cycle.

When the length of the cell cycle increases because of growth on a poor carbon source, the absolute length of S phase remains fairly constant (Barford and Hall 1976; Slater et al. 1977; Carter and Jagadish 1978), i.e., the S phase becomes a smaller fraction of the cell cycle. Most cell-cycle expansion occurs in the G_1 phase. However, a fourfold increase in cell-cycle length brought about by growth with a poor nitrogen source was accounted for by a proportional expansion of the three phases G_1, S, and G_2 (Rivin and Fangman 1980a). In another study, a strain growing with poor nitrogen sources exhibited expansion of the G_1 and S phases but not the G_2 phase (Johnston et al. 1980). It is clear that all phases of the cell cycle can expand, but no simple rules appear to govern phase expansion.

Timing of the cell-cycle phases can be referenced to the emergence of the daughter cell bud. In most diploid cells the bud emerges (becomes visible with the light microscope) at or about the time of entry into the S phase (Williamson 1965; Barford and Hall 1976; Slater et al. 1977; Johnston et al. 1980). However, in one diploid (B. Brewer and W. Fangman, unpubl.) and a haploid (Rivin and Fangman 1980a), the bud emerged about halfway through the S phase. Bud emergence, therefore, does not necessarily signal the beginning of the S phase. The lack of complete correlation between bud emergence and initiation of DNA synthesis is not surprising since temperature-sensitive mutations can uncouple these two processes (Hartwell 1978).

G_1-to-S-phase Transition

Cells are arrested in the G_1 phase by mating pheromones **a**-factor (Wilkinson and Pringle 1974) and α-factor (Hereford and Hartwell 1974), and by defects in several cell-cycle genes (Hereford and Hartwell 1974). That these genes act in a sequentially dependent pathway ($CDC28 \rightarrow CDC4 \rightarrow CDC7$) that leads to entry into the S phase (Fig. 4) was demonstrated by reciprocal shift experiments (transferring cultures to the restrictive temperature after arrest with mating pheromone, and vice versa) and by determining the arrest phenotype of cells containing mutations in two different *CDC* genes (Hereford and Hartwell 1974). **a**-Factor (Wilkinson and Pringle 1974) and α-factor (Hereford and Hartwell 1974) act at the same time as *CDC28*. Several additional *CDC* genes have been identified that appear to act at the same point in the cell cycle as *CDC28* (Fig. 4; Reed 1980). α-Factor has been shown to arrest cells close to the G_1 / S-phase boundary (Rivin and Fangman 1980a). Therefore, these mating pheromones and genes act during a short time interval at the end of the G_1 phase. The temporal sequence of gene action correlates with spindle pole body (SPB) phenotypes (Byers and Goetsch 1974): *cdc28* cells fail to duplicate the SPB, and *cdc4* cells do not separate the duplicated SPBs to form a complete spindle; with *cdc7* cells, SPB separation does occur, but final elongation of the spindle does not.

In contrast to other eukaryotes, proteins needed to complete the S phase appear to have been synthesized by the time DNA replication begins in yeast (Hereford and Hartwell 1973; Williamson 1973). The protein synthesis requirement is met prior to the *CDC7*-mediated step (Hereford and Hartwell 1974). Assuming that replication origins are activated throughout the S phase (Rivin and Fangman 1980b), these observations suggest that protein synthesis is not required for the activations. It is possible, however, that transit through the S phase requires continuous synthesis of specific proteins but that adequate levels of these proteins can be made even when the rate of synthesis is greatly reduced.

Histone synthesis may occur primarily during the S phase (Moll and Wintersberger 1976). Since cells can double their DNA in the absence of protein synthesis, either there is a pool of histones already available at the

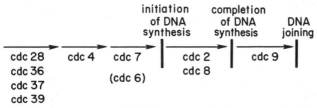

Figure 4   Sequence of *CDC*-gene-controlled steps as related to DNA replication. The sequence is based on the data of Hereford and Hartwell (1973) for *CDC4, CDC7,* and *CDC28*; those of Hartwell (1976) for *CDC2, CDC6, CDC8,* and *CDC9*; and those of Reed (1980) for *CDC36, CDC37,* and *CDC39*. The position of the *CDC6* step relative to the *CDC7* step as they relate to DNA replication is unknown. *CDC21*, involved in precursor biosynthesis, is not included.

onset of the S phase, or histone-deficient DNA is synthesized when protein synthesis is inhibited.

Molecular Parameters of the S Phase

The existence of multiple replication origins in yeast chromosomal DNA was first demonstrated by electron microscopy examination of DNA enriched in replicating structures either by culture-density transfer or by cell synchronization (Newlon et al. 1974). Replication from most origins is bidirectional (Petes and Williamson 1975a; Rivin and Fangman 1980b). Spacings between origins can be estimated to be about 100 kbp from DNA fiber autoradiography (Petes and Williamson 1975a; Rivin and Fangman 1980b) and about 60 kbp from electron microscopy (Newlon et al. 1974; Petes and Newlon 1974). Probably the best estimate of origin-to-origin spacings was obtained by electron microscopic analysis of DNA isolated from cells entering a synchronous S phase (Newlon and Burke 1980). The average spacing, determined for the smaller chromosomal DNAs as a function of increasing degree of replication, reached a minimum value of 36 kbp. If this spacing applies to large chromosomes, then an estimate of about 400 origins in the genome is obtained. Most adjacent replication origins are activated (start DNA synthesis) within a few minutes of each other (Newlon and Burke 1980; Rivin and Fangman 1980b), but some origins appear to be activated throughout the S phase (Rivin and Fangman 1980b). However, it has also been suggested that all origins are activated at the time DNA replication begins (Petes and Williamson 1975a) and that more direct evidence for the activation of origins throughout the S phase of yeast is needed.

The average replication fork rate has been estimated by DNA fiber autoradiography to be 2.1 kbp/min (24°C) in a diploid strain (Petes and Williamson 1975a), and 6.3 kbp/min (30°C) in a haploid strain (Rivin and Fangman 1980b). Sedimentation analysis of pulse-labeled DNA in another diploid gave a fork rate of 1.2 kbp/min (30°C) (Johnston and Williamson 1978). The differences in measured fork rate appear to be greater than can be accounted for by temperature or experimental procedure, but it is not known whether haploids always have higher fork rates than diploids. Replication fork rate can vary depending on the growth conditions. In cells growing slowly because of a limiting nitrogen source, a fourfold increase in S-phase length was completely accounted for by a fourfold decrease in fork rate (Rivin and Fangman 1980a,b).

REPLICATION GENES AND PROTEINS

Mutants

A large number of temperature-sensitive mutants that arrest at characteristic points in the cell cycle have been isolated (Hartwell et al. 1974; Hartwell

1978). Mutants defective in five cell-cycle genes (*CDC4, CDC7, CDC8, CDC21,* and *CDC28*) either fail to enter the S phase or stop DNA synthesis at restrictive temperatures (Hartwell 1973).

The products of *CDC28, CDC4,* and *CDC7* act sequentially prior to the initiation of DNA synthesis (Fig. 4), whereas the products of *CDC8* and *CDC21* are required continuously throughout the S phase (Hereford and Hartwell 1974; Hartwell 1976). The structure of DNA isolated from temperature-arrested cells is consistent with *cdc4* and *cdc7* mutants being defective in the initiation of DNA synthesis and *cdc8* mutants being defective in DNA chain elongation (Petes and Newlon 1974). Since the *CDC7*-mediated step is the last event identified before replication begins, it may be directly involved in the initiation of DNA synthesis. However, *CDC7* is apparently not essential for the initiation of premeiotic DNA synthesis (Schild and Byers 1978). *cdc8* mutants do not appear to be defective in precursor synthesis (Hereford and Hartwell 1971; Oertel and Goulian 1979) nor in DNA polymerase activities (Wintersberger et al. 1974; Prakash et al. 1979). DNA extracted from *cdc8* cells at restrictive temperatures contains single-strand bubbles of about 300 bases that disappear when the DNA is treated to remove protein (Klein and Byers 1978). The product of *CDC21* is thymidylate synthetase (Bisson and Thorner 1977), the enzyme that converts deoxyuridine-5'-monophosphate (dUMP) to thymidine-5'-monophosphate (TMP). The *CDC8-* and *CDC21*-gene products are both required for premeiotic DNA synthesis (Schild and Byers 1978).

In the original *cdc* mutant isolations, mutants defective in a number of genes were characterized as nuclear-division mutants because they appeared to complete DNA synthesis but were unable to undergo nuclear division (Culotti and Hartwell 1971; Hartwell 1973). The possible involvement of these genes in DNA synthesis was examined by asking whether the DNA made at restrictive temperature in these cells would allow them to divide at the permissive temperature when further DNA synthesis was blocked by hydroxyurea (Hartwell 1976). Mutants defective in two genes, *CDC2* and *CDC6*, failed to divide in this test. Therefore, the products of both of these genes appear to be required for the complete duplication of DNA. This interpretation depends upon the assumption that hydroxyurea blocks only DNA synthesis in yeast. These mutants could be defective in replication of a specific component of chromosomal DNA such as telomeres or centromeres. The *CDC6* gene appears to function prior to DNA synthesis, whereas *CDC2* acts sometime during S phase. DNA made in *cdc2* cells at restrictive temperatures contains small single-strand bubbles (M. Conrad and C. Newlon, unpubl.) similar to those in DNA from temperature-arrested *cdc8* cells (Klein and Byers 1978).

In contrast to *CDC2* and *CDC6*, most of the nuclear-division genes (*CDC5, CDC9, CDC13, CDC15, CDC16,* and *CDC23*) are not required to complete a normal S phase (Hartwell 1976). However, none of these genes can function until DNA synthesis is completed. One of the genes, *CDC9*, probably codes for a DNA ligase (Johnston and Nasmyth 1978): *cdc9* cells grown at restrictive temperatures accumulate nascent DNA with single-strand lengths of about 400 bases. However, no ligase activity was detected in *cdc9* cells at any temperature, although ligase activity is readily detected in wild-type cells. It has therefore been impossible to demonstrate that ligase activity itself is temperature-sensitive in *cdc9* cells. At restrictive temperatures *cdc9* cells are also defective in repair of DNA damaged either by ultraviolet light (Johnston and Nasmyth 1978) or methylmethanesulfonate (Johnston 1979). An increase in intragenic and intergenic recombination is observed after return to permissive temperatures (Fabre and Roman 1979; Game et al. 1979). By analogy with *CDC9*, the products of other nuclear-division *CDC* genes may be required for processes related to DNA synthesis.

Given the large number of proteins involved in replication of DNA chains in prokaryotes, it is surprising that so few genes necessary for DNA replication have been identified in yeast. There are several possible explanations for the small number of *cdc* mutants defective in DNA replication: (1) Some proteins required for DNA replication, such as a topoisomerase, also might play essential roles in transcription (Durnford and Champoux 1978). Mutants defective in such proteins would not necessarily arrest in the stage-specific manner required for selection as a *cdc* mutant. Temperature-sensitive mutants have been selected only on the basis of their having reduced rates of DNA synthesis (Johnston and Game 1978). Mutants in seven genes were also defective in RNA synthesis and did not arrest at specific stages in the cell cycle. Further studies are required to determine whether these mutants are defective in proteins directly involved in DNA synthesis. (2) Some proteins required for DNA synthesis might be present normally in such excess that temperature-sensitive mutants retain enough activity to remain phenotypically wild type. For example, *Escherichia coli* mutants with 1% wild-type levels of ligase are not defective in DNA synthesis (Gellert and Bullock 1970). The availability of temperature-sensitive suppressors in yeast (Rasse-Messenguy and Fink 1973) may make it possible to isolate nonsense mutations that eliminate residual enzyme activity. (3) Cells might contain several proteins capable of performing the same function (e.g., several proteins might provide energy for unwinding the DNA helix, for RNA priming of nascent chains, or for polymerizing deoxyribonucleotides). Defects in any one of these proteins might not be lethal.

The major attraction of yeast as a model system for studying eukaryotic DNA replication is that it is highly amenable to genetic manipulation. If its promise is to be realized, more genes with roles in DNA synthesis need to be identified. New selection procedures for isolating replication mutants need to be devised. Autonomously replicating plasmids derived from chromosomal DNA (see Autonomous Replicators) can be used for mutant isolations (B.-K. Tye, pers. comm.) and may be especially useful if different gene products are required to activate different initiation sites. For isolation of certain mutants, it may be necessary to screen large numbers of unselected colonies as was done to isolate the first *E. coli* DNA polymerase mutant (De Lucia and Cairns 1969).

Replication-related Proteins

Three DNA polymerase activities have been detected in yeast: two nuclear enzymes (A or I, B or II) and one mitochondrial enzyme (Wintersberger and Wintersberger 1970a,b; Chang 1977). All three polymerases are large (> 100,000 daltons) so that yeast, like many other lower eukaryotes and plants (Chang 1976), lacks an enzyme equivalent to the β DNA polymerase (30,000–50,000 daltons) found in multicellular animals (Wintersberger 1974a). DNA polymerase A accounts for most of the activity in the cell, and unlike DNA polymerase B, its activity increases during periods of DNA synthesis (Wintersberger and Wintersberger 1970a,b; Wintersberger 1974b; Chang 1977). However, because no mutants have been identified with defects in either polymerase, it is not known whether A, B, or an unidentified DNA polymerase is essential for chromosome replication. Only polymerase B has been shown to have an associated nuclease activity (3′ exonuclease) that could provide an error-correcting capability similar to that found in prokaryotic enzymes (Chang 1977; Wintersberger 1978). The absence of nuclease activities in DNA polymerase A is typical of other eukaryotic DNA polymerases.

A number of other proteins whose activities are compatible with roles in DNA replication have been isolated from yeast. These proteins include a topoisomerase capable of removing positive superhelical turns that could arise during replication (Durnford and Champoux 1978), RNase-H-like activities capable of removing RNA primers from DNA (Wyers et al. 1976), and a DNA-dependent ATPase (Plevani et al. 1980) and single-stranded DNA binding proteins (S. LaBonne and L. Dumas, in prep.; B. Crosby et al., pers. comm.), both of which may play roles in unwinding the DNA helix prior to its replication. However, no mutants have been identified with defects in any of these proteins; thus, none of them has been shown to be involved in DNA replication.

INITIATION OF DNA REPLICATION

It is not known whether initiation of replication occurs at specific sites on chromosomal DNA molecules. Data compatible with the existence of specific replication-origin sequences come from mutagenesis experiments with synchronized cells (Burke and Fangman 1975; Kee and Haber 1975). Different genes were shown to be maximally sensitive to mutagenesis by nitrosoguanidine, a mutagen that acts preferentially at replication forks during limited intervals of the S phase. Further information about initiation of replication comes from studies of chromosomal rRNA genes, the multiple-copy extrachromosomal 2μ DNA plasmid, and mitochondrial DNA. In addition, cloned chromosomal sequences that appear to promote autonomous replication of plasmids within yeast cells provide important new approaches to the study of replication initiation.

Autonomous Replicators

Yeast can be transformed with recombinant DNA plasmids containing a selectable yeast gene (Beggs 1978; Hinnen et al. 1978). Two fundamentally different types of transformation have been described (Struhl et al. 1979). Most plasmids with fragments of chromosomal DNA transform at low frequency (1–10 colonies/μg of DNA). The transformed cells contain the plasmid DNA integrated into the chromosome in a relatively stable manner (1% loss after 15 generations; Struhl et al. 1979). In contrast, high-frequency transformation (10^3–10^5 colonies/μg of DNA) is observed with some yeast sequences (Hsaio and Carbon 1979; Struhl et al. 1979). In these cases, the DNA is found as supercoiled, extrachromosomal circles that are highly unstable. For example, plasmids containing a 1.4-kbp EcoRI fragment with the TRP1 gene are lost from 18% of haploid (6% of diploid) cells after one generation of growth in nonselective media (Kingsman et al. 1979). To explain the properties of DNA sequences that transform at high frequency, it has been suggested that they contain replicator sequences (initiation sites for DNA synthesis) and are therefore capable of autonomous replication (Hsaio and Carbon 1979; Struhl et al. 1979). Alternate interpretations, such as integration into the chromosome followed by replication and excision, seem unnecessarily complex.

The following observations are consistent with the interpretation that sequences capable of high-frequency transformation are replicators:

1. Extrachromosomal copies of plasmids that transform at high frequency are always found in transformed cells.
2. A high frequency of transformation is also observed with fragments of 2μ DNA (Beggs 1978), a DNA clearly capable of self-replication in yeast.
3. Sequences capable of high-frequency transformation occur about once in 30–40 kbp of chromosomal DNA (Beach et al. 1980; Chan and Tye 1980). This frequency is very similar to the average spacing of replication origins (once in 36 kbp) detected in chromosomal DNA by electron microscopy (Newlon and Burke 1980).
4. The sequence associated with the TRP1 gene that promotes high-frequency transformation can act only in cis (Stinchcomb et al. 1979), does not replicate in the presence of α-factor nor in the absence of an active CDC7-gene product (R. Hice and W. Fangman, unpubl.), and is found in the nucleus (Kingsman et al. 1979). These properties are among those expected for a chromosomal origin of replication.

5. High-frequency transformation is conferred by a $\leq$ 800-bp region of the 1.4-kbp *TRP1* DNA fragment (Stinchcomb et al. 1979) and by $\leq$ 350 bp of the 6-kbp 2μ DNA (Broach and Hicks 1980). The subset of sequences with replicator activity in both DNAs is preferentially used for initiation of DNA synthesis in vitro (Scott 1980; S. Celniker and J. Campbell, unpubl.).

All of these observations strongly suggest that the sequences identified by high-frequency transformation act as origins for the replication of the plasmids containing them and, by inference, that initiation of replication in chromosomal DNA occurs at specific sites. However, it is important to demonstrate unequivocally that the sequences responsible for autonomous replication of recombinant DNA plasmids are the same sequences that function as origins of replication in intact chromosomal DNA molecules.

Plasmids that transform at high frequencies can contain either single-copy or repetitive yeast sequences. The replicator sequence in the *TRP1* plasmid is detected only once in the yeast genome. It is estimated that it could share a maximum of 20 contiguous bases with other origin sequences (Stinchcomb et al. 1979). In another study, six out of ten plasmids capable of high-freqency transformation contained only single-copy DNA (Chan and Tye 1980). However, four plasmids contained sequences present in multiple copies in the yeast genome. Three of these shared a common sequence that is present 10–15 times in yeast chromosomal DNA. This sequence correlates with replicator activity: Thirteen out of thirteen recombinant plasmids, chosen only by their ability to cross-hybridize with this sequence, were also capable of high-frequency transformation (Chan and Tye 1980). These results suggest that some chromosomal sequences responsible for high-frequency transformation are present in multiple copies in the genome and that others are either unique or share limited homology with other replicator sequences. Replicator-sequence variation may account for differences in efficiency of transformation, copy number, and plasmid stability exhibited by different plasmids. The presence of bacterial DNA and/or total plasmid size also may be important. A 1.4-kbp yeast plasmid has an increased copy number and greater stability in transformed cells compared to the 5.8-kbp bacteria-yeast hybrid plasmid from which it was derived (J. Scott, pers. comm.).

A fraction of DNA fragments from a wide variety of eukaryotes (*N. crassa, Dictyostelium discoideum, Caenorhabditis elegans, Drosophila melanogaster, Zea mays*), but not *E. coli*, also exhibit high-frequency transformation in yeast (Stinchcomb et al. 1980). These observations raise the possibility that replication origins are similar or identical in all eukaryotes. However, neither *Xenopus laevis* amplified rDNA, which presumably carries an origin for rolling-circle replication, nor the SV40 origin of replication can promote autonomous replication in yeast (Zakian 1981). These failures may be the consequence of atypical replication requirements for these two DNAs. In contrast, the origin regions from both the extrachromosomal rDNA of *Tetrahymena thermophila* (G. Kiss and R. Pearlman, in prep.) and mitochondrial DNA from *X. laevis* (Zakian 1981) do promote high-frequency transformation and extrachromosomal replication of plasmid DNAs.

The identification of sequences from yeast and other eukaryotes that are capable of autonomous replication in yeast opens up new approaches to the study of eukaryotic chromosomal DNA replication. For example, replicator sequences can be altered in vitro and reintroduced into yeast to locate the sequences necessary for proper initiation. Mutants in genes required for activation of specific chromosomal DNA replication origins might be identified by their inability to maintain these plasmids. Replicator plasmids can also serve as defined templates for in vitro replication studies. By analogy with work in prokaryotes, these small DNAs should be valuable for identifying and isolating factors required for chromosomal DNA replication. Finally, replicator sequences should be useful for learning about controls over initiation of DNA replication, because extra copies and novel arrangements of replicator sequences can be created. Transformed cells can contain many extra copies of a replicator sequence in extrachromosomal form or extra copies integrated into chromosomal DNA. Integration can occur into homologous DNA, producing a tandem duplication of the replicator sequence (Hsaio and Carbon 1979), and into nonhomologous DNA, producing an extra origin in a foreign location (Stinchcomb et al. 1979).

rRNA Genes

The rRNA genes are replicated continuously throughout the S phase in cells synchronized either by a starvation-refeeding procedure (Gimmler and Schweizer 1972) or by arrest in G_1 phase in a *cdc7* strain (Brewer et al. 1980). In contrast, single-copy genes appear to replicate during discrete, subintervals of the S phase (Burke and Fangman 1975; Kee and Haber 1975). The rDNA does not replicate in cells arrested in the G_1 phase of the cell cycle and appears to double by synthesis throughout the S phase (Brewer et al. 1980). Thus, replication of rDNA is probably limited to the S phase. At the replication fork rate observed in synchronous cultures produced by G_1 arrest (3.6 kbp/min; Rivin and Fangman 1980b), it would take about 2 minutes to replicate a single 8.4-kbp rDNA unit. Since S phase is about 75 minutes in these cells and rDNA replication continues throughout this entire period, rDNA replication cannot occur by the simultaneous activation of origins in each repeat unit. Staggered activation of origins in each repeat unit throughout the S phase is also incompatible with these data unless repeats also contain termination sites for DNA replication. Thus, it seems unlikely that rDNA replication occurs by activation of an origin in each repeat unit. A few origins activated early in S phase would be adequate for replication of the entire rRNA gene cluster. These origins could occur in flanking non-rDNA sequences and/or in one or a few repeat units.

In cells transformed with recombinant plasmids containing rDNA, the plasmid is always found integrated into the rDNA (Szostak and Wu 1979; Petes 1980). These plasmids transform at high frequency relative to nonreplicator unique-sequence DNA, as might be expected from the large amounts of homologous DNA provided by the 120 copies of the rRNA genes. However, the spacer (noncoding) region of at least one repeat unit transforms at higher frequencies than the rRNA coding region (10^4 vs. 10^2 transformants/μg of DNA; Szostak and Wu 1979). These transformants are transiently unstable (Szostak and Wu 1979), and extrachromosomal plasmid circles can be detected (J. Szostak, pers. comm.). Therefore, the high-frequency transformation of the spacer region is probably a consequence of its containing a replicator sequence. Another rDNA repeat does not appear to contain a replicator because extrachromosomal copies of the plasmid are never detected in transformed cells (T. Petes, pers. comm.). These data are consistent with some but not all repeat units having an origin for DNA replication.

2μ DNA

The plasmid called 2μ DNA is a 6-kbp circular molecule with two inverted repeats of 600 bp that are separated from each other by about 180 degrees (Beggs et al. 1976; Guerineau et al. 1976; Hollenberg et al. 1976; Cameron et al. 1977; Gubbins et al. 1977; Livingston and Klein 1977). There are approximately 50 copies per haploid genome ($\sim 2\%$ of cellular DNA) (Clark-Walker and Miklos 1974), but systematic studies of copy number in different strains or under different growth conditions have not been carried out. Although 2μ DNA has the same buoyant density in CsCl as chromosomal DNA, it is readily purified as a covalently closed circular molecule by density banding in CsCl-ethidium-bromide gradients (e.g., Zakian et al. 1979) and, because of its small size, can be selectively extracted from lysates (Livingston and Kupfer 1977). 2μ DNA is organized into nucleosomes (Livingston and Hahne 1979; Nelson and Fangman 1979) with a repeat length indistinguishable from that of chromosomal DNA (Nelson and Fangman 1979). Early results suggested that 2μ DNA had a cytoplasmic location (Clark-Walker 1972; Livingston 1977), but observations on the chromatin structure and replication properties of 2μ DNA (see below), as well as the direct analysis of nuclei (C. Saunders et al., pers. comm.), indicate that 2μ DNA is in the nucleus.

Control of the replication of the multiple-copy 2μ DNA plasmid is similar to that for chromosomal DNA. Its replication is blocked when cells are arrested in the G_1 phase with α-factor or by mutations in *CDC4, CDC7,* or *CDC28* (Livingston and Kupfer 1977). The gene products of *CDC8* and *CDC21* (Petes and Williamson 1975b; Livingston and Kupfer 1977) are also required for replication of 2μ DNA. In synchronized cells, replication of 2μ DNA is limited to the S phase, with the majority of replication occurring early in the synthetic period (Zakian et al. 1979). Density-transfer experiments show that each molecule of 2μ DNA replicates once and only once during the S phase (Zakian et al. 1979). The activation of origins once per cell cycle is similar to chromosomal DNA but contrasts with other multiple-copy plasmids. Bacterial plasmids (Rownd 1969; Bazaral and Helinski 1970; Andresdottier and Masters 1978; Gustafsson et al. 1978), mammalian mitochondrial DNA (Bogenhagen and Clayton 1977), and extrachromosomal rDNA molecules of *Physarum* (Zellweger et al. 1972; Newlon et al. 1973; Vogt and Braun 1977) are not limited to a single round of replication in a cell cycle, nor is replication of the eukaryotic DNAs limited to the S phase. Electron microscopy of replicating 2μ DNA has shown that replication begins only once on a DNA molecule (Petes and Williamson 1975b) and appears to be bidirectional since single-strand "whiskers" were seen at both ends of replication bubbles (V. Zakian, unpubl.; see Fig. 5).

Plasmids containing fragments of 2μ DNA transform cells at high frequency (Beggs 1978). The replicator acivity has been limited to a single *cis*-acting region of about 350 bp (Storms et al. 1979; Struhl et al. 1979; Broach and Hicks 1980). Since the initiation site has not been mapped on intact 2μ

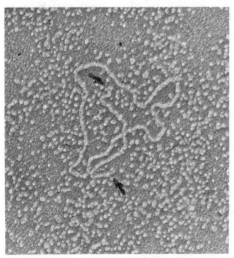

Figure 5 2μ DNA replication intermediate. Molecules were isolated from cells entering a synchronous S phase in the presence of hydroxyurea. The short branch at each fork (→) is interpreted to be a single strand of DNA produced by branch migration of a discontinuously replicated molecule. (Adapted from V. Zakian, unpubl.)

DNA, it is not known whether this sequence is the same as that normally used for replication. A second *trans*-acting region of the 2μ plasmid is also required for its efficient replication (Broach and Hicks 1980). In vitro replication systems using 2μ DNA as template (Jazwinski and Edelman 1979) should permit isolation of factors required for replication of both this plasmid and chromosomal DNA.

Mitochondrial DNA

On the basis of renaturation analysis and restriction endonuclease mapping, mitochondrial DNA is a molecule of approximately 75 kbp (see review by Locker et al. 1979). The physical map is circular, although only fragmented mitochondrial DNA is usually isolated. The number of mitochondrial DNA molecules per haploid genome ranges from about 10 to 40 (5–20% of cellular DNA). Copy number varies from strain to strain with carbon source and with growth conditions (Fukuhara 1969; Williamson 1970; Brewer and Fangman 1981). Since mitochondrial DNA is AT-rich (82%; Bernardi et al. 1972) compared with chromosomal DNA (60%), the two DNAs are readily separated by equilibrium density banding in CsCl ($\Delta\rho$ 0.016 g/cm³; Corneo et al. 1966) or in cesium gradients containing certain fluorescent dyes (Williamson and Fennell 1975). An abundant histonelike protein has been isolated from yeast mitochondria (Caron et al. 1979).

The replication of mitochondrial DNA can be uncoupled from that of chromosomal DNA in several ways. It is not limited to the S phase but occurs throughout the cell cycle (Williamson and Moustacchi 1971; Sena et al. 1975); it continues in the presence of α-factor (Petes and Fangman 1973; Cryer et al. 1974) and in the absence of both cytoplasmic and mitochondrial protein synthesis (Grossman et al. 1969). Unlike chromosomal DNA, replication of mitochondrial DNA does not require an active *CDC28-*, *CDC4-*, or *CDC7-* gene product (Cottrell et al. 1973; Newlon and Fangman 1975). However, two *CDC* genes essential for ongoing replication of chromosomal DNA (*CDC8* and *CDC21*) are also required for the synthesis of mitochondrial DNA (Wintersberger et al. 1974; Newlon and Fangman 1975).

Little is known at the molecular level about the replication of mitochondrial DNA. Since it is difficult to isolate intact molecules, it has been impossible to examine replicating intermediates by electron microscopy. A high rate of recombination (Dujon et al. 1974) has prevented determining whether all molecules replicate each cell cycle. A gradual shift in bouyant density rather than the appearance of discrete density classes is seen in density-transfer experiments (Williamson and Fennell 1974).

Inferences about replication origins have been made from the analysis of mitochondrial DNA from petite (respiratory-deficient) strains. Mitochondrial DNA from petite strains consists of tandem repeats of a subset (0.1–80%) of the sequences present in the grande (wild-type) mitochondrial genome (Borst and Grivell 1978; Locker et al. 1979). The amount of mitochondrial DNA in petite and wild-type cells is similar; thus, the concentration of the amplified sequence in a petite strain can be up to 10^3 times its concentration in wild-type cells. Hypersuppressive petites are strains in which the petite phenotype is selectively transmitted to $\geq 90\%$ of the vegetative diploid progeny after mating with wild-type cells. The DNAs from different hypersuppressive petites share a common sequence even though these petite strains can arise from different regions of the mitochondrial genome (De Zamaroczy et al. 1979; Blanc and Dujon 1980). It has been suggested that the common sequences in different hypersuppressive strains are origins for DNA replication (De Zamaroczy et al. 1979; Blanc and Dujon 1980). This interpretation would mean that mitochondrial DNA from hypersuppressive strains contains at least 100 origins per DNA molecule (Blanc and Dujon 1980). If multiple origins make it more likely that a DNA molecule will replicate, they could explain its selective transmission in crosses with wild-type cells. If the common sequences detected in hypersuppressive strains are origins of replication, wild-type mitochondrial DNA must contain at least three such sequences, since hypersuppressive strains can arise from at least three different regions of the mitochondrial genome. Furthermore, these sequences must be preferred but not required sites for initiation of replication since there are nonsuppressive petite strains whose DNAs lack them.

RADIOACTIVE LABELING, DENSITY LABELING, AND CELL SYNCHRONIZATION

Since yeast does not possess a thymidine kinase (Grivell and Jackson 1968), most DNA radioactive labeling, has employed the nonspecific nucleic acid precursors adenine and uracil. In spite of the 100-fold excess of RNA compared with DNA, this limitation has not been a serious one for many areas of study because RNA radioactive label can be efficiently removed by alkali or RNase treatments. However, the use of general nucleic acid precursors has limited studies on the details of in vivo replication, especially the detection of primers and discontinuous chains.

Mutants are available that permit specific radioactive labeling of yeast DNA. *tup* mutations allow yeast to take up TMP from the growth medium, and *tmp1* mutants, which are deficient in thymidylate synthetase (Bisson and Thorner 1977), have a requirement for TMP that can be met in the *tup* background (Brendel et al. 1975; Wickner 1975). The *tup* and *tmp1* mutations often result in the production of petite strains at high frequency and in poor cell viability. Moreover, incorporation of TMP by some *tup* strains occurs preferentially into mitochondrial DNA (Cryer et al. 1974). Therefore, these strains have not been used extensively for DNA structure and replication studies. However, a strain containing a mutation in the gene *TUP7* remains grande, has good viability, and incorporates TMP efficiently (50–60% substitution) (L. Bisson and J. Thorner, pers. comm.).

In vivo replication experiments involving dense isotope transfers have successfully employed $^{15}NH_4^+$ alone (e.g., Williamson and Fennell 1974; Sena et al. 1975), or $^{15}NH_4^+$ and [^{13}C]glucose (e.g., Newlon et al. 1974; Zakian et al. 1979). However, the ^{15}N-^{13}C combination gives a larger density difference in CsCl and allows excellent resolution of fully dense and hybrid density DNA with a verticle tube rotor (Brewer et al. 1980). 5-Bromo-2'-deoxyuridine-5'-monophosphate (BdUMP) has also been used with a *tup tmp1* strain (Leff and Eccleshall 1978).

Molecular analysis of the S phase depends upon the ability to isolate S-phase cells or to obtain homogeneous populations that undergo a synchronous S phase during subsequent incubation. Physical selection of a subset of cells from asynchronous populations has been accomplished by equilibrium density banding (Hartwell 1970; Shulman et al. 1973), zonal sedimentation (Sebastian et al. 1971), and centrifugal elutriation (Gordon and Elliott 1977; Elliott and McLaughlin 1978). Zonal sedimentation and centrifugal elutriation have the advantage that cells representing all stages of the cycle can be isolated. Physiological synchronization has been accomplished by cycles of carbon starvation and refeeding, followed by centrifugal enrichment for a population of small G_1-phase cells (Williamson and Scopes 1962; Williamson 1973). A double G1-phase block, involving arrest with α-factor followed by incubation at the restrictive temperature with a *cdc7* mutant, results in a synchronous S phase when cells are returned to the permissive temperature (Newlon et al. 1974; Zakian et al. 1979; Rivin and Fangman 1980a).

ACKNOWLEDGMENTS

We thank Dr. B. J. Brewer for useful suggestions about the contents and organization of this article. Our research is supported by grants from the National Institutes of Health and the American Cancer Society.

REFERENCES

Aarstad, K. and T.B Øyen. 1975. On the distribution of 5S RNA cistrons in the genome of *Saccharomyces cerevisiae. FEBS Lett.* **51**: 227.

Abelson, J.A. 1979. RNA processing and the intervening sequence problem. *Annu. Rev. Biochem.* **48**: 1035.

Andresdottier, V. and M. Masters. 1978. Evidence that F'lac replicates asynchronously during the cell cycle of *Escherichia coli* B/r. *Mol. Gen. Genet.* **163**: 205.

Barford, J.B. and R.J. Hall. 1976. Estimation of the length of cell cycle phases from asynchronous cultures of *Saccharomyces cerevisiae. Exp. Cell Res.* **102**: 276.

Bateman, A.J. 1975. Simplification of palindromic telomere theory. *Nature* **253**: 379.

Bazaral, M. and D.R. Helinski. 1970. Replication of a bacterial plasmid and an episome in *Escherichia coli. Biochemistry* **9**: 399.

Beach, D., M. Piper, and S. Shall. 1980. Isolation of chromosomal origins of replication in yeast. *Nature* **284**: 185.

Beckmann, J.S., P.F. Johnson, and J. Abelson. 1977. Cloning of yeast transfer RNA genes in *Escherichia coli*. *Science* **196**: 205.

Beggs, J.D. 1978. Transformation of yeast by a replicating hybrid plasmid. *Nature* **275**: 104.

Beggs, J.D., M. Guerineau, and J.F. Atkins. 1976. A map of the restriction targets in yeast 2 micron plasmid DNA cloned on bacteriophage lambda. *Mol. Gen. Genet.* **148**: 287.

Bell, G.I., L.J. De Gennaro, D.H. Gelfand, R.J. Bishop, P. Valenzuela, and W.J. Rutter. 1977. Ribosomal RNA genes of *Saccharomyces cerevisiae*. I. Physical map of the repeating unit and location of the regions coding for 5S, 5.8S, 18S, and 25S ribosomal RNAs. *J. Biol. Chem.* **252**: 8118.

Bernardi, G., G. Piperno, and G. Fonty. 1972. The mitochondrial genome of wild type yeast cells. I. Preparation and heterogeneity of mitochondrial DNA. *J. Mol. Biol.* **65**: 173.

Bisson, L. and J. Thorner. 1977. Thymidine 5′-monophosphate-requiring mutants of *Saccharomyces cerevisiae* are deficient in thymidylate synthetase. *J. Bacteriol.* **132**: 44.

Blamire, J., D.R. Cryer, D.B. Finkelstein, and J. Marmur. 1972. Sedimentation properties of yeast nuclear and mitochondrial DNA. *J. Mol. Biol.* **67**: 11.

Blanc, H. and B. Dujon. 1980. Replicator regions of the yeast mitochondrial DNA responsible for suppressiveness. *Proc. Natl. Acad. Sci.* **77**: 3942.

Bogenhagen, D. and D.A. Clayton. 1977. Mouse L cell mitochondrial DNA molecules are selected randomly for replication throughout the cell cycle. *Cell* **11**: 719.

Borst, P. and L.A. Grivell. 1978. The mitochondrial genome of yeast. *Cell* **15**: 705.

Brandriss, M.C., J.W. Stewart, F. Sherman, and D. Botstein. 1976. Substitution of serine caused by a recessive lethal suppressor in yeast. *J. Mol. Biol.* **102**: 467.

Brandt, W.F. and C. von Holt. 1976. The occurrence of histone H3 and H4 in yeast. *FEBS Lett.* **65**: 386.

Brendel, M., W.W. Fath, and W. Laskowski. 1975. Isolation and characterization of mutants of *Saccharomyces cerevisiae* able to grow after inhibition of dTMP synthesis. *Methods Cell Biol.* **11**: 287.

Brewer, B.J. and W.L. Fangman. 1980. Preferential inclusion of extrachromosomal genetic elements in yeast meiotic spores. *Proc. Natl. Acad. Sci.* **77**: 5380.

Brewer, B.J., V.A. Zakian, and W.L. Fangman. 1980. Replication and meiotic transmission of the yeast ribosomal RNA genes. *Proc. Natl. Acad. Sci.* **77**: 6739.

Broach, J.R. and J.B. Hicks. 1980. Replication and recombination functions associated with the yeast plasmid, 2μ circle. *Cell* **21**: 501.

Burke, W and W.L. Fangman. 1975. Temporal order in yeast chromosome replication. *Cell* **5**: 263.

Byers, B. and L. Goetsch. 1974. Duplication of spindle plaques and integration of the yeast cell cycle. *Cold Spring Harbor Symp. Quant. Biol.* **38**: 123.

———. 1975. Electron microscopic observations on the meiotic karyotype of diploid and tetraploid *Saccharomyces cerevisiae*. *Proc. Natl. Acad. Sci.* **72**: 5056.

Cameron, J.R., E.Y. Loh, and R.W. Davis. 1979. Evidence for transposition of dispersed repetitive DNA families in yeast. *Cell* **16**: 739.

Cameron, J.R., P. Philippsen, and R.W. Davis. 1977. Analysis of chromosmal integration and deletions of yeast plasmids. *Nucleic Acids Res.* **4**: 1429.

Caron, F., C. Jacq, and J. Rouvière-Yaniv. 1979. Characterization of a histone-like protein extracted from yeast mitochondria. *Proc. Natl. Acad. Sci.* **76**: 4265.

Carter, B.L.A. and M.N. Jagadish. 1978. Control of cell division in the yeast *Saccharomyces cerevisiae* cultured at different growth rates. *Exp. Cell Res.* **112**: 373.

Chan, C.S.M. and B.-K. Tye. 1980. Autonomously replicating sequences in *Saccharomyces cerevisiae*. *Proc. Natl. Acad. Sci.* **77**: 6329.

Chang, L.M.S. 1976. Phylogeny of DNA polymerase-β. *Science* **191**: 1183.

———. 1977. DNA polymerases from baker's yeast. *J. Biol. Chem.* **252**: 1873.

Cihlar, R.L. and P.S. Sypherd. 1980. The organization of the ribosomal RNA genes in the fungus *Mucor racemosus*. *Nucleic Acids Res.* **8**: 793.

Clark, L. and J. Carbon. 1980. Isolation of a yeast centromere and construction of functional small circular chromosomes. *Nature* **287:** 504.

Clark-Walker, G.D. 1972. Isolation of circular DNA from a mitochondrial fraction of yeast. *Proc. Natl. Acad. Sci.* **69:** 388.

Clark-Walker, G.D. and A.A. Azad, 1980. Hybridizable sequences between cytoplasmic ribosomal RNAs and 3 micron circular DNAs of *Saccharomyces cerevisiae* and *Torulopsis glabrata. Nucleic Acids Res.* **8:** 1009.

Clark-Walker, G.D. and G.L.G. Miklos. 1974. Localization and quantification of circular DNA in yeast. *Eur. J. Biochem.* **41:** 359.

Cockburn, A.F., M.J. Newkirk, and R.A. Firtel. 1976. Organization of the ribosomal RNA genes of *Dictyostelium discoideum*: Mapping of the nontranscribed spacer regions. *Cell* **9:** 605.

Corneo, G., C. Moore, D.R. Sanadi, L. Grossman, and J. Marmur. 1966. Mitochondrial DNA in yeast and some mammalian species. *Science* **151:** 687.

Cottrell, S., M. Rabinowitz, and G.S. Getz. 1973. Mitochondrial deoxyribonucleic acid synthesis in a temperature-sensitive mutant of deoxyribonucleic acid replication of *Saccharomyces cerevisiae. Biochemistry* **12:** 4374.

Cramer, J.H., M.M. Bhargava, and H.O. Halvorson. 1972. Isolation and characterization of γ DNA of *Saccharomyces cerevisiae. J. Mol. Biol.* **71:** 11.

Cramer, J.H., F.W. Farrelly, and R.H. Rownd. 1976. Restriction endonuclease analysis of ribosomal DNA from *Saccharomyces cerevisiae. Mol. Gen. Genet.* **148:** 233.

Cryer, D.R., C.D. Goldthwaite, S. Zinker, K.-B. Lam, E. Storm, R. Hirschberg, J. Blamire, D.B. Finkelstein, and J. Marmur. 1974. Studies on nuclear and mitochondrial DNA of *Saccharomyces cerevisiae. Cold Spring Harbor Symp. Quant. Biol.* **38:** 17.

Culotti, J. and L.H. Hartwell. 1971. Genetic control of the cell division cycle in yeast. III. Seven genes controlling nuclear division. *Exp. Cell Res.* **67:** 389.

De Lucia, P. and J. Cairns. 1969. Isolation of an *E. coli* strain with a mutation affecting DNA polymerase. *Nature* **224:** 1164.

De Zamaroczy, M., G. Baldacci, and G. Bernardi. 1979. Putative origins of replication in the mitochondrial genome of yeast. *FEBS Lett.* **108:** 429.

Dujon, B., P.P. Slonimski, and L. Weill. 1974. Mitochondrial genetics. IX. A model for recombination and segregation of mitochondrial genomes in *Saccharomyces cerevisiae. Genetics* **78:** 415.

Durnford, J.M. and J.J. Champoux. 1978. The DNA untwisting enzyme from *Saccharomyces cerevisiae. J. Biol. Chem.* **253:** 1086.

Elliott, S.G. and C.S. McLaughlin. 1978. Rate of macromolecular synthesis through the cell cycle of the yeast *Saccharomyces cerevisiae. Proc. Natl. Acad. Sci.* **75:** 4384.

Etcheverry, T., D. Colby, and C. Guthrie. 1979. A precursor to a minor species of yeast tRNA[Ser] contains an intervening sequence. *Cell* **18:** 11.

Fabre, F. and H. Roman. 1979. Evidence that a single DNA ligase is involved in replication and recombination in yeast. *Proc. Natl. Acad. Sci.* **76:** 4586.

Farabaugh, P.J. and G.R. Fink. 1980. Insertion of the eukaryotic transposable element Ty1 creates a 5-base pair duplication. *Nature* **286:** 352.

Federoff, N.V. 1979. On spacers. *Cell* **16:** 697.

Finkelstein, D.B., J. Blamire, and J. Marmur. 1972. Location of ribosomal RNA cistrons in yeast. *Nat. New Biol.* **240:** 279.

Finnegan, D.J., G.M. Rubin, M.W. Young, and D.S. Hogness. 1978. Repeated gene families in *Drosophila melanogaster. Cold Spring Harbor. Symp. Quant. Biol.* **42:** 1053.

Forte, M.A. and W.L. Fangman. 1976. Naturally occurring cross-links in yeast chromosomal DNA. *Cell* **8:** 425.

———. 1979. Yeast chromosomal DNA molecules have strands which are cross-linked at their termini. *Chromosoma* **72:** 131.

Franco, L., E.W. Johns, and J.M. Navlet. 1974. Histones from baker's yeast. Isolation and fractionation. *Eur. J. Biochem.* **45:** 83.

Fukuhara, H. 1969. Relative proportions of mitochondrial and nuclear DNA in yeast under various conditions of growth. *Eur. J. Biochem.* **11**: 135.

Gafner, J. and P. Philippsen. 1980. The yeast transposon Ty1 generates duplications of target DNA on insertion. *Nature* **286**: 414.

Game, J.C., L.H. Johnston, and R.C. von Borstel. 1979. Enhanced mitotic recombination in a ligase defective mutant of the yeast *Saccharomyces cerevisiae*. *Proc. Natl. Acad. Sci.* **76**: 4589.

Gellert, M. and M.L. Bullock. 1970. DNA ligase mutants of *Escherichia coli*. *Proc. Natl. Acad. Sci.* **67**: 1580.

Gimmler, G.M. and E. Schweizer. 1972. rDNA replication in a synchronized culture of *Saccharomyces cerevisiae*. *Biochem. Biophys. Res. Commun.* **46**: 143.

Goodman, H.M., M.V. Olson, and B.D. Hall. 1977. Nucleotide sequence of a mutant eukaryotic gene: The yeast tyrosine-inserting ochre suppressor SUP4-o. *Proc. Natl. Acad. Sci.* **74**: 5453.

Goodwin, G.H., L. Woodhead, and E.W. Johns. 1977. The presence of high mobility group non-histone chromatin proteins in isolated nucleosomes. *FEBS Lett.* **73**: 85.

Gordon, C.N. and S.G. Elliott. 1977. Fractionation of *Saccharomyces cerevisiae* cell populations by centrifugal elutriation. *J. Bacteriol.* **129**: 97.

Grivell, A.R. and J.F. Jackson. 1968. Thymidine kinase: Evidence for its absence from *Neurospora crassa* and some other microorganisms, and the relevance of this to the specific labelling of deoxyribonucleic acid. *J. Gen. Microbiol.* **54**: 307.

Grossman, L.J., E.S. Goldring, and J. Marmur. 1969. Preferential synthesis of yeast mitochondrial DNA in the absence of protein synthesis. *J. Mol. Biol.* **46**: 367.

Gubbins, E.J., C.S. Newlon, M.D. Kann, and J.E. Donelson. 1977. Sequence organization and expression of a yeast plasmid DNA. *Gene* **1**: 185.

Guerineau, M., C. Granchamp, and D. Slonimski. 1976. Circular DNA of a yeast episome with two inverted repeats: Structural analysis by a restriction enzyme and electron microscopy. *Proc. Natl. Acad. Sci.* **73**: 3030.

Gustafsson, P., K. Nordstrom, and J.W. Perram. 1978. Selection and timing of replication of plasmids Rldrd-19 and F' lac in *Escherichia coli*. *Plasmid* **1**: 187.

Guthrie, J. and J.A. Abelson. 1982. Organization and expression of tRNA genes in *Saccharomyces cerevisiae*. In *The molecular biology of the yeast* Saccharomyces. II. *Metabolism and gene expression* (ed. J. Strathern et al.), Cold Spring Harbor Laboratory, Cold Spring Harbor, New York. (In press.)

Hall, B.D. and A. Sentenac. 1982. RNA polymerases and transcription. In *The molecular biology of the yeast* Saccharomyces. II. *Metabolism and gene expression* (ed. J. Strathern et al.), Cold Spring Harbor Laboratory. Cold Spring Harbor, New York. (In press.)

Hartwell, L.H. 1970. Periodic density fluctuation during the yeast cell cycle and the selection of synchronous cultures. *J. Bacteriol.* **104**: 1280.

———. 1973. Three additional genes required for deoxyribonucleic acid synthesis in *Saccharomyces cerevisiae*. *J. Bacteriol.* **115**: 966.

———. 1976. Sequential function of gene products relative to DNA synthesis in the yeast cell cycle. *J. Mol. Biol.* **104**: 803.

———. 1978. Cell division from a genetic perspective. *J. Cell Biol.* **77**: 627.

Hartwell, L.H., J. Culotti, J.R. Pringle, and B.J. Reid. 1974. Genetic control of the cell division cycle in yeast. *Science* **183**: 46.

Hattman, S., C. Kenny, L. Berger, and K. Pratt. 1978. Comparative study of DNA methylation in three unicellular eucaryotes. *J Bacteriol.* **135**: 1156.

Hawthorne, D.C. and U. Leupold. 1974. Suppressor mutations in yeast. *Curr. Top. Microbiol. Immunol.* **64**: 1.

Hereford, L.H. and L.H. Hartwell. 1971. Defective DNA synthesis in permeabilized yeast mutants. *Nat. New Biol.* **234**: 171.

———. 1973. Role of protein synthesis in the replication of yeast DNA. *Nat. New Biol.* **244**: 129.

———. 1974. Sequential gene function in the initiation of *Saccharomyces cerevisiae* DNA synthesis. *J. Mol. Biol.* **84**: 445.

Hereford, L.M. and M. Rosbash. 1977. Number and distribution of polyadenylated RNA sequences in yeast. *Cell* **10**: 453.

Hereford, L.M., K. Fahrner, J. Woolford, Jr., M. Rosbash, and D.B. Kaback. 1979. Isolation of yeast histone genes H2A and H2B. *Cell* **18**: 1261.

Heywood, P. and P.T. Magee. 1976. Meiosis in protists. Some structural and physiological aspects of meiosis in algae, fungi, and protozoa. *Bacteriol. Rev.* **40**: 190.

Hinnen, A., J.B. Hicks, and G.R. Fink. 1978. Transformation of yeast. *Proc. Natl. Acad. Sci.* **75**: 1929.

Holland, J.P. and M.J. Holland. 1980. Structural comparison of two nontandemly repeated yeast glyceraldehyde-3-phosphate dehydrogenase genes. *J. Biol. Chem.* **255**: 2596.

Holland, M.J., G.L. Hager, and W.J. Rutter. 1977. Characterization of purified poly (adenylic acid)-containing messenger ribonucleic acid from *Saccharomyces cerevisiae*. *Biochemistry* **16**: 8.

Hollenberg, C.P., A. Degelmann, B. Kusterman-Kohn, and H.D. Royer. 1976. Characterization of 2- μm DNA of *Saccharomyces cerevisiae* by restriction enzyme analysis and integration in an *Escherichia coli* plasmid. *Proc. Natl. Acad. Sci.* **75**: 605.

Hsaio, C.-L. and J. Carbon. 1979. High frequency transformation of yeast by plasmids containing the cloned *ARG4* gene. *Proc. Natl. Acad. Sci.* **76**: 3829.

Javaherian, K., L.F. Liu, and J.C. Wang. 1978. Nonhistone proteins HMG1 and HMG2 change the DNA helical structure. *Science* **199**: 1345.

Jazwinski, S.M. and G.M. Edelman. 1979. Replication *in vitro* of the 2 μm DNA plasmid of yeast. *Proc. Natl. Acad. Sci.* **76**: 1223.

Johnston, G.C., R.A. Singer, S.O. Sharrow, and M.L. Slater. 1980. Cell division in the yeast *Saccharomyces cerevisiae* growing at different rates. *J. Gen. Microbiol.* **118**: 479.

Johnston, L.H. 1979. The DNA repair capability of *cdc9*, the *Saccharomyces cerevisiae* mutant defective in DNA ligase. *Mol. Gen. Genet.* **170**: 89.

Johnston, L.H. and J.C. Game. 1978. Mutants of yeast with depressed DNA synthesis. *Mol. Gen. Genet.* **161**: 205.

Johnston, L.H. and K.A. Nasmyth. 1978. *Saccharomyces cerevisiae* cell cycle mutant *cdc9* is defective in DNA ligase. *Nature* **274**: 891.

Johnston, L.H. and D.H. Williamson. 1978. An alkaline sucrose gradient analysis of the mechanism of nuclear DNA synthesis in the yeast *Saccharomyces cerevisiae*. *Mol. Gen. Genet.* **164**: 217.

Kaback, D.B. and N. Davidson. 1980. Organization of ribosomal RNA gene cluster in the yeast *Saccharomyces cerevisiae*. *J. Mol. Biol.* **138**: 745.

Kaback, D.B. and H.O. Halvorson. 1977. Magnification of genes coding for ribosomal RNA in *Saccharomyces cerevisiae*. *Proc. Natl. Acad. Sci.* **74**: 1177.

―――. 1978. Ribosomal DNA magnification in *Saccharomyces cerevisiae*. *J. Bacteriol.* **134**: 237.

Kaback, D.B., L.B. Angerer, and N. Davidson. 1979. Improved methods for the formation of R-loops. *Nucleic Acids Res.* **6**: 2499.

Kaback, D.B., M.M. Bhargava, and H.O. Halvoson. 1973. Location and arrangement of genes coding for ribosomal RNA in *Saccharomyces cerevisiae*. *J. Mol. Biol.* **79**: 735.

Kaback, D.B., H.O. Halvorson, and G.M. Rubin. 1976. Location and magnification of 5S RNA genes in *Saccharomyces cerevisiae*. *J. Mol. Biol.* **197**: 385.

Kedes, L.H. 1979. Histone genes and histone messengers. *Annu. Rev. Biochem.* **48**: 837.

Kee, S.G. and J.E. Haber. 1975. Cell cycle-dependent induction of mutations along a yeast chromosome. *Proc. Natl. Acad. Sci.* **72**: 1179.

Kingsman, A.J., L. Clarke, R.K. Mortimer, and J. Carbon. 1979. Replication in *Saccharomyces cerevisiae* of plasmid pBR313 carrying DNA from the yeast *trp-1* region. *Gene* **7**: 141.

Klein, H.L. and B. Byers. 1978. Stable denaturation of chromosomal DNA from *Saccharomyces cerevisiae* during meiosis. *J. Bacteriol.* **134**: 629.

Klein, H.L. and S.K. Welch. 1980. Inverted repeated sequences in yeast nuclear DNA. *Nucleic Acids Res.* **8**: 4651.

Knapp, G., R.C. Ogden, C.L. Peebles, and J. Abelson. 1979. Splicing of yeast tRNA precursors: Structure of the reaction intermediate. *Cell* **18**: 37.

Kramer, R.A., P. Philippsen, and R.W. Davis. 1978. Divergent transcription in the yeast ribosomal RNA coding region as shown by hybridization to separated strands and sequence analysis of cloned DNA. *J. Mol. Biol.* **123**: 405.

Lauer, G.O., T.M. Roberts, and L.C. Klotz. 1977. Determination of the nuclear DNA content of *Saccharomyces cerevisiae* and implications for the organization of DNA in yeast chromosomes. *J. Mol. Biol.* **114**: 507.

Leff, J. and T.R. Eccleshall. 1978. Replication of bromodeoxyuridylate-substituted mitochondrial DNA in yeast. *J. Bacteriol.* **135**: 436.

Lilley, D.M.J. and J.F. Pardon. 1979. Structure and function of chromatin. *Annu. Rev. Genet.* **13**: 197.

Livingston, D.M. 1977. Inheritance of the 2 μm DNA plasmid from *Saccharomyces. Genetics* **86**: 73.

Livingston, D.M. and S. Hahne. 1979. Isolation of a condensed, intracellular form of the 2 μm DNA plasmid of *Saccharomyces cerevisiae. Proc. Natl. Acad. Sci.* **76**: 3727.

Livingston, D.M. and H.L. Klein. 1977. DNA sequence organization of a yeast plasmid. *J. Bacteriol.* **129**: 472.

Livingston, D.M. and D.M. Kupfer. 1977. Control of *Saccharomyces cerevisiae* 2 μm DNA replication by cell division cycle genes that control nuclear DNA replication. *J. Mol. Biol.* **116**: 249.

Locker, J., A. Lewin, and M. Rabinowitz. 1979. The structure and organization of mitochondrial DNA from petite yeast. *Plasmid* **2**: 155.

Lohr, D. and L. Hereford. 1979. Yeast chromatin is uniformly digested by DNase I. *Proc. Natl. Acad. Sci.* **76**: 4285.

Lohr, D. and K.E. Van Holde. 1975. Yeast chromatin subunit structure. *Science* **188**: 165.

———. 1979. Organization of spacer DNA in chromatin. *Proc. Natl. Acad. Sci.* **76**: 6326.

Lohr, D., R.T. Kovacic, and K.E. Van Holde. 1977a. Quantitative analysis of the digestion of yeast chromatin by staphylococcal nuclease. *Biochemistry* **16**: 463.

Lohr, D., K. Tatchell, and K.E. Van Holde. 1977b. On the occurrence of nucleosome phasing in chromatin. *Cell* **12**: 829.

Lohr, D., J. Corden, K. Tatchell, R.T. Kovacic, and K.E. Van Holde. 1977c. Comparative subunit structure of HeLa, yeast, and chicken erythrocyte chromatin. *Proc. Natl. Acad. Sci.* **74**: 79.

Maizels, N. 1976. *Dictyostelium* 17S, 25S, and 5S rDNAs lie within a 38,000 base pair repeated unit. *Cell* **9**: 431.

Mardian, J.K.W. and I. Isenberg. 1978. Yeast inner histones and the evolutionary conservation of histone-histone interactions. *Biochemistry* **14**: 4304.

Meyerink, J.H., J. Klootwijk, R.J. Planta, A. van der Ende, and E.F.J. van Bruggen. 1979. Extrachromosomal circular ribosomal DNA in the yeast *Saccharomyces carlsbergensis. Nucleic Acids Res.* **7**: 69.

Moens, P.B. and E. Rapport. 1971. Spindles, spindle plaques, and meiosis in the yeast *Saccharomyces cerevisiae* (Hansen). *J. Cell Biol.* **50**: 344.

Moll, R. and E. Wintersberger. 1976. Synthesis of yeast histones in the cell cycle. *Proc. Natl. Acad. Sci.* **73**: 1863.

Morris, N.R. 1976. Nucleosome structure in *Aspergillus nidulans. Cell* **8**: 357.

Mortimer, R.K. and D.C. Hawthorne. 1975. Genetic mapping in yeast. *Methods Cell Biol.* **11**: 221.

Moustacchi, E. and D.H. Williamson. 1966. Physiological variations in satellite components of yeast DNA detected by density gradient centrifugation. *Biochem. Biophys. Res. Commun.* **23**: 56.

Nelson, D.A., W.R. Beltz, and R.L. Rill. 1977. Chromatin subunits from baker's yeast: Isolation and partial characterization. *Proc. Natl. Acad. Sci.* **74**: 1343.

Nelson, R.G. and W.L. Fangman. 1979. Nucleosome organization of the yeast 2 μm DNA plasmid: A eukaryotic minichromosome. *Proc. Natl. Acad. Sci.* **76**:6515.

Newlon, C.S. and W. Burke. 1980. Replication of small chromosomal DNAs in yeast. *ICN-UCLA Symp. Mol. Cell. Biol.* **19**:399.

Newlon, C.S. and W.L. Fangman. 1975. Mitochondrial DNA synthesis in cell cycle mutants of *Saccharomyces cerevisiae. Cell* **5**:423.

Newlon, C.S., G.E. Sonenshein, and C.E. Holt. 1973. Time of synthesis of genes for ribosomal ribonucleic acid in *Physarum. Biochemistry* **12**:2338.

Newlon, C.S., T.D. Petes, L.M. Hereford, and W.L. Fangman. 1974. Replication of yeast chromosomal DNA. *Nature* **247**:32.

Noll, M. 1976. Differences and similarities in chromatin structure of *Neurospora crassa* and higher eukaryotes. *Cell* **8**:349.

Oertel, W. and M. Goulian. 1979. Deoxyribonucleic acid synthesis in *Saccharomyces cerevisiae* cells permeabilized with ether. *J. Bacteriol.* **140**:333.

Olson, M.V., K. Loughney, and B.D. Hall. 1979a. Identification of the yeast DNA sequences that correspond to specific tyrosine-inserting nonsense suppressor loci. *J. Mol. Biol.* **132**:387.

Olson, M.V., B.D. Hall, J.R. Cameron, and R.W. Davis. 1979b. Cloning of the yeast tyrosine transfer RNA genes in bacteriophage lambda. *J. Mol. Biol.* **127**:285.

Øyen, T.D., G. Saelid, and G.V. Skuladottir. 1978. Study of a haploid strain with an unusually high rDNA content. III. Unequal meiotic segregation of the γ-DNA fraction. *Biochim. Biophys. Acta* **520**:88.

Petes, T.D. 1979. Yeast ribosomal DNA genes are located on chromosome XII. *Proc. Natl. Acad. Sci.* **76**:410.

———. 1980. Unequal mitotic recombination within tandem arrays of yeast ribosomal RNA genes. *Cell* **19**:765.

Petes, T.D. and D. Botstein. 1977. Simple Mendelian inheritance of the reiterated ribosomal DNA of yeast. *Proc. Natl. Acad. Sci.* **74**:5091.

Petes, T.D. and W.L. Fangman. 1972. Sedimentation properties of yeast chromosomal DNA. *Proc. Natl. Acad. Sci.* **69**:1188.

———. 1973. Preferential synthesis of yeast mitochondrial DNA in α-factor arrested cells. *Biochem. Biophys. Res. Commun.* **55**:603.

Petes, T.D. and C.S. Newlon. 1974. Structure of DNA in DNA replication mutants of yeast. *Nature* **251**:637.

Petes, T.D. and D.H. Williamson. 1975a. Fiber autoradiography of replicating yeast DNA. *Exp. Cell Res.* **95**:103.

———. 1975b. Replicating circular DNA molecules in yeast. *Cell* **4**:249.

Petes, T.D., B. Byers, and W.L. Fangman. 1974. Size and structure of yeast chromosmal DNA. *Proc. Natl. Acad. Sci.* **70**:3072.

Petes, T.D., L.M. Hereford, and K.G. Skryabin. 1978. Characterization of two types of yeast ribosomal RNA genes. *J. Bacteriol.* **134**:295.

Petes, T.D., S. Smolik-Utlaut, and T. Zamb. 1980. Genetic analysis of the repeating ribosomal genes of yeast. In *Molecular genetics in yeast, Alfred Benzon Symposium 16* (ed. D. von Wettstein et al.), Munksgaard, Copenhagen. (In press.)

Petes, T.D., C.S. Newlon, B. Byers, and W.L. Fangman. 1974. Yeast chromosomal DNA: Size, structure, and replication. *Cold Spring Harbor Symp. Quant. Biol.* **38**:9.

Philipps, G. and C. Gigot. 1977. DNA associated with nucleosomes in plants. *Nucleic Acids Res.* **4**:3617.

Philippsen, P., M.J. Thomas, R.A. Kramer, and R.W. Davis. 1978. Unique arrangement of coding sequences for 5S, 5.8S, 18S and 25S ribosomal RNA in *Saccharomyces cerevisiae* as determined by R-loop and hybridization analysis. *J. Mol. Biol.* **123**:387.

Piñon, R. 1978. Folded chromosomes in non-cycling yeast cells. Evidence for a characteristic g_0 form. *Chromosoma* **67**:263.

———. 1979a. Folded chromosomes in meiotic yeast cells: Analysis of early meiotic events. *J. Mol. Biol.* **129**:433.

————. 1979b. A probe into nuclear events during the cell cycle of *Saccharomyces cerevisiae:* Studies of folded chromosomes in *cdc* mutants which arrest in G1. *Chromosoma* **70:** 737.

Piñon, R. and D. Pratt. 1979. Folded chromosomes of mating-factor arrested yeast cells: Comparison with g_0 arrest. *Chromosoma* **73:** 117.

Piñon, R. and Y. Salts. 1977. Isolation of folded chromosomes from the yeast *Saccharomyces cerevisiae. Proc. Natl. Acad. Sci.* **74:** 2850.

Plevani, P., G. Badaracco, and L.M.S. Chang. 1980. Purification and characterization of two forms of DNA dependent ATPase from yeast. *J. Biol. Chem.* **255:** 4957.

Potter, S.S., W.J. Borein, Jr., P. Dunsmuir, and G.M. Rubin. 1979. Transposition of elements of the *412, copia* and *297* dispersed repeated gene families in *Drosophila. Cell* **17:** 415.

Prakash, L., D. Hinkle, and S. Prakash. 1979. Decreased UV mutagenesis in *cdc8*, a DNA replication mutant of *Saccharomyces cerevisiae. Mol. Gen. Genet.* **172:** 249.

Rasse-Messenguy, F. and G.R. Fink. 1973. Temperature-sensitive nonsense suppressors in yeast. *Genetics* **75:** 459.

Reed, S.I. 1980. The selection of *Saccharomyces cerevisiae* mutants defective in the start event of cell division. *Genetics* **95:** 561.

Rivin, C.J. and W.L. Fangman. 1980a. Cell cycle phase expansion in nitrogen-limited cultures of *Saccharomyces cerevisiae. J. Cell Biol.* **85:** 96.

————. 1980b. Replication fork rate and origin activation during the S phase of *Saccharomyces cerevisiae. J. Cell Biol.* **85:** 108.

Rothstein, R. 1979. Deletions of tyrosine tRNA gene in *S. cerevisiae. Cell* **17:** 185.

Rownd, R. 1969. Replication of a bacterial episome under relaxed control. *J. Mol. Biol.* **44:** 387.

Rubin, G.M. 1973. The nucleotide sequence of *Saccharomyces cerevisiae* 5.8S ribosomal ribonucleic acid. *J. Biol. Chem.* **248:** 3860.

Rubin, G.M. and J.E. Sulston. 1973. Physical linkage of the 5S cistrons to the 18S and 28S ribosomal RNA cistrons in *Saccharomyces cerevisiae. J. Mol. Biol.* **79:** 521.

Schild, D. and B. Byers. 1978. Meiotic effects of DNA-defective cell division cycle mutations of *Saccharomyces cerevisiae. Chromosoma* **70:** 109.

Schweizer, E., C. MacKechnie, and H.O. Halvorson. 1969. The redundancy of ribosomal and transfer RNA genes in *Saccharomyces cerevisiae. J. Mol. Biol.* **40:** 261.

Scott, J.F. 1980. Preferential utilization of a yeast chromosomal replication origin as template for enzymatic DNA synthesis. *ICN-UCLA Symp. Mol. Cell. Biol.* **19:** 379.

Sebastian, J., B.L.A. Carter, and H.O. Halvorson. 1971. Use of yeast populations fractionated by zonal centrifugation to study the cell cycle. *J. Bacteriol.* **108:** 1045.

Sena, E.P., J.W. Welch, H.O. Halvorson, and S. Fogel. 1975. Nuclear and mitochondrial deoxyribonucleic acid replication during mitosis in *Saccharomyces cerevisiae. J. Bacteriol.* **123:** 497.

Sherman, F. 1982. Suppression in the yeast *Saccharomyces cerevisiae.* In *The molecular biology of the yeast* Saccharomyces. II. *Metabolism and gene expression* (ed. J. Strathern et al.), Cold Spring Harbor Laboratory, Cold Spring Harbor, New York (In press.)

Shulman, R.W., L.H. Hartwell, and J.R. Warner. 1973. Synthesis of ribosomal proteins during the yeast cell cycle. *J. Mol. Biol.* **73:** 513.

Slater, M.L., S.O. Sharrow, and J.J. Gart. 1977. Cell cycle of *S. cerevisiae* in populations growing at different rates. *Proc. Natl. Acad. Sci.* **74:** 3850.

Sommer, A. 1978. Yeast chromatin: Search for histone H1. *Mol. Gen. Genet.* **161:** 323.

Spiker, S., J.K.W. Mardian, and I. Isenberg. 1978. Chromosomal HMG proteins occur in three eukaryotic kingdoms. *Biochem. Biophys. Res. Commun.* **82:** 129.

Stinchcomb, D.T., K. Struhl, and R.W. Davis. 1979. Isolation and characterization of a yeast chromosomal replicator. *Nature* **282:** 39.

Stinchcomb, D.T., M. Thomas, J. Kelly, E. Selker, and R.W. Davis. 1980. Eukaryotic DNA segments capable of autonomous replication in yeast. *Proc. Natl. Acad. Sci.* **77:** 4559.

Storms, R.K., J.G. McNeil, D.S. Khandekar, G. An, J. Parker, and J.D. Friesen. 1979. Chimeric plasmids for cloning of deoxyribonucleic acid sequences in *Saccharomyces cerevisiae. J. Bacteriol.* **140:** 73.

Strathern, J.N., C.S. Newlon, I. Herskowitz, and J.B. Hicks. 1979. Isolation of a circular derivative of yeast chromosome III: Implications for the mechanism of mating type interconversion. *Cell* **18:** 309.

Strobel, E., P. Dunsmuir, and G.M. Rubin. 1979. Polymorphism in the chromosomal locations of elements of *412, copia* and *297* dispersed repeated gene families in *Drosophila. Cell* **17:** 429.

Struhl, K., D.T. Stinchcomb, S. Scherer, and R.W. Davis. 1979. High-frequency transformation of yeast: Autonomous replication of hybrid DNA molecules. *Proc. Natl. Acad. Sci.* **76:** 1035.

Szostak, J.W. and R. Wu. 1979. Insertion of a genetic marker into the ribosomal DNA of yeast. *Plasmid* **2:** 536.

———. 1980. Unequal crossing over in the ribosomal DNA of *Saccharomyces cerevisiae. Nature* **284:** 426.

Thomas, J.G. and V. Furber. 1976. Yeast chromatin structure. *FEBS Lett.* **66:** 274.

Valenzuela, P., A. Venagas, F. Weinberg, R. Bishop, and W.J. Rutter. 1978. Structure of phenylalanine-tRNA genes: Intervening DNA segment within region coding for tRNA. *Proc. Natl. Acad. Sci.* **75:** 190.

Valenzuela, P., G.I. Bell, A. Venegas, E.T. Sewell, F.R. Masiary, L.J. DeGennaro, F. Weinberg, and W.J. Rutter. 1977. Ribosomal RNA genes of *Saccharomyces cerevisiae.* II. Physical map and nucleotide sequence of the 5S ribosomal RNA gene and adjacent intergenic regions. *J. Biol. Chem.* **252:** 8126.

Venegas, A., M. Quiroga, J. Zaldivar, W.J. Rutter, and P. Valenzuela. 1979. Isolation of yeast tRNA^{Leu} genes. DNA sequence of a cloned tRNA^{Leu} gene. *J. Biol. Chem.* **254:** 12306.

Vogt, V.M and R. Braun. 1977. The replication of ribosomal DNA in *Physarum polycephalum. Eur. J. Biochem.* **80:** 557.

Wallis, J.L., L.Hereford, and M. Grunstein. 1980. Histone H2b genes of yeast encode two different proteins. *Cell* **22:** 799.

Warner, J.R. 1982. The yeast ribosome: Structure, function, and synthesis. In *The molecular biology of the yeast* Saccharomyces. II. *Metabolism and gene expression* (ed. J. Strathern et al.), Cold Spring Harbor Laboratory, Cold Spring Harbor, New York. (In press.)

Weber, S. and I. Isenberg. 1980. HMG proteins of *Saccharomyces cerevisiae. Biochemistry* **19:** 2236.

Weintraub, H. and M. Groudine. 1976. Chromosomal subunits in active genes have altered conformation. *Science* **193:** 848.

Weisbrod, S., M. Groudine, and H. Weintraub. 1980. Interaction of HMG14 and 17 with actively transcribed genes. *Cell* **19:** 289.

Welch, S. 1978. "Structural genes and inverted repeats in the yeast genome." Ph.D. thesis, University of Washington, Seattle.

Whitney, P.A. and B.D. Hall. 1974. Repeated DNA in *Saccharomyces cerevisiae. Fed. Proc.* **33:** 1282.

Wickner, R.B. 1975. Mutants of *Saccharomyces cerevisiae* that incorporate deoxythymidine 5′-monophosphate into DNA *in vivo. Methods Cell Biol.* **11:** 295.

Wilkinson, L.E. and J.R. Pringle. 1974. Transient G1 arrest of *S. cerevisiae* cells of mating type α by a factor produced by cells of mating type **a.** *Exp. Cell Res.* **89:** 175.

Williamson, D.H. 1965. The timing of deoxyribonucleic acid synthesis in the cell cycle of *Saccharomyces cerevisiae. J. Cell Biol.* **25:** 517.

———. 1970. The effect of environmental and genetic factors on the replication of mitochondrial DNA in yeast. *Soc. Exp. Biol.* **24:** 247.

———. 1973. Replication of the nuclear genome in yeast does not require concomitant protein synthesis. *Biochem. Biophys. Res. Commun.* **52:** 731.

Williamson, D.H. and D.J. Fennell. 1974. Apparent dispersive replication of yeast mitochondrial DNA as revealed by density labelling experiments. *Mol. Gen. Genet.* **131:** 193.

———. 1975. The use of fluorescent DNA binding agent for detecting and separating yeast mitochondrial DNA. *Methods Cell Biol.* **12:** 335.

Williamson, D.H. and E. Moustacchi. 1971. The synthesis of mitochondrial DNA during the cell cycle in the yeast *Saccharomyces cerevisiae. Biochem. Biophys. Res. Commun.* **42**: 195.

Williamson, D.H. and A.W. Scopes. 1962. A rapid method for synchronizing division in the yeast, *Saccharomyces cerevisiae. Nature* **193**: 256.

Wintersberger, E. 1974a. Absence of a low molecular weight DNA polymerase from nuclei of the yeast, *Saccharomyces cerevisiae. Eur. J. Biochem.* **50**: 197.

———. 1974b. Deoxyribonucleic acid polymerase from yeast: Further purification and characterization of DNA-dependent DNA polymerase A and B. *Eur. J. Biochem.* **50**: 41.

———. 1978. Yeast DNA polymerase: Antigenic relationships, use of RNA primer and associated exonuclease activity. *Eur. J. Biochem.* **86**: 167.

Wintersberger, U. and E. Wintersberger. 1970a. Studies in deoxyribonucleic acid polymerases from yeast. I. Partial purification and properties of two DNA polymerases from mitochondria-free cell extracts. *Eur. J. Biochem.* **13**: 11.

———. 1970b. Studies in deoxyribonucleic acid polymerase from yeast. II. Partial purification and characterization of mitochondrial DNA polymerase from wild-type and respiration-deficient yeast cells. *Eur. J. Biochem.* **13**: 20.

Wintersberger, U., J. Hirsch, and A.M. Fink. 1974. Studies on nuclear and mitochondrial DNA replication in a temperature-sensitive mutant of *Saccharomyces cerevisiae. Mol. Gen. Genet.* **131**: 291.

Wintersberger, U., P. Smith, and K. Letnansky. 1973. Yeast chromatin. Preparation from isolated nuclei, histone composition and transcription capacity. *Eur. J. Biochem.* **33**: 123.

Woolford, J.L., Jr., L.M. Hereford, and M. Rosbash. 1979. Isolation of cloned DNA sequences containing ribosomal protein genes from *Saccharomyces cerevisiae. Cell* **18**: 1247.

Wyers, F., J. Huet, A. Sentenac, and P. Fromageot. 1976. Role of DNA-RNA hybrids in eukaryotes: Characterization of yeast ribonucleases H1 and H2. *Eur. J. Biochem.* **69**: 385.

Zakian, V.A. 1981. The origin of relication from *Xenopus laevis* mitochondrial DNA promotes high frequency transformation of yeast. *Proc. Natl. Acad. Sci.* (in press).

Zakian, V.A., B.J. Brewer, and W.L. Fangman. 1979. Replication of each copy of the yeast 2 micron DNA plasmid occurs during the S phase. *Cell* **17**: 923.

Zellweger, A., U. Ryser, and R. Braun. 1972. Ribosomal genes of *Physarum*: Their isolation and replication in the mitotic cycle. *J. Mol. Biol.* **64**: 681.

Cytology of the Yeast Life Cycle

Breck Byers
Department of Genetics
University of Washington
Seattle, Washington 98195

INTRODUCTION

The preceding papers in this volume clarify the major reasons that *Saccharomyces cerevisiae* is exceptionally well-suited to the analysis of cellular functions. The facility and rigor of its genetic analysis enable the researcher to apply the power of molecular genetics to the analysis of a wide variety of biochemical and cytological processes. One hopes, of course, that many of the processes revealed in yeast will provide clues to mechanisms operating in a broader range of eukaryotic organisms. Any similarities between processes occurring in yeasts and those in plant or animal cells may indicate useful experimental approaches for the latter organisms, which are less amenable to genetic analysis. In contrast, any differences should aid in defining those mechanisms that are fundamental to the eukaryotic mode of cellular function. Accordingly, this paper describes certain aspects of yeast cytology in comparison with our current understanding of other eukaryotic cells.

In keeping with the central importance of genetics to our study of yeast, I have chosen to stress cytological aspects that are most pertinent to the transmission of the genetic material: mitosis, conjugation, and meiosis. This approach will necessitate omission of a wealth of other cytological information. Many topics, including important early observations from freeze-fracture preparation for electron microscopy, are described in the seminal review by Matile et al. (1969). In more recent reviews, Hartwell (1974) and Pringle and Hartwell (this volume) discuss observations on cytological changes in the cell-division cycle, and Cabib (1975) and Ballou (1982) describe the structure and formation of the cell wall. Cytoplasmic membrane systems involved in secretion—the endoplasmic reticulum, Golgi apparatus, vacuoles, secretory vesicles, and so forth—are discussed by Schekman and Novick (1982). Yeast mitochondria, which display an intriguing tendency to become fused into one or more large aggregates, are described by Stevens (this volume). Meiosis and sporulation have been reviewed recently by Esposito and Esposito (1978) and Esposito and Klapholz (this volume).

VEGETATIVE GROWTH

Division by Budding

Typical electron microscopic views of the yeast cell and its organelles are depicted in Plates 1 and 2.[1] Nuclear changes occurring in the cell-division cycle and conjugation are diagramed in Figure 1 and shown as micrographs in Plates 3, 4, 5, and 6. Cultures growing in vegetative medium under conditions favoring either fermentation or oxidative metabolism are characterized by the presence of both unbudded and budded cells. The cells are unbudded in the early phase of the cell-division cycle, prior to the initiation of chromosomal replication. This population becomes more abundant as the culture medium becomes exhausted of essential nutrients and the culture approaches the stationary phase of growth. Unbudded cells from either growing or stationary cultures are slightly ovoid if haploid and are more strikingly ovoid, or elongate, if diploid. The surface contour is smoothly curved because intracellular osmotic pressure, or turgor, forces the plasma membrane firmly against the semirigid cell wall. As a cell enters a round of division, a portion of the wall gains a more acute arc of curvature. This blunt protuberance soon undergoes a striking transition of form to become nearly spherical, but it remains attached to the cell proper by a narrow neck. Further growth of the cell occurs largely in the bud, which eventually becomes about as large as the original cell. The original cell and the bud then become separated from each other, giving rise to the mother and daughter cells, respectively.

Cytological examinations of the cell, principally by electron microscopy, have revealed that the initiation of this asymmetric mode of cellular replication substantially precedes nuclear division. The mitotic spindle begins to form during the earliest phase of bud emergence, but the nucleus does not enter the neck of the bud until the bud has nearly completed its growth. Subsequently, the nucleus becomes much more elongate, eventually extending as far as the most distal cell surfaces at either extremity. Nuclear division is then accomplished when the elongate nucleus becomes pinched apart in the region passing through the neck. Throughout these phases of nuclear division, no condensation of the chromatin is noted by usual methods of specimen preparation for electron microscopy (Gordon 1977). Although condensed chromosomes have been described both by light microscopy (Wintersberger et al. 1975) and by electron microscopy (Williamson 1966),

[1]For Plates 1–12, see pp. 85–96.

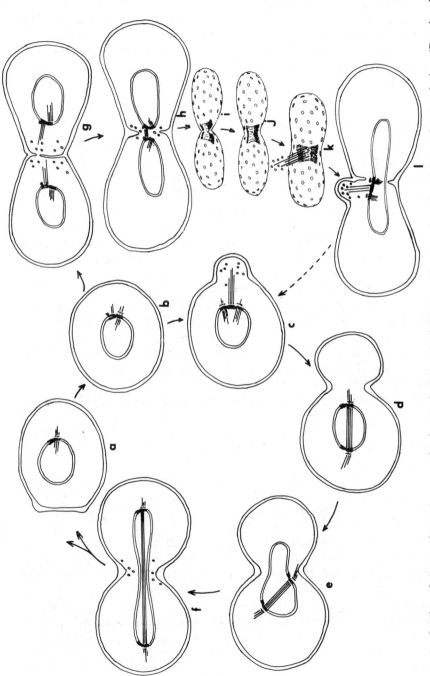

Figure 1 Outline of changes in nuclear organization during the cell-division cycle and conjugation. See text for description. (Reprinted, with permission, from Byers and Goetsch 1975a.)

the observed structures may have been artifactually condensed by the methods of specimen preparation. Yeast mitosis is also characterized by the persistence of the nuclear envelope, which undergoes dissolution during mitosis in higher eukaryotes. These distinctive features may limit the generality of information gleaned from the study of yeast, but they may also provide a useful comparison to focus our attention on the essential processes of cellular replication.

Nuclear Organization and Division

Much of the yeast nucleus consists largely of chromatin of homogeneous appearance, lacking any of the localized regions of condensation indicative of heterochromatin. This apparent lack of heterochromatin may derive merely from the paucity of highly redundant DNA (Bicknell and Douglas 1970), which occupies the constitutive heterochromatin of higher eukaryotes (Pardue and Gall 1970). In addition to chromatin, the yeast nucleus also contains a densely staining nucleolus (Plate 1b), which is crescent-shaped and closely applied to the nuclear envelope. Nuclear purification and subfractionation have confirmed the expected localization of the genes for ribosomal RNA in the nucleolar fraction (Smitt et al. 1973). It is interesting to note that the nucleolus occupies a greater proportion of the nuclear volume in yeast than in other eukaryotes; this may well reflect the greater requirement for ribosomal synthesis in an organism exhibiting a shorter generation time. During nuclear division, the nucleolus is partitioned between the two incipient daughter nuclei, retaining a texture similar to that seen at all other phases of the life cycle (Robinow and Marak 1966).

The nucleus is enclosed by a nuclear envelope similar to that seen in other eukaryotic cells. Detection of this structure by electron microscopy (Agar and Douglas 1955) firmly established the eukaryotic nature of yeast. The envelope bears a typical complement of nuclear pores and varies little in its appearance throughout the life cycle.

Although the persistence of the nuclear envelope first suggested that nuclear division might be amitotic, electron microscopy revealed that the yeast nucleus contains a mitotic spindle consisting of microtubules as in other eukaryotic cells (Robinow and Marak 1966). The two ends of the spindle consist of the spindle pole bodies (SPBs), which appear as discs of darkly staining amorphous material spanning the nuclear envelope (Plates 3 and 5). These organelles have also been termed "spindle plaques" (Moens and Rapport 1971a) but have been redesignated according to their homology with similar organelles in a variety of lower eukaryotes (Kubai 1975). Microtubules extend perpendicularly from the plane of the SPBs both into the nuclear interior and outward to the cytoplasm; the microtubules extending into the nucleus comprise the intranuclear spindle and are more numerous than the cytoplasmic ones. Both sets of microtubules have their proximal ends within relatively inconspicuous layers of material lying parallel to the more densely stained central layer of the SPB. Whereas the nuclear envelope surrounding the SPB is generally not well defined, the membrane of the envelope is quite darkly stained along one margin. This prominent sector of the surrounding envelope has been termed the half-bridge in order to differentiate it from the bridge, which is present after SPB duplication.

Duplication of the SPB is a prominent landmark in the pathways of events leading to both mitosis and meiosis (as described more fully below). Although the duplication mechanism is somewhat obscure, electron microscopy reveals two distinct stages, which are shown diagrammatically in Figure 2. In the first, a small patch of densely staining material similar in appearance to the SPB proper appears on the cytoplasmic surface of the half-bridge distal to the SPB itself. This satellite (Plate 4c) remains entirely outside the nucleus until the second phase, when the satellite is lost and two daughter SPBs of equal size are found within the nuclear envelope. Numerous observations on cells fixed near the time of transition from the first stage to the second have failed to reveal how this transition occurs. One possibility is that the satellite itself is transformed into one daughter SPB by insertion into the nuclear envelope. Another possibility is that the original SPB splits in two while the satellite is lost, perhaps contributing its substance to both daughter SPBs. Perhaps genetic dissection of the duplication process will be required to determine the actual mechanism. In any case, the daughter SPBs (Plates 4 and 5) flank a complete bridge, which appears to consist of two layers of darkly stained unit membrane interconnected by inflections adjacent to either SPB.

The duplicated SPBs remain paired for an extended period of time in both mitosis and meiosis. Byers and Goetsch (1975a) found that duplication in a growing culture was nearly coincident with the earliest phase of bud emergence and that the SPBs remained paired until the bud had enlarged to about one third of the diameter of the mother cell. During this period of bud

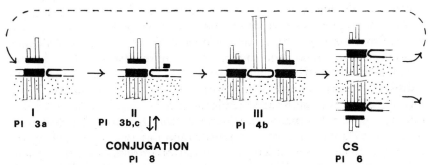

Figure 2 Stages of SPB development in the cell cycle: (I) single SPB; (II) modified single SPB; (III) duplicated SPB; (CS) complete spindle. The nucleoplasmic surface is stippled. Plate numbers refer to representative micrographs. (Adapted from Byers and Goetsch 1975a.)

growth, the duplicated SPBs remain on the same side of the nucleus as that which underlies the bud (as in Plate 4). Then, as the bud achieves a larger size, the fully formed spindle makes its appearance (Plate 6). Its formation must be quite rapid, because cytological observations have failed to reveal any intermediates in its formation, such as a stage after scission of the bridge but before aggregation of the two sets of microtubules into a single spindle. The intranuclear spindle consists of a bundle of microtubules ending on either of the two SPBs, which remain embedded within the nuclear envelope. The length of this original spindle is regularly about the diameter of the nucleus, so it extends from one side to the other without deforming the circular profile of the nuclear envelope. Two types of microtubules are evident within the spindle. One type is continuous across the nucleus from one SPB to the other, whereas the majority of microtubules are shorter, are attached at only one end to an SPB, and often diverge slightly from the axis defined by the continuous microtubules. Once the spindle has formed, it remains of constant length throughout the period of bud enlargement. When the nucleus migrates into the neck of the bud, the spindle does not immediately elongate further. Instead, it usually lies diagonal to the major axis of the nucleus, which has become elongated by its migration into the narrow neck of the bud. It is clear from these observations that this phase of nuclear elongation is not mediated by the spindle, which fails to extend into the extremities of the nucleus (Robinow and Marak 1966). Conversely, the failure of immediate spindle elongation upon nuclear elongation suggests that the length of the spindle is not determined by the passive migration of the SPBs along the nuclear envelope. The intrinsic stability of the shorter spindle length is confirmed by the terminal phenotypes of the *cdc* mutants defective in medial nuclear division; these mutants regularly undergo nuclear migration but maintain short spindles indefinitely at arrest (Byers and Goetsch 1974). In normal mitosis, spindle elongation occurs shortly after nuclear migration into the neck of the bud. The spindle poles move apart and come to lie at the extremities of the nucleus. The central bundle then becomes interrupted at its midpoint as the daughter nuclei pinch apart.

Valuable images of the mitotic process have been provided by Peterson and Ris (1976), who viewed thick sections by million-volt electron microscopy. This method permitted them to view the entire spindle in a single pair of stereoscopic views, thereby revealing the spatial association between the indistinct chromatin masses and the sparse array of microtubules. These images indicate that when the spindle is present, chromatin occupies a broad central zone near the center of the spindle and that the shorter (lateral) microtubules reach from either SPB to this chromatin-rich zone. Subsequently, the chromatin divides into two separate masses that move toward either spindle pole as the lateral microtubules shorten and the central microtubule bundle elongates. An obvious interpretation is that the lateral microtubules interconnect the chromatin masses with the SPBs. In many eukaryotic cells, differentially staining kinetochores appear to mediate the attachment of the chromosomes to microtubules extending toward the spindle poles. Kineto-chores have not been detected in yeast, but chromatin behavior suggests a similar mode of attachment. Depolymerization of the attached lateral micro-tubules would shorten them and thereby draw the chromosomes toward the pole. Because microtubule depolymerization is believed to occur at the ends of the polymer (Margolis et al. 1978), it is difficult to conceive a mechanism for permitting retention of attachments by both SPBs and kinetochores to the ends of microtubules during their depolymerization. This uncertainty is not limited to yeast cytology but is also an unsolved aspect of spindle mechanisms proposed for higher eukaryotes (Inoué and Sato 1967; Margolis et al. 1978).

The possible mechanisms for the concomitant elongation of the central microtubule bundle are numerous. It seems likely that the lengths of the microtubules themselves must increase in order to account for the extensive (up to eightfold) elongation of the whole spindle (Byers and Goetsch 1975a). Furthermore, it has recently been found that the antimicrotubule agent methyl benzimidazole-2-yl carbamate (MBC) causes loss of spindle micro-tubules (Quinlan et al. 1980). The occurrence of substantial chromosome loss in diploid strains subjected reversibly to MBC treatment (J. Wood, unpubl.) also points to an interaction with the spindle. The demonstrable association of MBC with the tubulin of *Aspergillus* (Davidse and Flach 1977) suggests that the loss of microtubules in yeast is caused by a shift in the equilibrium conditions required for maintenance of the polymerized state. Elongation might conversely be driven by an increase in the available pool of tubulin. But spindle elongation may also require the longitudinal sliding of the microtubules along one another. Strong evidence for a mechanism of the latter sort has been seen in the diatom nucleus, which is similar in its organization but contains many more microtubules. Pickett-Heaps and Tippit (1978) and McDonald et al. (1979) have demonstrated that the so-called "central spindle" of the diatom nucleus is composed of two sets of microtubules intermingled with one another in a highly ordered array. Each set is attached only to one pole of the spindle, so that each set is antiparallel to the other set with respect to their polar attachments. It seems probable that these microtubule sets are also antiparallel in the orientation of their con-stituent tubulin molecules. At anaphase, each set is translated in the direction of its polar attachment, thereby causing separation of the poles and elongation of the central spindle without any change in microtubule length. Reconstruction of images from serial sections revealed that each microtubule is most closely bounded by microtubules of the other set, so one can account for spindle elongation by the directional sliding forces between antiparallel microtubules. It has been proposed previously that sliding of this sort is mediated by discrete structures (presumably macromolecules) that are occasionally observed to interconnect adjacent microtubules (McIntosh 1974). A similar mechanism is found in the well-characterized bending mechanism of various eukaryotic cilia, where adjacent (though not antiparallel) microtubule doublets are forced to slide along one another by the energy derived from the adenosine triphosphatase activity of dynein (Gibbons 1977). Although probes for such activity are poorly developed, the sensitivity of spindle motion in permeabilized mammalian cells to vanadate ion is similar to that of dynein, thereby suggesting the essential activity of a similar protein (Cande and Wolniak 1979). The applicability of these mechanisms to the yeast spindle remains unknown, but one may hope that conditional mutants, such as the medial nuclear division mutants (Culotti and Hartwell 1971), will reveal the components essential to spindle function.

Implicit in the foregoing description is the assumption that the sister chromosomes are drawn apart by attachments to the shorter microtubules, which extend between the chromatin and either SPB. Although recognizable kinetochores have not been described in yeast, the unambiguous genetic demonstration of centromeres in both mitosis and meiosis (Mortimer and Hawthorne 1966) makes it clear that the segregation of yeast chromosomes must involve specific chromosomal sites. Bearing in mind the limitations of comparative observations, it is useful to consider the spindle behavior of the oomycete *Saprolegnia* (Heath 1980), which similarly maintains a closed nuclear envelope during mitosis but has detectable kinetochores. The kinetochores are individually borne on single microtubules and undergo duplication immediately before metaphase, when they come to lie in typical bipolar orientation between microtubules extending toward either pole. Although it is unclear why the kinetochores are not replicated before prophase, subsequent behavior is similar to that in higher eukaryotes. Nuclear division is preceded by a shortening of the kinetochore microtubules as the central spindle is elongated. Here, as in many other lower eukaryotes, no metaphase plate is formed; pairs of chromosomes are instead distributed widely throughout the length of the spindle. The broad distribution of yeast chromatin throughout the apparent time of metaphase might therefore derive simply from a similar metaphase organization in yeast.

Suggestive evidence for yeast kinetochores has been reported by Peterson and Ris (1976) from observations on lysed mitotic nuclei. The medial ends of some individual lateral microtubules are associated with chromatin strands in these views. These authors also note that there is only a sufficient number of lateral microtubules for each chromosome to be attached to a single microtubule, basing the chromosome number on the known number of genetic linkage groups. But other modes of spindle interaction with the chromosomes remain plausible. Pickett-Heaps and Tippit (1978) have argued from their analysis of the diatom spindle that its lateral microtubules do not impinge directly on the chromosomes. Instead, these microtubules appear to interact with a "collar" of densely staining material that surrounds the central spindle. This collar may serve as an attachment scaffold for the chromosomes. Disjunction in this case would result from the poleward separation of two collars rather than the separation of sets of individual chromosomes to either pole. In the absence of compelling evidence for yeast kinetochores, it remains possible that yeast disjunction also employs a mechanism of this sort. The demonstration by Williamson and Fennell (1981) that chromosome sets of similar replicative age cosegregate in mitosis certainly is consistent with a coordinate disjunctional mechanism for the entire genome.

Control of Spindle Formation and Function

Genetic analysis should facilitate approaches to understanding the molecular mechanisms required for mitosis in yeast. In particular, the cell-division-cycle (*cdc*) mutations (Hartwell et al. 1973) provide a valuable collection of variants within which we may expect to identify thermolabile components essential to processes such as microtubule polymerization, spindle formation, spindle elongation, and other aspects of nuclear division. But the primary defects of rather few *cdc* mutations have been identified, and none is known to cause a defect in a component of the spindle apparatus. Nevertheless, these mutations have provided considerable insight into the dependent interactions between various aspects of the cell-division cycle (see Pringle and Hartwell, this volume). With regard to spindle function, we have learned that nuclear elongation and the concomitant elongation of the spindle (Byers and Goetsch 1974) require the synthesis (Hartwell 1976) and ligation (Johnston and Nasmyth 1978) of the chromosomal DNA. The cytological manifestations of these mutants thereby reveal that the cell possesses a mechanism precluding spindle elongation until DNA replication is complete. By analogy, we may reasonably suspect that the primary defects of some mutants causing medial nuclear division arrest may reside in other DNA-independent processes rather than within the spindle apparatus per se.

In general, spindle development at the terminal phenotypes of the *cdc* mutants is appropriate to the stage of progression through the cell cycle (Byers and Goetsch 1974). Mutants with terminal phenotypes similar to stationary arrest of the cell cycle have a simple single SPB, whereas those arrested like wild-type cells exposed to a mating factor contain a satellite-bearing single SPB. Particularly intriguing is the arrest of two mutants, *cdc4* and *cdc34*, which contain a duplicated (but unseparated) SPB similar to that normally seen during bud emergence. These mutants regularly form new buds in the absence of nuclear DNA replication, which suggests that the cellular state associated with a duplicated SPB is competent in the initiation of bud emergence. Furthermore, each bud generally arises from the neck of the prior bud—a site immediately adjacent to the duplicated SPB. Moreover, in growing cells one sees cytoplasmic microtubules extending from the outer surface of the duplicated SPB into the base of the emerging bud. Taken together with the finding that all mutants that were budded at arrest had also undergone SPB duplication, these observations suggested an active role of the duplicated SPB in the induction of bud emergence. The mechanism of this induction is unknown, but the frequent occurrence of organelle displacement along parallel arrays of microtubules in other organisms (Porter 1976) suggests that vesicles that participate in bud development (Schekman and Novick 1982) are directed toward the budding site by the cytoplasmic microtubules extending outward from the duplicated SPB (Byers and Goetsch 1975a). If this proposal proves correct, we would reasonably expect the control of bud orientation to be determined by the orientation of the SPBs on the nuclear surface. Each new bud of a haploid cell typically arises adjacent to the bud of the previous cell cycle, whereas $\mathbf{a}/\alpha$ diploid cells ordinarily bud at the opposite pole (Freifelder 1960). One might postulate that the corresponding differences in SPB orientation result from intrinsic differences in nuclear rotation or migration imposed by the spindle mechanism in the preceding mitosis. Alternatively, the SPB-microtubule assembly may move freely with respect to the cellular cortex until a site receptive to stable interaction is encountered. In the latter case, the pertinent difference between haploid and diploid cells would reside in the cortical region rather than the nucleus.

Recent characterization of another strain, mutant in *cdc31*, requires reevaluation of the role played by the SPB (Byers 1981). This mutant superficially appears to be arrested similarly to other nuclear division mutants, but electron microscopy reveals that the spindle possesses an SPB at only one pole. Furthermore, both budding and DNA replication, which normally appear dependent on SPB duplication, occur in these cells as the culture undergoes arrest. Detailed observations reveal, however, that the single SPB differs from the normal type in that it is significantly larger and bears at least twice as many microtubules. As a consequence of this arrest, the ensuing mitosis directs all of the chromosomes toward a single pole, and the ploidy of the nucleate daughter cell is doubled (Schild et al. 1981). Although the phenotype of *cdc31* negates the dependency of bud emergence and DNA replication on SPB duplication per se, it remains tenable that the substance of the SPB must become doubled for the other events to occur. In any case, the SPB, though unduplicated, displays the typical spatial association with the emerging bud.

Although various roles of the SPB are suggested by *cdc* mutants, the most striking SPB-mediated process is the control of microtubule organization. The SPB represents one member of a large class of organelles that are found at the ends of microtubules and have been termed microtubule-organizing centers, or MTOCs (Pickett-Heaps 1969). The MTOCs of animal cells are frequently indistinct except where they are highly clustered around the centrioles. These constitute the centrosphere (Robbins and Gonatas 1964), which remains capable of inducing microtubule assembly in the absence of the centriole proper (Berns and Richardson 1977). In vitro experiments demonstrate that the centrosphere exerts its control over the distribution of microtubules by providing centers for the initiation of the polymerization of tubulin (Weisenberg and Rosenfeld 1975; Gould and Borisy 1977). In higher plants, which lack centrioles, the distribution of MTOCs is yet more dispersed and indistinct, whereas many algae and fungi bear microtubules on discrete MTOCs (Kubai 1975).

The exclusive localization of these sites in the SPBs renders yeast an invaluable organism for investigation of their function. The SPBs are detectable in lysates of vegetative yeast cells by their ability to initiate the in vitro assembly of microtubules (Hyams and Borisy 1978). This property is retained by the SPBs when they are separated from other components of the lysate by centrifugation in sucrose gradients (Byers et al. 1978). The latter study found the SPBs capable of nucleating a narrow range of microtubule numbers not significantly different from that found in vivo. Hyams and Borisy (1978), on the other hand, reported that the SPBs from rapidly growing cells bore a substantially lower capacity for microtubule initiation than did those from stationary cells. This difference was interpreted as an indication that sites for the initiation of microtubule polymerization are transferred from the SPB to the chromosomes as the cell progresses from stationary phase into mitosis. Nevertheless, the data are also consistent with the artifactual loss of microtubule-initiating capacity from the SPBs of growing cells, such as might occur if active microtubules together with attached initiating sites were forcibly disrupted from the SPB by chromatin dispersal during lysis.

Here, as in other systems, the mechanism of the initiation of polymerization remains obscure, but the morphology of yeast microtubules may provide a clue. When the spindles were viewed by negative staining for electron microscopy (Byers et al. 1978), the ends of microtubules attached to the SPB were found to differ from the distal ends. Whereas the central cavity of the microtubule is open to the external space at the distal end, the end proximal to the SPB is a closed surface. By analogy, the microtubule is topologically similar to an elongate test tube, with its open top situated distally and its bottom adjacent to the SPB. Because the closed end occupies the site where polymerization was initiated, the formation of this closed surface may be pivotal to the initiation of polymer growth. This possibility is favored by the further observation that microtubules induced in vitro also bear the closed end.

Cytoplasmic Features of Budding and Cytokinesis

Cytoplasmic aspects of the cell-division cycle include the formation of the bud and population of the bud with cytoplasmic organelles, in addition to the daughter nucleus, cytokinesis, and cell separation. Most pertinent features of cytoplasmic organization are amply described in the classic review by Matile et al. (1969), as well as in other chapters in this volume (see Introduction). During the early portion of the cell-division cycle, many small vesicles (ca.

30–40 nm in diameter) accumulate first at the site where the bud will emerge and are later seen (Plate 4) within the early bud (Matile et al. 1969). As noted earlier, cytoplasmic microtubules extending outward from the duplicated SPB are present in this same region (Byers and Goetsch 1974). Therefore, it is reasonable to imagine that the microtubules are responsible for the localization of the vesicles, which are in turn required for the remodeling of the overlying cell wall.

The cytoplasm also contains numerous mitochondria, which are of the typical form, with an inner membrane contorted into tightly paired involutions (cristae) and a more smoothly contoured outer membrane (Plate 1). The propensity for mitochondrial fusion during vegetative growth often results in the presence of a single major mitochondrial mass (plus a few separate small mitochondria) during most of the cell cycle (Stevens 1977 and this volume). It is difficult to determine whether this structural arrangement leads to population of the growing bud by a particular subset of mitochondria. Genetic analysis would require assaying the transmission of mitochondrial genomes from a cell containing a mixed population. When zygotes heterozygous for mitochondrial markers are analyzed (Wilkie and Thomas 1973; Dujon et al. 1974), it is found that one parental set predominates in the lineage from each zygote. The predominant parental set may represent the one most accessible to the emerging bud (Strausberg and Perlman 1978).

The cell-division cycle concludes with cytokinesis, which occurs at the neck of the bud shortly after nuclear division. The site of incipient division is occupied from the earliest stage of bud formation by a "filamentous ring" (Plate 7a,b), composed of a monolayer of circumferential filaments about 10 nm in diameter, regularly spaced 24 nm apart, immediately subjacent to the plasma membrane (Byers and Goetsch 1976a). During the same early period of bud emergence, chitin is deposited in a ring-shaped region of the cell wall in the neck of the bud (Hayashibe and Katohda 1973). This chitin ring is not located at the exact middle of the neck but lies at the side nearer the mother cell, to which it will remain attached as the bud scar after cytokinesis and cell separation. The respective roles of the filamentous ring and the chitin ring in cytokinesis have not been established, but cytological observations provide some clues. In vegetatively growing cells, the filamentous ring persists until immediately before cytokinesis, which appears to be accomplished by the fusion of small vesicles to produce a pair of membranes across the aperture enclosed by the chitin ring (Plate 7c). The apparent presence of chitin within the central portion of the bud scar suggests that the fusing vesicles are involved in chitin synthesis (Cabib et al. 1974). One cdc mutation, cdc24, renders the cell unable to form a bud, although nuclear division proceeds. In this mutant, chitin is found to be deposited widely throughout the cell wall (Sloat and Pringle 1978), so we know that chitin deposition per se is insufficient for bud formation. Normally, the filamentous ring might serve to suppress fusion of chitin-containing vesicles with the cytokinesis site until the bud is fully formed. But the filamentous ring must also play some positive role in cytokinesis because cells defective in genes essential for cytokinesis (*cdc3, cdc10, cdc11,* and *cdc12*) lack the filamentous ring that is possessed by all other mutants with a bud at the terminal phenotype (Byers and Goetsch 1976b).

CONJUGATION

Under the influence of the mating pheromones, yeast cells of opposite mating types depart from the growth cycle and undergo conjugation with one another to produce a zygote, which then buds to initiate the growth of the hybrid strain. The production and reception of pheromones has yielded little to microscopic analysis, but striking cytological changes are evident during the subsequent response (Byers and Goetsch 1975a). The cells of both mating types depart from the cell-division cycle in G_1 phase with a single SPB (as diagramed in Fig. 1). This SPB is modified, however, by the acquisition of a satellite like that seen in the later portion of G_1 during ongoing division (Fig. 2 and Plate 8). The mating cells accumulate agglutinins in their cell walls and may thereby eventually make intimate contact with appropriate mating partners. The walls of the fusing cells are gradually deformed to produce the smooth contour of the isthmus. Perforation of the paired cell walls then establishes cytoplasmic continuity between the two former cells, which now constitute the zygote. Karyogamy, the fusion of the two nuclei, generally begins as soon as perforation has occurred. Both nuclei, which had borne their satellite-bearing SPBs adjacent to the site of cell fusion, soon begin to migrate toward it. Because the extent of the perforation is slow to develop, the nuclei frequently make contact well before the aperture has enlarged fully. Consequently, neither nucleus appears able to pass completely through the aperture, and fusion of the nuclei occurs within the isthmus.

A pivotal feature of karyogamy in any eukaryote is the abrupt transition of ploidy when the nuclei fuse. In yeast, both haploid and diploid strains normally have one SPB per cell during G_1, so there must be a mechanism for the net reduction in SPBs during mating. That is, the diploid product of conjugation must come to possess one SPB, whereas two SPBs were provided by the two haploid parents. A hypothetical means by which this transition could occur is the direct formation of a structure equivalent to the duplicated SPB by fusion of the original SPBs, thereby bypassing duplication in the first diploid cell cycle; but this is not the mechanism. Instead, it has been found that the original diploid nucleus arises by fusion of the parental SPBs to form a larger, single SPB (Byers and Goetsch 1975a). This fusion (Fig. 1i,j) occurs along the lateral margins of the two SPBs in a manner such that each SPB proper is directly fused to the other and, similarly, the two satellites fuse with each other. This "conjugation SPB," when newly formed, is V-shaped in profile because it retains the angle described by the parental

SPBs at the time of their fusion. The V-shaped profile reverts to a planar form within an hour or two; henceforth, the diploid SPB in this cell and its descendants are distinguishable only by the greater size of the SPB. SPBs of diploid cells are found to occupy roughly twice as much volume as those of haploid cells when observed at equivalent stages of the cell-divison cycle. Tetraploid cells similarly display a further doubling in the volume of their SPBs in comparison with those in diploid cells. These differences in volume are due principally to differences in dimensions within the plane of the surrounding nuclear envelope; the dimension perpendicular to the envelope does not vary significantly with ploidy.

Soon after formation of the conjugation SPB, it undergoes duplication. At nearly the same time, DNA replication occurs (Sena 1973) and the first bud appears, usually near the midpoint of the isthmus (Fig. 1k,l). This localization of the first zygotic bud seems to depend upon the location of the duplicated SPB, which is borne on the zygotic nucleus trapped in the narrow aperture where the parental nuclei have fused. When SPB localization was experimentally disrupted by centrifugation, the first zygotic bud arose instead at a pole of the zygote (M. Hungate and B. Byers, unpubl.). Moreover, this pole was the same as that to which the nucleus had been displaced, again exemplifying the spatial influence of the SPB on bud emergence.

MEIOSIS AND SPORULATION

Although yeast meiosis and sporulation are readily amenable to genetic analysis (Esposito and Klapholz, this volume), cytological analysis has proven difficult. Pioneering cytological work by Mundkur (1961), Moor and Muhlethaler (1963), Engels and Croes (1968), and Moens and Rapport (1971a,b) has nonetheless demonstrated essential similarities to meiotic events in eukaryotes better suited to such observations. Diploid (*MATa/ MATα*) cells transferred to sporulation medium enter into a complex pathway of processes leading to ascospore production. Major processes in this pathway include chromosomal DNA replication, meiotic prophase with attendant genetic recombination, two successive meiotic divisions, and ascospore formation. One may note that the earlier events are similar to those in higher eukaryotes, whereas such similarity is not the case for all fungi. For example, a haplonic organism like the fission yeast *Schizosaccharomyces pombe* remains haploid until it is transferred to sporulation medium and then undergoes conjugation before premeiotic DNA replication begins (Egel and Egel-Mitani 1974). At least certain fungi, such as *Neottiella* (Rossen and Westergaard 1966) and *Coprinus* (Lu and Jeng 1975), delay karyogamy until the bulk of premeiotic DNA replication has occurred. In contrast, *S. cerevisiae* enters meiosis from the vegetative diplophase, as do higher plants and animals.

Meiotic Prophase

The entry of *S. cerevisiae* into the sporulation process can be induced by a variety of means, but all entail transfer into a sporulation medium, which contains an oxidizable source of energy, such as acetate, but is deficient in a metabolic nitrogen source. The induction of sporulation for cytological analysis may be accomplished as is commonly done for genetic analysis: A culture grown to stationary phase in a glucose-based medium is washed and transferred to sporulation medium. Greater synchrony has recently been achieved by effecting the transfer at a precisely timed late phase in the transition from logarithmic growth to stationary arrest (Petersen et al. 1978). In addition, numerous studies of yeast meiosis have employed a method of transfer from late logarithmic vegetative growth in acetate-based medium to sporulation medium (Roth and Halvorson 1969). This method permits meiosis to be completed more rapidly, but only certain strains are competent to make this transition. Although these and related methods (Fast 1973) have permitted many useful observations of meiosis, the lack of synchrony in yeast, relative to the elegantly developed synchrony in *Lilium* (Stern and Hotta 1977) and *Coprinus* (Lu 1967), continues to limit various analyses.

Regardless of the method of induction employed, yeast cells enter meiosis from the G_1 phase of the division cycle. Cells transferred from stationary phase are already largely in G_1 (or G_0), whereas those transferred from logarithmic growth complete cell division soon after transfer (see Hartwell 1974). Because the rate of bud growth is depressed in sporulation medium, the daughter cell formed by the bud is often quite small. These smaller cells lend to the asynchrony of sporulation by lagging substantially behind the larger cells; in addition, they often fail to complete sporulation (Haber and Halvorson 1972).

Electron microscopy of cells entering meiosis demonstrates that most nuclei bear a single, unduplicated SPB (Moens and Rapport 1971a). Any other stages of SPB duplication and spindle development seen at early times are probably attributable to the completion of mitotic cycles begun before transfer to sporulation medium. At later times, the SPBs undergo duplication and achieve a prolonged arrest of development at the duplicated, but unseparated, stage (Moens and Rapport 1971a). Then the SPBs undergo separation from one another to form the spindle for meiosis I.

These distinct alterations of SPB organization provide landmarks for the recognition of other processes occurring in meiotic prophase. Considering first the entry into meiosis, we note that the presence of a single SPB is consistent with the departure of the cell from mitosis in the prereplicative, or G_1, phase. As the culture begins premeiotic chromosomal DNA replication, the SPBs of many cells become duplicated, as they do at the analogous (S) stage of the cell-division cycle. Occasionally, however, a cell may enter meiosis directly from vegetative growth without regaining a single SPB. This exceptional behavior is indicated by the electron microscopic detection of an occasional meiotic cell bearing a very small anucleate bud, which is not pinched off from the cell proper (B. Byers and L. Goetsch, unpubl.). A cell of this sort had probably initiated bud emergence before reverting directly to the meiotic pathway. This pathway is consistent with the findings of Hirschberg and Simchen (1977), who tested the ability of many *cdc* mutants to enter meiosis after temperature-sensitive arrest of vegetative growth. None of the mutants with a complete spindle at vegetative arrest could enter meiosis without first completing the cell-division cycle. But cells defective in *cdc4*, which causes vegetative arrest with a duplicated SPB, could enter directly into meiosis, and electron microscopy (B. Byers and L. Goetsch, unpubl.) indicated that the duplicated SPB did not revert to a single form throughout this transition.

The SPB normally duplicates well after transfer to sporulation conditions and then persists in this form (Plate 10c) for a considerably longer period of time than in the cell-division cycle. The period of residence by the duplicated SPB appears to coincide with the periods of premeiotic DNA replication and meiotic recombination (Horesh et al. 1979), although the general asynchrony of meiotic cultures has prevented the precise timing of these events. It is perhaps more informative to inquire into the dependent relationships, as defined by the genetic dissection of the meiotic pathway described by Esposito and Klapholz (this volume). We may note here, however, that the inhibition of premeiotic DNA replication by hydroxyurea or by thermal inhibition of temperature-sensitive *cdc8* and *cdc21* mutants permits SPB duplication but not meiotic recombination or later stages of spindle development (Schild and Byers 1978). In contrast, similar inhibition of DNA replication in vegetative cells permits formation of the complete mitotic spindle but not its elongation (Byers and Goetsch 1974).

Meiotic prophase is of particular cytological interest because the organelles responsible for chromosomal synapsis are detectable. Early observations revealed regions of high electron density interlaced by flat zones, which were less electron dense and appeared similar to the central regions of the synaptonemal complex (SC) characteristic of other eukaryotes in meiosis (Engel and Croes 1968). Moens and Rapport (1971b) soon recognized that, in addition to this "polycomplex" body (Plate 9), there were bona fide SCs present in the nuclei of meiotic yeast (Plate 10a). The latter are elongate, smoothly curved structures composed of three principal layers: a central layer of weakly stained material, flanked by two darkly stained layers spaced about 100 nm apart. Later observations on partially synchronized cultures revealed a characteristic sequence of changes in the appearance of these structures (Zickler and Olson 1975). The darkly staining "dense body" was evident at early stages, frequently bearing short segments of the central elements of the SC in the manner originally seen by Engel and Croes (1968). Later, the lateral elements of the SC became faintly discernible as cells entered leptotene. The pairing of these structures in zygotene led to the presence of fully formed SC at pachytene. Then, in diplotene, the central elements of the SC appeared to lose their association with the chromosomes and often formed regular stacks similar to the so-called polycomplexes characteristic of this stage in many eukaryotes (Westergaard and von Wettstein 1972).

Besides permitting us to gain some understanding of the mechanisms involved in chromosomal synapsis, the visibility of the SC has provided a means for visualizing the yeast karyotype. Byers and Goetsch (1975b) employed serial sectioning of pachytene nuclei to reconstruct the courses followed by the chromosomes, which are paired along either side of the SC at this stage (Fig. 3). This work demonstrated that diploid cells contain about 17 bivalents, as predicted by genetic means. Individual chromosomes extended between telomeric attachments on the nuclear envelope (Plate 10d) and varied about tenfold in length. The entire karyotype totaled about 30 μm in length. Unfortunately, the lengths did not fall into classes of sufficient regularity that identification of specific members by length appeared likely.

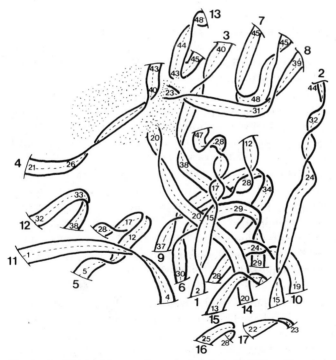

Figure 3 The pachytene karyotype of a diploid cell reconstructed from serial sections. The larger numbers represent rank-ordering by chromosome length, the smaller numbers are section numbers, and the stippled area represents the nucleolus. (Reprinted from Byers and Goetsch 1975b.)

Thus, although pachytene chromosomes are more readily distinguished than the amorphous circular structures identified as chromosomes in mitotic cells (Wintersberger et al. 1975; Galeotti and Williams 1978) and entire bivalents can be seen in lysates (Plate 11; B. Byers and L. Goetsch, unpubl.), the cytogenetics of yeast remains limited in its utility. In tetraploid nuclei, Byers and Goetsch (1975b) found that segments of SC of equal length were frequently associated at sites equidistant from either telomere of the paired members; these therefore appeared to represent the exchanges of pairing partners in quadrivalents expected from the pattern of recombination known in tetraploid yeast (Roman et al. 1955).

As in mitosis, the nucleolus remains prominent throughout meiotic prophase (Moens and Rapport 1971b; Zickler and Olson 1975). The karyotype of pachytene nuclei contained two segments of SC entering the nucleolar region (see Fig. 3), where their further continuity could not be established because of the dense staining of the nucleolus (Byers and Goetsch 1975b). These observations suggest that genes for ribosomal RNA (or other nucleolar functions) are borne either on telomeres of two bivalents or in an interior segment of one bivalent. The demonstration by Petes and Botstein (1977) that ordinary reciprocal recombination is unexpectedly rare in rRNA (rDNA) genes arranged in tandem suggests that synapsis, as mediated by the SC, is not only difficult to visualize but may actually be absent within the nucleolus. Petes (1979) has found, more recently, that all rDNA genes are clustered on chromosome XII, so the segments of SC associated with the nucleolus might represent proximal and distal portions of this chromosome.

As in other eukaryotes, the SC appears to participate in synapsis, which is probably a precondition for meiotic recombination, but it has proven difficult to establish any more direct role of the SC in recombination. An analysis of the time course of genetic recombination and cytological events by Olson and Zimmermann (1978) indicated that the bulk of reciprocal exchange was delayed until SC was readily detected but that some recombination (mostly conversion) occurred earlier during the premeiotic S phase. Genetic dissection of the meiotic process should provide for a more definitive assignment of interrelationships (Esposito and Klapholz, this volume). Particularly compelling in this regard is the meiotic behavior of diploids homozygous for cdc7, for which incubation in meiotic conditions at the restrictive temperature permits DNA replication but neither recombination nor formation of the SC (Schild and Byers 1978). Release of the temperature restraint permits both of the latter processes to proceed, so it is not unreasonable to infer a causal relationship between them.

Observations on pachytene nuclei have also revealed the presence of amorphous, densely staining bodies along the central region of the SC (Plate 10a,e) (Byers and Goetsch 1975b). These structures are similar in appearance to the "recombination nodules" described by Carpenter (1975) from *Drosophila* oocytes. The numbers and positions of recombination nodules are well correlated with sites of crossing-over, even when the distribution of crossovers is modified by genetic perturbations (Carpenter 1979). The recombination nodules in yeast are irregularly spaced along the SC at average intervals of 0.4 μm, as is consistent with the expected frequency of crossovers ($\sim$ 70 crossovers/30 μm of SC). Determination of the roles played by the recombination nodules in yeast, as in *Drosophila* and other eukaryotes, remains a major challenge of research into meiotic mechanisms.

Electron microscopy has also been applied to the study of yeast meiosis in searches for intermediates in the pairing and recombination of the chromosomal DNA. Such studies have been stimulated by the fact that the degree of recombination per length of DNA is particularly high in this species. Simchen and Friedmann (1975) have described possible intermediates in gene conversion; these appear to consist of associations between elongated doublestranded molecules and shorter pieces of single-stranded DNA. Klein and Byers (1978) found that DNA from cells in meiotic prophase contains clusters of small regions of strand separation, which might provide for intermolecular associations mediated by pairing of the exposed bases. More recently, Bell and Byers (1979) have found that the crossed strand-exchange form predicted by molecular models of recombination (Holliday 1964) arises within the repeated sequences of the 2-micron plasmid during meiosis. Further study will be required to determine whether similar structures are intermediate in chromosomal recombination.

Meiotic Division and Spore Formation

The SPBs remain paired throughout meiotic prophase and then undergo separation to form the spindle for meiosis I (Moens and Rapport 1971a). Although the absence of chromosomal condensation has precluded direct observation of chromosomal behavior, these investigators reasonably assume that disjunction occurs during the elongation of this spindle, which proceeds without the delay noted in mitosis. The conclusion of meiosis I is most notable for its failure to produce two separate nuclei; instead, the original nucleus remains single, gaining only a partial constriction about its circumference, perpendicular to the spindle. Then the two SPBs each undergo another duplication to produce two pairs of duplicated, but unseparated, SPBs. Separation of these pairs of SPBs then leads to formation of two spindles for meiosis II within a single nucleus. These two spindles frequently lie nearly perpendicular to one another and bear their poles on lobes extending outward from the center of the compound nucleus. Spindle elongation leads to the further protrusion of these lobes, but again nuclear division does not follow directly. The elongate spindles instead remain undivided, as does the overlying nuclear envelope, until the formation of spores is well advanced.

Spore formation is preceded by a striking modification in the structure of the SPBs. Whereas the SPBs present during meiosis I are structurally similar to those in mitotic cells, the SPB gains a wider and more prominent outer plaque during meiosis II (Moens and Rapport 1971a). It is against the cytoplasmic surface of this outer plaque that development of the prospore wall is initiated by the apparent coalescence of small vesicles into a flattened sac (Plate 12a). The extent of the prospore wall is initially limited to the width of the outer plaque, but lateral extensions then proceed along the nuclear surface until the prospore wall forms a cup nearly surrounding the nuclear lobe. The close apposition between the prospore wall and the nuclear envelope persists throughout this period of enlargement, but eventually they become separated from each other, permitting the invasion of mitochondria and other cytoplasmic components into the intervening space (Plate 12b). The entry of cytoplasm continues until the incipient spore has a diameter two to three times that of the nuclear lobe. Then nuclear division is completed by the interruption of the spindles and pinching apart of the nucleus. At nearly the same time, individualization of the presumptive spores takes place by the further extension of the edge of the prospore wall to complete its engulfment of the cytoplasmic mass.

The spore wall proper is then formed by the deposition of material between the two membranous layers of the prospore wall. Analysis of the macromolecular species present in mature spores suggests that glucan and mannose are principal constituents of the wall (Kane and Roth 1974). A dense outer layer, which arises late in wall formation (Lynn and Magee 1970), fails to appear in glucosamine auxotrophs (Whelan and Ballou 1975). The sensitivity of these auxotrophic spores to the β-glucanases employed for ascus dissection suggests that it is a glucosamine-containing constituent (perhaps chitin) of the dense outer layer that normally renders the spores resistant to the enzymes (Ballou et al. 1977).

One may note from the preceding outline of the spore-formation process that a considerable portion of the original cell does not become included in the four ascospores. The greatest exclusion is of cytoplasm, which consists of all detectable cytoplasmic components and remains in the ascal space between the prospore walls and the original cell wall. In addition, a small portion of the original nucleus fails to be incorporated into any of the four spores; this portion contains densely staining material that may include former components of the SC or nucleolus (Moens and Rapport 1971a). Brewer and Fangman (1980) and Brewer et al. (1980) have recently analyzed the propagation of several stable nucleic acids through the sporulation process and have found that none were transferred to spores without partial loss in relation to the transfer of chromosomal DNA. Mitochondrial DNA and rRNA entered the spores to about the same extent as that of mitochondrial and cytoplasmic volumes ($\sim 30\%$ each). Double-stranded RNA, 2-micron DNA, and ribosomal DNA entered the spores to about the same extent as chromosomal DNA, so it seems likely that they are associated with the nucleus (at least during meiosis II). The small nuclear lobe, which appears to be excluded from the spore nuclei (see above; Moens and Rapport 1971b), must therefore not contain a major fraction of these nucleic acids.

Genetic and natural variants of the meiotic processes have provided a wealth of information about the causal interrelationships in meiosis. Many of these have been reviewed by Esposito and Esposito (1978) and elsewhere in this volume, but certain observations pertaining to the relationship between SPB modification and prospore-wall formation are discussed below. One informative type of variation is seen in strains that regularly form only two diploid spores rather than four haploid ones (Grewal and Miller 1972). Moens (1974) found by electron microscopy that the SPBs of these strains gain the outer plaque normally characteristic of meiosis II during the sole meiotic division. Further analysis of newly formed hybrids from this genetic background demonstrated that DNA replication and a normal meiotic prophase, with coincident genetic recombination, occur before the single meiotic division (Moens et al. 1977). So it is evident that modification of the SPB occurs in the meiotic division immediately preceding spore formation. It was also noted that some hybrids underwent partial duplication of the SPBs at either pole of the spindle but again gained an outer plaque before prospore-wall formation. Schild and Byers (1980) found that some variants can proceed even farther through the normal sequence of meiotic events and yet form diploid spores. Simchen (1974) had found earlier that no spores were formed when homozygotes for cdc5 or cdc14 were challenged by their restrictive temperature soon after transfer to sporulation medium. However, Schild and Byers (1980) found that if the temperature was raised during meiosis II, the spindle did not elongate fully; nevertheless, an outer plaque appeared on each pole. Consequently, prospore-wall induction occurred on each separate SPB, but the two poles of each meiosis-II spindle remained so near to each other that their incipient prospore walls became fused together into a single prospore wall. Maturation of the fused wall then led to the production of a diploid spore.

More direct evidence for a positive role by the modified SPB is provided by the genetics and cytology of strains forming fewer than four haploid spores when sporulating under deleterious conditions. Light microscopy has shown that such strains generally complete both meiotic divisions but fail to encapsulate some of the haploid nuclei (Haber and Halvorson 1972). Genetic analysis of a wild-type strain subjected to an artificial prolongation of the meiotic process formed progressively fewer haploid spores unless the medium was replenished (Davidow et al. 1980). Analysis of centromere-linked markers demonstrated the predominant formation of nonsisters in two-spored asci. The cytological basis was found in the pattern of SPB modification at meiosis II: Sporulating cells destined to form two (nonsister) spores underwent modification of only one spindle pole, so only one product of each second meiotic division was encapsulated. This not only indicates the necessity of SPB modification for prospore-wall induction but also suggests that the starving meiotic cell may regulate the number of meiotic products to be encapsulated in order to maximize the successful use of limited resources.

PERSPECTIVES

The elegance of yeast genetics has stimulated research into a wide variety of cytological and biochemical processes in this organism. Implicit in this research has been the hope that the findings of this work would lead to a better understanding of biological mechanisms fundamental to higher eukaryotes as well. The overview gained by cytological methods indicates that our hopes will not be in vain, for many organelles, such as mitochondria and cytoplasmic membrane systems, appear quite similar to those in plants and animals. The particular attention paid to nuclear division mechanisms in this paper has likewise suggested basic similarities to the analogous mechanisms in higher eukaryotes. Despite the lack of nuclear envelope dispersal or any apparent chromatin condensation during mitosis, the mitotic spindle itself seems to share many features with that in other eukaryotes. In addition, the organization of the meiotic bivalent about the SC is clearly typical of eukaryotes in general.

In light of these similarities, it seems likely that the next decade of research will include fruitful structural and biochemical analyses of many specific organelle systems—the mitochondria, the mitotic and meiotic spindles, the nuclear envelope, the secretory membrane systems, the plasma membrane, the SC—in the exquisite detail attainable by molecular genetics. These findings should, in turn, stimulate a better understanding of cognate systems in higher eukaryotes.

Finally, it is perhaps reasonable to inquire whether the special advantage presented by genetic analysis of yeast will not be diminished by the burgeoning field of recombinant DNA research. The enormous potential presented by DNA cloning, in vitro mutagenesis, and DNA-mediated transformation in higher eukaryotes has made it evident (Brown 1981) that we shall soon enjoy a wealth of information about the control of transcription and translation of innumerable gene products, as identified by two-dimensional electrophoresis. We may note, on the other hand, that assessing the structural and enzymatic roles of these gene products is the venue of traditional genetics. The powerful inferences derived from basic genetic tools—conditional mutants, suppressors, epistatic interactions, and so forth—enable us to identify the multitude of genes involved in complex processes and to probe their interactions within the living cell. We may therefore confidently look forward to the further application of traditional genetics, bolstered by the power of genetic engineering, to the elucidation of mechanisms fundamental to the eukaryotic life cycle.

ACKNOWLEDGMENT

The author's research program is supported by the National Institutes of Health (grant GM-18541).

REFERENCES

Agar, H.D. and H.C. Douglas. 1955. Studies of budding and cell wall structure in yeast. *J. Bacteriol.* **70**:427.

Ballou, C.E. 1982. The yeast cell wall and cell surface. In *The molecular biology of the yeast* Saccharomyces. II. *Metabolism and gene expression* (ed. J. Strathern et al.), Cold Spring Harbor Laboratory, Cold Spring Harbor, New York (In press.)

Ballou, C.E., S.K. Maitra, J.W. Walker, and W.L. Whelan. 1977. Developmental defects associated with glucosamine auxotrophy in *Saccharomyces cerevisiae. Proc. Natl. Acad. Sci.* **74**:4351.

Bell, L. and B. Byers. 1979. Occurrence of crossed strand-exchange forms in yeast DNA during meiosis. *Proc. Natl. Acad. Sci.* **76**:3445.

Berns, M.W. and S.M. Richardson. 1977. Continuation of mitosis after selective laser microbeam destruction of the centriolar region. *J. Cell Biol.* **75**:977.

Bicknell, J.N. and H.C. Douglas. 1970. Nucleic acid homologies among species of *Saccharomyces. J. Bacteriol.* **101**:505.

Brewer, B.J. and W.L. Fangman. 1980. Preferential inclusion of extrachromosomal genetic elements in yeast meiotic spores. *Proc. Natl. Acad. Sci.* **77**:5380.

Brewer, B.J., V.A. Zakian, and W.L. Fangman. 1980. Replication and meiotic transmission of yeast ribosomal RNA genes. *Proc. Natl. Acad. Sci.* **77**:6739.

Brown, D.D. 1981. Gene expression in eukaryotes. *Science* **211**:667.

Byers, B. 1981. Multiple roles of the spindle pole bodies in the life cycle of *Saccharomyces cerevisiae.* In *Molecular genetics in yeast, Alfred Benzon Symp. 16* (ed. D. von Wettstein et al.), Munksgaard, Copenhagen (In press.)

Byers, B. and L. Goetsch. 1974. Duplication of spindle plaques and integration of the yeast cell cycle. *Cold Spring Harbor Symp. Quant. Biol.* **38**:123.

―――. 1975a. The behavior of spindles and spindle plaques in the cell cycle and conjugation of *Saccharomyces cerevisiae. J. Bacteriol.* **124**:511.

―――. 1975b. Electron microscopic observations on the meiotic karyotype of diploid and tetraploid *Saccharomyces cerevisiae. Proc. Natl. Acad. Sci.* **72**:5056.

―――. 1976a. A highly ordered ring of membrane-associated filaments in budding yeast. *J. Cell Biol.* **69**:717.

―――. 1976b. Loss of the filamentous ring in cytokinesis-defective mutants of budding yeast. *J. Cell Biol.* **70**:35a.

Byers, B., K. Shriver, and L. Goetsch. 1978. The role of spindle pole bodies and modified microtubule ends in the initiation of microtubule assembly in *Saccharomyces cerevisiae. J. Cell Sci.* **30**:331.

Cabib, E. 1975. Molecular aspects of yeast morphogenesis. *Annu. Rev. Microbiol.* **29**:191.

Cabib, E., R. Ulane, and B. Bowers. 1974. A molecular model for morphogenesis: The primary septum of yeast. *Curr. Top. Cell. Regul.* **8**:1.

Cande, W.Z. and S.M. Wolniak. 1979. Chromosome movement in lysed mitotic cells is inhibited by vanadate. *J. Cell Biol.* **79**:573.

Carpenter, A.T.C. 1975. Electron microscopy of meiosis in *Drosophila melanogaster* females. II. The recombination nodule—a recombination-associated structure of pachytene? *Proc. Natl. Acad. Sci.* **72**:3186.

―――. 1979. Recombination nodules and synaptonemal complex in recombination-defective females of *Drosophila melanogaster. Chromosoma* **75**:259.

Culotti, J. and L.H. Hartwell. 1971. Genetic control of the cell division cycle in yeast. *Exp. Cell Res.* **67**: 389.

Davidow, L.S., L. Goetsch, and B. Byers. 1980. Preferential occurrence of non-sister spores in two-spored asci of *Saccharomyces cerevisiae:* Evidence for regulation of spore-wall formation by the spindle pole body. *Genetics* **94**: 581.

Davidse, L.C. and W. Flach. 1977. Differential binding of methyl benzimidazole-2-yl carbamate to fungal tubulin as a mechanism of resistance to this antimitotic agent in mutant strains of *Aspergillus nidulans. J. Cell Biol.* **72**: 174.

Dujon, B., P.P. Slonimski, and L. Weill. 1974. Mitochondrial genetics. IX. A model for recombination and segregation of mitochondrial genomes in *Saccharomyces cerevisiae. Genetics* **78**: 415.

Egel, R. and M. Egel-Mitani. 1974. Premeiotic DNA synthesis in fission yeast. *Exp. Cell Res.* **88**: 127.

Engels, F.M. and A.F. Croes. 1968. The synaptinemal complex in yeast. *Chromosoma* **25**: 104.

Esposito, M.S. and R.E. Esposito. 1978. Aspects of the genetic control of meiosis and ascospore development inferred from the study of *spo* (sporulation-deficient) mutants of *Saccharomyces cerevisiae. Biol. Cell.* **33**: 93.

Fast, D. 1973. Sporulation synchrony of *Saccharomyces cerevisiae* grown in different carbon sources. *J. Bacteriol.* **116**: 925.

Freifelder, D. 1960. Bud position in *Saccharomyces cerevisiae. J. Bacteriol.* **80**: 567.

Galeotti, C.L. and K.L. Williams. 1978. Giemsa staining of mitotic chromosomes in *Kluyveromyces lactis* and *Saccharomyces cerevisiae. J. Gen. Microbiol.* **104**: 337.

Gibbons, I.R. 1977. Structure and function of flagellar microtubules. In *International cell biology 1976–1977* (ed. B.R. Brinkley and K.R. Porter), p. 348. Rockefeller University Press, New York.

Gordon, C.N. 1977. Chromatin behavior during the mitotic cell cycle of *Saccharomyces cerevisiae. J. Cell Sci.* **24**: 81.

Gould, R.R. and G.G. Borisy. 1977. The pericentriolar material in Chinese hamster cells nucleates microtubule formation. *J. Cell Biol.* **73**: 601.

Grewal, N.S. and J.J. Miller. 1972. Formation of asci with two diploid spores by diploid cells of *Saccharomyces. Can. J. Microbiol.* **18**: 1897.

Haber, J.E. and H.O. Halvorson. 1972. Cell cycle dependency of sporulation in *Saccharomyces cerevisiae. J. Bacteriol.* **109**: 1027.

Hartwell, L.H. 1974. *Saccharomyces cerevisiae* cell cycle. *Bacteriol. Rev.* **38**: 164.

――. 1976. Sequential function of gene products relative to DNA synthesis in the yeast cell cycle. *J. Mol. Biol.* **104**: 803.

Hartwell, L.H., R.K. Mortimer, J. Culotti, and M. Culotti. 1973. Genetic control of the cell division cycle in yeast. V. Genetic analysis of *cdc* mutants. *Genetics* **74**: 267.

Hayashibe, M. and S. Katohda. 1973. Initiation of budding and chitin ring. *J. Gen. Appl. Microbiol.* **19**: 23.

Heath, I.B. 1980. Behavior of kinetochores during mitosis in the fungus *Saprolegnia ferax. J. Cell Biol.* **84**: 531.

Hirschberg, J. and G. Simchen. 1977. Commitment to the mitotic cell cycle in yeast in relation to meiosis. *Exp. Cell Res.* **105**: 245.

Holliday, R. 1964. A mechanism for gene conversion in fungi. *Genet. Res.* **5**: 282.

Horesh, O., G. Simchen, and A. Friedmann. 1979. Morphogenesis of the synapton during yeast meiosis. *Chromosoma* **75**: 101.

Hyams, J.S. and G.G. Borisy. 1978. Nucleation of microtubules *in vitro* by isolated spindle pole bodies of the yeast *Saccharomyces cerevisiae. J. Cell Biol.* **78**: 401.

Inoué, S. and H. Sato. 1967. Cell motility and labile association of molecules. *J. Gen. Physiol. Suppl.* **50**: 259.

Johnston, L.H. and K.A. Nasmyth. 1978. *Saccharomyces cerevisiae* cell cycle mutant *cdc9* is defective in DNA ligase. *Nature* **274**: 891.

Kane, S.M. and R. Roth. 1974. Carbohydrate metabolism during ascospore development in yeast. *J. Bacteriol.* **118:** 8.

Klein, H.L. and B. Byers. 1978. Stable denaturation of chromosomal DNA from *Saccharomyces cerevisiae* during meiosis. *J. Bacteriol.* **134:** 629.

Kubai, D.F. 1975. The evolution of the mitotic spindle. *Int. Rev. Cytol.* **43:** 167.

Lu, B.C. 1967. Meiosis in *Coprinus lagopus:* A comparative study with light and electron microscopy. *J. Cell Sci.* **2:** 529.

Lu, B.C. and D.Y. Jeng 1975. Meiosis in *Coprinus.* VII. The prekaryogamy S-phase and the postkaryogamy DNA replication in *C. lagopus. J. Cell Sci.* **17:** 461.

Lynn, R.R. and P.T. Magee. 1970. Development of the spore wall during ascospore formation in *Saccharomyces cerevisiae. J. Cell Biol.* **44:** 688.

Margolis, R.L., L. Wilson, and B.I. Kiefer. 1978. Mitotic mechanism based on intrinsic microtubule behavior. *Nature* **272:** 450.

Matile, P., H. Moor, and C.F. Robinow. 1969. Yeast cytology. In *The yeasts* (ed. A.H. Rose and J.S. Harrison), vol. 1, p. 219. Academic Press, New York.

McDonald, K.L., M.K. Edwards, and J.R. McIntosh. 1979. Cross-sectional structure of the central mitotic spindle of *Diatoma vulgare.* Evidence for specific interactions between antiparallel microtobules. *J. Cell Biol.* **83:** 443.

McIntosh, J.R. 1974. Bridges between microtubules. *J. Cell Biol.* **61:** 166.

Moens, P.B. 1974. Modification of sporulation in yeast strains with two-spored asci (*Saccharomyces, Ascomycetes*). *J. Cell Sci.* **16:** 519.

Moens, P.B. and E. Rapport. 1971a. Spindles, spindle plaques, and meiosis in the yeast, *Saccharomyces cerevisiae* (Hansen). *J. Cell Biol.* **50:** 344.

――――. 1971b. Synaptic structures in the nuclei of sporulating yeast, *Saccharomyces cerevisiae* (Hansen). *J. Cell Sci.* **9:** 665.

Moens, P.B., M. Mowat, M.S. Esposito, and R.E. Esposito. 1977. Meiosis in a temperature-sensitive DNA synthesis mutant and in an apomictic yeast strain (*Saccharomyces cerevisiae*). *Phil. Trans. R. Soc. Lond. B* **277:** 351.

Moor, H. and K. Muhlethaler. 1963. Fine structure in frozen-etched yeast cells. *J. Cell Biol.* **17:** 609.

Mortimer, R.K. and D.C. Hawthorne. 1966. Genetic mapping in *Saccharomyces. Genetics* **53:** 165.

Mundkur, B. 1961. Electron microscopical studies of frozen dried yeast. III. Formation of the tetrad in *Saccharomyces. Exp. Cell Res.* **25:** 24.

Olson, L.W. and F.K. Zimmermann. 1978. Meiotic recombination and synaptonemal complexes in *Saccharomyces cerevisiae. Mol. Gen. Genet.* **166:** 151.

Pardue, M.L. and J.G. Gall. 1970. Chromosomal localization of mouse satellite DNA. *Science* **168:** 1356.

Petersen, J.G.L., L.W. Olson, and D. Zickler. 1978. Synchronous sporulation of *Saccharomyces cerevisiae* at high cell concentrations. *Carlsberg Res. Commun.* **43:** 241.

Peterson, J.B. and H. Ris. 1976. Electron microscopic study of the spindle and chromosome movement in the yeast *Saccharomyces cerevisiae. J. Cell Sci.* **22:** 219.

Petes, T.D. 1979. Yeast ribosomal DNA genes are located on chromosome XII. *Proc. Natl. Acad. Sci.* **76:** 410.

Petes, T.D. and D. Botstein. 1977. Simple Mendelian inheritance of the reiterated ribosomal DNA of yeast. *Proc. Natl. Acad. Sci.* **74:** 5091.

Pickett-Heaps, J.D. 1969. The evolution of the mitotic apparatus: An attempt at comparative ultrastructural cytology in dividing plant cells. *Cytobios* **3:** 257.

Pickett-Heaps, J.D. and D.H. Tippit. 1978. The diatom spindle in perspective. *Cell* **14:** 455.

Porter, K.R. 1976. Introduction: Motility in cells. In *Cell motility* (ed. R. Goldman et al.), vol. 3, p. 1. Cold Spring Harbor Laboratory, Cold Spring Harbor, New York.

Quinlan, R.A., C.I. Pogson, and K. Gull. 1980. The influence of the microtubule inhibitor, methyl benzimidazol-2-ylcarbamate (MBC) on nuclear division and the cell cycle in

84 B. Byers

Saccharomyces cerevisiae. J. Cell Sci. **46**: 341.

Robbins, E. and N.K. Gonatas. 1964. The ultrastructure of a mammalian cell during the mitotic cycle. *J. Cell. Biol.* **21**: 429.

Robinow, C.F. and J. Marak. 1966. A fiber apparatus in the nucleus of the yeast cell. *J. Cell Biol.* **29**: 129.

Roman, H., M.M. Phillips, and S.M. Sands. 1955. Studies of polyploid *Saccharomyces*. I. Tetraploid segregation. *Genetics* **40**: 546.

Rossen, J.M. and M. Westergaard. 1966. Studies on the mechanism of crossing over. II. Meiosis and the time of meiotic chromosome replication in the Ascomycete *Neottiella rutilans* (Fr.) Dennis. *C.R. Trav. Lab. Carlsberg* **35**: 233.

Roth, R. and H.O. Halvorson. 1969. Sporulation of yeast harvested during logarithmic growth. *J. Bacteriol.* **98**: 831.

Schekman R. and P. Novick. 1982. The secretory process and yeast cell surface assembly. In *The molecular biology of the yeast* Saccharomyces: *Metabolism and gene expression* (ed. J. Strathern et al.), Cold Spring Harbor Laboratory, Cold Spring Harbor, New York (In press.)

Schild, D. and B. Byers. 1978. Meiotic effects of DNA-defective cell division cycle mutations of *Saccharomyces cerevisiae. Chromosoma* **70**: 109.

———. 1980. Diploid spore formation and other meiotic effects of two cell-division-cycle mutations of *Saccharomyces cerevisiae. Genetics* **96**: 859.

Schild, D., H.N. Ananthaswamy, and R.K. Mortimer. 1981. An endomitotic effect of a cell cycle mutation of *Saccharomyces cerevisiae. Genetics* **97**: 551.

Sena, E.P., D.N. Radin, and S. Fogel. 1973. Synchronous mating in yeast. *Proc. Natl. Acad. Sci.* **70**: 1373.

Simchen, G. 1974. Are mitotic functions required in meiosis? *Genetics* **76**: 745.

Simchen, G. and A. Friedmann. 1975. Structure of DNA molecules in yeast meiosis. *Nature* **257**: 64.

Sloat, B.F. and J.R. Pringle. 1978. A mutant of yeast defective in cellular morphogenesis. *Science* **200**: 1171.

Smitt, W.W.S., J.M. Vlak, I. Molenaar, and T.H. Rozijn. 1973. Nucleolar function of the dense crescent in the yeast nucleus. *Exp. Cell. Res.* **80**: 313.

Stern, H. and Y. Hotta. 1977. Biochemistry of meiosis. *Phil. Trans. R. Soc. London B* **277**: 277.

Stevens, B.J. 1977. Variation in number and volume of the mitochondria in yeast according to growth conditions. A study based on serial sectioning and computer graphics reconstitution. *Biol. Cell.* **28**: 37.

Strausberg, R.L. and P.S. Perlman. 1978. The effect of zygotic bud position on the transmission of mitochondrial genomes in *Saccharomyces cerevisiae. Mol. Gen. Genet.* **163**: 131.

Weisenberg, R.C. and A.C. Rosenfeld. 1975. *In vitro* polymerization of microtubules into asters and spindles in homogenates of surf clam eggs. *J. Cell Biol.* **64**: 146.

Westergaard, M. and D. von Wettstein. 1972. The synaptinemal complex. *Annu. Rev. Genet.* **6**: 71.

Whelan, W.L. and C.E. Ballou. 1975. Sporulation in D-glucosamine auxotrophs of *Saccharomyces cerevisiae:* Meiosis with defective ascospore wall formation. *J. Bacteriol.* **124**: 1545.

Wilkie, D. and D.Y. Thomas. 1973. Mitochondrial genetic analysis by zygote cell lineages in *Saccharomyces cerevisiae. Genetics* **73**: 368.

Williamson, D.H. 1966. Nuclear events in synchronously dividing yeast cultures. In *Cell synchrony* (ed. I.L. Cameron and G.M. Padilla), p. 81. Academic Press, New York.

Williamson, D.H. and D.J. Fennell. 1981. Non-random assortment of sister chromatids in yeast mitosis. In *Molecular genetics in yeast, Alfred Benzon Symp. 16* (ed. D. von Wettstein et al.) Munksgaard, Copenhagen (In press.)

Wintersberger, U., M. Binder, and P. Fischer. 1975. Cytogenetic demonstration of mitotic chromosomes in the yeast *Saccharomyces cerevisiae. Mol. Gen. Genet.* **142**: 13.

Zickler, D. and L.W. Olson. 1975. The synaptonemal complex and the spindle plaque during meiosis in yeast. *Chromosoma* **50**: 1.

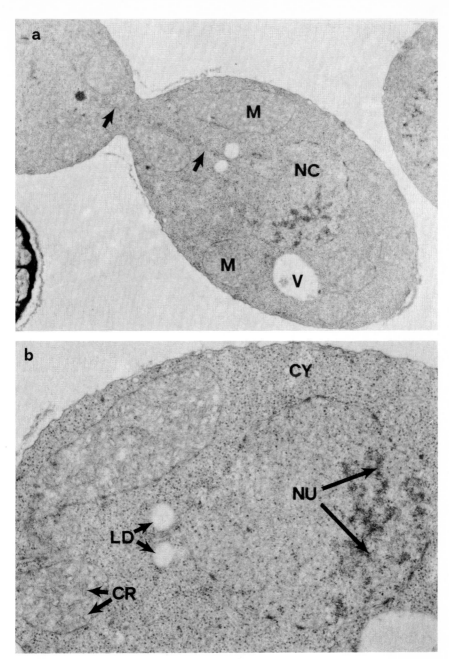

Plate 1 A budded yeast cell from which the wall was removed enzymatically after initial fixation. In *a*, the two poles of the nucleus (NC), seen late in division, are interconnected by a thin strand (arrows). A vacuole (V) and mitochondria (M) are indicated. Magnification, 15,000 ×. In *b*, a more detailed view of an adjacent serial section shows cristae (CR) in one of the mitochondria, lipid droplets (LD), the cytoplasm (CY) filled with ribosomes, and the crescent-shaped nucleolus (NU). Magnification, 30,000 ×.

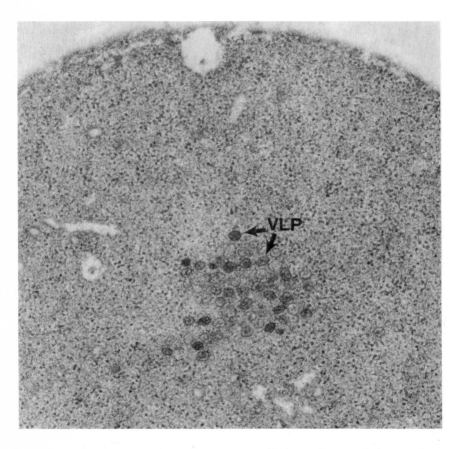

Plate 2   A cluster of viruslike particles (VLP) lies within the ribosome-laden cytoplasm of a vegetative yeast cell. Magnification, 50,000 ×.

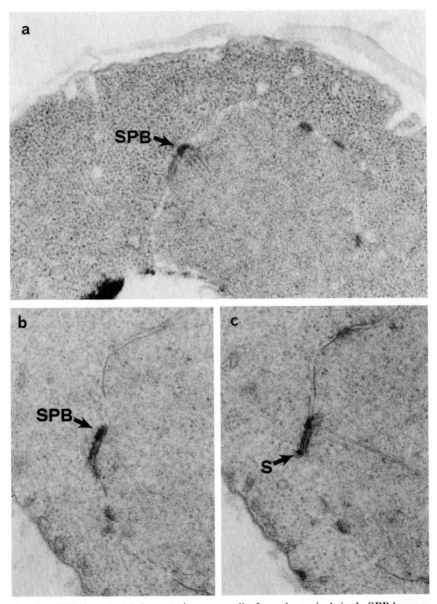

Plate 3 SPBs in unbudded vegetative yeast cells. In *a*, the typical single SPB bears a half-bridge to its lower left. In serial views *b* and *c* of another cell (at a later stage), a satellite (S) is present at the distal end of the half-bridge. Magnification, 50,000×.

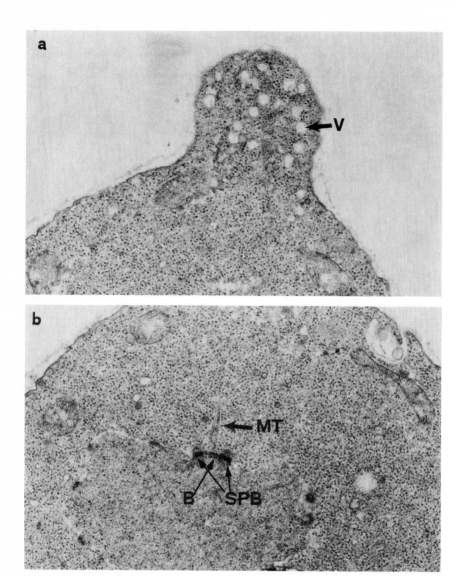

Plate 4 Two images from a serially sectioned vegetative cell with an early bud, which is filled with vesicles (V) shown in *a*. In *b*, the bridge (B) of the duplicated SPBs bears cytoplasmic microtubules (MT) directed toward the emerging bud. Magnification, 30,000 ×.

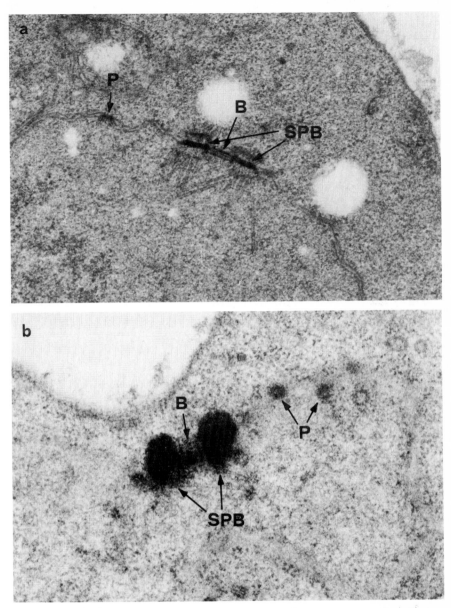

Plate 5 Duplicated SPBs in arrested *cdc4* mutant cells are seen here in two perpendicular orientations: (*a*) in transverse section and (*b*) in tangential section of the nuclear envelope. The discoidal structures of each SPB and of the intervening bridge (B) are evident. Note nuclear pores (P). Magnification, 50,000 ×.

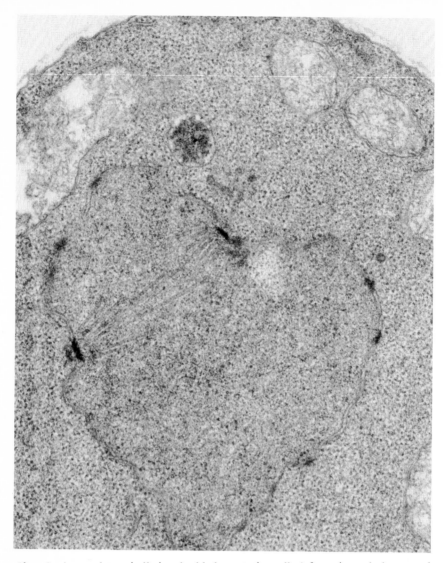

Plate 6   A complete spindle in a budded vegetative cell. A few microtubules extend between one SPB and the other; most are splayed out, presumably toward the kinetochores. Magnification, 50,000 ×.

Plate 7   Structures at the site of cytokinesis in vegetative cells. In *a*, a section passing just beneath the plasma membrane in the neck of the bud reveals the filamentous ring. Remnants of the chitin ring (CR), or incipient bud scar, are indicated. Magnification, 57,600 ×. In *b*, a section through the center of the neck displays the highly ordered filaments in end view as regularly spaced dots. Magnification, 48,000 ×. At a later stage, in *c* (modified from Byers and Goetsch 1976), the filamentous ring has been lost, as cytokinesis is effected by vesicle fusion (arrow). Magnification, 28,800 ×.

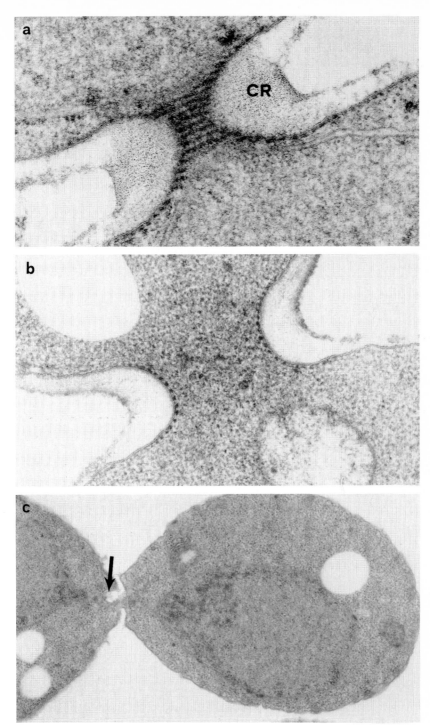

Plate 7 (See facing page for legend.)

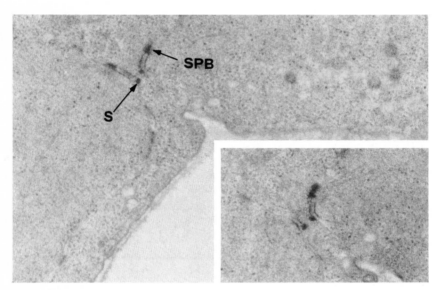

Plate 8   Two serial views of the satellite (S)-bearing SPBs beginning to fuse within a zygote at the initiation of karyogamy. Magnification, 32,400 ×.

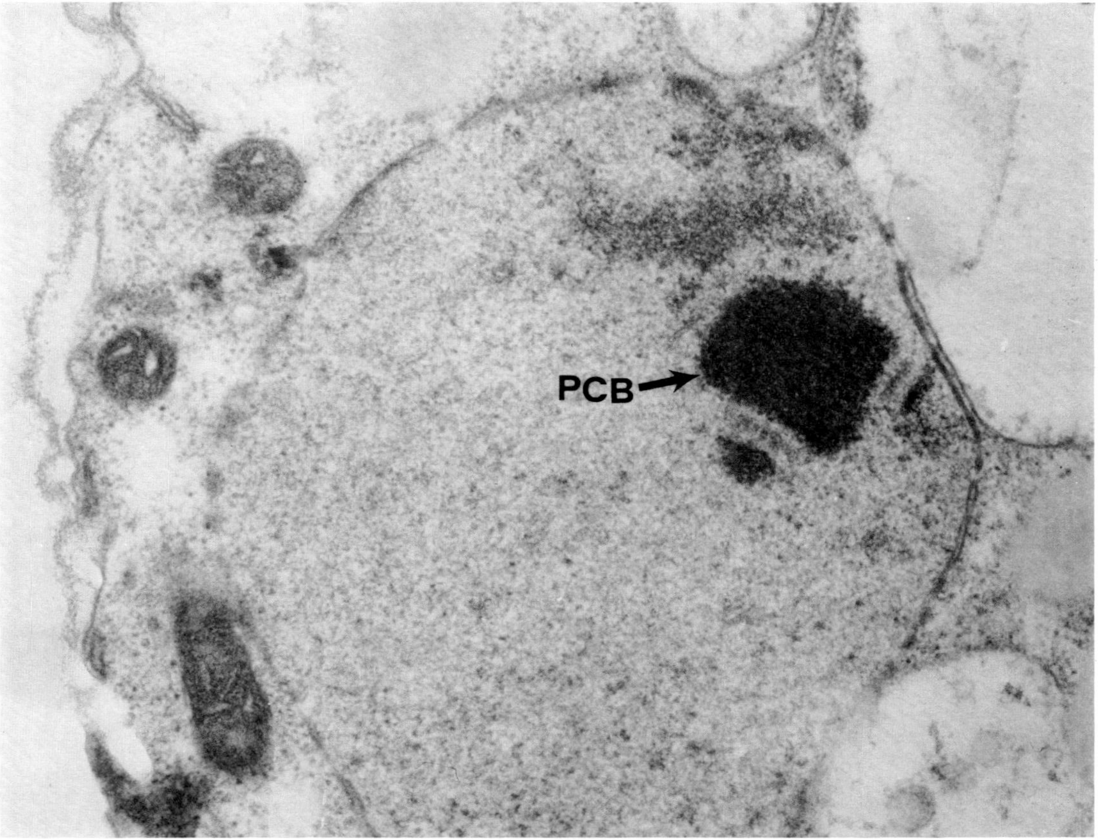

Plate 9 A sporulating cell seen early in prophase. The nucleus contains a prominent polycomplex body (PCB) traversed by central regions of the SC. Magnification, 39,500 ×.

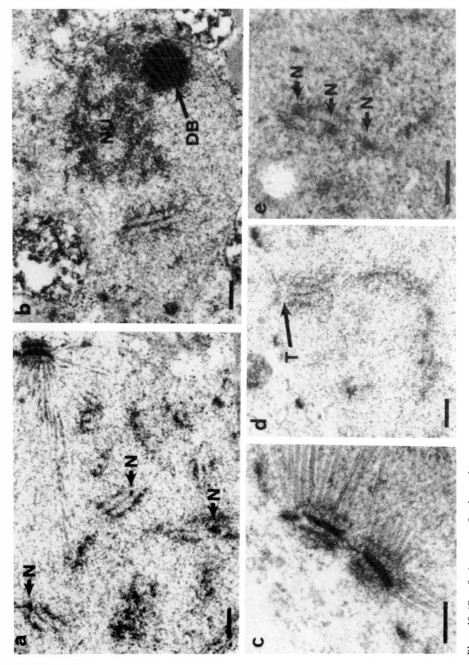

94

Plate 10 (See facing page for legend.)

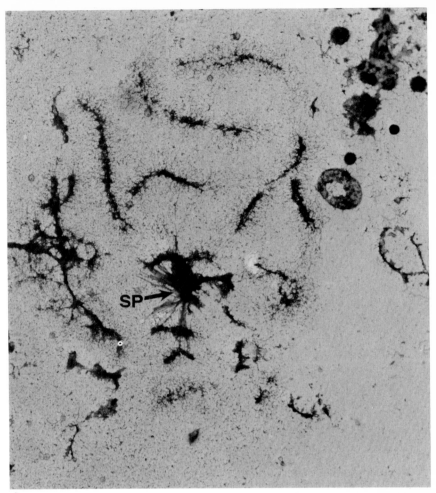

Plate 11 Pachytene chromosomes released on an aqueous surface and recovered on a plastic film for electron microscopy. The spindle pole (SP) and its microtubules are seen at the center. Magnification, 10,000×.

Plate 10 Nuclear structure during pachytene of meiosis: (*a*) Segments of SC with abundant recombination nodules (N) adjacent to the central element. (*b*) The nucleolus (NU) and dense body (DB). (*c*) The duplicated SPBs. (*d*) The attachment of a bivalent to the nuclear envelope by its telomere (T). (*e*) Three closely spaced recombination nodules (N). (Reprinted from Byers and Goetsch 1975b.) Bars represent 0.2 μm.

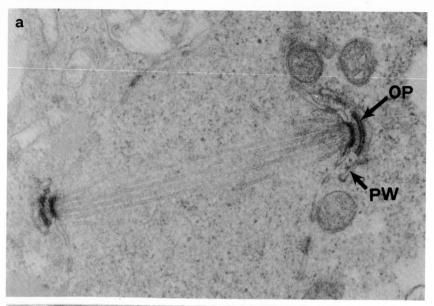

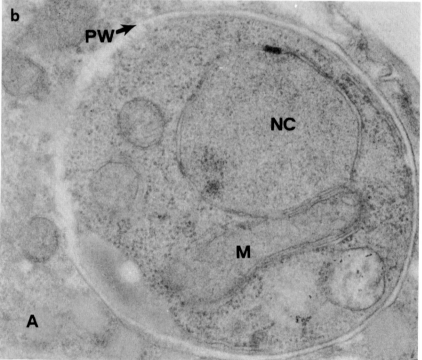

Plate 12 Two stages of the second meiotic division. In *a*, an early stage of prospore-wall (PW) formation has occurred adjacent to the modified outer plaque (OP) characteristic of this stage. At a later stage, in *b*, the expanded PW separates the nucleus (NC), mitochondria (M), and various other cytoplasmic components from the excluded portion of the ascus (A). Magnification, 40,000 ×.

The *Saccharomyces cerevisiae* Cell Cycle

John R. Pringle
Department of Cellular and Molecular Biology
Division of Biological Sciences, University of Michigan
Ann Arbor, Michigan 48109

Leland H. Hartwell
Department of Genetics, University of Washington
Seattle, Washington 98195

1. **How Does a Cell Carry Out a Cell Cycle?**
   A. Discontinuous (Stage-specific) Processes
   B. Continuous Processes
2. **How Is Cell Proliferation Controlled?**
   A. The Concept of Start
   B. Coordination of Successive Cell Cycles
   C. Coordination of Growth with Division
   D. Coordination of Cell Proliferation with the Availability of Essential Nutrients
   E. Coordination of Conjugation with Division
   F. Coordination of Meiosis with Mitosis
   G. Approaches to Start
   H. A Model for the Control of Start

INTRODUCTION

The cell cycle is the process of vegetative (asexual) cellular reproduction; in a normal cell cycle, one cell gives rise to two cells that are genetically identical to the original cell. Questions about the cell cycle can be conveniently divided into two categories. First, one can ask *how* a cell carries out a cell cycle, once it has undertaken to do so. Into this category fall questions about the morphological and biochemical aspects of cell-cycle events and about the mechanisms that ensure their temporal and functional coordination. Second, one can ask what determines *when* a cell will undertake a cell cycle, or how the overall control of cell proliferation is achieved. Into this category fall questions about the coordination of successive cell cycles, the coordination of growth with division, the coordination of cell proliferation with the availability of essential nutrients, and the selection of developmental alternatives. In the text that follows, we consider these two categories of questions in turn. Our bibliography is intended more as a guide to the literature than as a historically accurate record of the development of the field; we apologize to the earlier workers whose contributions thus get less explicit credit than they deserve.

HOW DOES A CELL CARRY OUT A CELL CYCLE?

As has often been noted, successful completion of a cell cycle requires a cell to integrate the processes that duplicate the cellular material with the processes that partition the duplicated material into two viable daughter cells. Another useful formulation of the problem is that successful cellular reproduction requires a cell to integrate the discontinuous (or stage-specific) events that occur once or a few times per cell cycle with the continuous processes of metabolism, maintenance, and growth that occur throughout the cell cycle and, indeed, even when the cell is not cycling. As discussed below, some of the processes of duplication are continuous and others discontinuous, whereas the processes of partitioning can probably all be regarded as discontinuous.

Discontinuous (Stage-specific) Processes

Landmark Events and a Temporal Map of the Cell Cycle

Among the many stage-specific processes of the cell cycle are some that can be monitored morphologically or biochemically with techniques that are presently available; these are referred to as landmark events (Hartwell 1974, 1978; Pringle 1978). A temporal map is simply a diagram that summarizes the temporal order of cell-cycle events. Methods for using asynchronous cultures, synchronous cultures, and single-cell studies to generate temporal maps of landmark events have been adequately reviewed elsewhere (Mitchison 1971; Hartwell 1974; Mitchison and Carter 1975; Pringle 1978) or illustrated by recent studies (Byers and Goetsch 1975; Hartwell and Unger 1977; Creanor and Mitchison 1979; Elliott and McLaughlin 1979a; Zakian et al. 1979; Rivin and Fangman 1980).

The major landmark events of the yeast cell cycle are diagramed and temporally mapped in Figure 1; individual events are described briefly below and in more detail elsewhere (Hartwell 1974; Byers and Goetsch 1975, 1976a; Petes 1980; Ballou 1982; Schekman and Novick 1982; Byers; Fangman and Zakian; both this volume). Like the cell cycles of other eukaryotes, the yeast cell cycle is conventionally divided into a G_1 phase, which precedes the initiation of chromosomal DNA replication; an S phase, during which chromosomal DNA is replicated; a subsequent G_2 phase; and an M phase, during which mitosis and nuclear division occur.

Late in the G_1 phase, the spindle-pole body ([SPB] the microtubule-organizing center in the nuclear envelope) develops a small satellite structure. Soon thereafter, the SPB becomes morphologically duplicated, and the initiation of chromosome replication occurs. At about this time, or somewhat later, events at the cell surface herald the appearance of the future daughter cell: A discrete ring of chitin appears in the largely nonchitinous wall of the mother cell, a small bud emerges within the confines of this chitin ring, and a ring of microfilaments (10 nm in dia.) appears in the cytoplasm adjacent to the cell membrane in the neck region connecting the mother to the

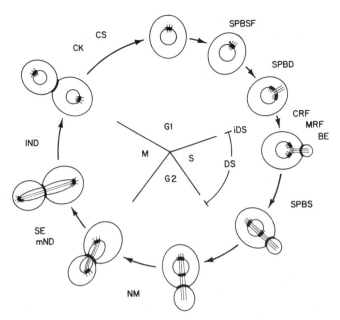

Figure 1 Major landmark events of the *S. cerevisiae* cell cycle. The diagram attempts to indicate the temporal order of events, but distances between events in the diagram are not necessarily proportional to the time intervals between these events; notably, the intervals from SPBSF to BE and from IND to CK may be substantially exaggerated in the diagram. In addition, there is in some cases uncertainty about the temporal order (see text). Abbreviations: SPBSF, spindle-pole-body satellite formation; SPBD, spindle-pole-body duplication; CRF, formation of the chitin ring (shown in the diagram as a heavy line at the mother-bud junction); MRF, formation of the microfilament ring (not shown in the diagram, but found adjacent to the cell membrane in the region of the mother-bud junction); BE, bud emergence; iDS, initiation of chromosomal DNA synthesis; DS, chromosomal DNA synthesis; SPBS, spindle-pole-body separation (and formation of a complete spindle); NM, nuclear migration; mND, medial stage of nuclear division; SE, spindle elongation (as indicated in the diagram, prior to mND, the spindle microtubules do not normally stretch the length of the nucleus; it is not clear whether mND and SE should be regarded as distinct events); IND, late stage of nuclear division; CK, cytokinesis; CS, cell separation (usually monitored after mild sonication, since the time interval from cytokinesis to natural cell separation is highly variable [Pringle and Mor 1975]). Landmark events not shown in the diagram include replication of the 2-micron plasmid DNA, which occurs during the S phase (Zakian et al. 1979); the periodic variations in vacuole structure and cell density (Hartwell 1970; Wiemken et al. 1970); and the various stages of cell separation (septum formation) as defined by Cabib and his colleagues (Molano et al. 1980, and references cited therein).

bud. In cells with small buds, the duplicated SPB and the extranuclear microtubules are clearly oriented toward the bud site; it is not certain whether this orientation is maintained throughout the cycle.

There also remain several points of uncertainty about the precise timing of these early cell-cycle events. (1) For some time, it was believed that bud emergence was dependent on (and thus, necessarily, subsequent to) the duplication of the SPB and that the initiation of DNA synthesis was dependent not only on SPB duplication but also on the subsequent separation of the duplicated structures (Byers and Goetsch 1974; Hartwell 1974, 1978; Hereford and Hartwell 1974). It is now clear (see below, Temporal and Functional Coordination of Stage-specific Events) that both bud emergence and the initiation of DNA synthesis can occur while the duplication of the SPB is blocked and that the initiation of DNA synthesis in fact can occur well in advance of SPB separation. It remains likely that SPB duplication precedes bud emergence and the initiation of DNA synthesis in normal cells, but since Byers and Goetsch (1975) observed no unbudded cells with duplicated SPBs, the time interval must be short. (2) Some studies have indicated that bud emergence and the initiation of DNA synthesis are nearly coincident (Hartwell 1974; Slater et al. 1977; Johnston et al. 1980), whereas other studies have indicated that bud emergence occurs midway through an S phase that lasts one third to one half of the cell cycle (Hartwell 1970; Rivin and Fangman 1980). Since bud emergence and the initiation of DNA synthesis appear to be independent of each other (see below, Temporal and Functional Coordination of Stage-specific Events), this apparent discrepancy may reflect genuine strain-dependent or growth-condition-dependent variations in the relative timing of these events. (3) A related puzzle is posed by the observation that essentially all of the budded cells in an exponentially growing population are able to divide in the presence of hydroxyurea (Slater 1973; Hartwell 1976). Since the drug seems to produce a rapid and specific inhibition of DNA synthesis and since the completion of DNA synthesis is a prerequisite for division (see below, Temporal and Functional Coordination of Stage-specific Events), this observation seems to imply that DNA synthesis is essentially complete by the time of bud emergence. Perhaps this was the case, or the S period was simply much shorter under the conditions of the hydroxyurea experiments than under the conditions of the experiments cited in relation to point 2. (4) Although it is clear that chitin-ring formation precedes bud emergence (Hayashibe and Katohda 1973; Cabib and Bowers 1975; Sloat and Pringle 1978), the timing of these events relative to microfilament-ring formation is uncertain. Cells with small buds display some microfilaments, but the ring is not fully developed (Byers and Goetsch 1976a). It seems possible that microfilament-ring formation is initiated prior to bud emergence or even prior to chitin-ring formation but that the initial filaments are difficult to visualize in unbudded cells.

Whatever the precise order of these early events, once the bud has emerged, it grows steadily as the cycle proceeds. Near the end of the S phase (the precise timing is again uncertain), separation of the SPBs occurs, with concomitant formation of a complete spindle. Somewhat later, the nucleus migrates to the neck connecting mother and bud; still later, the spindle quickly elongates and nuclear division occurs. As in other fungi, the nuclear envelope does not break down during nuclear division in yeast; this circumstance and the smallness of the individual yeast chromosomes have thus far precluded determination of when mitotic chromosome segregation occurs relative to the visible stages of nuclear division. Cytokinesis, or the complete separation of mother cytoplasm from daughter cytoplasm by a barrier of cell membrane, and cell separation, in which mother and daughter cells actually come apart, follow soon after the completion of nuclear division and seem to involve a second burst of chitin synthesis (Molano et al. 1980). Since, by definition, the G_1 phase begins at the end of nuclear division, cytokinesis and cell separation occur during early G_1.

In addition to the well-defined landmark events presented in Figure 1, a variety of metabolic and biosynthetic processes have been reported to be discontinuous (or, at least, to vary greatly in rate) during the cell cycle. For convenience, these are considered together with the continuous metabolic processes later in the text.

Definition of Stage-specific Functions by Mutations and Inhibitors

In addition to the known landmark events, many other stage-specific functions have been identified by the study of cell-division-cycle (*cdc*) mutants or, in a few cases, by the use of stage-specific inhibitors (notably α- and **a**-mating pheromones, as discussed in Coordination of Conjugation with Division).

The Nature of cdc *Mutations.*   By definition, a *cdc* mutation leads to a defect in a particular stage-specific function of the cell cycle (Hartwell 1974, 1978; Pringle 1978). The immediate effect of the mutation is said to be on the primary-defect event, which can be either synthesis or function of the *CDC*-gene product. However, the effects of the mutation can only be analyzed in terms of events that can be monitored biochemically or morphologically (i.e., landmark events). The first such event known to be affected by the mutation is called the diagnostic landmark. The diagnostic landmark may occur considerably later in the cycle than the primary-defect event and may require revision as new landmark events are recognized. For example, discovery of the microfilament ring (Byers and Goetsch 1976a) led to the recognition that mutants whose diagnostic landmark was originally cytokinesis (Hartwell 1971b) were also defective in the much earlier event of microfilament-ring formation (Byers and Goetsch 1976b).

In principle, a *cdc* mutation could affect a function that was either essential for division or only helpful (e.g., in increasing the fidelity of DNA replication or of mitotic chromosome segregation), and the mutation could either block completely, or only produce abnormalities in, the affected function. In practice, most *cdc* mutations in yeast studied to date produce complete (although not necessarily immediate) blockage of events that are essential for cell-cycle progress. Since such mutations are lethal, conditional mutants of some type must be used (Pringle 1975). To date, ordinary temperature-sensitive mutants (*ts*) have been used in most studies, but there have also been some recent successes in isolating cold-sensitive mutants (D. Moir and D. Botstein pers. comm.) and nonsense mutants suppressible by ordinary (Reed 1980b) or temperature-sensitive (Rai and Carter 1981) suppressors.

When conditional-lethal *cdc* mutants are placed under restrictive conditions, each cell ceases normal development at the same point in the cell cycle (the time at which the defective gene product would normally function). Thus, from an initially asynchronous population there develops a morphologically homogeneous population of cells exhibiting the characteristic terminal phenotype for the mutation (Hartwell 1974; Pringle 1978). This homogeneity has been the criterion by which *cdc* mutants have been identified among collections of conditional-lethal mutants (Hartwell et al. 1973). It must be emphasized that both continuous processes and discontinuous processes not dependent on the primary-defect event continue while the primary-defect event is blocked; thus, the terminal phenotype cannot be regarded simply as a normal stage of the cell cycle, and experiments involving the return of arrested cells to permissive conditions can be difficult to interpret (Pringle 1978).

It should be noted that the primary-defect event of a *cdc* mutation need not itself be a stage-specific function. For example, a mutation blocking the normally continuous synthesis of a gene product whose function is stage-specific will generally be identified as a *cdc* mutation. Also, the role of G_1 events in regulating the rate of cell proliferation (see below, How Is Cell Proliferation Controlled?) makes it clear that various mutations affecting continuous aspects of metabolism and growth can lead to G_1 arrest; the discovery that *cdc19* (Hartwell et al. 1973) is a temperature-sensitive pyruvate-kinase mutation (Kawasaki 1979) is a case in point, as is the G_1 arrest of the *tra3* mutant, whose primary defect seems to be in the regulation of amino acid metabolism (Wolfner et al. 1975).

Temporal Mapping of the Events Defined by cdc *Mutations.*   Information about the temporal order of primary-defect events is sometimes provided by studies of the functional relationships among these events (see below, Temporal and Functional Coordination of Stage-specific Events), since one event must follow another on which it is known to be dependent. In addition, it is sometimes possible to learn about the timing of gene-product synthesis and of gene-product function from the execution points that all conditional *cdc* mutants display. The execution point is the time in the cell cycle beyond which a shift from permissive to restrictive conditions can no longer prevent a mutant cell from successfully completing the cycle in question. Because of its dependence on whether the *cdc* mutation is conditional for synthesis or for function and on whether the mutation is tight or leaky, the execution point is clearly an allele-specific, rather than a gene-specific, parameter (see Hartwell et al. 1973, Table 3; Hartwell 1974, Fig. 3). Even allelic mutants that all display first-cycle arrest (i.e., each cell divides once or not at all after the shift to restrictive conditions) can have significantly different execution points. Thus, the extraction of useful information from execution-point data requires that we know the approximate rates at which gene-product activity is lost after a shift to restrictive conditions. Although it is difficult to judge these rates reliably in the absence of direct assays of the gene products, several useful indirect arguments are available; these arguments and other aspects of the gathering and interpretation of execution-point data have been considered in detail elsewhere (Pringle 1978, 1981).

Known CDC *Genes.* Approximately 50 yeast *CDC* genes have been defined on the basis of mutants whose diagnostic landmarks include formation of the SPB satellite, enlargement of the SPB, duplication of the SPB, the initiation of DNA synthesis, the propagation of DNA synthesis, medial nuclear division (or spindle elongation), late nuclear division, chitin-ring formation, and microfilament-ring formation. Table 1 presents a guide to the information available on these genes. The chromosomal location of each mapped gene is included in the table mainly to emphasize the increasing desirability and feasibility of mapping each newly identified *CDC* gene (Mortimer and Schild, this volume).

Total Number of CDC *Genes.* Recent attempts to identify new *CDC* genes using temperature-sensitive mutants have mostly yielded additional mutant alleles of genes already known. For example, the last 34 isolates arresting as multibudded, multinucleate cells have all carried alleles of the known genes *CDC3, CDC10, CDC11,* and *CDC12* (A. Adams and J. Pringle, unpubl.). From such results, it might be inferred that the 50 known *CDC* genes (Table 1) represent a majority of the total. However, a variety of arguments (reviewed in detail by Pringle [1981]) suggest that this conclusion is false and that the recent difficulties in extending the list of known *CDC* genes may be due instead to one or more of the following problems. (1) Different *CDC* genes vary greatly in their susceptibility to conventional temperature-sensitive mutations (Hartwell et al. 1973; Reed 1980a; J. Pringle et al. unpubl.). Thus, many *CDC* genes may be difficult or impossible to identify using conventional temperature-sensitive mutations exclusively. (2) Some yeast structural genes are represented by two or three nontandem copies (or near copies) per haploid genome (Hereford et al. 1979; Holland and Holland 1980). Any *CDC* genes in this category will be difficult to identify by recessive loss-of-function mutations, although they may be identifiable as extragenic suppressors of previously known *cdc* mutations (Jarvik and Botstein 1975; Morris et al. 1979). (3) *CDC* genes whose products are helpful, rather than essential, for cell-cycle progress would not have been detected by the screening procedures used to date, although they could presumably be studied using approaches modeled on those used successfully to study such mutations in *Drosophila* (reviewed by Hartwell 1978). (4) Genes whose

Table 1 Known *CDC* genes and their roles

Gene[a]	Terminal phenotype[b]	Diagnostic landmark[c]	Map position[d]	Basic characterization[e,f]	Remarks and/or additional references[e,f,g]
CDC1	A(?)	BE(?)	—	1,2,13	Many cells arrest with small buds; terminal phenotype also heterogeneous with respect to SPB; macromolecule synthesis shuts down quickly at 36°C; thus, probably not a good *cdc* mutant (14). Essential for mating (7). Also 5,10.
CDC2	H	DS(?)	4L	1,2,4,15	Makes DNA at 36°C but remains sensitive to hydroxyurea, which suggests that the DNA made is incomplete or defective (11), though direct evidence is lacking (16). Required for sporulation (5,17). Undergoes chromosome loss (9). Also 6,7,8,10.
CDC3	M	MRF	—	1,2,13,18	Originally described as defective in CK (13); later shown defective in MRF as well (18). Also 5,7,8,10,11.
CDC4	D	iDS (and SPBS)	6L	1,2,4,19,20	Defective in both of the mutually independent events iDS (4,11,19,20,21,22) and SPBS (2). Essential (probably as a prerequisite; see A Functional Sequence Map of the Yeast Cell Cycle) for 2-micron plasmid replication (23) but not for mitochondrial DNA replication (24). Essential for sporulation (5,25,26) and for karyogamy during mating (7,10). Possibly involved in folded-chromosome transformations (27). Cells released from a *cdc4* block can undergo meiosis directly (8; see Coordination of Meiosis with Mitosis). Also 6,28,29.
CDC5	K	NR(?)	13L	1,3,4	Since it terminates with two separate daughter nuclei (3), the diagnostic landmark is either CK or a late stage of ND involving reorganization of the daughter nuclei (tentatively denoted "nuclear reorganization"). Essential for sporulation (5,30) and for transfer of mitochondria during mating (7,10). Also 11.

CDC6	H	DS(?)	10L	1,2,4,15	Makes DNA at 36°C but remains sensitive to hydroxyurea, which suggests that the DNA made is incomplete or defective (11), though direct evidence is lacking (16). Apparently not required for sporulation (5), though leakiness may account for this result. Undergoes extensive chromosome loss (9). Also 7,8,10,12.
CDC7	H	iDS	4L	1,2,4,15,20	Essential for initiation of replication of chromosomal DNA (4,11,20,22,31) and of 2-micron plasmid DNA (23,32,33) but not of mitochondrial DNA (24). Essential for sporulation (5,17) but not for premeiotic DNA synthesis (17). Not required for development of a "g_1 folded chromosome" (27). Useful in synchronization of the nuclear pathway (33). Also 6,7,8,10,12,29,34,35.
CDC8	H	DS	10R	1,2,4,19	Essential for replication of chromosomal DNA (4,11,19,22,31,36), 2-micron plasmid DNA (23,32,34), and probably mitochondrial (24,31) DNA in vivo, as well as in permeabilized-cell (21,31) and in vitro (29,34) systems. Also essential for premeiotic DNA synthesis (5,17,26) and apparently for error-prone repair (37). Undergoes chromosome loss (9) and produces petites at high frequencies (68). Also 6,7,10.
CDC9	H	DS	4L	1,2,4,15,38	Synthesizes DNA (4,15) and becomes partially insensitive to hydroxyurea (11) at 36°C, but is defective in DNA ligase (38,39). Involved in recombination as well as replication (40) and is probably essential for sporulation (5). Undergoes chromosome loss (9). Also 8,10.
CDC10	M	MRF	3L	1,2,13,18	Originally described as defective in CK (13); later shown defective in MRF as well (18). Has been isolated on a recombinant plasmid (41). Also 5,7,8,10, 29,34.
CDC11	M	MRF	10R	1,2,13,18	Originally described as defective in CK (13); later shown defective in MRF as well (18). Cold-sensitive alleles have been isolated (42). Also 5,7,10,11,12.
CDC12	M	MRF	8R	1,3,13,18	Originally described as defective in CK (13); later shown defective in MRF as well (18). Nonleaky alleles are now available (43). Also 10.

Table 1 *(Continued)*

Gene[a]	Terminal phenotype[b]	Diagnostic landmark[c]	Map position[d]	Basic characterization[e,f]	Remarks and/or additional references[c,f,g]
CDC13	H	mND	—	1,2,4,15	No detectable defect in chromosomal (4,11,15) or plasmid (23) DNA synthesis. Necessary for sporulation (5). Undergoes chromosome loss (9). Also 6,7,8,10.
CDC14	J	IND	6R	1,2,4,15	No detectable defect in chromosomal (4,15) or mitochondrial (24) DNA synthesis. Necessary for sporulation (5,30). Undergoes chromosome loss (9). Also 6,7,8,10,12.
CDC15	J	IND	1R	1,2,4,15	No detectable defect in DNA synthesis (4,11,15). Apparently not required for sporulation (5). Also 7,8,10.
CDC16	H	mND	11L	1,2,4	No detectable defect in DNA synthesis (4,11). Probably necessary for sporulation (5). Possibly affected in synthesis of a microtubule-associated protein (44). Also 7,10,12.
CDC17	H	mND	—	1,2,4	No defect in DNA synthesis detected (4). Probably necessary for sporulation (5). Also 7,10.
CDC18	K	NR(?)	—	1,3	Terminal phenotype, diagnostic landmark like those of *cdc5* (q.v.). Not well studied, since no available allele gives first-cycle arrest (thus, all are presumed leaky). Also 10.
CDC19	A	SPBSF	1L	1,3,9	Apparently is the structural gene for pyruvate kinase (9,45). Also 10.
CDC20	H	mND	—	1,2,4	No defect in DNA synthesis detected (4). Probably necessary for sporulation (5). Possibly defective in the synthesis of a microtubule-associated protein (44). Also 10.
CDC21	H	DS	15R	1,2,4,46,47	The structural gene for thymidylate synthetase (46,47); essential for replication of chromosomal DNA (4,11), 2-micron plasmid DNA (32), and mitochondrial DNA (24) in vivo, though not in a permeabilized-cell system supplied with dTTP (31). Also essential for premeiotic DNA replication (17). Produces petites at high frequency (68). Also 7,10.

Gene					
CDC22	B(?)	SPBE(?)	—	1,2	Phenotype similar to that of *cdc28* mutants, but the original isolate (1) is a multiple mutant in which two mutations are required for the prototype phenotype (43). Noncomplementing isolates (12) also do not give the prototype phenotype. Also 10.
CDC23	H	mND	—	1,2,4	No detectable defect in chromosomal (4,11) or mitochondrial (24) DNA synthesis. Probably essential for sporulation (5). Also 10.
CDC24	C	CRF	—	1,2,14,48,49	Continues growth (6,48,49), DS (4,11,14), and ND (2,14) at 36°C (often for more than one cycle), but is defective in the apparently independent events (see A Functional Sequence Map of the Yeast Cell Cycle) of CRF and BE. Seems generally defective in the spatial localization of growth (48,49,50). Apparently required for mating (7). Also 5,10,51,52,53.
CDC25	A	SPBSF	—	1,3,6,54	Many mutant alleles, including several giving first-cycle arrest, have been isolated using procedures selective for G_1-arrest mutants (54,55). Apparently defective in forming "g_0 folded chromosomes" (27). Apparently not required for sporulation; indeed, mutant diploids sporulate in acetate-containing growth medium at 36°C (56). Also 10,51,52,53,57.
CDC26	H	mND(?)	6R	1,3	The available mutant does not give first-cycle arrest and has been little studied. It grows at 36°C on nonfermentable carbon sources (9). Also 10.
CDC27	J(?)	IND(?)	—	1,3	The available mutants do not give first-cycle arrest and have been little studied. Essential for transfer of mitochondria during mating (10). Undergoes chromosome loss (9).
CDC28	B	SPBE	2R	1,2,4,6,20	Essential (presumably as a prerequisite; see A Functional Sequence Map of the Yeast Cell Cycle) for replication of chromosomal DNA (4,20) and 2-micron plasmid DNA (23) but not of mitochondrial DNA (24). Apparently essential for sporulation (5,56) and for karyogamy during mating (7,10). A selection based on the ability of *cdc28*-arrested cells to conjugate (7) has yielded many additional alleles (58). Apparently defective in forming "g_0 folded chromosomes" (27). An amber mutant is available (59). Has been isolated on a recombinant plasmid (60). Also 29,51,52,61,62.

Table 1 (*Continued*)

Gene[a]	Terminal phenotype[b]	Diagnostic landmark[c]	Map position[d]	Basic characterization[e,f]	Remarks and/or additional references[e,f,g]
CDC29	A	SPBSF	9L	1,3	Little studied. Also 10.
CDC30	J	IND	—	1,3	The available mutants do not give first-cycle arrest and have been little studied. They grow at 36°C on nonfermentable carbon sources (9). Also 10.
CDC31	E	SPBD	—	1,3	For discussion of the significance of this mutant, see A Functional Sequence Map of the Yeast Cell Cycle. At intermediate temperatures, mutant strains undergo diploidization (63). Also 10.
CDC32	B	SPBE	—	1,3	Phenotype similar to that of *cdc28* mutants but not well characterized due to genetic intractability of the original mutant (1). A second allele has been isolated as an extragenic suppressor of a cold-sensitive *cdc* mutant (42). Also 10.
CDC33	A	SPBSF	—	3,6,64	Apparently essential for mating (7). Also 10.
CDC34	D	iDS(?) (and SPBS)	—	3,64	Phenotype apparently similar to that of *cdc4* mutants but not well characterized because the only available allele does not give first-cycle arrest (and thus is presumed to be leaky). Apparently essential for karyogamy (10).
CDC35	A	SPBSF	—	3,54	Numerous alleles, including some giving first-cycle arrest, have been isolated using a procedure selective for G_1-arrest mutants (54). Apparently not required for sporulation; indeed, mutant diploids sporulate in acetate-containing growth medium (56). Also 10,51,52.
CDC36	B	SPBE	4L	58	Phenotype of *MAT*a, *MAT*α, *MAT*a/*MAT*a, and *MAT*α/*MAT*α strains apparently similar to that of *cdc28* mutants (58,69); phenotype of *MAT*a/*MAT*α strains differs (58). Able to conjugate while arrested at 36°C (58). Also 10.
CDC37	B	SPBE	—	58	Phenotype apparently similar to that of *cdc28* mutants (58,69). Able to conjugate while arrested at 36°C (58) but defective in karyogamy (10).

CDC38	B(?)	—	—	—	Phenotype apparently similar to that of *cdc28* mutants. Not well characterized.
CDC39	B	SPBE	3R	58	As for *CDC36* (58,69). Also 10.
CDC40	H	DS(?)	—	65	Makes DNA at 36°C but remains sensitive to hydroxyurea, which suggests that the DNA made is incomplete or defective (65). Essential for sporulation (65).
CDC41	X	?	—	66	Has not been studied in detail. See footnote b. Also 10.
CDC42	C	CRF(?)	—	43	Phenotype apparently similar to that of *cdc24* mutants but has not been characterized in detail.
CDC43	C	CRF(?)	7L	43	As for *cdc42*.
CDC44	H	mND	—	42	Mutant csA10. Only a cold-sensitive allele available. Apparently completes one round of DNA synthesis and becomes insensitive to hydroxyurea at restrictive temperature.
CDC45	H	mND	—	42	Mutant csA18. Only a cold-sensitive allele available. Apparently completes one round of DNA synthesis and becomes insensitive to hydroxyurea at restrictive temperatures. Can be suppressed by certain temperature-sensitive alleles of *CDC46* and *CDC47*, which suggests that these gene products interact with the *CDC45*-gene product.
CDC46	H	?	—	42	Complementation group A (strains rA18-50, etc.). Only temperature-sensitive alleles available, isolated as extragenic suppressors of cold-sensitive mutations of *CDC45* and of a nonallelic cold-sensitive *cdc* mutation. No DNA measurements yet.
CDC47	H	?	—	42	Complementation group B (strain rA18-100). Only a temperature-sensitive allele available, isolated as an extragenic suppressor of a cold-sensitive mutation of *CDC45*. No DNA measurements yet.
CDC48	H	?	—	42	Mutant csE24 and related strains. A cold-sensitive and several temperature-sensitive alleles available. No DNA measurements yet.

Table 1 (Continued)

Gene[a]	Terminal phenotype[b]	Diagnostic landmark[c]	Map position[d]	Basic characterization[e,f]	Remarks and/or additional references[e,f,g]
CDC49	Y	?	—	42	Mutants csB71, etc. Only cold-sensitive alleles available. Like cdc50.
CDC50	Y	?	—	42	Mutant csF81. Only a cold-sensitive allele available. Like cdc49.
CDC51	H	?	—	42	Mutants csH80. Only a cold-sensitive allele available. No DNA measurements yet.
TRA3	X	?	15R	67	Isolated as a mutant derepressed for the enzymes of amino acid biosynthesis during growth at 23°C but displays a cdc phenotype at 36°C. See footnote b. Function is interdependent with the mating-pheromone-sensitive step (67; see A Functional Sequence Map of the Yeast Cell Cycle). Essential for sporulation (56). Also 52.

[a]Most of the genes recognized have been shown to be distinct both by complementation analysis and by linkage tests (either determination of map position or demonstration of nonlinkage to other *cdc* genes). In some cases, however, only complementation has been used or only mutants of similar phenotypes have been included in the analysis, or both.

[b]Although it seems possible that different mutant alleles of the same gene (e.g., conditional-for-function and conditional-for-synthesis alleles) could give different terminal phenotypes, so far no unequivocal case of this type is known. Symbols: (A) unbudded, uninucleate G_1 cells with no SPB satellite—cells do not continue growth under restrictive conditions; (B) unbudded, uninucleate G_1 cells with SPB satellites—cells continue growth and become misshapen "shmoo shapes" under restrictive conditions; (C) unbudded cells that are unable to form chitin rings but continue growth, DNA synthesis, and nuclear division under restrictive conditions; (D) multiply budded, growing, uninucleate G_1 cells whose SPBs are duplicated but not separated; (E) singly budded, growing, uninucleate G_2 cells whose SPBs seem approximately twice normal size but are not duplicated (resulting in a short, unipolar spindle); (H) singly budded, growing, uninucleate G_1 or G_2 cells with complete, short spindles in nuclei situated at the mother-bud neck; (J) singly budded, growing, uninucleate G_2 cells with complete, long spindles in nuclei extended through much of the length of mother and bud—mother and/or bud often develop elongations that may be abortive attempts at new budding; (K) singly budded, growing cells containing two nuclei, each with a G_1 DNA content and a single SPB—cells sometimes develop elongations like those described under J; (M) multiply budded, multinucleate growing cells; (X) unbudded, uninucleate G_1 (not definite for *cdc41*) cells that seem to continue growth under restrictive conditions without becoming misshapen (shmooing)—thus, seem not to fit in class A or B, but states of their SPBs are not known; (Y) uninucleate cells with small buds—cells seem to continue growth (without the buds increasing in size relative to the mother cells) under restrictive conditions—DNA contents and SPB morphologies not yet characterized.

[c]Abbreviations as given in the legend to Fig. 1, except for SPBE (spindle-pole-body enlargement), as distinct from SPBD (spindle-pole-body duplication); see A Functional Sequence Map of the Yeast Cell Cycle), and NR (nuclear reorganization; see remarks on *CDC5*).

[d]Data from Mortimer and Schild (this volume).

[a]References: (1) Hartwell et al. 1973; (2) Byers and Goetsch 1974; (3) Byers 1981; (4) Hartwell 1973b; (5) Simchen 1974; (6) Johnston et al. 1977a; (7) Reid and Hartwell 1977; (8) Hirschberg and Simchen 1977; (9) Kawasaki 1979; (10) Dutcher 1980; (11) Hartwell 1976; (12) Johnston and Game 1978; (13) Hartwell 1971b; (14) Hartwell et al. 1974; (15) Culotti and Hartwell 1971; (16) Johnston and Williamson 1978; (17) Schild and Byers 1978; (18) Byers and Goetsch 1976b; (19) Hartwell 1971a; (20) Hereford and Hartwell 1974; (21) Hereford and Hartwell 1971; (22) Petes and Newlon 1974; (23) Livingston and Kupfer 1977; (24) Newlon and Fangman 1974; (25) Simchen and Hirschberg 1977; (26) Zamb and Roth 1977; (27) Piñon 1979a; (28) Byers and Sowder 1980; (29) Jazwinski and Edelman 1976; (30) Schild and Byers (1980); (31) Oertel and Goulian 1979; (32) Petes and Williamson 1975; (33) Zakian et al. 1979; (34) Jazwinski and Edelman 1979; (35) Carter and Jagadish 1978b; (36) Klein and Byers 1978; (37) Prakash et al. 1979; (38) Johnston and Nasmyth 1978; (39) Game et al. 1979; (40) Fabre and Roman 1979; (41) Clarke and Carbon 1980; (42) D. Moir and D. Botstein, pers. comm.; (43) A. Adams and J. Pringle, unpubl.; (44) Shriver and Byers 1977; (45) G. Kawasaki, pers. comm.; (46) Game 1976; (47) Bisson and Thorner 1977; (48) Sloat and Pringle 1978; (49) Sloat et al. 1981; (50) Field and Schekman 1980; (51) Shilo et al. 1976; (52) Shilo et al. 1977; (53) Shilo et al. 1979; (54) Reid 1979; (55) S. Paris and J. Pringle, unpubl.; (56) V. Shilo et al. 1978; (57) B. Shilo et al. 1978; (58) Reed 1980a; (59) Reed 1980b; (60) Nasmyth and Reed 1980; (61) Jagadish and Carter 1977; (62) Taketo et al. 1980; (63) Schild et al. (1981); (64) Culotti 1974; (65) Kassir and Simchen 1978; (66) J. Culotti, pers. comm.; (67) Wolfner et al. 1975; (68) Newlon et al. 1979; (69) B. Byers and L. Goetsch, pers. comm.

[f]Some of the references listed are surveys for many mutants of genetic analyses (1); morphological properties (2,3); effects on DNA synthesis (4,11), growth (6), sporulation (5), or conjugation (7) and karyogamy (10); susceptibility to chromosome loss during mitosis after release from arrest at 36° C (9); or the ability of arrested mutants to proceed directly to meiosis without completing mitosis (8). For references 5–10, a specific comment is made only where an interesting effect was observed.

[g]For brevity, remarks refer indiscriminately to the apparent functions of the wild-type gene product or to the phenotypes of the corresponding mutants. Except as indicated, all mutants are conventional temperature-sensitive mutants for which 23° C is permissive temperature and 36° C is restrictive temperature. More information about some gene functions and their interrelations is provided in A Functional Sequence Map of the Yeast Cell Cycle.

111

products are responsible for two, or a few, essential stage-specific functions have probably been excluded from consideration by the definition and screening criterion used in isolating the existing set of *cdc* mutants, because such mutants would not have unique terminal phenotypes. (5) Similarly, if there is a flaw in the assumption that any mutation affecting a single, essential stage-specific function will yield a morphologically homogeneous terminal phenotype, some genes may have been excluded from consideration. However, the one attempt to test this assumption (by screening directly for mutants with defective DNA synthesis, without regard to cellular morphology) yielded only mutants in which RNA synthesis (a continuous process) as well as DNA synthesis was defective (Johnston and Game 1978). (6) Reactions catalyzed by distinct gene products in prokaryotes are often catalyzed by the separate domains of multifunctional polypeptides in eukaryotes (Kirschner and Bisswanger 1976; Pringle 1979). If the reactions catalyzed are sequential and if products are nondiffusible, an entire gene coding for a multifunctional protein should behave as a single complementation group. Thus, it is possible that the number of stage-specific functions is significantly greater than the number of *CDC* complementation groups.

Function of CDC *Genes in Mitosis and Meiosis.* The mitotic cell cycle occurs in *MATa* and *MATα* haploid cells, in *MATa/MATα* diploid cells, in zygotes producing their first daughter cells, in germinating spores, and in stationary-phase cells undertaking their first cycles in fresh medium. In all cases tested to date, the products of *CDC* genes seem to function in the mitotic cycle in all of these various circumstances (Hartwell 1974, 1978; L. Hartwell and J. Pringle, unpubl.). In addition, some, but not all, of the *CDC*-gene products also function during meiosis and sporulation (Table 1), as will be discussed further in Coordination of Meiosis with Mitosis.

Primary-defect Events and the Molecular Details of Cell-cycle Progress. One of the main incentives for isolating *cdc* mutants is their potential usefulness in dissecting the molecular mechanisms of cell-cycle events. Full realization of this potential depends on identification of the primary-defect events, an enterprise that has unfortunately proceeded only slowly. To date, in only three cases is the gene product affected by a *cdc* mutation known. There is good evidence that *CDC21* codes for thymidylate synthetase (Game 1976; Bisson and Thorner 1977), and at least one allele appears to be temperature-sensitive for function of the mutant gene product. Similarly, the *cdc19-1* allele codes for a temperature-labile pyruvate kinase (Kawasaki 1979, and pers. comm.). Finally, *cdc9-1* strains are deficient in DNA ligase activity both in vivo and in vitro (Johnston and Nasmyth 1978), although it is not yet clear if the temperature-sensitive mutation is in the structural gene or a controlling element.

In addition, some progress has been made toward elucidating the molecular details of the mutational blocks of the mutants defective in microfilament-ring formation and cytokinesis (Byers and Goetsch 1976b), in DNA synthesis (Petes 1980; Williamson and Johnston 1981; Fangman and Zakian, this volume), in chitin-ring formation and bud emergence (Sloat and Pringle 1978; Sloat et al. 1981), and in the behavior of the SPB (Byers and Goetsch 1975; Dutcher 1980; Byers 1981, and this volume).

Temporal and Functional Coordination of Stage-specific Events

It is clear that the stage-specific events of the cell cycle cannot occur in a random order if viable daughter cells are to be produced. A priori, it seems likely that the relative order of some events (e.g., DNA synthesis and nuclear division) must be invariant but that the relative order of other events (e.g., bud emergence and the stages of DNA synthesis) might well be flexible and thus could vary with strain and growth conditions. The available temporal-mapping data support these expectations (see above, Landmark Events and a Temporal Map of the Cell Cycle; Hartwell 1976; Hartwell and Unger 1977; Jagadish and Carter 1977; Slater et al. 1977; Carter and Jagadish 1978b; Johnston et al. 1980; Lord and Wheals 1980; Rivin and Fangman 1980).

There are at least two possible mechanisms by which the normal temporal order of cell-cycle events might be achieved (Mitchison 1971). First, a central timer or "clock" might trigger events at appropriate times. Second, the temporal order of events might be a consequence of their functional inter-relationships; i.e., the temporal order A, then B, might result from the functional dependence of event B upon the prior occurrence of event A.

The Possibility of Cell-cycle Clocks.   The periodic budding of *cdc4* mutants (Table 1) suggested the possibility of a central timer controlling bud emergence (Hartwell 1971a; Hartwell et al. 1974). However, the notion of budding as a simple periodic event is difficult to reconcile with the concept of control of cell-cycle initiation (and hence of budding) by the "Start" event, for which strong evidence exists (see below, How Is Cell Proliferation Controlled?). In addition, no other evidence has emerged for a central timer controlling cell-cycle events, and the known functional interrelationships of events seem adequate to explain what we know of their temporal order. Thus, we think it more likely that the behavior of *cdc4* mutants reflects their arrest at the position in the cycle at which budding is normally triggered, coupled with the gradual decay of a control element that normally restricts budding events to one per cycle.

Functional Sequence Maps.   A functional sequence map is a diagram summarizing the functional interrelatedness of cell-cycle events. In addition to suggesting how the normal temporal order of events may be achieved, a functional sequence map can provide important clues to the molecular nature of these events (Hartwell 1976).

Three principal methods are available for functional sequence mapping. The use of these methods and the interpretation of the resulting maps are subject to a variety of constraints and potential complications, as discussed in detail elsewhere (Pringle 1978, 1981). In the single-mutant method, strains carrying single *cdc* mutations are shifted to restrictive conditions, and the landmarks that do and do not occur are monitored. Comparison of different mutants can provide information about the functional relations both among landmark events and among the mutants' primary-defect events. In the double-mutant method, the terminal phenotypes of two single mutants are compared to that of the constructed double mutant. (Note that it must be possible to apply restrictive conditions for both mutations simultaneously.) The terminal phenotype of the double mutant reflects the functional relationship of the primary-defect events but yields no information on the functional relations of landmark events. The reciprocal-shift method requires two reversible blocks that can be applied independently and consists of two separate experiments. In the first experiment, block A alone is applied during a first incubation; when the cells are arrested, block A is removed and block B is applied, and the ability of the cells to complete some later landmark event during this second incubation is monitored. The second experiment is the reciprocal of the first: Block B alone is applied during the first incubation, and block A alone is applied during the second. The behavior of the cells during the two experiments reflects the functional relationship of the primary-defect events of the two blocks; as in the case of the double-mutant method, it yields no information on the functional relations of landmark events. The use of these methods is illustrated below.

A Functional Sequence Map of the Yeast Cell Cycle.   Figure 2 presents our current tentative picture of the functional organization of the yeast cell cycle. The following commentary refers to the letters attached to the individual steps in the map.

Steps A and B: Mutants of several genes arrest as unbudded, nongrowing, G_1 cells whose SPBs bear no satellites (Table 1), much as do nutrient-limited cells (Byers and Goetsch 1975; see below, Coordination of Cell Proliferation with the Availability of Essential Nutrients). In contrast, mutants of several other genes, like cells treated with an appropriate mating pheromone (see below, Coordination of Conjugation with Division), arrest as unbudded G_1 cells that continue to grow and whose SPBs bear satellites but do not enlarge appreciably or become duplicated (Table 1). Given these results, the fact that SPB satellite formation seems to be an antecedent of SPB duplication during the normal cell cycle (Byers and Goetsch 1975) and the interdependence of the steps controlled by *CDC28, CDC36, CDC37,* and *CDC39* with the mating-pheromone-sensitive step (Hereford and Hartwell 1974; Reed 1980a), it seems reasonable to postulate two distinct and sequential steps (A and B in Fig. 2). However, reciprocal-shift experiments have indicated that at least the *CDC25* and *CDC33* steps, and probably the *CDC35* step as well, are also interdependent with the mating-pheromone-sensitive step (Culotti 1974; Reid

1979). Thus, we remain uncertain as to how to view this complex of events at the beginning of the cycle, a point of considerable interest because of the importance of these events in the overall control of cell proliferation (see below, How Is Cell Proliferation Controlled?). In any case, it does seem clear that the complex of early events is prerequisite for most, if not all, subsequent steps of the cell cycle (Hartwell et al. 1974).

Steps C, D, and E: As indicated in Table 1, *cdc24* mutants are unable to bud but undergo SPB duplication and DNA synthesis; *cdc4* mutants are unable to initiate DNA synthesis but bud and duplicate their SPBs; and *cdc31* mutants bud and replicate their DNA but are unable to achieve duplication of their SPBs (although an apparent doubling in size of the SPBs does occur). From these facts, the single-mutant method allows the strong conclusion that steps C, D, and E are all independent of one other, as are the landmark events SPB duplication, initiation of DNA synthesis, and bud emergence. The lack of dependence of bud emergence and the initiation of DNA synthesis on the duplication of the SPB is a major revision of previous concepts (Byers and Goetsch 1974; Hartwell 1974; Hartwell 1978) necessitated by the recently discovered properties of *cdc31* mutants (Byers 1981).

Steps B, D, and F: The use of the double-mutant and reciprocal-shift methods to establish this dependent sequence has been described previously (Hereford and Hartwell 1974).

Step G: Although the *CDC4* and *CDC31* functions are independent (see above), both of these functions are necessary for SPB separation (Byers 1981). Since the *CDC31*- and *CDC4*-gene products must function well in advance of SPB separation in order to permit SPB duplication and the initiation of DNA synthesis (for the timing of these events, see above, Landmark Events and a Temporal Map of the Cell Cycle), it is unlikely that these gene products are the last whose function is required for SPB separation. However, no genes whose products function specifically in step G have been identified. Note that the removal of SPB separation from the B-D-F sequence (Byers and Goetsch 1974; Hartwell 1974, 1978) is justified not only by the behavior of *cdc31* mutants (Byers 1981) but also by the fact that this event occurs near the end, rather than near the beginning, of DNA synthesis.

Steps D, F, H, H', I, J, and K: Reciprocal-shift experiments have shown clearly that some gene-product-controlled steps (steps D and F) are prerequisites for, others (step H) are interdependent with, and still others (steps I, J, and K) are dependent on, the hydroxyurea-sensitive step (Hartwell 1976). The status of the *CDC9*-controlled step is less certain, but it is shown here as a separate step H' for two reasons. First, the results of reciprocal-shift experiments with hydroxyurea (Hartwell 1976), although most compatible with the dependent order shown, were in fact intermediate between the results expected from this order and those expected if the *CDC9*-controlled and hydroxyurea-sensitive steps were interdependent. Second, the finding that *cdc9* mutants are defective in DNA ligase (Johnston and Nasmyth 1978)

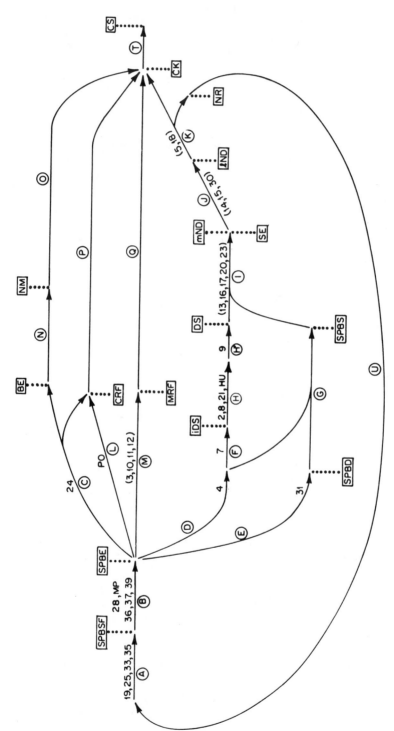

116

Figure 2 (See facing page for legend.)

makes it likely that the *CDC9*-controlled step really is dependent on the DNA-synthesis functions of step H and prerequisite for the nuclear-division functions of steps I, J, and K. (Note that step H presumably still lumps the chemically sequential steps of precursor synthesis and polymerization; the difficulty of resolving such sequences with the methods available for functional sequence mapping is expected [Pringle 1981].) The sequencing experiments with hydroxyurea did not allow any conclusions about the functional relations among the various events that were dependent on the hydroxyurea-sensitive step, and we have grouped these events into the dependent series I-J-K (1) on the very weak evidence (Pringle 1978) that their diagnostic landmarks (medial nuclear division or spindle elongation, late nuclear division, and the hypothetical nuclear reorganization [Table 1]) are sequential and (2) on the evidence from reciprocal-shift experiments that the hydroxyurea-sensitive step, the *CDC44* step, and the *CDC14* step form a dependent sequence (D. Moir and D. Botstein, pers. comm.). The simple series I-J-K is almost certainly an oversimplification, since the reciprocal-shift experiments of Moir and Botstein have also suggested that the functions comprising nuclear division lie in at least two parallel pathways. It does seem clear that these various events are all prerequisite for cytokinesis (Hartwell et

Figure 2   A tentative functional sequence map of the yeast cell cycle. Whereas some features of this map seem solidly established, other features are only weakly supported by the available evidence, and some features are frankly speculative (see text). In this map, if two events are *interdependent* (i.e., neither can be completed while the other is blocked), their symbols are associated with the same arrow (A,B). If event B is dependent upon event A (i.e., A can be completed while B is blocked, but not vice versa), their symbols are associated with arrows connected head-to-tail, or by an unbroken sequence of arrows connected head-to-tail (A B or A X Y B; the latter example summarizes a dependent series, or dependent sequence, of B upon Y upon X upon A). If two events are independent (i.e., either can be completed while the other is blocked), their symbols are associated with arrows in parallel (i.e., arrows that cannot be connected by an unbroken sequence of arrows connected head-to-tail). For reasons discussed elsewhere (Pringle 1981), only events defined by *cdc* mutations or stage-specific inhibitors are associated directly with the arrows, whereas each landmark event is shown in a balloon attached to the head of the arrow(s) representing the last gene-product-function step(s) known to be necessary for completion of the landmark event. The positioning of landmarks from left to right corresponds approximately to their temporal order, except that SPBE and SPBD seem to be essentially coincident in normal cells. Encircled letters are reference symbols for the discussion in the text; *CDC* genes are indicated by their numbers; MP, PO, and HU indicate the stage-specific inhibitors mating pheromone, polyoxin D, and hydroxyurea, respectively; abbreviations for landmark events are given in the legend to Fig. 1, except for SPBE (spindle-pole-body enlargement; see text and Table 1), and NR (nuclear reorganization; see Table 1). The functional relations among events whose symbols are grouped within parentheses are not known.

al. 1974). It is uncertain whether step E and the putative step G are in fact prerequisite for the gene-product functions presently grouped as step I (as shown in Fig. 2) or are simply an independent prerequisite for medial nuclear division and spindle elongation.

Steps C, L, and M: It is clear that the CDC24-controlled function (step C) is essential both for bud emergence and for chitin-ring formation (Sloat and Pringle 1978), as shown in Figure 2. Indeed, since the CDC24-gene product seems generally involved in generating cellular spatial organization (Field and Schekman 1980; Sloat et al. 1981), it seems likely that its function is also necessary for normal microfilament-ring formation. However, the difficulty of looking for the microfilament ring in unbudded cdc24 cells has thus far precluded a direct test of this possibility (B. Byers, pers. comm.), and we have not indicated such a relationship in the map. In addition, it is possible that the CDC24-gene product acts independently of step B (as opposed to the dependent relationship shown in Fig. 2), since cdc24 cells arrested with mating pheromone at restrictive temperature do not show the normal localization of growth to the shmoo tip (Field and Schekman 1980). Also uncertain is the status of step L. This step is defined by the observation that polyoxin D, a potent inhibitor of fungal chitin synthetases in vitro, can apparently block chitin synthesis (and, eventually, normal septation) without blocking budding or nuclear division (Cabib and Bowers 1975). Similar results were obtained by starving a glucosamine auxotroph for glucosamine (Ballou et al. 1977). Although it is not clear whether the inhibition of chitin synthesis in vivo by these methods is either complete or entirely specific (Ballou et al. 1977; Sloat and Pringle 1978), the most likely interpretation of the results available is that chitin synthesis is a step (L) that is a prerequisite for cytokinesis but not for steps C, D, or E. Since, in addition, mutants defective in genes CDC24, CDC4, and CDC31 all synthesize chitin at restrictive temperature (Sloat and Pringle 1978; Sloat et al. 1981; A. Adams and J. Pringle, unpubl.), the single-mutant method leads to the conclusion that step L is independent of steps C, D, and E. A further uncertainty is whether step L is really dependent on step B (as shown in Fig. 2), since cells arrested by mating pheromone do deposit chitin in their growing tips (Schekman and Brawley 1979). Even more problematic is the status of step M. Since mutants defective in CDC3, CDC10, CDC11, or CDC12 undergo DNA synthesis and SPB duplication, whereas mutants defective in CDC4 or CDC31 form microfilament rings (Byers and Goetsch 1976b), the single-mutant method establishes that the CDC3, CDC10, CDC11, and CDC12 functions are independent of steps D and E. Reciprocal-shift experiments with mating pheromone have established that at least the CDC3 and CDC11 functions are dependent on step B (Bücking-Throm et al. 1973; CDC10 and CDC12 were not tested). However, since the relations among these four functions are not known, it is not clear whether a single step M or a complex of steps is required. Moreover, since it is not known whether cdc24 mutants, polyoxin-treated cells, or glucosamine auxotrophs form microfilament rings, whether *cdc3,* *cdc10, cdc11,* and *cdc12* mutants form chitin rings, or when exactly the microfilament ring is formed relative to bud emergence and chitin-ring formation, the relationship between the step or steps M and steps C and L is unknown. Thus, the treatment of step M in Figure 2 is justified less by evidence than by its potential value in focusing attention on the unanswered questions involving this part of the functional-sequence map.

Steps N, O, P, and Q: Since steps C, L, and M are required for early landmark events and also for the late event of cytokinesis, it seems likely that gene products functioning in the presently hypothetical steps N, O, P, and Q will eventually be discovered. The modest justification for positioning the landmark event nuclear migration in this part of the map has been given previously (Hartwell et al. 1974).

Step T: Although this step has not been studied in detail, it is clear that a variety of gene products function in the step or steps leading from cytokinesis to the actual separation of mother and daughter cells (Hartwell et al. 1974; Pringle and Mor 1975).

Step U: The coordination of successive cycles is considered later in the text.

Interpretation and Implications of the Functional Sequence Map. As Figure 2 and the associated discussion show, although many details remain obscure, the broad outlines of the functional organization of the yeast cell cycle seem reasonably clear. First, the discontinuous aspects of cellular reproduction seem to occur as a well-defined program of discrete, gene-controlled events, each of which occurs at a specific point in the program. Second, this program of stage-specific events includes some but not all (see below, Continuous Processes) aspects of the essential duplication of cellular materials, as well as all known aspects of the partitioning of the duplicated materials into daughter cells. Third, the program contains dependent sequences of events but also incorporates branch points (where several independent events depend on a common precursor event), convergence points (where one event depends on several independent precursor events), and parallel pathways of mutually independent events. Fourth, the program seems to have a definite beginning (in the event or complex of events denoted A and B in Figure 2) and end (in cytokinesis and cell separation). A cell that has completed division but has not yet carried out steps A and B is, at least temporarily, not engaged in the program of stage-specific events and might sensibly be said to be noncycling, even if continuous processes of metabolism and growth are continuing at a high rate. As will be made clear later in the text, cells generally carry out steps A and B and thus undertake the program of stage-specific events only when they will be able to complete the entire program and divide. Fifth, the functional interrelations among the stage-specific events seem to provide an adequate explanation for what is known of their temporal relations (see above, Landmark Events and a Temporal Map of the Cell Cycle; Temporal and Functional Coordination of Stage-specific Events). Some groups of events (e.g., those involved in DNA synthesis and nuclear division) occur in rigidly prescribed temporal orders (which are probably essential for successful reproduction) because they form dependent sequences. In contrast, the relative timing of events lying in parallel pathways (e.g., functions involved in bud emergence or the initiation of DNA synthesis) may vary with strain and growth conditions, but the pathways cannot get grossly out of phase because they are all dependent on the same event(s) at the beginning of the cycle and are all prerequisite for completion of the cycle by cytokinesis. Sixth, it should be remembered that many of the stage-specific events must be localized in space as well as in time (Sloat et al. 1981; Schekman and Novick 1982; and see below, the discussion of *CDC8* and *CDC21*).

Finally, it is possible to address the question of whether progress through the program of stage-specific events is controlled by the sequential synthesis of the appropriate gene products, by the sequential activation of gene products that are present continuously, or both. Analysis of this question would be greatly facilitated if it were known which *cdc* mutants are temperature-sensitive for synthesis and which are temperature-sensitive for function of the affected gene products. Unfortunately, direct information allowing this discrimination is available only for a few *cdc* mutants (see above, Primary-defect Events and the Molecular Details of Cell-cycle Progress).

Another approach to these questions is provided by the experiments of Byers and Sowder (1980), who formed hybrid cells containing temperature-sensitive *cdc* nuclei in cytoplasms containing normal levels of the wild-type products of the genes in question. For all *cdc* mutants tested, except *cdc4*, these hybrid cells could complete several cell cycles at the restrictive temperature, which suggests either that the wild-type gene products are normally present in considerable excess over what is required for completion of a cell cycle or that the cells normally contain excess amounts of exceptionally stable mRNAs for these gene products. The former (and seemingly more likely) interpretation has two corollaries. First, most *cdc* mutants showing first-cycle arrest (such as those used in developing the functional sequence map shown in Fig. 2) must be temperature-sensitive for function of the mutant gene products. Thus, the map provides no information about possible dependency relationships involving gene-product synthesis. Second, even if many *CDC*-gene products are synthesized discontinuously (in contrast to most cellular proteins—see below, Protein Synthesis), most steps in the cell cycle must be triggered by the stage-specific activation of gene products or by the stage-specific availability of substrates, rather than by the stage-specific synthesis of gene products already present in excess. This conclusion is consistent with the observation that mitochondrial DNA synthesis is continuous during the cell cycle (see below, Synthesis of Mitochondrial DNA and Double-stranded RNA). Since this process depends on the *CDC8*- and *CDC21*-gene products (Newlon and Fangman 1975), it seems that these gene products must be present continuously but activated in a temporally and spatially localized fashion during the cell cycle.

On the other hand, if the results of Byers and Sowder (1980) are due to the presence of unusually stable mRNAs, then stage-specific translation of these mRNAs would remain a possible mechanism for controlling progress through the program of stage-specific events. In any case, however, it seems unlikely that stage-specific transcription of *CDC* genes plays a very general role in this control. In presenting this view of the program of stage-specific events as predominantly a sequence of gene-product activations, we do not mean to imply that stage-specific transcription or translation plays no role in the cell cycle. Since the budded interval of the cell cycle increases slightly in length at slow growth rates, it is likely that some stage-specific components must be synthesized as the cycle proceeds (see below, Coordination of Growth with Division), a conclusion supported by the observation that a complete blockage of protein synthesis can prevent cells from traversing most intervals of the cell cycle (see below, Coordination of Cell Proliferation with the Availability of Essential Nutrients). Moreover, histone protein and histone mRNA do seem to appear in a stage-specific fashion (see below, Protein Synthesis). Finally, the fact that the *CDC4*-gene product does not appear to be present in excess (Byers and Sowder 1980) leaves open the possibility that stage-specific transcription or translation of this and a few other key *CDC*-gene products is important in controlling cell-cycle events.

Continuous Processes

The continuous processes occur throughout the program of stage-specific events leading to division (Figs. 1 and 2) and also, for the most part, in cells that have not undertaken this program or that are blocked in its execution.

Maintenance, Intermediary Metabolism, and Overall Growth

Whether or not environmental conditions permit net growth (i.e., a balanced increase in the components of cell mass) and cell proliferation, a cell must continuously expend energy (Ball and Atkinson 1975; Lagunas 1976; Lillie and Pringle 1980) for such purposes as protein turnover (Bakalkin et al. 1976; Betz 1976; Sumrada and Cooper 1978b; Holzer and Heinrich 1980) and the maintenance of intracellular pH (Kotyk 1963; Navon et al. 1979) and ionic composition (Watson 1970). The amounts of energy required can be substantial (Watson 1970) and may be greater in growing than in nongrowing (stationary-phase) cells (Lagunas 1976).

When net growth is possible, generally it seems to proceed continuously at whatever rate is allowed by the supply of energy and raw materials. Small daughter cells increase steadily in volume prior to initiating the program of stage-specific events (Hartwell and Unger 1977; Johnston et al. 1977a), and total cell mass seems to increase steadily during the cell cycle (Mitchison 1971). Moreover, growth continues unabated when cell-cycle progress is blocked by any of a variety of *cdc* mutations or stage-specific inhibitors (Table 1; see below, Coordination of Growth with Division; Johnston et al. 1977a). (The mutants with terminal phenotype A [Table 1] seem, at first glance, to be exceptions to this rule; however, it is likely that the G_1 arrest of these mutants is secondary to a primary lesion affecting growth [Johnston et al. 1977a].) The only convincing exception to the rule of continuous growth is the "lag phase" that stationary-phase cells undergo before beginning growth after exposure to fresh medium (Scopes and Williamson 1964; Williamson 1964), although there may also be culture conditions under which major fluctuations in the rate of growth occur during the cell cycle (von Meyenburg 1969; Wiemken et al. 1970).

A steady increase in total cell mass probably reflects a more-or-less continuous generation of ATP and formation of biosynthetic precursors, as well as a continuous accumulation of macromolecules (see below). Surprisingly, however, some cultures in which total cell mass and protein content increased continuously showed stepwise increases both in total nitrogen (Williamson 1964) and in the rate of oxygen consumption (Scopes and Williamson 1964; Williamson 1964). The meaning of these observations is not clear, and it is possible that they are merely artifacts of the synchronization procedure. However, pronounced cyclic changes in fermentation and respiration rates have also been observed with cells cycling synchronously under very different conditions (Küenzi and Fiechter 1969; von Meyenburg 1969), where the changes in energy metabolism have been related to cyclic changes in reserve-carbohydrate levels (Küenzi and Fiechter 1969, 1972). In addition, stepwise changes in rates of nutrient uptake have been observed using a technique that seems to avoid synchronization artifacts (Kubitschek and Edvenson 1977).

Components of Growth

Synthesis of Cell-wall Polysaccharides.   Chitin synthesis has been treated as a discontinuous process because it apparently occurs mainly in one or two bursts (see above, Landmark Events and a Temporal Map of the Cell Cycle). However, chitin synthesis does occur in cells arrested by mating pheromone (Schekman and Brawley 1979) and may also occur at a low rate during much of the cell cycle (Cabib and Bowers 1975; Sloat et al. 1981). In contrast, synthesis of the major cell-wall polysaccharides (glucan and mannan) is clearly continuous in small cells growing prior to budding (Katohda et al. 1976), in cells arrested by mating pheromone (Lipke et al. 1976), and in budded cells (Hayashibe et al. 1977), although there are indications of a decrease in rate near the time of division (Wiemken et al. 1970; Hayashibe et al. 1977). In budded cells, the deposition of new glucan and mannan is confined to the bud and, during most of the cycle, to its growing tip (Sloat et al. 1981; Schekman and Novick 1982).

RNA Synthesis. A variety of studies have agreed that total RNA and the major classes of RNA are synthesized steadily during the cell cycle, but they have not agreed on the precise pattern of this synthesis. The recent study of Elliott and McLaughlin (1979b) seems to have avoided most potential artifacts by (1) labeling cells during balanced exponential growth and then separating them at 2° C according to position in the cell cycle and (2) using a dual-label (pulse and long-term) technique to look at rates of synthesis and accumulation. This study concluded that the several rRNA species, tRNA, and poly(A)-containing mRNA are all accumulated continuously and synthesized at exponentially increasing rates as the cell cycle proceeds. It is also clear that total RNA accumulation continues unabated in cells whose cell-cycle progress has been blocked by mating pheromones (Throm and Duntze 1970; Wilkinson and Pringle 1974), hydroxyurea (Slater 1973), or various *cdc* mutations (Culotti and Hartwell 1971; Hartwell 1971a, 1973b).

Since each of the major classes of RNA is contributed by the transcription of many genes, the question of how the transcription of individual genes occurs during the cell cycle remains unresolved. Now that clones of many individual genes are available, this issue should soon be clarified. Indeed, studies with cloned histone genes have already suggested that the corresponding mRNA is present only during S phase (Hereford et al. 1981), although the data on protein synthesis (see below) suggest that this is a special case.

Protein Synthesis. It seems clear that total protein is accumulated continuously and synthesized at an exponentially increasing rate during the cell cycle (Mitchison 1971; Elliott and McLaughlin 1978). Moreover, it now seems that this pattern of synthesis also applies to nearly all individual proteins. This conclusion is based primarily on studies using two-dimensional gel electrophoresis in an approach similar to that used by Elliott and McLaughlin (1979b) to study RNA synthesis (see above). Forty-nine ribosomal proteins and 130 other proteins were examined quantitatively; none showed any significant deviation from an exponentially increasing rate of synthesis during the cell cycle, and only two showed any indication of periodic degradation or modification (Elliott and McLaughlin 1978, 1979a; Elliott et al. 1979). Qualitative examination of an additional 400 protein spots did not reveal any obvious cases of synthesis restricted to a discrete interval of the cell cycle (Elliott and McLaughlin 1978). Only for histones is there good evidence of discontinuous synthesis during the cell cycle (Moll and Wintersberger 1976), and even here the possibility of an artifact cannot at present be ruled out (Groppi and Coffino 1980).

Since the studies just discussed all have dealt necessarily only with proteins of moderate to high abundance, it remains conceivable that many low-abundance proteins are synthesized discontinuously. However, the results of

Byers and Sowder (1980; see above, Interpretation and Implications of the Functional Sequence Map) with regard to *CDC*-gene products (at least some of which are presumably in the low-abundance class and which seem a priori the most likely candidates for periodic synthesis) do not encourage this idea.

The conclusion that nearly all proteins are synthesized continuously is difficult to reconcile with the many studies indicating that the activities of most enzymes undergo stepwise increases during the cell cycle (Halvorson et al. 1971; Mitchison 1971, 1977; Yashphe and Halvorson 1976; del Rey et al. 1979). Although it is not certain, it seems very likely that some of the enzymes studied are included among the spots visualized on the two-dimensional gels. The apparent discrepancy could be resolved by supposing that the enzymes are synthesized continuously but that newly made proteins are activated only during discrete intervals of the cell cycle. However, there is no evidence for such a control mechanism, and its purpose is obscure. It is likely that many of the apparent steps in enzyme activity are synchronization artifacts (Mitchison 1977; Creanor and Mitchison 1979). However, some of the later studies were carried out in ways that seem to avoid such problems (Yashphe and Halvorson 1976; del Rey et al. 1979), so that a puzzle clearly remains.

Synthesis of Mitochondrial DNA and Double-stranded RNA. Although the issue has been controversial, it now seems clear that mitochondrial DNA synthesis is continuous through the cell cycle (Sena et al. 1975) and continues for some time both in mating-pheromone-arrested cells (Petes and Fangman 1973) and in arrested *cdc* mutants not specifically defective in DNA synthesis (Newlon and Fangman 1975). The "killer" double-stranded RNAs can be replicated in G_1-arrested cells but are not made during S phase (Zakian and Fangman 1979).

Integration of Continuous and Discontinuous Processes

Although some uncertainties remain, the general picture that has emerged is of a continuous background of maintenance, metabolism, and (the environment permitting) net growth, on which can be superimposed the program of discontinuous events (including some special aspects of growth) that leads to cell division. The problems of coordination facing a cell are basically two: (1) It must coordinate the various continuous processes so that growth is balanced, and (2) it must divide approximately once per doubling in mass achieved by balanced growth. The former topic is outside the scope of this review (see, however, Ludwig et al. 1977; Warner and Gorenstein 1978; Swedes et al. 1979; Warner 1982); note that the coordination of mitochondrial DNA replication and of double-stranded RNA replication with other aspects of growth are especially interesting problems. The latter topic is considered below, in Coordination of Growth with Division.

HOW IS CELL PROLIFERATION CONTROLLED?

The Concept of Start

Many observations to be reviewed in this section suggest the existence of a unique control point in the cell cycle at which mating pheromones, growth, nutrient availability, and events of the previous cell cycle exert control over cellular reproduction. This control step has been termed Start (Hartwell et al. 1974), since its completion marks a commitment by the cell under normal conditions to complete the program of stage-specific events leading to cell division. Start is a prerequisite for the initiation of DNA synthesis, bud emergence, and all other known stage-specific events of the cell cycle and is operationally defined as the position of arrest by growth and nutrient limitation, by mating-pheromone exposure, and by certain *cdc* mutations (*cdc25, cdc28, cdc33, cdc36, cdc37, cdc39*) at the restrictive temperature. Thus, Start corresponds to steps A and B in Figure 2.

Coordination of Successive Cell Cycles

Yeast cells arrested in nuclear division by any of a variety of *cdc* mutations or inhibitors neither put out new buds nor initiate new rounds of DNA synthesis, whereas mutants in which cytokinesis is blocked undergo repeated cycles of budding, DNA synthesis, and nuclear division (Hartwell et al. 1974). These observations suggest that the Start of cell cycle n depends upon the completion of the nuclear division but not necessarily of the cytokinesis of cell cycle $n-1$ (cf. step U in Fig. 2).

Coordination of Growth with Division

During balanced growth, populations of cells maintain a constant size distribution for countless generations. Control mechanisms must exist that coordinate the continuous accumulation of most cellular constitutents with the distribution of the accumulated material to daughter cells at division. Indeed, the existence of homeostatic mechanisms that maintain cell size is evident in *Saccharomyces cerevisiae* (Johnston et al. 1977a) and has been rigorously demonstrated in the fission yeast *Schizosaccharomyces pombe* (Fantes 1977).

A model to explain the coordination between growth and division in *S. cerevisiae* was proposed by Johnston et al. (1977a), developed further by Hartwell and Unger (1977), and has subsequently received considerable support (Adams 1977; Jagadish and Carter 1977; Jagadish et al. 1977; Johnston 1977; Johnston et al. 1977b; Slater et al. 1977; Carter and Jagadish 1978a; Carter et al. 1978; Tyson et al. 1979; Lord and Wheals 1980). The model consists of two proposals. First, it proposes that growth, rather than progress through the program of stage-specific events (Fig. 2), is normally rate-limiting for cellular proliferation, i.e., that the cell can normally complete the program of events from Start through cytokinesis in less time than it requires for a complete doubling of its macromolecular constituents. Second, the model proposes that growth to a critical size is a prerequisite for the completion of one and only one step in the cell cycle, namely Start. The term size is employed deliberately for its vagueness, since we do not know how the cell monitors its size. However, it seems that the cell monitors some parameter, probably a specific macromolecule, whose amount is reasonably closely coupled to cell volume, mass, protein content, and RNA content.

Two tenets of this model were experimentally verified by Johnston et al. (1977a): (1) Completion of stage-specific events subsequent to Start does not require significant growth, and (2) completion of Start does require growth to a critical size. Upon nitrogen starvation, cells that had passed Start completed division with less than a 10% increase in protein and produced abnormally small daughters as small as one-tenth the volume of the mother cells. Upon reinoculation into fresh medium, these small cells budded after intervals of time that were inversely related to their initial volumes. The small cells were located at or before Start, and they did not execute Start until they had attained a size that was about the same as that at which mother cells (those with one or more bud scars) executed Start.

If, as the model proposes, growth is normally rate-limiting, then control mechanisms that limit growth in the absence of division should not be necessary. Indeed, there appears to be no immediate feedback from the program of stage-specific events to growth, since cdc mutants blocked at various stages of the cell cycle are able to continue growth, attaining cell volumes, masses, and protein contents two to three times those of cells in balanced growth (Johnston et al. 1977a).

The mode of coordination between growth and division is more dramatically evident in S. cerevisiae than in most organisms as a consequence of the fact that the sibling cells at division are generally unequal in size (Hartwell and Unger 1977; Johnston et al. 1977a; Carter and Jagadish 1978a; Lord and Wheals 1980). The mother cell (identified by time-lapse photography or by the permanent bud scar on its surface) is usually larger than the daughter cell it produces. The discrepancy in size is greater at slower growth rates when growth rate is limited either by carbon source (Carter and Jagadish 1978a; Lord and Wheals 1980) or by limiting concentrations of cycloheximide (Hartwell and Unger 1977). The larger mother cell has a shorter subsequent cell cycle than the daughter cell, and the difference in cycle times is accounted for by the longer G_1 period of the daughter cell (Fig. 3).

The difference between the cell cycles of the mother and daughter is adequately accounted for by the model of Johnston et al. (1977a). The mother cell proceeds from budding to division without a doubling of its mass when growth is rate-limiting because no stage-specific event subsequent to Start

Figure 3   Asymmetric division in *S. cerevisiae*. Parent and daughter cells are usually unequal in size at division. The daughter cell (upper cell at division in the figure) grows more and takes longer to pass Start than does the parent cell.

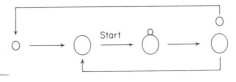

This asymmetric mode of division has implications for the population age distribution that must be taken into account in making calculations on the basis of cell-cycle data. Appropriate equations have been derived describing the cell-cycle age distribution (Hartwell and Unger 1977), the genealogical age distribution (Lord and Wheals 1980), and the mean cell age and cell volume (Tyson et al. 1979).

depends upon the attainment of a specific increment of growth. Since the mother cell wall is rigid, most of the mass remains with the mother cell at division. Both mother and daughter grow subsequent to division, but since the daughter is smaller than the mother, it must grow more and, hence, wait a longer time, before completing Start.

The model of Johnston et al. (1977a) specifies only a critical size necessary for the Start event and that the mass-doubling time normally exceeds the time required to complete the program of stage-specific cell-cycle events. Recent observations suggest that more parameters will be necessary to account for all aspects of the coordination between growth and division in *S. cerevisiae*. A moderate increase in the interval of the cell cycle from bud emergence to division is evident at slow growth rates, which is not predicted by the model (von Meyenburg 1968; Hartwell and Unger 1977; Jagadish and Carter 1977, 1978; Slater et al. 1977; Carter and Jagadish 1978a; Tyson et al. 1979; Lord and Wheals 1980). The dependence of the budded interval upon growth rate suggests that some stage-specific components are not present in excess (see above, Interpretation and Implications of the Functional Sequence Map) and, hence, that their rate of accumulation can be rate-limiting under conditions of slow growth. Furthermore, the volume at which cells complete Start is not constant but varies systematically both with growth rate and with cell age. At slower growth rates, cells initiated buds at smaller volumes than did cells at rapid growth rates, when growth rate was controlled by varying the rate of supply of glucose in a chemostat, by changing the carbon source, or by changing the nitrogen source (Johnston et al. 1979; Lorincz and Carter 1979; Tyson et al. 1979; Lord and Wheals 1980). The mother-cell volume at bud emergence varies with the genealogical age of the mother, increasing about 20 percent each generation (Hartwell and Unger 1977; Johnston et al. 1979; Lorincz and Carter 1979). Since mother cells are apparently the same volume at division as they were at the previous Start event, it is noteworthy that even the mother cell exhibits a growth requirement for the next Start. This fact and the variability of volume at Start with growth rate suggest that the cell does not monitor volume itself but, rather, some parameter only loosely correlated with cell volume. Furthermore, were it not the case that the mother cell had a growth requirement for Start, the mother cell might not be able to control its division in response to nutritional deprivation. (The next section will argue that nutritional control of division acts through the critical size requirement at Start.)

The point in the cell cycle at which cells become committed to complete Start at a given size was investigated by shifting cells from a medium in which Start normally occurs at a smaller size to one in which Start normally occurs at a larger size and vice versa (Johnston et al. 1979; Lorincz and Carter 1979). The results indicate that adjustment to the size characteristic of the new medium is made by cells in G_1 rather close to the time at which Start occurs.

Coordination of Cell Proliferation with the Availability of Essential Nutrients

Cellular reproduction responds in an orderly manner to the availability of required nutrients. Starvation of a population of prototrophic cells for carbon and energy, ammonia, sulfate, phosphate, biotin, or potassium leads to uniform arrest of the cells in the unbudded phase of the cell cycle (Hartwell 1974; Pringle and Mor 1975; Unger and Hartwell 1976; Johnston et al. 1977a; Piñon 1978; Sumrada and Cooper 1978b; Piñon and Pratt 1979; Lillie and Pringle 1980). In most if not all cases, the unbudded cells are arrested at or before Start, since supply of the missing nutrient under conditions restrictive for Start does not lead to budding or division.

Growth-rate limitation by nutrients produces a characteristic change in the intervals of the cell cycle, which also suggests a major control point at Start. A variety of different measurements concur in finding that the interval of time from some event early in the cell cycle to the time of division is relatively constant, increasing only modestly at very slow growth rates. Many observations are consistent with the notion that most of the increase in cell-cycle time that occurs when cells are placed under conditions of slower growth, either by glucose limitation in a chemostat (von Meyenburg 1968; Jagadish and Carter 1977, 1978; Carter and Jagadish 1978b) or by changes in carbon source (Barford and Hall 1976; Jagadish and Carter 1977, 1978; Slater et al. 1977; Carter and Jagadish 1978b; Lord and Wheals 1980), occurs in the interval from division to Start. Thus, the interval from bud emergence to division (von Meyenburg 1968; Slater et al. 1977; Lord and Wheals 1980), the interval from the end of DNA synthesis to division (Jagadish and Carter 1978), the interval from the initiation of DNA synthesis to division (Barford and Hall 1976; Slater et al. 1977), the interval from the execution of the *CDC7* function to division (Carter and Jagadish 1978b), the interval from the execution of the *CDC28* function to division (Jagadish and Carter 1977), and the interval from the execution point for mating-factor sensitivity to division (Jagadish and Carter 1977) are relatively constant with generation times ranging from about 2 to 6 hours. The largest variation in the budded interval was reported in a study comparing many carbon sources in three different media (Tyson et al. 1979), but these authors too found the greatest variation to occur in the interval from division to bud emergence. Only one report disagrees with these studies, finding a proportionate increase in G_1, S, and G_2 at two different growth rates achieved by changes in nitrogen source (Rivin and Fangman 1980); possibly some nitrogen limitations produce a different response, or this study may have been complicated by the phenomenon of basic amino acid inhibition of DNA synthesis (Sumrada and Cooper 1978a), since the strain utilized required lysine. Another recent study, employing a prototrophic strain, did find a selective expansion of G_1 when growth rate was altered by varying the nitrogen source (Johnston et al. 1980). In summary, many nutrients control the same step in the cell cycle, Start, and may work by a common mechanism.

Two conditions must be met for a population of cells to arrest at Start during nutritional starvation. At some point in the history of the culture, the declining level of the limiting nutrient must be inadequate for cells that have not completed Start to do so (first condition), yet still adequate for cells that have just completed Start to finish the remaining events of the cycle and divide (second condition). Since nutrients are necessary for macromolecular synthesis, the most parsimonious explanation of how the first condition is met is that the failure to complete Start is a direct consequence of the cells' inability to satisfy the critical size requirement for Start. However, it is also possible that nutrients control the cell cycle via some metabolic signal. This possibility was critically tested for the sulfate-assimilation pathway, and the results demonstrated that no intermediate of sulfate assimilation between sulfate uptake and the synthesis of methionyl tRNA was a positive regulator of Start (Unger and Hartwell 1976).

If nutrients control the cell cycle through the growth requirement at Start, then all nutritional starvations should satisfy the first of the two conditions mentioned above. However, not all starvations permit uniform arrest of the starved populations of cells at Start. Starvation for magnesium, many auxotrophic starvations, and even abrupt withdrawal of glucose can result in populations of cells arrested at various positions in the cell cycle (Cooper et al. 1979; J. Pringle and L. Hartwell, unpubl.). This fact can be accounted for by supposing that only certain limitations permit the cells to satisfy the second condition, which may require the synthesis of certain stage-specific proteins. It is clear that some proteins must be accumulated during the cell cycle, since cells make little progress in the mitotic cycle (with the exception of completing DNA synthesis; Hereford and Hartwell 1973; Williamson 1973) when protein synthesis is arrested with inhibitors (Slater 1974; Unger and Hartwell 1976; Johnston et al. 1977b). Nevertheless, cells progress through the events from Start to division with little delay when protein accumulation is severely limited (though not prevented) by nitrogen starvation (Johnston et al. 1977a, b). These observations suggest that some proteins are either present in limiting concentrations or synthesized at discrete times in the cell cycle and that the synthesis of these proteins is regulated by growth rate: Their differential rates of synthesis must be increased at slow growth rates and during the onset of starvation conditions. Such proteins would define a sub-set of *CDC* proteins, since most *CDC*-gene products appear to be present in excess (see above, Interpretation and Implications of the Functional Sequence Map) and since most proteins seem to be synthesized continuously (see above, Protein Synthesis).

S. cerevisiae cells that have arrested at or before Start in response to nu-tritional starvation are in a physiological state distinct from that of cells that are at or before Start in growing cultures. Cells in populations that have entered stationary phase because of any of a variety of nutritional limitations display resistance to killing by heat, by cell-wall-degrading enzymes, and by various chemicals; such resistance is orders of magnitude greater than that of the G_1 cells in actively growing populations (Schenberg-Frascino and Moustacchi 1972; Deutch and Parry 1974; Parry et al. 1976; Piedra and Herrera 1976; S. Paris and J. Pringle, unpubl.). Also, at least two types of nutritional limitation yield stationary-phase populations whose cells contain a characteristic "g_0 folded chromosome" (Piñon 1978, 1979a,b; Piñon and Pratt 1979). The stationary-phase *S. cerevisiae* cell may be analogous to the G_0 animal cell. However, the existence of a distinct G_0 stage of the cell cycle would be more strongly evidenced by the isolation of mutations in genes whose products function in the transition to or from the G_0 state but have no role in actively proliferating cells.

Coordination of Conjugation with Division

In addition to the mitotic cell cycle and the putative G_0 state, conjugation (mating) between cells of opposite mating types constitutes yet a third program in the *S. cerevisiae* life cycle (Thorner, this volume). When actively growing cultures of opposite mating types are mixed, the cell cycles of conjugating cells become synchronized prior to cell fusion (Hartwell 1973a) by the mating pheromones that are constitutively produced and secreted by cells of each mating type (Stötzler et al. 1976; Betz and Duntze 1979). Each pheromone specifically arrests cells of opposite mating types at Start, prior to bud emer-gence, the initiation of DNA synthesis, and SPB duplication (Throm and Duntze 1970; Bücking-Throm et al. 1973; Hartwell et al. 1974; Hereford and Hartwell 1974; Wilkinson and Pringle 1974; Byers and Goetsch 1975). During conjugation, cell agglutination (Campbell 1973) is followed by cell fusion and nuclear fusion (Byers and Goetsch 1975). The resulting zygote then embarks on a mitotic cell cycle that employs the program of cell-cycle events discussed above (see How Does a Cell Carry Out a Cell Cycle?).

The synchronization of cells of opposite mating types is of obvious value, if not necessity, when the conjoined elements are to embark upon a mitotic program. However, it is not obvious a priori that a particular site in the cycle would be better than another and, hence, it was of interest to determine whether conjugation was restricted to the Start step or merely facilitated at this step. Under normal conditions, the mating pheromones assure synchronization of the cells at this point and preclude challenge at other steps. However, when *cdc* mutants were presynchronized at various steps in the cycle by incubation at the restrictive temperature and then challenged to mate, it was found that only mutants arrested at the *CDC28* step (Start) were able to mate (Reid and Hartwell 1977).

Coordination of Meiosis with Mitosis

During meiosis, events similar to those in the mitotic cell cycle occur but achieve a different outcome. DNA synthesis is followed by extensive recombination involving a structure unique to meiotic cells, the synaptonemal complex, and chromosome segregation occurs twice. The meiotic segregations of chromosomes differ from the mitotic in that, in the first segregation, homologous centromeres separate but sister centromeres do not, and, in the second, segregation is not preceded by DNA replication. Furthermore, cytokinesis and cell division in meiotic *S. cerevisiae* cells are accomplished by the growth of new cell wall and cell membrane within the cytoplasm of the mother cell, in contrast to the septum formation of mitotic cells (Moens 1971). It is of considerable interest to determine whether the cell calls upon an entirely new genetic program during meiosis or merely alters the mitotic one.

This question was answered for *S. cerevisiae* by testing the ability of the temperature-sensitive *cdc* mutants to complete meiosis at the restrictive temperature (Simchen 1974; Table 1). Nearly all of the genes that are known to control nuclear events during mitosis—SPB duplication and segregation, DNA replication, chromosome segregation—were found to be essential for meiosis as well. In contrast, four of the genes (*CDC24, CDC3, CDC10, CDC11*) that control the parallel pathway(s) in the mitotic cycle that leads to budding and cytokinesis (Fig. 2) were found not to be essential for meiosis and spore formation. Clearly, the meiotic program is achieved through modification of the mitotic one. It will be of interest to determine the functional sequence map for *CDC*-gene-product functions during meiosis, since some of the unique outcomes of meiosis might be achieved merely by changing the order of mitotic gene expression. Genes that function uniquely in *S. cerevisiae* meiosis have also been identified previously (reviewed by Baker et al. 1976; Esposito and Klapholz, this volume). It should be possible to locate the positions of the meiosis-specific functions within the altered mitotic program.

Cells are induced to switch from the mitotic to the meiotic program by starvation for nitrogen in the presence of a nonfermentable carbon source (Esposito and Klapholz, this volume). A shift from growth medium to sporulation medium results in most cells completing the mitotic cell cycle and arresting in G_1 (probably at Start) before they embark upon the meiotic program (see Hirschberg and Simchen 1977). Is it necessary for meiosis to begin from the Start position of the mitotic program (as is the case for conjugation), or can cells begin meiosis from other points? This question was answered by first arresting *cdc* mutants at the restrictive temperature in growth medium and then shifting them to the permissive temperature in sporulation medium (Hirschberg and Simchen 1977). Most *cdc* mutants finished the mitotic cycle in sporulation medium before undergoing meiosis, but *cdc4* cells proceeded directly through meiosis. Thus, the meiotic program can apparently be initiated from the *CDC4* step in the mitotic cycle, at least one step after Start. Similarly, complete asci that have small, attached, anucleate buds are occasionally observed in wild-type cultures (Byers, this volume). These cells may have been able to enter meiosis after bud emergence in the mitotic cycle.

Approaches to Start

Appreciation of the significance of Start as a control point for *S. cerevisiae* cell division has led to further attempts to clarify the nature of this step. This section will enumerate some promising approaches.

The kinetics with which cells in a population complete Start have been examined and found to approximate first order, leading to the suggestion that the rate-limiting event at Start is "probabilistic" (Shilo et al. 1976). This idea is difficult to reconcile with the overwhelming evidence (reviewed above, in Coordination of Growth with Division) that attainment of the critical cell size is normally the rate-limiting "deterministic" event at Start, and the issue has been debated (Nurse and Fantes 1977; Shilo et al. 1977; Wheals 1977). Recently, Nurse (1980) has suggested the likely interpretation that both deterministic and probabilistic events occur at Start, and A. Wheals (pers. comm.) has proposed a "sloppy size control model" for Start that well accounts for all observations by incorporating a strongly deterministic parameter relating to cell size together with a sloppy size-monitoring mechanism that permits a degree of randomness in the behavior of cells with apparently identical sizes.

The nature of the substance that the cell monitors as a metric of size has been sought by exposing cells to various inhibitors and perturbations of the growth medium; the results have been interpreted as suggesting that the accumulation of stable RNA may be more closely coordinated with the completion of Start than is the accumulation of protein (Bedard et al. 1980).

Our current understanding of Start suggests specific phenotypes for mutants altered in gene products that mediate this step, and a number of the expected mutants have been found. A selection procedure for *cdc* mutants that arrest specifically at Start and maintain the capacity to conjugate has identified 33 mutants in four genes: *CDC28, CDC36, CDC37,* and *CDC39* (Reed 1980a). Mutants resistant to division arrest at Start by mating pheromone have been sought and found in ten different genes: *STE2, STE3, STE4, STE5, STE6, STE7, STE8, STE9, STE11,* and *STE12* (MacKay and Manney 1974a,b; Manney and Woods 1976; Hartwell 1980). Most of these mutants are pleiotropic for other sex-related phenotypes, and further understanding will be dependent upon sorting out those that are specific for division arrest from the others. Mutants with altered size control have recently been reported; one (*whil*) executes Start at a smaller than normal size, and another (*whi2*) fails to arrest at Start when growth is limited by nutritional deprivation (Sudbery et al. 1980).

Since the SPB is implicated in the control of division at Start by virtue of its enlargement coincident with Start (Fig. 2), insight into the nature of this structure is highly desirable. No substantial progress has been made in purifying the SPB as a first step in analyzing its structure. However, it has been demonstrated that SPBs in vitro act as nucleation sites for the polymerization of microtubules (Byers et al. 1978; Hyams and Borisy 1978), an activity that should aid ultimately in their purification and functional analysis.

Another exciting recent development is the suggestion of an approach to the genetics of the SPB. It appears that mutants with karyogamy defects may have lesions in gene products that are or act upon structural components of the SPB. Mutations producing defects in karyogamy during conjugation have the unusual property that they behave as if dominant to their wild-type alleles in heterokaryons but are recessive in diploids. This behavior is consistent with the mutant gene products being located in a nuclear structure (Conde and Fink 1976; Fink and Conde 1977), since it has been shown that nuclei fuse during conjugation at their SPBs (Byers and Goetsch 1975). Furthermore, of 40 *cdc* mutants that were tested for karyogamy defects, only three (*cdc4, cdc28,* and *cdc37*) were found to be defective (Dutcher 1980); all three are mutants whose cell-division defects may have to do with early steps in the SPB cycle. Finally, and perhaps most suggestively, in heterokaryons, the *cdc4* mutations are dominant to the *CDC4* allele for the karyogamy defect but recessive for the cell-division defect. This paradoxical behavior is nicely accounted for by the hypothesis that the *CDC4* product is a component of the SPB (for explanation, see Dutcher 1980).

A Model for the Control of Start

How cells coordinate their growth and division to maintain a constant size is a question that has received considerable experimental and theoretical attention. Recently, models put forward by various investigators have been classified, compared, and evaluated (Fantes et al. 1975). Models differ in whether the concentration or the absolute amount of a critical substance is important, in whether the substance acts to inhibit or to activate division, in the time course of accumulation of the substance, in whether or not the substance is incorporated into a structural component of the cell, and in the point in the cell cycle at which control is exerted. Since the critical substance has not been identified in any type of cell, few constraints exist to limit tenable models.

The data just reviewed for *S. cerevisiae* localize the point of control to a specific stage of the cell cycle, implicate the duplication of a specific nuclear structure as the controlling step, identify several gene products essential for this step, and identify other gene products that integrate completion of this step with growth. Although these facts do not distinguish among many of the proposed formal models for the integration of growth and division, they do permit the construction and testing of models with concrete rather than hypothetical elements, examples of which are suggested below.

We propose that doubling of the mass of the SPB is the event that commits the cell to the division cycle and marks the completion of Start. The duplicated SPB then activates the other genetic programs leading to budding, cytokinesis, DNA replication, etc. (Fig. 2). The SPB is potentially capable of communicating with the cell surface to initiate formation of the chitin ring, bud, and microfilament ring, since microtubules emanate from the SPB toward the cell surface at this stage of the cell cycle. The SPB is also capable of communicating with the chromosomes to initiate replication, since it may be connected to them either by microtubules or by a direct interaction of the chromosomes with the appropriate region of the nuclear envelope. The transmission of a signal from the SPB upon its duplication to these other cellular structures might result either from a duplication in the number of microtubules or from an allosteric change in the microtubule polymer.

Start itself could be regulated either by controls over the accumulation of SPB subunits or by controls over the assembly of subunits previously accumulated. In the simplest model (adapted from the model elaborated by Donachie [1968] for *E. coli*), some limiting subunit would be accumulated in constant proportion to total cellular protein, and Start would occur when the critical subunit had accumulated to an *amount* sufficient to duplicate the mass of the SPB. The simplest way to maintain a constant differential rate of synthesis of the critical SPB subunit is by some form of autogenous control; however, since the concentration of the free subunit itself would not be constant (it would fall to zero at each Start event), another component must be postulated (e.g., a fragment clipped off the critical subunit before its assembly) that actually controls synthesis autogenously (see Sompayrac and Maaløe 1973). The time course of accumulation of the limiting subunit may be more complicated (involving, e.g., an unstable subunit). Under this model, the products of the genes *CDC28, CDC36, CDC37,* and *CDC39* could be subunits (although not necessarily the limiting subunit) of the SPB itself or proteins that are necessary for its assembly; consistent with this idea is the fact that *cdc28* and *cdc37* mutants are pleiotropic, exhibiting a karyogamy defect during conjugation (Dutcher 1980). *whi1* mutants, which reduce the size of the cell at Start, could identify a site or product that controls the accumulation of the limiting SPB subunit. The mating pheromones could arrest cells at Start either by inhibiting the synthesis of the limiting subunit (or, for that matter, of any essential subunit) or by inhibiting the assembly of the subunits into a SPB. The fact that diploid cells are roughly twice the size of haploid cells at Start would be accounted for by the fact that the SPB is roughly twice as big in diploids as in haploids (see Byers, this volume), so that twice the protein accumulation would be necessary to provide enough of the limiting SPB subunit.

Additional controls over the accumulation of the limiting SPB subunit are suggested by the fact that cells complete Start at different sizes in different media. Furthermore, the fact that cells can adjust to a new size at Start rather soon after a shift of medium suggests that the amount of the limiting SPB subunit can be changed quickly in response to nutritional clues. Finally, the fact that the *whi2* mutant alters the size requirement for Start specifically during nutritional starvation also suggests additional control mechanisms.

ACKNOWLEDGMENTS

We thank the numerous colleagues who have shared their unpublished results with us. Unpublished work from our own laboratories has been supported by National Institutes of Health grants GM-17709 (to L.H.H.) and GM-23936 (to J.R.P.) and National Science Foundation grant PCM 78-25607 (to Barbara F. Sloat and J.R.P.).

Note Added in Proof

Conflicting data have now appeared on the timing of double-stranded RNA replication during the cell cycle. Compare A. M. Newman et al. 1981. *J. Virol.* **38:** 263 with V. A. Zakian et al. 1981. *Mol. Cell. Biol.* **1:** 673.

REFERENCES

Adams, J. 1977. The interrelationship of cell growth and division in haploid and diploid cells of *Saccharomyces cerevisiae*. *Exp. Cell Res.* **106:** 267.

Bakalkin, G.Y., S.L. Kalnov, A.S. Zubatov, and V.N. Luzikov. 1976. Degradation of total protein at different stages of *Saccharomyces cerevisiae* yeast growth. *FEBS Lett.* **63:** 218.

Baker, B.S., A.T.C. Carpenter, M.S. Esposito, R.E. Esposito, and L. Sandler. 1976. The genetic control of meiosis. *Annu. Rev. Genet.* **10:** 53.

Ball, W.J., Jr. and D.E. Atkinson. 1975. Adenylate energy charge in *Saccharomyces cerevisiae* during starvation. *J. Bacteriol.* **121:** 975.

Ballou, C. 1982. The yeast cell wall and cell surface. In *The molecular biology of the yeast*

Saccharomyces. II. *Metabolism and gene expression* (ed. J. Strathern et al.), Cold Spring Harbor Laboratory, Cold Spring Harbor, New York. (In press.)

Ballou, C.E., S.K. Maitra, J.W. Walker, and W.L. Whelan. 1977. Developmental defects associated with glucosamine auxotrophy in *Saccharomyces cerevisiae. Proc. Natl. Acad. Sci.* **74**: 4351.

Barford, J.P. and R.J. Hall. 1976. Estimation of the length of cell cycle phases from asynchronous cultures of *Saccharomyces cerevisiae. Exp. Cell Res.* **102**: 276.

Bedard, D.P., R.A. Singer, and G.C. Johnston. 1980. Transient cell cycle arrest of *Saccharomyces cerevisiae* by amino acid analog β-2-DL-thienylalanine. *J. Bacteriol.* **141**: 100.

Betz, H. 1976. Inhibition of protein synthesis stimulates intracellular protein degradation in growing yeast cells. *Biochem. Biophys. Res. Commun.* **72**: 121.

Betz, R. and W. Duntze. 1979. Purification and partial characterization of a factor, a mating hormone produced by mating-type-a cells from *Saccharomyces cerevisiae. Eur. J. Biochem.* **95**: 469.

Bisson, L. and J. Thorner. 1977. Thymidine 5'-monophosphate-requiring mutants of *Saccharomyces cerevisiae* are deficient in thymidylate synthetase. *J. Bacteriol.* **132**: 44.

Bücking-Throm, E., W. Duntze, L.H. Hartwell, and T.R. Manney. 1973. Reversible arrest of haploid yeast cells at the initiation of DNA synthesis by a diffusible sex factor. *Exp. Cell Res.* **76**: 99.

Byers, B. 1981. Multiple roles of the spindle pole bodies in the life cycle of *Saccharomyces cerevisiae.* In *Molecular genetics in yeast, Alfred Benzon Symp. 16* (ed. D. von Wettstein et al.). Munksgaard, Copenhagen. (In press.)

Byers, B. and L. Goetsch. 1974. Duplication of spindle plaques and integration of the yeast cell cycle. *Cold Spring Harbor Symp. Quant. Biol.* **38**: 123.

———. 1975. Behavior of spindles and spindle plaques in the cell cycle and conjugation of *Saccharomyces cerevisiae. J. Bacteriol.* **124**: 511.

———. 1976a. A highly ordered ring of membrane-associated filaments in budding yeast. *J. Cell Biol.* **69**: 717.

———. 1976b. Loss of the filamentous ring in cytokinesis-defective mutants of budding yeast. *J. Cell Biol.* **70**: 35a.

Byers, B. and L. Sowder. 1980. Gene expression in the yeast cell cycle. *J. Cell Biol.* **87**: 6a.

Byers, B., K. Shriver, and L. Goetsch. 1978. The role of spindle pole bodies and modified microtubule ends in the initiation of microtubule assembly in *Saccharomyces cerevisiae. J. Cell Sci.* **30**: 331.

Cabib, E. and B. Bowers. 1975. Timing and function of chitin synthesis in yeast. *J. Bacteriol.* **124**: 1586.

Campbell, D.A. 1973. Kinetics of the mating-specific aggregation in *Saccharomyces cerevisiae. J. Bacteriol.* **116**: 323.

Carter, B.L.A. and M.N. Jagadish. 1978a. The relationship between cell size and cell division in the yeast *Saccharomyces cerevisiae. Exp. Cell Res.* **112**: 15.

———. 1978b. Control of cell division in the yeast *Saccharomyces cerevisiae* cultured at different growth rates. *Exp. Cell Res.* **112**: 373.

Carter, B.L.A., A. Lorincz, and G.C. Johnston. 1978. Protein synthesis, cell division and the cell cycle in *Saccharomyces cerevisiae* following a shift to a richer medium. *J. Gen. Microbiol.* **106**: 222.

Clarke, L. and J. Carbon. 1980. Isolation of a yeast centromere and construction of functional small circular chromosomes. *Nature* **287**: 504.

Conde, J. and G.R. Fink. 1976. A mutant of *Saccharomyces cerevisiae* defective for nuclear fusion. *Proc. Natl. Acad. Sci.* **73**: 3651.

Cooper, T.G., C. Britton, L. Brand, and R. Sumrada. 1979. Addition of basic amino acids prevents G-1 arrest of nitrogen-starved cultures of *Saccharomyces cerevisiae. J. Bacteriol.* **137**: 1447.

Creanor, J. and J.M. Mitchison. 1979. Reduction of perturbations in leucine incorporation in synchronous cultures of *Schizosaccharomyces pombe* made by elutriation. *J. Gen. Microbiol.* **112**: 385.

Culotti, J. 1974. "Coordination of cell cycle events in *Saccharomyces cerevisiae*." Ph.D. thesis, University of Washington, Seattle.

Culotti, J. and L.H. Hartwell. 1971. Genetic control of the cell division cycle in yeast. III. Seven genes controlling nuclear division. *Exp. Cell Res.* **67**: 389.

del Rey, F., T. Santos, I. Garcia-Acha, and C. Nombela. 1979. Synthesis of 1,3- β-glucanases in *Saccharomyces cerevisiae* during the mitotic cycle, mating, and sporulation. *J. Bacteriol.* **139**: 924.

Deutch, C.E. and J.M. Parry. 1974. Sphaeroplast formation in yeast during the transition from exponential phase to stationary phase. *J. Gen. Microbiol.* **80**: 259.

Donachie, W.D. 1968. Relationship between cell size and time of initiation of DNA replication. *Nature* **219**: 1077.

Dutcher, S.K. 1980. "Genetic control of karyogamy in *Saccharomyces cerevisiae*." Ph.D. thesis, University of Washington, Seattle.

Elliott, S.G. and C.S. McLaughlin. 1978. Rate of macromolecular synthesis through the cell cycle of the yeast *Saccharomyces cerevisiae. Proc. Natl. Acad. Sci.* **75**: 4384.

———. 1979a. Synthesis and modification of proteins during the cell cycle of the yeast *Saccharomyces cerevisiae. J. Bacteriol.* **137**: 1185.

———. 1979b. Regulation of RNA synthesis in yeast. III. Synthesis during the cell cycle. *Mol. Gen. Genet.* **169**: 237.

Elliott, S.G., J.R. Warner, and C.S. McLaughlin. 1979. Synthesis of ribosomal proteins during the cell cycle of the yeast *Saccharomyces cerevisiae. J. Bacteriol.* **137**: 1048.

Fabre, F. and H. Roman. 1979. Evidence that a single DNA ligase is involved in replication and recombination in yeast. *Proc. Natl. Acad. Sci.* **76**: 4586.

Fantes, P.A. 1977. Control of cell size and cycle time in *Schizosaccharomyces pombe. J. Cell Sci.* **24**: 51.

Fantes, P.A., W.D. Grant, R.H. Pritchard, P.E. Sudbery, and A.E. Wheals. 1975. The regulation of cell size and the control of mitosis. *J. Theoret. Biol.* **50**: 213.

Field, C. and R. Schekman. 1980. Localized secretion of acid phosphatase reflects the pattern of cell surface growth in *Saccharomyces cerevisiae. J. Cell Biol.* **86**: 123.

Fink, G.R. and J. Conde. 1977. Studies on *KAR1*, a gene required for nuclear fusion in yeast. In *International Cell Biology, 1976–1977* (ed. B.R. Brinkley and K.R. Porter), p. 414. Rockefeller University Press, New York.

Game, J.C. 1976. Yeast cell-cycle mutant *cdc21* is a temperature-sensitive thymidylate auxotroph. *Mol. Gen. Genet.* **146**: 313.

Game, J.C., L.H. Johnston, and R.C. von Borstel. 1979. Enhanced mitotic recombination in a ligase-defective mutant of the yeast *Saccharomyces cerevisiae. Proc. Natl. Acad. Sci.* **76**: 4589.

Groppi, V.E., Jr. and P. Coffino. 1980. G_1 and S phase mammalian cells synthesize histones at equivalent rates. *Cell* **21**: 195.

Halvorson, H.O., B.L.A. Carter, and P. Tauro. 1971. Synthesis of enzymes during the cell cycle. *Adv. Microb. Physiol.* **6**: 47.

Hartwell, L.H. 1970. Periodic density fluctuation during the yeast cell cycle and the selection of synchronous cultures. *J. Bacteriol.* **104**: 1280.

———. 1971a. Genetic control of the cell division cycle in yeast. II. Genes controlling DNA replication and its initiation. *J. Mol. Biol.* **59**: 183.

———. 1971b. Genetic control of the cell division cycle in yeast. IV. Genes controlling bud emergence and cytokinesis. *Exp. Cell Res.* **69**: 265.

———. 1973a. Synchronization of haploid yeast cell cycles, a prelude to conjugation. *Exp. Cell Res.* **76**: 111.

———. 1973b. Three additional genes required for deoxyribonucleic acid synthesis in *Saccharomyces cerevisiae. J. Bacteriol.* **115**: 966.

———. 1974. *Saccharomyces cerevisiae* cell cycle. *Bacteriol. Rev.* **38**: 164.

————. 1976. Sequential function of gene products relative to DNA synthesis in the yeast cell cycle. *J. Mol. Biol.* **104**: 803.

————. 1978. Cell division from a genetic perspective. *J. Cell Biol.* **77**: 627.

————. 1980. Mutants of *Saccharomyces cerevisiae* unresponsive to cell division control by polypeptide mating hormone. *J. Cell Biol.* **85**: 811.

Hartwell, L.H. and M.W. Unger. 1977. Unequal division in *Saccharomyces cerevisiae* and its implications for the control of cell division. *J. Cell Biol.* **75**: 422.

Hartwell, L.H., J. Culotti, J.R. Pringle, and B.J. Reid. 1974. Genetic control of the cell division cycle in yeast. *Science* **183**: 46.

Hartwell, L.H., R.K. Mortimer, J. Culotti, and M. Culotti. 1973. Genetic control of the cell division cycle in yeast: V. Genetic analysis of *cdc* mutants. *Genetics* **74**: 267.

Hayashibe, M. and S. Katohda. 1973. Initiation of budding and chitin-ring. *J. Gen. Appl. Microbiol.* **19**: 23.

Hayashibe, M., N. Abe, and M. Matsui. 1977. Mode of increase in cell wall polysaccharides in synchronously growing *Saccharomyces cerevisiae*. *Arch. Microbiol.* **114**: 91.

Hereford, L.M. and L.H. Hartwell. 1971. Defective DNA synthesis in permeabilized yeast mutants. *Nat. New Biol.* **234**: 171.

————. 1973. Role of protein synthesis in the replication of yeast DNA. *Nat. New Biol.* **244**: 129.

————. 1974. Sequential gene function in the initiation of *Saccharomyces cerevisiae* DNA synthesis. *J. Mol. Biol.* **84**: 445.

Hereford, L.M., M.A. Osley, J.R. Ludwig, Jr., and C.S. McLaughlin. 1981. Cell-cycle regulation of yeast histone mRNA. *Cell* **24**: 367.

Hereford, L.M., K. Fahrner, J. Woolford, Jr., M. Rosbash, and D.B. Kaback. 1979. Isolation of yeast histone genes H2A and H2B. *Cell* **18**: 1261.

Hirschberg, J. and G. Simchen. 1977. Commitment to the mitotic cell cycle in yeast in relation to meiosis. *Exp. Cell Res.* **105**: 245.

Holland, J.P. and M.J. Holland. 1980. Structural comparison of two non-tandemly repeated yeast glyceraldehyde-3-phosphate dehydrogenase genes. *J. Biol. Chem.* **255**: 2596.

Holzer, H. and P.C. Heinrich. 1980. Control of proteolysis. *Annu. Rev. Biochem.* **49**: 63.

Hyams, J.S. and G.G. Borisy. 1978. Nucleation of microtubules in vitro by isolated spindle pole bodies of the yeast *Saccharomyces cerevisiae*. *J. Cell Biol.* **78**: 401.

Jagadish, M.N. and B.L.A. Carter. 1977. Genetic control of cell division in yeast cultured at different growth rates. *Nature* **269**: 145.

————. 1978. Effects of temperature and nutritional conditions on the mitotic cell cycle of *Saccharomyces cerevisiae*. *J. Cell Sci.* **31**: 71.

Jagadish, M.N., A. Lorincz, and B.L.A. Carter. 1977. Cell size and cell division in yeast cultured at different growth rates. *FEMS Microbiol. Lett.* **2**: 235.

Jarvik, J. and D. Botstein. 1975. Conditional-lethal mutations that suppress genetic defects in morphogenesis by altering structural proteins. *Proc. Natl. Acad. Sci.* **72**: 2738.

Jazwinski, S.M. and G.M. Edelman. 1976. Activity of yeast extracts in cell-free stimulation of DNA replication. *Proc. Natl. Acad. Sci.* **73**: 3933.

————. 1979. Replication *in vitro* of the 2-μm DNA plasmid of yeast. *Proc. Natl. Acad. Sci.* **76**: 1223.

Johnston, G.C. 1977. Cell size and budding during starvation of the yeast *Saccharomyces cerevisiae*. *J. Bacteriol.* **132**: 738.

Johnston, G.C., J.R. Pringle, and L.H. Hartwell. 1977a. Coordination of growth with cell division in the yeast *Saccharomyces cerevisiae*. *Exp. Cell Res.* **105**: 79.

Johnston, G.C., R.A. Singer, and E.S. McFarlane. 1977b. Growth and cell division during nitrogen starvation of the yeast *Saccharomyces cerevisiae*. *J. Bacteriol.* **132**: 723.

Johnston, G.C., C.W. Ehrhardt, A. Lorincz, and B.L.A. Carter. 1979. Regulation of cell size in the yeast *Saccharomyces cerevisiae*. *J. Bacteriol.* **137**: 1.

Johnston, G.C., R.A. Singer, S.O. Sharrow, and M.L. Slater. 1980. Cell division in the yeast *Saccharomyces cerevisiae* growing at different rates. *J. Gen. Microbiol.* **118**: 479.

Johnston, L.H. and J.C. Game. 1978. Mutants of yeast with depressed DNA synthesis. *Mol. Gen. Genet.* **161**: 205.

Johnston, L.H. and K.A. Nasmyth. 1978. *Saccharomyces cerevisiae* cell cycle mutant *cdc9* is defective in DNA ligase. *Nature* **274**: 891.

Johnston, L.H. and D.H. Williamson. 1978. An alkaline sucrose gradient analysis of the mechanism of nuclear DNA synthesis in the yeast *Saccharomyces cerevisiae. Mol. Gen. Genet.* **164**: 217.

Kassir, Y. and G. Simchen. 1978. Meiotic recombination and DNA synthesis in a new cell cycle mutant of *Saccharomyces cerevisiae. Genetics* **90**: 49.

Katohda, S., N. Abe, M. Matsui, and M. Hayashibe. 1976. Polysaccharide composition of the cell wall of baker's yeast with special reference to cell walls obtained from large- and small-sized cells. *Plant Cell Physiol.* **17**: 909.

Kawasaki, G. 1979. "Karyotypic instability and carbon source effects in cell cycle mutants of *Saccharomyces cerevisiae.*" Ph.D. thesis, University of Washington, Seattle.

Kirschner, K. and H. Bisswanger. 1976. Multifunctional proteins. *Annu. Rev. Biochem.* **45**: 143.

Klein, H.L. and B. Byers. 1978. Stable denaturation of chromosomal DNA from *Saccharomyces cerevisiae* during meiosis. *J. Bacteriol.* **134**: 629.

Kotyk, A. 1963. Intracellular pH of baker's yeast. *Folia Microbiol.* **8**: 27.

Kubitschek, H.E. and R.W. Edvenson. 1977. Midcycle doubling of uptake rates of adenine and serine in *Saccharomyces cerevisiae. Biophys. J.* **20**: 15.

Küenzi, M.T. and A. Fiechter. 1969. Changes in carbohydrate composition and trehalase-activity during the budding cycle of *Saccharomyces cerevisiae. Arch. Mikrobiol.* **64**: 396.

———. 1972. Regulation of carbohydrate composition of *Saccharomyces cerevisiae* under growth limitation. *Arch. Mikrobiol.* **84**: 254.

Lagunas, R. 1976. Energy metabolism of *Saccharomyces cerevisiae*. Discrepancy between ATP balance and known metabolic functions. *Biochim. Biophys. Acta* **440**: 661.

Lillie, S.H. and J.R. Pringle. 1980. Reserve carbohydrate metabolism in *Saccharomyces cerevisiae*: Responses to nutrient limitation. *J. Bacteriol.* **143**: 1384.

Lipke, P.N., A. Taylor, and C.E. Ballou. 1976. Morphogenic effects of α-factor on *Saccharomyces cerevisiae* **a** cells. *J. Bacteriol.* **127**: 610.

Livingston, D.M. and D.M. Kupfer. 1977. Control of *Saccharomyces cerevisiae* 2μm DNA replication by cell division cycle genes that control nuclear DNA replication. *J. Mol. Biol.* **116**: 249.

Lord, P.G. and A.E. Wheals. 1980. Asymmetrical division of *Saccharomyces cerevisiae. J. Bacteriol.* **142**: 808.

Lorincz, A. and B.L.A. Carter. 1979. Control of cell size at bud initiation in *Saccharomyces cerevisiae. J. Gen. Microbiol.* **113**: 287.

Ludwig, J.R. II, S.G. Oliver, and C.S. McLaughlin. 1977. The regulation of RNA synthesis in yeast. II. Amino acids shift-up experiments. *Mol. Gen. Genet.* **158**: 117.

MacKay, V. and T.R. Manney. 1974a. Mutations affecting sexual conjugation and related processes in *Saccharomyces cerevisiae*. I. Isolation and phenotypic characterization of non-mating mutants. *Genetics* **76**: 255.

———. 1974b. Mutations affecting sexual conjugation and related processes in *Saccharomyces cerevisiae*. II. Genetic analysis of nonmating mutants. *Genetics* **76**: 273.

Manney, T.R. and V. Woods. 1976. Mutants of *Saccharomyces cerevisiae* resistant to the α mating-type factor. *Genetics* **82**: 639.

Mitchison, J.M. 1971. *The biology of the cell cycle.* Cambridge University Press, Cambridge, England.

———. 1977. Enzyme synthesis during the cell cycle. In *Cell differentiation in microorganisms, plants and animals* (ed. L. Nover and K. Mothes), p. 377. Gustav Fischer Verlag, Jena.

Mitchison, J.M. and B.L.A. Carter. 1975. Cell cycle analysis. *Methods Cell Biol.* **11**: 201.

Moens, P.B. 1971. Fine structure of ascospore development in the yeast *Saccharomyces cerevisiae. Can. J. Microbiol.* **17**: 507.

Molano, J., B. Bowers, and E. Cabib. 1980. Distribution of chitin in the yeast cell wall. An ultrastructural and chemical study. *J. Cell Biol.* **85**: 199.

Moll, R. and E. Wintersberger. 1976. Synthesis of yeast histones in the cell cycle. *Proc. Natl. Acad. Sci.* **73**: 1863.

Morris, N.R., M.H. Lai, and C.E. Oakley. 1979. Identification of a gene for α-tubulin in *Aspergillus nidulans. Cell* **16**: 437.

Nasmyth, K.A. and S.I. Reed. 1980. Isolation of genes by complementation in yeast: Molecular cloning of a cell-cycle gene. *Proc. Natl. Acad. Sci.* **77**: 2119.

Navon, G., R.G. Shulman, T. Yamane, T.R. Eccleshall, K.-B. Lam, J.J. Baronofsky, and J. Marmur. 1979. Phosphorus-31 nuclear magnetic resonance studies of wild-type and glycolytic pathway mutants of *Saccharomyces cerevisiae. Biochemistry* **18**: 4487.

Newlon, C.S. and W.L. Fangman. 1975. Mitochondrial DNA synthesis in cell cycle mutants of *Saccharomyces cerevisiae. Cell* **5**: 423.

Newlon, C.S., R.D. Ludescher, and S.K. Walter. 1979. Production of petites by cell cycle mutants of *Saccharomyces cerevisiae* defective in DNA synthesis. *Mol. Gen. Genet.* **169**: 189.

Nurse, P. 1980. Cell cycle control—Both deterministic and probabilistic? *Nature* **286**: 9.

Nurse, P. and P. Fantes. 1977. Transition probability and cell-cycle initiation in yeast. *Nature* **267**: 647.

Oertel, W. and M. Goulian. 1979. Deoxyribonucleic acid synthesis in *Saccharomyces cerevisiae* cells permeabilized with ether. *J. Bacteriol.* **140**: 333.

Parry, J.M., P.J. Davies, and W.E. Evans. 1976. The effects of "cell age" upon the lethal effects of physical and chemical mutagens in the yeast, *Saccharomyces cerevisiae. Mol. Gen. Genet.* **146**: 27.

Petes, T.D. 1980. Molecular genetics of yeast. *Annu. Rev. Biochem.* **49**: 845.

Petes, T.D. and W.L. Fangman. 1973. Preferential synthesis of yeast mitochondrial DNA in α factor-arrested cells. *Biochem. Biophys. Res. Commun.* **55**: 603.

Petes, T.D. and C.S. Newlon. 1974. Structure of DNA in DNA replication mutants of yeast. *Nature* **251**: 637.

Petes, T.D. and D.H. Williamson. 1975. Replicating circular DNA molecules in yeast. *Cell* **4**: 249.

Piedra, D. and L. Herrera. 1976. Selection of auxotrophic mutants in *Saccharomyces cerevisiae* by a snail enzyme digestion method. *Folia Microbiol.* **21**: 337.

Piñon, R. 1978. Folded chromosomes in non-cycling yeast cells. Evidence for a characteristic g_0 form. *Chromosoma* **67**: 263.

———. 1979a. A probe into nuclear events during the cell cycle of *Saccharomyces cerevisiae*: Studies of folded chromosomes in *cdc* mutants which arrest in G_1. *Chromosoma* **70**: 337.

———. 1979b. Folded chromosomes in meiotic yeast cells: Analysis of early meiotic events. *J. Mol. Biol.* **129**: 433.

Piñon, R. and D. Pratt. 1979. Folded chromosomes of mating-factor arrested yeast cells: Comparison with G_0 arrest. *Chromosoma* **73**: 117.

Prakash, L., D. Hinkle, and S. Prakash. 1979. Decreased UV mutagenesis in *cdc8*, a DNA replication mutant of *Saccharomyces cerevisiae. Mol. Gen. Genet.* **172**: 249.

Pringle, J.R. 1975. Induction, selection, and experimental uses of temperature-sensitive and other conditional mutants of yeast. *Methods Cell Biol.* **12**: 233.

———. 1978. The use of conditional lethal cell cycle mutants for temporal and functional sequence mapping of cell cycle events. *J. Cell. Physiol.* **95**: 393.

———. 1979. Proteolytic artifacts in biochemistry. In *Limited proteolysis in microorganisms* (ed. G.N. Cohen and H. Holzer), p. 191. United States Government Printing Office, Washington, D.C.

———. 1981. Genetic approaches to study of the cell cycle. In *Cellular dynamics: Mitosis/cytokinesis* (ed. A.M. Zimmerman and A. Forer). Academic Press, New York. (In press.)

Pringle, J.R. and J.-R. Mor. 1975. Methods for monitoring the growth of yeast cultures and for dealing with the clumping problem. *Methods Cell Biol.* **11**: 131.

Rai, R. and B.L.A. Carter. 1981. The isolation of nonsense mutation in cell division cycle genes of the yeast *Saccharomyces cerevisiae*. *Mol. Gen. Genet.* **181**: 556.

Reed, S.I. 1980a. The selection of *S. cerevisiae* mutants defective in the start event of cell division. *Genetics* **95**: 561.

————. 1980b. The selection of amber mutations in genes required for completion of start, the controlling event of the cell division cycle of *S. cerevisiae*. *Genetics* **95**: 579.

Reid, B.J. 1979. "Regulation and integration of the cell division cycle and the mating reaction in the yeast *Saccharomyces cerevisiae*." Ph.D. thesis, University of Washington, Seattle.

Reid, B.J. and L.H. Hartwell. 1977. Regulation of mating in the cell cycle of *Saccharomyces cerevisiae*. *J. Cell Biol.* **75**: 355.

Rivin, C.J. and W.L. Fangman. 1980. Cell cycle phase expansion in nitrogen-limited cultures of *Saccharomyces cerevisiae*. *J. Cell Biol.* **85**: 96.

Schekman, R. and V. Brawley. 1979. Localized deposition of chitin on the yeast cell surface in response to mating pheromone. *Proc. Natl. Acad. Sci.* **76**: 645.

Schekman, R. and P. Novick. 1982. The secretory process and yeast cell surface assembly. In *The molecular biology of the yeast* Saccharomyces. II. *Metabolism and gene expression* (ed. J. Strathern et al.), Cold Spring Harbor Laboratory, Cold Spring Harbor, New York. (In press.)

Schenberg-Frascino, A. and E. Moustacchi. 1972. Lethal and mutagenic effects of elevated temperature on haploid yeast. I. Variations in sensitivity during the cell cycle. *Mol. Gen. Genet.* **115**: 243.

Schild, D. and B. Byers. 1978. Meiotic effects of DNA-defective cell division cycle mutations of *Saccharomyces cerevisiae*. *Chromosoma* **70**: 109.

Schild, D. and B. Byers. 1980. Diploid spore formation and other meiotic effects of two cell-division-cycle mutations of *Saccharomyces cerevisiae*. *Genetics* **96**: 859.

Schild, D., H.N. Ananthaswamy, and R.K. Mortimer. 1981. An endomitotic effect of a cell-cycle mutation of *Saccharomyces cerevisiae*. *Genetics* (in press).

Scopes, A.W. and D.H. Williamson. 1964. The growth and oxygen uptake of synchronously dividing cultures of *Saccharomyces cerevisiae*. *Exp. Cell Res.* **35**: 361.

Sena, E.P., J.W. Welch, H.O. Halvorson, and S. Fogel. 1975. Nuclear and mitochondrial deoxyribonucleic acid replication during mitosis in *Saccharomyces cerevisiae*. *J. Bacteriol.* **123**: 497.

Shilo, B., V.G.H. Riddle, and A.B. Pardee. 1979. Protein turnover and cell-cycle initiation in yeast. *Exp. Cell Res.* **123**: 221.

Shilo, B., V. Shilo, and G. Simchen. 1976. Cell-cycle initiation in yeast follows first-order kinetics. *Nature* **264**: 767.

————. 1977. Transition probability and cell-cycle initiation in yeast. *Nature* **267**: 648.

Shilo, B., G. Simchen, and A.B. Pardee. 1978. Regulation of cell-cycle initiation in yeast by nutrients and protein synthesis. *J. Cell. Physiol.* **97**: 177.

Shilo, V., G. Simchen, and B. Shilo. 1978. Initiation of meiosis in cell cycle initiation mutants of *Saccharomyces cerevisiae*. *Exp. Cell Res.* **112**: 241.

Shriver, K. and B. Byers. 1977. Yeast microtubules: Constituent proteins and their synthesis. *J. Cell Biol.* **75**: 297a.

Simchen, G. 1974. Are mitotic functions required in meiosis? *Genetics* **76**: 745.

Simchen, G. and J. Hirschberg. 1977. Effects of the mitotic cell-cycle mutation *cdc4* on yeast meiosis. *Genetics* **86**: 57.

Slater, M.L. 1973. Effect of reversible inhibition of deoxyribonucleic acid synthesis on the yeast cell cycle. *J. Bacteriol.* **113**: 263.

————. 1974. Recovery of yeast from transient inhibition of DNA synthesis. *Nature* **247**: 275.

Slater, M.L., S.O. Sharrow, and J.J. Gart. 1977. Cell cycle of *Saccharomyces cerevisiae* in populations growing at different rates. *Proc. Natl. Acad. Sci.* **74**: 3850.

Sloat, B.F. and J.R. Pringle. 1978. A mutant of yeast defective in cellular morphogenesis. *Science* **200**: 1171.

Sloat, B.F., A. Adams, and J.R. Pringle. 1981. Roles of the *CDC24* gene product in cellular morphogenesis during the *Saccharomyces cerevisiae* cell cycle. *J. Cell Biol.* **89**: 395.

Sompayrac, L. and O. Maaløe. 1973. Autorepressor model for control of DNA replication. *Nat. New Biol.* **241**: 133.

Stötzler, D., H.-H. Kiltz, and W. Duntze. 1976. Primary structure of α-factor peptides from *Saccharomyces cerevisiae. Eur. J. Biochem.* **69**: 397.

Sudbery, P.E., A.R. Goodey, and B.L.A. Carter. 1980. Genes which control cell proliferation in the yeast *Saccharomyces cerevisiae. Nature* **288**: 401.

Sumrada, R. and T.G. Cooper. 1978a. Basic amino acid inhibition of cell division and macromolecular synthesis in *Saccharomyces cerevisiae. J. Gen. Microbiol.* **108**: 45.

———. 1978b. Control of vacuole permeability and protein degradation by the cell cycle arrest signal in *Saccharomyces cerevisiae. J. Bacteriol.* **136**: 234.

Swedes, J.S., M.E. Dial, and C.S. McLaughlin. 1979. Regulation of protein synthesis during energy limitation of *Saccharomyces cerevisiae. J. Bacteriol.* **138**: 162.

Taketo, M., S.M. Jazwinski, and G.M. Edelman. 1980. Association of the 2-μm DNA plasmid with yeast folded chromosomes. *Proc. Natl. Acad. Sci.* **77**: 3144.

Throm, E. and W. Duntze. 1970. Mating-type-dependent inhibition of deoxyribonucleic acid synthesis in *Saccharomyces cerevisiae. J. Bacteriol.* **104**: 1388.

Tyson, C.B., P.G. Lord, and A.E. Wheals. 1979. Dependency of size of *Saccharomyces cerevisiae* cells on growth rate. *J. Bacteriol.* **138**: 92.

Unger, M.W. and L.H. Hartwell. 1976. Control of cell division in *Saccharomyces cerevisiae* by methionyl-tRNA. *Proc. Natl. Acad. Sci.* **73**: 1664.

von Meyenburg, H.K. 1968. Der Sprossungszyklus von *Saccharomyces cerevisiae. Pathol. Microbiol.* **31**: 117.

———. 1969. Energetics of the budding cycle of *Saccharomyces cerevisiae* during glucose limited aerobic growth. *Arch. Mikrobiol.* **66**: 289.

Warner, J.R. 1981. The yeast ribosome: Structure, function, and synthesis. In *The molecular biology of the yeast* Saccharomyces. II. *Metabolism and gene expression* (ed. J. Strathern et al.), Cold Spring Harbor Laboratory, Cold Spring Harbor, New York. (In press.)

Warner, J.R. and C. Gorenstein. 1978. Yeast has a true stringent response. *Nature* **275**: 338.

Watson, T.G. 1970. Effects of sodium chloride on steady-state growth and metabolism of *Saccharomyces cerevisiae. J. Gen. Microbiol.* **64**: 91.

Wheals, A.E. 1977. Transition probability and cell-cycle initiation in yeast. *Nature* **267**: 647.

Wiemken, A., P. Matile, and H. Moor. 1970. Vacuolar dynamics in synchronously budding yeast. *Arch. Mikrobiol.* **70**: 89.

Wilkinson, L.E. and J.R. Pringle. 1974. Transient G1 arrest of *S. cerevisiae* cells of mating type α by a factor produced by cells of mating type **a**. *Exp. Cell Res.* **89**: 175.

Williamson, D.H. 1964. Division synchrony in yeasts. In *Synchrony in cell division and growth* (ed. E. Zeuthen), p. 351. Interscience, New York.

———. 1973. Replication of the nuclear genome in yeast does not require concomitant protein synthesis. *Biochem. Biophys. Res. Comm.* **52**: 731.

Williamson, D.H. and L.H. Johnston. 1981. The synthesis of chromosomal replicons in brewer's yeast. In *The fungal nucleus* (ed. K. Gull and S.G. Oliver), Cambridge University Press, Cambridge, England. (In press.)

Wolfner, M., D. Yep, F. Messenguy, and G.R. Fink. 1975. Integration of amino acid biosynthesis into the cell cycle of *Saccharomyces cerevisiae. J. Mol. Biol.* **96**: 273.

Yashphe, J. and H.O. Halvorson. 1976. β-D-galactosidase activity in single yeast cells during cell cycle of *Saccharomyces lactis. Science* **191**: 1283.

Zakian, V.A. and W.L. Fangman. 1979. Synthesis of yeast double-stranded RNAs during the cell cycle. *J. Cell Biol.* **83**: 9a.

Zakian, V.A., B.J. Brewer, and W.L. Fangman. 1979. Replication of each copy of the yeast 2 micron DNA plasmid occurs during the S phase. *Cell* **17**: 923.

Zamb, T.J. and R. Roth. 1977. Role of mitotic replication genes in chromosome duplication during meiosis. *Proc. Natl. Acad. Sci.* **74**: 3951.

Pheromonal Regulation of Development in *Saccharomyces cerevisiae*

Jeremy Thorner
Department of Microbiology and Immunology
University of California
Berkeley, California 94720

INTRODUCTION

Mating of the two different haploid cell types of baker's yeast *Saccharomyces cerevisiae* to form a diploid provides a relatively simple system for studying the biochemical basis of intercellular signaling and developmental control in eukaryotic cells. Conjugation occurs efficiently when cultures of opposite mating type are mixed but only exceedingly rarely when cultures of the same haploid cell type are mixed. Thus, there must be specific biochemical features that distinguish and define the mating type of a yeast cell and that mediate the intercellular interactions occurring between them to permit conjugation. Indeed, substantial progress has been made over the last 10 years in describing events in the mating process at the molecular level (for reviews, see Crandall 1977, 1978; Crandall et al. 1977; Manney and Meade 1977; MacKay 1978; Goodenough 1980; Thorner 1980; Manney et al. 1981). One exciting outcome of these investigations is the demonstration that reciprocal exchange of diffusible oligopeptide pheromones is the intercellular stimulus that initiates the physiological changes in haploid cells that culminate in mating. In this paper, mating will be discussed in light of recent advances in our understanding of the cellular and molecular biology of yeast. Where possible, the identification and biochemical mode of action of cell-type-specific gene products that function in this developmental process will be emphasized, particularly the pivotal role of the mating pheromones.

AN OVERVIEW OF MATING EVENTS

A diagrammatic representation of the conjugation process is shown in Figure 1 and is based on the observations of a large number of laboratories. Several groups have devised methods for achieving reasonably synchronous mating of mass populations and have followed the kinetics of mating in such systems (Bilinski et al. 1975; Sena et al. 1975; Rodgers and Bussey 1978). There are a number of discrete stages in the formation of a zygote that can be observed in the light microscope. These events are accompanied by underlying changes in the ultrastructure of the mating cells, which are discussed later (see also Byers, this volume).

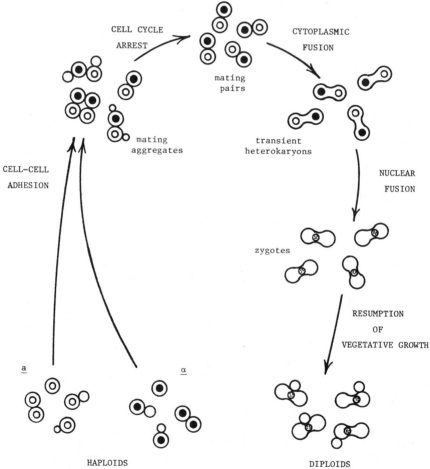

Figure 1 Schematic representation of the conjugation process in *S. cerevisiae.* (Reprinted, with permission, from Thorner 1980.)

Upon mixing liquid cultures of haploids of opposite mating type together, the cells begin to adhere to one another. Some strains of *S. cerevisiae* display an intrinsically high degree of agglutinability toward cells of opposite mating type and start to agglomerate immediately. For the majority of strains, however, this specific intercellular adhesion is an inducible function. Usually, therefore, most of the **a** and α cells become tightly agglutinated to one another, pairwise and in larger aggregates, only after 30–60 minutes following admixture. Naturally, the time course of this and the ensuing events depends upon the particular culture conditions (medium, temperature, cell density, etc.).

By approximately one doubling time (90–120 minutes), the agglutinated cells have all halted growth in the G_1 phase of their cell-division cycle (Pringle and Hartwell, this volume), since sonic disruption of the aggregates at this stage reveals them to be a collection of unbudded cells almost exclusively.

Once in firm contact and with their cell cycles synchronized, autolytic activities remove the cell-wall and plasma-membrane integuments between cell pairs in the appropriate orientation, resulting in cytoplasmic continuity by about 180 minutes after mixing. About this same time, a few free morphologically elongated cells, markedly pear-shaped ("shmoos") or distended to form even longer and more bizarre forms (conjugation-tube projections or copulatory-process extensions), can be found in mating mixtures (Fig. 2). Such directional cell-surface growth can be seen even more dramatically when cells of opposite mating type are placed on agar medium about 10–15 μm apart but not touching (Levi 1956; Herman 1971; MacKay and Manney 1974a,b). Obviously, in successfully mated pairs, cell-envelope dissolution and resynthesis must be conducted in a reasonably controlled fashion to permit cell fusion but to avoid cell lysis.

Zygote formation is completed by the union of the two haploid nuclei to form a single diploid nucleus. Under growth conditions, the zygotes can begin to form daughter diploid buds by about 4 hours after the initial mixing of the haploid cultures (Fig. 2). Under conditions of nutrient limitation, the diploid zygotes can commence meiosis and sporulation.

GENETIC CONTROL OF CONJUGATION: THE MATING-TYPE LOCUS

The genetic composition of a yeast cell at the mating-type locus (*MAT*), a region mapping on the right arm of chromosome III, determines its cell type and, hence, its developmental capabilities (Herskowitz and Oshima, this volume). Haploid cells of opposite mating type, **a** and α (determined by the *MAT***a** and *MAT*α loci, respectively), mate efficiently with each other but not with cells of like mating type. In contrast, normal **a**/α diploid cells (*MAT***a**/*MAT*α) do not mate with either **a** or α haploids nor with other heterozygous diploids (Mortimer and Hawthorne 1969).

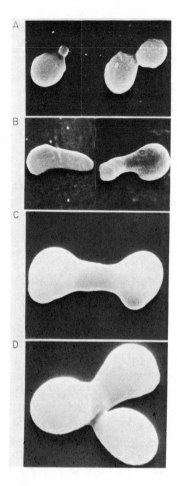

Figure 2 Morphology of yeast cells during various stages in conjugation. (*A*) Vegetatively growing haploids. (*B*) Pheromone-treated haploids. (*C*) Early zygote. (*D*) Budded zygote. (Scanning electron micrographs kindly provided by M. Davis, and by E. Schabtach and I. Herskowitz.)

MAT also controls other aspects of the life cycle. In contrast to mating proficiency, meiosis and sporulation (Esposito and Klapholz, this volume) are normally observed only in cells containing at least one representative of each *MAT* allele. *MAT* even seems to exert some morphogenic influence during vegetative growth. Haploid cells usually produce new buds in the region immediately surrounding their original birth scar; whereas, heterozygous diploids tend to bud first at the cell pole opposite their birth scar and then alternately from one pole and then from the other (or sometimes from the medial region of the cell) and, hence, display a much more random budding pattern (Freifelder 1960; Streiblova 1970). *MAT* also seems to govern the degree of resistance of yeast cells to DNA-damaging agents (Haynes and Kunz, this volume). For example, it has been reported that $\mathbf{a}/\alpha$ diploids are better able to survive after treatment with a chemical mutagen than are $\mathbf{a}/\mathbf{a}$ or α/α homozygous diploids (Livi and MacKay 1980).

The fact that cells heterozygous for *MAT* have distinctly different properties from cells with only one kind of MAT information originally suggested that $MATa$ and $MAT\alpha$ are codominant; i.e., one MAT locus is not merely the absence of the other. The properties of mutations mapping in $MATa$ and $MAT\alpha$ (Sprague et al. 1981; Herskowitz and Oshima, this volume) also indicated that they are not simply alternative alleles but contain distinct, functionally nonequivalent blocks of DNA. Indeed, direct isolation and analysis of the $MATa$ and $MAT\alpha$ regions by recombinant DNA methodology (Hicks et al. 1979; Nasmyth and Tatchell 1980; Strathern et al. 1980) have confirmed that although the $MATa$ and MATα segments contain some homology in flanking regions, they differ from each other by substitutions of completely nonhomologous DNA sequences of 650 and 750 bp, respectively. It has also been found (Klar et al. 1981; Nasmyth et al. 1981a,b) that the $MATa$ and $MAT\alpha$ regions each specify two distinct divergent transcripts.

The most attractive current picture for the function of the $MATa$ and $MAT\alpha$ loci is that their specific transcripts encode separate sets of regulatory proteins (Strathern et al. 1981; Tatchell et al. 1981; Herskowitz and Oshima, this volume). The regulatory proteins encoded by a particular MAT locus are essential for controlling the expression of other genes elsewhere in the yeast genome, which determine many of the cell's characteristics (its mating ability, its budding pattern, its capacity for sporulation, etc.). Such a model would predict that **a** and α haploids differ detectably in the types of poly(A)-containing mRNA sequences that they express, although this has not yet been demonstrated directly, except for the transcripts from the MAT regions themselves. Of special interest here is that the particular genes at MAT direct whether or not a cell will elaborate a mating pheromone, and if the cell does so, which pheromone it will produce.

MATING PHEROMONES

Each haploid cell type secretes into the culture medium a specific oligopeptide pheromone, or mating factor. These extracellular signals trigger biochemical changes reciprocally in their respective target cells: haploids of the opposite mating type. The responses elicited include, in order of their appearance, (1) cell-surface alterations that enhance the strength and selectivity of cell-cell contacts; (2) transient arrest of cell growth, specifically in G_1 phase of the cell cycle, that stays nuclear DNA synthesis and, in effect, synchronizes the growth stages of the mating partners; and (3) new wall and membrane synthesis that is appropriate to cell fusion rather than to normal budding. Heterozygous diploids (**a**/α cells) neither produce pheromones nor do they respond to them.

α-Factor

Structure and Properties

The α-mating-type character is correlated with production by the cell of the oligopeptide pheromone called α-factor (Levi 1956; Duntze et al. 1970). The biochemical and morphological changes listed above that are initiated upon exposure of **a** cells to α-factor are presumably preparatory to conjugation, although they occur even in the absence of α cells themselves. The **a**-cell responses known to be elicited by α-factor provide the bases for three types of bioassays for routine detection of the pheromone.

First, because the presence of α-factor reversibly arrests the growth of **a** cells, specifically in the early G_1 period of the cell cycle, one can monitor (1) the accumulation of unbudded cells in the population as observed in the light microscope (Bucking-Throm et al. 1973), (2) an inhibition of the increase in total cell number as quantitated in an automatic electronic particle counter (Chan 1977), or (3) the presence of an area of nongrowth in a lawn of responsive **a** cells as detected visually (Fig. 3). Second, since an eventual manifestation of α-factor action on **a** cells is their pronounced and characteristic elongation, one can score the appearance of morphologically aberrant cells (shmoos) under the microscope (Duntze et al. 1973). Third, the ability of a pheromone to induce increased agglutination between responsive tester cells and cells of the opposite mating type can be measured (1) spectrophotometrically either as an increase in light-scattering particles due to clumping (Yanagishima et al. 1977b) or as a decrease in absorbance due to the settling out of aggregated cells (Hartwell 1980), (2) as an increase in the average size of aggregates as determined by the difference in total cell number quantitated in an electronic particle counter before and after sonication (Betz et al. 1978), or (3) by following the increase in frequency of zygote formation directly (Fehrenbacher et al. 1978).

The α-factor is produced constitutively by α cells, upon which it exerts no evident effect. This pheromone was first purified from filtrates of liquid cultures of α haploids and characterized by Duntze and his collaborators (Duntze et al. 1973; Stötzler et al. 1976; Stötzler and Duntze 1976). The α-factor molecules obtained by various laboratories using several different *S. cerevisiae* α strains are linear tridecapeptides or dodecapeptides with the same amino acid sequence (Table 1). (Although one laboratory reported that a hexapeptide that possessed agglutination-inducing activity could be isolated from a particular *S. cerevisiae* α strain [Table 1], this material showed such activity only under one set of assay conditions [Sakurai et al. 1975] and was much less efficacious than α-factor itself assayed under the same conditions. Furthermore, in subsequent work, authentic α-factor could be purified from the same strain [A. Sakurai, pers. comm.].)

Perhaps remarkably, the pheromone produced by the α-like haploids of a different species (*S. kluyveri*) can, at high concentration, cause G_1 arrest and cell-shape changes in *S. cerevisiae* **a** cells (McCullough and Herskowitz 1979). This molecule, therefore, is at least functionally similar to *S. cerevisiae* α-factor. That the *S. kluyveri* α-factor is indeed structurally related to the *S. cerevisiae* α pheromone is suggested by the fact that it cross-reacts with antibody directed against the carboxyterminal end of *S. cerevisiae* α-factor and by the fact that its amino acid composition is quite similar to that of the *S. cerevisiae* α pheromone (Table 1).

Figure 3 Inhibition of **a**-cell growth by α-factor pheromone. A sample (50 μl) of either 90% methanol alone (*left column*), or 90% methanol containing 0.5 μg (*middle column*) or 1.0 μg (*right column*) of pure α-factor was spotted against the top edge of plates containing minimal medium and was allowed to soak into the agar. When dry, the plates were sprayed very briefly with suspensions of wild-type **a** cells (strain X2180-1A) at 10^5/ml (*top row*), 10^6/ml (*second row*), 10^7/ml (*third row*), and 10^8/ml (*bottom row*). Note that (1) the growth inhibitory effects of α-factor can be readily detected on agar plates, (2) the response of cells to small changes in α-factor concentration (in this case, twofold) can be easily measured, and (3) the ability to detect α-factor in this way is strongly dependent on the population density. (Photograph kindly provided by Y. Jones-Brown.)

S. cerevisiae α-factor is composed entirely of unmodified L-amino acids, and both its amino and carboxyl termini are unblocked. Varying amounts of des-Trp α-factor (α-factor lacking the aminoterminal Trp residue) and molecules that have suffered oxidation damage (the Met residue has been converted to methionine sulfoxide) are also found in α-factor preparations. The des-Trp α-factor is fully biologically active, but the oxidized species show about a tenfold reduction in specific activity. The carboxyterminal half of the pheromone molecule (-Pro-Gly-Gln-Pro-Met-Tyr-COOH) resembles a portion of the polyproline-II-like chain segment seen in the three-dimensional structure of the gastroenteric hormone, avian pancreatic polypeptide (-Pro-Ser-Gln-Pro-Thr-Tyr-) (Blundell and Humbel 1980). This region may confer some specific and relatively stable conformation on the pheromone, despite its relatively short overall chain length.

α-Factor is a relatively hydrophobic peptide, based on its content of nonpolar residues (Meek 1980). In fact, adsorption of α-factor to hydrophobic matrices, like octyl-Sepharose (Ciejek et al. 1977) and polystyrene beads (J. Strazdis and V. MacKay, pers. comm.), has been used as a purification procedure. Purified α-factor is readily soluble in the milligram-per-milliliter range in solvents containing 80–100% methanol; whereas in completely aqueous buffers near neutral pH, solutions of only 100 μg/ml or less can be prepared (E. Ciejek, unpubl.). In addition, at alkaline pH, purified

Table 1 Biologically active peptides produced by *Saccharomyces* species

Organism	Mating type	Primary structure[a]	Reference[b]
S. cerevisiae X2180-1B	α	Trp-His-Trp-Leu-Gln-Leu-Lys-Pro-Gly-Gln-Pro-Met-Tyr His-Trp-Leu-Gln-Leu-Lys-Pro-Gly-Gln-Pro-Met-Tyr	1,2,3
S. cerevisiae T22	α	His-Trp-Leu-Gln-Leu-Lys-Pro-Gly-Gln-Pro-Met-Tyr	4
S. cerevisiae H15	α	Trp-His-Trp-Leu-Gln-Leu-Lys-Pro-Gly-Gln-Pro-Met-Tyr Arg-Gly-Pro-Phe-Pro-Ile	5
S. kluyveri IF01894	α	? ? ? (His-Trp-Phe-Ser-Leu-Lys-Ser-Gly-Gln-Pro-Met-Tyr)	6
S. cerevisiae X2180-1A	a	Tyr-Ile-Ile-Lys-Gly-Val-Phe-Trp-Ala-Asx-Pro Tyr-Ile-Ile-Lys-Gly-Leu-Phe-Trp-Ala-Asx-Pro Tyr-Ile-Ile-Lys-Gly-Val-Phe-Trp-Ala-Asp?-Pro	7

[a]The inferred sequence (in parenthesis) and the most likely positions (indicated by ?) for the amino acid replacements in the *S. kluyveri* α phermone are based on the limited differences in its amino acid composition from that of *S. cerevisiae* α-factor, on the finding that the *S. kluyveri* molecule cross-reacts with antibody directed against the carboxyterminal portion of *S. cerevisiae* α-factor (Y. Jones-Brown et al., in prep.), and on the codon changes allowed by single-base (Leu → Phe and Pro → Ser) and double-base (Gln → Ser) changes, as determined from the genetic code.

[b](1) Stötzler and Duntze (1976); (2) Ciejek et al. (1977); (3) Tanaka et al. (1977b); (4) Sakurai et al. (1976a); (5) Sakurai et al. (1976b), and A. Sakurai (pers. comm.); (6) Sakurai et al. (1980); (7) Betz and Duntze (1979), and W. Duntze (pers. comm.).

α-factor has a pronounced tendency to partition into the nonaqueous phase of isobutanol-water mixtures (J. Thorner, unpubl.).

Both des-Trp and intact α-factor have been prepared using chemical means, either by solid-phase peptide synthesis (Ciejek et al. 1977) or by conventional solution methods for peptide synthesis (Masui et al. 1977). These chemically constructed oligopeptides display the entire array of biological activities that have been ascribed to the natural pheromone. The fact that the completely synthetic molecules elicit all of the responses typically observed in **a** cells that have been exposed to natural α-factor alone, or to α cells in mating mixtures, provides unambiguous proof that α-factor is indeed the sole primary signal for inducing the mating program in **a** cells.

Attempts have been made to define the portions of the α-factor primary structure that are essential for its biological activity by preparing altered forms of the molecule chemically. These alterations have included (1) synthesis of truncated molecules lacking certain individual residues or blocks of residues (Ciejek et al. 1977; Masui et al. 1977; Stötzler et al. 1977; Tanaka et al. 1977a; Ciejek and Thorner 1979; Ciejek 1980), (2) replacement of various amino acids at one or more positions (Masui et al. 1979; Samokhin et al. 1979), and (3) covalent modification of either terminus or reactive amino acid side chains (Lipke 1976; Stötzler et al. 1976; Maness and Edelman 1978; Tanaka and Kita 1978; Finkelstein and Strausberg 1979; E. Ciejek et al., unpubl.). The major conclusions of such structure-function studies are that His 2, Leu 6, and Lys 7 are residues especially crucial for biological activity (Table 1). On the other hand, every alteration of the primary sequence of the α-factor molecule that has been attempted to date, except removal or manipulation of Trp 1 (and possibly Tyr 13), results in a drastic reduction ($\geq 10^4$) of specific biological activity.

Mode of Action: Pheromone Binding

On the basis of analogy with the features of peptide hormone action in higher eukaryotic systems (Cuatrecasas and Hollenberg 1976), it could be argued that α-factor initially elicits its effects by binding to a specific receptor protein on the surface of **a** haploids. In accordance with this view, the dose dependence of **a**-cell response to the pheromone follows saturation kinetics and is first-order with respect to α-factor concentration, whether measured by the appearance of elongated cells (Udden and Finkelstein 1978), by the induction of G_1 arrest (Samokhin et al. 1981; S. Moore, pers. comm.), or by the enhancement of agglutinability (S. Moore, pers. comm.). These observations are compatible with the existence on **a** cells of identical and noninteracting receptor molecules for α-factor, each of which can bind only a single molecule of the pheromone. It is not possible to determine from these kinetic experiments whether only one such receptor molecule need be occupied at any given time to trigger the mating program. The pheromone concentration required to induce G_1 arrest (Chan 1977), aberrant **a**-cell morphology (Udden and Finkelstein 1978), or increased adhesiveness (S. Moore, pers. comm.) is strongly dependent on the **a**-cell population density (Fig. 4). This reflects, in part, the fact that **a**-cell-bound proteases degrade α-factor quite rapidly (see below, Mode of Action: Pheromone Proteolysis). Thus, at high **a**-cell concentrations, the concentration of α-factor required to elicit biological responses must be correspondingly high to insure that sufficient intact pheromone is present. At very low **a**-cell density (10^2/ml), where the rate of loss of α-factor due to proteolysis is greatly reduced (and where titration of the pheromone by the **a** cells themselves should not significantly effect the free α-factor concentration), the K_m for causing G_1 arrest and for inducing increased agglutinability is quite low, about 10^{-10} M (S. Moore, pers. comm.). Although this value is a phenomenological constant (since ligand binding has not been measured directly), it may nevertheless provide an estimate for the actual dissociation constant of α-factor from its initial binding site. The true dissociation constant may be different from the apparent K_m measured biologically due to the contribution of factors that would influence this kinetic parameter but whose effects are currently unknown; e.g., the rates of entry and exit of α-factor through the cell wall, the rate of new receptor synthesis and the rate of turnover (or recycling), if any, of α-factor-receptor complexes, and the percentage of the population of receptor molecules that must be occupied for a given time to ultimately elicit a particular biological response.

Although kinetic evidence is consistent with the presence of a specific α-factor receptor on **a** cells, direct measurements of the binding of radioactively

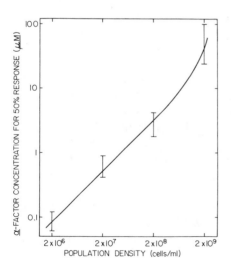

Figure 4 Induction of aberrant **a**-cell morphology by α-factor pheromone. Cells from an exponentially growing culture of wild-type **a** cells (strain X2180-1A) were collected by centrifugation, washed with fresh minimal medium buffered at pH 4.8 with 50 mM Na-succinate, and resuspended at the indicated densities. Samples of each cell suspension were exposed to different concentrations of pure α-factor using the microtiter dish assay described previously (Ciejek and Thorner 1979). The range of pheromone concentration required to cause about 50% of the tester cells to undergo a pronounced elongation (shmoo formation) was determined by scoring each sample in the phase-contrast microscope. (Data kindly provided by R. Kunisawa.)

labeled α-factor by **a** and α cells have failed to reveal dramatic preferential binding to intact **a** haploids (Maness and Edelman 1978; Thorner 1980; D. Finkelstein, pers. comm.). There are several factors that could explain the difficulty in demonstrating specific binding of α-factor to **a** cells, even if it exists.

First, it seems likely that the unusual properties of the yeast cell surface itself may cause a high degree of nonspecific association of the pheromone, which interferes with detection of its interaction with a true receptor. The outermost cell-wall layer of most *S. cerevisiae* strains is a complex phosphomannoprotein polymer (Ballou 1982). Because of the high density of phosphate groups carried by this surface-glycoprotein complex, yeast cells bear a net negative charge and can be stained by cationic dyes such as alcian blue. The α-factor has an overall positive charge at all pH values in the physiological range (3–7). Hence, nonspecific electrostatic interactions between α-factor and phosphomannan might contribute in a major way to the binding of the pheromone by yeast cells. Another property of yeast mannan also poses a second potential problem in demonstrating specific binding of α-factor to **a** cells. It has been shown by Parks and his co-workers that the mannan layer has a high affinity for lipophilic compounds such as sterols (Thompson et al. 1973). Since the pheromone is quite a nonpolar molecule, as reflected by its solubility characteristics discussed above, noncovalent hydrophobic interactions with the mannan might also contribute to nonspecific association of the pheromone with the yeast cell surface. Despite attempts to avoid these particular problems by choosing appropriate mannan mutants (*mnn* strains) lacking phosphate moieties (Ballou 1982) and special solvent conditions with which to measure the binding of ^{3}H-labeled α-factor, there was only threefold-to-tenfold more pheromone bound to **a** cells than to α cells (Thorner 1980). This amount of cell-associated radioactivity corresponded to 10^4–10^5 molecules of the pheromone bound per **a** cell. Unfortunately, the low specific radioactivity of the label employed in these experiments necessitated the use of high cell concentrations and a concentration of ^{3}H-labeled α-factor far exceeding that which would be required to saturate a specific receptor with the affinity suggested by the biologically measured K_m value. Therefore, the major component of the binding observed in this (Thorner 1980) and in earlier studies (Maness and Edelman 1978; D. Finkelstein, pers. comm.) is probably nonspecific association of the pheromone with low affinity sites that are not physiologically relevant. On the other hand, responsive mammalian cells appear to have quite a large number ($\sim 10^5$) of specific receptors for binding glucagon, even when such measurements are made at concentrations of the hormone clearly in the physiological range (10^{-10}–10^{-8} M). What is needed for yeast is a highly radioactive and fully biologically active α-factor derivative, such that reliable binding studies can be performed at moderate cell densities and with pheromone concentrations in the physiological range (10^{-11}–10^{-8} M). Although radioiodination would seem the method of choice for preparing such a labeled derivative, it has yet to be directly demonstrated that

^{125}I-labeled α-factor, made by any of the available procedures and then freed of unreacted pheromone by some separation method, does indeed retain biological activity (Maness and Edelman 1978; Finkelstein and Strausberg 1979; G. Samokhin, pers. comm.).

If a cells do have specific receptors for α-factor, then one might expect to find receptor-defective strains among a-cell mutants selected as nonresponsive to α-factor. If this receptor is specific in the sense that it recognizes only α-factor, then receptor defects might be expected to affect the pheromone-responsiveness of a cells only. Indeed, Hartwell (1980) has recently identified a number of different genes that, when defective, cause a cells to be both α-factor-resistant and sterile, i.e., mating-deficient. One of these mutations, *ste2*, is mating-type-specific, in that it confers pheromone insensitivity and sterility to a haploids only. Perhaps this locus represents the structural gene for α-factor receptor or a component thereof. Obviously, binding of radioactively labeled α-factor to *ste2* strains should be examined.

Mode of Action: Pheromone Proteolysis

A second source of interference in detecting interaction of α-factor with a putative receptor is that a cells are able to rapidly destroy the pheromone, as well as respond to it. Throm and Duntze (1970) originally found that the effects of α-factor are transient. Even when α-factor is not removed from treated cultures by a change of medium, the a cells eventually recover from growth arrest by the pheromone and resume normal vegetative multiplication. At a constant input cell density, the higher the initial pheromone concentration is, the longer the period of arrest; and at a constant initial α-factor concentration, the higher the input cell density is, the shorter the period of arrest (Chan 1977; Stötzler et al. 1977; Manney et al. 1981). Studies by Hicks and Herskowitz (1976) and Chan (1977) were the first to suggest that this reversal or recovery is due to destruction of α-factor activity by the a cells. However, their physiological experiments could not distinguish between enzymatic inactivation of the pheromone and removal of the α-factor from solution (either by adsorption to an extracellular inhibitor [Yanagishima et al. 1977a] or through binding and internalization of the molecule by the a cells themselves). By using ^{125}I-labeled α-factor, Maness and Edelman (1978) were able to demonstrate that the pheromone is in fact chemically altered in the presence of a cells. Since only the carboxyterminal Tyr residue was labeled, they could not determine the precise nature of this alteration. Nonetheless, these investigators suggested that this chemical alteration represented limited and relatively specific proteolytic cleavage of α-factor. They further proposed that this proteolysis was involved in the mechanism of action of the pheromone since preincubation of a cells with the protease inhibitor Trasylol appeared to prevent α-factor-induced G_1 arrest. However, this inhibitor was the only one out of five tested that showed this effect. They also hypothesized that the requirement for proteolysis might be to produce fragments of the pheromone, which then penetrate the cell membrane and act intracellularly.

Independent investigations of the metabolism of α-factor by yeast cells (Ciejek and Thorner 1979; Finkelstein and Strausberg 1979; H. Kita, pers. comm.) have demonstrated unequivocally that the chemical alteration detected by Maness and Edelman is indeed proteolytic cleavage of the α-factor molecule. The rate of degradation is quite rapid. At a-cell densities of 10^6–10^7/ml and pheromone concentrations of approximately 10^{-6} M, the rate of α-factor proteolysis at 25°C is approximately 10^7 pheromone molecules cleaved per minute per a cell (Ciejek and Thorner 1979; Finkelstein and Strausberg 1979). The primary scission occurs between Leu 6 and Lys 7 (Ciejek and Thorner 1979) (Table 1). In contrast to the suggestions of Maness and Edelman, however, these studies indicate that only the intact pheromone is the biologically active agent and that the observed proteolytic breakdown is one mechanism by which a cells inactivate α-factor in order to recover from G_1 arrest. The results that lead to these conclusions are summarized as follows.

First, whether the pheromone is radioactively tagged at its carboxyl terminus or throughout the molecule, essentially all of the label is recovered in the culture medium, even after prolonged incubation with a cells. The fact that proteolytic fragments are recovered essentially quantitatively in the culture fluid does not, in itself, completely rule out an essential role for α-factor cleavage in its biochemical mode of action. Perhaps proteolysis occurs concomitantly during a very transient interaction of α-factor with its receptor; perhaps a protease is part of the receptor; or perhaps a cells rapidly internalize the pheromone, degrade it intracellularly to active fragments, and then secrete the breakdown products. However, the fragments that are produced are not biologically active themselves; and it should be recalled that all partial segments of the pheromone prepared to date by chemical or enzymatic means have specific biological activities that are considerably reduced or undetectable.

Second, the most pronounced effect of agents or conditions that inhibit proteolysis of the molecule is not to prevent response to the pheromone but rather to impede recovery from its effects. For example, the presence of protease inhibitors that reduced the rate of degradation of ^{3}H-labeled α-factor markedly prolonged the period of pheromone-induced cell-cycle arrest of a cells (Ciejek and Thorner 1979). The effect of Trasylol observed by Maness and Edelman was probably an artifact due to nonspecific growth inhibition by benzyl alcohol, an antimicrobial agent present in all commercial preparations of this protease inhibitor.

Finally, the apparent K_m for proteolytic destruction of the pheromone is at least one to two orders of magnitude higher than the α-factor concentration required to induce cell-cycle arrest when both processes are measured at the same cell density (Finkelstein and Strausberg 1979). The disparity between the K_m for proteolytic degradation and the K_m for biological activity of the pheromone makes it unlikely that proteolytic cleavage is a direct step in α-factor action.

Work with mutants further supports the conclusion that proteolysis is not an obligatory part of the mechanism of α-factor action. By screening for mutants that were "super-sensitive" to the growth-inhibitory effects of α-factor, R. Chan and C. Otte (in prep.) were able to identify at least two genes (*SST1* and *SST2*) that, when defective, *enhance* **a**-cell response to α-factor. Cells of the **a** mating type carrying an *sst1* mutation degrade ^{3}H-labeled α-factor much more slowly than their normal parent strain and appear to be deficient in a membrane-bound peptidase (Ciejek 1980; E. Ciejek et al., in prep.). Thus, it is clear that mutants defective in their ability to cleave α-factor proteolytically show no impairment in their response to the pheromone whatsoever. Strains bearing *sst1* mutations may be useful in attempts to measure accurately the binding of radioactively labeled α-factor to **a** cells.

Despite the fact that normal **a** haploids can destroy α-factor by proteolysis, it is clear that **a** cells can nevertheless begin to divide in the presence of the pheromone even under conditions where the rate of α-factor degradation is insignificant. This effect was first observed by Manney et al. (1981) who carefully examined the kinetics of the resumption of **a**-cell budding at concentrations of α-factor well below the K_m for the proteolysis reaction. This finding has been amply confirmed under other experimental conditions where destruction of α-factor by proteolytic cleavage is negligible, such as when a protease-resistant α-factor analog is used (Samokhin et al. 1981), in cultures of *sst1* mutants (R. Chan and C. Otte, in prep.), at low **a**-cell density (S. Moore, pers. comm.), or when **a** cells are constantly perfused with fresh α-factor-containing medium (S. Moore, pers. comm.). In other words, the **a** haploids eventually become desensitized or habituated to a given concentration of α-factor. Such a process is well known in other hormonal systems (Cuatrecasas and Hollenberg 1976). Whether this second recovery process is due to "down regulation" of putative α-factor-receptor complexes by endocytosis or whether it is the result of slow regeneration in the cells of some key metabolite is not yet known. In this regard, however, the *sst2* mutation identified by R. Chan and C. Otte (in prep.) is especially noteworthy. Strains bearing the *sst2* lesion, although apparently able to degrade free α-factor (L. Blair and J. Thorner, unpubl.), nevertheless require an exceedingly long time before commencing the cell cycle after having been exposed to the pheromone. Even when removed from the presence of α-factor by medium exchange, *sst2* mutant cells seem to remain "locked" in G_1 because they fail to divide (but continue to increase enormously in volume) and only begin to bud again normally after an extended period of time (R. Chan and C. Otte, in prep.).

Unlike the target **a** cells, α cells growing exponentially do not appreciably degrade their pheromone (and do not elaborate an extracellular inhibitor that prevents **a** cells from doing so [Finkelstein and Strausberg 1979]). Hence, the ability to destroy α-factor by proteolysis appears to define another apparent distinction between the two mating types. However, the biochemical basis for this difference is not completely clear. Extracts from α cells (Finkelstein and Strausberg 1979), spheroplasts of α cells (Maness and Edelman 1978), and even old stationary-phase α cells (Tanaka and Kita 1977; Ciejek 1980) do attack α-factor and cleave the molecule in a manner similar to that observed using whole **a** cells. Since the spectrum of membrane-bound peptidases in **a** and α cells appears very similar, if not identical (Ciejek 1980), it seems that what distinguishes **a** and α haploids with respect to α-factor proteolysis is the ease of access of the pheromone to the degradative system.

In fact, several additional lines of evidence suggest that **a** and α cells may differ in their ability to respond to the pheromone mainly because of the degree to which their cell surfaces exclude α-factor (due, perhaps, to the presence or absence of specific receptors). In this regard, it was initially reported that spheroplasts of **a** cells could still respond to α-factor, as judged by the induction of nonround forms (Maness and Edelman 1978). Since other work has demonstrated that the pheromone-induced morphological change of intact **a** cells is manifest mainly at the level of chitin synthesis (Schekman and Brawley 1979) and other cell-wall components (Lipke et al. 1976), it is difficult to assess the significance of the observed shape changes in the putative spheroplasts. Indeed, the response of **a** spheroplasts to α-factor has recently been reinvestigated, and a preliminary communication reports that **a**-cell spheroplasts do not respond to the pheromone, as indicated by their continued synthesis of DNA in the presence of α-factor (Polster and MacKay 1980). One interpretation of this result is that a specific α-factor receptor may reside, at least in part, in the cell wall. On the other hand, the lack of response may be due to disruption of proper α-factor-receptor association under the unusual osmotic conditions required to stabilize the spheroplasts or to rampant degradation of the pheromone as a result of increased exposure of membrane peptidases because the cell wall has been removed.

In another attempt to localize the site of pheromone action and thereby to cast some light on its mode of action, Tanaka and Kita (1978) prepared a fluorescent derivative of α-factor. They found that only **a** cells, and not α cells and **a**/α diploids, bound the fluorescent material. For at least 2 hours, the label remained at the cell periphery. By 5 hours, diffuse fluorescence appeared throughout the cell, and by 8 hours, fluorescence seemed to be localized over the nucleus. Unfortunately, these experiments are hard to interpret because they were not well controlled. For example, the specificity of the observed effects was not checked by incubating **a** cells with fluorescent α-factor in the presence of a large excess of competing unlabeled pheromone. Furthermore, the fluorescent pheromone was prepared by reacting the available amino groups of α-factor (α-NH_2 of Trp 1 and ε-NH_2 of Lys 7 [Table 1] with fluorescein isothiocyanate (FITC). Since reaction of these same amino groups with acetic anhydride (Lipke 1976), fluorescamine, or dansylchloride (Maness and Edelman 1978) completely abolishes the biological activity of α-factor, it is hard to understand how a derivative containing the considerably bulkier FITC group could retain biological activity. In fact, another laboratory claims that FITC-α-factor is not biologically active (G. Samokhin, pers. comm.). Since Tanaka and Kita did not purify the fluorescent derivative away from unreacted pheromone, its specific biological activity is unknown. Also, even if it is true that only **a** cells permit the passage of FITC-α-factor through their cell walls, the observed "binding" of the fluorescent pheromone at the cell surface may still be an artifact. The highly hydrophobic FITC-α-factor may partition into the hydrocarbon interior of the lipid bilayer of the plasma membrane, rather than interact with a specific receptor. No experiments with spheroplasts and FITC-α-factor were reported. In addition, since some of the effects of normal α-factor can be detected within as short a time as 15 minutes following admixture with **a** cells (Fehrenbacher et al. 1978) and since conjugation is normally completed within 4 hours, the time course seen for FITC-α-factor by Tanaka and Kita is somewhat disturbing. Thus, the "intracellular accumulation" observed after very prolonged incubation (5–8 hr) may represent the sequestering of metabolically inert FITC-Trp and/or FITC-Lys, or peptides containing these modified amino acids, produced by proteolysis of the fluorescent pheromone.

Mode of Action: A Second Messenger Hypothesis

Many peptide hormones in other systems appear to influence the physiology of their target cells by either stimulating or inhibiting the activity of membrane-bound adenylate cyclase and, hence, modulating the intracellular level of 3'-, 5'-cyclic AMP (cAMP). For this reason, we recently examined (Liao and Thorner 1980) the effects of α-factor on the activity of the membrane-bound adenylate cyclase of *S. cerevisiae* (Varimo and Londesborough 1976). When enzyme activity in plasma-membrane preparations was measured by a sensitive and specific radiochemical assay (Salomon et al. 1974), addition of α-factor markedly inhibited adenylate cyclase activity in a dose-dependent manner. The integrity of the cell membrane appeared essential for this effect since solubilization of the enzyme with the detergent Triton X-100 completely abolished inhibition by α-factor. The pheromone had no detectable effect on the activity of the cAMP phosphodiesterase present in yeast cell extracts. The in vitro inhibition of adenylate cyclase activity by α-factor was apparently specific; and the following observations show that the effect of α-factor on the membrane-bound adenylate cyclase in vitro is physiologically relevant to the mode of action of the pheromone in vivo.

First, partial sequences of the α-factor molecule that are biologically inactive in vivo (e.g., Pro-Gly-Gln-Pro-Met-Tyr [Ciejek and Thorner 1979]) did not inhibit adenylate cyclase activity in vitro; conversely, peptide portions of α-factor that do have some detectable effects on **a** cells in vivo (e.g., Leu-Lys-Pro [Ciejek 1980]) did show some inhibitory action on the adenylate cyclase in vitro. Second, heterologous peptides, some of which are hormones known to alter the activity of adenylate cyclases in the membranes of other organisms, had little or no effect on the yeast enzyme. Third, artificially elevating the intracellular concentration of cAMP antagonized α-factor action. For example, the presence of cAMP (1–10 mM) in the culture medium markedly speeded the recovery of **a** haploids from α-factor-induced G_1 arrest. Even more strikingly, the presence of inhibitors of cAMP-phosphodiesterase activity in the culture medium could prevent almost completely the α-factor-induced cell-cycle arrest of **a** cells (H. Liao and J. Thorner, in prep.). Finally, one class (*ste5*) of the pheromone-insensitive **a**-cell mutants described earlier (which were selected to be resistant at 34°C but not at 23°C to the growth stasis caused by α-factor [Hartwell 1980]) had adenylate cyclase activity that was insensitive to α-factor inhibition when their plasma membranes were assayed at the high temperature. The adenylate-cyclase activity was fully sensitive to pheromone inhibition if the *ste5* membranes were assayed at the lower temperature (Liao and Thorner 1980). Therefore, at least part of the mechanism of action of α-factor on **a** cells appears to be mediated through an inhibition of adenylate cyclase activity, which would lower the intracellular level of cAMP. How could a lowering of cAMP level perturb or disrupt normal vegetative multiplication and divert cells into their mating program?

It has been amply borne out that many of the responses of mammalian cells to changes in cAMP concentration are mediated through changes in the enzymic activity of cAMP-dependent protein kinases (Greengard 1978; Gottesman 1980). Like higher eukaryotes, yeast cells seem to have a complete cAMP-based regulatory system, which, in addition to adenylate cyclase, includes two (high and low K_m) cyclic nucleotide phosphodiesterases (Fujimoto et al. 1974; Londesborough 1977), protein kinase activities (Takai et al. 1974; Lerch et al. 1975; Kudlicki et al. 1978), and a cAMP-binding protein that is apparently a protein kinase regulatory subunit (Dery et al. 1979; Jaynes et al. 1979; Hixson and Krebs 1980). Moreover, Ca^{++}-dependent regulatory components that interlock with cyclic nucleotide-based control systems (Cheung 1980) have been demonstrated in other lower eukaryotes. For example, the protozoan *Tetrahymena pyriformis* and the cellular slime mold *Dictyostelium discoideum* have calmodulin proteins with properties very similar to those found in bovine brain (Jamieson et al. 1979; Clarke et al. 1980). A similar protein seems to be present in *S. cerevisiae* (G. Stetler and J. Thorner, unpubl.).

The lowering of intracellular cAMP concentration presumably caused by α-factor would reduce the activity of cAMP-dependent protein kinase(s) (or other specific cAMP-dependent functions) and therefore would lower the rate of phosphorylation of certain target proteins. A number of yeast proteins are already known that undergo reversible phosphorylation and dephosphorylation, e.g., ribosomal proteins (Becker-Ursic and Davies 1976; Zinker and Warner 1976), RNA polymerase subunits (Bell et al. 1977), and nicotinamide adenine dinucleotide (NAD)-glutamate dehydrogenase (Hemmings 1980). In other words, in yeast, control of cell division and regulation of the gene expression and other metabolic functions required for conjugation may involve proteins whose activities or structural assembly are affected by their degree of phosphorylation. Thus, we propose specifically that the presence of α-factor may prevent the phosphorylation (by cAMP-dependent protein kinases) of certain key target proteins that we postulate are required, in their phosphorylated form, to allow progression through G_1 and to repress the expression of conjugation-specific functions.

Although only **a** cells respond to the presence of α-factor in vivo, inhibition by α-factor of adenylate cyclase activity in vitro was observed with membrane preparations from α haploids and $\mathbf{a}/\alpha$ diploids as well. However, it should be stressed that these plasma-membrane fractions were prepared from spheroplasts produced by extensive digestion of the cell wall with Zymolyase 60,000 (a commercially available lytic-enzyme preparation containing potent $\beta(1 \rightarrow 3)$-glucanase activity and other hydrolytic enzymes). Therefore, these findings provide an additional indirect indication that, in vivo, the pheromone-insensitive α haploids and $\mathbf{a}/\alpha$ diploids may differ from the target **a** cells mainly in their ability to exclude the entry of α-factor from its site of action.

a-Factor

Structure and Properties

When **a** and α cells are mixed for mating, approximately equal numbers of unbudded cells of both mating types are found, indicating some reciprocal effect of **a** cells on α cells (Hartwell 1973). Indeed, an activity was initially detected in concentrated filtrates from **a**-cell cultures that was able to arrest the growth of α cells transiently (Wilkinson and Pringle 1974). This material was, however, much less efficacious (on a per milligram of protein basis) than similarly concentrated α-cell-culture medium and did not appear to induce elongation of α cells. Subsequent attempts to purify and characterize this material yielded ambiguous results as to its precise chemical nature (Betz et al. 1977). Recently though, Duntze's laboratory has succeeded in purifying to homogeneity two related undecapeptides and a modified form of one of them (Table 1) from **a**-cell-culture medium. All three species have the biological activities expected of an **a**-cell pheromone, but the Val-containing molecule has a markedly higher specific activity than the others (Betz and Duntze 1979). These substances arrest α cells as unbudded cells in the G_1 phase, increase the agglutinability of α cells toward **a** cells, and eventually cause the pronounced elongation of α cells. These oligopeptides have none of these effects on **a** cells or on $\mathbf{a}/\alpha$ diploids.

It is clear from their amino acid composition and their behavior during purification that the undecapeptide **a**-factors are exceedingly hydrophobic, which undoubtedly accounts for the difficulties experienced in earlier attempts to isolate and characterize the **a**-cell pheromones. The implications of the apparent differences between these molecules remain to be determined. For example, do the different molecules represent the products of processing a single larger protein, or are there in fact two expressed **a**-factor genes? If the latter is true, it may explain the observation that **a**-factor production seems to be enhanced by the presence of α-factor (V. MacKay, pers. comm.). In other words, one **a**-factor may be expressed constitutively, and the other may be inducible by α-factor.

By using the super-sensitive mutation *sst2*, discussed above (R. Chan and C. Otte, in prep.), it is possible to construct α-cell tester strains that respond dramatically to the growth-inhibitory action of the **a** pheromones. By using such an assay system, it has been found that some α haploids, but not the majority of α strains, when interposed as a barrier (Hicks and Herskowitz 1976) between an **a**-factor-releasing strain and responsive α cells, are able to attenuate markedly the effects of the **a**-cell pheromone (L. Blair and J. Thorner, in prep.). Whether this inactivation of **a**-factor activity is actually due to proteolysis of the molecule, analogous to the degradation of α-factor by **a** cells, is not yet known. It is clear, however, that α cells bearing the *sst1* mutation are *not* super-sensitive to **a**-factor (R. Chan and C. Otte, in prep.); therefore, *SST1* is probably not involved in the inactivation of **a**-factor activity by certain α strains.

Pheromone Biosynthesis

It is clear from the DNA sequence of *MATα* (Astell et al. 1981; Nasmyth et al. 1981a) that the information for the structure of α-factor is not coded, at least in any straightforward way, at the mating-type locus itself. An α-cell-specific mutation (*ste13-1*) that causes mating deficiency (G. Sprague et al., pers. comm.) also appears to cause a direct alteration in the primary structure of the α-factor molecule that is secreted. This mutation segregates like a single nuclear gene but is genetically unlinked to *MAT*. The variant molecule produced can be resolved from normal α-factor by high-pressure liquid chromatography on RP-C18 or by gel filtration through Sephadex G-25. The nature of the separations observed indicates that the variant is somewhat larger and less hydrophobic than wild-type α-factor, which suggests the presence of additional polar residues (D. Julius et al., in prep.). Furthermore, functionally α/α diploids bearing both the mutant allele and a normal α-factor gene appear to produce only the normal form of the pheromone. This lack of codominance suggests that the mutation represents a lesion in a gene that results in altered processing of an otherwise normal pheromone precursor rather than a change within the DNA sequence of the pheromone structural gene itself. Strains producing this and other mutant forms of α-

factor were identified among a collection of α-cell mutants that were apparently pheromone-deficient when tested for biological activity but that nevertheless secreted normal levels of pheromone-related peptides when examined immunologically with anti-α-factor antibody (L. Blair et al.; Y. Jones-Brown et al.; both in prep.). Mutants producing such altered forms of α-factor are not mating-defective. However, mutants identified in the same screen, but producing no pheromone detectable either immunologically or biologically, are sterile. At least some of the latter mutations are not at $MAT\alpha$, and may not be allelic to the mutations that cause the production of altered α-factors. Obviously, regulation of the expression of the α-factor gene is of great interest, and attempts to clone this sequence are in progress in several laboratories. Such cloned segments should also greatly aid study of the precise mechanism of α-factor synthesis.

Because α-factor is normally present in α-cell-culture medium at low concentration, 10^{-7}–10^{-8} M, as determined by a radioimmunoassay (Y. Jones-Brown et al., in prep.), and because the total weight of the pheromone recovered from large-scale (200-liter) preparations is only 10–20 mg (Ciejek et al. 1977; Ciejek 1980), there may be a misleading impression that α-factor production represents only a tiny fraction of the synthetic capacity of an α haploid. This yield is deceptively small because of the low molecular weight of the peptide. By comparison, about as much α-factor is made, on a molar basis, by a yeast cell as phosphoglycerate kinase, which is a major yeast glycolytic enzyme (Scopes 1973).

The appearance of α-factor in extracellular culture fluid is rapidly inhibited by the addition of cycloheximide or by shifting α cells carrying temperature-sensitive mutations that block RNA and/or protein synthesis to the nonpermissive temperature (Scherer et al. 1974). Such observations suggest that α-factor synthesis does involve ribosomal translation of a specific mRNA, as opposed to construction on a multienzyme complex like the peptide antibiotics of bacteria (Lipmann 1971). In addition, the small size of the α pheromone itself and the finding that α strains auxotrophic for amino acids other than those found in the α-factor molecule ceased producing the pheromone when deprived of the required amino acids (Scherer et al. 1974) suggest further that α-factor is produced by the processing of a larger precursor polypeptide, as has been found for many peptide hormones in higher eukaryotes (e.g., Lomedico et al. 1977). Indeed, Tanaka and Kita (1977) have reported the presence in α-cell-culture medium of a small amount of an α-factor-related polypeptide, for which they determined the total amino acid composition, as well as a partial sequence for a few residues at the amino terminus. Their results indicate that this molecule may have the following structure: Trp-His-Trp-Leu-Gln-Leu-Lys-Pro-Gly-Gln-Pro-Met-Tyr-(-Ala, Asx, Gly, Ile, Lys, Val, Ala, Asx, Glx, Ser, Thr). This material apparently represents an incompletely processed form of an α-factor precursor, wherein the additional amino acids have not yet been removed from the carboxyl terminus of the pheromone. This putative precursor was biologically inactive, and none of the limited chemical and enzymatic treatments attempted converted it to a functional pheromone (Tanaka and Kita 1977). Perhaps, among α-cell mutants that are unable to release active α-factor will be those that are defective in the protease(s) responsible for processing pheromone precursor. Such mutants might accumulate substrate amounts of precursor forms that could be used to study the mechanism of processing in vitro.

It should be noted that α-factor, the putative pro-α-factor, and the **a**-factors all have distributions of amino acid residues that closely resemble those seen in the "signal sequences" of the precursor forms of many proteins that are inserted in or secreted through cell membranes (Schekman and Novick 1982). Thus, it is possible that production of a particular pheromone by a given mating type may be the natural outcome of the process of insertion of a membrane protein or the process of export of a periplasmic or mannan-bound polypeptide normally present at the yeast cell surface. The *kex2* mutation is perhaps significant in this regard. The *kex2* mutation prevents secretion of killer toxin (Wickner, this volume) and affects the processing and secretion of many other extracellular proteins and glycoproteins by yeast cells (Rodgers et al. 1979). Furthermore, the *kex2* lesion causes α cells but not **a** cells to be mating-deficient (Leibowitz and Wickner 1976). Most notably, the *kex2* defect also prevents α haploids from releasing any detectable α-factor activity or immunologically cross-reacting material (Y. Jones-Brown et al., in prep.). However, the *kex2* mutation does not prevent **a** cells from releasing active **a**-factor (L. Blair, unpubl.).

Obviously, elucidation of the detailed mechanism of biosynthesis of the pheromone molecules awaits further identification, purification, and characterization of precursor form(s), as well as the isolation of the structural gene(s) coding for these precursors through the use of recombinant DNA methodology.

CYTOLOGICAL AND PHYSIOLOGICAL EVENTS

Intercellular Signaling

Clearly, pheromone production and response are correlated with the mating process. In support of their important role in the physiology of mating are the findings that many mating-defective haploids do not produce their particular pheromone (MacKay and Manney 1974a,b; Y. Jones-Brown et al., in prep.) and that **a**-cell mutants selected as resistant to α-factor are almost invariably mating-deficient (Manney and Woods 1976; Hartwell 1980).

However, if a sporulated diploid of *S. cerevisiae* is placed on nutrient medium in the laboratory, the tetrad of spores will germinate within the ascus and, once metabolically active, these haploids will immediately fuse pairwise to reform two diploid cells. Nevertheless, the fact that *S. cerevisiae* has evolved a system involving diffusible pheromones for inducing the mating program suggests that, in nature, individual spores are frequently released from asci. In other words, it seems likely that haplophase for heterothallic strains is not necessarily a transient stage of the life cycle in the wild. Indeed, many other microorganisms, both eukaryotic and prokaryotic, appear to release extracellular signals that trigger a conjugation process similar to that found in *S. cerevisiae*. In particular, many systems have evolved a diffusible peptide or a molecule containing a peptide moiety as the intercellular signal. Examples include other yeast species (*S. kluyveri* [McCullough and Herskowitz 1979], *Rhodosporidium toruloides,* and *Tremella mesenterica* [Tsuchiya et al. 1978]); protozoa (*Tetrahymena thermophila* [Adair et al. 1978]); slime molds (*Polysphondylium violaceum* and *P. pallidum* [Wurster et al. 1976]); and Gram-positive bacteria (*Streptococcus faecalis* [Dunny et al 1978]).

Because the mating pheromones of *S. cerevisiae* mutually arrest vegetative growth of **a** and α haploids, it has been suggested that this response represents a "phasing" mechanism, whereby the **a**- and α-cell cycles are synchronized, and is a prerequisite for mating (Bucking-Throm et al. 1973; Hartwell 1973). Indeed, the elegant work of Hartwell and his collaborators has demonstrated that the presence of α-factor stays the cell-division cycle of **a** cells at a specific point in very early G_1 phase. The α-factor block is prior to both duplication of the nuclear spindle pole body (Byers and Goetsch 1974, 1975) and the initiation of chromosomal DNA replication, and prior to all but a very few of the cell-divison-cycle functions (*CDC* genes) that have been defined in yeast to date by the use of temperature-sensitive mutations (Pringle and Hartwell this volume). In fact, it appears that only cells arrested in G_1 phase mate efficiently, regardless of whether they are held at that stage by α-factor (Bucking-Throm et al. 1973), by shift of a temperature-sensitive mutant (*cdc28*) to nonpermissive temperature (Reid and Hartwell 1977), by nutritional deprivation (Mortimer 1955), or by direct isolation of the unbudded fraction of normal populations (Sena et al. 1975).

Whatever the molecular mechanism of pheromone action, it is clear from reciprocal shift experiments (Hereford and Hartwell 1974; Hartwell 1976) that α-factor prevents the initiation of a cell cycle by the target **a** cells if they are already in early G_1; but once an **a** cell has started through its cell cycle, it will complete that one cycle even if α-factor is present. In other words, α-factor appears to cause **a** cells to "idle" in G_1, as opposed to other treatments (like starvation for phosphate or sulfate, addition of ethionine [Singer et al. 1978], or shift of a temperature-sensitive pyruvate kinase mutant [*cdc19*] to a nonpermissive condition), which presumably cause **a** cells to arrest as unbudded cells because of inhibition of metabolism or limitation for some metabolite required for subsequent growth. In support of this view, measurements of macromolecular synthesis indicate that in **a** cells exposed to α-factor, the rates of bulk RNA, protein, polysaccharide synthesis (Throm and Duntze 1970; Sumrada and Cooper 1978; Schekman and Brawley 1979), and even mitochondrial DNA synthesis (Cryer et al. 1974) are relatively unaffected, despite the fact that chromosomal DNA replication and normal bud emergence are clearly prevented. Because cell-wall and protein synthesis continue in the presence of α-factor while cell division is inhibited, a cells expand in volume and increase in length, which accounts for the eventual formation of morphologically distinct cells (shmoos). Also consistent with the view that α-factor-induced arrest is different from simple nutritional deprivation is the observation that the degree of chromosome "folding" in pheromone-treated a cells is distinguishable from that of either starved a cells or exponentially growing a cells (Piñon and Pratt 1979).

Certain other a-cell responses to the presence of the α pheromone, like release into the cytoplasm of vacuolar constituents (at least, arginine) (Sumrada and Cooper 1978) and the lack of accumulation of mRNAs for histones H2A and H2B (Hereford et al. 1980), are observed after other treatments or under other conditions that arrest haploids in G_1. In fact, the apparent reduction in protein methylation observed after exposure of a cells to α-factor (D. Finkelstein, pers. comm.), may represent dilution of the exogenous $[CH_3\text{-}^3H]$methionine label used to make such measurements due to release of the large pool of S-adenosylmethionine that is normally stored in the vacuole (Nakamura and Schlenck 1974). On the other hand, enzymatic methylation of phospholipids appears to play some role in the transduction through plasma membranes of a variety of receptor-mediated signals in a number of higher eukaryotic systems (Hirata and Axelrod 1980).

Cell-Cell Contact

Although G_1 arrest facilitated by the pheromones appears to be a necessary condition for conjugation, understandably the opposite mating types must be brought into close proximity to permit their fusion. Since yeast cells are nonmotile, the efficiency of mating depends on the opportunity for cell-cell contact provided by random collision. Depending upon the particular technique used in the laboratory to bring mating partners into juxtaposition, anywhere from 10–20% (gentle swirling in liquid cultures) to nearly 100% (direct pairing of a and α cells by micromanipulation) of the haploids of a given mating type can go on to yield viable zygotes with cells of the opposite mating type.

Because successful mating requires close cell-cell contact, it is very significant that the earliest detectable response of a cells to exposure to α-factor is some cell-surface alteration that markedly increases their agglutinability toward α cells (Fehrenbacher et al. 1978; Shimoda et al. 1978). This increase in intercellular adhesion presumably increases the probability that opposite mating types will remain in close contact after they have met by chance collisions.

When they are mixed under nongrowing conditions, a and α cells that have not been exposed to each other previously (so-called naive cells) bind to each other only weakly, although they do interact with each other more strongly than they cohere to cells of like mating type (Campbell 1973; Fehrenbacher et al. 1978), which suggests that they normally express only a basal level of mating-specific cell-surface agglutinins. In contrast, within 20–60 minutes after incubation of mating mixtures in growth medium, or after incubation of each haploid with the appropriate purified pheromone in growth medium, cells of opposite mating type (so-called preconditioned cells) adhere to each other quite strongly (Yanagishima et al. 1977b; Betz et al. 1978; Fehrenbacher et al. 1978), which suggests that expression of mating agglutinins has been highly induced. Initially, two classes had been observed (Sakai and Yanagishima 1972) among various available **a**-cell stocks with regard to this particular response: (1) those **a** strains that agglomerated with α cells immediately upon admixture (constitutive strains) and (2) those that required a period of mutual incubation with α cells before a strong agglutination reaction was obtained (inducible strains). The competence of most α haploids to agglutinate is enhanced much less dramatically, or not at all, by prior exposure to **a** cells or to **a**-factor (Betz et al. 1978; Fehrenbacher et al. 1978; L. Hartwell, pers. comm.). It has been found that the high degree of intrinsic agglutinability of many constitutive **a** strains observed at 23–28°C can be eliminated by growth of the cells at 34–36°C (a physiologically reasonable temperature). After culturing in this high-temperature range, the formerly constitutive **a** strains behave like inducible cells, i.e., they are only weakly agglutinative when naive, but when preconditioned with α-factor, they display markedly increased agglutinability toward α cells (Doi and Yoshimura 1978; Doi et al. 1979; Tohoyama et al. 1979).

One possible interpretation of these results is that intrinsic agglutinability and inducible mating-specific adhesion are separable functions carried out by different macromolecules that differ in their thermostability. This seems unlikely because glycoprotein fractions (which appear to mediate mating-type-specific intercellular recognition, see below) partially purified from constitutive **a** strains and from preconditioned inducible **a** strains seem to have very similar properties (Tohoyama et al. 1979); furthermore, the agglutinin made by constitutive strains at low temperature is lost only slowly, by growth dilution, after shift to 36°C (Doi and Yoshimura 1978).

An alternative possibility is that in inducible **a** cells, synthesis of mating-specific agglutinin is normally repressed at all temperatures by a negative regulatory element. However, the action of this repressor can be prevented somehow by exposure of the cells to α-factor, which results in induction of the synthesis of mating agglutinin. On the basis of this model, in the so-called constitutive **a** strains, expression of mating-specific agglutinin is derepressed at 23–28°C because the repressor bears a defect that makes its function (or synthesis) cold-sensitive. Hence, at lower temperatures, where the repressor is inactive (or not present), mating agglutinin is made whether or not cells are exposed to α-factor (constitutive phenotype); but at higher temperatures, where the repressor is operative, mating agglutinin can be made only if α-factor is present (inducible phenotype). This model predicts that total elimination of the repressor (a null mutation) would make an **a** cell constitutive for expression of mating agglutinin at all temperatures. Indeed, a mutation (*sag1*) conferring such a completely constitutive phenotype has been found among **a**-cell stocks (Doi and Yoshimura 1978). If the *sag1* mutation represents the loss of a negative regulatory element that operates in *trans*, then its constitutivity should be recessive to inducibility. This was, in fact, found to be the case by constructing **a**/**a** diploids through protoplast fusion of an inducible **a** strain and the *sag1* constitutive **a** mutant. Moreover, the *sag1* mutation segregated like a single nuclear gene in crosses to an inducible α strain and was inseparable from the *MAT***a** locus in all 59 tetrads examined. These results were confirmed in a related study done in a converse fashion by another group. Yanagishima and Nakagawa (1980) have reported that the inducible character of certain natural **a** stocks is dominant to the constitutive character of certain other **a** strains in **a**/**a** diploids constructed by protoplast fusion. Furthermore, inducibility of such **a** strains segregated like a single Mendelian trait and was linked to *MAT***a**. (However, other findings by these same workers suggest that the control of agglutination may be more complex than this. They also examined certain other inducible **a** variants, derived by spontaneous mutation at 28°C from constitutive strains and identified by a rough-to-smooth change in colony morphology. The inducible character of these revertants was recessive to constitutivity; and the lesion conferring inducibility [*saa1*] was apparently completely unlinked to the mating-type locus by tetrad analysis. The inducible revertants were also peculiar because they became constitutive again when cultured at 22°C and because they had a tendency to flocculate, which required the addition of EDTA during tests for mating-specific agglutination and, perhaps, made scorings ambiguous.)

It has been suggested (J. Shuster, pers. comm.) that the *sag1* mutation may define a regulatory role for the protein encoded by the *a2* transcript of *MAT***a**, a gene previously without an identifiable function (Herskowitz and Oshima, this volume). In contrast, the *saa1* mutation is difficult to interpret at this time.

At the chemical level, the cell-wall synthesis that occurs in **a** cells in the presence of α-factor seems altered in several ways. The newly made wall material contains more glucan and less mannan than does an unexposed **a** cell wall (Lipke et al. 1976). Furthermore, the mannan of **a**-cell shmoos contains an increased proportion of shorter oligosaccharide side chains and unsubstituted $\alpha(1 \rightarrow 6)$ backbone, which probably explains their greater binding of fluorescently labeled concanavalin A (Tkacz and MacKay 1979) and antibody directed against $\alpha(1 \rightarrow 6)$-mannosyl haptens (Lipke and Ballou 1980). Also, cells treated with α-factor have higher chitin synthetase activity than untreated cells and appear to deposit chitin diffusely throughout the elongating portion of the cells (Schekman and Brawley 1979) rather than in the narrow ring found at the division septum in normal cells (Byers, this volume). Because of all of these cell-wall changes, a-cell shmoos are more susceptible to glucanase digestion than normal cells. Electron micrographs reveal that the protuberance that forms on elongating a cells in fact has a thinner wall layer with a fuzzy exocellular coat and contains subtending small vesicles (Lipke et al. 1976). Such small vesicles have been observed at the presumptive bud site and in enlarging buds in vegetative cells (Sentandreau and Northcote 1969) and presumably carry the enzyme activities and precursors for cell-wall and membrane growth (Schekman and Novick 1982).

Conceivably, cell-surface alterations like those discussed above may reflect or be part of the synthesis, activation, or exposure of cell-surface-associated agglutinins responsible for mating-specific intercellular adhesion. In any event, the pheromone-induced acquisition of increased agglutinability does require protein synthesis (Betz et al. 1978; Fehrenbacher et al. 1978). However, inhibitors of polysaccharide and glycoprotein biosynthesis (including tunicamycin, 2-deoxyglucose, and amphomycin) fail to prevent α-factor-induced enhancement of agglutination, even at concentrations that would be inhibitory to subsequent growth. Hence, the protein synthesis that is required does not appear to involve de novo glycoprotein assembly (C. Atcheson et al., in prep.).

Attempts have been made to isolate cell-surface molecules from haploid cells that may correspond to mating agglutinins. For example, by treating a cells or α cells with snail digestive juice (a mixture of some 30 different hydrolytic enzymes that degrades yeast cell walls) (Shimoda et al. 1975) or by autoclaving haploid cells (Yoshida et al. 1976; Hagiya et al. 1977), glycoprotein fractions are released, which appear to act as univalent blockers of the aggregation of a and α cells. These molecules are only incompletely characterized as yet. (It should be mentioned here that cellular aggregation in another yeast species, *Hansenula wingei*, does appear to be mediated by cell-wall-bound glycoproteins [Crandall 1977]. These components have been characterized extensively [Yen and Ballou 1973, 1974; Burke et al. 1980], despite the fact that, in order to release them from the cell wall, treatments must be used that destroy portions of the molecules to some degree. Such cell-surface glycoproteins also seem to be involved in mating-type-specific agglutinations in at least two other species of ascomycetous yeasts [Burke et al. 1980].)

Some mutations that alter *S. cerevisiae* mannan structure, like *mnn2*, (Ballou 1982) do seem to elevate the intrinsic adhesiveness of haploids toward cells of the opposite mating type (Radin 1976). However, all of these mutants still show at least some enhancement of agglutinability after exposure to pheromones (G. Fehrenbacher et al., unpubl.). Strains bearing other mannan alterations (A364A, e.g., which is *mnn1*) show a prototypically inducible phenotype for agglutinability (Fehrenbacher et al. 1978).

It is clear that only unbudded cells actually conjugate. Similarly, it appears that the unbudded fractions of naive cell cultures of opposite mating type stick together best, albeit weakly (Campbell 1973). However, in contrast to division arrest by the pheromones, induction of agglutinability by pheromones can occur at any stage in the cell cycle, not just G_1. For example, after preconditioning with pheromones, haploids at any stage of the cell-division cycle are found in mating aggregates (Yanagishima et al. 1977b; Fehrenbacher et al. 1978), and preconditioned haploids at various stages of the cell cycle obtained by density-gradient fractionation of pheromone-treated cultures are all agglutination-competent (Tohoyama et al. 1979). Furthermore, budded cells blocked in S phase by hydroxyurea treatment (Doi and Yoshimura 1978), or cells blocked at many stages of the cell cycle by raising *cdc* mutants to a nonpermissive temperature (L. Hartwell, pers. comm.), can be fully induced for agglutination by mating pheromones. These findings agree with the fact that, kinetically, the induction of increased agglutinability is an early response to pheromone treatment, detectable after 15 minutes and maximal by 45–60 minutes. In fact, if the unbudded fraction of a vegetative population of **a** haploids is isolated, the cells will not agglutinate tightly with unbudded α haploids and subsequently yield zygotes unless the unbudded **a** cells have first been incubated with α-factor (or with the α cells themselves) for 20–60 minutes (Radin 1976). This indicates that merely being in G_1 is not sufficient to confer mating competence, at least in **a** cells, and suggests therefore that α-factor is essential for efficient mating.

These results, taken together, indicate the following time course for initial mating events. First, the pheromones induce cell-surface changes that increase and stabilize cell-cell contacts. While being held in close apposition in mating aggregates, the population accumulates in G_1 because the presence of the pheromones also prevents the cells from progressing through the cell-division cycle. Concomitantly, subsequent functions required for cell fusion (or to circumvent budding) are called into play, again presumably because of the continued presence of the pheromones. Whether the primary signal generated by a pheromone (e.g., lowering the intracellular level of cAMP) has many separate effects, or sets off a cascade of dependent steps, is unknown.

Cell Fusion

Even less is understood at the biochemical level about the actual steps of zygote formation. Once in firm contact, the cell walls and plasma membranes separating the two G_1-arrested cells must be removed to permit cell fusion (plasmogamy). The initial breakdown of the wall and membrane is presumably accomplished by autolytic enzymes (Shimoda and Yanagishima 1972). The endogenous glucanases (and mannanases) of yeast, their physiological roles during normal growth, and their regulation by pheromones during mating, have been studied to only a limited extent (del Rey et al.

1979). Both endo-β(1 → 3)- and exo-β(1 → 3)-glucanase activities have been demonstrated in *S. cerevisiae* (for review, see Phaff 1977). Both enzymes have been reported to be glycoproteins that are released into the medium upon spheroplasting of yeast cells (Biely et al. 1976). However, at least a significant fraction (~ 10%) of the total activities of these enzymes has been found associated with endoplasmic reticulum-derived small vesicles (Matile et al. 1971; Cortat et al. 1972), presumably those involved in normal cell-wall and membrane growth at the bud site (Schekman and Novick 1982).

The frequency with which a haploid will fuse with more than one other cell of the opposite mating type is very low (< 1 multiple mating/ 10^4 matings that result in normal diploids), even under conditions where it can be shown that a haploid cell is in intimate contact with as many as six cells of the opposite mating type (Rodgers and Bussey 1978). Presumably, some other process distinct from simple cell-cell contact restricts cell fusion to the appropriate cell pair. Perhaps fusion is restricted because of the requirement for some site or organelle to be in proper orientation in both haploids.

In electron micrographs of conjugating pairs (Osumi et al. 1974), the glucan layers of opposing parental cell walls do seem to become progressively thinner, similar to what is observed for **a** cells treated with α-factor (Lipke et al. 1976). Fragmentation of the plasma membrane by vesiculation occurs concomitantly until, finally, both the wall material and the membrane become sufficiently open to allow cytoplasmic continuity of the apposed haploid partners. As might be anticipated, cell permeability is altered at this stage of conjugation (Shimoda and Yanagishima 1972). Electron microscopy observations (Byers and Goetsch 1974, 1975) also show that during the time of cell fusion, microtubules that radiate from the outer surface of the spindle pole body of each haploid nucleus often appear to reach the region of contact between the two cells and frequently pass through the aperture formed once the two membranes have merged with each other. Like **a** cells arrested in G_1 by α-factor, or like the temperature-sensitive cell-division-cycle mutant with the earliest known block in G_1 (*cdc28*), the nuclear spindle pole body of a conjugating haploid is essentially single but with an additional half-bridge or satellite structure (Byers, this volume). It appears that the extranuclear microtubules that project off of the cytoplasmic face of each of the spindle pole bodies lengthen and eventually interconnect the spindle pole bodies of the two nuclei. Therefore, the site for cell fusion and, hence, the restriction on partner formation, may be directed by the microtubules that radiate from the spindle pole body toward a specific place on the cell membrane. In this regard, it should be recalled that during vegetative growth, haploids always tend to bud from the apex of the cell near the birth scar. Indeed, scanning electron microscopy reveals that the abnormal wall growth that occurs in **a** cells treated with α-factor also seems to extend from this same cell pole (Thorner 1980). In accordance with this finding, it has been reported recently that the fewer the number of buds the haploid cell has already produced, the greater the mating potential of the cell (Müller 1980). Perhaps, too many bud scars occlude the construction of a site for cell fusion. Equally striking is the observation (J. Rine, pers. comm.) that on a solid surface, the cell-wall outgrowth of α-factor-treated **a** cells tends to form most readily when the birth scar pole is facing the source of the pheromone. In the brown algae *Fucus* and *Pelvetia*, it has been reported that a gradient of calcium ionophore will cause eggs to send out their rhizoidal processes from the side of the cell exposed to the highest Ca^{++} concentration (Robinson and Cone 1980). Similarly, in the water mold *Blastocladiella*, outgrowth of the rhizoidal tips can be directed toward the highest concentrations of either synthetic proton conductors or the proton-conducting ionophore gramicidin D (Harold and Harold 1980; Stump et al. 1980). Perhaps α-factor treatment results in the generation of such ion currents that serve to set the direction of assembly or orientation of morphogenic determinants in the yeast cell (like the spindle-pole-body-associated microtubules).

In addition to cell-wall and membrane breakdown, continued resynthesis of all cell-envelope components must be occurring in order to prevent lysis of the mating pairs. Although the biosyntheses of the complex macromolecules involved, including glucan (Shematek et al. 1980), mannan (Ballou 1982), chitin (Cabib 1975; Farkas 1979), and phospholipids (Henry 1981), are being investigated in some detail in vegetatively growing cells, very little is known about the restructuring of these components in conjugating cells. The bizarre shapes eventually formed by **a** cells in the presence of α pheromone but in the absence of α cells probably reflects a chronic condition of being unable to complete the normal agglutination and fusion process.

Perhaps mutants that still respond to pheromones by arresting in G_1 but are nevertheless mating-deficient (MacKay and Manney 1974a,b; J. Rine, pers. comm.; L. Blair, unpubl.) are blocked in other aspects or at a late stage of conjugation. For example, such mutants may be defective in agglutinin synthesis or in functions necessary for the cell-wall dissolution (or cell-wall resynthesis) required for conjugation. It has been shown, however, that a mutant (*exb1-1*) apparently lacking exo-β(1 → 3)-glucanase activity shows no growth impairment and is not detectably mating-deficient (Santos et al. 1979).

Nuclear Fusion

The last step of conjugation, nuclear fusion (karyogamy), is initiated specifically at the distal ends of the half-bridges of the spindle pole bodies on each haploid nucleus (Byers and Goetsch 1974, 1975). These fuse, resulting in the formation of a single, large satellite-bearing spindle pole body (Byers, this volume). Hence, after the two haploid nuclei merge, the new diploid nucleus has a complete spindle pole body and is therefore ready to resume a normal vegetative cell cycle or to commence meiosis. Just how the two nuclear envelopes sort themselves out is not known, although in higher eukaryotes nuclear membrane dissolution and reformation has been correlated with the phosphorylation and dephosphorylation of the nuclear-membrane-associated lamina proteins (Gerace and Blobel 1980; Shelton et al. 1980). Indeed, several "nuclear-limited" mutations of *S. cerevisiae* have been identified that prevent nuclear fusion (Conde and Fink 1976; Dutcher 1980). Thus, the consequence of mating a strain carrying such a mutation is the formation of a zygote that remains binucleate, i.e., a heterokaryon. These mutations have been called *kar* (*kar*yogamy-defective). Since such a zygote is identified by its ability to segregate both heterokaryons and haploid heteroplasmons during its subsequent growth and division, obviously nuclear fusion is not a necessary outcome of cell fusion, nor is nuclear fusion required for viability. It is interesting that haploids bearing one of several *cdc* mutations, including *cdc4, 28, 34,* and *37,* also show the Kar⁻ phenotype (Byers; Pringle and Hartwell; both this volume). The behavior of microtubules, spindle pole bodies, and nuclear-membrane-associated proteins in all of these mutants is of immense interest.

PROSPECTUS

Clearly, the information that has been obtained about the control of cell type in yeast and about the mating process and its regulation is still rudimentary. However, the opportunities are now available to answer many of these questions in detail at the molecular level.

Since it has been shown both genetically (Herskowitz and Oshima, this volume) and biochemically (Nasmyth et al. 1981a; Tatchell et al. 1981) that *MAT* codes for some sort of master regulatory proteins, it should become possible to study how these polypeptides control the expression of cell-type-specific and mating-type-specific functions both in vivo and in vitro once such genes have been isolated by recombinant DNA techniques. In this regard, it is absolutely essential to begin to pinpoint the precise biochemical defects caused by mutations that result both in mating-type-specific sterility and in mating deficiency that is not cell-type-specific. For example, do **a**-cell-specific sterile mutations truly represent lesions in components like an α-factor receptor or an **a**-cell-surface agglutinin? Once such defects are defined, and convenient assays are devised to detect the presence of the wild-type gene products, direct transformation of these mutants with hybrid vectors containing yeast DNA will allow the isolation of these genes. Conversely, such genes can be isolated by the ability of the wild-type alleles on a plasmid vector to restore mating competence to strains bearing sterile mutations at the locus of interest. The cloned DNA segments could then be used to identify mRNA and protein products, which may help define the function of these genes in the mating process.

It should be stressed that the yeast pheromones, whose expression is presumably under the direct control of the mating-type locus, are themselves regulatory proteins. These intercellular signals may have important effects on the pattern of transcription in their target cells, as suggested by the apparent absence of histone H2A and H2B mRNAs in α-factor-treated **a** cells, and this question should be investigated further.

The initial mode of action of α-factor in lowering the rate of synthesis of cAMP and thus lowering the level of this intracellular second messenger suggests that the yeast system could be used to elucidate the properties of a eukaryotic effector-responsive adenylate cyclase system. If **a**-factor exerts its effects on α cells through a similar pathway, then defects in components of an entire cAMP-dependent regulatory cascade might exist among mutations that cause pheromone-insensitivity and sterility but that are not cell-type-specific. Thus, abnormalities in functions like adenylate cyclase itself and its regulatory subunit, cAMP phosphodiesterase and a calmodulinlike protein, or cAMP-dependent protein kinase and its regulatory subunit, may be uncovered. Mutants with such lesions might be expected to show pleiotropic phenotypes in that other cellular processes, e.g., carbon metabolism (Fraenkel 1982), might also be affected.

ACKNOWLEDGMENTS

I express my deepest appreciation to present and former members of my laboratory who conducted much of the work described in this review, in particular Peter Baum, Buff Blair, Tony Brake, David Julius, Riyo Kunisawa, Hans Liao, June Lugovoy, Gary Stetler, and also Cathy Atcheson, Linda Bisson, Elena Ciejek, Mike Davis, George Fehrenbacher, Y'Vonne Jones-Brown, and Karen Perry. I am also very grateful for the constant stimulation and generous assistance provided by Lee Hartwell, Ira Herskowitz, David Botstein, Beth Jones, George Sprague, and Jasper Rine. I thank Jeff Shuster and Duane Jenness for their critical reading of the final manuscript. Studies reported here were mainly supported by U.S. Public Health Service grant GM21841 (to J.T.), and also by postdoctoral research fellowships from the Jane Coffin Childs Memorial Fund for Medical Research (to Y'Vonne Jones-Brown), from the National Institutes of Health (to Hans Liao), from the Damon Runyon–Walter Winchell Cancer Fund (to Tony Brake), and from the American Cancer Society (to Buff Blair and Gary Stetler).

REFERENCES

Adair, W.S., R. Barker, R.S. Turner, and J. Wolfe. 1978. Demonstration of a cell-free factor involved in cell interactions during mating in *Tetrahymena*. *Nature* **274:** 54.

Astell, C.R., L. Ahlstrom-Jonasson, M. Smith, K. Tatchell, K.A. Nasmyth, and B.D. Hall. 1981. The sequence of the DNAs coding for the mating type genes of *Saccharomyces cerevisiae*. *Cell* (in press).

Ballou, C.E. 1982. The yeast cell wall and cell surface. In *The molecular biology of the yeast* Saccharomyces. II. *Metabolism and gene expression* (ed. J. Strathern et al.), Cold Spring Harbor Laboratory, Cold Spring Harbor, New York. (In press.)

Becker-Ursic, D. and J. Davies. 1976. *In vivo* and *in vitro* phosphorylation of ribosomal proteins by protein kinases from *Saccharomyces cerevisiae. Biochemistry* **15:** 2289.

Bell, G.I., P. Valenzuela, and W.J. Rutter. 1977. Phosphorylation of yeast DNA-dependent RNA polymerases *in vivo* and *in vitro*: Isolation of enzymes and identification of phosphorylated subunits. *J. Biol. Chem.* **252:** 3082.

Betz, R. and W. Duntze. 1979. Purification and partial characterization of a-factor, a mating hormone produced by mating-type a cells from *Saccharomyces cerevisiae. Eur. J. Biochem.* **95:** 469.

Betz, R., W. Duntze, and T.R. Manney. 1978. Mating factor-mediated sexual agglutination in *Saccharomyces cerevisiae. FEMS Lett.* **4:** 469.

Betz, R., V.L. MacKay, and W. Duntze. 1977. a-Factor from *Saccharomyces cerevisiae*: Partial characterization of a mating hormone produced by cells of mating type a. *J. Bacteriol.* **132:** 462.

Biely, P., Z. Kratky, and S. Bauer. 1976. Interaction of concanavalin A with external mannan proteins of *S. cerevisiae*: Glycoprotein nature of β-glucanases. *Eur. J. Biochem.* **70:** 75.

Bilinski, T., J. Litwinska, J. Zuk, and W. Gajewski. 1975. Synchronous zygote formation in yeasts. *Methods Cell Biol.* **11:** 89.

Blundell, T.L. and R.E. Humbel. 1980. Hormone families: Pancreatic hormones and homologous growth factors. *Nature* **287:** 781.

Bucking-Throm, E., W. Duntze, L.H. Hartwell, and T.R. Manney. 1973. Reversible arrest of haploid yeast cells at the initiation of DNA synthesis by a diffusible sex factor. *Exp. Cell Res.* **76:** 99.

Burke, D., L. Mendonca-Prevaito and C.E. Ballou. 1980. Cell-cell recognition in yeast: Purification of *Hansenula wingei* 21-cell sexual agglutinin and comparison of the factors from three genera. *Proc. Natl. Acad. Sci.* **77:** 318.

Byers, B. and L. Goetsch. 1974. Duplication of spindle plaques and integration of the yeast cell cycle. *Cold Spring Harbor Symp. Quant. Biol.* **38:** 123.

―――. 1975. Behavior of spindles and spindle plaques in the cell cycle and conjugation of *Saccharomyces cerevisiae. J. Bacteriol.* **124:** 511.

Cabib, E. 1975. Molecular aspects of yeast morphogenesis. *Annu. Rev. Microbiol.* **29:** 191.

Campbell, D. 1973. Kinetics of mating-specific aggregation in *Saccharomyces cerevisiae. J. Bacteriol.* **116:** 323.

Chan, R. 1977. Recovery of *Saccharomyces cerevisiae* mating type a cells from G1 arrest by α-factor. *J. Bacteriol.* **130:** 766.

Cheung, W.Y. 1980. Calmodulin plays a pivotal role in cellular regulation. *Science* **207:** 19.

Ciejek, E.M. 1980. "α-Factor, an oligopeptide mating pheromone from the yeast *Saccharomyces cerevisiae*: Purification, chemical synthesis, and cell-mediated proteolysis." Ph.D. thesis, University of California, Berkeley.

Ciejek, E. and J. Thorner. 1979. Recovery of *S. cerevisiae* a cells from G1 arrest by α-factor pheromone requires endopeptidase action. *Cell* **18:** 623.

Ciejek, E., J. Thorner, and M. Geier. 1977. Solid phase peptide synthesis of α-factor, a yeast mating pheromone. *Biochem. Biophys. Res. Commun.* **78:** 952.

Clarke, M., W.L. Bazari, and S.C. Kayman. 1980. Isolation and properties of calmodulin from *Dictyostelium discoideum. J. Bacteriol.* **141:** 397.

Conde, J. and G.R. Fink. 1976. A mutant of *Saccharomyces cerevisiae* defective for nuclear fusion. *Proc. Natl. Acad. Sci.* **73:** 3651.

Cortat, M., P. Matile, and A. Wiemken. 1972. Isolation of glucanase-containing vesicles from budding yeast. *Arch. Microbiol.* **82:** 189.

Crandall, M. 1977. Mating-type interactions in microorganisms. In *Receptors and recognition* (ed. P. Cuatrecasas and M.F. Greaves), vol. 3, p. 45. Wiley, New York.

————. 1978. Mating-type interactions in yeasts. *Symp. Soc. Exp. Biol.* **32:** 105.

Crandall, M., R. Egel, and V.L. MacKay. 1977. Physiology of mating in three yeasts. *Adv. Microb. Physiol.* **15:** 307.

Cryer, D.R., C.D. Goldthwaite, S. Zinker, K.B. Lam, E. Storm, R. Hirschberg, J. Blamire, D.B. Finkelstein, and J. Marmur. 1974. Studies on nuclear and mitochondrial DNA of *Saccharomyces cerevisiae. Cold Spring Harbor Symp. Quant. Biol.* **38:** 17.

Cuatrecasas, P. and M.D. Hollenberg. 1976. Membrane receptors and hormone action. *Adv. Protein Chem.* **30:** 251.

del Rey, F., T. Santos, I. Carcia-Acha, and C. Nombela. 1979. Synthesis of 1,3-β-glucanases in *S. cerevisiae* during the mitotic cycle, mating, and sporulation. *J. Bacteriol.* **139:** 924.

Dery, C., S. Cooper, M.A. Savageau, and S. Scanlon. 1979. Identification and characterization of the cAMP-binding proteins of yeast by photoaffinity labelling. *Biochem. Biophys. Res. Commun.* **90:** 933.

Doi, S. and M. Yoshimura. 1978. Temperature-dependent conversion of sexual agglutinability in *Saccharomyces cerevisiae. Mol. Gen. Genet.* **162:** 251.

Doi, S., Y. Suzuki, and M. Yoshimura. 1979. Induction of sexual cell agglutinability of **a** mating type cells by α-factor in *Saccharomyces cerevisiae. Biochem. Biophys. Res. Commun.* **91:** 849.

Dunney, G.M., B.L. Brown, and D.B. Clewell. 1978. Induced cell aggregation and mating in *Streptococcus faecalis*: Evidence for a bacterial sex pheromone. *Proc. Natl. Acad. Sci.* **75:** 3479.

Duntze, W., V.L. MacKay, and T.R. Manney. 1970. *Saccharomyces cerevisiae*: A diffusible sex factor. *Science* **168:** 1472.

Duntze, W., D. Stötzler, E. Bucking-Throm, and S. Kalbitzer. 1973. Purification and partial characterization of α-factor, a mating type-specific inhibitor of cell reproduction from *S. cerevisiae. Eur. J. Biochem.* **35:** 357.

Dutcher, S. 1980. "Genetic control of karyogamy in *Saccharomyces cerevisiae.*" Ph.D. thesis, University of Washington, Seattle.

Farkas, V. 1979. Biosynthesis of cell walls of fungi. *Microbiol. Rev.* **43:** 117.

Fehrenbacher, G., K. Perry, and J. Thorner. 1978. Cell-cell recognition in *Saccharomyces cerevisiae*: Regulation of mating-specific adhesion. *J. Bacteriol.* **134:** 893.

Finkelstein, D.B. and S. Strausberg. 1979. Metabolism of α-factor by **a** mating type cells of *Saccharomyces cerevisiae. J. Biol. Chem.* **254:** 796.

Fraenkel, D.G. 1982. Carbohydrate metabolism. In *The molecular biology of the yeast* Saccharomyces. II. *Metabolism and gene expression* (ed. J. Strathern et al.), Cold Spring Harbor Laboratory, Cold Spring Harbor, New York. (In press.)

Freifelder. D. 1960. Bud position in *Saccharomyces cerevisiae. J. Bacteriol.* **80:** 567.

Fujimoto, F., A. Tchikawa, and K. Tomita. 1974. Purification and properties of adenosine-3',5'-monophosphate phosphodiesterase from baker's yeast. *Arch. Biochem. Biophys.* **161:** 54.

Gerace, L. and G. Blobel. 1980. The nuclear envelope lamina is reversibly depolymerized during mitosis. *Cell* **19:** 277.

Goodenough, U. 1980. Sexual microbiology: The mating reactions of *Chlamydomonas, Tetrahymena,* and *Saccharomyces* and their genetic control. *Symp. Soc. Gen. Microbiol.* **30:** 301.

Gottesman, M.M. 1980. Genetic approaches to cyclic AMP effects in cultured mammalian cells. *Cell* **22:** 329.

Greengard, P. 1978. Phosphorylated proteins as physiological effectors. *Science* **199:** 146.

Hagiya, M., K. Yoshida, and N. Yanagishima. 1977. The release of sex-specific substances responsible for sexual agglutination from haploid cells of *S. cerevisiae. Exp. Cell Res.* **104:** 263.

Harold, R.L. and F.M. Harold. 1980. Oriented growth of *Blastocladiella emersonii* in gradients of ionophores and inhibitors. *J. Bacteriol.* **144:** 1159.

Hartwell, L.H. 1973. Synchronization of haploid yeast cell cycles, a prelude to conjugation. *Exp. Cell Res.* **76:** 111.

―――. 1976. Sequential function of gene products relative to DNA synthesis in the yeast cell cycle. *J. Mol. Biol.* **104**: 803.

―――. 1980. Mutants of *Saccharomyces cerevisiae* unresponsive to cell division control by polypeptide hormones. *J. Cell Biol.* **85**: 811.

Hemmings, B.A. 1980. Purification and properties of the phospho- and dephospho-forms of yeast NAD-dependent glutamate dehydrogenase. *J. Biol. Chem.* **255**: 7925.

Henry, S.A. 1982. The membrane lipids of yeast: Biochemical and genetic studies. In *The molecular biology of the yeast* Saccharomyces. II. *Metabolism and gene expression* (ed. J. Strathern et al.), Cold Spring Harbor Laboratory, Cold Spring Harbor, New York. (In press.)

Hereford, L.M. and L.H. Hartwell. 1974. Sequential gene function in the initiation of *S. cerevisiae* DNA synthesis. *J. Mol. Biol.* **84**: 445.

Hereford, L.M., K. Fahrner, and M.A. Osley. 1980. Replication of histone H2A and H2B expression. *10th International Conference on Yeast Genetics and Molecular Biology (1980),* (ed. A. Goffeau and J.M. Wiame), p. 78 (Abstr.), Université Catholique de Louvain, Louvain-la-Neuve, Belgium.

Herman, A.I. 1971. Sex-specific growth responses in yeasts. *Antonie van Leeuwenhoek J. Microbiol. Serol.* **34**: 379.

Hicks, J.B. and I. Herskowitz. 1976. Evidence for a new diffusible element of mating pheromones in yeast. *Nature* **260**: 246.

Hicks, J.B., J. Strathern, and A. Klar. 1979. Transposable mating type genes in *Saccharomyces cerevisiae. Nature* **282**: 478.

Hirata, F. and J. Axelrod. 1980. Phospholipid methylation and biological signal transmission. *Science* **209**: 1082.

Hixson, C.S. and E.G. Krebs. 1980. Characterization of a cyclic AMP-binding protein from baker's yeast: Identification as a regulatory subunit of cyclic AMP-dependent protein kinase. *J. Biol. Chem.* **255**: 2137.

Jamieson, G., T.C. Vanaman, and J.J. Blum. 1979. Presence of calmodulin in *Tetrahymena. Proc. Natl. Acad. Sci.* **76**: 6471.

Jaynes, P., J. McDonough, and H.R. Mahler. 1979. Cyclic AMP binding and protein kinase activity associated with plasma membranes of *Saccharomyces cerevisiae. J. Cell Biol.* **83**: 290 (Abstr.).

Klar, A.J.S., J.N. Strathern, J.R. Broach, and J.B. Hicks. 1981. Regulation of transcription in expressed and unexpressed mating type cassettes of yeast. *Nature* **289**: 239.

Kudlicki, W., N. Grankowski, and E. Gasior. 1978. Isolation and properties of two protein kinases from yeast which phosphorylate casein and some ribosomal proteins. *Eur. J. Biochem.* **84**: 493.

Leibowitz, M. and R.B. Wickner. 1976. A chromosomal gene required for killer plasmid expression, mating, and sporulation in *S. cerevisiae. Proc. Natl. Acad. Sci.* **73**: 2061.

Lerch, K., L. Muir, and E.H. Fischer. 1975. Purification and properties of a yeast protein kinase. *Biochemistry* **14**: 2015.

Levi, J.D. 1956. Mating reaction in yeast. *Nature* **177**: 753.

Liao, H. and J. Thorner. 1980. Yeast mating pheromone α-factor inhibits adenylate cyclase. *Proc. Natl. Acad. Sci.* **77**: 1898.

Lipke, P.N. 1976. "Morphogenetic effects of α-factor on *S. cerevisiae* **a** cells." Ph.D. thesis, University of California, Berkeley.

Lipke, P.N. and C.E. Ballou. 1980. Altered immunochemical reactivity of *S. cerevisiae* **a** cells after α-factor-induced morphogenesis. *J. Bacteriol.* **141**: 1170.

Lipke, P.N., A. Taylor, and C.E. Ballou. 1976. Morphogenic effects of α-factor on *Saccharomyces cerevisiae* **a** cells. *J. Bacteriol.* **127**: 610.

Lipmann, F. 1971. Attempts to map a process evolution of peptide biosynthesis. *Science* **173**: 875.

Livi, G.P. and V.L. MacKay. 1980. Mating-type regulation of methyl methanesulfonate sensitivity in *Saccharomyces cerevisiae. Genetics* **95**: 259.

Lomedico, P., S.J. Chan, D.F. Steiner, and G.F. Saunders. 1977. Immunological and chemical characterization of bovine preproinsulin. *J. Biol. Chem.* **252**: 7971.

Londesborough, J. 1977. Characterization of an adenosine-3',5'-cyclic monophosphate phosphodiesterase from baker's yeast. *Biochem. J.* **163**: 467.

MacKay, V.L. 1978. Mating type-specific pheromones as mediators of sexual conjugation in yeast. In *Molecular control of proliferation and differentiation* (ed. J. Papaconstantinous and W.J. Rutter), p. 243. Academic Press, New York.

MacKay, V.L. and T.R. Manney. 1974a. Mutations affecting sexual conjugation and related processes in *S. cerevisiae*. I. Isolation and phenotypic characterization of non-mating mutants. *Genetics* **76**: 255.

―――. 1974b. Mutations affecting sexual conjugation and related processes in *S. cerevisiae*: II. Genetic analysis of non-mating mutants. *Genetics* **76**: 273.

Maness, P.F. and G.M. Edelman. 1978. Inactivation and chemical alteration of mating factor α by cells and spheroplasts of yeast. *Proc. Natl. Acad. Sci.* **75**: 1304.

Manney, T.R. and J. Meade. 1977. Cell-cell interactions during mating in *S. cerevisiae*. In *Receptors and recognition* (ed. J.L. Reissig), vol. 3, p. 281. Wiley, New York.

Manney, T.R. and V. Woods. 1976. Mutants of *S. cerevisiae* resistant to the α mating type factor. *Genetics* **82**: 639.

Manney, T.R., W. Duntze, and R. Betz. 1981. The isolation, characterization, and physiological effects of the *S. cerevisiae* sex pheromones. In *Sexual interactions in eukaryotic microbes* (ed. D.H. O'Day and P.A. Horgan), p. 21. Academic Press, New York.

Masui, Y., N. Chino, S. Sakakibara, T. Tanaka, T. Murakami, and H. Kita. 1977. Synthesis of the mating factor of *S. cerevisiae* and its truncated peptides: The structure-activity relationship. *Biochem. Biophys. Res. Commun.* **78**: 534.

―――. 1979. Amino acid substitution of mating factor of *S. cerevisiae*: Structure-activity relationship. *Biochem. Biophys. Res. Commun.* **86**: 982.

Matile, P., M. Cortat, A. Wiemken, and A. Frey-Wyssling. 1971. Isolation of glucanase-containing particles from budding *Saccharomyces cerevisiae*. *Proc. Natl. Acad. Sci.* **68**: 636.

McCullough, J. and I. Herskowitz. 1979. Mating pheromones of *Saccharomyces kluyveri*: Pheromone interactions between *S. kluyveri* and *S. cerevisiae*. *J. Bacteriol.* **138**: 146.

Meek, J.L. 1980. Prediction of peptide retention times in high-pressure liquid chromatography on the basis of amino acid composition. *Proc. Natl. Acad. Sci.* **77**: 1632.

Mortimer, R.K. 1955. Evidence for two types of X-ray-induced lethal damage in *Saccharomyces cervisiae*. *Radiat. Res.* **2**: 361.

Mortimer, R.K. and D.C. Hawthorne. 1969. Yeast genetics. In *The yeasts* (ed. A.H. Rose and J.S. Harrison), vol 1, p.385. Academic Press, New York.

Müller, I. 1980. Parental age and the life span of zygotes of *Saccharomyces cerevisiae*. *10th International Conference on Yeast Genetics and Molecular Biology (1980)*, (ed. A. Goffeau and J.M. Wiame), p. 55 (Abstr.), Université Catholique de Louvain, Louvain-la-Neuve, Belgium.

Nakamura, K.D. and F. Schlenk. 1974. Active transport of exogenous S-adenosyl-methionine and related compounds into cells and vacuoles of *Saccharomyces cerevisiae*. *J. Bacteriol.* **120**: 482.

Nasmyth, K.A. and K. Tatchell. 1980. The structure of transposable yeast mating type loci. *Cell* **19**: 753.

Nasmyth, K.A., K. Tatchell, B.D. Hall, C. Astell, and M. Smith. 1981a. A physical analysis of mating type loci. *Cold Spring Harbor Symp. Quant. Biol.* **45**: 961.

―――. 1981b. A position effect on the control of transcription at yeast mating type loci. *Nature* **289**: 244.

Osumi, M., C. Shimoda, and N. Yanagishima. 1974. Mating reaction in *S. cerevisiae*: V. Changes in the fine structure during the mating reaction. *Arch. Microbiol.* **97**: 27.

Phaff, H.J. 1977. Enzymatic yeast cell wall degradation. *Adv. Chem.* **160**: 244.

Piñon, R. and D. Pratt. 1979. Folded chromosomes of mating factor-arrested yeast cells: Comparison with G_0 arrest. *Chromosoma* **73**: 117.

Polster, R.K. and V.L. MacKay. 1980. Requirement of intact cell wall for mating hormone response in *Saccharomyces cerevisiae*. *Annual Meeting of the American Society for Microbiology*, p. 123. (Abstr.), American Society for Microbiology, Washington, D.C.

Radin, D.N. 1976. "Genetics and physiology of mating in *Saccharomyces cerevisiae*." Ph.D. thesis, University of California, Berkeley.

Reid, B. and L.H. Hartwell. 1977. Regulation of mating in the cell cycle of *S. cerevisiae*. *J. Cell Biol.* **75**: 355.

Robinson, K.R. and R. Cone. 1980. Polarization of fucoid eggs by a calcium ionophore gradient. *Science* **207**: 77.

Rodgers, D.T. and H. Bussey. 1978. Fidelity of conjugation in *Saccharomyces cerevisiae*. *Mol. Gen. Genet.* **162**: 173.

Rodgers, D.T., D. Saville, and H. Bussey. 1979. *S. cerevisiae* killer expression mutant *kex2* has altered secretory proteins and glycoproteins. *Biochem. Biophys. Res. Commun.* **90**: 187.

Sakai, K. and N. Yanagishima. 1972. Mating regulation in *S. cerevisiae*. II. Hormonal regulation of agglutinability of **a** type cells. *Arch. Microbiol.* **84**: 191.

Sakurai, A., S. Tamura, N. Yanagishima, and C. Shimoda. 1975. Isolation of a peptide factor controlling sexual agglutination in *S. cerevisiae*. *Proc. Natl. Acad. Sci.* **51**: 291.

Sakurai, A., S. Tamura, N. Yanagishima, and C. Shimoda. 1976a. Structure of the peptidyl factor inducing sexual agglutination in *S. cerevisiae*. *Agric. Biol. Chem.* **40**: 1057.

Sakurai, A., K. Sakata, S. Tamura, K. Aizawa, N. Yanagishima, and C. Shimoda. 1976. Isolation and structure elucidation of α substance-IB, a hexapeptide inducing sexual agglutination in *S. cerevisiae*. *Agric. Biol. Chem.* **40**: 1451.

Sakurai, A., Y. Sato, K.H. Park, N. Takahashi, N. Yanagishima, and I. Banno. 1980. Isolation and chemical characterization of a mating pheromone produced by *Saccharomyces kluyveri*. *Agric. Biol. Chem.* **44**: 1451.

Salomon, Y., C. Landos, and M. Rodbell. 1974. A highly sensitive adenylate cyclase assay. *Anal. Biochem.* **58**: 541.

Samokhin, G.P., L.V. Lizlova, J.D. Bespalova, M.I. Titov, and V.N. Smirnov. 1979. Substitution of Lys7 by Arg does not affect biological activity of α-factor, a yeast mating pheromone. *FEMS Lett.* **15**: 435.

———. 1981. The effect of α-factor on the rate of cell-cycle initiation in *Saccharomyces cerevisiae*. *Exp. Cell Res.* **131**: 267.

Santos, T., F. Del Rey, J. Conde, J. Villanueva, and C. Nombela. 1979. *Saccharomyces cerevisiae* mutant defective in exo-1,3-β-glucanase production. *J. Bacteriol.* **139**: 333.

Schekman, R. and V.L. Brawley. 1979. Localized deposition of chitin on the yeast cell surface in response to mating pheromone. *Proc. Natl. Acad. Sci.* **76**: 645.

Schekman, R. and P. Novick. 1982. The secretory process and yeast cell-surface assembly. In *The molecular biology of the yeast* Saccharomyces. II. *Metabolism and gene expression* (ed. J. Strathern et al.), Cold Spring Harbor Laboratory, Cold Spring Harbor, New York (In press.)

Scherer, G., G. Haag, and W. Duntze. 1974. Mechanism of α-factor biosynthesis in *Saccharomyces cerevisiae*. *J. Bacteriol.* **119**: 386.

Scopes, R.K. 1973. 3-Phosphoglycerate kinase. In *The enzymes* (ed. P.D. Boyer), p. 335. Academic Press, New York.

Sena, E.P., D.N. Radin, J. Welch, and S. Fogel. 1975. Synchronous mating in yeasts. *Methods Cell Biol.* **11**: 71.

Sentandreau, R. and D.R. Northcote. 1969. The formation of buds in yeast. *J. Gen. Microbiol.* **55**: 393.

Shelton, K.R., V.H. Guthrie, and D.C. Cochran. 1980. On the variation of the major nuclear envelope (lamina) polypeptides. *Biochem. Biophys. Res. Commun.* **93**: 867.

Shematek, E., J.A. Braatz, and E. Cabib. 1980. Biosynthesis of the yeast cell wall. I. Preparation and properties of β-(1 $\rightarrow$ 3)glucan synthetase. *J. Biol. Chem.* **255**: 888.

Shimoda, C. and N. Yanagishima. 1972. Mating reaction in *S. cerevisiae*. III. Changes in autolytic activities. *Arch. Microbiol.* **85**: 310.

Shimoda, C., S. Kitano, and N. Yanagishima. 1975. Mating reaction in *S. cerevisiae*. VII. Effect of proteolytic enzymes on sexual agglutinability and isolation of crude sex-specific substances responsible for sexual agglutination. *Antonie van Leeuwenhoek J. Microbiol. Serol.* **41**: 513.

Shimoda, C., N. Yanagishima, A. Sakurai, and S. Tamura. 1978. Induction of sexual agglutinability of **a** mating type cells as the primary action of the peptidyl sex factor from α mating type cells in *S. cerevisiae*. *Plant Cell Physiol.* **19**: 513.

Singer, R.A., G.C. Johnston, and D. Bedard. 1978. Methionine analogs and cell division regulation in the yeast *Saccharomyces cerevisiae*. *Proc. Natl. Acad. Sci.* **75**: 6083.

Sprague, G., J. Rine, and I. Herskowitz. 1981. Homology and non-homology at the yeast mating type locus. *Nature* **289**: 250.

Stötzler, D. and W. Duntze. 1976. Isolation and characterization of four related peptides exhibiting α-factor activity from *Saccharomyces cerevisiae*. *Eur. J. Biochem.* **65**: 257.

Stötzler, D., R. Betz, and W. Duntze. 1977. Stimulation of yeast mating hormone activity by synthetic oligopeptides. *J. Bacteriol.* **132**: 28.

Stötzler, D., H. Kiltz, and W. Duntze. 1976. Primary structure of α-factor peptides from *Saccharomyces cerevisiae*. *Eur. J. Biochem.* **69**: 397.

Strathern, J.N., J.B. Hicks, and I. Herskowitz. 1981. Control of cell type by the mating type locus: The $\alpha1$-$\alpha2$ hypothesis. *J. Mol. Biol.* **147**: 357.

Strathern, J., E. Spatola, C. McGill, and J.B. Hicks. 1980. The structure and organization of transposable mating type cassettes in *Saccharomyces* species. *Proc. Natl. Acad. Sci.* **77**: 2829.

Streiblova, E. 1970. Study of scar formation in the life cycle of heterothallic *Saccharomyces cerevisiae*. *Can. J. Microbiol.* **16**: 827.

Stump, R.F., K.R. Robinson, R.L. Harold, and F.M. Harold. 1980. Endogenous electrical currents in the water mold *Blastocladiella emersonii* during growth and sporulation. *Proc. Natl. Acad. Sci.* **77**: 6673.

Sumrada, R. and T.G. Cooper. 1978. Control of vacuole permeability and protein degradation by the cell cycle arrest signal in *Saccharomyces cerevisiae*. *J. Bacteriol.* **136**: 234.

Takai, Y., H. Yamahura, and Y. Nishizuka. 1974. Adenosine-3′,5′-monophosphate-dependent protein kinase from *Saccharomyces cerevisiae*. *J. Biol. Chem.* **244**: 530.

Tanaka, T. and H. Kita. 1977. Degradation of mating factor by α-mating type cells. *J. Biochem.* **82**: 1689.

———. 1978. Site of action of mating factor in **a** mating type cell of *Saccharomyces cerevisiae*. *Biochem. Biophys. Res. Commun.* **83**: 1319.

Tanaka, T., H. Kita, and K. Narita 1977a. Purification and its structure-activity relationship of mating factor of *Saccharomyces cerevisiae*. *Proc. Natl. Acad. Sci.* **53**: 67.

Tanaka, T., H. Kita, T. Murakami, and K. Narita. 1977b. Purification and amino acid sequence of mating factor from *Saccharomyces cerevisiae*. *J. Biochem.* **82**: 1681.

Tatchell, K., K.A. Nasmyth, B.D. Hall, C. Astell, and M. Smith. 1981. *In vitro* mutation analysis of the mating type locus in yeast. *Cell* (in press).

Thompson, E.D., B.A. Knights, and L.W. Parks. 1973. Identification and properties of a sterol-binding polysaccharide isolated from *S. cerevisiae*. *Biochim. Biophys. Acta* **304**: 132.

Thorner, J. 1980. Intercellular interactions of the yeast *Saccharomyces cerevisiae*. In *Molecular genetics of development: An introduction to recent research on experimental systems* (ed. T.J. Leighton and W.A. Loomis, Jr.), p. 119. Academic Press, New York.

Throm, E. and W. Duntze. 1970. Mating type-dependent inhibition of deoxyribonucleic acid synthesis in *S. cerevisiae*. *J. Bacteriol.* **104**: 1388.

Tkacz, J.S. and V.L. MacKay. 1979. Sexual conjugation in yeast: Cell surface changes in response to the action of mating horomones. *J. Cell Biol.* **80**: 326.

Tohoyama, H., M. Hagiya, K. Yoshida, and N. Yanagishima. 1979. Regulation of the production of the agglutination substances responsible for sexual agglutination in *S. cerevisiae*: Changes associated with conjugation and temperature. *Mol. Gen. Genet.* **174**: 269.

Tsuchiya, E., S. Fukui, Y. Kamiya, Y. Sakagami, and M. Fujino. 1978. Requirements of chemical structure for hormonal activity of lipopeptidyl factors inducing sexual differentiation in vegetative cells of heterobasidiomycetous yeasts. *Biochem. Biophys. Res. Commun.* **85**: 459.

Udden, M.M. and D.B. Finkelstein. 1978. Reaction order of *S. cerevisiae* alpha-factor-mediated cell cycle arrest and mating inhibition. *J. Bacteriol.* **133**: 1501.

Varimo, K. and J. Londesborough. 1976. Solubilization and other studies on adenylate cyclase of baker's yeast. *Biochemical J.* **159**: 363.

Wilkinson, L.E., and J.R. Pringle. 1974. Transient G1 arrest of *S. cerevisiae* cells of mating type α by a factor produced by cells of mating type **a**. *Exp. Cell Res.* **89**: 175.

Wurster, B., P. Pan, G.G. Tyan, and J.T. Bonner. 1976. Preliminary characterization of the acrasin of the cellular slime mold *Polysphondylium violaceum*. *Proc. Natl. Acad. Sci.* **73**: 795.

Yanagishima, N. and Y. Nakagawa. 1980. Mutants inducible for sexual agglutination in *Saccharomyces cerevisiae*. *Mol. Gen. Genet.* **178**: 241.

Yanagishima, N., T. Shimizu, M. Hagiya, K. Yoshida, A. Sakurai, and S. Tamura. 1977a. Physiological detection of a binding substance for the agglutinability-inducing pheromone, α substance-I, in *S. cerevisiae*. *Plant Cell Physiol.* **18**: 1181.

Yanagishima, N., K. Yoshida, M. Hagiya, Y. Kawanabe, C. Shimoda, A. Sakurai, S. Tamura, and M. Osumi. 1977b. Sexual cell agglutination in *Saccharomyces cerevisiae*. In *Growth and differentiation in microorganisms* (ed. T. Ishikawa), p. 193. University Park Press, Baltimore.

Yen, P.H. and C.E. Ballou. 1973. Composition of a specific intercellular agglutination factor. *J. Biol. Chem.* **248**: 8316.

———. 1974. Partial characterization of the sexual agglutination factor from *Hansenula wingei* Y-2340 Type 5 cells. *Biochemistry* **13**: 2428.

Yoshida, K., M. Hagiya, and N. Yanagishima. 1976. Isolation and purification of the sexual agglutination substance of mating type **a** cells in *S. cerevisiae*. *Biochem. Biophys. Res. Commun.* **71**: 1085.

Zinker, S. and J.R. Warner. 1976. The ribosomal proteins of *Saccharomyces cerevisiae*: Phosphorylated and exchangeable proteins. *J. Biol. Chem.* **251**: 1799.

Control of Cell Type in
Saccharomyces cerevisiae:
Mating Type and
Mating-type Interconversion

Ira Herskowitz*
Department of Biology and
Institute of Molecular Biology
University of Oregon
Eugene, Oregon 97403

Yasuji Oshima
Department of Fermentation Technology
Osaka University
Osaka, Japan 565

INTRODUCTION

The study of yeast mating and mating-type interconversion has attracted considerable interest for several reasons. (1) The mating process involves oligopeptide pheromones, specific for each haploid cell type (mating types **a** and α), which act specifically to arrest the cells of opposite mating types in the G_1 phase of the cell cycle. Studies of the action of these pheromones may well

*Present address: Department of Biochemistry and Biophysics, University of California, San Francisco, California 94143.

shed light on the mechanism of growth control. Such studies are described elsewhere in this volume. (2) The mating type of a cell is determined by alleles of the mating-type locus *MAT*a or *MAT*α, which in some manner controls a variety of cellular processes: pheromone production and response, agglutination, mating, and sporulation. A working hypothesis is that the mating-type locus codes for regulatory proteins that control expression of unlinked genes. Regulatory proteins are well known in prokaryotes as agents for controlling protein synthesis and act in many cases to control transcription initiation and termination. One of the central questions in gene control and in eukaryotic development is to understand the mechanism by which differentiated cells express a distinctive set of proteins. Analysis of how the mating-type locus determines cell type in yeast should provide information on gene control in yeast, e.g., on the nature of regulatory proteins, and provide an intellectual model for analogous processes in higher eukaryotes. Yeast regulatory proteins are also discussed by Jones and Fink (1982), Fraenkel (1982), Oshima (1982), and Sherman (1982). (3) Yeast cells can exhibit either a heterothallic life cycle, in which the mating type of a cell is quite stable, or a homothallic life cycle, in which the mating type of a cell is unstable. The interconversion of mating types is a genetic switch—a change of the alleles of the mating-type locus—and is under a variety of genetic and physiological controls. A genetic switch poses an intriguing problem to the geneticist for two reasons. First, the switch involves a change between states that are quite different from each other; thus, a simple mutational event such as a single point mutation does not seem to be an adequate explanation. Second, the switch occurs at a frequency much higher than that expected for a simple mutational event. The switch between mating types in yeast has been found to occur by a novel mechanism involving rearrangement of the genome: Transposable elements (controlling elements or cassettes) coding for a and α regulators become activated by movement to the mating-type locus. Extrapolations of this mode of gene control provide a model for events occurring during development and differentiation in higher eukaryotes. (4) Many yeast strains of industrial importance are polyploid, e.g., triploid or tetraploid. The production of such strains by breeding offers a special challenge for the yeast geneticist. An understanding of the genetic determinants governing mating and mating-type interconversion has provided important new avenues for construction of such strains.

This paper reviews our present understanding of control of cell type by the mating-type locus and of the mechanism of mating-type interconversion.

CONTROL OF CELL TYPE

Genes Required for Mating

The two haploid cell types of *Saccharomyces cerevisiae* are determined by alleles of the mating-type locus *MAT*a for a cells and *MAT*α for α cells. Cells of opposite mating types mate efficiently, whereas those of like mating types mate only rarely. The **a**/α diploid formed by the mating of **a** and α cells is a third cell type. In contrast to **a**/**a** and α/α cells, which behave as their respective haploids, **a**/α cells are unable to mate but can be induced to undergo meiosis and sporulation.

In addition to mating and sporulation, several other processes are controlled by the mating-type locus (reviewed by Thorner, this volume). **a** and α cells each produce a pheromone (**a**-factor and α-factor, respectively) that acts specifically on cells of the other mating type to cause cell-cycle arrest. Likewise, **a** and α cells agglutinate with each other. Finally, **a** cells have a mechanism for inactivating α-factor (Hicks and Herskowitz 1976a; Chan 1977). **a**/α cells do not produce or respond to the pheromones and do not agglutinate; in contrast to **a**/**a** and α/α cells, **a**/α cells undergo premeiotic DNA synthesis when induced to sporulate, and they exhibit increased X-ray and methylmethanesulfonate resistance during vegetative growth (reviewed by Lemontt 1980). The mating-type locus also controls the budding pattern: **a**/α cells exhibit a polar budding pattern, and **a**/**a** or α/α cells exhibit a medial pattern (Hicks et al. 1977b). For the phenotypes described above, the dosage of the mating-type loci is unimportant. **a**/**a** strains behave the same as **a** strains; α/α strains behave as α, and **a**/α, **a**/**a**/α, **a**/α/α, and **a**/**a**/α/α strains all behave the same. Thus, homoallelic strains exhibit mating and associated processes, and heteroallelic strains exhibit sporulation and associated processes.

How does the mating-type locus control all of the different processes described above? MacKay and Manney (1974b) proposed that the mating-type loci code for regulatory proteins that control expression of other genes coding for mating- and sporulation-specific processes. This view derived from the identification of genes required for mating via isolation of mating-deficient mutants (MacKay and Manney 1974a,b). Some mutants (obtained from α but not **a** strains) carry mutations of the mating-type locus itself. The α-mating-type locus has at least two complementation groups, called *MATα1* and *MATα2* (described below). The other mutations are unlinked to the mating-type locus and are of three types (MacKay and Manney 1974a,b): those affecting mating only by α strains (which define α-specific genes, αsg), those affecting mating only by **a** strains (which define **a**-specific genes, **a**sg), and those that lead to defective mating in both **a** and α cells (nonspecific genes, nsg). Additional mutant isolation and complementation analysis (Hartwell 1980) indicates that there are at least two **a**sg *(ste2* and *ste6)* and six nsg genes. Four αsg genes have been identified: *ste3* (MacKay and Manney 1974a,b); *ste13* (J. Rine, pers. comm.); *tup1* (Wickner 1974); and *kex2* (Leibowitz and Wickner 1976). It should be noted that no structural genes for mating-specific functions have yet been identified. Thus, it is not known whether the structural gene for α-factor is at the mating-type locus or whether it is, for example, one of the unlinked α-specific genes.

Mutations of the **a**-mating-type locus were not found among mating-defective mutants but were identified by other phenotypes (Kassir and Simchen 1976; Klar et al. 1979b). These mutations (originally called **a*** or **a**° and now called *mata1*) do not affect mating ability, but they do affect the properties of diploid cells: *mata1*/*MATα* cells maintain the ability to mate as an α and cannot be induced to sporulate.

Structure of the Mating-type Locus

Our present understanding of the structure of the mating-type locus and the relationship between **a** and α alleles comes from genetic analysis of mutations of the mating-type locus and from physical studies of the cloned *MATa* and *MATα* alleles. MacKay and Manney (1974b) observed that *MATa*/*matα1-5* diploids were sporulation-proficient and yielded 2 **a**:2 nm (nonmater) asci. Recent analysis has extended these observations: *MATa*/*matα1-5* and *MATa*/*matα1-2am* strains do not yield *MATα*[+] recombinants ($< 10^{-5}$; Sprague et al. 1981). In contrast, *MATa*/*matα2-1* and *MATa*/*matα2-4* strains yield *MATα*[+] recombinants at a frequency of approximately 10^{-3} (Sprague et al. 1981). Recombination between *matα1* and *matα2* occurs with a frequency of approximately 10^{-2} (Sprague et al. 1981; Strathern et al. 1981). The results with the *matα1* mutants suggest that *MATa* and the segment of *MATα* coding for *MATα1* are nonhomologous blocks of DNA. The results with *matα2* mutants suggest that part of the *MATα2* gene may be present at the **a**-mating-type locus.

The recent cloning of the **a**- and α-mating-type loci (see below, Tests of the Cassette Model, Physical Tests) has allowed a direct analysis of the relationship between the alleles of the mating-type locus and has provided information on transcription of the mating-type locus (Hicks et al. 1979; Nasmyth and Tatchell 1980; Strathern et al. 1980). Restriction endonuclease DNA fragments containing *MATa* and *MATα* are heterologous for approximately 750 bp (as determined by heteroduplex analysis) and differ in size, with the *MATa* fragment being slightly smaller than the *MATα* fragment (as determined by electrophoresis). Nucleotide sequencing shows that *MATa* and *MATα* are heterologous for 642 bp and 747 bp, respectively (Nasmyth et al. 1981).

Two transcripts, both approximately 850 bases in length, are initiated from *MATα* (from the Yα region; see Fig. 5). The transcript initiated on the left side of Yα continues leftward, and the transcript initiated on the right side of Yα continues rightward into regions X and Z, respectively. The leftward transcript appears to correspond to *MATα2*, and the rightward transcript to *MATα1* (Klar et al. 1981; Nasmyth et al. 1981). *MATa* produces two transcripts with similar orientations, only one of which corresponds to a known *MATa* function.

A Hypothesis for Control of Cell Type

As noted previously, the α-mating-type locus has two complementation groups: *MATα1* and *MATα2*. Although both *matα1* and *matα2* mutants were isolated as mating-deficient mutants, they have striking differences.

Mating and Associated Phenotypes

matα1 mutants exhibit reduced efficiency of mating with both **a** and α cells and do not exhibit α-specific phenotypes such as α-factor production or agglutination with **a** cells. *matα2* mutants, likewise, are defective in mating with **a** cells but exhibit two phenotypes of **a** cells: (1) They exhibit **a**-specific degradation of α-factor (Bar function; Hicks and Herskowitz 1976a) and respond to α-factor (Tkacz and MacKay 1979); (2) they mate as **a** cells with an efficiency that is 1% that of standard **a** cells (Hicks 1975; Strathern et al. 1981).

Sporulation

MATa/matα1 strains give normal sporulation, whereas *MATa/matα2* strains are deficient in sporulation. The only mutations of the **a**-mating-type locus *mata1* do not affect mating but lead to a defect in sporulation by *mata1/MATα* diploids (Kassir and Simchen 1976; Klar et al. 1979b). Because all of these mutations are recessive to their wild-type, mating-type-locus alleles, the mutant behavior is presumably due to loss of functions normally expressed by **a**- and α-mating-type loci. These observations have led to the following hypothesis (the α1-α2 model for control of cell type; Strathern et al. 1981; Fig. 1). (1)*MATα1* codes for the α1 product, a positive regulator of α-specific functions, which are presumed to be coded by the αsg genes. *matα1* mutants are defective in mating because they do not express these genes. (2) *MATα2* codes for the α2 product, a negative regulator of **a**-specific functions, which are presumed to be coded by the **a**sg genes. *matα2* mutants are defective in mating because they express **a**-specific and α-specific functions simultaneously. (3) Each mating-type locus must code for a sporulation regulator function necessary to induce sporulation since *MATa* and *MATα* are codominant: *MATa/MATα* strains differ from *MATa/MATa* and *MATα/MATα* strains both in mating and in sporulation. The sporulation regulator produced by the α-mating-type locus is coded by *MATα2*, and that of the **a**-mating-type locus (**a**1 product) is coded by *MATa1*. Because nonsense mutations exist for *MATα1* (Herskowitz et al. 1980a) and *MATa1* (Klar 1980), it is likely that they code for proteins.

According to this hypothesis, a *MATα* cell expresses α phenotypes because α1 function (the product of *MATα1*) induces α-specific functions and because α2 function (the product of *MATα2*) inhibits expression of **a**-specific functions. A *MATa* cell is constitutive for **a**-specific functions. Sporulation occurs only in *MATa/MATα* cells because both **a**1 (the product of *MATa*)

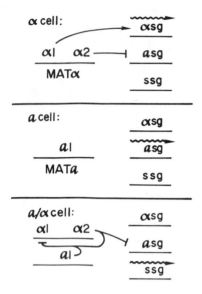

Figure 1 A hypothesis for control of cell type by the mating-type locus. Shown are genes expressed in α, **a**, and **a**/α cells, respectively. (*αsg*) α-specific genes; (**a***sg*) **a**-specific genes; (*ssg*) sporulation-specific genes and other genes expressed in **a**/α cells. (—) Expression of the genes; (→) stimulation of expression; (———┤) inhibition of expression. Expression of *ssg* may be controlled by a negative regulator (e.g., *RME, CSP, SCA*) whose synthesis or activity is inhibited in **a**/α cells.

and α2 functions are present. The manner in which **a**1-α2 stimulates sporulation is not clear. Recessive mutations unlinked to the mating-type locus (*sca, csp, rme*) that allow sporulation of **a**/**a** and α/α cells have been found (Gerlach 1974; Hopper and Hall 1975a; Kassir and Simchen 1976). One possibility is that the concerted action of **a**1 and α2 inactivates the wild-type *SCA-, CSP-,* and *RME*-gene products and thereby promotes sporulation. The manner in which mating is inhibited in **a**/α cells is likewise unclear. It is assumed that **a**/α cells produce neither **a**- nor α-specific functions. One possible mechanism for inhibition of mating in **a**/α cells is that α2 inhibits **a**-specific functions, and a complex of **a**1 and α2 inhibits synthesis of α1. This hypothesis is supported by the observation that α1 message is absent in **a**/α diploids (Klar et al. 1981; Nasmyth et al. 1981).

The α1-α2 hypothesis makes several explicit predictions that have been confirmed: (1) A *matα1 matα2* double mutant should be unable to turn on α-specific functions and be unable to turn off **a**-specific functions. It should thus mate as an **a**, but the resulting *matα1 matα2/MATα* diploid should mate as an α and be unable to sporulate. This prediction has been confirmed (Strathern 1977; Strathern et al. 1981, G. F. Sprague, pers. comm.). (2) A *matα2* mutant that also carries a defect in degradation of α-factor should produce α-factor. This prediction has been confirmed by the finding that *matα2* mutants able to produce α-factor have a defect in the **a**-specific Bar function (G. F. Sprague and I. Herskowitz, in prep.). Furthermore, additional mutations of *MAT***a** and *MATα* have recently been obtained by in vitro mutagenesis of cloned *MAT***a** and *MATα* (Nasmyth et al. 1981). In general, these mutations behave as expected for *mat**a***, *matα1*, and *matα2* mutations characterized earlier.

Before leaving this topic, we note that the basic premise that the mating-type locus codes for regulators of expression of other genes has not been demonstrated directly. The ability to clone **a***sg* and *αsg* will make it possible to determine whether the mating-type locus controls transcription of these genes.

MATING-TYPE INTERCONVERSION

Discovery and Definition of Homothallism

As described above, the life cycle of *Saccharomyces* yeasts includes both diploid and haploid phases. When diploid cells are placed on a sporulation medium, vegetative growth ceases and the cells become committed to meiosis and sporulation, in which each diploid cell is transformed into an ascus usually consisting of four haploid spores. The ascospores and haploid cells arising from them after germination constitute the haplophase. As described below, yeasts differ in the manner by which the diplophase is reconstituted. In the first few cell-division cycles after germination, significant developmental differences occur, depending on whether the yeast is homothallic or heterothallic.

In a pioneering study of yeasts, Winge (1935) observed homothallism in *S. cerevisiae* var. *ellipsoideus* and later in *S. chevalieri* (Winge and Roberts 1949). Diploid cells were observed in a culture grown from a single spore; these cells could sporulate but did not have mating ability. Heterothallism in *Saccharomyces* yeasts was first described by Lindegren and Lindegren (1943), who observed single-spore cultures that grew into haploid clones having a stable mating type, either **a** or α, and were unable to sporulate. When two haploids of complementary mating types were mixed together, there arose a diploid zygote that gave rise to a diploid clone capable of sporulation. **a** and α mating types always segregated 2 **a**:2 α in each ascus.

The definitions of homothallism and heterothallism used above follow the definitions conceptualized in Mucorales by Blakeslee (1904a,b): Homothallism denotes self-fertility (ability to form diploids from a single-spore culture); heterothallism denotes obligatory cross-fertility (ability to form diploids only between cells of different mating types that are derived from separate single-spore cultures). In practice, homothallism and heterothallism in yeasts are determined by procedures similar to those used by Winge and Lindegren, such as inspecting cell shape or, more generally, testing cells in a single-spore culture for sporulation and for mating ability (or other haploid phenotypes, such as response to α-factor). The terms homothallism and diploidization are used interchangeably in this paper.

Genetic Control of Homothallism

Discovery of Homothallic Genes

The different behaviors of the Winge and Lindegren strains were due to the presence of the D gene (for diploidization), now called HO, in the Winge strains; the Lindegren strains contained the d allele, now called ho, an inactive allele of D (Winge and Roberts 1949). The HO gene proved to be epistatic to the $MATa$ and $MAT\alpha$ alleles and segregated independently from the mating-type locus. Studies described below revealed that there are three genes (HO, HML, and HMR) that play prominent roles in homothallism; the HO/ho (D/d) diploids of Winge and Roberts (1949) were homozygous for two of these ($HML\alpha$ and $HMRa$).[1]

The first evidence of multiple genes controlling homothallism came from the work of Takahashi (1958), who proposed the existence of three loci. The relationship between these loci and HO, HML, and HMR is discussed in detail by Takano and Oshima (1970a). During a genetic study on film formation with a sherry yeast, *S. oviformis*, a diploid was obtained with a distinctive segregation pattern (Takano and Oshima 1967). This diploid produced asci showing a 2 homothallic:2 **a** segregation in each tetrad. The diploid homothallic segregants again showed 2 homothallic:2 **a** segregation. This type of segregation (later called Hq segregation by Santa Maria and Vidal [1970]) is to be contrasted with the 4 homothallic:0 heterothallic segregation (Ho segregation) seen in the classical D strain. The Hq segregation indicated that these strains were homozygous for an allele that is unable to promote diploidization by **a** cells. The locus defined by this analysis was originally designated HM, subsequently HMa, and is now called HML (Takano and Oshima 1967; Oshima and Takano 1972; Harashima et al. 1974). The HML allele present in the Ho strain that allows **a** cells to diploidize is designated $HML\alpha$; the HML allele in the Hq strain is $HMLa$ (see Table 1).

Another type of homothallic diploid strain, a strain of *S. norbensis* described by Santa Maria and Vidal (1970), produced asci showing 2 homothallic diploid:2 α segregation in each tetrad. Upon sporulation, the diploid homothallic segregants exhibited the same segregation as the original strain. This type of segregation (Hp) is strikingly similar to the Hq segregation of the *S. oviformis* strain, with the difference being in the mating type of the stable haploid segregants. The Hp segregation indicated that the *S. norbensis* strain was homozygous for an allele that is unable to promote diploidization by α cells. The locus defined by this analysis was originally designated $HM\alpha$ (Oshima and Takano 1972; Harashima et al. 1974) and is now called HMR. The HMR allele present in the Ho strain that allows α cells to diploidize is designated $HMRa$; the allele in *S. norbensis* (Hp) is $HMR\alpha$.

---

[1]The nomenclature conventions of the 1978 International Congress on Yeast Genetics at Rochester, New York, are used in this paper. The relationship to the previous nomenclature is given in Table 1. Justifications for these changes are given below.

Table 1   Genetic control of the yeast life cycle

| Diploid[a] | Segregation pattern | Genotype of *HO/HO* strains | |
		Harashima et al.[b]	revised nomenclature
Ho type I	4 homothallic (Ho type I)	*hm*a *MAT*a *hm*α	*HML*a *MAT*a *HMR*α
		*hm*a *MAT*α *hm*α	*HML*a *MAT*α *HMR*α
Ho type II	4 homothallic (Ho type II)	*HM*a *MAT*a *HM*α	*HML*α *MAT*a *HMR*a
		*HM*a *MAT*α *HM*α	*HML*α *MAT*α *HMR*a
Hp	2 homothallic (Hp):	*HM*a *MAT*a *hm*α	*HML*α *MAT*a *HMR*α
	2 heterothallic (α)	*HM*a *MAT*α *hm*α	*HML*α *MAT*α *HMR*α
Hq	2 homothallic (Hq):	*hm*a *MAT*a *HM*α	*HML*a *MAT*a *HMR*a
	2 heterothallic (a)	*hm*a *MAT*α *HM*α	*HML*a *MAT*α *HMR*a

[a]Ho type I is a perfect homothallic strain derived from a cross between an α segregant from an Hp strain and an **a** segregant from an Hq strain; Ho type II is *S. chevalieri* and original homothallic *S. cerevisiae*; Hp is *S. norbensis*; Hq is *S. oviformis*. Additional information is given in the text.
[b]Nomenclature of Harashima et al. (1974).

In summary, as a result of crosses among standard heterothallic strains, fully homothallic (Ho type) strains, and segregants from semihomothallic strains (Hp and Hq types), three loci were identified as being involved in homothallism: The *HO* locus is required for diploidization of both **a** and α haploid strains; all *ho* strains are heterothallic. As noted above, *HMR*a allows α cells to diploidize, and *HML*α allows **a** cells to diploidize. Thus, *HO HMR*a (*HML*a or *HML*α) *MAT*α and *HO HML*α (*HMR*a or *HMR*α) *MAT*a cells diploidize, whereas *HO HMR*a *HML*a *MAT*a cells are stable **a** haploids and *HO HMR*α *HML*α *MAT*α cells are stable α haploids.

An initially surprising observation (see Naumov and Tolstorukov 1973; Harashima et al. 1974) is that both *HO MAT*a and *HO MAT*α strains with the nonstandard *HML* and *HMR* alleles (*HML*a *HMR*α) are able to diploidize (Table 1).

Dominance Relationships

The above results led to the hypothesis that *HMR*α (originally *hm*α) and *HML*a (originally *hm*a) are not simply inactive versions of *HMR*a and *HML*α. The first suggestion that this might be the case came from segregation of various ascus types with respect to homothallism from *MAT*a *HO HML*a *HMR*a × *MAT*α *HO HML*α *HMR*α crosses, in which one possible explanation for the observed results was that some of the homothallism genes were duplicated (Oshima and Takano 1972). Subsequent analysis led to the specific proposal that *HMR*α is functionally equivalent to *HML*α and that *HML*a is functionally equivalent to *HMR*a (Naumov and Tolstorukov 1973; Harashima et al. 1974). Thus, the duplicate *HML* and *HMR* genes in the analysis of Oshima and Takano (1972) were *HMR*α and *HML*a themselves.

According to the equivalence hypothesis, $HMRa$ and $HMR\alpha$ (originally $HM\alpha$ and $hm\alpha$, respectively) should be codominant, and $HML\alpha$ and $HMLa$ (HMa and hma) should be codominant. Evidence that $HMLa$ is not simply a recessive defect in $HML\alpha$ came from the observation that $MAT\alpha/$ $MAT\alpha$ sectors generated from an $HML\alpha/HMLa$ $MAT\alpha/MATa$ $HMR\alpha/$ $HMR\alpha$ HO/ho strain could give rise to a/a and a/α cells (Klar and Fogel 1977). Demonstration of codominance for $HML\alpha$ and $HMLa$ and for $HMRa$ and $HMR\alpha$ came from studies of diploids constructed by protoplast fusion (Arima and Takano 1979) and from statistical analysis of segregants from tetraploids (Harashima and Oshima 1980).

Dominance of HO to ho has been demonstrated in a variety of studies (Hopper and Hall 1975b; Hicks et al. 1977b; Klar and Fogel 1977).

Gene Functions

Several possibilities were enumerated by Winge and Roberts (1958) for diploidization. Diploids might be formed by (1) endomitosis, i.e., absence of nuclear division after chromosome duplication; (2) direct diploidization, i.e., fusion of mother and daughter nuclei before cytokinesis; or (3) fusion of two separate cells. Although Winge and Laustsen (1937) suggested that direct diploidization occurs in a Danish baking yeast, Winge (1935) had proposed earlier that diploidization occurs most often by fusion between two cells in a single-spore culture of a homothallic strain. This view was further strengthened by observing that cell fusion occurs between two cells within a few generations after germination, in general after the second cell division in single-spore cultures (see, e.g., Takano and Oshima 1967). Subsequent studies (Hicks and Herskowitz 1976b; Strathern and Herskowitz 1979) showed that mating never occurs before the second cell division after germination or between mother and daughter cells produced by a single cell division (see Control of Mating-type Interconversion).

How do two cells fuse to produce the zygote? Winge and Roberts (1958) suggested that cell fusion occurs between cells of the same mating type in a homothallic yeast, unlike matings of heterothallic yeast, which must be of complementary mating types. Hawthorne (1963b) suggested that the $HO(D)$ gene acts by causing mutation of one mating-type allele to the other during cell growth; hence, diploidization would occur in homothallic strains just as in heterothallic strains, by fusion of cells of opposite mating types. This view was confirmed by genetic examination of **a** cells that were derived from a $MAT\alpha$ HO $HMRa$ $HMLa$ spore (Takano and Oshima 1970b). The genetic determinant for the **a** mating type in these cells was mapped to the same site as the mating-type locus of standard heterothallic strains (with a small discrepancy resulting from sampling fluctuation). These **a** clones were phenotypically indistinguishable from standard heterothallic **a** cells in their mating ability, in their ability to promote sporulation in a/α diploids, and in their response to the homothallism genes. Similarly, an α clone derived from a homothallic **a** spore did not show any differences from a standard heterothallic α strain; it carried a normal α allele of the mating-type locus (Oshima and Takano 1971; see also Hicks and Herskowitz 1976b, 1977).

These studies led to the important conclusion that diploidization results from the interconversion of a cell's mating type and that the change in mating type is due to a change of the mating-type locus itself. The *HO* gene promotes the high frequency of switching between mating types, but it is not necessary for the maintenance of the new mating type; i.e., the new mating-type allele is stable after *HO* is replaced by *ho* (Takano and Oshima 1970b; Oshima and Takano 1971; Hicks and Herskowitz 1976b). An *HMRa* or *HMLa* allele is necessary for switching from *MATα* to *MATa*, whereas an *HMRα* or *HMLα* allele is necessary for switching from *MATa* to *MATα*.

The frequency and pattern of the mating-type switch have been demonstrated on descendants of single homothallic spores by monitoring the sensitivity of α-factor (Hicks and Herskowitz 1976b; Strathern and Herskowitz 1979) and are discussed in detail below.

Mapping

The map position of the *HO* gene has recently been found to be 27 centiMorgans (cM) distal to the *cdc9* locus on chromosome IV (G. Kawasaki, pers. comm.). Loose linkage between both *HML* and *HMR* with *MAT* was detectable from pooled tetrad data. Further information on the positions of *HML* and *HMR* involved analysis of segregants from a diploid heterozygous at three loci: *HMRa/HMRα* (or *HMLα/HMLa*), *MATa/MATα*, and for a standard chromosome III marker. To analyze such data required development of a novel algebraic procedure (Harashima and Oshima 1976, 1977) that allowed placement of the *HMR* locus at 65 cM from the centromere on the right arm of chromosome III and placement of the *HML* locus at 64 cM from the centromere on the left arm of chromosome III (Fig. 2; Harashima and Oshima 1976). These map positions (*HML* on the left arm and *HMR* on the right arm of chromosome III) and the codominance and functional

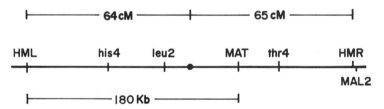

Figure 2 Genetic map of chromosome III showing selected markers. Map positions are drawn to approximate genetic scale according to Harashima and Oshima (1976) and the map compiled by Mortimer and Schild (this volume). The distance between *HML* and *MAT* is ~180 kb (Strathern et al. 1979b). There are no known markers distal to *HML* and *MAL2*.

equivalence relationships described previously provide the basis for the current nomenclature used for the homothallism genes (see Table 1; also Arima and Takano 1979; Harashima and Oshima 1980).

Controlling Element Model

In summary of the above, the *MAT*a and *MAT*α alleles of the mating-type locus are interchangeable with each other in *HO* cells at high frequency by specific action of the homothallism genes. Conversion from *MAT*α to *MAT*a requires both *HO* and *HMR*a (or *HML*a), whereas *HO* and *HML*α (or *HMR*α) are required for conversion from *MAT*a to *MAT*α. Two general categories of molecular mechanisms for mating-type interconversion have been proposed that differ from each other in the stucture of the mating-type locus (Fig. 3). In one case, the information for both *MAT*a and *MAT*α alleles is present at the mating-type locus; DNA modification (Holliday and Pugh 1975; D. C. Hawthorne, pers. comm.) or intramolecular recombination (Brown 1976; Hicks and Herskowitz 1977) at a hypothetical control site at the mating-type locus allows expression of either the *MAT*a or *MAT*α allele (so-called flip-flop models; Hicks et al. 1977a). In such a scheme, *HO*, *HML*, and *HMR* would code for enzymes involved in modification or rearrangement of the mating-type-locus DNA. The second type of hypothesis, which can be called an insertion model, was originally proposed by Oshima and Takano

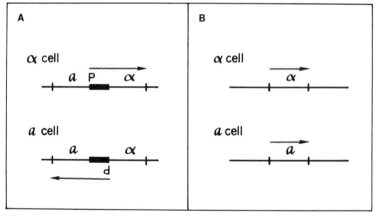

Figure 3 Two types of models for mating-type interconversion. The mating-type locus is shown for α and a cells. (a, α) Genes determining cell type; (———▶) expression. (*A*) Both a and α regulators are present at the mating-type locus (flip-flop model). In this particular model, a region between a and α regulators is invertible and contains an essential controlling site (such as a promoter, P). (*B*) The mating-type locus contains either a or α regulators but not both (insertion model). Insertion of an *HML*α or *HMR*α element at the mating-type locus confers the α cell type; insertion of an *HML*a or *HMR*a element confers the a cell type.

(1971) and modified by Harashima et al. (1974). In this case, the elementary structure of the mating-type locus for both the *MATa* and *MATα* alleles is essentially the same. The *HMR* and *HML* genes produce some kind of transposable genetic element, a controlling element analogous to those proposed in maize by McClintock (1957), which have a specific affinity for the mating-type locus. The association of the *HMRa* or *HMLa* element with the mating-type locus would form the *MATa* allele, and the association of an *HMLα* or *HMRα* element would form the *MATα* allele. The *HO* gene would control the association and removal (i.e., the insertion and excision) of these controlling elements at the mating-type locus; the *ho* allele is unable to promote these events. Confirmation and elaboration of the controlling-element model are discussed in later sections.

THE CONTROLLING ELEMENTS OF MATING-TYPE INTERCONVERSION ARE GENETIC CASSETTES

Healing of *mat* Mutations by Mating-type Interconversion

A remarkable property of the mating–type interconversion process is that mutations of the mating-type locus are restored to function, i.e., they are healed, as a result of mating-type interconversion. Spores that initially are *matα1-2* or *matα1-5* and carry *HO*, *HMLα*, and *HMRa* give rise to cells with a functional *MATα*-mating-type locus within three or four cell divisions after germination (Hicks and Herskowitz 1977; D. C. Hawthorne, pers. comm.). Healing of these mutations also occurs in *ho* strains. In contrast, mutations in **a**sg and αsg are not healed in *HO* strains. These observations led to the proposal that yeast cells contain additional, unexpressed *MATα* information in the genome at the *HMLα* (or *HMRα*) locus that can replace defective *matα* information (Hicks and Herskowitz 1977; Hicks et al. 1977a). By extension, the yeast genome was also proposed to contain unexpressed *MATa* information at the *HMRa* (or *HMLa*) locus. Consistent with this proposal is the observation that all mutations affecting functions of the mating-type locus are healable. Not only *matα1⁻* mutations but also *matα2* and *mata1* mutations are healed (Klar et al. 1979b; Strathern et al. 1979a). The observation that some of the healable *mat* mutations are nonsense (*matα1-2*, Herskowitz et al. 1980a; *matα-ochre*, D. C. Hawthorne, pers. comm.) indicates that *HMLα* is not simply a transposable regulatory site but contains structural gene information. (The ability to activate information at *HMLα* and *HMRa* in *ho* strains as a result of *sir*, *mar*, and *cmt* mutations further supports this view and is described below.)

These observations led to a specific form of the controlling element model, the cassette model. In this hypothesis, the *MATa* and *MATα* loci are nonhomologous blocks of DNA that code for regulators of cell type (as described earlier). The mating type of a cell is determined by insertion of one of the two blocks (cassettes) into a site on chromosome III (the mating-type locus) where the cassette is expressed, e.g., because it is adjacent to an active promoter or other essential controlling site. This site is analogous to the playback head of a tape recorder, and the **a** or α information is analogous to a tape cassette that is expressed when plugged into position. The yeast genome contains silent **a** and α information at *HML* and *HMR* that is not expressed, because these cassettes are not adjacent to a functional controlling site. Interconversion of mating types occurs by replacement of the cassette at the mating-type locus by a copy of the information at *HML* or *HMR*, a process catalyzed by some action of the *HO* gene (Fig. 4).

Tests of the Cassette Model

The cassette model leads to two specific predictions: (1) *HML*α and *HMR***a** are unexpressed copies of α and **a** information, respectively. (2) Mutations in *HML*α and *HMR***a** should give rise to mutations at the mating-type locus as a result of mating-type interconversion. These contentions are supported by both genetic and physical observations.

Mating-type Interconversion by Gross Rearrangement of Chromosome III

Circumstantial evidence for *HML*α and *HMR***a** being silent α and **a** information comes from studies of two types of chromosomal rearrangements observed in *ho* strains. *ho* strains switch mating types at a frequency of approximately 10^{-6}. In approximately 10% of these switches, the new mating-type determinant is inseparable from an extensive deletion or other rearrangement removing essential genes (Hawthorne 1963a; Strathern et al. 1979b). These **a**-lethal and α-lethal mutations are maintained as heterozygotes.

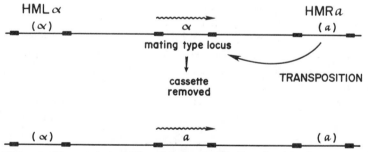

Figure 4 Cassette/controlling element model of mating-type interconversion. (*Top*) Structure of chromosome III of an α cell; (*bottom*) structure of chromosome III after interconversion to become an **a** cell. (——➤) Expression of the cassette at *MAT*; parentheses indicate that the cassette is silent. (▬) Sites that may be involved in the transposition process (see Fig. 5).

The **a**-lethal mutation, obtained in switches from α to **a**, is a deletion on the right arm of chromosome III, extending from the mating-type locus and ending between *thr4* and *MAL2* (Hawthorne 1963a). On the basis of the map position of *HMR*a, it was proposed that this deletion activates **a** information at *HMR*a by fusing it to a site at the mating-type locus (Hicks et al. 1977a). The involvement of *HMR*a in such events is confirmed by the observation that such deletions isolated in strains with mutations at the *HMR*a locus lead to defective **a**-mating-type loci (Kushner 1979; Klar 1980).

The α-lethal mutation, obtained in switches from **a** to α, removes the *thr4* gene and alters the linkage relationships between markers on chromosome III (such as *his4* and *leu2*) (Strathern et al. 1979b; R. Mortimer and D. C. Hawthorne, pers. comm.). On the basis of the map position of *HML*α, it was proposed that α information at this locus may become fused with an essential site at the mating-type locus by the formation of a circular chromosome. This interpretation is supported by the finding that strains carrying the α-lethal mutation contain a 63μ circular derivative of chromosome III (Strathern et al. 1979b; see Fig. 2).

Physical studies described below show that *HML* and *HMR* share homologous DNA sequences with the mating-type locus, which are presumably involved in formation of these rearrangements.

Functional Equivalence of HMLα *and* MATα

Mutations unlinked to the mating-type locus have been found in *ho* strains, which suppress the defects of mating-type-locus mutations. The ability of one such suppressor, *sir1-1*, to suppress the mating defect of *mat*α*1-5* requires *HML*α: *mat*α*1-5 ho sir1-1 HML*α *HMR*a strains mate, whereas *mat*α*1-5 ho sir1-1 HML*a *HMR*a strains do not. This observation and others described later in the text indicate that *HML*α codes for the same function as *MAT*α and, by extension, that *HMR*a codes for the same functions as *MAT*a.

"Wounding" of the Mating-type Locus in HMLα *and* HMRa *Mutants*

The cassette model predicts that a mutation within the structural genes of *HML*α or *HMR*a should give rise to a mutation of the mating-type locus as a result of mating-type interconversion. Mutations in *HML*α and *HMR*a have been obtained by two different methods and have been found to give rise to mutations of the mating-type locus, i.e., to wounding of the mating-type locus. For example, an *HO MAT*a *HMR*a strain with a mutation in *HML*α (*hml*α*-66*) switches back and forth between two cell types: an **a** cell and a cell defective in mating. Likewise, an *HO hml*α*-66 HMR*a cell, initially with a functional *MAT*α, first switches to *MAT*a and then to *mat*α⁻. The defect in mating by cells of the latter type is due to a mutation of the mating-type locus that has the properties of a *mat*α*1* mutation (Kushner et al. 1979). Similar results have been obtained for *hmr*a mutations (Blair et al. 1979; Klar and Fogel 1979; Klar 1980). A particularly striking observation is that *hmr*a-nonsense mutations give rise to nonsense mutations of the **a**-mating-type locus, *mata*-nonsense (Klar 1980).

Physical Tests

The **a**- and α-mating-type loci and the *HMLα* and *HMRa* loci have been cloned by a two-step procedure. First, yeast DNA sequences were identified that could allow *mata⁻* or (a presumed) *matαl⁻* mutation to mate as α. This complementation assay allowed isolation of *HMLα* (Hicks et al. 1979) and *MATα* (Nasmyth and Tatchell 1980). Since these cloned DNA segments hybridized to three different genomic DNA fragments that were shown to be *HML*, *HMR*, and *MAT*, the second step was to clone these cross-hybridizing DNA fragments. The cloned *HMLα*, *HMRa*, *MATa*, and *MATα* segments have been used for several incisive studies: (1) By using the *HMLα* segment as probe, it was shown that mating-type interconversion (in both *ho* and *HO* strains) was accompanied by a physical change in the yeast genome. A change from α to **a** resulted in the change of only a single restriction endonuclease fragment, the fragment carrying the mating-type locus (Hicks et al. 1979; Strathern et al. 1980). The fragment carrying *MATa* is approximately 150 bp smaller than the fragment carrying *MATα*. (2) By using *HMLα* (and other) probes, it was shown that strains with **a**-lethal and α-lethal chromosomes contain novel restriction endonuclease fragments that hybridized to the probe. These novel restriction fragments are those predicted for fusions between *MAT* and *HMRa* or between *MAT* and *HMLα* (Strathern et al. 1980). (3) DNA heteroduplexes have been formed between all combinations of *HMLα*, *HMRa*, *MATα*, and *MATa* and reveal several important points (Nasmyth and Tatchell 1980; Strathern et al. 1980). First, *MATa* and *MATα* DNA fragments are nonhomologous for approximately 750 bp. The distinctive sequence present at *MATa* is found at *HMRa* but not at *HMLα*. Likewise, the distinctive sequence present at *MATα* is found at *HMLα* but not at *HMRa*. These results provide a physical demonstration that *MATα* and *HMLα* are homologous, as are *MATa* and *HMRa*. Second, the **a**- and α-specific regions present at *HML*, *HMR*, and *MAT* are flanked by regions of homology present at all three loci. These relationships are summarized in Figure 5, in which it is seen that *HML* and *HMR* both contain homologous sequences flanking the mating-type locus, with the homology extending

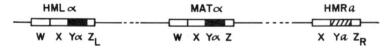

Figure 5 Homologous and nonhomologous sequences at *HMLα*, *MAT*, and *HMRa* determined by heteroduplex analysis (Nasmyth and Tatchell 1980; Strathern et al. 1980). Orientation and nomenclature are from Strathern et al. (1980). Approximate segment lengths (in bp): (W) 700; (X) 700; (Ya) 600; (Yα) 750; (Z_L) 300; (Z_R) 250.

further for *HML* than for *HMR*. The significance of this difference is not known. The existence and position of these homologous segments provide a likely explanation for the formation of **a**-lethal and α-lethal rearrangements: Recombination between these sequences produces a deletion or circle. The possible role of the homologous sequences in the mating-type interconversion process is discussed below.

Mechanics of Mating-type Interconversion

In this section experiments are described bearing on the mechanism by which information from *HML* and *HMR* becomes transposed to the mating-type locus and the functions and sites involved in this process.

Nonreciprocality

The substitution of one cassette by another involves a nonreciprocal transfer of information from *HML* or *HMR* to the mating-type locus. When a strain that is *HO HMLα MATα HMR*a switches to *MAT*a, it remains *HMLα HMR*a (Harashima et al. 1974). Thus, the information at *HML* and *HMR* is not "used up" as a result of mating-type interconversion but remains intact. These observations indicate that a copy of the information from *HML* or *HMR* is transposed to the mating-type locus.

Fate of the Exiting Cassette

What happens to the α information at the mating-type locus when an α strain switches to **a**? Because *HO HMLα MATα HMR*a cells give rise to *HO HMLα MAT*a *HMR*a cells rather than to *HO HMLα MAT*a *HMR*α, it is clear that the α information previously at *MAT* is not transferred to the *HMR* locus. To determine whether the excised information is stored somewhere else in the genome and whether it ever returns to the mating-type locus, the switching behavior of *HO HML*a *MAT*a *HMR*a strains that were derived from an *HO HML*a *MAT*α *HMR*a strain by mating-type interconversion has been analyzed (Rine et al. 1981). In nearly 10⁵ colonies derived from such an *HO HML*a *MAT*a *HMR*a strain, all remained *MAT*a. Similar observations have been made with an *HO HMLα MATα HMR*α strain derived by switching from an *HO HMLα MAT*a *HMR*α strain. Thus, the cassette previously at the mating-type locus is not readily retrieved after it leaves the mating-type locus. The exiting cassette may be degraded or diluted out during growth of cells after being excised from the chromosome.

Sites and Functions Necessary for Switching

The allele of the α-mating-type locus (*MATα-inc*, "inconvertible") from the yeast *S. diastaticus* behaves as if it were insensitive to switching. *HO HMLα MATα-inc HMR*a strains switch to *MAT*a at low frequency (Takano et al. 1973). Once a *MATα-inc* strain has switched to *MAT*a, subsequent interconversions occur at normal frequency. Because $MAT\alpha$-*inc* has a *cis*-dominant defect in switching (Strathern et al. 1979a; Takano and Arima 1979), this natural variant allele may identify a site essential for removal of the α cassette. The observation that the defect in switching by $MAT\alpha$-*inc* strains is healable indicates that the *inc* mutation affects part of the information of the mating-type locus that is removed during the interconversion process. Similar mutations have been found that affect the **a**-mating-type locus (MAT**a**-*inc*; Mascioli and Haber 1980). Another type of mutation (STK, which is presumably dominant) is tightly linked to the mating-type locus and also appears insensitive to switching, but it is not healable. Both *inc* and STK mutations appear to provoke changes from $HML\alpha$ to HML**a** and from HMR**a** to $HMR\alpha$ (Haber et al. 1980a,b). These mutations presumably identify sites at the mating-type locus that are necessary for switching and that may be recognition sites for the machinery of mating-type interconversion.

In addition to HO, two other genes have been found that are necessary for mating-type interconversion. The $SWI1$ gene is necessary for switching both from α to **a** and from **a** to α in HO $HML\alpha$ HMR**a** strains. Segregation data indicate that this gene is distinct from HO, HML, HMR, and MAT (Haber and Garvik 1977). Several other mutations have been found to affect mating-type interconversion, but their relationship to *swi1* remains to be determined (Blair 1979; Oshima and Takano 1980). The $RAD52$ gene, which is required for mitotic and meiotic general recombination, also affects mating-type interconversion: HO $HML\alpha$ HMR**a** *rad52* strains are inviable (Malone and Esposito 1980; R. E. Malone and J. E. Haber, pers. comm.). The observation that HO $HML\alpha$ $MAT\alpha$-*inc* HMR**a** *rad52* strains are viable indicates that the lethality of the *rad52* mutation is due to abortive mating-type interconversion and suggests that the $RAD52$-gene product may be needed to complete a step in mating-type interconversion.

Molecular Mechanism of Mating-type Interconversion

Although several models can be proposed for mating-type interconversion, an asymmetric gene conversion model may be most likely based on circumstantial evidence and biological precedent. In such a model, HO, $SWI1$, and $RAD52$ code for (or control) the protein machinery that carries out the steps described above. This model accounts for the points mentioned previously: the nonreciprocality of transfer of information from HML or HMR to MAT; the permanent loss of information displaced from MAT; the extensive homologous sequences flanking HML, HMR, and MAT; and the involvement of $RAD52$ in mating-type interconversion.

Consider a switch from $MAT\alpha$ to MAT**a**. The first step involves synapsis of HMR**a** and $MAT\alpha$, perhaps facilitated by a protein that recognizes homologies at the mating-type locus and HMR**a**. The second step is the invasion of a DNA strand from HMR**a** into the $MAT\alpha$ duplex, followed by degradation of the $MAT\alpha$ strand and its replacement by the invading **a** strand. The **a**-α heteroduplex at the mating-type locus is then resolved using the **a** information as template, and the gap at *HMR***a** is filled.

Other types of models involve freely diffusing intermediates, e.g., copies of **a** information from *HMR***a**. Initial attempts to find such intermediates by Southern blotting have been unsuccessful (Hicks et al. 1979).

An interesting puzzle associated with mating-type interconversion is the position of the silent cassettes relative to the mating-type locus: *HML* and *HMR* are physically linked to the mating-type locus but are only loosely linked genetically. Whether there is any functional significance to this arrangement should soon be resolved by studying switching behavior of strains with translocations involving *HML*, *HMR*, and the mating-type locus.

Control of Expression of Information at *HML* and *HMR*

Mutations Allowing In Situ Expression

Because the cassette at the mating-type locus rather than at *HML* or at *HMR* confers a cell's phenotype, the information at *HML* and *HMR* must not be expressed. This conclusion is reinforced by the existence of mutations of the mating-type locus that have recessive defects in mating and sporulation. There are several a priori mechanisms that might explain the differences in expression of information at *MAT* versus the information at *HML* and *HMR*. One extreme hypothesis is that *HML* and *HMR* information is transcribed and translated but does not contain the complete information for **a**1, α1, and α2 functions. If *HML*α, for example, were to code for only part of α1 and α2, then transcription and translation of *HML*α would not provide α1 and α2 functions to the cell. In another hypothesis, *HML* and *HMR* contain complete information for **a**1, α1, and α2, but these functions are not produced because *HML* and *HMR* lack essential sites necessary for transcription or translation. Studies of mutations that allow expression of *HML*α and *HMR***a** in *ho* strains in situ, i.e., without genetic rearrangement, demonstrate that the **a** and α information at *HML* and *HMR* is an intact version of the **a** and α information of the mating-type locus and that *HML* and *HMR* expression is under some kind of negative control.

The *sir1-1* mutation, selected as a suppressor of the *matα1-5* mutation, suppresses the mating and sporulation defects of all mutations of the mating-type locus (Herskowitz et al. 1977; Rine et al. 1979) and allows *MAT***a**/*MAT***a** and *MATα*/*MATα* strains to sporulate efficiently. The ability of *sir1-1* to suppress the mating defect of *matα1-5* is dependent on the presence of *HMLα* information: *matα1-5 ho sir1-1 HMLα HMR***a** strains mate, whereas *matα1-5 ho sir1-1 HML***a** *HMR***a** or *ho sir1-1 hmlα-66 HMR***a** strains do not (Rine et al. 1979; L. C. Blair and J. D. Rine, pers. comm.).

The *mar1-1* mutation was identified by its ability to allow *MAT***a**/*MAT***a** strains to sporulate and because it led to a mating defect in *MATa* and *MATα* cells (Klar et al. 1979a). These behaviors were explained by proposing that *mar1-1* allows expression of information at *HMLα* and *HMRa*, thereby giving *MATa/MATa*, *MATa*, and *MATα* strains the phenotypes of *MATa/MATα* cells with respect to mating and sporulation. This hypothesis was supported by the observation that the mating defect of *mar1-1* strains is relieved by substitution of the standard *HML* and *HMR* alleles with natural variant or mutant alleles (Klar et al. 1979a; Klar 1980). Haber and George (1979) have shown that the *cmt* mutation of Hopper and Hall (1975b) is analogous to *mar1-1*.

Rine (1979) has performed a systematic isolation of mutations allowing expression of *HMLα* or *HMRα* by isolating derivatives of *HMLα matal⁻ HMRα ho* strains that mate as α. Thus far, four complementation groups of *SIR* (*silent information regulator*) have been found: *SIR1* (defined by *sir1-1*), *SIR2* (which includes *mar1-1*), *SIR3* (which includes *cmt*), and *SIR4*. All of these mutations are recessive to wild type, all lead to expression of both *HMLα* and *HMRa*, and all are unlinked to one another. Because most *sir* mutations (including an *ochre* mutation in *SIR1*; J. D. Rine, pers. comm.) lead to the nonmating phenotype of *MATa/MATα* cells, it is not surprising that *sir* mutants also have been obtained as α-factor-resistant mutants (Hartwell 1980; J. McCullough, pers. comm.): *ste8* and *ste9* (Hartwell 1980) are in complementation groups *SIR3* and *SIR4*, respectively (Rine 1979).

Since in situ expression of *HMLα* and *HMRa* can supply **a**1, α1, and α2 functions, *HMLα* and *HMRa* must contain complete coding information for **a**1, α1, and α2 and be kept silent in wild-type strains by controlling synthesis of these functions. Recent results indicate that stable transcripts of information from *HML* and *HMR* are not observed in wild-type strains (Klar et al. 1980; Nasmyth et al. 1981). The *SIR* genes may thus control transcription or RNA stability. Whether the four *SIR* genes are concerned exclusively with controlling expression of *HML* and *HMR* is not known; it is clear that *sir* mutants are not grossly pleiotropic.

Because the cassette at the mating-type locus is expressed even in *SIR⁺* strains, the mating-type locus presumably lacks a site of Sir action. Transposition of cassettes from *HML* or *HMR* to the mating-type locus thus appears to activate this information by removing it from a site of Sir action.

SAD, an **a** Cassette at a New Position

The *SAD* mutation was identified by its ability to allow efficient sporulation by *MATα/MATα* strains (Hopper and MacKay 1980). *SAD* is dominant to wild type (*sad⁺*), is unstable, and is located on the right arm of chromosome III, between *thr4* and *HMR* (Hopper and MacKay 1980; Kassir and Herskowitz 1980). Because *SAD* allows sporulation not only by *MATα/ MATα* strains but also by *matal/MATα* strains, it may act by alleviating the requirement of **a**1 product for sporulation or by supplying **a**1 function itself.

Two observations support the latter hypothesis (Y. Kassir et al., in prep.): (1) *HO HMLα MATα SAD HMRα* strains can switch to *MATa*. (2) *SAD* strains contain a new DNA fragment that hybridizes with a *MATa* probe. Although *SAD* is able to provide **a**l function in *SIR⁺* strains, *MATα SAD* strains mate normally (i.e., they do not behave as **a**/α). Hence, expression of **a**l from *SAD* may be low per se or may still be partially sensitive to Sir.

CONTROL OF MATING-TYPE INTERCONVERSION

Switching Pattern—Asymmetry

Mating-type interconversion is under a variety of controls that lead to the production of specific cell lineages during the growth of a clone of homothallic cells. As noted above, in Mating-type Interconversion, switching of a homothallic cell from one mating type to another can be followed microscopically by observing changes from α-factor resistance (in *MATα* cells) to α-factor sensitivity (in *MATa* cells) (Hicks and Herskowitz 1976b). Two "rules" of switching can account for the observed lineages (Fig. 6; Hicks et al. 1977b; Strathern and Herskowitz 1979): (1) Both cells of any cell division are always of the same mating type. For example, α cells produce two α progeny or two **a** progeny, never one **a** and one α. (2) Only cells that have budded at least once ("experienced cells"; cells S [2], S [4], and D1 [4] in Fig. 6) are competent

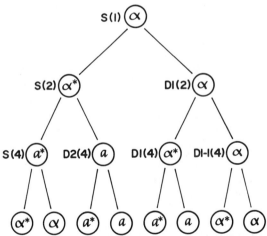

Figure 6   Pattern of switching by a homothallic cell. Mother and daughter cells are separated by micromanipulation; **a** and α cells are distinguished by their response to α-factor. (S) spore cell; (D1) the spore's first daughter; (D2) the spore's second daughter; (D1-1) the first daughter of D1. Numbers in parentheses indicate the stage (1 cell, 2 cells, 4 cells). (*) Experienced cells. This lineage shows the pattern produced when S (2), S (4), and D1 (4) have given rise to switched cells (see Strathern and Herskowitz 1979).

to switch mating types. Cells that have not budded ("inexperienced cells"; cells S [1], D1 [2], etc., in Fig. 6) are rarely if ever observed to switch. A clone of homothallic cells is thus a mixed population of cells that differ in competence to switch.

Cells are able to switch mating types in successive cell divisions but do so only after they have become experienced. (Examples of such cells are S [2] and S [4]; Fig. 6.) Switching continues at high frequency for many generations (at least 12) after spore germination (Klar and Fogel 1977; see also Strathern and Herskowitz 1979).

The observation that sister cells of any cell division are always identical indicates that the event that changes cell type occurs early in the cell-division cycle. In particular, it has been proposed that the cassette at the mating-type locus is replaced sometime between late G_1 (after the point in the cell cycle at which mating pheromones arrest cells) and early S (before the mating-type locus is replicated). Because cells can switch cell types in successive cell cycles, the cellular apparatus necessary for one mating type must be replaced by that of the other mating type. This turnover may reflect destruction and resynthesis of such components in all cells, or it may be a special property of *HO* cells. If indeed the mating-type locus is changed early in the cell cycle, then the cell has until its next G_1 to "change its clothing."

Experienced cells give rise to switched cells in 73% of their cell divisions; inexperienced cells give rise to switched cells in 0.1% (or less) of their cell divisions (Strathern and Herskowitz 1979). Hence, there is an asymmetric distribution of potential for mating-type interconversion at each cell division. Why a cell must bud at least once in order to acquire competence is unknown. The difference between mother and daughter cells may reflect asymmetric distribution of a factor that is necessary to activate the switching machinery in the mother cell (or that inhibits switching in the daughter cell). Another possibility is that the mating-type-locus DNA must acquire modifications (such as methylation) before it can be a substrate for the interconversion machinery. According to this explanation, the DNA strands of the mating-type locus must be distributed in a specific manner (Hicks and Herskowitz 1976b). In any event, lineages of homothallic cells are analogous to stem-cell lineages. Each cell division always produces one cell incapable of switching and a second cell that has the competence to switch.

Control of Switching by the Mating-type Locus—Directionality of Switching and Inhibition in a/α Cells

Analysis of the switching pattern also reveals that the switching process has directionality: Experienced cells switch to the opposite mating type at a frequency greater than 50% (73% of experienced cells switch). In other words, the cassette that becomes inserted into the mating-type locus is not chosen at random. Rather, the cassette at the mating-type locus directs the insertion of the heterologous cassette from *HML* or *HMR*. Two kinds of mechanisms can be imagined for this directionality. The mating-type locus itself might code for a function that specifically promotes replacement by the heterologous cassette or inhibits replacement by the homologous cassette. Another possibility is that the structure of the cassettes at the mating-type locus and at *HML* and *HMR* regulate which cassette is chosen. For example, the cassettes or regions around the mating-type locus, *HML,* or *HMR* might contain recognition regions for the transposition machinery analogous to the attachment sites of lambdoid prophages. It should be noted that directionality may not be absolute, since homologous cassette substitution does occur: *HMLα matα1-5 HMRα HO* strains do switch to *MATα* (Rine et al. 1981). Whether homologous cassette substitution occurs in the 27% of the cases when heterologous cassette substitution does not occur is unclear.

The mating-type locus does not affect the frequency of switching in haploid cells; switching from **a** to *α* occurs at the same frequency as from *α* to **a** (Strathern and Herskowitz 1979). In **a**/*α* diploids, however, interconversion does not occur (Takano and Oshima 1970b; Hicks et al. 1977b; Klar and Fogel 1977). The turnoff of switching is not due to diploidy per se since *MATa/MATa* and *MATα/MATα* diploids exhibit normal high-frequency mating-type interconversion. The same functions of the mating-type loci involved in turning on sporulation by *MATa/MATα* diploids and in turning off mating phenotypes are also involved in turning off switching in *MATa/MATα* diploids: *MATa/matα2* and *mata1⁻/MATα* diploids continue to switch mating types (Klar et al. 1979b; Strathern et al. 1979a). A unifying explanation for directionality and inhibition in **a**/*α* diploids would be for the *α*2 and **a**1 functions to inhibit transposition of the silent *α* and **a** cassettes, respectively.

CONCLUDING REMARKS: APPLICATIONS, COMPARISONS, AND EXTRAPOLATIONS

Breeding Polyploid Yeasts

Breeding triploid or tetraploid strains of yeast requires matings between diploids and other diploids or haploids. Understanding how the mating-type locus controls mating ability and how the yeast life cycle is determined provides information useful for constructing polyploid yeasts. For example, tetraploids can be formed efficiently from homothallic **a**/*α* diploids if these strains are first converted to **a**/**a** or *α*/*α* strains by mitotic recombination (Klar and Fogel 1977; Takano et al. 1977). Isogenic strains (differing only at the mating-type locus) with ploidies of 1n, 2n, 3n, 4n, etc., can likewise be constructed utilizing homothallic strains. For example, a diploid of genotype *HO/HO HMLα MATa HMRα/HMLα MATα HMRα* (formed by diploidization of an *HO HMLα MATa HMRα* cell) yields a stable *α* haploid upon sporulation. This diploid can also yield a stable $MAT\alpha/MAT\alpha$ diploid by mitotic recombination, which can mate with the $HO\ HML\alpha\ MAT\mathbf{a}\ HMR\alpha$ progenitor to form a stable isogenic triploid strain. A stable $\alpha/\alpha/\alpha$ strain can then be derived from this triploid by mitotic recombination and can be mated with the **a** progenitor by spore-to-cell mating to form a stable tetraploid strain. This set of strains would then be available for a definitive study of the effects of ploidy on fermentation and other behaviors.

Comparisons with Other Yeasts

The mechanism of homothallism for two yeasts that do not interbreed with *S. cerevisiae*, the yeasts *Kluyveromyces lactis* and *Schizosaccharomyces pombe*, may be similar to that of *S. cerevisiae*. *K. lactis* has two mating types, **a** and α, which can be interconverted by action of two genes. *H*α, which is only 9 cM from the mating-type locus, is necessary for conversion from α to **a**; *H*a, which is unlinked to the mating-type locus, is necessary for conversion from **a** to α (Herman and Roman 1966). By analogy with *S. cerevisiae*, the *H*α and *H*a loci may correspond to *HMR*a and *HML*α, i.e., they may be silent mating-type-locus information.

In *S. pombe*, crosses among heterothallic strains (mating types plus and minus) and homothallic (*h*90) strains revealed that the region controlling the life cycle has two parts, *mat1* and *mat2* (Leupold 1959; Egel 1977). According to one hypothesis, *mat1* contains information (*M*) for the minus mating type, and *mat2* contains information (*P*) for the plus mating type; an invertible region between *mat1* and *mat2* leads to mutually exclusive expression of either *M* or *P* (Egel 1977). In a more recent hypothesis (reviewed by Leupold 1980; R. Egel, pers. comm.), *mat2* is analogous to one of the *HM* loci of *S. cerevisiae*, in that it carries silent *P* information. When this information is transposed to *mat1* (a "playback locus" analogous to *MAT* of *S. cerevisiae*), a cell expresses the plus mating type. According to this hypothesis, the minus mating type might be determined by insertion of *M* information from a hitherto unrecognized silent locus; alternatively, *M* information is already present at *mat1*.

More information on *K. lactis* and *S. pombe* is clearly required in order to know whether homothallism occurs as in *S. cerevisiae*. It seems likely that these yeasts interconvert mating types by rearrangement of their genomes, utilizing controlling elements or cassettes. It will be most interesting to understand the mechanism of homothallism in organisms more distantly related to *S. cerevisiae* than *K. lactis* and *S. pombe*, e.g., in the homothallic alga *Chlamydomonas monoica*.

Extrapolations to Multicassette Systems

A cassette mechanism provides a means by which a cell can switch to any of several different states. For example, consider a cell whose genome contains several silent cassettes, *A, B, C...*, and a playback locus (analogous to *MAT* in *S. cerevisiae*) that allows expression of whichever cassette is inserted at this position. A cell with an *A* cassette at the playback locus exhibits an A phenotype and can switch to another phenotype by substitution of a different cassette.

Antigenic variation of trypanosomes may occur by such a mechanism. Silent genes coding for different forms of the major cell-surface antigen appear to be activated by transposition of these genes to a site where they are expressed (Hoeijmakers et al. 1980). A cassette mechanism can also be invoked to account for the sequential appearance of different cell types during development in multicellular eukaryotes. As described above, the yeast switching process exhibits directionality: Cells competent to switch mating types switch preferentially to the opposite mating type. In a cell with multiple cassettes, one can imagine that an active cassette promotes its own replacement by another cassette. Consequently, cells of type A would appear before cells of type B, etc. (Strathern and Herskowitz 1979; Herskowitz et al. 1980b).

ACKNOWLEDGMENTS

We thank Jasper Rine, George Sprague, Jr., Jeff Strathern, and Beth Jones for comments on the manuscript; J. Rine, G. Sprague, J. Strathern, Jim Hicks, Kim Nasmyth, Kelly Tatchell, Isumu Takano, and others for communication of unpublished results; Kerrie Rine for figures; and Julie Dunn and Nancy Caretto for preparation of the manuscript. Work from I.H.'s laboratory was supported by research grants and a Career Development Award from the U.S. Public Health Service. Work from Y.O.'s laboratory was supported by research grants from the Ministry of Education, Science, and Culture of Japan.

REFERENCES

Arima, K. and I. Takano. 1979. Evidence for co-dominance of the homothallic genes, *HMα/hmα* and *HMa/hma*, in *Saccharomyces* yeasts. *Genetics* **93**: 1.

Blair, L.C. 1979. "Genetic analysis of mating type switching in yeast." Ph.D. thesis, University of Oregon, Eugene.

Blair, L.C., P.J. Kushner, and I. Herskowitz. 1979. Mutations of the *HMa* and *HMα* loci and their bearing on the cassette model of mating type interconversion in yeast. *ICN-UCLA Symp. Mol. Cell. Biol.* **14**: 13.

Blakeslee, A.F. 1904a. Zygospore formation, a sexual process. *Science* **19**: 864.

———. 1904b. Sexual reproduction in the Mucorineae. *Proc. Am. Acad. Arts Sci.* **40**: 205.

Brown, W.S. 1976. A cross-over shunt model for alternate potentiation of yeast mating-type alleles. *J. Genet.* **62**: 81.

Chan, R.K. 1977. Recovery of *Saccharomyces cerevisiae* mating-type **a** cells from G1 arrest by α factor. *J. Bacteriol.* **130**: 766.

Egel, R. 1977. "Flip-flop" control and transposition of mating-type genes in fission yeast. In

DNA insertion elements, plasmids, and episomes (ed. A. Bukhari et al.), p. 447. Cold Spring Harbor Laboratory, Cold Spring Harbor, New York.

Fraenkel, D.G. 1982. Carbohydrate metabolism. In *The molecular biology of the yeast* Saccharomyces. II. *Metabolism and gene expression.* (ed. J. Strathern et al.), Cold Spring Harbor Laboratory, Cold Spring Harbor, New York. (In press.)

Gerlach, W.L. 1974. Sporulation in mating-type homozygotes of *Saccharomyces cerevisiae.* *Heredity* **32:** 241.

Haber, J.E. and B. Garvik. 1977. A new gene affecting the efficiency of mating type interconversions in homothallic strains of *Saccharomyces cerevisiae.* *Genetics* **87:** 33.

Haber, J.E. and J.P. George. 1979. A mutation that permits the expression of normally silent copies of mating-type information in *Saccharomyces cerevisiae.* *Genetics* **93:** 13.

Haber, J.E., D.W. Mascioli, and D.T. Rogers. 1980a. Illegal transposition of mating type genes in yeast. *Cell* **20:** 519.

Haber, J.E., W.T. Savage, S.M. Raposa, B. Weiffenbach, and L.B. Rowe. 1980b. Genetic identification of sequences adjacent to the yeast mating type locus that are essential for transposition of new mating type alleles. *Proc. Natl. Acad. Sci.* **77:** 2824.

Harashima, S. and Y. Oshima. 1976. Mapping of the homothallic genes, $HM\alpha$ and $HM\mathbf{a}$, in *Saccharomyces* yeasts. *Genetics* **84:** 437.

———. 1977. Frequencies of twelve ascus-types and arrangement of three genes from tetrad data. *Genetics* **86:** 535.

———. 1980. Functional equivalence and co-dominance of homothallic genes, $HM\alpha/hm\alpha$ and $HM\mathbf{a}/hm\mathbf{a}$, in *Saccharomyces* yeasts. *Genetics* **95:** 819.

Harashima, S., Y Nogi, and Y. Oshima. 1974. The genetic system controlling homothallism in *Saccharomyces* yeasts. *Genetics* **77:** 639.

Hartwell, L.H. 1980. Mutants of *S. cerevisiae* unresponsive to cell division control by polypeptide mating hormones. *J. Cell. Biol.* **85:** 811.

Hawthorne, D.C. 1963a. A deletion in yeast and its bearing on the structure of the mating type locus. *Genetics* **48:** 1727.

———. 1963b. Directed mutation of the mating type alleles as an explanation of homothallism in yeast. *Proc. Int. Congr. Genet.* **1:** 34.

Herman, A. and H. Roman. 1966. Allele specific determinants of homothallism in *Saccharomyces lactis.* *Genetics* **53:** 727.

Herskowitz, I., J. Rine, G. Sprague, Jr., and R. Jensen. 1980a. Control of cell type in yeast by genetic cassettes. *Miami Winter Symp.* **17:** 133.

Herskowitz, I., J.N. Strathern, J.B. Hicks, and J. Rine. 1977. Mating type interconversion in yeast and its relationship to development in higher eukaryotes. *ICN-UCLA Symp. Mol. Cell. Biol.* **8:** 193.

Herskowitz, I., L. Blair, D. Forbes, J. Hicks, Y. Kassir, P. Kushner, J. Rine, G. Sprague, Jr., and J. Strathern. 1980b. Control of cell type in the yeast *Saccharomyces cerevisiae* and a hypothesis for development in higher eukaryotes. In *The molecular genetics of development* (ed. W. Loomis and T. Leighton), p. 79. Academic Press, New York.

Hicks, J.B. 1975. "Interconversion of mating types in yeast." Ph.D. thesis, University of Oregon, Eugene.

Hicks, J.B. and I. Herskowitz. 1976a. Evidence for a new diffusible element of mating pheromones in yeast. *Nature* **260:** 246.

———. 1976b. Interconversion of yeast mating types. I. Direct observations of the action of the homothallism (*HO*) gene. *Genetics* **83:** 245.

———. 1977. Interconversion of yeast mating types. II. Restoration of mating ability to sterile mutants in homothallic and heterothallic strains. *Genetics* **85:** 373.

Hicks, J.B., J.N. Strathern, and I. Herskowitz. 1977a. The cassette model of mating-type interconversion. In *DNA insertion elements, plasmids, and episomes* (ed. A.I. Bukhari et al.), p. 457. Cold Spring Harbor Laboratory, Cold Spring Harbor, New York.

———. 1977b. Interconversion of yeast mating types. III. Action of the homothallism (*HO*) gene in cells homozygous for the mating type locus. *Genetics* **85:** 395.

Hicks, J.B., J.N. Strathern, and A.J.S. Klar. 1979. Transposable mating type genes in *Saccharomyces cerevisiae*. *Nature* **282**: 478.

Hoeijmakers, J.H.J., A.C.C. Frasch, A. Bernards, P. Borst, and G.A.M. Cross. 1980. Novel expression-linked copies of the genes for variant surface antigens in trypanosomes. *Nature* **284**: 78.

Holliday, R. and J.E. Pugh. 1975. DNA modification mechanisms and gene activity during development. *Science* **187**: 226.

Hopper, A.K. and B.D. Hall. 1975a. Mating type and sporulation in yeast. I. Mutations which alter mating-type control over sporulation. *Genetics* **80**: 41.

———. 1975b. Mutation of a heterothallic strain to homothallism. *Genetics* **80**: 77.

Hopper, A.K. and V.L. MacKay. 1980. Control of sporulation in yeast: *SAD1*—A mating-type specific, unstable alteration that uncouples sporulation from mating-type control. *Mol. Gen. Genet.* **180**: 301.

Jones, E. and G. Fink. 1982. The regulation of amino acid and nucleotide biosynthesis in yeast. In *The molecular biology of the yeast* Saccharomyces. II. *Metabolism and gene expression*. (ed. J. Strathern et al.), Cold Spring Harbor Laboratory, Cold Spring Harbor, New York. (In press.)

Kassir, Y. and I. Herskowitz. 1980. A dominant mutation (*SAD*) bypassing the requirement for the **a** mating type locus in yeast sporulation. *Mol. Gen. Genet.* **180**: 315.

Kassir, Y. and G. Simchen. 1976. Regulation of mating and meiosis in yeast by the mating-type region. *Genetics* **82**: 187.

Klar, A.J.S. 1980. Interconversion of yeast cell types by transposable genes. *Genetics* **95**: 631.

Klar, A.J.S. and S. Fogel. 1977. The action of homothallism genes in *Saccharomyces* diploids during vegetative growth and the equivalence of *hm***a** and *HM*α loci functions. *Genetics* **85**: 407.

———. 1979. Activation of mating type genes by transposition in *Saccharomyces cerevisiae*. *Proc. Natl. Acad. Sci.* **76**: 4539.

Klar, A.J.S., S. Fogel, and K. MacLeod. 1979a. *MAR1*—A regulator of *HM***a** and *HM*α loci in *Saccharomyces cerevisiae*. *Genetics* **93**: 37.

Klar, A.J.S., S. Fogel, and D.N. Radin. 1979b. Switching of a mating-type **a** mutant allele in budding yeast *Saccharomyces cerevisiae*. *Genetics* **92**: 759.

Klar, A.J.S., J.N. Strathern, J.R. Broach, and J.B. Hicks. 1981. Regulation of transcription of expressed and unexpressed mating type cassettes of yeast. *Nature* (in press).

Kushner, P.J. 1979. "Control of cell type by mobile genes in yeast." Ph.D. thesis, University of Oregon, Eugene.

Kushner, P.J., L.C. Blair, and I. Herskowitz. 1979. Control of yeast cell types by mobile genes—A test. *Proc. Natl. Acad. Sci.* **76**: 5264.

Leibowitz, M.J. and R.B. Wickner. 1976. A chromosomal gene required for killer plasmid expression, mating, and spore maturation in *Saccharomyces cerevisiae*. *Proc. Natl. Acad. Sci.* **73**: 2061.

Lemontt, J.F. 1980. Genetic and physiological factors affecting repair and mutagenesis in yeast. In *DNA repair and mutagenesis in eukaryotes* (ed. F.J. deSerres et al.), p. 85. Plenum Press, New York.

Leupold, U. 1959. Studies on recombination in *Schizosaccharomyces pombe*. *Cold Spring Harbor Symp. Quant. Biol.* **23**: 161.

———. 1980. Transposable mating-type genes in yeasts. *Nature* **283**: 811.

Lindegren, C.C. and G. Lindegren. 1943. A new method for hybridizing yeast. *Proc. Natl. Acad. Sci.* **29**: 306.

MacKay, V.L. and T.R. Manney. 1974a. Mutations affecting sexual conjugation and related processes in *Saccharomyces cerevisiae*. Isolation and phenotypic characterization of nonmating mutants. *Genetics* **76**: 255.

———. 1974b. Mutations affecting sexual conjugation and related processes in *Saccharomyces cerevisiae*. II. Genetic analysis of nonmating mutants. *Genetics* **76**: 273.

Malone, R.E. and R.E. Esposito. 1980. The *RAD52* gene is required for homothallic interconversion of mating types and spontaneous mitotic recombination in yeast. *Proc. Natl. Acad. Sci.* **77**: 503.

Mascioli, D.W. and J.E. Haber. 1980. A *cis*-acting mutation within the *MAT*a locus of *Saccharomyces cerevisiae* that prevents efficient homothallic mating type switching. *Genetics* **94**: 341.

McClintock, B. 1957. Controlling elements and the gene. *Cold Spring Harbor Symp. Quant. Biol.* **21**: 197.

Nasmyth, K.A. and K. Tatchell. 1980. The structure of transposable yeast mating type loci. *Cell* **19**: 753.

Nasmyth, K.A., K. Tatchell, B.D. Hall, C. Astell, and M. Smith. 1981. A physical analysis of mating type loci in *Saccharomyces cerevisiae*. *Cold Spring Harbor Symp. Quant. Biol.* **45**: 961.

Naumov, G.I. and I.I. Tolstorukov. 1973. Comparative genetics of yeast. X. Reidentification of mutators of mating types in *Saccharomyces*. *Genetika* **9**: 82.

Oshima, T. and I. Takano. 1980. Mutants showing heterothallism from a homothallic strain of *Saccharomyces cerevisiae*. *Genetics* **94**: 841.

Oshima, Y. 1982. Regulatory circuits for gene expression: The metabolism of galactose and phosphate. In *The molecular biology of the yeast* Saccharomyces. II. *Metabolism and gene expression* (ed. J. Strathern et al.), Cold Spring Harbor Laboratory, Cold Spring Harbor, New York (In press.)

Oshima, Y. and I. Takano. 1971. Mating types in *Saccharomyces*: Their convertibility and homothallism. *Genetics* **67**: 327.

―――. 1972. Genetic controlling system for homothallism and a novel method for breeding triploid cells in *Saccharomyces*. In *Fermentation technology today* (ed. G. Terui), p. 847. Society of Fermentation Technology, Osaka, Japan.

Rine, J.D. 1979. "Regulation and transposition of cryptic mating type genes in *Saccharomyces cerevisiae*." Ph.D. thesis, University of Oregon, Eugene.

Rine, J.D., J.N. Strathern, J.B. Hicks, and I. Herskowitz. 1979. A suppressor of mating type locus mutations in *Saccharomyces cerevisiae*: Evidence for and identification of cryptic mating type loci. *Genetics* **93**: 877.

Rine, J., R. Jensen, D. Hagen, L. Blair, and I. Herskowitz. 1981. Pattern of switching and fate of the replaced cassette in yeast mating-type interconversion. *Cold Spring Harbor Symp. Quant. Biol.* **45**: 95.

Sherman, F. 1982. Supression in the yeast *Saccharomyces cerevisiae*. In *The molecular biology of the yeast* Saccharomyces. II. *Metabolism and gene expression* (ed. J. Strathern et al.), Cold Spring Harbor Laboratory, Cold Spring Harbor, New York (In press.)

Sprague, G.F.J., J. Rine, and I. Herskowitz. 1981. Homology and non-homology at the yeast mating type locus. *Nature* **289**: 250.

Santa Maria, J. and D. Vidal. 1970. Segregación anormal del "mating type" en *Saccharomyces*. *Inst. Nac. Invest. Agron. Conf.* **30**: 1.

Strathern, J.N. 1977. "Regulation of cell type in *Saccharomyces cerevisiae*." Ph.D. thesis, University of Oregon, Eugene.

Strathern, J.N. and I. Herskowitz. 1979. Asymmetry and directionality in production of new cell types during clonal growth: The switching pattern of homothallic yeast. *Cell* **17**: 371.

Strathern, J.N., L.C. Blair, and I. Herskowitz. 1979a. Healing of *mat* mutations and control of mating type interconversion by the mating type locus in *Saccharomyces cerevisiae*. *Proc. Natl. Acad. Sci.* **76**: 3425.

Strathern, J.N., J.B. Hicks, and I. Herskowitz. 1981. Control of cell type by the mating type locus: The α1-α2 hypothesis. *J. Mol. Biol.* **147**: 357.

Strathern, J.N., C.S. Newlon, I. Herskowitz, and J.B. Hicks. 1979b. Isolation of a circular derivative of yeast chromosome III: Implications for the mechanism of mating type interconversion. *Cell* **18**: 309.

Strathern, J.N., E. Spatola, C. McGill, and J.B. Hicks. 1980. The structure and organization of the transposable mating type cassettes in *Saccharomyces*. *Proc. Natl. Acad. Sci.* **77**: 2829.

Takahashi, T. 1958. Complementary genes controlling homothallism in *Saccharomyces*. *Genetics* **43**: 705.

Takano, I. and K. Arima. 1979. Evidence of insensitivity of the α-*inc* allele to the function of the homothallic genes in Saccharomyces yeasts. *Genetics* **91**: 245.

Takano, I. and Y. Oshima. 1967. An allele specific and a complementary determinant controlling homothallism in *Saccharomyces oviformis*. *Genetics* **57**: 875.

————. 1970a. Allelism tests among various homothallism-controlling genes and gene systems in Saccharomyces. *Genetics* **64**: 229.

————. 1970b. Mutational nature of an allele-specific conversion of the mating type by the homothallic gene *HOα* in Saccharomyces. *Genetics* **65**: 421.

Takano, I., T. Kusumi, and Y. Oshima. 1973. An α mating-type allele insensitive to the mutagenic action of the homothallic gene system in *Saccharomyces diastaticus*. *Mol. Gen. Genet.* **126**: 19.

Takano, I., T. Oshima, S. Harashima, and Y. Oshima. 1977. Tetraploid formation through the conversion of the mating-type alleles by the action of homothallic genes in the diploid cells of *Saccharomyces* yeasts. *J. Ferment. Technol.* **55**: 1.

Tkacz, J.S and V.L. MacKay. 1979. Sexual conjugation in yeast. Cell surface changes in respon the action of mating hormones. *J. Cell Biol.* **80**: 326.

Wickner, R.B. 1974. Mutants of *Saccharomyces cerevisiae* that incorporate deoxythymidine-5'-monophosphate into deoxyribonucleic acid *in vivo*. *J. Bacteriol.* **117**: 252.

Winge, Ö. 1935. On haplophase and diplophase in some Saccharomyces. *C. R. Trav. Lab. Carlsberg Ser. Physiol.* **21**: 77.

Winge, Ö. and O. Laustsen. 1937. On two types of spore germination, and on genetic segregation in *Saccharomyces*, demonstrated through single-spore cultures. *C. R. Trav. Lab. Carlsberg Ser. Physiol.* **22**: 99.

Winge, Ö. and C. Roberts. 1949. A gene for diploidization in yeast. *C. R. Trav. Lab. Carlsberg Ser. Physiol.* **24**: 341.

————. 1958. Life history and cytology of yeasts. In *The chemistry and biology of yeasts* (ed A.H. Cook), p. 93. Academic Press. New York.

Meiosis and Ascospore Development

Rochelle Easton Esposito and Sue Klapholz
Department of Biology
The University of Chicago
Chicago, Illinois 60637

INTRODUCTION

Sporulation in yeast includes meiosis and ascospore development. It has been the focus of numerous studies for two primary reasons. First, it provides a relatively simple model system for the investigation of eukaryotic differentiation and the manner in which a complex series of biochemical, morphological, and genetic events are coordinated into a successful developmental pathway. Second, two events of major genetic consequence occur during meiosis: genetic recombination and chromosome segregation. Both of these events play a profound role in the generation of new genotypes and euploid genomes during sexual reproduction. Despite the central importance of the meiotic process, specific knowledge of the genetic and biochemical control of gametogenesis in eukaryotic organisms is very limited.

Utility of Yeast For Studies of Sporulation

As an experimental system, yeast presents the opportunity to study meiosis and gamete development in an organism that possesses all of the technical advantages of microbial systems, while exhibiting chromosome behavior typical of higher eukaryotic cells. It has a number of attractive features that make it particularly well-suited for an analysis of meiotic cell differentiation: (1) Yeast has well-developed genetics and is readily manipulated biochemically. (2) Large numbers of single cells can be stimulated to undergo meiosis in a defined medium. (3) Meiosis can be interrupted and viable cells recovered at various stages of development. (4) All of the meiotic products of a given meiosis can be recovered in association with one another, permitting precise reconstruction of exchange and segregation events. (5) Aspects of the life cycle and availability of selective systems permit isolation of mutants defective in the formation of final meiotic products, as well as mutants specifically altered in exchange and chromosome disjunction. (6) The presence of well-characterized mitotic recombination events, e.g., spontaneous and induced mitotic recombination, genetic transformation, and mating-type interconversion, provide useful probes to analyze the interactions among gene functions involved in meiotic recombination and mitotic processes.

There are also several limitations in the manipulation of yeast that have slowed progress in defining the parameters of sporulation. One is that cells become relatively impermeable to the uptake of labeled precursors as sporulation progresses, due to an increase in the pH of the medium. Although improved isotopic labeling can be achieved by lowering the pH, this procedure reduces sporulation and alters the kinetics of macromolecular synthesis. The extensive RNA and protein degradation that occurs during sporulation further complicates quantitation of the rates of macromolecular synthesis, because both the cellular pools and uptake of precursors are continually changing. In addition, not all of the cells in a population complete sporulation. Even in the most competent genetic backgrounds selected for high sporulation proficiency, and under optimum sporulation conditions, 15–25% of cells do not sporulate. These may never enter the sporulation process or may be arrested at various stages and eventually undergo cell lysis. Those cells that do sporulate do so asynchronously, usually taking between 12 and

24 hours from the time spores are first observed until maximum spore production has been achieved. Once spore-wall formation has been initiated during meiosis II, disruption of the cells becomes more difficult. Finally, individual chromosomes are not visible by light or electron microscopy, and their behavior can only be inferred from genetic experiments and cytological observations of synaptonemal complex (SC) formation in prophase of meiosis I. Efforts to overcome these difficulties have met with some success. Furthermore, it is clear that even with the limitations described above, the specific advantages of yeast have made it presently one of the more useful systems for analysis of the genetic and biochemical control of gametogenesis in eukaryotes.

Rationales for and Approaches to the Study of Sporulation

Studies of yeast meiosis and spore formation have been concerned with the following long-range goals: (1) determination of the morphogenic, biochemical, and genetic events that occur in wild-type cells during sporulation; (2) identification of genes that are indispensable for sporulation and of the roles of meiotic functions in other cellular processes; (3) determination of the stage(s) at which these genes have their effects and the dependency relationships among them; (4) distinction between events that are both necessary and unique to sporulating cells and those that are required, but not sporulation-specific; (5) identification of sporulation-specific gene products and their functions during sporulation; (6) definition of the events that trigger meiotic development and the mechanisms that control gene expression during the sporulation process.

The overall strategies in attaining these goals are based on several complementary experimental approaches. The first includes characterization of the normal sequence of landmark events of the sporulation process by cytological, biochemical, and genetic studies of wild-type cells as they proceed through meiosis and spore formation. The second involves perturbation of the process by physiological or genetic methods through the use of inhibitors, change of culture conditions, or mutations. The third entails analysis of the end products of meiotic development in order to infer events that have occurred in their formation. The fourth involves the cloning of genes required for sporulation to understand the level of regulation of gene expression during development and, ultimately, to relate specific gene products to their function(s) during meiosis.

The key aim in all of the studies performed thus far has been to develop appropriate criteria for defining sporulation-specific functions in biochemical terms. This is a crucial point in evaluating whether there are genes or gene products that function only during sporulation, independent of the cellular response to the starvation conditions that stimulate meiotic development, and in understanding the regulation of this process. In most cases, controls have involved comparing the behavior of a/α, a/a, and α/α cells during sporulation. Events occurring only in a/α cells are considered to be sporulation-specific. Those present in all cell types are viewed as nonspecific and probably related to starvation. This assumption is based on the fact that a/a and α/α cells do not undergo premeiotic DNA replication and are presumed to be blocked early in the process. Studies of sporulation mutants have indicated, however, that a defect early in meiosis need not result in a complete block of subsequent events. Hence, the dismissal of events observed in all cell types during sporulation as nonspecific may be inappropriate. Moreover, events identified as a/α-specific need not be confined to the sporulation process but may be under mating-type control during vegetative growth as well.

An alternative approach has been to select for mutations in sporulation-specific genes by screening for mutants that sporulate under nutritional conditions or in genetic backgrounds normally not conducive to spore formation. This rationale is based on the expectation that mutations would be recovered that "turn on" sporulation-specific gene functions. However, the same problems in evaluating the specificity of the gene functions apply, e.g., such a mutation may simply act by inducing a starvation response. Additionally, its wild-type allele may participate in other aspects of cell metabolism. Developing methodologies to identify gene functions unique to the sporulation process therefore remains an outstanding problem at the present time.

Overview of Sporulation

Meiosis and spore formation in yeast are generally triggered by starvation conditions. Typically, diploids and cells of higher ploidy undergo sporulation when both $MATa$- and $MAT\alpha$-mating-type information is present. In certain genetic backgrounds, however, both haploids and mating-type homozygotes can complete meiotic development. Following entry into meiosis from the G_1 stage of the mitotic cell cycle, chromosomes undergo premeiotic DNA synthesis, pairing, genetic recombination, and reductional and equational segregation. The morphological changes associated with chromosome behavior during meiosis center on development of the SC and assembly of the spindle apparatus (see Byers, this volume). This includes spindle pole body (SPB) duplication and separation and microtubule assembly in the two divisions. The SPBs become modified in meiosis II and serve to nucleate spore-wall formation. As the spore walls grow and close, they envelop four haploid genomes. The spore walls mature, and asci become visible in the light microscope. Spore number per ascus and spore ploidy depend upon nutritional conditions and cell genotype.

During sporulation, considerable macromolecular synthesis and degradation occur. Although sporulation-specific transcription has been detected, the identification of sporulation-specific proteins has been a difficult problem. Mutant analysis indicates that most of the gene functions required for the mitotic cell-division cycle (*CDC* genes) are also required for meiosis, except those involved in bud initiation and cytokinesis (Pringle and Hartwell, this volume). A number of other genes involved in aspects of vegetative metabolism, such as DNA repair and respiration, are also required for sporulation. In addition, a systematic search for conditional mutants defective in spore production suggests that there are 50–100 genes that are indispensable for sporulation but nonessential for mitotic cell division.

This paper is intended to be a comprehensive review of progress in understanding both the biochemistry and genetic control of sporulation in yeast over the last 10 years. We will not attempt to cover early work on sporulation, which dealt largely with developing conditions to maximize spore production (for reviews, see Miller 1959; Phaff et al. 1966; Fowell 1969, 1975). Several other reviews have been written on the genetics and physiology of sporulation in yeast (Haber and Halvorson 1972b, 1975; Tingle et al. 1973; M. S. Esposito and R. E. Esposito 1974, 1975, 1978; Haber et al. 1975, 1977; for general reviews of meiosis, see also Baker et al. 1976; Heywood and Magee 1976).

ENTRY INTO MEIOSIS

Conditions for Sporulation

Sporulation generally requires the presence of both *MAT***a**- and *MAT*α-mating-type alleles (Roman and Sands 1953). It occurs during nitrogen deprivation, in the presence of a nonfermentable carbon source such as acetate, in cells adapted to respiration. The conditions for sporulation in yeast are similar to those that promote spore formation in bacteria (see reviews by Fowell 1975; Haber and Halvorson 1975).

Depending upon the strain and culture conditions, a low to moderate level of asci may be formed in stationary phase, particularly when acetate or glycerol is used as a carbon source (Dawes 1975; Fast 1978). Maximal sporulation, however, occurs when cells are transferred from growth medium to sporulation medium. The sporulation medium most commonly used contains simply 1% potassium acetate and the auxotrophic requirements of the strain, at neutral or near-neutral pH. Under these conditions, no cell division occurs other than completion of the mitotic cycle initiated in the presporulation growth medium (Croes 1967a). Alternatively, a small amount of glucose (0.1%) and yeast extract (0.25%) may be added to permit a few cell divisions prior to sporulation (McClary et al. 1959; Croes 1967a). Sporulation occurs over a broad range of temperatures with an optimum at approximately 30°C;

ascus production is poor above 35°C and reduced below 25°C (Esposito and Esposito 1969).

The effects of various carbon and nitrogen sources on sporulation were examined in detail by J. J. Miller and co-workers. They demonstrated both glucose and nitrogen repression of sporulation and maximum spore formation in the presence of acetate (Adams and Miller 1956; Miller and Halpern 1956; Miller 1957, 1963a,b, 1964; Miller and Hoffmann-Ostenhof 1964). Glucose inhibition of sporulation has been interpreted as being due to repression of TCA-cycle enzymes required for acetate utilization (Miyake et al. 1971). The effect may also be due to its known inhibition of gluconeogenic enzymes required for the synthesis of storage carbohydrates, which accumulate during sporulation (see Fraenkel 1982). In addition to glucose, high levels of other sugars (e.g., fructose, maltose, mannose, sucrose), as well as CO_2, also inhibit sporulation (Adams and Miller 1954; Miller 1957).

Nitrogen repression of sporulation is less well understood and appears to be under complex genetic control (Piñon 1977; Vezinhet et al. 1979; see also Cooper 1982). The inhibitory effect of ammonium salts on sporulation is similar to observations made in other organisms where NH_4^+ has been shown to block cell development, e.g., gametogenesis in *Chlamydomonas reinhardii* (Sager and Granick 1954), heterocyst formation in blue-green algae (Fleming and Haselkorn 1974), and sporulation in *Bacillus megaterium* (Schaeffer et al. 1965). A level of 2–4 mM $(NH_4)_2SO_4$, approximately threefold less than the amount added to vegetative growth medium, generally inhibits sporulation. Piñon (1977) has demonstrated that spore production is maximally inhibited when NH_4^+ is added during premeiotic S and prophase of meiosis I. Although there is some dispute as to whether premeiotic DNA replication initiates in the presence of NH_4^+ (Piñon 1977; Croes et al. 1978b), many other processes are inhibited that normally occur during sporulation. These include the induction of glyoxylate-cycle enzymes and adaptation to acetate utilization (Gosling and Duggan 1971), RNA and protein synthesis (Durieu-Trautmann and Delavier-Klutchko 1977), protein degradation (Croes et al. 1978b; Opheim 1979), glycogen degradation (Fonzi et al. 1979), and the production of trinucleate and tetranucleate cells (Miller 1963a). NH_4^+ has no effect once cells have become committed to spore formation (Piñon 1977). Although the basis of the nitrogen effect is presently unclear, it has been suggested that the inhibition of sporulation is due directly to the presence of ammonium ions rather than to derivatives synthesized from it. Among other nitrogen-containing compounds examined, only glutamine and methylamine, a nonmetabolizable analog of ammonia, showed similar repression effects; glutamate had no effect (Piñon 1977). The possibility that NADH-dependent glutamic dehydrogenase (GDH) is involved has been ruled out since nonsense mutations in the gene coding for GDH have no effect on nitrogen sensitivity (Newlon 1979), and this enzyme is degraded early during sporulation (Betz 1977).

Is acetate necessary for sporulation? The available data suggest that it is, although its presence is not required continuously during the process. Acetate carbon is utilized in macromolecular synthesis throughout sporulation and, under limiting acetate conditions, the number of spores per ascus is reduced (Esposito et al. 1969; Fast 1973). However, the fact that exposure to acetate for brief periods (3-7 hr), followed by incubation in water, will promote sporulation indicates that exogenous acetate is not required throughout sporulation; the yield of asci in this case depends upon the length of the incubation period in acetate (Darland 1969; Simchen et al. 1972).

Sporulation Competency and Cell-cycle Dependency

The conditions required to sporulate cells indicate that respiratory competency and sufficient macromolecular reserves are important factors in inducing spore formation. Two procedures have been used to optimize ascus production. One employs glucose-grown cells harvested from early stationary phase during the shift from fermentation to respiratory metabolism (Croes 1967b). The other employs respiratory-competent cells grown in acetate medium and harvested during exponential phase (Roth and Halvorson 1969). In the former regimen, the ability of cells to undergo meiosis is directly proportional to oxygen consumption (QO_2) upon transfer to sporulation medium and inversely proportional to the amount of glucose remaining in the medium at the time of shift (Croes 1967b). At early stationary phase, cells respire the acetate and evolve CO_2 without any apparent lag (Esposito et al. 1969). Sporulation competency also depends upon the RNA and protein content per cell, and decreases in stationary phase when turnover of these macromolecules occurs (Halvorson 1958a,b; Croes 1967b). The latter regimen employs a steady-state log culture and, hence, a more uniform cell population and usually results in more extensive and synchronous sporulation. Generally, as cells approach stationary phase, sporulation competency declines. This decline is most likely due to turnover of macromolecules, as has been shown in stationary-phase, glucose-grown cells. The precise period during growth of a culture when sporulation competency is maximal by either procedure is highly strain-dependent and not necessarily correlated with growth rate in a particular carbon source. The sporulation-negative phenotype of some strains can in fact be reversed by simply altering the time at which cells are transferred to sporulation medium (see *SUP3*, Table 1). The yield of asci also depends upon cell density and acetate concentration (Fowell 1975). Generally, 1×10^7 cells/ml to 1×10^8 cells/ml can be sporulated efficiently in 1% KAc.

In most strains, sporulation is not synchronous. The time of initiation of spore production is similar, regardless of the method used to sporulate cells, and occurs approximately 10 hours after exposure to sporulation medium. In the glucose growth–acetate sporulation procedure, maximum ascus production is achieved by 36–48 hours, whereas in the acetate growth–acetate sporulation regimen, final yields of mature spores are reached by 24 hours. No major differences have been observed in the time of initiation of various other biochemical, genetic, and cytological landmark events using these two procedures (Hopper et al. 1974). The primary effects seem to be restricted to the synchrony with which these events occur. Stationary-phase cells grown in a 10% glucose medium, followed by sporulation at high cell density (2×10^8 cells/ml to 3×10^8 cells/ml), has also been used as a method to improve sporulation synchrony (Petersen et al. 1978). The most rapid sporulation, however, has been observed in strain SK1, in which spore formation occurs over a 2-hour period (Kane and Roth 1974). This property can be transmitted in crosses and should provide a useful tool for obtaining more synchronous populations to precisely define the time and duration of various landmark events.

Studies employing both homothallic and heterothallic diploid strains have demonstrated that cells enter meiosis from the G_1 stage of the mitotic cell cycle (M.S. Esposito and R.E. Esposito 1974; Hartwell 1974; Shilo et al. 1978). One report suggests that cells can also enter meiosis after the S phase without undergoing premeiotic DNA synthesis, but this has not yet been substantiated (Croes et al. 1976). Within a short period of time following introduction of cells into sporulation medium, the proportion of unbudded cells increases to nearly 100%. Budded cells that contain an ascus and a bud without a nucleus are very rare, indicating that cells distributed throughout the mitotic cell cycle usually complete mitosis before they sporulate (M.S. Esposito and R.E. Esposito 1974). Several investigators have also shown that homogeneous cell fractions sporulate in a more synchronous fashion and that sporulation competency varies in cells at different stages of the cell cycle. Maximum sporulation competency is observed in mother cells at cell separation. The mother cells at this time are usually larger and sporulate to a greater extent than the buds (Yanagita et al. 1970; Haber and Halvorson 1972a; Sando et al. 1973; C. Milne, cited in Hartwell 1974; Tsuboi and Yanagishima 1974).

The sporulation process rarely occurs with 100% efficiency. Some cells may not form asci at all, whereas others may produce asci containing fewer than four spores. The proportion of cells containing zero, one, two, three, or four ascospores is dependent upon both physiological factors (noted below) and genetic factors (see below, Genetic Control of Meiosis and Spore Formation). The physiological factors include (1) the presporulation (growth) and sporulation media composition, (2) cell age, (3) respiratory sufficiency, and (4) cell density during sporulation. For example, short exposures to acetate followed by incubation in water can result in nearly 100% two-spored asci among sporulated cells (S. Klapholz, unpubl.). Increased levels of two-spored asci also occur in cells exposed to glucose during sporulation (Miller and

Halpern 1956), in small cells or buds that exhibit overall poorer sporulation than larger mother cells (Haber and Halvorson 1972a), in sporulating cells lacking mitochondrial DNA due to exposure to ethidium bromide during the last one or two divisions in acetate presporulation medium (Kuenzi et al. 1974), and in populations sporulated at high cell density (Davidow et al. 1980). In contrast, under certain conditions, the number of spores per ascus can also be increased. For example, in a natural variant called 19e1, which normally forms asci containing only two spores, Bilinski and Miller (1980) made the intriguing discovery that pregrowth in the presence of high levels of glucose (6%), followed by sporulation in acetate medium, resulted in the production of up to 20% three- and four-spored asci among total asci (see *spo12*, Table 1). A further increase to over 70% three- and four-spored asci was achieved when zinc sulfate was added to both presporulation and sporulation media. The basis of the zinc effect is unkown.

Commitment Stages and Trigger Events

Commitment to sporulation is generally defined by interrupting the normal process of development by changing culture conditions (e.g., media, temperature, addition of inhibitors, etc.) to those not usually conducive to ascospore formation. The ability of a cell to complete sporulation after the point of disruption is taken as an indication that a prior event(s) committing the cell to meiosis has occurred. Completion of the sporulation process is monitored in two ways: (1) by microscopic examination of cultures for the presence of ascospores and (2) by plating on media to detect the formation of haploid meiotic products, e.g., by assaying the segregation of a heterozygous recessive drug-resistance marker such as *can1*.

Thus far, three developmental stages have been described with respect to commitment to sporulation (Simchen et al. 1972). The first stage, termed "readiness," is defined as the period when cells after exposure to sporulation medium are able to sporulate in water without further stimulus by the medium. Readiness begins after 3 hours of exposure to sporulation medium and lasts until about 7 hours. If cells are returned to growth medium before 7 hours, they undergo mitosis without forming spores. At this point, the cells are not yet irreversibly committed to haploidization and spore formation. Simchen et al. (1972) have suggested that readiness is followed by a second stage, designated "precommitment," characterized by a brief loss of plating efficiency on growth medium. At this stage cells exhibit readiness but are thought to lose their ability to return to vegetative growth in nutrient medium, prior to acquiring the capability of completing sporulation in such medium. The third stage, "commitment to sporulation," was previously defined by Kirsop (1954) and Ganesan et al. (1958). At this stage, cells form asci even when challenged with growth medium. Irreversible commitment to sporulation occurs after premeiotic DNA replication and initiation of commitment to genetic recombination, coincident with SPB separation at pachytene of meiosis I (R. Esposito and M. Esposito 1974; Horesh et al. 1979). Commitment to completion of the mitotic cell cycle also occurs at a time associated with SPB separation, although in mitosis this takes place prior to the S phase (Hirschberg and Simchen 1977). These results suggest that SPB separation or an event closely associated with spindle development may be involved in irreversibly fixing the mode of cell division.

Is there a single event that commits cells to the entire process of meiosis and ascospore development? Two lines of evidence argue against this view. First, there appears to be a series of individual commitments to different landmark events, rather than one early commitment event for the whole process, as has also been determined for bacterial sporulation (Mandelstam 1971). It has been possible, for example, to separate commitment to the two primary genetic events of meiosis, genetic recombination and chromosome segregation, by either genetic means or perturbation of wild-type meiosis (Sherman and Roman 1963; Esposito et al. 1974; Klapholz and Esposito 1980b). Second, the stage of commitment to a particular event depends upon the composition of the sporulation medium and the medium to which cells are transferred (Freese et al. 1975). For example, the behavior of sporulating yeast cells transferred to water (Simchen et al. 1972) and various growth media (S. Klapholz, unpubl.) supports the general conclusion that cells can reverse spore development and resume cell division more readily in a rich growth medium than in a poor one. The observations described above suggest that commitment to a specific landmark is a complex multistep process, involving (1) development of a competent state in which preconditions, necessary but not sufficient for the landmark to be executed, are satisfied, followed by (2) one or more reversible gene functions directly involved in execution of the landmark and, finally, (3) an irreversible event that commits the cell to complete the landmark. Commitment to at least two landmarks, genetic recombination and spore formation, has been shown to include both reversible and irreversible substages (see below, Genetic Recombination).

What are the events that trigger the initiation of sporulation? At present, little is known about the nature of these events other than that cells initiate meiotic development from G_1 of the cell cycle. An early proposal, that the absence of nitrogen triggers sporulation (Miller 1963a,b), does not appear to be correct, as the presence of nitrogen during the first 4 hours has no effect on the further progress of sporulation. Croes (1967b) has suggested that the absence of nitrogen combined with incubation in a poor carbon source triggers meiotic development by causing a state of metabolic unbalance in which the glyoxylate cycle fails to produce sufficient levels of amino acids for protein synthesis. This insufficiency was postulated to stimulate proteolysis which, in turn, would trigger other events required for sporulation. The fact that the addition of glyoxylate enhances the level of sporulation (Bettleheim and Gay 1963) has been interpreted as support for this view (Croes 1967b).

However, it should be noted that glyoxylate also promotes mitotic division in sporulation medium, perhaps increasing the proportion of cells capable of entering meiosis from G_1.

MORPHOLOGICAL DEVELOPMENT

Chromosome Behavior

Morphological development during sporulation of yeast has been followed by light microscopy (Fig. 1), fluorescence microscopy (Fig. 2), and electron microscopy of thin-sectioned or freeze-etched material (see Byers, this volume). The segregation of chromatin during the two meiotic divisions is readily visualized by light microscopy, employing a DNA stain such as Giemsa (Ganesan 1959; Pontefract and Miller 1962), and by fluorescence microscopy using mithramycin (Slater 1976), 33258 Hoechst (Lemke et al. 1978), or diamidino-2-phenylindole (DAPI) (Davidow et al. 1980). These procedures permit classification of cells into several stages: mononucleate, first nuclear division, binucleate, second nuclear division, and tetranucleate (Fig. 2).

Electron microscopy studies reveal that in *Saccharomyces cerevisiae*, as in other Ascomycetes (Zickler 1970), chromatin segregation takes place without dissolution (Hashimoto et al. 1960; Mundkur 1961) or division (Moens and Rapport 1971b; Guth et al. 1972) of the parental nucleus. Toward the end of the second meiotic division, the nuclear membrane "buds" at each of the four poles, and a prospore wall initiates development at each nuclear lobe (Moens 1971; Moens and Rapport 1971b). At prospore-wall closure, the four asco-

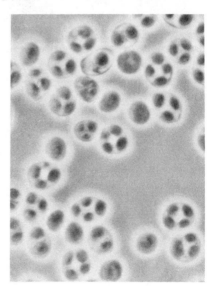

Figure 1 A wild-type diploid (K65-3D) after 72 hr in sporulation medium. Both small unsporulated cells and mature asci are present. Magnification, 1450 ×, with phase optics. (Courtesy of J. Wagstaff and S. Klapholz).

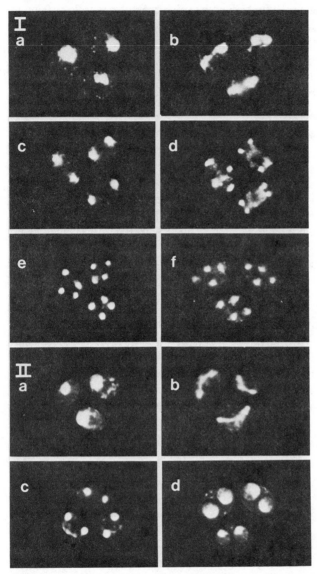

Figure 2 Sporulating cells of *S. cerevisiae* stained with DAPI, a DNA-specific fluorescent dye. *I a–f* show the stages of sporulation in tetrad-producing strains: (*a*) mononucleate; (*b*) first nuclear division; (*c*) binucleate; (*d*) second nuclear division; (*e*) tetranucleate; (*f*) asci. *II a–d* show the stages of sporulation in a dyad-producing strain (ATCC4117): (*a*) mononucleate; (*b*) nuclear division; (*c*) binucleate; (*d*) asci. Electron microscopy studies indicate that the nuclear envelope does not break down and does not actually divide until the completion of the second meiotic division when prospore-wall closure occurs (see text for details). Brightly staining dots in the cytoplasm represent mitochondrial DNA (Williamson and Fennell 1975). (Courtesy of S. Klapholz.)

spore nuclei are pinched off from the parental nucleus. The morphological stages observed by electron microscopy are summarized in Figure 3.

There have been no convincing demonstrations of individual condensed chromosomes with either light or electron microscopy. Our understanding of bivalent formation during meiosis-I prophase has been obtained solely from electron microscopy studies of SC morphogenesis. A number of these studies

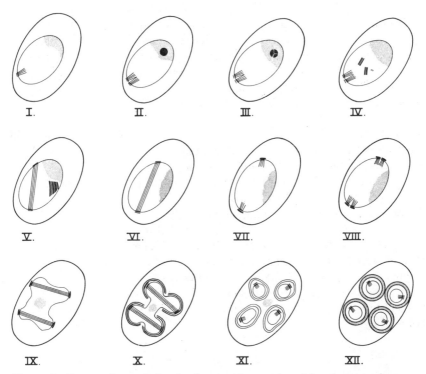

Figure 3 Stages of sporulation in *S. cerevisiae* as viewed by electron microscopy. Details are presented in the text. (*I*) A cell with a nucleus containing a single unduplicated SPB, with its associated microtubules radiating into the nucleoplasm. The stippled region represents the nucleolus. (*II*) SPB duplication and the appearance of the dense body (DB). (*III*) Appearance of SC-like elements within the DB. (*IV*) Disappearance of the DB and appearance of SCs in the nucleoplasm. (*V*) Formation of an intranuclear meiosis-I spindle, disappearance of the SCs, and appearance of a single polycomplex body (PCB). (*VI*) Disappearance of the PCB. (*VII*) Enlargement of the outer layer (plaque) of each SPB and breakdown of the meiosis-I spindle. (*VIII*) Duplication of the two SPBs. (*IX*) Formation of two intranuclear meiosis-II spindles. (*X*) Prospore-wall growth around each nuclear bud, initiated at the outer plaque of each SPB. (*XI*) Prospore-wall closure, separating the nucleus into four separate haploid nuclei. The nucleolus is not incorporated into the spores. (*XII*) Insertion of spore-wall material between unit membranes of the prospore wall. Spores become spherical and mature.

have been facilitated by the use of temperature-sensitive *cdc* mutants that arrest during sporulation at a restrictive temperature (33.5°C or 34°C) and sporulate at a pace slower than normal at a permissive temperature (20°C) (Byers and Goetsch 1975; Schild and Byers 1978; Horesh et al. 1979).

The classical stages of meiotic prophase have been observed in *S. cerevisiae:* (1) leptotene (unpaired lateral elements), (2) zygotene (short pieces of completed SCs and lengths of unpaired lateral elements), and (3) pachytene (extensive lengths of SCs) (Zickler and Olson 1975). Byers and Goetsch (1975) traced SCs through the nuclei of serially sectioned cells in order to reconstruct the meiotic karyotype. Their cytological findings are consistent with current genetic data for 17 linkage groups (Mortimer and Schild 1980). Following pachytene, the individual complexes disappear, and a single polysynaptonemal complex (polycomplex body) is observed in the nucleus (Moens and Rapport 1971b; Zickler and Olson 1975; Horesh et al. 1979).

Prior to the appearance of SCs, a darkly staining spherical structure, the "dense body," is observed in the nucleus (Engles and Croes 1968; Moens and Rapport 1971a; Zickler and Olson 1975; Horesh et al. 1979). It is often, but not always, found in association with the nucleolus. The dense body is one of the earliest cytological landmarks of sporulation, visible prior to the onset of the premeiotic S phase (Zickler and Olson 1975; Horesh et al. 1979). At a slightly later stage, the dense body houses structures that look like SC central elements (Engles and Croes 1968; Moens and Rapport 1971a; Zickler and Olson 1975; Horesh et al. 1979). When SCs become visible in the nucleoplasm, the dense body disappears (Zickler and Olson 1975) or may remain, shrunken and devoid of the central elementlike structures (Horesh et al. 1979). Horesh et al. (1979) postulate that the central elements of the SCs are formed in the dense body and are then transported to lie between the paired homologs (see Fig. 3).

Spindle Development

When cells are introduced into sporulation medium, a single SPB is generally observed, situated within a discontinuity in the nuclear envelope (Moens and Rapport 1971b; Peterson et al. 1972; Horesh et al. 1979). The SPB then duplicates, and the two SPBs remain in a side-by-side configuration (Moens and Rapport 1971b) until after the SCs are no longer visible (Zickler and Olson 1975; Horesh et al. 1979). The SPBs are next observed at opposite sides of the nucleus with the meiosis-I spindle running between them (Moens and Rapport 1971b; Peterson et al. 1972; Zickler and Olson 1975; Horesh et al. 1979). After the spindle has reached its maximal length, the outer plaque of each SPB enlarges. The SPBs then duplicate and separate, forming the two intranuclear meiosis-II spindles. The mechanisms of SPB duplication and separation remain obscure.

Spore-wall Synthesis and Maturation

The development of the ascospore walls has been investigated by a number of workers employing thin-sectioning or freeze-etching techniques for electron microscopy (Hashimoto et al. 1960; Mundkur 1961; Marquardt 1963; Osumi et al. 1966; Lynn and Magee 1970; Moens 1971; Moens and Rapport 1971b; Guth et al. 1972; Beckett et al. 1973; Illingworth et al. 1973; Davidow et al. 1980). The prospore wall, a double membrane structure, initiates development at the cytoplasmic side of each SPB during the second meiotic division (Moens 1971; Moens and Rapport 1971b). Davidow et al. (1980) provide strong evidence that the modified (enlarged) outer plaque of the meiosis-II SPB is necessary for prospore-wall formation. The investigators observed that under conditions of high-temperature arrest (36°C), followed by downshift to 23°C, two-spored asci are formed at elevated levels. Genetic and cytological analyses demonstrated that the two spores contain nonsister haploid nuclei, resulting from failure of two of the four meiosis-II products to be packaged into ascospores. Electron microscopy revealed cells in which two SPBs (always nonsisters) did not have enlarged outer plaques, whereas the other two were normal in morphology. Prospore-wall formation was initiated at the two normal SPBs but not at the two that were unmodified.

As the prospore wall grows and encloses the nuclear lobe, cytoplasmic material and organelles flow under its rim (Lynn and Magee 1970; Moens 1971; Guth et al. 1972). After closure, spore-wall growth occurs by the insertion of material between the two unit membranes of the prospore wall (Marquardt 1963; Lynn and Magee 1970; Beckett et al. 1973; Illingworth et al. 1973). Light microscopy examination of fixed cells stained with Sudan black B demonstrates that fat globules accumulate in the ascus cytoplasm during sporulation (Pontefract and Miller 1962). As spore-wall formation begins, masses of fat can be seen beside the maturing spores; in cells with newly formed spore walls, fat deposits appear to be in association with both the inner and outer spore-wall surfaces. Electron microscopy studies have extended this observation, demonstrating that lipid vesicles become associated with the outer spore-wall membrane and appear to be involved in spore-wall growth (Lynn and Magee 1970; Beckett et al. 1973; Illingworth et al. 1973). The inner membrane of the prospore wall becomes the spore plasmalemma; the outer membrane apparently contributes to the surface protein layer of the spore (Briley et al. 1970; Beckett et al. 1973; Illingworth et al. 1973).

Mitochondrial Behavior

During sporulation, there is a single mitochondrion per cell (Stevens 1978; see also Stevens, this volume). This organelle is highly branched and extends throughout much of the cytoplasm. By serial-section reconstructions of sporulating cells, Stevens (1978) observed that during prophase of meiosis I, most mitochondrial profiles are located close to the cell periphery. During the two meiotic divisions, they change in distribution and are largely found in proximity to the dividing nucleus. As prospore-wall formation begins, the single mitochondrion, which surrounds the parental nucleus (Stevens 1978; B. Byers and L. Goetsch, cited in Brewer and Fangman 1980), is distributed into the newly developing ascospores (Moens and Rapport 1971b; Zickler and Olson 1975; Stevens 1978). A significant portion of the mitochondrion, however, is not incorporated into the ascospores and remains behind in the ascus cytoplasm (Stevens 1978; Brewer and Fangman 1980). The distribution of the mitochondrion into the ascospores is apparently a nonrandom process; although the four spores (not including spore walls) take up 30% of the ascus space, 52% of the mitochondrial mass is enclosed within them (Brewer and Fangman 1980). Mature ascospores generally contain a single branched mitochondrion (Stevens 1978; Brewer and Fangman 1980).

MACROMOLECULAR SYNTHESIS

Permeability

During sporulation, there is a characteristic rise in medium pH, from pH 7 to approximately pH 9. The rise in pH is thought to be due to the accumulation of bicarbonate and carbonate ions as cells respire acetate and evolve CO_2 (Fowell 1969). The increase begins immediately after transfer of cells to sporulation medium and reaches a maximum after 7 hours, when commitment to the reductional division is first detected (Croes 1967a; Esposito et al. 1969; Mills 1972). Concomitant with the increase in medium pH, cells become relatively impermeable to the uptake of amino acids and nucleotide precursors (Mills 1972). Mills and co-workers attempted to increase cell permeability by lowering the pH to 6.0 during brief pulses with labeled precursors (Mills 1972; Curiale and Mills 1975; Curiale et al. 1976). This protocol dramatically improves label incorporaton and has minimal effects on ascus formation and the recovery of intragenic recombinants (Mills 1972; Hopper et al. 1974; Kuenzi and Roth 1974). However, it causes an increase in the rate of RNA and protein synthesis (Wejksnora and Haber 1976). To avoid such artifacts, McCusker and Haber (1977) developed a method of sporulating cells at a constant pH of 5.5 in medium buffered with succinic acid. The general applicability of this procedure remains to be determined, since substantial reductions in ascus production had been previously reported for a number of strains sporulated in buffered medium (Mills 1972; Arnaud and Galzy 1973; Mills et al. 1977). Alternative conditions have been used to lengthen the period in which cells are permeable by adjusting initial pH, acetate concentration, and cell density (Mills et al. 1977). Another approach has been to employ a transport mutant in which uptake is increased and does not limit incorporation of precursors into macromolecules (Croes et al. 1978a). Despite the difficulties in labeling cells, the various methods used to measure macromolecular synthesis have nevertheless led to rather similar conclusions about the timing of landmark biochemical events, which are summarized in Figure 4.

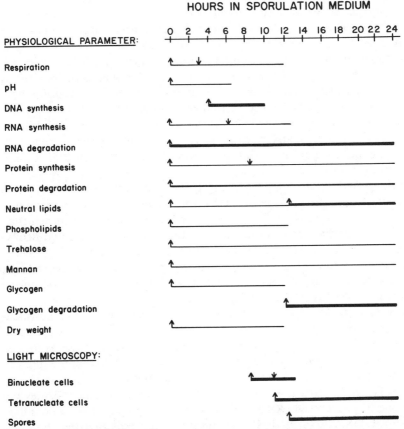

Figure 4 Timeline of physiological and light microscopy parameters. Thin lines indicate the period during which an event occurs in sporulating a/α cells, as well as in nonsporulating a/a or α/α cells incubated in sporulation medium. Bold dark lines indicate the period during which an event occurs uniquely in a/α cells. The intermediate line indicates a conflict in the literature regarding the mating-type specificity of the event. The timeline represents a composite of data from a number of independent studies on sporulation (cited in the text). It should be emphasized that both strain differences and sporulation conditions influence the specific kinetics of each of these events. Modified from Tingle et al. (1973), Hopper et al. (1974), and Fast (1978).

DNA Metabolism

Kinetics of Premeiotic DNA Replication

DNA synthesis during sporulation has been monitored by several procedures: the diphenylamine reaction, the DABA fluorometric test, and isotope incorporation (reviewed by Haber and Halvorson 1975). When logarithmic acetate-grown cells are shifted to sporulation medium, there is a small initial increase in incorporation of labeled precursors into DNA during the first 2 hours as cells complete mitotic rounds of replication. The DNA content per sporulating cell then doubles in a discrete period beginning 4–6 hours and ending 8–10 hours after transfer to sporulation medium (Croes 1966, 1967a; Esposito et al. 1969; Roth and Lusnak 1970; Simchen et al. 1972). Generally, the increase in DNA content in a culture is proportional to the number of cells that sporulate. Premeiotic DNA synthesis requires the presence of both a- and α-mating-type alleles, except in certain mutant genetic backgrounds (Roth and Lusnak 1970; see below, Genetic Control of Meiosis and Spore Formation). It also occurs in n + 1 $MATa/MAT\alpha$ cells, which are unable to complete meiosis and form viable ascospores (Roth and Fogel 1971; Kuenzi and Roth 1974).

Cells prelabeled during growth and transferred to unlabeled sporulation medium incorporate the label into DNA to the same extent and in a temporal pattern similar to cells placed in labeled sporulation medium (Simchen et al. 1972). This finding, and the observation that deoxynucleotide and deoxynucleoside pools decrease during the first 12 hours of sporulation (Sando and Miyaki 1971), has led to the suggestion that premeiotic replication utilizes precursors derived primarily from the extensive turnover of RNA during sporulation (Simchen et al. 1972).

Recently, the length of the premeiotic S phase was measured in individual cells by autoradiography (Williamson et al. 1980). The length of the premeiotic S phase was calculated at 65 ± 12 minutes, approximately twice as long as that in vegetative cells. In a number of other systems, the duration of premeiotic S has also been found to be longer than the mitotic S phase (for review, see Baker et al. 1976). In *Triturus*, this is due to fewer initiation sites per genome in meiotic cells (Callan 1972). In yeast, however, fiber autoradiography of DNA isolated during premeiotic S has indicated that both the replicon size (84 kb) and rate of fork migration (2 kb/min) are the same as in mitotic cells (Johnston and Johns 1980; Williamson et al. 1980; see also Fangman and Zakian, this volume). To reconcile these differences, Williamson et al. (1980) suggested that the longer S phase in meiosis may be due to staggering in the initiation of synthesis in different replicons. The use of specific cloned genes as probes to order the time of replication of different regions of the genome in mitosis and meiosis in highly synchronous populations may help to resolve this question. To better understand the relationship between DNA replication and the recombination process and to determine whether the findings in *Lilium* of a small amount of DNA replication during zygotene and pachytene (Stern and Hotta 1977, 1978) apply to other meiotic systems, a more detailed analysis of the timing of DNA synthesis in yeast relative to the cytological stages of meiosis defined by electron microscopy is also required.

Replication of Mitochondrial DNA

The presence of a mechanism coordinating the overall duplication of the mitochondrial and nuclear genomes during sporulation is suggested by data demonstrating a constant proportion of mitochondrial DNA in vegetative cells and haploid ascospores (10–11% of total DNA) (Kuenzi and Roth 1974; Kuenzi et al. 1974; Tingle et al. 1974). However, two lines of evidence indicate that mitochondrial DNA and nuclear DNA synthesis must also, to some extent, be independently regulated during meiosis as they have been shown to be in mitosis. First, in diploids and $n+1$ $MATa/MAT\alpha$ disomic strains, mitochondrial DNA content increases, prior to chromosomal replication, during the first 4 hours of sporulation (Kuenzi and Roth 1974; Piñon et al. 1974). Second, in mutant strain 132, mitochondrial DNA synthesis occurs in the absence of nuclear synthesis (Piñon et al. 1974). It should be noted that radioactive precursors are also incorporated into mitochondrial DNA well after the main round of premeiotic replication. This occurs in the absence of a net increase and may represent repair synthesis (Kuenzi and Roth 1974).

Relationship of Premeiotic DNA Synthesis to Other Meiotic Events

It has long been thought that the timing and quality of premeiotic replication differs from mitotic replication in creating conditions that lead to chromosome pairing and genetic exchange. This view predicts that premeiotic replication utilizes at least some unique gene functions that are nonessential for mitotic replication. Such functions should be indispensable not only for the process of meiotic recombination but for subsequent meiotic events dependent upon recombination.

Evidence that premeiotic DNA synthesis is required for meiosis and spore formation is based on studies with inhibitors of DNA synthesis, such as hydroxyurea (HU) and 8-hydroxyquinoline (Silva-Lopez et al. 1975; Simchen et al. 1976; Mills 1978), and on the behavior of mutants defective in sporulation (reviewed by Esposito and Esposito 1978). Short pulses or continuous treatment with 40–100 mM HU within the first 8 hours of sporulation inhibit ascus production (Silva-Lopez et al. 1975; Simchen et al. 1976). However, if cells are removed from sporulation medium containing HU and plated on growth medium, high viability is retained except for a brief period between 7 hours and 7½ hours, indicating that the effect of HU is generally reversible. The HU-sensitive period with respect to viability coincides with precommitment (see above, Entry into Meiosis) and a period of UV sensitivity (Simchen et al. 1973, 1976; Salts et al. 1976; Hottinguer-de-Margerie and Moustacchi 1979). Simchen et al. (1976) have suggested that it reflects a stage at which lesions are not effectively repaired. Premeiotic replication has also been shown to be essential for commitment to recombination. Studies with HU support the general view that preconditions promoting exchange are established during replication (see below, Genetic Control of Meiosis and Spore Formation).

Evidence exists that mitotic and meiotic DNA replication are under separate genetic control but share some functions in common from studies with *cdc* and sporulation-defective strains (Table 1). Several mutants that fail to exhibit mitotic DNA synthesis undergo premeiotic DNA replication (e.g., *cdc2, cdc7*). Other mutants that fail to exhibit premeiotic DNA replication undergo mitotic DNA synthesis (e.g., *spo7, spo8, spo9, mei1, mei2,3*). Still others are blocked in replication in both processes (e.g., *cdc4, cdc8, cdc21*). In addition, the association of replication with other landmark events also differs in mitosis and meiosis. For example, in mitosis, DNA synthesis depends upon SPB duplication; but in meiosis, SPB duplication and DNA replication are independent events (see below, Genetic Control of Meiosis and Spore Formation). Although the mutations in the first two groups described above result in blocks in either mitotic or meiotic replication, their wild-type alleles may also play a role in other aspects of DNA metabolism. Compare, for example, the behavior of *cdc7* and *spo7* (Table 1). The *CDC7* gene is required for mitotic, but not meiotic, DNA replication; nevertheless, it appears to be necessary for genetic exchange in meiosis (Schild and Byers 1978). The *SPO7* gene, on the other hand, is required for meiotic, but not mitotic, replication (M. S. Esposito and R. E. Esposito 1974). Since the *spo7-1* mutant exhibits antimutator activity in mitosis, its wild-type allele clearly functions during vegetative growth, perhaps in DNA repair synthesis (Esposito et al. 1975). These data emphasize the complexity of the interrelationships between mitosis and meiosis and caution against the oversimplification of the specificity of functions of these genes.

DNA Structure

One approach to understanding the behavior of chromosomes during meiosis and mitosis has been to isolate DNA-protein complexes by sucrose gradient centrifugation, following lysis of spheroplasts in a nonionic detergent. Four fast sedimenting chromatin complexes have been reported in diploid cells at different stages of the mitotic cell cycle (Piñon and Salts 1977; Piñon 1979a). One of them, the slowest sedimenting species, accumulates only during stationary phase, nitrogen starvation, and within the first 3 hours of sporulation (Piñon 1978, 1979b). Another, found in mitotic cells prior to replication, is lacking in meiotic cells. It has been proposed that the latter complex represents an association between unreplicated chromatin and duplicated SPBs and/or components of the spindle. If this interpretation is correct, the failure to detect this complex in meiosis is of interest and may be related to differences in the relationship of DNA replication to SPB duplication and separation in mitosis and meiosis. Mitotic replication requires SPB duplication and separation, whereas premeiotic replication occurs prior to, and does not require SPB separation. Additional studies in this area should be of value in elucidating chromosome-spindle interactions during mitosis and meiosis.

Molecular weight changes in meiotic DNA have also been examined, following gentle lysis of spheroplasts in alkaline and neutral sucrose gradients. Although no double-strand breaks have been observed, single-strand scissions in both template and newly synthesized DNA have been detected during premeiotic S phase, coinciding with the time of commitment to recombination (Jacobson et al. 1975; Kassir and Simchen 1980). Approximately 70–200 single-strand breaks were found per meiotic cell. Jacobson et al. (1975) suggested that the accumulation of these breaks may be related to commitment to recombination, since previous calculations indicate that approximately 150 recombination events occur per meiotic cell (Hurst et al. 1972). However, in a recent study employing the highly synchronously sporulating strain SK1, few if any single-strand nicks were observed during sporulation (Resnick et al. 1981). This paradox may be attributed to the rapid formation and repair of breaks in SK1 or, alternatively, to nicks incurred during extraction in the former studies. One approach in evaluating the significance of these single-strand lesions in meiotic cells is through the analysis of mutants defective in meiotic recombination. Two mutants that fail to exhibit commitment to recombination, *cdc40* and *rad52-1*, also accumulate single-strand scissions during incubation in sporulation medium (Kassir and Simchen 1980; Resnick et al. 1981). On the basis of the *cdc40* phenotype, Kassir and Simchen (1980) proposed that the accumulation of nicks is related to DNA replication rather than to the recombination process. However, it may also be argued that single-strand scissions are associated with exchange and that both the *cdc40* and *rad52-1* mutants are defective in a stage of the recombination process that occurs after nick formation.

Nuclear DNA from sporulating cells has been further analyzed by electron microscopy, with the primary aim of detecting recombination intermediates. Simchen and Friedman (1975) examined DNA prepared from lysed spheroplasts and CsCl density gradient centrifugation. They observed linear molecules up to 300–400 μm in length. Duplex molecules that appear to be paired in short regions flanked by single-stranded tails were observed from 6 hours onward. Klein and Byers (1978) examined DNA derived from spheroplasts lysed in sucrose gradients and recovered linear molecules, some of which appeared to be replication intermediates containing bubbles and terminal forks. An unusual type of molecule, containing small, clustered, denatured regions of about 300 bases in length, spaced 1.5–3 kb apart was also observed between 6 and 8 hours of sporulation. The relationship of the structures described above to premeiotic replication and recombination remains to be determined.

RNA Metabolism

RNA Synthesis

RNA synthesis increases immediately after induction of sporulation. Most of the synthesis occurs between 4 hours and 10 hours, reaching a maximum at 6 hours, regardless of the sporulation regimen (Esposito et al. 1969, 1970; Sando and Miyake 1971; Mills 1972; Sogin et al. 1972; Chaffin et al. 1974; Hopper et al. 1974; Wejksnora and Haber 1974). In some studies, an increase of approximately 50% and then a decline in total RNA per cell has been detected (Esposito et al. 1969; Sando and Miyake 1971), whereas in others, the level of RNA per cell appears nearly constant or declines to 50% of the initial value by the time of maximum spore formation (Croes 1967a; Hopper et al. 1974; Wejksnora and Haber 1974). These differences may reflect variations in the extent of RNA degradation in different strains. Although it is known that protein synthesis is required until the late stages of spore formation, it has not yet been demonstrated that RNA synthesis is necessary for sporulation, due to the lack of studies with inhibitors of RNA synthesis.

All classes of RNA are made during sporulation. The distribution of various classes of RNA in sporulating cells generally resembles that observed in vegetative populations with the following exception. A new stable unmethylated 20S RNA species has been detected specifically during sporulation. This RNA is found only in a/α cells and comprises 15% of the stable RNA made (Kadowaki and Halvorson 1971a,b; Sogin et al. 1972; Wejksnora and Haber 1974). It is unrelated to a 20S methylated rRNA precursor found in vegetative cells and is located in a 32S RNP particle, whose presence during sporulation depends upon new protein synthesis (Wejksnora and Haber 1978). It is not present in all strains. Crosses between those that are $20S^+$ and $20S^-$ indicate that it accumulates under the genetic control of a cytoplasmic element, independent of mitochondrial DNA, 2-micron DNA, ψ, *URE3*, and double-stranded killer RNA (Garvic and Haber 1978). The function of this unmethylated 20S RNA during meiosis is not understood. It has been suggested that its accumulation is part of a nitrogen-starvation response, coincident with, but not required for, sporulation, since strains that do not have the 20S unmethylated RNA nevertheless sporulate (Garvic and Haber 1978).

Other differences have been described between vegetative and sporulation RNA metabolism. The rate of rRNA synthesis is slower during sporulation and less sensitive to cycloheximide inhibition (Chaffin et al. 1974; Magee and Hopper 1974; Wejksnora and Haber 1974). The *rna2* mutant reduces ribosomal protein synthesis in mitosis but not meiosis (Pearson and Haber 1980).

Altered RNA polymerase activities have also been reported (Magee 1974). The most extensively studied differences, however, are described in detail below.

RNA Degradation

RNA synthesis is accompanied by extensive RNA degradation (Croes 1967a; Esposito et al. 1969; Chaffin et al. 1974; Hopper et al. 1974). Approximately 50–70% of total preexisting RNA is degraded in a/α cells and about half that amount is broken down in a/a and α/α cells. These data suggest that at least a portion of the RNA degradation is sporulation-specific (Hopper et al. 1974). A rise in the level of RNase within the first 8–12 hours after inoculation into sporulation medium, specific for a/α sporulation-competent diploids, has also been reported (Tsuboi 1976; Tsuboi and Yanagishima 1976). Preliminary kinetic studies suggest that the predominant RNase activity in sporulating cells differs from that in vegetative cells (Tsuboi and Yanagishima 1976).

Active Ribosomes

Within the first 12 hours of sporulation, the total number of ribosomes declines by about 50–65%. Approximately half of the remaining ribosomes are newly synthesizsed (Hopper et al. 1974). There have been conflicting reports regarding the proportion of ribosomes involved in protein synthesis during sporulation. Frank and Mills (1978) have found that the proportion of ribosomes in polysomes depends on the presporulation carbon source and the growth phase of the cells prior to transfer to sporulation medium (see Mills 1974). They observed a constant level of 65–85% active ribosomes in acetate-grown logarithmic cells and in sporulating populations, using the method of Martin (1973). This procedure measures active monosomes, as well as active ribosomes in larger polysomes. Cycloheximide was added in these studies to prevent runoff of ribosomes from mRNA, but this treatment can also lead to polysome loading. Hence, the actual percentage of active ribosomes during sporulation may have been overestimated. In contrast, Kraig and Haber (1980) reported a similar level of ribosomes in polysomes in acetate-grown logarithmic cells (75–80%), but they noted a threefold decrease in the percentage of ribosomes in polysomes and an increase in the proportion of monosomes in sporulating cells. The monosomes were not irreversibly inactivated, as shown by their reincorporation into polysomes in return-to-growth studies. Since active monosomes were not measured in this study, the percentage of ribosomes engaged in protein synthesis during sporulation is still an open issue.

The average half-life of mRNA (20 min) and the proportion of poly(A) RNA (1.3–1.4%) has been shown to be similar in vegetative and sporulating cells (Kraig and Haber 1980). However, the level of mRNA adenylation is about twofold lower during sporulation, possibly indicating shorter tracts of poly(A) (C. Saunders, cited in Kraig and Haber 1980). Kraig and Haber (1980) have proposed that the loss of polysomes in their experiments is not due to mRNA limitation but rather to a decline in mRNA translation (perhaps related to the lower level of adenylation) under nitrogen-starvation conditions. Supporting this view was their observation of a gradual decrease in in vitro translation of mRNA from meiotic cells. However, this decrease was insufficient to account for the very rapid loss of polysomes detected during sporulation in their studies. Turnover of specific messages, the loss of polysomes, and initiation of mRNA translation are important areas that require further study.

Sporulation-specific mRNA Synthesis

Identification of sporulation-specific gene products has been a primary goal in understanding the regulation of gene expression during sporulation. Recently, Mills (1980) reported the first evidence for sporulation-specific mRNA populations. He isolated poly(A) RNA and constructed cDNA probes from both vegetative and sporulating cells. The ability of the poly(A) RNA to protect the various cDNA probes from nuclease digestion was examined. About 7% of the cDNA made from sporulating cells after 12 hours of sporulation does not hybridize to vegetative RNA. Furthermore, about 5–7% of the cDNA made from poly(A) RNA from a/α cells after 12 hours of sporulation does not completely hybridize to poly(A) RNA from α/α cells. These results suggest a sporulation-specific mRNA population of approximately 7%. Thus, at least one aspect of gene regulation during meiosis and spore development appears to be at the level of transcription.

Protein Metabolism

General Protein Synthesis

Protein synthesis is essential for sporulation; the addition of an inhibitor of protein synthesis, such as cycloheximide or ethionine, to a sporulating culture at any time up to the appearance of mature asci has an inhibitory effect (Sando 1960; Croes 1967a; Esposito et al. 1969; G. Darland 1969, cited in Tingle et al. 1973; Magee and Hopper 1974). Magee and Hopper (1974) demonstrated that most major events of sporulation—DNA synthesis, meiosis-I and meiosis-II chromosome segregation, recombination, RNA degradation, glycogen synthesis and breakdown, ascus formation and, to some extent, RNA synthesis—require cellular protein synthesis.

Mitochondrial protein synthesis may also be required for sporulation, since erythromycin inhibits sporulation in erythromycin-sensitive, but not -resistant, strains (Puglisi and Zennaro 1971). In addition, exposure of cells to ethidium bromide during the first 4 hours of sporulation greatly reduces the level of ascus formation (Newlon and Hall 1978). Ethidium bromide causes destruction of mitochondrial DNA (Goldring et al. 1970; Perlmann and Mahler 1971) and inhibits mitochondrial transcription and, consequently, protein synthesis (Fukuhara and Kujawa 1970). In apparent contradiction to these results is the observation that sporulation can occur in cells rendered petite by exposure to ethidium bromide shortly before transfer to sporulation medium (Kuenzi et al. 1974; Newlon and Hall 1978; Dujon, this volume).

The net amount of cellular protein increases 10–35% during the first 6–10 hours of sporulation and declines thereafter (Croes 1967a; Esposito et al. 1969; Sando and Miyake 1971; Hopper et al. 1974; Magee and Hopper 1974; Vezinhet et al. 1974; Croes et al. 1978a; Peterson et al. 1979). The increase in protein content is not specific to a/α cells; Hopper et al. (1974) observed essentially the same kinetics of label incorporation into protein in a/α, a/a, and α/α diploids incubated in sporulation medium.

General Protein Degradation

Under the conditions of nitrogen starvation required for sporulation, protein synthesis depends upon the internal amino acid pool and supply of nitrogen compounds from protein degradation. Upon transfer of cells to sporulation medium, the amino acid pool size begins to decrease, declining to approximately one-third that of vegetative cells (Ramirez and Miller 1964). The concentration of all amino acids decreases, except for proline (Ramirez and Miller 1964) and glutamic acid (P. Rousseau 1972, cited in Tingle et al. 1973) which, for reasons unknown, increase.

Extensive degradation of vegetative proteins occurs during sporulation (Esposito et al. 1969; Hopper et al. 1974; Klar and Halvorson 1975; Betz and Weiser 1976a; Betz 1977). A similar rate of degradation has been observed for proteins synthesized during sporulation itself (Betz and Weiser 1976a). The amount of breakdown is generally measured as the loss of vegetatively labeled protein and is a minimum estimate since breakdown products may be reutilized. The minimum net percentage of vegetative protein that is degraded in a/α cells during sporulation ranges from about 25% to 70% in different studies (Esposito et al. 1969; Hopper et al. 1974; Klar and Halvorson 1975; Betz and Weiser 1976a). Variation in strains, growth and sporulation procedures, sporulation kinetics, and labeled precursor may be responsible for the differences among these estimates.

It is presently unclear whether extensive protein breakdown upon transfer to sporulation medium is specific to a/α cells or occurs to an equivalent extent and at the same rate in a/a and α/α cells. Hopper et al. (1974) observed that [35S]methionine prelabeled cultures of a/α cells exhibited 25–30% protein breakdown, whereas near isogenic a/a and α/α cells exhibited only 0–10% breakdown. Betz and Weiser (1976a) reported that 60–70% of vegetative proteins labeled with [3H]leucine were turned over in an a/α diploid at a rate of 2.5–3%/hr. A related haploid exhibited less protein degradation (~50%) at a somewhat slower rate than the diploid. Zubenko and Jones (1981), however, found similar levels of degradation of vegetative proteins (50–75%, depending on the label) in near isogenic $\mathbf{a}/\alpha$ and $\mathbf{a}/\mathbf{a}$ diploids. In addition, they found that both $\mathbf{a}/\alpha$ and $\mathbf{a}/\mathbf{a}$ diploids homozygous for a mutation affecting one or more of the three major proteinases exhibited the same reduced rates and extents of protein breakdown.

During sporulation, there is an increase in the activities of three proteinases: proteinase A (Chen and Miller 1968; Klar and Halvorson 1975; Betz and Weiser 1976a), proteinase B (Klar and Halvorson 1975; Betz and Weiser 1976a), and proteinase C (carboxypeptidase Y) (Klar and Halvorson 1975). In another study, the increase in proteinase-C activity was not considered significant (Betz and Weiser 1976a). No novel proteinase activities are detected during sporulation (Klar and Halvorson 1975; Betz and Weiser 1976a). In addition, the proteinases isolated from sporulating and vegetatively growing cells exhibit identical DEAE-Sephadex elution patterns and are indistinguishable in their sensitivities to proteinase inhibitors (Klar and Halvorson 1975; Betz and Weiser 1976a). As has been demonstrated for RNase activity (see above; Tsuboi 1976; Tsuboi and Yanagishima 1976), the elevation in proteinase activity during sporulation occurs only in $\mathbf{a}/\alpha$ cells and is prevented by the addition of cycloheximide to the sporulation medium (Klar and Halvorson 1975).

Mutants have been isolated that exhibit reduced or no detectable activity of one or more of the three proteinases (Betz 1975; Wolf and Fink 1975; Jones 1977; Zubenko et al. 1979). Diploids homozygous for a mutation that results in a low level of proteinase-A activity are unable to complete sporulation (Betz 1975). Diploids homozygous for a mutation at the *prb1* locus, the structural gene for proteinase B, complete sporulation but form asci that are abnormal in appearance (Zubenko et al. 1979; Zubenko and Jones 1981). On the other hand, diploids homozygous for the *prc1-1* mutation lack proteinase-C activity but sporulate normally (Wolf and Fink 1975). Zubenko and Jones (1981) have shown, by the analysis of proteinase-defective mutants, that proteinases B and C are responsible for approximately half the total amount of protein degradation during sporulation.

Both proteinase and ribonuclease activities have been localized to the yeast vacuole (Matile and Wiemken 1967; Cabib et al. 1973; Hasilik et al. 1974), which undergoes extensive fragmentation during sporulation (Svihla et al. 1964). Trew et al. (1979) have recently found that $\mathbf{a}/\mathbf{a}$ and α/α diploids also undergo vacuolar fragmentation when incubated in sporulation medium, which suggests that any release of RNase and proteinases from the vacuole would not be specific to $\mathbf{a}/\alpha$ cells.

Specific Protein Synthesis, Degradation, and Expression

To address the question of whether individual differences exist between sporogenous and asporogenous cell types, one-dimensional and two-dimensional gel comparisons of labeled proteins were performed (Hopper et al. 1974; Petersen et al. 1979; Trew et al. 1979; Wright and Dawes 1979; Kraig and Haber 1980). Although differences in the protein patterns of vegetative and sporulating cells were detected, few, if any, differences were seen when $\mathbf{a}/\alpha$ and $\mathbf{a}/\mathbf{a}$ or α/α cell types were compared. In most of these studies, proteins were labeled by short (5 min) (Kraig and Haber 1980) or long ($\frac{1}{2}$–5 hr) (Petersen et al. 1979; Trew et al. 1979; Kraig and Haber 1980) pulses with [35S]methionine during growth or sporulation. In others, proteins were prelabeled during vegetative growth with [35S]sulfate (Wright and Dawes 1979) or [35S]methionine (Kraig and Haber 1980) and sampled during incubation in unlabeled sporulation medium. Protein patterns were resolved by either Coomassie-blue staining, autoradiography, or fluorography.

The protein patterns produced by $\mathbf{a}/\alpha$ cells indicated that both the growth phase and the composition of the growth medium have a significant effect on the protein patterns generated by a particular strain. A number of differences have been demonstrated between exponential-phase, acetate-grown cells and sporulating cells (Petersen et al. 1979; Trew et al. 1979; Kraig and Haber 1980). For example, Trew et al. (1979) reported 9 proteins specific to vegetative cells during exponential growth and 11 proteins specific to cells incubated in sporulation medium, using a two-dimensional gel system that resolved some 700 individual polypeptides. However, in each study, the same protein patterns were also generated by asporogenous diploid or haploid cell types (Petersen et al. 1979; Trew et al. 1979; Kraig and Haber 1980). Moreover, no qualitative differences have been detected between stationary-phase, glucose-grown cells and sporulating cells (Petersen et al. 1979), which suggests that many of the changes noted above are likely to be associated with a starvation response. One study has presented evidence that sporulation-specific processing may take place (Wright and Dawes 1979). By prelabeling galactose-grown cells, a number of qualitative and quantitative changes in protein patterns were observed during a 24-hour period in sporulation medium; approximately 20 of these changes were found in $\mathbf{a}/\alpha$, but not in $\mathbf{a}/\mathbf{a}$, cells.

A number of explanations have been proposed to explain why, except for the study of Wright and Dawes (1979), no $\mathbf{a}/\alpha$-specific proteins have been detected during sporulation. Hopper et al. (1974) and Hopper and Hall (1975) suggest that there may exist only a few crucial genes that require both $MAT\mathbf{a}$ and $MAT\alpha$ for expression and that their products may be too few or too low in concentration to detect, even with two-dimensional gels. Alternatively, such proteins may escape detection due to poor solubility, extremely low isoelectric points, or low molecular weight (Petersen et al. 1979).

The activities of individual proteins during the course of sporulation have been examined by a number of investigators (Chen and Miller 1968; Miyake et al. 1971; Vezinhet et al. 1974; Klar and Halvorson 1975; Betz and Weiser 1976a,b; Klar et al. 1976; Ballou et al. 1977; Colonna and Magee 1978; Matur and Berry 1978; Fonzi et al. 1979; Opheim 1979; del Rey et al. 1980; Pearson and Haber 1980). The timing of the increases(s) and/or decrease(s) in the level of a particular protein may serve as a useful landmark in the sporulation process. One such example is 2-amino-2-deoxy-D-glucose-6-phosphate ketol-isomerase, an enzyme required for the synthesis of D-glucosamine, which is thought to be a component of the spore wall (Ballou et al. 1977). This enzyme is not detected in $\mathbf{a}/\alpha$ cells until late in meiosis when tetranucleate cells are abundant and, thus, serves as a landmark for ascospore maturation. Mutants at the *GCN1* locus, which lack this enzyme, produce abnormal asci (Whelan and Ballou 1976; Ballou et al. 1977; see Table 1).

In most cases studied, the same changes in the activity or abundance of a particular protein found in $\mathbf{a}/\alpha$ cells are observed in asporogenous cells incubated in sporulation medium (Vezinhet et al. 1974; Betz and Weiser 1976b; Fonzi et al. 1979; Pearson and Haber 1980). A considerable $\mathbf{a}/\alpha$-specific increase in the level of three proteinases (Klar and Halvorson 1975) and ribonuclease activity (Tsuboi 1976) during sporulation has been reported. However, these activities are also present at reduced levels in both vegetative cells and asporogenous cells under sporulation conditions. One activity that may truly be unique to $\mathbf{a}/\alpha$ sporulating cells is an α-1,4-glucosidase (Colonna and Magee 1978). Colonna and Magee (1978) found no activity in vegetative cells or in sporulating $\mathbf{a}/\alpha$ cells prior to 8 hours and only trace levels of activity in α/α cells at any time during incubation in sporulation medium. In $\mathbf{a}/\alpha$ cells, the activity rose from 8 hours to 18 hours and remained constant until at least 24 hours. In addition, there has been a recent report of a second peak of 1,3-β-glucanase activity, which begins a few hours before mature asci are formed in $\mathbf{a}/\alpha$ cells and is not detected in α/α cells (del Rey et al. 1980). This activity appears to be due to a new 1,3-β-glucanase, not present in vegetative cells.

Carbohydrate Metabolism

During sporulation, most of the assimilated acetate (62%) is respired as CO_2, approximately 16% is incorporated into macromolecules, and the remainder is present in the stable pool (Esposito et al. 1969). A large portion of the acetate incorporated into macromolecules can be accounted for by the synthesis of carbohydrates, which increase approximately sixfold during sporulation (Kane and Roth 1974). These carbohydrates are utilized in spore-wall construction and are thought to serve as energy sources during ascus development and spore germination.

A dramatic increase (70%) in dry weight occurs early in sporulation (Croes 1967a; Esposito et al. 1969), due largely to the accumulation of the storage carbohydrates trehalose and glycogen and, to a lesser extent, the cell-wall components glucan and mannan (Roth 1970; Kane and Roth 1974). Trehalose is synthesized continuously during sporulation, rising to a level approximately tenfold higher than that in vegetative cells (Roth 1970; Kane and Roth 1974). As in other fungi, it is found localized in ascospores. The glycogen-glucan content increases approximately eightfold, and the mannan content doubles within the first 12 hours in sporulation medium, prior to the appearance of mature asci. The synthesis of all of the carbohydrates described above occurs in **a**/**a** and α/α cells and is considered to be a response to starvation conditions (Kane and Roth 1974). At present, only trehalose accumulation has been shown to be required for ascus production, as mutants defective in its accumulation do not sporulate (Lillie and Pringle 1980).

Substantial degradation of glycogen specific to **a**/α cells takes place late in sporulation, during the appearance of asci (Hopper et al. 1974; Kane and Roth 1974). During this same period in **a**/α cells, a 10-fold increase in non-specific alkaline phosphatase activity occurs, and the level of intracellular glucose increases 30- to 40-fold (Fonzi et al. 1979). Glycogen degradation and/or the increase in alkaline phosphatase activity (which may be degrading phosphorylated sugars to glucose) are thought to be responsible for the accumulation of glucose during the late stages of sporulation.

The breakdown of glycogen has been studied in some detail, since it is one of the few biochemical parameters considered to be sporulation-specific. Glycogen degradation depends upon new protein synthesis during sporulation (Magee and Hopper 1974) and is inhibited by the addition of nitrogen (Opheim 1979). A number of glycolytic intermediates and enzyme activities have also been examined during sporulation to understand the regulation of glycogen synthesis and degradation (Fonzi et al. 1979). No substantial changes in glycolytic intermediates or enzyme activities, other than a 50% decline in fructose biphosphatase activity and a fourfold increase in the level of fructose 1,6-biphosphate (which may activate glycogen synthesis), were observed. Furthermore, no changes in ATP levels, energy charge, or cAMP, which correlate with glycogen synthesis or breakdown, were detected. Recently, Colonna and Magee (1978) discovered a glycogenolytic enzyme activity, absent in vegetative cells, that appears coincidently with the initiation of glycogen degradation and the appearance of asci. The glycogen degradation complex has been separated into two activities: One acts on α-1,4 bonds (probably a glucamylase), and the other acts on α-1,6 bonds (a debranching enzyme). As might be expected, the activity of the whole complex is absent in cells blocked early in meiosis, during premeiotic DNA synthesis and/or recombination (Clancy et al. 1980). This was shown by the addition of the inhibitors HU or sulfanilamide (Colonna et al. 1977) and, genetically, by analysis of the *spo7-1* and *cdc4* mutants. Glycogen degrading activity is also absent in wild-type cells arrested at pachytene by high temperature but is present in the *spo3-1* mutant, defective late in sporulation during packaging of haploid nuclei into spores. Recently, another **a**/α-specific activity, a 1,3-β-glucanase, was detected after 7 hours of sporulation (del Rey et al.

1980). Thus far, the study of carbohydrate metabolism has provided the best documented examples of sporulation-specific enzyme activities.

Lipid Metabolism

Cytological studies demonstrate the accumulation of lipid vesicles in the cytoplasm during sporulation and suggest that they play a crucial role in spore-wall development (see above, Morphological Development). Biochemical studies have also shown that extensive lipid synthesis takes place during sporulation (Esposito et al. 1969; Henry and Halvorson 1973; Illingworth et al. 1973; Vezinhet et al. 1974). The bulk of lipids synthesized during sporulation are triglycerides and sterol esters and, to a lesser extent, phospholipids (Henry and Halvorson 1973; Illingworth et al. 1973). Label-incorporation studies with [14C]acetate reveal two major periods of synthesis (Esposito et al. 1969; Henry and Halvorson 1973; Illingworth et al. 1973), resulting in a fourfold increase in cellular lipid content (Illingworth et al. 1973). The first synthetic period occurs between 0 and 10 to 18 hours in sporulation medium (apparently depending upon the strain and presporulation-sporulation regimen). It is observed in haploid, as well as in diploid, cells (Henry and Halvorson 1973). Almost all of the phospholipids and some neutral lipids are synthesized during this period, in both cell types. The kinetics of ^{32}P incorporation into three classes of phospholipids further indicate no differences between the sporulating and nonsporulating cells. The start of the second phase of lipid synthesis, which lasts 5–6 hours, coincides with the time of the first appearance of mature asci (between 20 and 24 hr in sporulation medium). This synthetic period is specific to a/α cells and consists almost entirely of neutral lipid synthesis.

GENETIC RECOMBINATION

Most of the information about the properties of meiotic recombination comes from the analysis of the end products of meiotic development. A number of rules of recombination have emerged from these studies concerning the distribution and frequency of exchange events and the characteristics of recombined regions. Knowledge of these rules has contributed to the development of a general model of the mechanism of homologous exchange. This model involves single-strand scission, strand invasion, heteroduplex formation, mismatch correction, and the possibility of associated exchange of outside markers (for reviews, see Fogel et al. 1979 and Fogel et al., this volume). The essential features of the interaction of DNA molecules appear to be similar in mitosis and meiosis (see Esposito and Wagstaff, this volume). In addition, detailed analysis of the properties of recombination in both mitosis and meiosis suggest that the individual steps are part of a highly regulated process. The presence of specific genetic mechanisms controlling both the timing and chromosomal location of recombination events may be inferred from (1) the nonrandom distribution of meiotic exchange, (2) the association of high levels of recombination with reductional chromosome segregation in meiosis, (3) the rare occurrence of exchange between homologous chromosomes in mitosis, (4) the absence of a direct relationship of mitotic and meiotic recombination frequencies to physical distance between genetic markers, and (5) differences in the time at which recombination takes place relative to DNA replication in mitosis and meiosis.

These observations raise a number of questions regarding the development and regulation of high-recombination potential during meiosis, which are not easily answered by standard genetic analysis of meiotic products. Some questions of interest are: What are the signals at the cellular level to begin and complete exchange during meiosis? Does exchange begin and end in all regions of the genome at the same time? What events commit a cell to high levels of exchange? What factors control the overall distribution of exchange, and why do they differ in meiosis and mitosis? Does commitment to meiotic levels of exchange also commit a cell to reductional segregation? How are genetic recombination and chromosome segregation genetically coordinated? What phenotypes are expected of mutants defective in various aspects of the exchange process, and how can such mutants be recovered and analyzed?

Methods of Analysis

The questions raised above have not been readily addressed by analysis of meiotic products for two reasons. First, the events of interest are completed prior to the formation of the final product. Second, the utility of Rec⁻ mutants to probe the factors required for the development of high-recombination potential is limited by a low recovery of viable gametes. This is due to the fact that meiotic recombination is required for proper segregation of homologous chromosomes at meiosis I. Where it has been studied, mutants defective in exchange exhibit a high level of nondisjunction and form aneuploid meiotic products (Baker et al. 1976). In yeast, these gametes are nearly always inviable. As a consequence of the high chromosome number, the probability of recovering spores that are not lacking one or more chromosomes and that are not multiply disomic for others is extremely low. Several procedures that circumvent these problems have been employed to analyze exchange and to characterize Rec⁻ mutants. These are described below.

"Return-to-growth" Protocol

The temporal properties of meiotic recombination may be analyzed by taking advantage of the fact that yeast cells can be interrupted during meiotic development and returned to vegetative growth without loss of cell viability. When diploid yeast are removed from a meiosis-inducing environment at an early stage of development, the cells revert to mitotic division on growth medium. With increasing time in sporulation medium, the cells become irreversibly committed to the completion of meiosis and form haploid spores before resuming vegetative growth (see above, Entry into Meiosis). Sherman and Roman (1963) first utilized the return-to-growth protocol to examine the progress of intragenic recombination during meiosis. R. E. Esposito and M. S. Esposito (1974) then extended this procedure to analyze commitment to intergenic exchange during meiosis. These studies demonstrated that when appropriately marked strains are returned to growth, commitment to both intragenic and intergenic recombination can be detected in cells prior to commitment to the reductional division. Recombinant genotypes are recovered initially in diploids that remain heterozygous for markers closely linked to their centromeres. With increasing time of incubation in sporulation medium, recombinants are recovered as haploids.

Intragenic recombination is measured by an increase in prototroph production when diploids heteroallelic for auxotrophic loci are plated on selective omission media. Intergenic recombination, on the other hand, is generally monitored among cells initially selected for an intragenic exchange event, usually on an independent chromosome. Intergenic recombination is detected by homozygosis of recessive markers originally in heterozygous condition, in a manner analogous to the detection of mitotic crossing-over. Intergenic exchange assayed among prototrophs derived from the growth of haploid spores is detected by changes in linkage relationships in a manner similar to random spore analysis. The rationale for examining intergenic events among cells committed to an intragenic event is based on the following considerations: (1) Selection for at least one other recombination event identifies the sporulation-competent fraction of the population, since nearly all recombinants at the end of sporulation are present as haploids. This facilitates comparisons of intergenic exchange in different strains, in different regions of the genome, and at various times during sporulation in a more uniform population. (2) Selection for a prototrophic recombinant, which generally results from a gene-conversion event, allows one to study the clonal derivatives of a single cell. Early in meiosis, when sporulating cells are capable of returning directly to mitotic division, the prototrophic chromatid is present in heterozygous condition in a diploid cell. Following commitment to sporulation, the prototrophic chromatid is recovered in the clonal products of a single haploid spore. In the latter case, the other members of the tetrad are unable to grow or mate with the prototrophic spore on the selective medium.

It is difficult to determine when the actual exchange event occurs in the return-to-growth protocol. It may be completed in the sporulation medium, prior to, or at the time of, plating or on the selective-growth medium. To avoid unnecessary assumptions, the initial enhancement in the level of recombinant colonies has therefore been called "commitment to recombination." It may, in fact, reflect completion of the exchange process or represent an event that commits the cell to exchange (R. E. Esposito and M. S. Esposito 1974). That the latter possibility is likely to be the case is suggested by studies of cells arrested in pachytene and the effects of HU on recombination (described later in this section). For convenience, the cells that are induced to enter meiosis by exposure to sporulation medium but that revert to mitosis when transferred to growth medium are referred to as "meiototic" cells (Esposito and Esposito 1978). Under the return-to-growth conditions, such cells experience aspects of both meiotic prophase and mitotic cell division.

The return-to-growth protocol has been employed in extensive studies of the temporal properties of commitment to exchange in different regions of the genome in wild-type cells as they proceed through meiosis. It has also been the primary method used to identify putative Rec⁻ mutants. Such mutants have been diagnosed by their failure to produce prototrophic colonies resulting from gene conversion of heteroallelic markers (see below, Genetic Control of Meiosis and Spore Formation). An advantage to using this method is that the return to mitotic growth permits Rec⁻ cells to be rescued into viable euploid products, since they proceed directly to an equational, rather than a reductional, chromosome division after prophase. This analysis, however, is limited by the fact that cells in sporulation medium eventually become committed to reductional segregation. After this point, they are incapable of returning to mitotic growth without first completing sporulation. Hence, a population of mutant cells that cannot complete a normal reductional division in the absence of exchange, and are past the point where they can return to mitotic growth, will ultimately yield a limited and selected sample of viable cells for analysis. Furthermore, the absence of prototrophic colonies may be due to factors other than a specific Rec⁻ defect. For example, cells may be blocked in a function required for the return to vegetative growth. Nevertheless, the meiototic system has been a useful probe to study the progress of commitment to exchange in wild-type cells during meiosis and in the identification of putative Rec⁻ mutants.

The spo13 Bypass System

A recently developed method for the positive identification of recombination-defective mutations involves the use of the *spo13-1* mutation (Klapholz 1980; Klapholz and Esposito 1980c; Malone and Esposito 1981). In strains containing this mutation, meiotic recombination is followed by a single meiosis-II-like division. The meiosis-I reductional segregation is "bypassed," and two diploid spores are formed. This system has features in common with the return-to-growth protocol, in that cells become induced for meiotic exchange and then directly undergo an equational centromere disjunction. The particular utility of this mutant, however, is that viable spores can be recovered in the absence of recombination in *spo13-1 rec* diploid genetic backgrounds. Consequently, biases in the recovery of viable products after commitment to spore formation and ambiguities in interpretation of the absence of recombinant colonies can be eliminated. The *spo13* bypass system has been used to identify *spo11-1* and *rad50-1* as bona fide meiotic Rec⁻ mutations (for further details, see below, Genetic Control of Meiosis and Spore Formation). Other putative Rec⁻ mutants, identified as such by the return-to-growth protocol, are not "rescued" by *spo13-1*. These include *rad6-1* and *rad52-1* (Malone and Esposito 1981). At present, it is unclear why only certain *spo13-1 rec* double-mutant combinations yield viable meiotic products. (One possibility is that the mutants that are not rescued by *spo13-1* have additional defects that block the progress of meiosis.) Nevertheless, this finding has provided a convenient way to order gene functions involved in recombination (described below, in Genetic Control of Meiosis and Spore Formation). Because *spo13-1* undergoes a single equational division, it has also been possible to develop conditions that allow haploids (n or n + 1) to sporulate. Cells enter prophase of meiosis I, become induced for meiotic recombination (detected on the disome), undergo equational division, and form viable haploid two-spored asci (see below, Chromosome Segregation). This system will hopefully provide a new useful tool with which to obtain recessive Rec⁻ mutations.

Detection of Recombined DNA Molecules

Recently, L. S. Davidow et al. (pers. comm.) have devised a method for the detection of recombined DNA molecules, which should permit a precise definition of the time of genetic recombination relative to commitment to exchange and other landmark meiotic events. This method takes advantage of a diploid strain possessing two adjacent restriction site polymorphisms in the *SUP4* region. Recombined DNA molecules can be recognized by the appearance of novel restriction fragments using appropriate probes. Recombinant molecules have been observed between 4 and 6 hours of sporulation by this procedure. Another approach to the detection of recombined molecules, through the visualization of recombination intermediates by electron microscopy of DNA during sporulation, has already been discussed above, in Macromolecular Synthesis.

Temporal Properties of Commitment to Exchange

Analysis of the kinetics and quality of recombination events detected by the return-to-growth protocol has led to several conclusions, which are described below.

Commitment to Intragenic and Intergenic Exchange Occurs
Prior to Commitment to Meiosis-I Segregation

In all of the studies reported to date, commitment to intragenic recombination, monitored at a number of independent loci and in unrelated strains, is initiated coincidently with, or shortly after, the initiation of premeiotic DNA synthesis (Sherman and Roman 1963; R. E. Esposito and M. S. Esposito 1974; Hopper et al. 1974, 1975; Roth 1974; Silva-Lopez et al. 1975; Kassir and Simchen 1978; Plotkin 1978; Plotkin and Esposito 1978; Schild and Byers 1978). An increase in prototrophic colonies is first detected between 4 hours and 6 hours. Maximum levels are reached by 12–18 hours after sporulation induction. Recombinants are initially recovered in diploid cells; after 7 hours they begin to be recovered as haploid meiotic products. This time coincides with commitment to reductional segregation and haploidization, determined by cytological studies of the appearance of binucleate cells and the segregation of recessive drug-resistance markers.

Studies of meiototic cells exposed to sporulation medium for 9 hours revealed that the level of intergenic recombination in different regions of the genome varied from 0–100% of the final meiotic value. This observation led to the suggestion that there may be an internal cellular program in which different regions of the genome become committed to exchange at different times (R. E. Esposito and M. S. Esposito 1974). Furthermore, it raised the question of whether commitment to full levels of meiotic exchange was completed in all regions prior to commitment to the reductional division. A detailed kinetic analysis of the progress of intergenic exchange in two intervals on chromosome VII was undertaken to answer this question (Esposito et al. 1974). Intergenic exchange on chromosome VII was monitored among prototrophs selected for a gene-conversion event on chromosome II. Prototrophs recovered throughout meiosis were partitioned into two classes: those that returned to growth (meiototic cells) and those that were committed to the completion of meiosis. Cells committed to meiosis, recovered at any time during sporulation, exhibited full meiotic exchange, as expected if cells complete meiosis normally on the selective medium. On the other hand, prototrophic cells capable of returning directly to mitosis initially exhibited no intergenic exchange. However, with increasing time of incubation in sporulation medium, they displayed full meiotic levels of recombination (30% and 16% recombination in the centromere-*ade6* and centromere-*trp5* regions, respectively). Thus, although genetic recombination appears to be necessary for proper chromosome disjunction at meiosis I, meiotic levels of exchange do not commit cells to a reductional segregation of centromeres or to haploidization. More extensive studies of both intragenic and intergenic recombination in 19 intervals on chromosomes II and VII indicate that induction of commitment to meiotic recombination in all intervals occurs prior to the reductional division of homologs (Plotkin 1978; Plotkin and

Esposito 1978). Hence, commitment to genetic recombination and commitment to reductional segregation are separable events, most likely residing on independent but coordinated pathways (see also Genetic Control of Meiosis and Spore Formation).

Recombinational Events Detected in Meiotic Cells Are Qualitatively Similar to Exchange Events in Meiotic Cells

The polarity of gene conversion has been examined in mitotic, meiotic, and meiotic cells to determine if meiotic and meiotic events are similar (Esposito and Plotkin 1978). Meiotic tetrad analysis indicated that single-site conversion at the *leu1* locus, in a diploid containing *leu1-c/leu1-12*, occurred in 3% of all tetrads; the *leu1-c* allele was converted in 97% of these cases. The meiotic $LEU1^+$ cells similarly exhibited highly polar conversion (94% were converted for *leu1-c*). In a control study, $LEU1^+$ mitotic cells exhibited non-polar conversion (50% were converted for *leu1-c*), a feature of gene conversion typical of mitotic cells (Esposito and Wagstaff, this volume).

Evidence for a Temporal Program of Commitment to Genetic Recombination during Meiosis

Evidence for the existence of a temporal program governing commitment to exchange during meiosis is based upon three observations: (1) variations in the extent of intergenic recombination in different genetic regions relative to the final meiotic yield, within a given class of prototrophs; (2) differences in total recombination capacity among prototrophs selected at independent loci; and (3) an indication of different initiation times in the recovery of different classes of prototrophs by kinetic analysis.

The first two points are illustrated in Figure 5 (from the data of Plotkin 1978). Comparisons of the meiotic and meiotic maps on chromosome VII among $LYS2^+$ and $HIS7^+$ intragenic recombinants, recovered after 14 hours of incubation in sporulation medium, indicate that nearly full meiotic levels of exchange have occurred in the two prototrophic classes. The meiotic map of chromosome VII among $LYS2^+$ cells is 55% of the meiotic map (i.e., 85.6/156.9 centiMorgans [cM]), constructed from control prototrophs committed to the completion of meiosis. The meiotic map among $HIS7^+$ cells is 67% of the meiotic map (103.1/153.0 cM). The meiotic maps of the $LYS2^+$ and $HIS7^+$ prototrophs are nearly identical (156.9 cM and 153.0 cM, respectively), as expected. Inspection of individual prototrophic classes demonstrates that in each case, individual regions exhibit different proportions of the final meiotic yield of recombination. This suggests that these regions either initiate exchange at different times or exhibit region-specific rates of exchange that are not simply proportional to map distance and may be under an additional type of regulation. Kinetic studies suggest that both of these are likely to be the case (Plotkin 1978).

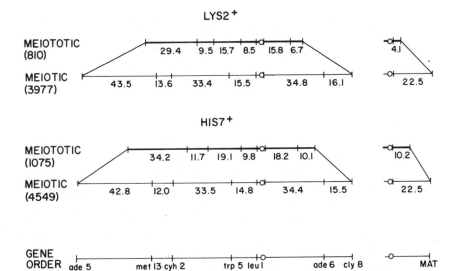

Figure 5 Comparison of genetic maps in sporulating cells returned to mitotic growth medium after 14 hr in sporulation medium. A heterothallic diploid, Z270, was used in this study. The relevant genotype of this strain is

MA*T*a	*lys2-2*	*his7-1*	*ADE5*	*met13-c*	*cyh2*	*trp5-c*	*leu1-c*	*ade6*	*cly8*
MA*T*α	*lys2-1*	*his7-2*	*ade5*	*met13-d*	*CYH2*	*trp5-d*	*leu1-d*	*ADE6*	*CLY8*

Intergenic map distances on chromosomes VII and III were determined among different protrophic subpopulations, selected for an intragenic recombination event in either the *lys2* or *his7* locus, located on the right arm of chromosome II, 72 cM and 123 cM, respectively, from the centromere. Prototrophic colonies derived from cells that remain diploid and return directly to mitotic division and that are not yet committed to haploidization are termed meiotic. Those derived from cells committed to the completion of meiosis and that have become haploid are designated meiotic. See text for details. (Courtesy of D. Plotkin.)

Comparison of the two meiotic maps of chromosome VII indicates that the *LYS2*⁺ map is 83% of the *HIS7*⁺ map. This suggests that the *HIS7*⁺ population is a biologically older one with respect to genetic exchange, since its total recombination capacity is higher. Hence, the selection for intragenic recombinational events at different loci appears to permit an asynchronously sporulating population to be fractionated into subclasses of various recombinational age, on the basis of the level of commitment to exchange elsewhere in the genome. Comparisons of the total level of intergenic exchange on other independent chromosomes and among other prototroph classes indicate that total recombination capacity may be used to determine the order in which commitment to recombination takes place in the intragenic regions, independent of the kinetics of prototroph production.

Finally, these experiments indicate that the level of commitment to meiosis among $LYS2^+$ and $HIS7^+$ prototrophs is the same (83% and 81%, respectively), even though the total recombination capacity of each population differs. These data provide further support for the view that commitment to recombination and commitment to meiotic segregation are on independent pathways.

The Actual Exchange Event Probably Occurs after Premeiotic S Phase

Two arguments have been raised in support of the view that although events during DNA synthesis commit cells to recombination, the exchange events themselves occur after replication, most likely at pachytene. The first involves studies with the DNA synthesis inhibitor HU. When HU is added prior to replication, both DNA synthesis and recombination are blocked (Silva-Lopez et al. 1975). However, when it is added progressively during replication, depending upon the time of addition, an increasingly higher proportion of cells complete replication in sporulation medium and form asci. This fraction is in great excess of the fraction that forms stable recombinants, which led to the suggestion that recombination events occur after replication. The asci after HU treatment were dissected and found to have high viability, but they were not analyzed for their exchange levels. This presents a paradox, because the data suggesting that viable haploid genomes are formed in the absence of recombination are contrary to most observations on the dependency of normal chromosome segregation on genetic recombination. In another study with HU, Simchen et al. (1976) found that short pulses of HU during premeiotic S phase actually increased exchange, which suggests that preconditions that promote exchange were established during DNA synthesis.

The strongest argument, however, that exchange occurs after DNA replication, comes from the behavior of a strain that undergoes a reversible thermal arrest at pachytene at 36°C (Davidow et al. 1980). A substantial enhancement of recombination frequencies has been found in such arrested cells returned to the permissive temperature and allowed to form asci (~ 10-fold to 20-fold for gene conversion, 2-fold for reciprocal recombination, and 40-fold to 50-fold for postmeiotic segregation) (L.S. Davidow, pers. comm.). These results support the classical view that exchange occurs at pachytene. Studies involving the detection of recombinant DNA molecules during meiosis should further resolve this question.

CHROMOSOME SEGREGATION

Haploid Meiosis

Haploids disomic for chromosome III and heterozygous at the mating-type locus are able to undergo a number of sporulation-specific events. These include premeiotic DNA synthesis, commitment to recombination (Roth and Fogel 1971; Roth 1973), and RNA synthesis (Kadowaki and Halvorson 1971a). Asci containing the outlines of spores, designated incipient asci (Roth and Fogel 1971), are sometimes observed when $MATa/MAT\alpha$ disomic haploids are sporulated, which indicates that both meiotic divisions and spore-wall formation occur in a fraction of the cells. However, mature asci containing viable spores are not formed. Since spore maturation does not take place in disomic haploids, it apparently cannot occur with the multiply aneuploid products that are formed (as expected from the segregation of a haploid genome through two meiotic divisions). Similarly, in haploid plants that undergo meiosis, the aneuploid products rarely mature (see Ivanov 1938).

Occasionally, in haploid plants, viable haploid products are formed when only one of the two meiotic divisions takes place (Belling and Blakeslee 1927; Lesley and Frost 1928). In yeast, the *spo13-1* mutation, which causes a/α diploids to undergo a single, generally equational division during sporulation (Klapholz and Esposito 1980b), also permits sporulation of $MATa/MAT\alpha$ disomic haploids (Klapholz 1980). The level of sporulation (two-spored asci) in these haploids can be as high as 20%. The spores exhibit 50% viability and are generally n + 1 in ploidy. Preliminary studies show that normal levels of recombination take place between *leu2* and *MAT* on the disome, and aberrant segregation of chromosome III, yielding a haploid (**a** or α) and a non-mating (a/α, $a/\alpha/\alpha$, or $a/a/\alpha$) spore pair, occurs approximately 15% of the time. Completely haploid *spo13-1* strains carrying the *mar1* mutation, which allows expression of silent mating-type information (Klar et al. 1979), sporulate with the same efficiency as a/α *spo13-1* strains (Klapholz 1980). The introduction of a meiotic hypo-recombination mutation, *spo11-1*, into the disomic *spo13-1* background eliminates recombination and aberrant segregation of chromosome III and greatly increases sporulation efficiency and spore viability (Klapholz 1980). These findings raise the possibility that recombination functions, perhaps nonhomologous exchange, contribute to the low efficiency of ascus production, spore inviability, and aberrant segregation observed during sporulation of *spo13-1* haploids (see below, Genetic Control of Meiosis and Spore Formation).

Another apparent case of haploid sporulation has been reported by Klar (1980). When **a** or α *kar1* (Conde and Fink 1976) haploids are mated with α/α or a/a *KAR1* diploids, respectively, and then plated on sporulation medium, unbudded zygotes containing six ascospores are frequently observed (Klar 1980; R. Malone, pers. comm.). Genetic analysis of the hexad products revealed that two of the spores had the same genotype as the haploid parent (Klar 1980). Using a $MATa/MATa$ disomic haploid *kar1* strain with non-complementing *leu2* and *his4* heteroalleles on the disome, Klar (1980) demonstrated the occurrence of high levels of recombination, a landmark event of meiosis, in the *kar1* nucleus. Whether the haploid *kar1* nucleus undergoes meiosis with an abortive reductional division, or a mitotic division with the enclosure of the two nuclei into spores due to the influence of the diploid nucleus (Klar 1980), remains an unresolved and intriguing question.

Trisomy and Triploidy

Two modes of chromosome pairing are possible for trisomic chromosomes: trivalent and bivalent-univalent. Most studies are consistent with the view that trivalent chromosome associations with random two-from-one disjunction predominate in singly trisomic strains. This has been shown in strains trisomic for chromosome I (Hawthorne and Mortimer 1960; Cox and Bevan 1962), chromosome III (Shaffer et al. 1971; Haber 1974), and chromosome XI (Culbertson and Henry 1973). The data from these analyses suggest that trivalent pairing may be as frequent as 100% for chromosome III (Shaffer et al. 1971) and 86% for chromosome XI (Culbertson and Henry 1973). In one study involving tetraploid strains that are trisomic for chromosome III, evidence that all three chromosomes can participate in a double exchange event without interference is presented (Riley and Manney 1978). The authors found that the level of exchange between *leu2* and its centromere (6.5 cM) was approximately three times that observed in normal diploids, proportional to the increase in available nonsister chromatid pairs from 4 to 12.

In another study of trisomic segregation involving chromosome I, James and Inhaber (1975) proposed bivalent-univalent chromosome association with preferential pairing of "like" homologs to account for their data. However, ascal types, which could either be due to a single gene-conversion event or trisomic pairing with random segregation, were not genetically analyzed to distinguish between the two possiblities. The authors also pointed out that other studies of trisomic segregation for chromosome I (Hawthorne and Mortimer 1960; Cox and Bevan 1962) do not suggest preferential pairing and that their own use of a homothallic strain as the origin of the like homologs (*ADE1*) and an unrelated heterothallic strain as the origin of the unlike chromosome (*ade1*) may have influenced the mode of pairing. Similar observations of preferential pairing have been made in crosses of α/α disomic haploids to a genetically unrelated **a** haploid derived from strain SK1 (J. Game, pers. comm.).

Although triploids sporulate at normal levels, their products are highly inviable; generally only 10–25% of spores germinate (Mortimer and Hawthorne 1969; Parry and Cox 1970; S. Klapholz, unpubl.). The surviving aneuploid products tend to be disomic for certain chromosomes more frequently than for others (Parry and Cox 1970). Furthermore, disomy for a particular chromosome is frequently coincident with disomy for others. In an early study, two triploids exhibited uncharacteristically high spore viability (60–80%), and one proved capable of generating asci containing two diploid plus two haploid spores (Pomper et al. 1953). This unusual behavior suggested that a nonrandom segregation mechanism might operate in certain triploid strains (Pomper et al. 1953).

Tetrasomy and Tetraploidy

Tetrasomic segregation for chromosome I has been examined in $2n + 2$ diploids (Cox and Bevan 1962; James and Inhaber 1975). Both studies indicated a predominance of bivalent pairing. In addition, James and Inhaber (1975) concluded that there was preferential pairing of "like" homologs in an $ADE1/ADE1/ade1/ade1$ tetrasomic. Roman et al. (1955) examined the segregation of markers in $+/+/-/-$ and $+/-/-/-$ configurations in tetraploids during sporulation. The observed ascus types for three loci, including MAT on chromosome III, suggest that both bivalent and tetravalent pairing occur and that the relative frequency of each may depend upon the particular chromosome.

Nondisjunction

The rare occurrence of nondisjunction has been reported in tetraploids, where it results in monosomic and trisomic spores (Roman et al. 1955; Bruenn and Mortimer 1970). Upon sporulation, monosomics yield two live (haploid) and two inviable (presumably nullosomic) spores. Presumptive monosomics for large chromosomes, such as VII, grow and sporulate poorly (Bruenn and Mortimer 1970; Parry and Zimmerman 1976). Most monosomics are unstable, giving rise to mixed clones containing both 2n and $2n - 1$ cells (Bruenn and Mortimer 1970). However, stable monosomics for chromosomes I and VI have been obtained (Strömnaes 1968; Bruenn and Mortimer 1970). Nondisjunction in diploids during meiosis, resulting in disomic and presumably nullosomic (inviable) spores, has been infrequently observed and has not been quantitated. Disomic strains have been recovered as spontaneous meiotic segregants from tetrads (Rodarte et al. 1968; Mortimer and Hawthorne 1973). Stable chromosome-III disomics have also been obtained from a sporulated diploid by selection (Shaffer et al. 1971).

A high frequency of nondisjunction ($\sim 5\%$) has been observed among the viable meiotic products of a diploid heterozygous for a ring chromosome (Strathern et al. 1981). This diploid contains one normal chromosome III and a circular derivative of chromosome III that lacks much of the right arm. Along with the increased nondisjunction frequency for chromosome III (> 10 times that of a normal diploid), a decrease in chromosome III meiotic map values, expected from the lethality associated with single exchange events between a circular and a linear chromosome, was found.

GENETIC CONTROL OF MEIOSIS AND SPORE FORMATION

This section addresses the functions of genes required for sporulation and the coordination of the various landmark events into an ordered sequence. Genetic dissection of cell differentiation by analysis of mutations that terminate or cause abnormal patterns of behavior has been extremely useful in reducing complex developmental systems into series of discrete steps more amenable to study. From the pleiotropic nature of the defects caused by the presence of such mutations, inferences can be made about the dependency relationships among various events and, in certain cases, gene functions can be deduced that were nc⁺ previously recognized by studies of wild-type cells. The utility of this approach is that in many cases, these inferences can be made in the absence of specific knowledge about the biochemical lesion caused by the mutation (see also Pringle and Hartwell, this volume).

Detection of Mutants

Three broad approaches have been taken to recover mutations causing defects in gamete formation (reviewed by Esposito and Esposito 1975, 1978; Baker et al. 1976). The first involves inducing and specifically selecting mutations affecting gametogenesis. The second involves screening for variants occurring in nature. The third entails analysis of mutants previously isolated for defects in other aspects of cell development for their behavior during gametogenesis. These procedures are expected to yield mutations in genes specifying enzymatic activities and/or structural components required for the execution of landmark events and mutations in genes that regulate the expression of the first class of genes. At present, it is a difficult matter to distinguish between structural and regulatory mutations, since there is almost a complete lack of knowledge of the specific enzymatic activities required for sporulation. One might expect, however, that recessive mutations would be found in both classes, whereas dominant mutations are more likely to be associated with regulatory functions. To gain a comprehensive view of the genetic control of meiosis and spore formation, it is thus desirable to develop procedures that permit the detection and recovery of both recessive and dominant mutations. Mutagenesis of diploid cells or cells of higher ploidy, in which sporulation normally occurs, a priori, is expected to yield a low recovery of recessive alleles. In yeast, this problem has been overcome in two ways: (1) by the use of homothallic strains to diploidize both recessive and dominant mutations induced in haploid spores (Esposito and Esposito 1969) and (2) by mutagenesis of n + 1 **a**/α disomic strains, which undergo a number of meiotic landmarks, such as DNA replication and commitment to recombination, but do not form viable ascospores (Roth and Fogel 1971; Roth 1976). Heterothallic haploid (Simchem et al. 1972) and diploid cells (Hopper and Hall 1975) have also been mutagenized to recover dominant mutations. Conditional and nonconditional sporulation-defective mutations have been isolated from sporulation-proficient wild-type strains employing these methods. In addition, mutations have been sought that suppress the inability of diploid cells to sporulate under nutritional conditions or in genetic backgrounds that do not normally permit ascus production.

The following specific criteria have been utilized to detect mutant strains: (1) absence of sporulation, (2) reduction in spore number per ascus, (3) spore inviability, (4) absence of intragenic recombination during meiosis, (5) sporulation of mating-type homozygotes, and (6) sporulation of glucose-grown cells in stationary phase. The rationales for selecting these criteria are briefly described below. The detailed procedures for the detection and genetic analysis of mutants recovered by the methods outlined above have been reviewed previously and will not be dealt with here (Esposito and Esposito 1975).

Schemes for the Selection and Recovery of Mutants

Taking advantage of the homothallic life cycle, Esposito and Esposito (1969) mutagenized haploid ascospores derived from a sporulation-proficient strain bearing the *HO* gene. Diploid survivors, resulting from mating-type switching and mating within the ascosporal clones, were screened for failure to produce mature asci at one or more temperatures. Strains that were respiratory-deficient and unable to metabolize acetate, or temperature-sensitive for mating ability or vegetative growth, were excluded from analysis. The rationale for this approach was based on the assumption that selection for defects in the production of mature ascospores would result in the unbiased recovery of mutations in gene functions required throughout meiotic development. It was anticipated that the exclusion of cell-cycle mutants would focus attention on gene functions uniquely required for meiosis. Conditional mutants were isolated to facilitate genetic and biochemical analysis (Esposito et al. 1970). Recessive mutations in 11 independent loci (*spo1* through *spo11*) and three dominant mutations were recovered by this procedure. The number of indispensable gene functions required for ascus production and nonessential for vegetative growth was estimated at 48 ± 27 on the basis of the number of repeat mutations within the same gene (Esposito et al. 1972). Temperature-shift experiments combined with genetic, cytological, and biochemical analyses indicate that these loci function at various stages throughout sporulation. In a second mutant hunt employing similar procedures, ten additional nuclear genes were identified, as well as a collection of conditional mutants that sporulate but form inviable ascospores when sporulated at a restrictive temperature (R. E. Esposito and M. S. Esposito, unpubl.). The latter class of mutants was sought to recover lesions resulting in aneuploid spore production due to defects in genetic recombination and/or chromosome segregation. Mutations that affect the viability of the ascospores often reduce ascus yield; many of these mutations may thus be in loci already defined by *spo* alleles.

A *MATa/MATα* n + 1 disomic strain was utilized by Fogel and Roth (1971) and Roth (1973, 1976) to specifically isolate mutations in genes required for premeiotic DNA replication and/or genetic recombination. The disomic chromosome pair was marked with *leu2* heteroalleles. Survivors of mutagenesis were replica-plated to sporulation medium to induce sporulation and, subsequently, to leucineless medium to detect recombinants. The rationale for this approach is based on the premise that selection of mutants in a near-haploid strain will permit recovery of recessive (as well as dominant) mutations on all chromosomes except for the disome. This procedure also permits a specific enrichment for mutants blocked early in meiotic events, at or prior to commitment to recombination. Variants defective in both DNA replication and recombination were distinguished from those blocked only in prototroph production by screening for the occurrence of premeiotic S phase. The mutations *mei1*, *mei2*, and *mei3* were isolated by this procedure (Roth 1973). An additional 48 temperature-sensitive mutants blocked in prototroph production were recovered, 10 of which were also blocked in premeiotic S phase. The number of independent loci among these mutants has not yet been determined.

Another novel approach has been to screen for mutations specifically altering mating-type control of sporulation, which permits **a**/**a** and α/α diploids to sporulate (Hopper and Hall 1975; Hopper et al. 1975). The rationale for this mutant hunt was based on the assumption that the mating-type locus controls the initial cellular response to the nutritional conditions that induce sporulation. This interpretation depends on the observation that mating-type homozygotes do not execute premeiotic S phase, an early landmark event. It was anticipated that the selection of mutants with altered mating-type control would therefore provide information on how this control operates. Mutants were selected in an α/α diploid heterozygous for two recessive drug-resistance markers, *cyh2* and *can1*. The formation of haploid meiotic products was monitored by selection for the simultaneous segregation of both drug-resistance markers into the same cell. The segregation of two markers was employed to reduce the probability of recovering drug-resistant diploid clones due to mitotic recombination. Among several mutants isolated by this procedure, the *csp1* mutant, studied most extensively, was later found to be identical to *rme1*, isolated by Kassir and Simchen (1976).

Dawes (1975) used an analogous rationale to isolate mutants that sporulate under nutritional conditions in which the wild type does not. The condition chosen was stationary phase following growth in glucose. By this approach, he also hoped to isolate mutations in genes that control the initiation of sporulation. Furthermore, as mentioned earlier, it was anticipated that some of the difficulties in distinguishing a sporulation-specific function from a starvation response might be avoided by selecting for a "turn-on" of gene functions under starvation or mild starvation conditions. Mutations were induced by the procedure of Esposito and Esposito (1969), employing the same parental strain used to isolate *spo* mutants (Dawes 1975). Mutagenized ascospores were germinated, grown to stationary phase, and treated with ether to selectively kill vegetative cells (Dawes and Hardie 1974). After repeating this protocol a number of times, survivors were screened for the presence of asci in stationary phase. Mutations in several independent genes have been isolated by this method. Strains carrying these lesions generally do not grow well on a wide range of nonfermentable carbon sources. Reversions of one of the mutations, *spd1*, were selected by their ability to grow normally in glycerol medium. It was expected that such reversions might lead to the recovery of *spo* mutations causing defects in the initiation of meiosis, since they modify the expression of another mutation thought to act early in the process. Indeed, *spo* mutations were recovered. It is not yet known how many independent genes are represented by these *spo* mutations (Dawes et al. 1980).

Screening of Wild-type Strains

Grewal and Miller (1972) surveyed a large number of sporogenic strains of *Saccharomyces* and found that approximately 10% of them (15 of 140) produced predominantly two-spored asci. Among these natural variants, chromatin staining revealed that three strains of *S. cerevisiae* (var. *ellipsodeus*), ATCC4098, ATCC4117, and 19e1, formed asci with only two nuclei. The remaining strains produced asci with four nuclei, only two of which were present in mature ascospores. The DNA content of the vegetative cells and ascospores of ATCC4117 and 19e1 proved to be diploid (Grewel and Miller 1972), and only one nuclear division during sporulation was observed by Giemsa staining (C. Robinow, cited in Grewal and Miller 1972). Electron microscopy studies of sporulation in these two variants revealed many features in common with standard tetrad-producing meiosis (Moens 1974; Moens et al. 1977). These include landmarks of meiosis-I prophase (Moens 1974; Moens et al. 1977) and a single spindle that is initially meiosis-I-like (or mitotic), in character and later resembles that of meiosis II (Moens 1974).

Genetic analysis of ATCC4117 led to the isolation from it of two recessive mutations, designated *spo12-1* and *spo13-1*, which were briefly described in the section on Genetic Recombination (Klapholz and Esposito 1980a; see Table 1). Each mutation alone causes cells to undergo a single, generally equational division during sporulation to produce asci with two diploid or near-diploid spores (Klapholz and Esposito 1980a,b). Recombination precedes the single division, at typically high meiotic levels and, occasionally, reductional or aberrant segregation for one or more chromosomes takes place (Klapholz and Esposito 1980b). The number of different genes involved in the two-diploid-spored phenotype of natural variants appears to be small.

The 19e1 strain has recently been shown to be *spo12 SPO13*, and ATCC4098, *spo12 spo13*, in genotype by complementation analysis (S. Klapholz, unpubl.). Surprisingly, *spo13-1* was also discovered to be an ocher nonsense mutation, which indicates that this allele is not a wild-type polymorphism. These mutations are discussed in more detail later in this section.

Analyses of two-spored asci in other strains produced under standard sporulation conditions have demonstrated that, in most cases, the ascospores are the haploid products of two meiotic divisions (Bevan 1953; Magni 1958; Takahashi 1962; Grewal and Miller 1972; Esposito et al. 1974; James 1974). Haploid two-spored asci may contain either randomly or nonrandomly packaged meiotic products. If the two spores contain a random selection of two of the four meiosis-II nuclei, then approximately 33% will contain sister nuclei and be alike for centromere-linked markers, and 67% will contain nonsister nuclei. In one study of a strain that produced 44% two-spored asci (Takahashi 1962), individual centromere-marker segregation patterns indicated random inclusion of two of the four haploid genomes into ascospores. However, as pointed out by Davidow et al. (1980), the number of tetratype asci for two centromere markers was in vast excess of that expected, making it impossible to interpret whether the two haploid spores truly contained a random assortment of genomes. James (1974) genetically analyzed several strains that produced a large proportion of asci with fewer than four haploid spores (~80% of the total). The two-spored asci of one strain contained random meiotic products, whereas those formed by three other strains contained pairs of nonsister nuclei at least 90% of the time. In another study, the former strain produced almost exclusively nonsister spore pairs (see Discussion in Davidow et al. 1980); the reason for the discrepancy between the two studies is not clear.

Analysis of Mutants Defective in Other Aspects of Cell Development

All of the screenings described previously had as their primary goal the detection and recovery of sporulation-defective mutants. Sporulation-defective variants have also been identified by indirect means, by screening mutants affecting other aspects of cell development. The classes that have received the most attention have been the *cdc, rad,* and *mat* mutants. Sporulation-defective mutations have also been revealed by genetic analysis of strains bearing mutations in genes that, a priori, would not be expected to be required for sporulation (e.g., *kex2, tup1, spe2, SUP3*; see Table 1).

The *cdc* mutants were examined by Simchen (1974), with the expectation that some of the functions required for DNA replication and nuclear division in mitosis might also be utilized in meiosis. He found that all of the *cdc* mutants, other than those defective in bud emergence and cytokinesis, are also defective in meiosis (Table 1). The *rad* mutants, particularly those involved in the repair of X-ray damage, were examined by several groups because of their involvement in mitotic recombination and repair. The expectation was that meiosis and mitosis would require some common recombination functions (Game et al. 1980; Prakash et al. 1980; Malone and Esposito 1981). This was found to be the case by the analysis of a number of *rad* mutants, discussed later in this section and in Table 1. Finally, *mat* mutants were examined because of the requirement for sporulation of heterozygosity at the mating-type locus (MacKay and Manney 1974a,b). Such studies have led to the conclusion that the *MATa*- and *MATα2*-gene functions are necessary for sporulation, since the *mat*a* mutation (Kassir and Simchen 1976) and lesions in the *MATα2*, but not *MATα1*, complementation group (Sprague et al. 1981; Strathern et al. 1981) prevent sporulation. *MATa/matα2* and *mata*/MATα* diploids thus lack the functional *MAT* heterozygosity required for sporulation (see review by Herskowitz and Oshima, this volume).

Terminal or Abnormal Mutant Phenotypes

Mutations affecting the process of sporulation can be divided into four general categories: (1) mutations that result in reduced ability or failure to complete sporulation, (2) mutations that affect the viability and/or ploidy of the final meiotic products, (3) mutations that permit sporulation in genetic backgrounds not normally permitting sporulation, and (4) mutations that allow sporulation under nutritional conditions normally not conducive to sporulation. The phenotypes of mutants whose defective sporulation behavior has been attributed to one or, in a few cases, two nuclear genes are summarized in Table 1.

A number of other nuclear genes in addition to those listed in Table 1 are required for sporulation. These include the numerous *PET* loci required for mitochondrial function, loci required for fatty acid metabolism (Keith et al. 1969), and loci required for the TCA cycle (Ogur et al. 1965). Mutations in all of these loci render the cells respiratory-deficient and, hence, incapable of sporulation. Several loci involved in amino acid biosynthesis are also required for sporulation (Wejksnora and Haber 1974; Lucchini et al. 1978). For example, *met2* and *met13* mutations result in an imbalance in different rRNA species and a lack of sporulation. These defects are reversible by the addition of methionine to the sporulation medium (Wejksnora and Haber 1974). Similarly, *aro1* and *aro2* mutations result in an almost complete absence of sporulation, which can be reversed by supplementing the sporulation medium with three aromatic amino acids. The *aro1* mutation acts early, preventing premeiotic DNA replication (Lucchini et al. 1978).

All cytoplasmic petite mutants and many, but not all, *mit⁻* mutants are unable to sporulate (Ephrussi and Hottinguer 1952; Sherman and Slonimski 1964; Pratje et al. 1979). In addition, a cytoplasmic condition, AM (abnormal meiosis), induced by ^{32}P decay, causes an extra round of DNA synthesis

Table 1 Phenotypes of Mutants Affecting Sporulation

Mutant locus	Map position	Description of mutant sporulation phenotype	Reference
cdc2	IV, L	The *cdc2-1* homozygote sporulates normally at 21°C. At 33.5°C, 0–14% asci are formed. Premeiotic DNA synthesis occurs. Commitment to intragenic recombination is reduced to 7% of the 21°C level. No SCs are observed. Cells arrest after SPBI duplication or separation. No complete spindles are formed.	Simchen (1974); Schild and Byers (1978)
cdc4	VI, L	*cdc4-1* and *cdc4-3* homozygotes sporulate normally at 25°C. At 34°C, <0.5% asci are formed. In *cdc4-3* at 34°C, no haploidization occurs. There is a 20–40% increase in DNA content, half of which is due to mitochondrial replication. (At 36°C, an increase in DNA of <10% is observed.) Commitment to intragenic recombination occurs at 50% of the wild-type level. SCs accumulate during arrest in ~15% of cells. Glycogen degrading activity is absent.	Simchen (1974); Byers and Goetsch (1975); Simchen and Hirschberg (1977); Zamb and Roth (1977); Horesh et al. (1979); Clancy et al. (1980)
cdc5	XIII, L	The *cdc5-1* homozygote sporulates at near normal levels at 25°C. At 34°C, no asci are formed. No haploidization occurs. Premeiotic DNA synthesis takes place. Commitment to intragenic recombination occurs at wild-type levels. SCs are formed, and paired lateral elements devoid of the central region accumulate during arrest.	Simchen (1974); Horesh et al. (1979)
cdc7	IV, L	*cdc7-1* and *cdc7-4* homozygotes sporulate normally at 21°C. At 34°C, 0–1% asci are formed. Premeiotic DNA synthesis occurs. No SCs are formed. Commitment to intragenic and intergenic recombination is absent. Cells arrest after SPBI duplication. The SPBs eventually enlarge, invaginate from the nuclear envelope into the center of the nucleus, and may fragment into three or four smaller SPBs.	Simchen (1974); Schild and Byers (1978); Horesh et al. (1979)
cdc8	X, R	The *cdc8-2*, *cdc8-3*, and *cdc8-4* homozygotes exhibit reduced sporulation at 25°C. *cdc8-1* sporulates at a reduced level at 21°C (18% asci), and normally at 17°C (55% asci). At 33.5°C, no asci are formed in any of the mutants. In *cdc8-1*, premeiotic DNA synthesis does not occur. Temperature-shift experiments demonstrate that DNA elongation is blocked. Cells arrest after SPBI duplication and do not form SCs.	Simchen (1974); Zamb and Roth (1977); Schild and Byers (1978)

cdc9	IV, L	The cdc9-2 homozygote exhibits reduced sporulation at 25°C and 33.5°C. At 33.5°C haploidization is not observed. Commitment to intragenic recombination occurs at normal levels. (cdc9-1 haploids are deficient in DNA ligase activity.)	Simchen (1974); Johnston and Nasmyth (1978)
cdc13		The cdc13-1 homozygote sporulates normally at 25°C. No asci are formed at 33.5°C.	Simchen (1974)
cdc14	VI, R	cdc14-1 and cdc14-2 homozygotes sporulate normally at 25°C. No asci are formed at 33.5°C.	Simchen (1974)
cdc16	XI, L	The cdc16-1 homozygote sporulates at a greatly reduced level at 25°C. No asci are formed at 33.5°C.	Simchen (1974)
cdc17		The cdc17-1 homozygote sporulates at a reduced level at 25°C and 33.5°C.	Simchen (1974)
cdc20		The cdc20-1 homozygote exhibits greatly reduced or no sporulation at 25°C. No asci are formed at 33.5°C.	Simchen (1974)
cdc21	XV, R	Homozygotes for a mutation in the cdc21 locus (the structural gene for thymidylate synthetase), cdc21-1, sporulate normally at 21°C. No asci are formed at 33.5°C. No premeiotic DNA synthesis occurs, due to a defect in elongation. Commitment to intragenic recombination is absent. No SCs are formed. Cells arrest with duplicated SPBIs.	Game (1976); Bisson and Thorner (1977); Schild and Byers (1978)
cdc23		The cdc23-1 homozygote does not sporulate at 25°C or 33.5°C.	Simchen (1974)
cdc25		The cdc25-2 homozygote sporulates normally in acetate sporulation medium at 25°C and 34°C. It also sporulates in rich or defined acetate growth medium at 34°C.	Shilo et al. (1978)
cdc28	II, R	The cdc28 homozygote exhibits greatly reduced sporulation at 25°C and 34°C.	Simchen (1974); Shilo et al. (1978)
cdc35		The cdc35-1 homozygote sporulates normally in acetate sporulation medium at 25°C and 34°C. It also sporulates in acetate-based rich or defined growth medium at 34°C.	Shilo et al. (1978)

Table 1 *(Continued)*

Mutant locus	Map position	Description of mutant sporulation phenotype	Reference
cdc40		The *cdc40* homozygote sporulates at 25°C. No asci are formed at 34°C. Premeiotic DNA synthesis takes place. Haploidization does not occur. Commitment to intragenic recombination is absent. Cells arrest prior to the first nuclear division.	Kassir and Simchen (1978)
chl	XVI, L	The *chl* homozygote sporulates at the normal level but produces spores with reduced viability. Intergenic recombination on chromosome III is reduced to half the wild-type level.	Liras et al. (1978)
con1		The *con1* homozygote forms <1% asci. Premeiotic DNA synthesis occurs. Commitment to intragenic recombination is absent.	Fogel and Roth (1974)
con2		The *con2* homozygote forms 6% asci containing inviable ascospores. Premeiotic DNA synthesis occurs. Commitment to intragenic recombination is absent.	Fogel and Roth (1974)
con3		The *con3* homozygote forms 24% asci containing inviable ascospores. Premeiotic DNA synthesis occurs. Commitment to intragenic recombination is absent.	Fogel and Roth (1974)
DSM1	VII, R	See *spo8, DSM1*.	
gcn1		*gcn1* (*1-1, 1-5, 1-9*) homozygotes (D-glucosamine auxotrophs) sporulate normally until ascospore maturation, with or without D-glucosamine in the sporulation medium. Haploidization occurs at the wild-type level. Intergenic exchange is normal among the haploids. Commitment to intragenic recombination occurs at wild-type levels. *gcn1-1* diploids form 25–60% abnormal asci; spores lack refractility and are easily digestible by external β-glucanases. The mutant spores lack the outer surface layer.	Whelan and Ballou (1976); Ballou et al. (1977)
kex2	XIV, R	*kex2* homozygotes (11 alleles tested) form 0–1% asci. Premeiotic DNA synthesis occurs. Commitment to intragenic recombination occurs at wild-type level. Approximately 30% of cells reach multinucleate stages of sporulation.	Liebowitz and Wickner (1976)

Mutant	Location	Description	References
mei1		The *mei1* homozygote does not sporulate. Premeiotic DNA synthesis does not occur. Commitment to intragenic recombination occurs at < 1% of the wild-type level. Carbohydrate, protein, and RNA synthesis occur as in wild type.	Roth (1973)
mei2 mei3		The *mei2 mei3* double mutant diploid does not sporulate. Premeiotic DNA synthesis does not occur. Commitment to intragenic recombination is absent. Carbohydrate and protein synthesis occur normally. RNA synthesis may be slightly defective. The phenotypes of the single *mei2* and *mei3* homozygotes have not been described.	Roth (1973)
mutant 17017		Strain 17017 and its meiotic derivatives are incapable of sporulation. This is due to a mutation that causes a 5-fold reduction in the level of proteinase-A activity. Proteinase-A activity rises 0–2-fold during sporulation in the homozygous mutant diploids, compared to over 3.5-fold in the wild type.	Betz (1975)
pep4	V, L	The *pep4-3* homozygote does not sporulate. No haploidization occurs. It exhibits a 60–70% reduction in total protein degradation during sporulation.	Zubenko and Jones (1981)
prb1		Homozygotes for mutations in *prb1* (the structural gene for proteinase B) form asci containing abnormal spores. The spores are smaller than usual and are embedded in a thick matrix. Haploidization is normal. Protein degradation is reduced to 50% of the wild-type level during sporulation.	Zubenko et al. (1979); Zubenko and Jones (1981)
prc1		*prc1-1* and *prc1-125* homozygotes (defective in the structural gene for proteinase C) sporulate at a normal level. Haploidization occurs normally. Protein degradation is reduced to 80% of the wild-type level during sporulation.	Wolf and Fink (1975); Zubenko and Jones (1981)
rad6	VII, L	The *rad6-1* homozygote does not sporulate. No haploidization takes place. Pre-meiotic DNA synthesis occurs. Commitment to intragenic recombination is at 9% of the wild-type level. Viability remains high throughout sporulation.	Parry and Cox (1968); Game and Mortimer (1974); Game et al. (1980)
rad50	IV, R	The *rad50-1* homozygote sporulates at a reduced level (40% asci). No haploidization occurs. Viability declines rapidly starting at or prior to S. No viable ascospores are produced. Premeiotic DNA synthesis occurs. Commitment to intragenic recombination is absent.	Game and Mortimer (1974); Game et al. (1980); Malone and Esposito (1981)

Table 1 (Continued)

Mutant locus	Map position	Description of mutant sporulation phenotype	Reference
rad51	V, R	rad51-1 and mut5-1 (both alleles of the rad51 gene) homozygotes sporulate at a reduced level. mut5-1 diploids form 7–10% asci, but viable spores are found at a frequency of only 10⁻⁴. Premeiotic DNA synthesis occurs. Commitment to intragenic recombination is at 10% of the wild-type level. Among rare viable spores, intragenic and intergenic exchanges are at wild-type levels.	Game and Mortimer (1974); Morrison and Hastings (1979)
rad52	XIII, L	The rad52-1 homozygote produces <30% asci. No haploidization occurs. Viability rapidly declines, as in rad50-1. No viable spores are produced. Premeiotic DNA synthesis occurs. Commitment to intragenic and intergenic recombination is absent. By 48 hours, 21% binucleate and 14% tetranucleate cells are formed.	Game and Mortimer (1974); Game et al. (1980); Prakash et al. (1980)
rad53		The rad53-1 homozygote sporulates at a normal level. Only 10% of spores are viable.	Game and Mortimer (1974)
rad55	IV, R	The rad55-1 homozygote exhibits reduced sporulation. Less than 25% of spores are viable.	Game and Mortimer (1974)
rad57	IV, R	The rad57-1 homozygote sporulates normally at 34.5°C. At 23°C, sporulation is reduced (40% asci). No haploidization occurs. Viability rapidly declines, as in rad50-1. No viable spores are produced. Commitment to intragenic recombination is absent.	Game and Mortimer (1974); Game et al. (1980)
rec2		The rec2 homozygote produces 1-2% asci. Less than 10% of spores are viable.	Rodarte-Ramon (1972)
rec3		The rec3 homozygote does not sporulate.	Rodarte-Ramon (1972)
rec4		The rec4 homozygote sporulates at a normal level (60%) and exhibits high spore viability (90%). Single-site conversion at the arg4 locus is drastically reduced, but exchange in other regions of the genome is normal.	Rodarte-Ramon (1972); Sanfilippo (1976)
rme1	VII, R	α/α and a/a and a/a* rme1 homozygotes exhibit reduced sporulation (30% asci). Premeiotic DNA synthesis occurs. Commitment to intragenic recombination is	Hopper et al. (1975); Kassir and Simchen (1976)

reduced to 25–30% of the wild-type level. Among spores, however, recombination levels are normal. Haploid *rme1* strains also undergo premeiotic DNA replication but do not complete sporulation (*rme* was formerly called *CSP* by Hopper et al. [1975]).

Gene	Location	Description	References
sca		**a**/**a** and α/α *sca* homozygotes are capable of sporulation.	Gerlach (1974)
spd1	XV, L	*spd1-1* and *spd1-2* homozygotes form asci on rich medium containing acetate or glycerol as the carbon source. They do not exhibit nitrogen repression of sporulation.	Dawes (1975); Vezinhet et al. (1979)
spe2	XV, L	The *spe2-2/spe2-4* diploid (deficient in S-adenosylmethionine decarboxylase activity and lacking spermine and spermidine) is unable to sporulate in the absence of exogenous spermine or spermidine. No haploidization occurs.	Cohn et al. (1978)
spo1	XIV, R	The *spo1-1* homozygote sporulates normally at 25°C. At 34°C, <1% asci are formed. Premeiotic DNA synthesis occurs. Commitment to intragenic and intergenic recombination is normal. SCs are formed. Cells arrest during SPBI duplication.	Esposito et al. (1970); R. Esposito and M. Esposito (1974); Moens et al. (1974); Klapholz (1980)
spo2		The *spo2-1* homozygote sporulates normally at 25°C. At 34°C, <1% asci are formed. Premeiotic DNA synthesis takes place. Commitment to intragenic and intergenic recombination occurs. The nucleus divides in two at meiosis I and again at meiosis II, yielding four separate nuclei. Prospore-wall development is abnormal, and cells that reach the tetranucleate stage arrest with small anucleate immature ascospores and free unenclosed nuclei.	Esposito et al. (1970); R. Esposito and M. Esposito (1974); Moens et al. (1974)
spo3		The *spo3-1* homozygote sporulates at near normal levels at 20°C and 25°C. At 30°C, 4% asci are formed, containing one or two randomly packaged haploid spores. At 34°C, no asci are formed. Premeiotic DNA synthesis takes place. Commitment to intragenic and intergenic recombination occurs. Meiosis-II prospore-wall closure precedes the completion of the meiosis-II division, resulting in unenclosed nuclei and anucleate and aneuploid spores that fail to mature. By 36 hours, 17% of cells are binucleate and 6% are trinucleate or tetranucleate.	Esposito et al. (1970); R. Esposito and M. Esposito (1974); M. Esposito et al. (1974); Moens et al. (1974)

Table 1 (Continued)

Mutant locus	Map position	Description of mutant sporulation phenotype	Reference
spo4		The spo4-1 homozygote sporulates normally at 20°C. At 34°C, development precedes essentially as described for spo3-1. Less than 1% asci are formed.	Esposito et al. (1972); M. Esposito and R. Esposito (1974); Esposito and Esposito (1978); P. Moens (unpubl.)
spo5		The spo5-1 homozygote sporulates normally at 20°C. At 34°C, development precedes essentially as described for spo3-1. No asci are produced.	Esposito et al. (1972); M. Esposito and R. Esposito (1974); Esposito and Esposito (1978); P. Moens (unpubl.)
spo7		The spo7-1 diploid produces < 1% asci at 20°C, 2% asci at 30°C, and no asci at 34°C. At 34°C, premeiotic DNA synthesis does not occur. Commitment to intragenic and intergenic recombination is absent. Cells generally arrest prior to SPB1 duplication. By 36 hours, 90% of cells are mononucleate, 6% are binucleate, and 3% are trinucleate or tetranucleate. Glycogen degrading activity is absent.	Esposito et al. (1972); M. Esposito and R. Esposito (1974); Esposito et al. (1975); Clancy et al. (1980); P. Moens (unpubl.)
spo8		The spo8-1 diploid produces < 20% asci at 25°C. At 34°C, no asci are formed. Premeiotic DNA synthesis is reduced at both 25°C and 34°C (max. increase, 30%). Commitment to intragenic recombination is reduced to 10–20% of the wild-type level. Commitment to intergenic recombination is absent. Cells arrest prior to SPB1 duplication.	Esposito et al. (1972); M. Esposito and R. Esposito (1974); Fast (1978); S. Klapholz (unpubl.); P. Moens (unpubl.)
spo8 DSM1	VII, R	The spo8-1 DSM1 double mutant sporulates normally at 25°C. At 34°C, no asci are formed. Haploidization occurs at reduced levels. Premeiotic DNA synthesis takes place. Commitment to intragenic recombination occurs. Commitment to intergenic exchange is reduced to 30% of the wild-type level. Among haploidized cells, normal levels of intergenic recombination are observed. DSM1 has no apparent affect on Spo⁺ strains.	Fast (1978)
spo9		spo9-1 and spo9-2 diploids sporulate normally at 20°C. At 34°C, no asci are formed. spo9-1 homozygotes do not undergo premeiotic DNA synthesis. Most cells arrest prior to SPB1 duplication.	Esposito et al. (1972); M. Esposito and R. Esposito (1974); P. Moens (unpubl.)

Gene	Map position	Description	References
spo10		The *spo10-1* homozygote produces ~1% asci at 20°C, 24% at 30°C, and 1% at 34°C. At 34°C, premeiotic DNA synthesis occurs. Cells arrest prior to SPBI duplication and accumulate aggregates of SCs.	Esposito et al. (1972); M. Esposito and R. Esposito (1974); Esposito and Esposito (1978); S. Klapholz (unpubl.)
spo11	VIII, L	The *spo11-1* homozygote sporulates at a reduced level at 25°C (<30% asci), and <10% of spores are viable. At 25°C, premeiotic DNA synthesis occurs. At 34°C, 0–10% asci are formed. Viability declines rapidly to 2% by 48 hours. The original *spo11-1* isolate (M90) shows a 10% increase in DNA synthesis at 34°C. No SCs are formed, but unpaired lateral elements are observed. Both meiotic divisions occur in 25% of cells, and one or more immature ascospores are present in ~ 30% of arrested cells. Recent studies indicate that premeiotic DNA synthesis occurs at the normal level in some *spo11-1* diploids at 34°C. In these strains, haploidization for all or many chromosomes occurs at a frequency of ~1%. Commitment to intragenic recombination is absent. In addition, no intragenic or intergenic recombination is detected among the rare haploidized products.	Esposito et al. (1972); M. Esposito and R. Esposito (1974); Moens et al. (1977); Fast (1978); Klapholz (1980); Klapholz and Esposito (1980c); C. Giroux (unpubl.)
spo12		The *spo12-1* homozygote sporulates at normal levels, producing asci containing only two spores. The spores have diploid or near-diploid genomes and exhibit normal or slightly reduced (80–90%) viability. Intergenic and intragenic recombination occur at wild-type or higher levels. 95% of asci result from a single equational division; the remaining 5% result from a single reductional or mixed division. Aberrant chromosome segregation yielding monosomic and trisomic spore pairs also takes place (at a frequency of ~10% for chromosome III).	Klapholz and Esposito (1980a,b)
spo13	VIII, R	The *spo13-1* (an ocher allele) homozygote sporulates at normal levels, producing asci containing only two spores. The spores have diploid or near-diploid genomes and exhibit reduced (<75%) viability. Intergenic and intragenic recombination occur at wild-type or higher levels. 95% of asci result from a single equational division; the remaining 5% result from a single reductional or mixed division. Aberrant chromosome segregation yielding monosomic and trisomic spore pairs also takes place (at a frequency of ~15% for chromosome III).	Klapholz and Esposito (1981a,b)

265

Table 1 (Continued)

Mutant locus	Map position	Description of mutant sporulation phenotype	Reference
SUP3	XV, L	The *SUP3-o* (ocher suppressor) homozygote sporulates at a reduced level ($<10\%$ asci). Premeiotic DNA synthesis occurs at ~40% of the wild-type level. One or both nuclear divisions occur in 20% of cells. By inoculating cells into sporulation medium from logarithmic growth in glucose-based rich medium, normal levels of sporulation can be achieved. (Other ocher suppressors affect sporulation to varying extents.)	Rothstein et al. (1977)
tra3	XV, R	The *tra3* homozygote sporulates normally and exhibits readiness at 25°C. At 34°C, no sporulation or readiness occurs.	Shilo et al. (1978)
tup1	III, R	*tup1-1*, *cyc9-1*, *flk1*, and *umr7* (all alleles of *tup1*) homozygotes exhibit reduced or no sporulation: 20% for *flk1/flk1*; 0–4% for *tup1/tup1*; 0% for both *cyc9-1/cyc9-1* and *umr7/umr7*.	Wickner (1974); Lemontt et al. (1980); Rothstein and Sherman (1980); Stark et al. (1980)

Abbreviations: SC, synaptonemal complex; SPBI, meiosis-I SPB.

during sporulation, which results in abnormal marker segregation patterns (Hottinguer-de-Margerie 1967; Moustacchi et al. 1967).

Coordination of Events: Relationships Between Mitosis and Meiosis

Entry into Meiosis Requires "Start" Functions

Initiation of the cell cycle, which begins at a point in G_1 termed Start (Hartwell 1971), requires several gene functions, including *CDC28, CDC25, CDC35,* and *TRA3* (Hereford and Hartwell 1974; Wolfner et al. 1975; Johnson et al. 1977). To determine whether any Start functions are required for entry into meiosis, which also occurs from G_1, Shilo et al. (1978) undertook a study of four temperature-sensitive Start mutants. They found that the *CDC28* and *TRA3* loci are required for both initiation of the cell cycle and entry into meiosis (see Table 1). The *cdc25* and *cdc35* mutations had no effect on normal sporulation. However, unlike the wild type, diploids carrying either of these mutations were capable of sporulation in acetate-based growth medium at their restrictive temperatures (see Table 1). Thus, the authors proposed that the *CDC25* and *CDC35* loci may regulate the choice between meiosis and mitosis in response to nutritional stimuli.

Nonconditional *spd1* mutations also permit sporulation of diploid cells in rich medium and prevent growth of haploid and diploid cells in a wide array of nonfermentable carbon sources (Dawes 1975; Vezinhet et al. 1979). The mutant cells arrest during G_1, at or prior to the *cdc28-1* arrest point (Start), when incubated in defined glycerol medium (Vezinhet et al. 1979). The *spd1* mutation may induce sporulation as a consequence of inhibiting growth; alternatively, it may turn on sporulation-specific functions that are epistatic to the normal growth functions.

Premeiotic DNA Synthesis and Mitotic DNA Synthesis Require Some Gene Functions in Common

A number of mutants have been described that fail to undergo a complete round of premeiotic DNA synthesis. These include *cdc4-3, cdc8-1, cdc21-1, mei1, mei2, mei3, spo7-1, spo8-1,* and *spo9-1* (Table 1). The three thermal-sensitive *cdc* mutants and the *spo7-1* mutant are also defective in some aspects(s) of mitotic DNA metabolism. During the cell cycle, *cdc4-3* arrests at SPB separation, prior to the initiation of S phase, and *cdc8-1* and *cdc21-1* arrest during the elongation of DNA synthesis (Hartwell 1971, 1973, 1976; Byers and Goetsch 1974). The *spo7-1* mutation causes antimutator activity for certain classes of spontaneous mutations (Esposito et al. 1975).

Not all mutations that prevent mitotic DNA replication block premeiotic S phase. For example, thermal-sensitive mutations in the *CDC7* gene, required for the initiation of DNA synthesis during the cell cycle (Hartwell 1973, 1976), do not prevent premeiotic DNA synthesis (Schild and Byers 1978).

However, although the mutant cells appear to undergo the normal level of DNA synthesis during sporulation at 33.5°C, they do not exhibit commitment to recombination or form SCs. Upon release to 21°C, SC formation and commitment to recombination occurs. Release is sometimes accompanied by a small (10–20%) HU-resistant increase in DNA content (Schild and Byers 1978). These findings indicate that the *CDC7*-gene product plays a different role in mitosis and meiosis but, in each case, is required for some aspect of DNA metabolism. Schild and Byers (1978) suggest that the *CDC7*-gene product may be required for pachytene repair synthesis which, in other systems, is insensitive to HU (Hotta et al. 1977; Stern and Hotta 1977).

Meiotic and Mitotic Recombination Share Common Gene Functions but Are Also under Separate Genetic Control

Three sporulation-defective mutations, *rad50-1, rad52-1* and *spo11-1*, have been examined in detail for their effects on recombination during mitosis and meiosis. All three abolish meiotic exchange (Game et al. 1980; Klapholz 1980; Prakash et al. 1980; Malone and Esposito 1981). Two of them, *rad50-1* and *rad52-1* also affect mitotic exchange. The *rad50-1* mutation causes a mitotic hyper-recombination phenotype (Malone and Esposito 1981), whereas *rad52-1* results in a deficiency in spontaneous and induced homologous exchange and mating-type switching (Saeki et al. 1974; Resnick 1975; Malone and Esposito 1980). Only *spo11-1* appears to be meiosis-specific. It has no detectable effect on homologous mitotic exchange or mating-type switching (Esposito and Esposito 1969; Klapholz and Esposito 1980c). In contrast to the behavior of *spo11-1*, another mutation, *rem1-1*, which causes elevated exchange in mitosis, has no effect in meiosis (Golin and Esposito 1977). Hence, *rem1-1* is mitosis-specific. The behavior of these mutants thus provides evidence for both common and independent gene functions governing the recombination processes in mitosis and meiosis (for general reviews of mitotic vs. meiotic recombination, see Malone et al. [1980] and Esposito and Wagstaff [this volume]).

Coordination of Events: Integration of Meiotic Landmarks

Premeiotic DNA Synthesis Is Required for Sporulation

Without exception, all strains that fail to undergo premeiotic DNA replication do not sporulate. These include sporulation-defective mutants listed in Table 1 and strains that lack functional alleles of both *MATa* and *MATα*. This latter class includes **a**/**a** and α/α-mating-type homozygotes (Roth and Lusnak 1970) and **a**/α strains bearing mutations in *MATa* (Kassir and Simchen 1976). Diploid **a**/α strains that contain mutations in the *MATα2* complementation group (Strathern et al. 1981) also do not sporulate, presumably because they too are blocked in premeiotic S phase. Two other arguments can be made to support the view that premeiotic S is required for sporulation. First, data described in the next section indicate that premeiotic S is necessary for commitment to recombination, normally a prerequisite for the formation of euploid genomes. Second, modifiers (*DSM1, rme1, sca*; see Table 1) that permit DNA replication in genetic backgrounds that normally do not allow sporulation also restore ascus production.

The behavior of modifiers has been examined in detail in the interactions involving *rme1* and mating-type control of sporulation and *spo8-1* and *DSM1*. The *rme1* allele allows a full round of DNA replication and partially restores commitment to intragenic recombination and ascus production in **a**/**a** and α/α strains (Hopper et al. 1975) and in **a**/α diploids carrying the *mat***a*** mutation (Kassir and Simchen 1976). The interaction of *spo8-1* and *DSM1* is more complex. In a *spo8-1 dsm1* strain, premeiotic S phase, recombination, and ascus production are severely reduced at 25°C and absent at 34°C. When the *DSM1* allele is present, DNA synthesis, recombination, and ascus formation are restored to near wild-type levels at 25°C; however, only DNA synthesis is restored at 34°C (Fast 1978; Fast et al. 1978). This temperature-sensitive effect may be due to the specific interaction of the *spo8-1*- and *DSM1*-gene products. The *rme1* and *DSM1* modifiers are therefore similar in that under appropriate conditions they both suppress defects in DNA synthesis, recombination, and ascus production. Moreover, each is a cryptic allele, having little or no detectable effect on sporulation in the absence of the lesions conferred by mating-type homozygosis or the *spo8-1* mutation, respectively. It is of interest that the recessive *rme1* allele and the partially dominant *DSM1* allele reside at the same chromosomal location and, thus, may be alleles of a single gene or part of a complex locus.

Commitment to Recombination Requires at Least the Initiation of Premeiotic S

All of the mutations that block premeiotic DNA synthesis prevent cells from achieving commitment to recombination, with the exception of *cdc4-3* (see Table 1). In the case of *cdc4-3*, there is a 20–40% increase in DNA content at 34°C (half of which appears to be mitochondrial by CsCl gradient analysis) and 50% of the wild-type level of commitment to intragenic recombination (Simchen and Hirschberg 1977). At 36°C, a negligible ($<10\%$) increase in DNA content occurs and there is no commitment to intragenic recombination. The authors conclude that only the initiation of DNA synthesis is required for commitment to recombination. They argue that a low level of synthesis (10–20%) occurs in all cells at 34°C, as opposed to a full round of replication occurring in a small fraction of cells, since almost all cells appear to arrest at the mononucleate stage by chromatin staining (Simchen and Hirschberg 1977).

These data, however, may be interpreted in another way. It has been shown by electron microscopy that approximately 15% of *cdc4-3* diploid cells accumulate SCs during arrest at 34°C (Byers and Goetsch 1975; Horesh et al. 1979), which suggests that duplicated chromosomes are present in at least 15% of the mutant cells. Thus, it seems more likely that the recombination observed in *cdc4-3*, sporulated at 34°C, is due to pachytene arrest in a fraction of cells, resulting in an elevated level of exchange upon their return to vegetative growth. Such behavior has been reported for wild-type cells arrested in pachytene (L. Davidow, pers. comm.) and could account for the observations of Simchen and Hirschberg (1977), if one assumes that cells that pass the initial DNA-synthesis block incurred by *cdc4-3* are blocked prior to the meiosis-I division. In support of this argument is the demonstration by temperature-shift experiments that *cdc4-3* is required during several stages of sporulation (Simchen and Hirschberg 1977). Thus, the general rule that completion of premeiotic DNA synthesis is required for commitment to recombination may, in fact, be valid. However, the completion of premeiotic S phase is not in itself sufficient for the occurrence of recombination. Numerous mutations that do not appear to affect premeiotic DNA synthesis, block commitment to recombination (e.g., *cdc2-1*, *cdc40*, *con* mutants, *rad* mutants; see Table 1).

Meiosis-I SPB Duplication and Premeiotic DNA Replication Are Independent Events

Mutants that do not undergo premeiotic DNA synthesis arrest either prior to meiosis-I SPB (SPBI) duplication (*spo7-1*, *spo8-1*, *spo9-1*) or after SPBI duplication but prior to separation (*cdc8-1*, *cdc21-1*) (Schild and Byers 1978; S. Klapholz, unpubl.; P. Moens, unpubl.). Cells exposed to HU during sporulation also arrest prior to SPB separation (Schild and Byers 1978). These results demonstrate that SPBI duplication is not dependent upon premeiotic DNA replication. However, SPB separation and the subsequent formation of the meiosis-I spindle may be dependent upon the completion of DNA synthesis.

During the vegetative cell cycle, the initiation of S phase is dependent upon SPB duplication and separation, indicating coordinate control of these events (Byers and Goetsch 1974). This does not appear to be the case during meiosis, since the *spo1-1* mutant is unable to complete SPBI duplication (Moens et al. 1974) but undergoes premeiotic S (Esposito et al. 1970).

Commitment to Recombination Is Not Required for Meiosis-I Spindle Formation to Occur, but Both Events Share Common Gene Functions

The behavior of some mutants suggests that recombination is required for meiosis-I spindle formation, whereas the behavior of others clearly contradicts this notion. Several mutants that undergo premeiotic DNA replication but do not exhibit commitment to recombination (*cdc2-1*, *cdc7-1*, *cdc7-4*, and *cdc40*) arrest prior to the first meiotic division (Kassir and Simchen 1978; Schild and Byers 1978). This suggests that commitment to recombination, as well as the ability to undergo SPBI duplication and separation, is required for meiosis-I spindle formation. However, other evidence indicates that both meiotic divisions can occur in the absence of recombination. For example, the *spo11-1* mutant, which is completely recombination-deficient (Klapholz 1980; Klapholz and Esposito 1980c), is capable of forming a normal meiosis-I spindle and undergoing both meiotic divisions (Moens et al. 1977). It should be pointed out, however, that the products of *spo11-1* meiosis are largely inviable and that the rare viable products are generally aneuploid (Klapholz and Esposito 1980c). Similarly, the *rad52-1* mutant does not exhibit commitment to recombination but forms a reduced level of tetranucleate cells and asci containing inviable spores (Game et al. 1980; Prakash et al. 1980). The behavior of the latter two mutants suggests that both divisions can take place without commitment to recombination but that viable euploid products cannot be formed.

To reconcile all of these results with the view that meiosis-I spindle formation is not dependent upon commitment to recombination, one could argue that the *cdc2, cdc7*, and *cdc40* mutants undergo an incomplete or abnormal round of premeiotic DNA replication or are defective in another event required for both recombination and spindle formation. The work of Schild and Byers (1978; described earlier in this section) suggests that at least the *cdc7* mutants may be aberrant in some aspect of meiotic DNA replication. It has also been proposed that the *spo11-1* mutant uncouples the normal sequence of events, permitting haploidization in the absence of recombination (see Esposito and Esposito 1978). However, it seems unlikely that *spo11-1, rad52-1*, and the many other sporulation mutants that form asci, but fail to exhibit commitment to recombination (see Table 1), are all regulatory mutants of this type. A unifying hypothesis for these data is the following: Meiosis-I spindle development is not dependent upon commitment to recombination, but both events are dependent upon some common gene functions (see also Fig. 6).

Commitment to Meiotic Levels of Recombination Is Not Sufficient for Haploidization

Studies of both wild-type and mutant strains indicate that commitment to recombination and commitment to the reductional division are independent events. It is clear from the return-to-growth experiments discussed above (in Genetic Recombination) that exposure to sporulation medium can stimulate cells to undergo meiotic levels of exchange in the absence of reductional chromosome segregation. In addition, the *spo1-1, cdc5-1,* and *cdc9-2* mutants exhibit commitment to recombination without commitment to meiosis (R. E. Esposito and M. S. Esposito 1974; Simchen 1974; Horesh et al. 1979). In the case of *spo1-1*, failure to undergo haploidization is due to a defect in SPBI duplication (Moens et al. 1974). The *spo12-1* and *spo13-1* mutants exhibit meiotic levels of exchange but undergo only one, generally equational, division, yielding diploid or near-diploid ascospores (Klapholz and Esposito 1980b). These mutants appear to be defective in the completion of the first meiotic division (Klapholz 1980; Klapholz and Esposito 1980b). Hence, although recombination may be necessary for proper meiosis-I segregation, it does not commit cells to the formation of haploid meiotic products.

Completion of Meiosis I Is Not Required for Meiosis II

Approximately 95% of the two-spored asci produced by *spo12-1* or *spo13-1* diploids results from a single equational division of chromosomes (Klapholz and Esposito 1980b). These data indicate that the completion of the meiosis-I reductional division is not necessary for the occurrence of meiosis II (for discussion, see Klapholz and Esposito 1980b). However, aberrant segregation of one or more chromosomes (e.g., 12% for chromosome III) takes place during the single division (Klapholz and Esposito 1980b). This abnormal behavior is correlated with the presence of meiotic levels of exchange; the coupling of *spo13-1* with a meiotic Rec⁻ mutation (*spo11-1* or *rad50-1*) concurrently abolishes or reduces both recombination and aberrant segregation (see further in this section). Hence, although prior reductional division is not required for the occurrence of meiosis II, it may be necessary to ensure a proper equational division of chromosomes at meiosis II after recombination has taken place.

Haploidization Is Not Sufficient for Ascus Maturation

A number of mutants complete the two meiotic divisions but are defective in some aspect of ascus maturation (e.g., *gcn1, kex2, prb1*; see Table 1). In strains homozygous for a mutation in the *gcn1* locus, haploidization occurs at wild-type levels, but refractile ascospores are not produced due to a defect in the outer layer of the spores (Whelan and Ballou 1976; Ballou et al. 1977). A number of other mutants reach the tetranucleate stage of meiosis but do not yield mature ascospores, possibly due to the highly aneuploid nature of the final products or to abnormal interactions between the nuclear and prospore-wall membranes (e.g., *spo2-1, spo3-1, spo4-1, spo5-1*; see Table 1). For example, in the *spo3-1* mutant, prospore-wall closure precedes the completion of the second meiotic division, resulting in unenclosed nuclei and aneuploid ascospores that fail to mature (Moens et al. 1974). The *spo2-1* mutant undergoes precocious division of the nuclear envelope at meiosis I and again at meiosis II (Moens et al. 1974). Subsequent spore-wall development is abormal, resulting in immature anucleate ascospores and unenclosed nuclei. A study of cells induced to form elevated levels of two-spored asci by reversible high-temperature arrest during pachytene has demonstrated that modification of the SPBIIs is required for the initiation of prospore-wall development (Davidow et al. 1980; see above, Morphological Development). The mutants described above demonstrate that the completion of normal spore-wall development depends upon a number of gene functions involved in nuclear segregation.

Ascospore Maturation and the Formation of Viable Meiotic Products Are Independent Events

As illustrated above, the recovery of viable haploid meiotic products is not dependent upon the formation of refractile ascospores. It has also been shown that the formation of mature asci does not insure that the meiotic products are viable. For example, monosomic diploids form mature asci that contain two inviable nullosomic spores and two viable haploid spores (Bruenn and Mortimer 1970). Triploid strains produce highly inviable aneuploid ascospores but sporulate at normal levels (see above, Chromosome Segregation). A number of meiotic recombination-defective mutants also form a moderate level of mature asci containing inviable, presumably aneuploid, spores (e.g., *rad50-1, rad57-1, con3*; see Table 1). The extent to which different Rec⁻ mutant strains produce mature asci (e.g., c.f. *spo11-1* and *rad50-1*; Table 1) may reflect the specific nature of the mutant lesion or, perhaps, the degree to which the ascospores are aneuploid.

Ordering Meiotic Gene Functions by Multiple-mutant Interactions

The analysis of multiply mutant sporulation-defective strains has only recently been undertaken in an attempt to investigate the number of independent sequential pathways of events involved in the process of sporulation and to order genes that act on the same dependency pathway. These results are summarized below and in Figure 6.

Genes Involved in Meiotic Recombination Act on a Different Sequential Pathway than Do Genes Required for Meiosis-I Segregation

Three genes required for meiotic recombination and viable spore formation, *RAD50* (Game et al. 1980), *RAD52* (Game et al. 1980; Prakash et al. 1980), and *SPO11* (Klapholz 1980; Klapholz and Esposito 1980c), act on a different

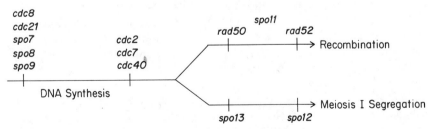

Figure 6 A simple dependency scheme for genes required for meiotic recombination and/or meiosis-I chromosome segregation. This model is based on studies of single- and multiple-mutant strains (described in the text and in Table 1). The order of *SPO11*, with respect to *RAD50* and *RAD52*, has not been determined.

sequential pathway than do *SPO12* and *SPO13*, which are required for meiosis-I segregation (Klapholz and Esposito 1980a,b). This was deduced from studies of double mutant diploids containing a Rec⁻ mutation: *rad50-1, rad52-1* or *spo11-1*, and either *spo12-1* or *spo13-1* (see Table 1). In each case, the double mutant exhibited a novel sporulation phenotype, combining features present in each single mutant strain. For example, *spo11-1 spo13-1* and *rad50-1 spo13-1* diploids produce asci containing two viable diploid spores (like *spo13-1*) but do not undergo meiotic recombination (like *spo11-1* and *rad50-1*) (Klapholz 1980; Klapholz and Esposito 1980c; Malone and Esposito 1981). The *rad52-1 spo13-1* double mutant forms dyads (like *spo13-1*) that contain inviable spores (like *rad52-1*) (Malone and Esposito 1981). In each instance, the lack of epistasis between the two mutations indicates that *RAD50, RAD52*, and *SPO11* do not act on the same sequential pathway as *SPO12* or *SPO13*. It should be emphasized that although these genes act in different sequential (i.e., independent) pathways, genetic evidence suggests that these pathways diverged from a common pathway (see Fig. 6) and must eventually converge at or prior to ascus formation.

Genes That Act on the Same Sequential Pathway Can Be Ordered with the Aid of a Third Mutation in a Gene on an Independent Pathway

It has not been possible to order genes that act on the same sequential pathway by simply comparing double and single mutant strains, when both single mutants express the same phenotype. However, it has been possible to show that *RAD50* acts prior to *RAD52* on one sequential pathway during sporulation and that *SPO13* acts before *SPO12* on a second pathway by comparing the appropriate double and triple mutant strains (see Fig. 6). In the first instance, the *rad50-1* mutation was shown to be epistatic to *rad52-1* by the finding that *rad50-1 rad52-1 spo13-1* triple mutants behave similarly to the *rad50-1 spo13-1* double mutant in forming viable recombinationless dyads and different from the *rad52-1 spo13-1* double mutant, which forms inviable dyads (Malone and Esposito 1981).

Similar analysis indicates that the *SPO13* gene acts prior to *SPO12* on a sequential pathway. Single *spo12-1* and *spo13-1* and double *spo12-1 spo13-1* mutant diploids have almost identical sporulation phenotypes (Klapholz 1980; Klapholz and Esposito 1980a,b). However, the phenotype of each single mutant in the presence of *spo11-1*, which acts on an independent pathway, is strikingly different. The *spo11-1 spo12-1* double mutant, like *spo11-1*, produces a reduced level of asci at 25°C and only a few percent asci at 34°C. The spores formed at 25°C exhibit poor (10%) viability (Klapholz 1980; Klapholz and Esposito 1980c). However, like *spo12-1*, the double mutant produces only two-spored asci, and the rare viable spores are diploid or near diploid in ploidy. In contrast, the *spo11-1 spo13-1* double mutant (described in the preceding section) produces two viable diploid spores per ascus. At 34°C no recombination is detected, whereas at 25°C it occurs at a greatly reduced level (Klapholz and Esposito 1980c). The *spo11-1 spo12-1 spo13-1* triple mutant behaves like the *spo11-1 spo13-1* double mutant, demonstrating that *spo13-1* is epistatic to *spo12-1* and, hence, that *SPO13* acts before *SPO12*.

Further support of this conclusion comes from the finding that **a**/α *spo13-1* and **a**/α *spo12-1 spo13-1* disomic haploids sporulate at an appreciable level (see above, Chromosome Segregation), whereas **a**/α *spo12-1* disomic haploids do not (Klapholz 1980; Klapholz and Esposito 1980c). In this analysis, the presence of an additional sporulation-defective mutation was not required to show that *spo13-1* is epistatic to *spo12-1*. Instead, this study exploited a unique phenotypic difference between *spo12-1* and *spo13-1* involving another meiotic parameter, the normal requirement for homologous pairs of chromosomes (i.e., a diploid or near-diploid state) for the formation of viable spores. Like wild-type strains, the *spo12-1* mutant requires the presence of homologous chromosomes to form viable meiotic products. On the other hand, both the single *spo13-1* and the double *spo12-1 spo13-1* mutants can sporulate in a haploid or near-haploid background (see above, Chromosome Segregation). Thus, *spo13-1* is epistatic to *spo12-1* by an independent criterion—the ability to utilize univalents in meiosis. This may also account for the ability of *spo13-1* to "rescue" certain Rec⁻ mutants, which may contain unpaired or univalent chromosomes due to defects in chromosome pairing and/or exchange (see above, Genetic Recombination).

The studies described above and in earlier parts of this section point out that the analysis of a mutation in more than one genetic background can often facilitate its characterization. In addition, such an analysis may provide important clues regarding the function of its wild-type allele. For example, the analysis of *spo11-1* and *rad50-1* in the presence of *spo13-1* provided an independent confirmation of their putative Rec⁻ phenotypes, ruling out the possibility that they were lesions in genes that affect the survival and/or detection of recombinants. Models for the coordination of landmark events of sporulation have been proposed, taking into account the phenotypes of various single mutants (Moens et al. 1977; Esposito and Esposito 1978; Schild and Byers 1978; Horesh et al. 1979). The analysis of multiple-mutant interactions, which has only recently been initiated, has already proved useful in testing aspects of these models and in ordering meiotic gene functions (Fig. 6). Further work in this area, complemented by molecular studies of cloned sporulation-specific genes, should provide important new insights into the genetic control of meiosis and ascospore formation.

ACKNOWLEDGMENTS

This review is dedicated to Herschel Roman, whose seminal work on genetic recombination has provided the inspiration for much of our own research.

We are extremely grateful to our many colleagues who have contributed preprints, unpublished data, and lively discussion. Special thanks are due to Terrance Cooper, Lance Davidow, Ian Dawes, Michael Esposito, Dale Fast, John Game, Craig Giroux, James Haber, Elizabeth Jones, Pete Magee, Robert Malone, Dallice Mills, Peter Moens, Diane Plotkin, John Pringle, and Giora Simchen. We are also particularly indebted to Joseph Wagstaff, Patrick Brown, and Terence Martin for helpful comments on this review. The assistance of Mary Haynes, Esther Lacey, Candace Waddell, and Peter Gauchier in the preparation of the manuscript is gratefully acknowledged. This work was supported by National Institutes of Health grants GM-23277, GM-29182, and CA-19265 (project 508) (U.S. Public Health Service), and postdoctoral training grant CA-09273 (U.S. Public Health Service) awarded to S.K.

REFERENCES

Adams, A.M. and J.J. Miller. 1954. Effect of gaseous environment and temperature on ascospore formation in *Saccharomyces cerevisiae* Hansen. *Can. J. Bot.* **32**: 320.

Arnaud, A. and P. Galzy. 1973. On the role of pH in the sporulation of *Saccharomyces cerevisiae*. *Folia Microbiol.* **18**: 281.

Baker, B.S., A.T.C. Carpenter, M.S. Esposito, R.E. Esposito, and L. Sandler. 1976. The genetic control of meiosis. *Annu. Rev. Genet.* **10**: 53.

Ballou, C., S. Maitra, J. Walker, and W. Whelan. 1977. Developmental defects associated with glucosamine auxotrophy in *Saccharomyces cerevisiae*. *Proc. Natl. Acad. Sci.* **74**: 4351.

Beckett, A., R.F. Illingworth, and A.H. Rose. 1973. Ascospore wall development in *Saccharomyces cerevisiae*. *J. Bacteriol.* **113**: 1054.

Belling, J. and A.F. Blakeslee. 1927. The assortment of chromosomes in haploid *Daturas*. *Cellule* **37**: 355.

Bettleheim, K.A. and J.L. Gay. 1963. Acetate-glyoxylate medium for sporulation of *Saccharomyces cerevisiae*. *J. Appl. Bacteriol.* **26**: 224.

Betz, H. 1975. Loss of sporulation ability in a yeast mutant with low proteinase A levels. *FEBS Lett.* **100**: 171.

——. 1977. Protein degradation and sporulation in yeast. In *Cell differentiation in microorganisms, plants and animals* (ed. L. Nover and K. Mothes), p. 243. Gustav Fischer, Jena.

Betz, H. and U. Weiser. 1976a. Protein degradation and proteinases during yeast sporulation. *Eur. J. Biochem.* **62**: 65.

——. 1976b. Protein degradation during yeast sporulation: Enzyme and cytochrome patterns. *Eur. J. Biochem.* **70**: 385.

Bevan, E.A. 1953. Genetic analysis of intact incomplete asci of yeast. *Nature* **171**: 576.

Bilinski, C.A. and J.J. Miller. 1980. Induction of normal ascosporogenesis in two-spored *Saccharomyces cerevisiae* by glucose, acetate and zinc. *J. Bacteriol.* **143**: 343.

Bisson, L. and J. Thorner. 1977. Thymidine-5′-monophosphate-requiring mutants of *Saccharomyces*. *J. Bacteriol.* **132**: 44.

Brewer, B. and W. Fangman. 1980. Preferential inclusion of extrachromosomal genetic elements in yeast meiotic spores. *Proc. Natl. Acad. Sci.* **77**: 6739.

Briley, M.S., R.F. Illingworth, A.H. Rose, and D.J. Fisher. 1970. Evidence for a surface protein layer on the *Saccharomyces cerevisiae* ascospore. *J. Bacteriol.* **104**: 588.

Bruenn, J. and R.K. Mortimer. 1970. Isolation of monosomics in yeast. *J. Bacteriol.* **102**: 548.

Byers, B. and L. Goetsch. 1974. Duplication of spindle plaques and integration of the yeast cell cycle. *Cold Spring Harbor Symp. Quant. Biol.* **38**: 123.

———. 1975. Electron microscopic observations on the meiotic karyotype of diploid and tetraploid *Saccharomyces cerevisiae. Proc. Natl. Acad. Sci.* **72**: 5056.

Cabib, E., R. Ulane, and B. Bowers. 1973. Yeast chitin synthetase: Separation of the zymogen from its activating factor and recovery of the latter in the vacuole fraction. *J. Biol. Chem.* **248**: 1451.

Callan, H.G. 1972. Replication of DNA in the chromosomes of eukaryotes. *Proc. R. Soc. Lond.* B **181**: 19.

Chaffin, W.L., S.J. Sogin, and H.O. Halvorson. 1974. Nature of ribonucleic acid synthesis during early sporulation in *Saccharomyces cerevisiae. J. Bacteriol.* **120**: 872.

Chen, A.W.-C. and J.J. Miller. 1968. Proteolytic activity of intact yeast cells during sporulation. *Can. J. Microbiol.* **14**: 957.

Clancy, M.J., L. Smith, and P.T. Magee. 1980. Regulation of a sporulation-specific enzyme activity. In *Tenth International Conference on Yeast Genetics and Molecular Biology, Louvain-la-Neuve, Belgium.* p. 152.

Cohn, M.S., C.W. Tabor, and H. Tabor. 1978. Isolation and characterization of *Saccharomyces cerevisiae* mutants deficient in *S*-adenosylmethionine decarboxylase, spermidine and spermine. *J. Bacteriol.* **134**: 208.

Colonna, J., J.M. Gentile, and P.T. Mayer. 1977. Inhibition by sulfanilamide of sporulation in *Saccharomyces cerevisiae. Can. J. Microbiol.* **23**: 659.

Colonna, W.Y. and P.T. Magee. 1978. Glycogenolytic enzymes in sporulating yeast. *J. Bacteriol.* **134**: 844.

Conde, J. and G. Fink. 1976. A mutant of *Saccharomyces cerevisiae* defective for nuclear fusion. *Proc. Natl. Acad. Sci.* **73**: 3651.

Cooper, T. 1982. Transport in *Saccharomyces cerevisiae.* In *The molecular biology of the yeast* Saccharomyces. II. *Metabolism and gene expression* (ed. J. Strathern et al.), Cold Spring Harbor Laboratory, Cold Spring Harbor, New York. (In press.)

Cox, B.S. and E.A. Bevan. 1962. Aneuploidy in yeast. *New Phytol.* **61**: 342.

Croes, A.F. 1966. Duplication of DNA during meiosis in baker's yeast. *Exp. Cell Res.* **41**: 452.

———. 1967a. Induction of meiosis in yeast. I. Timing of cytological and biochemical events. *Planta* **76**: 209.

———. 1967b. Induction of meiosis in yeast. II. Metabolic factors leading to meiosis. *Planta* **76**: 227.

Croes, A.F., G.J. DeVries, and J.M. Steijns. 1978a. Amino acid uptake and protein synthesis during early meiosis in *Saccharomyces cerevisiae. Planta* **139**: 93.

Croes, A.F., H.J. Dodemont, and C. Stumm. 1976. Induction of meiosis in *Saccharomyces cerevisiae* at different points in the mitotic cycle. *Planta* **130**: 131.

Croes, A.F., J.M. Steijns, G. DeVries, and T.M. Van der Putte. 1978b. Inhibition of meiosis in *Saccharomyces cerevisiae* by ammonium ions: Interference of ammonia with protein metabolism. *Planta* **141**: 205.

Culbertson, M.R. and S.A. Henry. 1973. Genetic analysis of hybrid strains trisomic for the chromosome containing a fatty acid synthetase gene complex (*fas1*) in yeast. *Genetics* **75**: 441.

Curiale, M.S. and D. Mills. 1975. Characterization of pulse-labeled ribonucleic acid in polysomes of sporulating yeast cells. In *Spores* (ed. P. Gerhardt et al.), vol. 6, p. 165. American Society for Microbiology, East Lansing, Michigan.

Curiale, M.S., M.M. Petryna, and D. Mills. 1976. Effect of culture medium pH. *J. Bacteriol.* **126**: 661.

Darland, G.K. 1969. "The physiology of sporulation in *Saccharomyces cerevisiae.*" Ph.D. thesis, University of Washington, Seattle.

Davidow, L.S., L. Goetsch, and B. Byers. 1980. Preferential occurrence of non-sister spores in two spored asci of *Saccharomyces cerevisiae*—Evidence of regulation of spore wall formation by the spindle pole body. *Genetics* **94**: 581.

Dawes, I.W. 1975. Study of cell development using derepressed mutations. *Nature* **255**: 707.

Dawes, I.W. and I.H. Hardie. 1974. Selective killing of vegetative cells in sporulated yeast cultures by exposure to diethyl ether. *Mol. Gen. Genet.* **131**: 281.

Dawes, I.W., G.R. Calvert, and T.A. Martin. 1980. Initiation of meiosis and sporulation in *Saccharomyces cerevisiae:* Regulation of protein synthesis in mutants altered in initiation. *Tenth International Conference on Yeast Genetics and Molecular Biology. Louvain-La-Neuve, Belgium.*

del Rey, F., T. Santos, I. Garcia-Acha, and C. Nombela. 1980. Synthesis of β-glucanases during sporulation in *Saccharomyces cerevisiae*: Formation of a new sporulation-specific 1, 3-β-glucanase. *J. Bacteriol.* **143**: 621.

Durieu-Trautmann, O. and C. Delavier-Klutchko. 1977. Effect of ammonia and glutamine on macromolecular synthesis and breakdown during sporulation of *Saccharomyces cerevisiae*. *Biochem. Biophys. Res. Commun.* **79**: 438.

Engles, F.M. and A.F. Croes. 1968. The synaptinemal complex in yeast. *Chromosoma* **25**: 104.

Ephrussi, B. and H. Hottinguer. 1952. On an unstable cell state in yeast . *Cold Spring Harbor Symp. Quant. Biol.* **16**: 75.

Esposito, M.S. and R.E. Esposito. 1969. The genetic control of sporulation in *Saccharomyces*. I. The isolation of temperature-sensitive sporulation-deficient mutants. *Genetics* **61**: 79.

―――. 1974. Genes controlling meiosis and spore formation in yeast. *Genetics* **78**: 215.

―――. 1975. Mutants of meiosis and ascospore formation. *Methods Cell Biol.* **11**: 303.

―――. 1978. Aspects of the genetic control of meiosis and ascospore development inferred from the study of *spo* (sporulation-deficient) mutants of *Saccharomyces cerevisiae*. *Biol. Cell.* **33**: 93.

Esposito, M.S., M. Bolotin-Fukuhara, and R.E. Esposito. 1975. Antimutator activity during mitosis by a meiotic mutant of yeast. *Mol. Gen. Genet.* **139**: 9.

Esposito, M.S., R.E. Esposito, and P.B. Moens. 1974. Genetic analysis of two spored asci produced by the *spo3* mutant of *Saccharomyces*. *Mol. Gen. Genet.* **135**: 91.

Esposito, M.S., R.E. Esposito, M. Arnaud, and H.O. Halvorson. 1969. Acetate utilization and macromolecular synthesis during sporulation of yeast. *J. Bacteriol.* **100**: 180.

―――. 1970. Conditional mutants of meiosis in yeast. *J. Bacteriol.* **104**: 202.

Esposito, R.E. and M.S. Esposito. 1974. Genetic recombination and commitment to meiosis in *Saccharomyces*. *Proc. Natl. Acad. Sci.* **71**: 3172.

Esposito, R.E. and D. Plotkin. 1978. A comparison of the properties of intragenic recombination in mitotic, meiototic, and meiotic populations. In *Ninth International Conference on Yeast Genetics and Molecular Biology; Rochester, New York.* p. 80.

Esposito, R.E., D.J. Plotkin, and M.S. Esposito. 1974. The relationship between genetic recombination and commitment to chromosomal segregation at meiosis. In *Mechanisms in recombination* (ed. R.F. Grell), p. 277. Plenum Press, New York .

Esposito, R.E., N. Frink, P. Bernstein, and M.S. Esposito. 1972. The genetic control of sporulation in *Saccharomyces*. II. Dominance and complementation of mutants of meiosis and spore formation. *Mol. Gen. Genet.* **114**: 241.

Fast, D. 1973. Sporulation synchrony of *Saccharomyces cerevisiae* grown in various carbon sources. *J. Bacteriol.* **116**: 925.

―――. 1978. "Premeiotic DNA synthesis and recombination in the yeast *Saccharomyces*." Ph.D. thesis, University of Chicago, Illinois.

Fast, D., R.E. Esposito, and M.S. Esposito. 1978. Evidence that the *SPO8* gene product mediates premeiotic DNA synthesis and intergenic recombination during sporulation of *Saccharomyces cerevisiae*. In *Ninth International Conference on Yeast Genetics and Molecular Biology, Rochester, New York*, p. 80.

Fleming, H. and R. Haselkorn. 1974. The program of protein synthesis during heterocyst differentiation in nitrogen-fixing blue-green algae. *Cell* **3**: 159.

Fogel, S. and R. Roth. 1974. Mutations affecting meiotic gene conversion in yeast. *Mol. Gen. Genet.* **130**: 189.

Fogel, R. Mortimer, K. Lusnak, and F. Tavares. 1979. Meiotic gene conversion: A signal of the basic recombination event in yeast. *Cold Spring Harbor Symp. Quant. Biol.* **43:** 1325.

Fonzi, W.A., M. Shanley, and D.J. Opheim. 1979. Relationship of glycolytic intermediates, glycolytic enzymes and ammonia to glycogen metabolism during sporulation in the yeast *Saccharomyces cerevisiae. J. Bacteriol.* **137:** 285.

Fowell, R.R. 1969. Sporulation and hybridization of yeasts. In *The yeasts* (ed. A.H. Rose and J.S. Harrison), vol. I, p. 303. Academic Press, New York.

———. 1975. Ascospores of yeast. In *Spores* (ed. P. Gerhart et al.), vol. 6, p. 124. American Society for Microbiology, Washington.

Fraenkel, D.G. 1982. Carbohydrate metabolism. In *The molecular biology of the yeast* Saccharomyces. II. *Metabolism and gene expression* (ed. J. Strathern et al.), Cold Spring Harbor Laboratory, Cold Spring Harbor, New York. (In press.)

Frank, K.R. and D. Mills. 1978. Ribosome activity and degradation in meiotic cells of *Saccharomyces cerevisiae. Mol. Gen. Genet.* **160:** 59.

Freese, E.B., P. Cooney, and E. Freese. 1975. Conditions controlling commitment of differentiation in *Bacillus megaterium. Proc. Natl. Acad. Sci.* **72:** 4037.

Fukuhara, H. and C. Kujawa. 1970. Selective inhibition of the *in vivo* transcription of mitochondrial DNA by ethidium bromide and by acriflavine. *Biochem. Biophys. Res. Commun.* **41:** 1002.

Game, J. 1976. Yeast cell-cycle mutant *cdc21* is a temperature-sensitive thymidylate auxotroph. *Mol. Gen. Genet.* **146:** 313.

Game, J.C. and R.K. Mortimer. 1974. A genetic study of X-ray sensitive mutants in yeast. *Mutat. Res.* **24:** 281.

Game, J.C., T.J. Zamb, R.J. Braun, M. Resnick, and R.M. Roth. 1980. The role of radiation (*rad*) genes in meiotic recombination in yeast. *Genetics* **94:** 51.

Ganesan, A.T. 1959. The cytology of *Saccharomyces. C. R. Lab. Carlsburg* **31:** 149.

Ganesan, A.T., H. Holter, and C. Roberts. 1958. Some observations on sporulation in *Saccharomyces. C. R. Lab. Carlsberg* **13:** 1.

Garvik, B. and J. Haber. 1978. New cytoplasmic genetic element that controls 20S RNA synthesis during sporulation in yeast. *J. Bacteriol.* **134:** 261.

Gerlach, W.L. 1974. Sporulation in mating type homozygotes of *Saccharomyces cerevisiae. Heredity* **32:** 241.

Goldring, E.S., L.I. Grossman, D. Krupnick, D. Cryer, and J. Marmur. 1970. The petite mutation in yeast: Loss of mitochondrial deoxyribonucleic acid during induction of petites with ethidium bromide. *J. Mol. Biol.* **52:** 323.

Golin, J.E. and M.S. Esposito. 1977. Evidence for joint genic control of spontaneous mutation and genetic recombination during mitosis in *Saccharomyces. Mol. Gen. Genet.* **150:** 127.

Gosling, J.P. and P.F. Duggan. 1971. Activities of tricarboxylic acid cycle enzymes, glyoxylate cycle enzymes and fructose diphosphate in baker's yeast during adaptation to acetate oxidation. *J. Bacteriol.* **106:** 980.

Grewal, N.S. and J.J. Miller. 1972. Formation of asci with two diploid spores by diploid cells of *Saccharomyces. Can. J. Microbiol.* **18:** 1897.

Guth, E., T. Hashimoto, and S.F. Conti. 1972. Morphogenesis of ascospores in *Saccharomyces cerevisiae. J. Bacteriol.* **109:** 869.

Haber, J.E. 1974. Bisexual mating behavior in a diploid of *Saccharomyces cerevisiae*: Evidence for genetically controlled non-random chromosome loss during vegetative growth. *Genetics* **78:** 843.

Haber, J.E. and H.O. Halvorson. 1972a. Cell cycle dependency of sporulation in *Saccharomyces cerevisiae. J. Bacteriol.* **109:** 1027.

———. 1972b. Regulation of sporulation in yeast. *Curr. Top. Dev. Biol.* **7:** 61.

———. 1975. Methods in sporulation and germination of yeasts. *Methods Cell Biol.* **11:** 46.

Haber, J.E., M.S. Esposito, P.T. Magee, and R.E. Esposito. 1975. In *Current trends in genetic and biochemical study of yeast sporulation. Spores* (ed. P. Gerhardt et al.), vol. 6, p. 132. American Society for Microbiology, Washington.

Haber, J.E., P.J. Wejksnora, D.D. Wygal, and E.Y. Lai. 1977. Controls of sporulation in *Saccharomyces cerevisiae.* In *Eukaryotic microbes as model developmental systems* (ed. D.H. O'Day and P.A. Horgen), vol. 2, p. 129. Marcel Dekker, New York.

Halvorson, H.O. 1958a. Intracellular protein and nucleic acid turnover in resting yeast cells. *Biochim. Biophys. Acta* **27**: 255.

———. 1958b. Studies on protein and nucleic acid turnover in growing cultures of yeast. *Biochim. Biophys. Acta* **27**: 267.

Hartwell, L. 1971. Genetic control of cell division cycle in yeast. II. Genes controlling DNA replication and its initiation. *J. Molec. Biol.* **59**: 183.

———. 1973. Three additional genes required for deoxyribonucleic acid synthesis in *Saccharomyces cerevisiae. J. Bacteriol.* **115**: 966.

———. 1974. *Saccharomyces cerevisiae* cell cycle. *Bacteriol. Rev.* **38**: 164.

———. 1976. Sequential function of gene products relative to DNA synthesis in the yeast cell cycle. *J. Molec. Biol.* **104**: 803.

Hashimoto, T., P. Gerhardt, S.F. Conti, and H.B. Naylor. 1960. Studies of the fine structure of microorganisms. V. Morphogenesis of nuclear and membrane structures during ascospore formation in yeast. *J. Biophys. Biochem. Cytol.* **7**: 305.

Hasilik, A., H. Muller, and H. Holzer. 1974. Compartmentation of the tryptophan-synthase-proteolyzing system in *Saccharomyces cerevisiae. Eur. J. Biochem.* **48**: 111.

Hawthorne, D.C. and R.K. Mortimer. 1960. Chromosome mapping in *Saccharomyces*: Centromere-linked genes. *Genetics* **45**: 1085.

Henry, S.A. and H.O. Halvorson. 1973. Lipid synthesis during sporulation of *Saccharomyces cerevisiae. J. Bacteriol.* **114**: 1158.

Hereford, L. and L.H. Hartwell. 1974. Sequential gene function in the initiation of *S. cerevisiae* DNA synthesis. *J. Mol. Biol.* **84**: 445.

Heywood, P. and P.T. Magee. 1976. Meiosis in protists. *Bacteriol. Rev.* **40**: 190.

Hirschberg, J. and G. Simchen. 1977. Commitment to mitotic cell cycle in yeast in relation to meiosis. *Exp. Cell Res.* **105**: 245.

Hopper, A.K. and B.D. Hall. 1975. Mating-type and sporulation in yeast. I. Mutations which alter mating-type control over sporulation. *Genetics* **80**: 41.

Hopper, A.K., J. Kirsch, and B.D. Hall. 1975. Mating-type and sporulation in yeast. II. Meiosis, recombination, and radiation sensitivity in an α/α diploid with altered sporulation control. *Genetics.* **80**: 61.

Hopper, A.K., P.T. Magee, S.K. Welch, M. Friedman, and B.D. Hall. 1974. Macromolecule synthesis and breakdown in relation to sporulation and meiosis in yeast. *J. Bacteriol.* **119**: 619.

Horesh, O., G. Simchen, and A. Friedmann. 1979. Morphogenesis of the synapton during yeast meiosis. *Chromosoma* **75**: 101.

Hotta, Y., A.C. Chandley, and H. Stern. 1977. Biochemical analysis of meiosis in the male mouse. II. DNA metabolism at pachytene. *Chromosoma* **62**: 255.

Hottinguer-de-Margerie, H. 1967. "Determinisme genetique d'un type d'anomalies meiotiques et mitotique induites par les desintegrations du ³²P chez la levure." Ph.D. thesis, Faculte des Sciences, Paris.

Hottinguer-de-Margerie, H. and E. Moustacchi. 1979. Abolition of the cyclic variations in radiosensitivity in a sporulation mutant blocked in premeiotic DNA synthesis. *Mol. Gen. Genet.* **175**: 259.

Hurst, D.D., S. Fogel, and R.K. Mortimer. 1972. Conversion associated recombination in yeast. *Proc. Natl. Acad. Sci.* **69**: 101.

Illingworth, R.F., A.H. Rose, and A. Beckett. 1973. Changes in the lipid composition and fine structure of *Saccharomyces cerevisiae* during ascus formation. *J. Bacteriol.* **113**: 373.

Ivanov, A. 1938. Experimental production of haploids in *Nicotiana rustica* L. *Genetica* **20**: 295.

Jacobson, G., R. Piñon, R.E. Esposito, and M.S. Esposito. 1975. Single strand scissions of chromosomal DNA during commitment to recombination at meiosis. *Proc. Natl. Acad. Sci.* **72:** 1887.

James, A.P. 1974. A new method of detecting centromere linkage in homothallic yeast. *Genet. Res.* **23:** 201.

James, A.P. and E.R. Inhaber. 1975. Evidence of preferential pairing of chromosomes at meiosis in aneuploid yeast *Genetics* **79:** 561.

Johnson, G.C., J.R. Pringle, and L.H. Hartwell. 1977. Coordination of growth with cell division in the yeast *Saccharomyces cerevisiae. Exp. Cell Res.* **105:** 79.

Johnston, L.H. and A.L. Johns. 1980. An analysis of the mechanism of premeiotic DNA synthesis in yeast. In *Tenth International Conference on Yeast Genetics and Molecular Biology, Louvain-la-Neuve, Belgium.* p. 54.

Johnston, L.H. and K. Nasmyth. 1978. *Saccharomyces cerevisiae* cell cycle mutant *cdc9* is defective in DNA ligase. *Nature* **274:** 891.

Jones, E.W. 1977. Proteinase mutants of *Saccharomyces cerevisiae. Genetics* **85:** 23.

Kadowaki, K. and H.O. Halvorson. 1971a. Appearance of a new species of ribonucleic acid during sporulation in *Saccharomyces cerevisiae. J. Bacteriol.* **105:** 826.

―――. 1971b. Isolation and properties of a new species of ribonucleic acid synthesized in sporulating cells of *Saccharomyces cerevisiae. J. Bacteriol.* **105:** 831.

Kane, S. and R. Roth. 1974. Carbohydrate metabolism during ascospore development in yeast. *J. Bacteriol.* **118:** 8.

Kassir, Y. and G. Simchen. 1976. Regulation of mating and meiosis in yeast by the mating type region. *Genetics* **82:** 187.

―――. 1978. Meiotic recombination and DNA synthesis in a new cell cycle mutant of *Saccharomyces cerevisiae. Genetics* **90:** 49.

―――. 1980. Single-strand scissions of nuclear yeast DNA occur without meiotic recombination. *Curr. Genet.* **2:** 79.

Keith, A.D., M. Resnick, and A.B. Haley. 1969. Fatty acid denaturase mutants of *Saccharomyces cerevisiae. J. Bacteriol.* **98:** 415.

Kirsop, B.H. 1954. Studies in yeast sporulation. I. Some factors influencing sporulation. *J. Inst. Brew.* **60:** 393.

Klapholz, S. 1980. "The genetic control of chromosome segregation during meiosis in yeast." Ph.D. thesis, University of Chicago, Illinois.

Klapholz, S. and R.E. Esposito. 1980a. Isolation of *spo12-1* and *spo13-1* from a natural variant of yeast that undergoes a single meiotic division. *Genetics* **96:** 567.

―――. 1980b. Recombination and chromosome segregation during the single division meiosis in *spo12-1* and *spo13-1* diploids. *Genetics* **96:** 589.

―――. 1980c. Interactions of genes controlling recombination and chromosome segregation in meiosis. In *Tenth International Conference on Yeast Genetics and Molecular Biology, Louvaine-la-Neuve, Belgium.* p. 153.

Klar, A.J.S. 1980. Mating-type functions for meiosis and sporulation in yeast act through cytoplasm. *Genetics* **94:** 597.

Klar, A.J.S. and H.O. Halvorson. 1975. Proteinase activities of *Saccharomyces cerevisiae* during sporulation. *J. Bacteriol.* **124:** 863.

Klar, A.J.S., A. Cohen, and H.O. Halvorson. 1976. Control of enzyme synthesis and stability during sporulation of *Saccharomyces cerevisiae. Biochimie* **58:** 219.

Klar, A.J.S., S. Fogel, and K. MacCleod. 1979. *MAR1*, a regulator of the *HM*a and *HM*α loci in *Saccharomyces cerevisiae. Genetics* **92:** 759.

Klein, H. and B. Byers. 1978. Stable denaturation of chromosomal DNA from *Saccharomyces cerevisiae* during meiosis. *J. Bacteriol.* **134:** 629.

Kraig, E. and J.E. Haber. 1980. Messenger ribonucleic acid and protein metabolism during sporulation of *Saccharomyces cerevisiae. J. Bacteriol.* **144:** 1098.

Kuenzi, M.T. and R. Roth. 1974. Timing of mitochondrial DNA synthesis during meiosis in *Saccharomyces cerevisiae. Exp. Cell Res.* **85:** 377.

Kuenzi, M.T., M.A. Tingle, and H.O. Halvorson. 1974. Sporulation of *Saccharomyces cerevisiae* in the absence of a functional mitochondrial genome. *J. Bacteriol.* **117**: 80.

Lemke, P.A., B. Kugelman, H. Morimoto, E.C. Jacobs, and J.R. Ellison. 1978. Fluorescent staining of fungal nuclei with a benzimidazol derivative. *J. Cell Sci.* **29**: 77.

Lemontt, J.F., D.R. Fugit, and V.L. MacKay. 1980. Pleiotropic mutations at the *tup1* locus that affect the expression of mating-type-dependent functions in *Saccharomyces cerevisiae*. *Genetics* **94**: 899.

Lesley, M.M. and H.B. Frost. 1928. Two extreme "small" matthiola plants: A haploid with one and a diploid with two additional chromosome fragments. *Am. Nat.* **62**: 22.

Liebowitz, M. and R. Wickner. 1976. A chromosomal gene required for killer plasmid expression, mating and spore maturation in *Saccharomyces cerevisiae*. *Proc. Natl. Acad. Sci.* **73**: 2061.

Lillie, S.H. and J.R. Pringle. 1980. Reserve carbohydrate metabolism in *Saccharomyces cerevisiae*: Responses to nutrient limitation. *J. Bacteriol.* **143**: 1384.

Liras, P., J. McCusker, S. Mascioli, and J.E. Haber. 1978. Characterization of a mutation in yeast causing nonrandom chromosome loss during mitosis. *Genetics* **88**: 651.

Lucchini, G., A. Biraghi, M.L. Carbone, A. DeScrilli, and G.E. Magni. 1978. Effect of mutation in the aromatic amino acid pathway on sporulation of *Saccharomyces cerevisiae*. *J. Bacteriol.* **136**: 55.

Lynn, R.R. and P.T. Magee. 1970. Development of the spore wall during ascospore formation in *Saccharomyces cerevisiae*. *J. Cell Biol.* **44**: 688.

MacKay, V. and T.R. Manney. 1974a. Mutations affecting sexual conjugation and related processes in *Saccharomyces cerevisiae*. I. Isolation and phenotypic characterization of nonmating mutants. *Genetics* **76**: 255.

———. 1974b. Mutations affecting sexual conjugation and related processes in *Saccharomyces cerevisiae*. II. Genetic analysis of nonmating mutants. *Genetics* **76**: 273.

Magee, P.T. 1974. Changes in DNA dependent RNA polymerase in sporulating yeast. *Mol. Biol. Rep.* **1**: 275.

Magee, P.T. and A.K. Hopper. 1974. Protein synthesis in relation to sporulation and meiosis in yeast. *J. Bacteriol.* **119**: 952.

Magni, G.E. 1958. Changes in radiosensitivity during meiosis in *Saccharomyces cerevisiae*. II. *U.N. Int. Conf. Peaceful Uses of Atomic Energy* **22**: 427.

Malone, R.E. and R.E. Esposito. 1980. The *RAD52* gene is required for homothallic interconversion of mating types and spontaneous mitotic recombination in yeast. *Proc. Natl. Acad. Sci.* **77**: 503.

———. 1981. Recombinationless meiosis in *Saccharomyces cerevisiae*. *Mol. Cell Biol.* **1**: 89.

Malone, R.E., J.E. Golin, and M.S. Esposito. 1980. Mitotic versus meiotic recombination in *Saccharomyces cerevisiae*. *Curr. Genet.* **1**: 241.

Mandelstam, J. 1971. Recurring patterns during development in primitive organisms. *Symp. Soc. Exp. Biol.* **25**: 1.

Marquardt, H. 1963. Elektronenoptische untersuchungen uber die ascosporenbildung bie *Saccharomyces cerevisiae* unter cytologischem und cytogenetischem aspekt. *Arch. Mikrobiol.* **46**: 308.

Martin, T.E. 1973. A simple and general method to determine the proportion of active ribosomes in eucaryotic cells. *Exp. Cell Res.* **80**: 496.

Matile, P. and A. Wiemken. 1967. The vacuole as the lysosome of the cell. *Arch. Mikrobiol.* **56**: 148.

Matur, A. and D.R. Berry. 1978. The use of step enzymes as markers during meiosis and ascospore formation in *Saccharomyces cerevisiae*. *J. Gen. Microbiol.* **109**: 205.

McClary, D.O., W.L. Nutley, and G. R. Miller. 1959. Effect of potassium versus sodium in the sporulation of *Saccharomyces*. *J. Bacteriol.* **78**: 362.

McCusker, J.H. and J.E. Haber. 1977. Efficient sporulation of yeast in media buffered near pH 6. *J. Bacteriol.* **132**: 180.

Miller, J.J. 1957. The metabolism of yeast sporulation. II. Stimulation and inhibition by monosaccharides. *Can. J. Microbiol.* **3**: 81.

————. 1959. A comparison of the sporulation physiology of yeast and aerobic *Bacilli. Wallerstein Lab Commun.* **22**: 267.

————. 1963a. Determination by ammonium of the manner of yeast nuclear division. *Nature* **198**: 214.

————. 1963b. The metabolism of yeast sporulation. V. Stimulation and inhibition of sporulation and growth by nitrogen compounds. *Can. J. Microbiol.* **8**: 259.

————. 1964. A comparison of the effects of several nutrients and inhibitors on yeast meiosis and mitosis. *Exp. Cell Res.* **33**: 46.

Miller, J.J. and C. Halpern. 1956. The metabolism of yeast sporulation. I. Effect of certain metabolites and inhibitors. *Can. J. Microbiol.* **2**: 519.

Miller, J.J. and O. Hoffmann-Ostenhof. 1964. Spore formation and germination in *Saccharomyces. Allg. Mikrobiol. Morphol. Physiol. Oekol. Mikroorg.* **4**: 273.

Mills, D. 1972. Effect of pH on adenine and amino acid uptake during sporulation in *Saccharomyces cerevisiae. J. Bacteriol.* **112**: 519.

————. 1974. Isolation of polyribosomes from yeast during sporulation and vegetative growth. *Appl. Microbiol.* **27**: 944.

————. 1978. 8-Hydroxyquinoline inhibition of DNA synthesis and intragenic recombination during yeast meiosis. *Mol. Gen. Genet.* **162**: 221.

————. 1980. Quantitative and qualitative analysis of polyadenylated RNA sequences during meiosis of *Saccharomyces cerevisiae.* In *Tenth International Conference on Yeast Genetics and Molecular Biology, Louvain-la-Neuve, Belgium.* p. 55.

Mills, D., M. Schiller, and M.M. Petryna. 1977. Characteristics of pulse-labeled RNA in relation to medium pH and acetate concentration during meiosis of yeast. *Mol. Gen. Genet.* **156**: 71.

Miyake, S., N. Sando, and S. Sato. 1971. Biochemical changes in yeast during sporulation. II. Acetate metabolism. *Dev. Growth Differ.* **12**: 285.

Moens, P.B. 1971. Fine structure of ascospore development in the yeast *Saccharomyces cerevisiae. Can. J. Microbiol.* **17**: 507.

————. 1974. Modification of sporulation in yeast strains with two-spored asci (*Saccharomyces,* Ascomycetes). *J. Cell Sci.* **16**: 519.

Moens, P. B. and E. Rapport. 1971a. Synaptic structures in the nuclei of sporulating yeast, *Saccharomyces cerevisiae.* (Hansen). *J. Cell Sci.* **9**: 665.

————. 1971b. Spindles, spindle plaques and intranuclear meiosis in the yeast *Saccharomyces cerevisiae. J. Cell Biol.* **50**: 344.

Moens, P.B., R.E. Esposito, and M.S. Esposito. 1974. Aberrant nuclear behavior at meiosis and annucleate spore formation by sporulation-deficient (*spo*) mutants of *Saccharomyces cerevisiae. Exp. Cell Res.* **83**: 166.

Moens, P.B., M. Mowat, M.S. Esposito, and R.E. Esposito. 1977. Meiosis in a temperature-sensitive DNA synthesis mutant and in an apomictic yeast strain (*Saccharomyces cerevisiae*). *Philos. Trans. R. Soc. Lond. B* **277**: 351.

Morrison, D.P. and P.J. Hastings. 1979. Characterization of the mutator mutation *mut5-1. Mol. Gen. Genet.* **175**: 57.

Mortimer, R.K. and D.C. Hawthorne. 1969. Yeast genetics. In *The yeasts* (ed. A.H. Rose and J.S. Harrison), vol. I, p. 386. Academic Press, New York.

————. 1973. Genetic mapping in *Saccharomyces.* IV. Mapping of temperature-sensitive genes and the use of disomic strains in localizing genes. *Genetics* **74**: 33.

Mortimer, R.K. and D. Schild. 1980. Genetic map of *Saccharomyces cerevisiae. Microbiol. Rev.* **44**: 519.

Moustacchi, E., H. Hottinguer-de-Margerie, and F. Fabre. 1967. A novel character induced in yeast by ^{32}P decay: The ability to manifest high frequencies of abnormal meiotic segregations. *Genetics* **57**: 909.

Mundkur, R. 1961. Electron microscopy of frozen-dried yeast. III. Formation of tetrad. *Exp. Cell Res.* **25**: 24.

Newlon, M.C. 1979. NADP-specific glutamate dehydrogenase is not involved in repression of yeast sporulation by ammonia. *Mol. Gen. Genet.* **176**: 297.

Newlon, M.C. and B.D. Hall. 1978. Inhibition of yeast sporulation by ethidium bromide. *Mol. Gen. Genet.* **165**: 113.

Ogur, M., A. Roshanmanesh, and S. Ogur. 1965. Tricarboxylic acid cycle mutants in *Saccharomyces*: Comparison of independently derived mutants. *Science* **147**: 1590.

Opheim, D.J. 1979. Effects of ammonium ions on activity of hydrolytic enzymes during sporulation of yeast. *J. Bacteriol.* **138**: 1022.

Osumi, M., N. Sando, and S. Miyake. 1966. Morphological changes in yeast cell during sporogenesis. *J. Women's Univ. J.* **13**: 70.

Parry, E.M. and B.S. Cox. 1970. The tolerance of aneuploidy in yeast. *Genet. Res.* **16**: 333.

Parry, J.M. and F.K. Zimmerman. 1976. The detection of monosomic colonies produced by mitotic chromosome nondisjunction in the yeast *Saccharomyces cerevisiae*. *Mutat. Res.* **36**: 49.

Pearson, N.J. and J.E. Haber. 1980. Changes in the regulation of ribosomal protein synthesis during vegetative growth and sporulation in yeast. *J. Bacteriol.* **143**: 1411.

Perlmann, P.S. and H.R. Mahler. 1971. Molecular consequences of ethidium bromide mutagenesis. *Nat. New Biol.* **231**: 12.

Petersen, J.G., M.C. Kielland-Brandt, and T. Nilsson-Tillgren. 1979. Protein patterns of yeast during sporulation. *Carlsberg Res. Commun.* **44**: 149.

Petersen, J.G., L. Olson, and D. Zickler. 1978. Synchronous sporulation of *Saccharomyces cerevisiae* at high cell concentrations. *Carlsberg Res. Commun.* **43**: 241.

Peterson, J.B., R.H. Gray, and H. Ris. 1972. Meiotic spindle plaques in *Saccharomyces cerevisiae*. *J. Cell Biol.* **53**: 837.

Phaff, H.J., M.W. Miller, and E.M. Mrak. 1966. *The life of yeasts.* Harvard University Press, Cambridge, Massachusetts.

Piñon, R. 1977. Effects of ammonium ions on sporulation of *Saccharomyces cerevisiae*. *Exp. Cell Res.* **105**: 367.

———. 1978. Folded chromosomes in non-cycling yeast cells. *Chromosoma* **67**: 263.

———. 1979a. A probe into nuclear events during the cell cycle of *Saccharomyces cerevisiae*: Studies of folded chromosomes in *cdc* mutants which arrest in G1. *Chromosoma* **70**: 337.

———. 1979b. Folded chromosomes in meiotic yeast cells: Analysis of early meiotic events. *J. Mol. Biol.* **129**: 433.

Piñon, R. and Y. Salts. 1977. Isolation of folded chromosomes from the yeast *Saccharomyces cerevisiae*. *Proc. Natl. Acad. Sci.* **74**: 2850.

Piñon, R., Y. Salts, and G. Simchen. 1974. Nuclear and mitochondrial DNA synthesis during yeast sporulation. *Exp. Cell Res.* **83**: 231.

Plotkin, D. 1978. "Commitment to meiotic recombination: A temporal analysis." Ph.D. thesis, University of Chicago, Illinois.

Plotkin, D. and R.E. Esposito. 1978. Temporal studies of recombination during sporulation in yeast. In *Ninth International Conference on Yeast Genetics and Molecular Biology, Rochester, New York.* p. 81.

Pomper, S., K.M. Daniels, and D.W. McKee. 1953. Genetic analysis of polyploid yeast. *Genetics* **39**: 343.

Pontefract, R.D. and J.J. Miller. 1962. The metabolism of yeast sporulation. IV. Cytological and physiological changes in sporulating cells. *Can. J. Microbiol.* **8**: 573.

Prakash, S., L. Prakash, W. Burke, and B.A. Montelone. 1980. Effects of the *RAD52* gene on recombination in *Saccharomyces cerevisiae*. *Genetics* **94**: 31.

Pratje, E., R. Schulz, S. Schnierer, and G. Michaelis. 1979. Sporulation of mitochondrial respiratory deficient *mit⁻* mutants of *Saccharomyces cerevisiae*. *Mol. Gen. Genet.* **176**: 411.

Puglisi, P.P. and E. Zennaro. 1971. Erythromycin inhibition of sporulation in *Saccharomyces cerevisiae*. *Experientia* **27**: 963.

Ramirez, C. and J.J. Miller. 1964. The metabolism of yeast sporulation. VI. Changes in amino acid content during sporogenesis. *Can. J. Microbiol.* **10**: 623.

Resnick, M.A. 1975. The repair of double-strand breaks in chromosomal DNA of yeast. In *Molecular mechanisms for repair of DNA* (ed. P. Hanawalt and R. Setlow), part B, p. 549. Plenum Press, New York.

Resnick, M.A., J.M. Kasimos, J.G. Game, R.J. Braun, and R.M. Roth. 1981. DNA changes during meiosis in a repair-deficient mutant (*rad52*) of yeast. *Science* **212**: 543.

Riley, M.I. and T.R. Manney. 1978. Tetraploid strains of *Saccharomyces cerevisiae* that are trisomic for chromosome III. *Genetics* **89**: 667.

Rodarte, U., S. Fogel, and R.K. Mortimer. 1968. Detection of recombination defective mutants in *Saccharomyces cerevisiae*. *Genetics* **60**: 216.

Rodarte-Ramon, U.S. 1972. Radiation induced recombination in *Saccharomyces:* The genetic control of recombination in mitosis and meiosis. *Radiat. Res.* **49**: 148.

Roman, H. and S.M. Sands. 1953. Heterogeneity of clones of *Saccharomyces* derived from haploid ascospores. *Proc. Natl. Acad. Sci.* **39**: 171.

Roman, H., M.M. Philips, and S.M. Sands. 1955. Studies of polyploid *Saccharomyces*. I. Tetraploid segregation. *Genetics* **40**: 546.

Roth, R. 1970. Carbohydrate accumulation during the sporulation of yeast. *J. Bacteriol.* **101**: 53.

―――. 1973. Chromosome replication during meiosis: Identification of gene functions required for premeiotic DNA synthesis. *Proc. Natl. Acad. Sci.* **70**: 3087.

―――. 1976. Temperature-sensitive yeast mutants defective in meiotic recombination and replication. *Genetics* **83**: 675.

Roth, R. and S. Fogel. 1971. A selective system for yeast mutants deficient in meiotic recombination. *Mol. Gen. Genet.* **112**: 295.

Roth, R. and H.O. Halvorson. 1969. Sporulation of yeast harvested during logarithmic growth. *J. Bacteriol.* **98**: 831.

Roth, R. and K. Lusnak. 1970. DNA synthesis during yeast sporulation: Genetic control of an early developmental event. *Science* **168**: 493.

Rothstein, R.J. and F. Sherman. 1980. Genes affecting the expression of cytochrome *c* in yeast: Genetic mapping and genetic interactions. *Genetics* **94**: 871.

Rothstein, R., R.E. Esposito, and M.S. Esposito. 1977. The effect of ochre suppression on meiosis and ascospore formation in *Saccharomyces*. *Genetics* **85**: 35.

Saeki, T., I. Machida, and S. Nakai. 1974. Split-dose recovery controlled by XS1 gene in yeast. *Radiat. Res.* **59**: 95.

Sager, R. and S. Granick. 1954. Nutritional control of sexuality in *Chlamydomonas reinhardi*. *J. Gen. Physiol.* **37**: 729.

Salts, Y., G. Simchen, and R. Piñon. 1976. DNA degradation and reduced recombination following UV irradiation during meiosis in yeast *(Saccharomyces)*. *Mol. Gen. Genet.* **146**: 55.

Sando, N. 1960. Ascospore formation of yeast. IV. Effects of metabolic inhibitors. *Sci. Rep. Tohoku Univ. Fourth Ser. (Biol.)* **26**: 157.

Sando, N. and S. Miyake. 1971. Biochemical changes in yeast during sporulation. I. Fate of nucleic acids and related compounds. *Dev. Growth Differ.* **12**: 273.

Sando, N., M. Maeda, T. Endo, R. Oka, and M. Hayashibe. 1973. Induction of meiosis and sporulation in differently aged cells of *Saccharomyces cerevisiae*. *J. Gen. Appl. Microbiol.* **19**: 359.

Sanfilippo, D. 1976. "Studies on the relationship between gene conversion and reciprocal recombination in heteroallelic reversion-deficient strains of yeast, *Saccharomyces cerevisiae*." Ph.D. thesis, University of California, Berkeley.

Schaeffer, P., J. Millet, and J.P. Aubert. 1965. Catabolic repression of bacterial sporulation. *Proc. Natl. Acad. Sci.* **54**: 704.

Schild, D. and B. Byers. 1978. Meiotic effects of DNA-defective cell division cycle mutations of *Saccharomyces cerevisiae. Chromosoma* **70:** 109.

Shaffer, B., I. Brearley, R. Littlewood, and G.R. Fink. 1971. A stable aneuploid of *Saccharomyces cerevisiae. Genetics* **67:** 483.

Sherman, F. and H. Roman. 1963. Evidence for two types of allelic recombination in yeast. *Genetics* **48:** 255.

Sherman, F. and P. Slonimski. 1964. Respiration-deficient mutants of yeast. II. Biochemistry. *Biochim. Biophys. Acta* **90:** 1.

Shilo, V., G. Simchen, and B. Shilo. 1978. Initiation of meiosis in cell-cycle initiation mutants of *Saccharomyces cerevisiae. Exp. Cell Res.* **112:** 241.

Silva-Lopez, E., T.J. Zamb, and R. Roth. 1975. Role of premeiotic replication in gene conversion in yeast. *Nature* **253:** 212.

Simchen, G. 1974. Are mitotic functions required in meiosis? *Genetics* **76:** 745.

———. 1978. Cell cycle mutants. *Annu. Rev. Genet.* **12:** 161.

Simchen, G. and A. Friedmann. 1975. Structure of DNA molecules in yeast meiosis. *Nature* **257:** 64.

Simchen, G. and J. Hirschberg. 1977. Effects of mitotic cell-cycle mutation *cdc4* on yeast meiosis. *Genetics* **86:** 57.

Simchen, G., D. Idar, and Y. Kassir. 1976. Recombination and hydroxyurea inhibition of DNA synthesis in yeast meiosis. *Mol. Gen. Genet.* **144:** 21.

Simchen, G., R. Piñon, and Y. Salts. 1972. Sporulation in *Saccharomyces cerevisiae:* Premeiotic DNA synthesis, readiness and commitment. *Exp. Cell Res.* **75:** 207.

Simchen, G., Y. Salts, and R. Piñon. 1973. Sensitivity of meiotic yeast cells to ultraviolet light. *Genetics* **73:** 531.

Slater, M.L. 1976. Rapid nuclear staining method for *Saccharomyces cerevisiae. J. Bacteriol.* **126:** 1339.

Sogin, S., J.E. Haber, and H.O. Halvorson. 1972. Relationship between sporulation-specific 20S ribonucleic acid and ribosomal ribonucleic acid processing in *Saccharomyces cerevisiae. J. Bacteriol.* **112:** 806.

Stark, H.C., D. Fugit, and D.B. Mowshowitz. 1980. Pleiotropic properties of a yeast mutant insensitive to catabolite repression. *Genetics* **94:** 921.

Stern, H. and Y. Hotta. 1977. Biochemistry of meiosis. *Phil. Trans. R. Soc. Lond. B* **277:** 277.

———. 1978. Regulatory mechanisms in meiotic crossing-over. *Annu. Rev. Plant Physiol.* **29:** 415.

Stevens, B.J. 1978. Behavior of mitochondria during sporulation in yeast. In *Electron microscopy* (ed. J.M. Sturgess), vol. 2, p. 406. Microscopical Society of Canada, Toronto.

Strathern, J.N., J.B. Hicks, and I. Hershkowitz. 1981. Control of cell type in yeast by the mating-type locus. The $\alpha1$-$\alpha2$ hypothesis. *J. Mol. Biol.* **147:** 357.

Strathern, J., C. Newlon, I. Herskowitz, and J.B. Hicks. 1979. Isolation of a circular derivative of yeast chromosome III: Implications for the mechanism of mating type interconversion. *Cell* **18:** 309.

Strömnaes, O. 1968. Genetic changes in *Saccharomyces cerevisiae* grown on media containing dl-para-fluoro-phenylalanine. *Hereditas* **59:** 197.

Svihla, G., J.L. Dainko, and F. Schlenk. 1964. Ultraviolet microscopy of the vacuole of *Saccharomyces cerevisiae* during sporulation. *J. Bacteriol.* **88:** 449.

Takahashi, T. 1962. Genetic segregation of two-spored asci in *Saccharomyces. Bull. Brew. Sci.* **8:** 1.

Tingle, M.A., M.T. Kuenzi, and H.O. Halvorson. 1974. Germination of yeast spores lacking mitochondrial deoxyribonucleic acid. *J. Bacteriol.* **117:** 89.

Tingle, M., A.J. Singh-Klar, S.A. Henry, and H.O. Halvorson. 1973. Ascospore formation in yeast. *Symp. Soc. Gen. Microbiol.* **23:** 209.

Trew, B.J., J.D. Friesen, and P.B. Moens. 1979. Two-dimensional protein patterns during growth and sporulation in *Saccharomyces cerevisiae. J. Bacteriol.* **138:** 60.

Tsuboi, M. 1976. Correlation among turnover of nucleic acids, ribonuclease activity, and sporulation ability of *Saccharomyces cerevisiae. Arch. Microbiol.* **111**: 13.

Tsuboi, M. and N. Yanagishima. 1974. Changes in sporulation ability during the vegetative cell cycle in *Saccharomyces cerevisiae. Bot. Mag. Tokyo* **87**: 183.

———. 1976. Promotion of sporulation by caffeine pretreatment in *Saccharomyces cerevisiae.* II. Changes in ribonuclease activity during sporulation. *Arch. Microbiol.* **108**: 149.

Vezinhet, F., J.H. Kinnaird, and I.W. Dawes. 1979. The physiology of mutants derepressed for sporulation in *Saccharomyces cerevisiae. J. Gen. Microbiol.* **115**: 391.

Vezinhet, V., M. Roger, M. Pellecuer, and P. Galzy. 1974. Genetic control of some metabolic modifications during the sporulation of *Saccharomyces cerevisiae* Hansen. *J. Gen. Microbiol.* **81**: 373.

Wejksnora, P.J. and J.E. Haber. 1974. Methionine-dependent synthesis of ribosomal ribonucleic acid during sporulation and vegetative growth of *Saccharomyces cerevisiae. J. Bacteriol.* **120**: 1344.

———. 1976. Influence of pH on the rate of ribosomal ribonucleic acid synthesis during sporulation in *Saccharomyces cerevisiae. J. Bacteriol.* **127**: 128.

———. 1978. Ribonucleoprotein particle appearing during sporulation in yeast. *J. Bacteriol.* **134**: 246.

Whelan, W.L. and C.E. Ballou. 1976. Sporulation in D-glucosamine auxotrophs of *Saccharomyces cerevisiae*: Meiosis with defective ascospore wall formation. *J. Bacteriol.* **124**: 1545.

Wickner, R.B. 1974. Mutants of *Saccharomyces cerevisiae* that incorporate deoxythymidine-5'-monophosphate into deoxyribonucleic acid *in vivo. J. Bacteriol.* **117**: 252.

Williamson, D.H. and D.J. Fennell. 1975. The use of fluorescent DNA-binding agent for detecting and separating yeast mitochondrial DNA. *Methods Cell Biol.* **12**: 335.

Williamson, D.H., L.H. Johnston, D.J. Fennell, and G. Simchen. 1980. Premeiotic DNA synthesis in *Saccharomyces cerevisiae* replicon size and rate of replication fork movement. In *Tenth International Conference on Yeast Genetics and Molecular Biology, Louvain-la-Neuve, Belgium.* p. 181.

Wolf, D. and G. Fink. 1975. Proteinase C (carboxypeptidase Y) mutant of yeast. *J. Bacteriol.* **123**: 1150.

Wolfner, M., D. Yep, F. Messenguy, and G.R. Fink. 1975. Integration of amino acid biosynthesis into the cell cycle of *S. cerevisiae. J. Mol. Biol.* **96**: 273.

Wright, J.F. and I.W. Dawes. 1979. Sporulation-specific protein changes in yeast. *FEBS Lett.* **104**: 183.

Yanagita, T., M. Yagishawa, S. Oishi, N. Sando, and T. Suto. 1970. Sporogenic activities of mother and daughter cells in *Saccharomyces cerevisiae. J. Gen. Appl. Microbiol.* **16**: 347.

Zamb, T.J. and R. Roth. 1977. Role of mitotic replication genes in chromosome duplication during meiosis. *Proc. Natl. Acad. Sci.* **74**: 3951.

Zickler, D. 1970. Division spindle and centrosomal plaques during mitosis and meiosis in some Ascomycetes. *Chromosoma* **30**: 287.

Zickler, D. and L.W. Olson. 1975. The synaptonemal complex and the spindle plaque during meiosis in yeast. *Chromosoma* **50**: 1.

Zubenko, G.S. and E.W. Jones. 1981. Protein degradation, meiosis and sporulation in proteinase-deficient mutants of *Saccharomyces cerevisiae. Genetics* **97**: 45.

Zubenko, G.S., A.P. Mitchell, and E.W. Jones. 1979. Septum formation cell division and sporulation in mutants of yeast deficient in proteinase B. *Proc. Natl. Acad. Sci.* **76**: 2395.

Mechanisms of Meiotic Gene Conversion, or "Wanderings on a Foreign Strand"[1]

Seymour Fogel,* Robert K. Mortimer,† and Karin Lusnak*
*Department of Genetics and
†Department of Biophysics and Medical Physics,
University of California
Berkeley, California 94720

INTRODUCTION

Over the last decade and a half, studies in organisms amenable to tetrad (or octad) analysis have provided a reasonably complete descriptive inventory of intragenic recombination (Fogel et al. 1979; Nicolas 1979; Rossignol et al. 1979; Sang and Whitehouse 1979). This paper aims to provide an overview of the relevant data. At the same time it attempts a synthesis of the new information in the expectation that a scaffold will emerge from which testable molecular hypotheses concerning generalized and intragenic recombination can be constructed. A comprehensive review of the literature concerning meiotic recombination will not be attempted. Much of the data have been cogently summarized in texts by Esser and Kunnen (1967), Whitehouse (1969), Grell (1974), Kushev (1974), Catcheside (1977), Fincham et al. (1979), Stahl (1979b) and in several critical reviews by Emerson (1969), Mortimer

[1] Sir Walter Scott, "The Lay of the Last Minstrel."

and Hawthorne (1969), Fogel and Mortimer (1971), Radding (1973), Hotch-kiss (1974), Hastings (1975), Esposito and Esposito (1977), Pukkila (1977), Resnick (1979), Stahl (1979a), and Petes (1980b).

Here, we approach the problem of intragenic meiotic recombination or recombinagenic events within short DNA segments by considering four questions.

1. What is the phenomenon of gene conversion?
2. What are the methods that can be employed to study this phenomenon?
3. What are the salient features of the conversion process?
4. Can methodologies of recombinant DNA technology be adapted to advance the analysis of intragenic recombination from the genetic level to the molecular level?

We may begin by asking, What is meiotic gene conversion? In organisms subject to octad analysis, irregular segregations or gene conversions are detected as departures from the $4^+:4^-$ intraascal segregations expected for a single heterozygous site. Thus, the appearance of asci displaying spore or spore-clone patterns of $6^+:2^-$, $2^+:6^-$, $5^+:3^-$, $3^+:5^-$, aberrant $4^+:4^-$, and others constitute primary exceptions to the fundamental Mendelian segregation principle and, for this reason, they have continued to attract the fungal geneticists' attention over the last 50 years.

At this juncture, some remarks on the classification of yeast tetrads are appropriate. Unless noted otherwise, the standard culture and handling procedures described by Mortimer and Hawthorne (1969) and Sherman and Lawrence (1974) were employed. Sporulated yeast cultures typically contain asci with four, three, or two spores. Ordinarily, only four-spored asci are taken for dissection by micromanipulation. In the studies reported here, several new techniques were devised and utilized. One involved a modified ascus-dissection procedure designed to permit the reliable detection of post-meiotic segregation (pms). Twenty asci were dissected in a 6 × 6-mm grid directly onto the surface of nutrient agar contained in an inverted 10-cm petri dish. After 3–4 days of incubation at 30°C, the resultant master was replica-plated to a series of plates containing various media diagnostic for each of the segregating markers. This protocol maintained the spatial organization of the spore progenies and permitted the detection of pms as sectoring for the growth response on one or more diagnostic plates. We have designated the protocol given above as the plate-dissection method.[2]

In effect, the plate-dissection method converts tetrads into octads and, hence, we are able to record for any single marker the 12 ascus segregation classes shown in groups a, b, and c (Fig. 1). Of these, only the first $4^+:4^-$ pattern is normal; the remaining patterns represent segregations that imply the gain or loss of informational copies by one chromosome at the expense of

---

[2]It is a pleasure to acknowledge the contribution of D.A. Campbell (University of California, Berkeley) in the development of the method and the final design of the plate carrier.

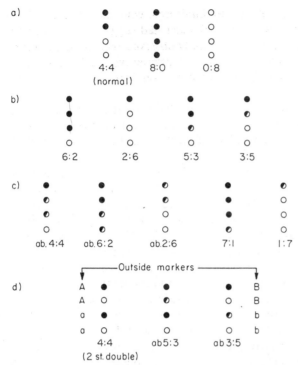

Figure 1   Diagram of various segregation classes of yeast tetrads. The ability to detect sectored ascosporal clones, in effect, converts tetrads into octads. (*a, b, c*) Normal segregation and 11 aberrant segregation classes. Classes shown in *d* are detected only if outside markers are available.

corresponding information contained in the nonsister homolog. Such events are presumed to occur during a four-strand stage at some point in early meiotic prophase after the major round of meiotic DNA replication has occurred (Fogel et al. 1979). Ratios of the type $6^+:2^-$, $6^-:2^+$, and two-strand double $4^+:4^-$ are assumed to reflect enzymatic repair of transient heteroduplex (h) DNA (an assumption addressed below); whereas segregations displaying one or more sectored ascosporal clones (pms) are taken to indicate persistent base-pair mismatches in the hDNA that failed to be repaired and were subsequently resolved in a single postmeiotic replication cycle. Three additional classes (as in Fig. 1d) may be identified when the central heterozygosity is flanked by heterozygous marker genes A/a and B/b.

Irregular segregations were reported in fungi and mosses by Wettstein (1924), Brunswick (1926), and Kniep (1928). Similar unusual ascal types in yeast were noted later by Lindegren (1949), Mundkur (1949), Winge and Roberts (1950), and Roman (1957). Such aberrant segregations were designated as "allele-induced mutations" by Lindegren, who later adopted Winkler's (1930) term of "gene conversion." Winkler had suggested that not only did gene conversion account for the regularly observed aberrant segregations but, in addition, gene conversion provided an alternative to the chiasmatype theory as a means to account for recombination between linked genetic factors.

Over the intervening years, it became abundantly clear that a wide variety of conventional genetic events could generate data superficially indistinguishable from gene conversion, not all of which are true gene conversions. A partial listing is given in Table 1. To distinguish gene conversions from other genetic events, one normally works with a diploid strain heterozygous for a comparatively large number of sites, say 15–20. Normal chromosome distribution is indicated by the regular $4^+:4^-$ segregation pattern for each marker, and exceptional conversion asci typically display aberrant segregations for a single marker. In the case where a central marker flanked by two closely linked outside markers manifests an irregular segregation, e.g., $6^+:2^-$, and the proximal and distal flanking markers display normal $4^+:4^-$ parental configurations, we would record a gene-conversion event for only the central marker. Frequently, however, the outside markers in such conversional asci are reciprocally recombined. Thus, the central marker exhibits nonreciprocal behavior, whereas the outside markers may retain the parental array or recombine reciprocally with approximately equal likelihood. The pronounced correlation between conversion and crossing-over led to the notion that the two processes were mechanistically linked (Mitchell 1955; Fogel and Hurst 1967). Moreover, because conversion and crossing-over are correlated, they may reflect two alternative aspects of a single basic event or process. Thus, we proposed that gene conversion is a signal of the basic meiotic recombination event in yeast (Fogel et al. 1979).

MECHANISTIC INTERPRETATIONS

Over the past 15 years, a plethora of molecular models has been advanced to account for both gene conversion and crossing-over or reciprocal recombination, in general. Convenient summaries and comparisons between the various models may be found in Stahl (1969, 1979b), Paszewski (1970), Hotchkiss (1974), and Catcheside (1977). As will become apparent, these models have different predictions as to the distribution of the segregation classes depicted in Figure 1. Here, we present Holliday's original model (1964), which provided signal impetus to development in this arena. The model is given in Figure 2. He was among the first to suggest a scheme that would account, in molecular terms, for the known behavior of various genetic markers in hybrids. His original model posited the formation of symmetrical hybrid or hDNA on each of the two interacting nonsister chromatids with two different associated base-pair mismatches at the included points of heterozygosity. Base-pair mismatch correction was assumed to be mediated by two distinct

Table 1 Causes of irregular segregation

Event	Affected loci	Segregation pattern
Chromosomal		
polyploidy	most genes	$8^+{:}0^-$, $6^+{:}2^-$, and $4^+{:}4^-$
polysomy	genes on one chromosome	$8^+{:}0^-$, $6^+{:}2^-$, and $4^+{:}4^-$
meiotic nondisjunction	genes on one chromosome	occasional $8^+{:}0^-$, $6^+{:}2^-$, and $4^+{:}4^-$
mitotic crossing-over before meiosis	genes distal to crossover	$8^+{:}0^-$ or $0^+{:}8^-$
Multiple gene control		
Polymeric	single traits	$8^+{:}0^-$, $6^+{:}2^-$, and $4^+{:}4^-$
Complementary	single traits	$4^+{:}4^-$, $2^+{:}6^-$, and $0^+{:}8^-$
Suppressors	suppressible traits	$8^+{:}0^-$, $6^+{:}2^-$, and $4^+{:}4^-$
False tetrad		
Four spores from more than one meiosis	most genes	irregular segregation
Binucleate spores	most genes	irregular segregation
Gene conversion	any gene	occasional $6^+{:}2^-$ and $2^+{:}6^-$

Tetrads displaying pms are not listed.

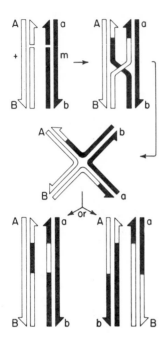

Figure 2 The Holliday model for recombination. At homologous special sites (recombinators), a bivalent enzyme nicks chains of the same polarity. Over part of their length, the nicked chains exchange partners and are ligated in their new positions. The resulting Holliday structure can be displayed as a Chi form, which illustrates the structural equivalence of its two modes of resolution. A vertical cut yields patched chromatids, which are not crossovers for flanking markers. In this model, conversion results when a patch or splice falls on a marked site and one of the resulting heteroduplexes ($+/m$) enjoys mismatch correction, or both heteroduplexes are corrected to the same information content. In the former case, the resulting meiotic tetrad is a $5^+:3^-$ or a $5^-:3^+$; in the latter case, a $6^+:2^-$ or $6^-:2^+$ tetrad results. When correction fails on both chromatids, the resulting tetrad is an aberrant 4:4. (Reprinted, with permission, from Stahl 1978.)

enzymatic processes, namely excision (by a nuclease) and replacement (by a polymerase) of a modal length of single DNA chains. Thus, $6^+:2^-$ segregations would eventuate if the strands bearing mutant alleles were excised, whereas $2^+:6^-$ segregations would eventuate if the strands bearing wild-type alleles were excised. These represent corrections in the same direction, i.e., to mutant or to wild type. Repair of only a single mismatch in the symmetrical pair (i.e., failure to correct one mismatch) generates pms of the $5^+:3^-$ or $3^+:5^-$ type. However, if both members of the pair remain uncorrected and instead undergo resolution by a single postmeiotic replication cycle, aberrant $4^+:4^-$ segregations ensue (Kitani et al. 1962; Kitani and Olive 1967; Kitani 1978). In yeast, this distinctive category (see Fig. 1c) is marked by a tetrad yielding two sectored ascosporal clones accompanied by a clone of each parental type. Also, it should be apparent that correction in opposite directions might result in either a normal $4^+:4^-$ restoration (Hastings et al. 1980) or, if close flanking markers are present, a two-strand double crossover. Clearly, such two-strand doubles arise as a consequence of random correction rather than classical crossing-over, and they should add to the ranks of conventional two-strand doubles. But other segregations, including $ab6^+:2^-$, $ab2^+:6^-$, $7^+:1^-$, $1^+:7^-$, $ab5^+:3^-$, and $ab3^+:5^-$, are not readily accommodated within the Holliday model. These can be accounted for by the involvement of all four chromatids in producing two segments of symmetric or asymmetric hybrid DNA (Lamb and Wickramaratne 1973; Lamb 1975). Essential to Holliday's model is a cross-stranded structure or half chiasma, generated by the nicking, separating, and reannealing of single DNA strands derived from the interacting parental chromatids. The chromatids may be considered as single duplex DNA molecules, and the half-chiasma is easily resolvable with or without outside marker recombination (Fig. 2).

As indicated above, the Holliday model carries implicit, clear testable predictions of two sorts. The first requires a rather high frequency of aberrant $4^+:4^-$ segregations—a symptom of repair failure or noncorrection in symmetrical hDNA segments spanning a heterozygous marker site. The second requires a much enhanced frequency of observed two-strand double crossovers among nonconversion asci. These arise from random corrections of the paired symmetrical heteroduplexes flanked by closely linked outside markers. The conspicuous paucity of supportive data in both categories, first reported by us at the European Molecular Biology Organization (EMBO) workshop on recombination (held in 1973 at Aviemore, Scotland), prompted the formulations of the new recombination model of Meselson and Radding (1975) shown in Figure 3. The Meselson-Radding model posits an initial asymmetric phase of hybrid DNA formation as a recombination intermediate. The Holliday-like structure may isomerize and subsequently yield symmetrical hybrid DNA segments. The two phases may differ in extent in different fungi. Yeast appears to generate very little or no symmetrical hDNA, whereas intragenic recombination events in *Ascobolus* typically initiate asymmetrically but often terminate in a symmetrical phase. Except for certain data bearing on the position of conversion-associated exchanges

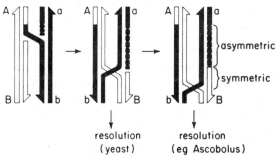

Figure 3   The Meselson-Radding model for recombination. DNA synthesis, initiated at a nick, displaces a chain, which then is taken up by the homologous chromatid. Coordinate chain extension and hydrolysis elongate the region of hybrid DNA formation. When the enzymes fall off, the structure can isomerize, and branch migration leads to hybrid DNA on both chromatids. This structure can then be visualized in a Chi form like that in Fig. 2; resolution can be to a partially asymmetric pair of either patches or splices. (Reprinted, with permission, from Stahl 1978.)

in yeast, virtually all of the data from both of these fungi are accommodated by the Meselson-Radding model.

PROPERTIES OF GENE CONVERSION

Recombination between homologous DNA segments can be studied as spontaneous or induced events in mitotic and meiotic systems. Mitotic recombination is covered by Esposito and Wagstaff (this volume) and by Roman (1980) and will not be considered here. Instead, our major focus is on the meiotic aspect of the conversion phenomenon. Primarily, we were interested in assessing whether recombination in yeast involves homoduplex DNA or heteroduplex DNA and, if the latter is involved, whether it is mainly symmetrical or asymmetrical. In addition, we wished to investigate the role of specific base-pair mismatches, the effects of comparatively long insertions and deletions, and the position and frequency of conversion-associated outside-marker recombination events.

Necessary and essential to any critical test of such questions is the ability to detect pms in unselected tetrads. Clearly, pms comprises the principal and, perhaps, the sole genetic evidence for the occurrence of hybrid DNA. It should be noted that our earlier yeast studies (Fogel and Mortimer 1969; Fogel et al. 1971; Hurst et al. 1972; Mortimer and Fogel 1974) were conducted using state-of-the-art conventional techniques of patching ascosporal clones via sterile toothpick transfers—a technique that efficiently obscured the reliable observation of these critical segregations, though they were already known in *Neurospora* (Case and Giles 1964), *Ascobolus* (Leblon 1972), and *Saccharomyces* (Esposito 1971). This paper (and our review in Fogel et al. [1979]) rectifies this shortcoming and describes a reexamination of gene conversion and its salient features on the basis of the plate-dissection method.

Gene conversion may be investigated according to several different protocols: selected recombinant spores, selected tetrads in which one spore is recombinant, and unselected tetrads. Principal among these are selective techniques that permit detection and enumeration of "recombinant" wild-type spores or asci produced by heteroallelic hybrids. Thus, from the general type cross $m1+ \times +m2$, where $m1+$ and $+m2$ represent independently derived alleles, one can assay the frequency of prototrophic (++) spores in a random spore population or the frequency of prototroph-containing asci among sporulated cells by conventional plating-dilution methods. Rapid and relatively easy to perform, these techniques yield data that permit the construction of genetic fine-structure maps. Outstanding examples of such maps have been published for the *ade8* locus (by Esposito 1967) and for the *his1* locus (by Korch and Snow 1973). Although they depict a spatial distribution of the various alleles within a single gene, they convey no information that bears on the underlying mechanism that led to the production of the "wild-type recombinants."

Polarity and Associated Crossing-over

Typically, when heteroallelic crosses in repulsion ($m1+/+m2$) are studied, additional information on the mechanism of recombination can be gathered from the analysis of intact asci selected to contain a prototrophic spore (Fogel and Hurst 1967). In heteroallelic hybrids, prototrophic spores arise from $6^+:2^-$ (or $5^+:3^-$) segregations of either allele or from reciprocal recombination between the mutant alleles. In the latter instance, the ascus would contain four spores, as follows: ++, $m1m2$, $m1+$, and $+m2$. Thus, among 1081 tetrads selected on the basis that they contained a prototrophic spore, about 90% displayed a $6^+:2^-$ conversion of one allele or the other, and only 10% arose as a consequence of an apparently reciprocal event between the two mutated sites (Fogel and Hurst 1967). Accordingly, for the most part, the appearance of a "wild-type" or prototrophic recombinant in heteroallelic crosses is dependent on a nonreciprocal event. In contrast, for markers that are further apart, as in intergenic events, recombinants typically manifest a reciprocal character.

Two additional primary features of the conversion process became evident from this and from comparable studies on selected tetrads. First is the feature designated as polarity. As shown in Table 2, alleles widely separated on a fine-structure map differ considerably in their conversion propensities. The data in Tables 2 and 3 were derived from three hybrids involving, respectively, *his1-7 × his1-1*, *his1-315 × his1-1*, and *his1-315 × his1-204*. Starting at the centromere proximal end, the order of the mutant sites in the *his1* locus is *315, 7, 204, 1*. Heterozygous flanking markers were also present: *thr3*, which is about 2.5 map units proximal to *his1-315*, and *arg6*, which is about 10 map units distal to *his1-1*. Several conclusions may be drawn from the tabulated data: The crosses differ significantly with respect to the proportion of asci showing reciprocal recombination, but the ratio of proximal conversions to distal conversions is nearly constant, i.e., about 6–7 times as many conver-

Table 2 Recombination events at the *his1* locus of yeast

Cross	Prototrophs (per 10^5 asci)	Reciprocals (% of aberrant asci)	Allele converted		
			proximal(P)	distal(D)	P/D
his1-7 × his1-1	293	12.6 ± 1.4	452	75	6.02
his1-315 × his1-1	416	7.6 ± 1.6	234	34	6.8
his1-315 × his1-204	344	1.6 (± 0.9)	161	24	6.3

(Adapted from Fogel and Hurst 1967.)

Table 3   Distribution of combinations of flanking marker-gene associations among *HIS1*⁺ recombinant ascospores in the yeast cross *THR3 his1-x arg6* × *thr3 his1-y ARG6*[a]

Type of *HIS*⁺ recombinant	*THR3 arg6*	*thr3 ARG6*	*thr3 arg6*	*THR3 ARG6*
reciprocal	4	0	97	0
proximal convert	474	5	265	103
distal convert	0	49	77	7
Totals	478	54	439	110

[a]Adapted from $\dfrac{THR3 \quad his1\text{-}x \quad arg6}{thr3 \quad his1\text{-}y \quad ARG6}$ (Fogel and Hurst 1967).

sions are found at the proximal site compared with events at the distal site. From these results and from studies on other loci (see below, Analysis of Gene Conversion at *ARG4*; Fig. 4), it is evident that there is a gradient of conversion frequency of alleles across genes. This polarity suggests specific sites involved in the origin or resolution of conversion events.

The second feature of the conversion process that is evident from the studies on selected tetrads is the association of outside marker recombination. About half of the prototrophic spores are recombined for outside markers. For example, the data on recombination of outside markers for the *his1* crosses discussed above (Table 2) are pooled and displayed in Table 3. We note that among the strands bearing a prototrophic recombinant at *HIS1*, 528 convertant strands retain the input parental arrays (474 + 49 + 5), whereas a nearly equivalent number of strands, 452 (265 + 103 + 77 + 7), display an outside marker pattern indicative of associated crossing-over. A very different distribution is found among the reciprocal recombinants between the alleles (Table 3). Here, with minor exceptions, a single class of outside marker combination is observed—one that is unequivocally diagnostic for the sequence of the mutant sites within the locus. The fact that the four potential arrays of outside markers are not present in equal numbers—one parental and one recombinant class predominate (see Table 3, Totals)— reflects the different conversion rates of proximal and distal alleles of *his1*, i.e., polarity. This provides an independent basis for assaying polarity relations among the various mutant sites within a genetic locus.

Coconversions

As informative as the selective tetrads have been, a considerable measure of even greater detail can be gleaned from the complete analysis of unselected tetrads.

Our experience with a large sample of nearly 25,000 unselected tetrads from two-point heteroallelic crosses reveals additional features of the conversion phenomenon. Eight major ascal types are regularly encountered. Ascal types including one or more pms will not be considered at this juncture. Their significance and special importance will be discussed in a subsequent section. Table 4 displays normal segregations, single-site conversions, symmetrical double-site conversions, and reciprocal recombinants. Single-site conversions (types b, c, d, and e) are marked by $6^+:2^-$ or $2^+:6^-$ segregations for either the proximal or distal allele accompanied by a normal $4^+:4^-$ segregation of the adjacent allele. Reciprocal recombinants (type h) exhibit all of the features expected to ensue from conventional reciprocal exchange between the mutant sites. With rare exceptions, symmetrical double-site conversions (types f and g, the novel category found only among unselected tetrads), fall into two subsets that involve a $6^+:2^-$ segregation of one allele accompanied by a symmetrical $2^+:6^-$ segregation of the second allele. We have referred to this process as informational transfer (Fogel and Mortimer 1969) but, following a suggestion by D. Stadler (pers. comm.), we have designated the double-site symmetrical event as coconversion. Coconversions are most simply explained as the consequence of a single event involving the nonreciprocal replacement of a segment of information in one homolog with information identical to that contained in the corresponding section of a nonsister homolog.

If it is assumed that coconversions arise not as single events but via independent conversional events at the two intragenic sites, then, in addition to the coconversion classes, six additional ascal types may be predicted to occur in a sample of unselected tetrads, which are displayed in Table 5 (Table 4f,g equals Table 5g,h). On the assumption that double-site conversions represent coincident events, we would expect, as a first approximation, all categories to occur with equal frequency. However, in our study of 5 isogenic hybrids involving 6500 unselected tetrads subjected to complete ascertainment by

Table 4 Main types of asci recoverable from heteroallelic diploids

Normal segregations	Single-site conversions				Symmetrical double-site conversions		Reciprocal recombinants
(a)	(b)[a]	(c)[a]	(d)[b]	(e)[b]	(f)[c]	(g)[c]	(h)
m1 +	*m1* +	*m1* +	*m1* +	*m1* +	*m1* +	*m1* +	*m1* +
m1 +	*m1* +	*m1* +	*m1* +	*m1* +	*m1* +	*m1* +	*m1* +
m1 +	+ +	*m1* +	*m1* +	*m1 m2*	+ *m2*	*m1* +	*m1 m2*
m1 +	+ +	*m1* +	*m1* +	*m1 m2*	+ *m2*	*m1* +	*m1 m2*
+ *m2*	+ *m2*	*m1 m2*	+ +	+ *m2*	+ *m2*	*m1* +	+ +
+ *m2*	+ *m2*	*m1 m2*	+ +	+ *m2*	+ *m2*	*m1* +	+ +
+ *m2*	+ *m2*	+ *m2*	+ *m2*	+ *m2*	+ *m2*	+ *m2*	+ *m2*
+ *m2*	+ *m2*	+ *m2*	+ *m2*	+ *m2*	+ *m2*	+ *m2*	+ *m2*

[a]Conversion of +/*m1*.
[b]Conversion of +/*m2*.
[c]Conversion of +/*m1* and +/*m2*.

Table 5 Ascal types expected if heteroallelic sites convert independently

(a)	(b)	(c)	(d)	(e)	(f)	(g)	(h)
+ +	m1 m2	+ +	m1 m2	+ +	+ +	+ m2	+ m2
+ +	m1 m2	+ +	m1 m2	+ +	+ +	+ m2	+ m2
+ +	m1 m2	+ +	m1 m2	+ m2	m1 +	m1 +	m1 +
+ +	m1 m2	+ +	m1 m2	+ m2	m1 +	m1 +	m1 +
+ +	m1 m2	m1 +	m1 +	+ m2	m1 +	m1 +	+ m2
+ +	m1 m2	m1 +	m1 +	+ m2	m1 +	m1 +	+ m2
m1 m2	+ +	+ m2	+ m2	m1 m2	m1 m2	m1 +	+ m2
m1 m2	+ +	+ m2	+ m2	m1 m2	m1 m2	m1 +	+ m2

backcrossing, such coincident conversion events were conspicuously rare. In fact, their frequencies are far below the predicted values obtained by multiplying the appropriate probabilities. Hence, we are led to the view that coconversion represents a single primary event in DNA.

On the basis of the above argument, we infer that conversion usually involves a sizable DNA segment rather than a narrowly restricted point. Widely separated alleles convert independently. As the distance between the mutant sites is decreased, the frequency of coconversion increases at the expense of single-site conversions. The converted segment has a variable length, but modally it is on the order of several hundred nucleotides, as judged from our earlier studies using the mitotic X-ray mapping procedures developed by Manney and Mortimer (1964) and a calibration factor derived from investigations on the genetics of cytochrome c by Parker and Sherman (1969) and Moore and Sherman (1974, 1975). This estimate is also consistent with recently published physical data on cloned segments bearing the *ARG4* locus (Clarke and Carbon 1978) and our own cloning studies (S. Fogel, unpubl.).

As pointed out above, prototrophs among the meiotic products of heteroallelic diploids derive mainly from $6^+:2^-$ segregations of either parental site and, to a much lesser extent, from reciprocal recombination between the mutated sites. Yet, the frequency of such wild-type recombinants may be used as a measure of the distance between the input sites, and reasonably additive maps can be obtained. We may summarize the probable genetic basis of this result as follows: For alleles in close proximity, conversion of one allele is usually associated with conversion of the other and, consequently, for heteroallelic combinations in a *trans* array, no prototrophs are generated. As the distance between the alleles increases, the likelihood of coconversion decreases, and the prototroph frequency increases. For alleles *arg4-4* and *arg4-17,* internal markers lying at opposite ends of the structural gene specifying arginino-succinase, only 3 among 58 exceptional tetrads (5%) in a sample of

697 asci were coconversions. This contrasts with the finding for the closely spaced alleles *arg4-2* and *arg4-17*. Here, of 36 exceptional tetrads in a sample of 544 unselected asci, 27, or 75%, were diagnosed as coconversional events. Coconversion has been observed in numerous other instances at other loci on many different chromosomes. In addition, it has been demonstrated that the coconverted segment may extend from one gene into or across an adjoining gene (Bassel and Mortimer 1971; Fogel et al. 1971). It is entirely reasonable to suppose that coconversion, reported in *Neurospora, Schizosaccharomyces, Ascobolus, Drosophila,* and maize, is a general phenomenon that can be detected following the interaction of homologous DNA segments either interchromosomally (Scherer and Davis 1980) or intrachromosomally (Klein and Petes 1980).

The gene-conversion phenomenon is characterized by several additional meiotic features, and these will be discussed here. In the sections that follow, pms will be considered in some detail.

pms and Mismatch Repair

Unselected tetrads (23,135) drawn from 20 closely related hybrids were scored for their segregation patterns at all heterozygous sites (Fogel et al. 1979). In all, 30 markers were scored in conjunction with the plate-dissection method for pms detection, and each hybrid contained 8–13 uniquely testable markers. The tetrads were assigned as normal $4^+:4^-$ segregations or to 11 other categories of aberrant segregations in Table 6. Of these, seven display one or more sectored ascosporal clones indicative of pms.

All 30 heterozygous sites show aberrant segregations at frequencies ranging from 0.6–18%. The $8^+:0^-$ and $0^+:8^-$ segregations are excluded. In yeast these surely reflect mitotic events initiated prior to the onset of meiosis (Roman 1957). However, such segregations in *Sordaria* and *Ascobolus,* where karyogamy immediately precedes meiosis, undoubtedly have a meiotic origin. Moreover, as pointed out by Lamb (1972) and Lamb and Wickramaratne (1973), the rare $8^+:0$ and $0^+:8^-$ segregations closely approximate the $7^+:1^-$, $1^+:7^-$, $ab6^+:2^-$ and $ab2^+:6^-$ classes. Quite plausibly, these wider ratio segregations are accounted for by assuming the involvement of all four chromatids at a single site with the occurrence of symmetrical hybrid DNA in each of the interacting sets. Here, we may note that in yeast, apart from the $8^+:0^-$ and $0^+:8^-$ asci, only 11 such wide ratio segregations were found in the 23,135 unselected asci. Nicolas (1979) recently reported on the conversion patterns of 80 mutants in 17 ascospore characters. This comprehensive study notes conspicuous similarities and differences among various fungi.

Table 7 displays a similar set of data drawn from five isogenic diploids. These are uniformly heterozygous at ten miscellaneous sites distributed throughout the genome, and they are variously heterozygous at one, two, or three sites within the *arg4* locus. The five hybrids were derived by single or

Table 6 Distributions of aberrant segregations for 30 heterozygous sites

Gene	Segregations	Hybrids[a]	8:0	0:8	6:2	2:6	5:3	3:5	ab4:4	ab6:2	ab2:6	7:1	1:7	ρ Events (%)	σp	χ² heterogeneity[b]	pms/total (%)
pet1	4,924	4	1	38	4	27	0	0	0	0	0	0	0	0.63	0.35	*	0
trp1	20,826	17	8	39	62	47	12	8	0	0	0	0	1	0.63	0.29	*	16.2
mat1	23,135	20	0	8	93	104	0	0	0	0	0	0	0	0.85	0.34	*	0
ura3	2,315	3	0	0	10	9	2	1	0	0	0	0	0	0.95	0.84		13.6
ade6	1,589	1	0	3	4	9	3	3	0	0	0	0	0	1.20	—	—	31.6
his5-2	2,315	3	2	2	14	10	4	1	0	0	0	0	0	1.26	0.17		17.2
tyr1	8,391	7	7	6	59	44	5	1	0	0	0	0	0	1.30	0.41		5.5
CUP1	18,016	15	20	16	123	124	6	3	0	0	0	0	0	1.42	0.52	*	3.5
gal2	2,416	2	1	3	36	17	0	0	0	0	0	0	0	1.78	0.09		0
leu2-1	3,203	3	2	2	41	25	0	0	0	0	0	0	0	2.06	0.65		0
trp5-48	2,315	3	3	1	23	31	1	0	0	0	0	0	0	2.38	0.56		2
met1	1,589	1	1	1	18	17	18	6	0	0	0	0	0	3.72	—	—	40.7
met10	892	1	0	0	17	16	1	0	0	0	0	0	0	3.82	—	—	2.9
ura1	15,014	13	29	17	331	275	11	15	1	0	0	0	0	4.23	1.48	**	4.3
ilv3	9,487	9	1	2	230	239	9	6	0	0	0	0	0	5.10	0.72		3.1
lys1-1	2,315	3	3	1	51	78	3	7	0	0	0	0	0	5.76	0.20		3.0

SUP6	892	1	0	0	22	33	0	0	0	0	0	0	6.16	—		0
thr1	21,220	18	13	15	691	594	14	22	0	0	0	0	6.23	1.15	**	2.7
his4-4	12,533	11	11	12	411	303	39	41	1	0	1	0	6.36	1.16		10.3
ade8-10	1,118	1	2	64	48	30	0	0	0	0	0	0	7.41	—	**	0
met13	10,454	10	151	15	459	474	1	0	0	0	3	0	9.07	2.15		0.1
ade8-18	15,480	13	29	2221	351	239	375	294	1	1	0	2	9.72	0.89		54.1
cdc14	892	1	1	0	48	39	1	1	10	0	0	0	9.99	—		2.2
ade7	1,028	1	3	159	47	27	11	8	0	0	0	0	10.7	—		20.4
his2	2,481	2	5	10	215	231	2	0	0	0	0	0	18.2	3.74	**	0.5
arg4-4	1,188	1	0	0	5	13	0	1	0	0	0	0	1.6	—		5.2
arg4-3	2,405	2	0	0	28	23	5	3	0	0	0	0	2.45	0.26		1.4
arg4-19	5,352	4	5	5	87	68	2	6	0	0	0	0	3.04	1.16	**	4.9
arg4-16	14,490	11	8	11	508	302	95	289	4	1	1	1	8.29	1.58	**	32.5
arg4-17	13,476	13	18	11	485	551	38	20	0	0	0	0	8.12	1.79	**	5.3
Total	221,751	324	2662	4521	3999	657	740	16	1	4	3	3	4.54			3.5 (median)
Percent				2.06	1.83	0.30	0.34	0.0073								

[a] Data pooled from segregations of hybrids G12, M141, M150, X3653, S204, 5246, 5275, 5276, 5276-3, 5308, 5310, 5313, 5420, 5420-1, 5420-1-1, 5462, 5463, 5475, and 5497.

[b] * and ** indicate significance at the 5% and 1% levels, respectively.

Table 7 Basic conversion percentage at miscellaneous heterozygous sites in five isogenic diploids

hybrids	Alleles at arg4	no. of tetrads	Percentage of total events												
			trp1	pet1	arg4-3	arg4-16	arg4-36	thr1	cup1	ade8-18	his4-Δ26	ilv3	lys1	met13	a/α
5559	$\dfrac{3+36}{+16+}$	1146	0.2	1.0	2.7	8.6	9.2	7.5	1.1	9.9	4.5	3.4	4.9	11.2	0.6
5571	$\dfrac{3++}{+16+}$	1072	0	0.8	1.8	7.0	—	5.8	0.75	8.1	4.9	2.8	2.6	9.3	0.6
5574	$\dfrac{3+36}{+++}$	1073	0.4	0.8	2.1	—	7.4	6.3	1.3	8.5	5.0	4.0	3.4	11.8	1.4
5577	$\dfrac{+++}{+16\,36}$	908	0.6	1.5	—	7.8	9.3	5.6	0.6	8.5	4.2	3.8	4.0	9.0	1.1
5576	$\dfrac{+++}{+16+}$	863	0.8	0.9	—	5.6	—	5.7	1.6	8.1	3.8	4.1	5.0	9.5	0.7
Total		5062	Av. 0.4	1.0	2.2	7.3	8.6	6.2	1.1	8.6	4.5	3.7	4.0	10.2	0.9

successive mitotic conversions from the three-point cross 5559. Tabulated values represent basic conversion frequencies on a percentage basis, i.e., $6^+:2^- + 2^+:6^- + 5^+:3^- + 3^+:5^-$/total tetrads × 100. The average value for each heterozygous site is in close agreement with the corresponding value in Table 6. These range from about 0.6% to about 10%, or nearly 20-fold. Polarity relations among the *arg4* alleles are evident. The most distally located site, *arg4-36*, converts almost four times as frequently as the most proximal *arg4-3* allele. *arg4-36* and *arg4-16* are closely adjacent nonsense mutants. The former is an ocher mutant and exhibits only occasional pms, whereas the latter is a complementing mutant that exhibits a high pms frequency.

pms is also recorded in Table 6. This comprises the primary evidence for the presence of uncorrected hDNA segments. In turn, these include nucleotide base mispairings or nonpairings when the heteroduplex subtends the mutant site. If the faulty pairings are repaired by enzymatic excision-resynthesis systems, gene conversions of the $6^+:2^-$ or $2^+:6^-$ variety result. Of the 30 miscellaneous heterozygous sites, 24 displayed pms with frequencies ranging up to 54% of total aberrant segregations. In a single instance—the *MAT* locus—where the sample size (23,135 asci) was sufficiently large, pms failed to occur, though 197 conversions were detected. The *MAT*a and *MAT*α alleles differ by a large nonhomology (Hicks et al. 1979; Nasmyth and Tatchell 1980). The lack of pms might be construed to imply that a heteroduplex spanning the mating-type locus produces a highly distorted DNA configuration that is recognized by the cell's excision-repair system and corrected with high efficiency. On the other hand, the conversion of substitutions (and deletions or insertions) may occur by a mechanism that does not involve heteroduplex formations.

An important test of the generalized notion that gene conversions (or restorations) and pms, respectively, represent corrections and noncorrections of an underlying hDNA structure emerges from studies by Williamson and Fogel (1980). Pleiotropic recessive mutants at four different *cor* (correction-deficient) loci were reported. Isolated as meiotic "hyper rec" mutants, these strains displayed elevated prototroph frequencies when meiotic intragenic recombination was assessed. In addition, they were mitotic mutators. The spontaneous forward mutation rate to canavanine resistance (can^R) in *cor1-1* and *cor2* strains, respectively, is 6.68×10^{-6} and 1.82×10^{-6}/cell/division, i.e., two orders of magnitude above the control strain's value of 1.76×10^{-8}/cell/division.

Genetic analyses involving these *cor* strains revealed that map distances for several gene-centromere and gene-gene intervals are essentially unchanged. On the other hand, the ratio of gene conversions ($6^+:2^-$ and $2^+:6^-$) to pms events ($5^+:3^-$ and $3^+:5^-$) is different in wild-type and *cor*⁻ strains (Table 8). If gene conversions represent correction of a heteroduplex and *cor*⁻ strains are defective in the correction of such structures, the increased pms frequency in *cor*⁻ strains should be accompanied by a corresponding decline in the $6^+:2^-$

and $2^+:6^-$ gene-conversion classes. In general, the data for *arg4-16 thr1 lys1-1 met13* and *his2* fit this model. For *his4-4* the total number of events is significantly higher in the *cor1 cor3* and *cor2 cor3* backgrounds, and we pose no explanation at this time. The *ade8-18* allele has a high pms frequency in wild-type strains. The observation that it is not particularly sensitive to the *cor⁻* mutations may indicate that heteroduplexes involving this allele are substrates for a different correction mechanism.

Our studies with strains doubly homozygous for two different *cor* mutations reveal that the *cor2 cor3* and *cor1 cor3* combinations may have additive effects. The pms frequency for most heterozygous markers is greater in the double mutant than the largest pms effect for either single mutant. Table 8 presents representative data for the *cor2/cor2 cor3/cor3* double mutant. These results support a multiple system hypothesis for correction of hDNA in meiosis.

We analyzed a *cor1-1* homozygote triply heteroallelic at the *arg4* locus and also studied all four *cor* homozygotes carrying heteroalleles at *his4*. These studies revealed that when correction does occur, the repair tracts in mutant *cor* strains are often shorter than in wild type. Certain tetrad classes are encountered that are not observed in wild type: for example, tetrads with a spore clone that segregates postmeiotically for the two alleles *arg4-3* and *arg4-16* and where segregation of the *arg4-36* allele is $2^+:6^-$. In addition, an apparent increase in single- and double-site pms events at the expense of coconversion tetrad types is found, which could result from restoration to the parental marker configuration at one or two of the three sites included within the heteroduplex. The finding that correction tracts are shorter may be sufficient to account for the pms effect.

The impairment of heteroduplex correction throughout the genome in *cor* strains allows us to assess the occurrence of symmetrical hybrid DNA as a meiotic recombination intermediate in yeast. If symmetrical initiation occurred often, aberrant $4^+:4^-$ tetrads would comprise about 25% of all aberrant segregations, i.e., when correction is 50%. Clearly, this is not found. Thus, only 2 of 98 irregular segregations were aberrant $4^+:4^-$'s at *his2* in *cor1-1/cor1-1* strains. In addition, when data for all sites in the various mutants are collected, aberrant $4^+:4^-$ segregations are as rare as aberrant 6^+2^- and $2^+:6^-$ tetrads. Moreover, their frequency of occurrence is equal to the product of the frequency of independent initiations. Thus, such tetrads probably arise from two independent asymmetric events.

Mutants of the *cor* type will undoubtedly play a major role in disentangling the molecular genetics and biochemistry of recombination into those reactions that impinge on hybrid DNA formation and those involving the recognition and repair of distortions within hDNA. Similar conclusions apply to the *pac1* mutant reported by L. S. Davidow et al. (pers. comm.) and *spo11* and *spo13* studied by Klapholz and Esposito (1980a,b).

Table 8 Percentage gene conversions and pms for various heterozygous markers in mutant cor strains

Marker	+/+		cor1/cor1		cor2/cor2		cor3/cor3		cor2 cor3 / cor2 cor3	
	gc[a]	pms[b]	gc	pms	gc	pms	gc	pms	gc	pms
his4-4	7.2	0	6.6	12.0	3.0	2.4	13.6	1.2	6.3	14.2
arg4-16	6.1	3.0	4.5	6.8		—		—		—
thr1	4.5	0.6	5.2	4.5	4.2	4.2	2.4	8.4	2.1	12.5
lys1-1	7.6	0.2	1.2	13.9	2.9	0.6	1.4	5.5	2.1	6.2
met13	9.8	0.1	7.6	2.5	6.4	4.1	7.4	12.1	3.8	16.5
ade8-18	3.7	5.7	5.1	6.0	3.8	3.2	4.0	4.8	3.7	7.3
his2	12.9	0	9.3	11.3	8.5	1.2	12.6	5.1	7.5	11.0

[a] Gene conversion frequency × 100 denoted as gc.
[b] Postmeiotic segregation frequency × 100 denoted as pms.

Parity

Given the totals for each locus in Table 6 and similar data from our earlier studies on unselected tetrads (Fogel and Mortimer 1969), a parity principle is evidently operative in yeast—the likelihood of conversions in either direction is equal, i.e., $6^+:2^-$ segregations occur as often as $2^+:6^-$ segregations. Similar studies in *Ascobolus* and *Sordaria* (Leblon 1972; Yu-Sun et al. 1977; Nicolas 1979) indicate extreme dissymmetries for these and other ascal classes. Some dissymmetries also occur in yeast. However, the magnitudes of the dissymmetry coefficients $(6^+:2^-/2^+:6^-)$ range from 0.38–2.12 in yeast, compared with values of 0.01–100 in other Ascomycetes. In *Ascobolus,* the most pronounced disparities are encountered for ICR-170-induced mutations—presumptive single-base addition or deletion alterations (Rossignol et al. 1979). In *Ascobolus,* base substitution mutants exhibit coefficients closer to unity. The yeast mutants shown in Table 6 were induced with UV or ethylmethanesulfonate (EMS). At least half are judged to be base-substitution mutants on the criteria of suppressibility, complementation, or temperature sensitivity. To date, only a single frameshift mutant, *his4-519* (kindly provided by G. Fink (Cornell University, Ithaca, New York)), has been studied in this laboratory (M. Williamsson, pers. comm.), with respect to gene conversion. Among 978 unselected asci, she found 27 and 43 asci of the $6^+:2^-$ and $2^+:6^-$ varieties, respectively, along with a single $3^+:5^-$ ascus. These data fall within the sampling limits of a 1:1 expectation and thus are consistent with overall parity.

Among the 30 sites examined in Table 6, 12 show statistically significant departures from parity among the conversion events, i.e., $6^+:2^-$ does not equal $2^+:6^-$ segregations. For two sites, *arg4-16* and *arg4-17,* with significant disparities of 1.7 and 0.9, respectively, it may be noted that the sum of $6^+:2^-$ and $5^+:3^-$ is about equal to the sum of $2^+:6^-$ and $3^+:5^-$ segregations. This finding suggests that pms and conversion are genetic end products that originate in hDNA depending on whether or not it has experienced enzymatic repair. The variability in total events (Table 6) may represent site- or even region-specific effects similar to those reported in *Neurospora* and yeast (Catcheside 1975; Lawrence et al. 1975). Rossignol et al. (1979) emphasize that small differences in the direction of correction, i.e., mutant to wild type or vice versa, and the occurrence of symmetrical hDNA can account for the observed parity departures.

Fidelity of Conversion

Since conversion is most simply detected as $6^+:2^-$ or $2^+:6^-$ segregations in otherwise normal tetrads, it might be assumed that gene conversion is a process akin to spontaneous mutation and therefore represents a rich source of allelic diversity. What can be said of the identity or nonidentity between parental and converted alleles? In three separate studies it was suggested that the additional minus spore in a $2^+:6^-$ tetrad carried a mutation identical to the parent. Roman and Jacobs (1958) established this by an analysis of mitotic reversion rates, a sensitive allele-specific phenomenon, of the parental and conversion-derived alleles. Zimmerman (1968) examined the gene products of the derived alleles. Finally, Fogel and Mortimer (1968, 1970) utilized nonsense mutations of the amber and ocher types to demonstrate that the conversion-derived minus alleles contained precisely the parental defect. These collective findings lead us to the view that gene conversion is essentially a conservative process that neither creates nor destroys genetic information. Rather, it may be inferred that gene conversion operates with virtually complete fidelity and involves replacement of the genetic information in the relevant DNA segment with information that is identical to that carried in the corresponding homologous segment.

It may be noted in passing that the hybrid DNA repair model proposed by Holliday (1964, 1974) suggests the possibility of generating new alleles via gene conversion. This expectation is not consonant with the data presented above. Admittedly, this feature is not central to Holliday's model nor is it to those of Stadler and Towe (1963), Whitehouse (1963), Emerson (1969), and others listed by Hotchkiss (1974), which attribute considerable significance to the roles and functions of the mismatched base pairs in hDNA.

Conversion of Deletions and Insertions

Since the conversion of large deletions or insertions by the same mechanism as that postulated for conversion of single-base substitutions would involve a heteroduplex structure with an extensive single-strand loop, one might expect that conversion of such structures would show disparity. Studies on gene conversion of large deletions in yeast have been reported by Fink (1974), Fink and Styles (1974), and Lawrence et al. (1975). Taken collectively, these studies and ours established that deletions (1) convert in either direction with fidelity and approximate parity; (2) coconvert with other sites in the same gene or adjacent genes, and (3) display normal associated outside marker exchange frequencies. We undertook a more detailed study of two heterozygous deletions, *his4-Δ15* and *his4-Δ26,* studied by Fink. From direct measurements on the gene product, he estimated the respective deletions to be 400 bp and 300 bp in length (G.R. Fink, pers. comm.). In our study, pms and gene conversion were routinely scored by the plate-dissection method, a procedure not employed in the studies noted earlier. The data are displayed in Table 9. Data for the ocher mutant *his4-4,* which falls within the $\Delta 15$ deletion, are included for comparison. Significant excesses of $6^+:2^-/2^+:6^-$ are found for the point mutation *his4-4* and the $\Delta 15$. The $\Delta 26$ deletion displays a modest disparity in the opposite direction. Both deletions fail to display any pms among a total of 364 conversions in 8238 unselected tetrads. On the assumption that the ratio of pms to total events is the same for the deletions

Table 9 Conversion of deletions

his4 A B C	Segregations	Aberrant segregations					Total events (%)	pms/events (%)
		6:2	2:6	5:3	3:5	other		
4 —X—	12,510	411	303	39	41	2	6.36	10.3
15 ▰▬	1,965	55	21	0	0	0	3.66	0
26 ▰▬	6,273	124	164	0	0	0	4.59	0

and *his4-4,* we would have predicted 80 pms events. Thus, we conclude that deletions convert at a near normal frequency and with approximate parity but do not display pms. Limited results with conversion of insertions have yielded essentially equivalent data. A. Hinnen (unpubl.) observed $6^+{:}2^-$ and $2^+{:}6^-$ conversions of a 10-kbp insertion at the *LEU2* locus. Similarly, S. Fogel (unpubl.) observed two conversions in 360 tetrads of an 11-kpb plasmid integrated at *ADE8*. Furthermore, it is again worth noting that the substitution difference between the *MATa* and *MATα* alleles showed no pms in over 23,000 tetrads, although the conversion rate was about 1% and exhibited parity. The observed conversional behavior of deletions, insertions, and substitutions poses special problems for any molecular model that purports to account for gene conversion and crossing-over. Whereas single-site conversion data yield minimal estimates of pairing frequency in particular regions, coconversions and conversions of deletions and insertions impinge on lengths of pairing segments. Accordingly, gene-conversion analyses of insertions and deletions can provide valuable information concerning the topography of chromatid pairing at meiosis, structure of the synaptonemal complexes, and the frequency and length of pairings at any single site.

In view of this data, Radding (1979) developed a model that does not invoke excision repair. The unpaired loop is first replicated, and then a recombinational event results in either the insertion of the duplex bubble or the release of the bubble. Stahl (1979b) proposed a model in which the formation of a replication bubble ("sex circle") precedes the recombination events. Thus, an incoming duplex DNA may be involved in the conversion process. Such a duplex DNA could span the deletion or insertion and, hence, no single-stranded bubble need exist as a molecular intermediate in the conversion process. We will return to a discussion of the usefulness of deletions and insertions as tools in the analysis of gene conversion.

Effect of Specific Mismatch on Conversion and pms

The 30 different heterozygous sites in Table 6 display uncorrelated pms and conversion frequencies. Most likely, the various heterozygosities represent a miscellaneous collection of potential base-pair mismatches, and these may determine the observed pattern differences in conversion and pms frequencies. To evaluate the role of a specific mismatch in conversion, we synthesized a diploid hybrid that carried an ocher allele from one parent and an amber allele from the other parent in the same codon. The various nonsense mutants at *trp1, his4,* and *arg4* were derived from each other by EMS mutagenesis, according to Hawthorne's (1969) protocol. Heterozygous for *amber/ochre* at three centromere-linked but independently assorting sites on chromosomes IV, III, and VIII, the diploid could generate two possible mismatches at each site—AC and TG—during meiosis. Extensive comparable data for *+/amber* or *+/ochre* combinations are also given in Table 10. Scoring for ochre or amber in the 1118 unselected tetrads was simplified by carrying an amber suppressor homozygous in the diploid. From this study we conclude that given identical mismatches at three different genomic sites, each nonetheless displays a distinctive conversion and pms profile specific for the site and not for the potential mismatch. If the specific mismatch determined either the frequency of heteroduplex formation or correction, we would have expected that the *ochre/amber* combinations at each locus would be indistinguishable in conversional behavior. Conversion patterns, however, appear to be determined more by the location of a heterozygous site in the chromosome than by the potential mismatch. Effectively, this same conclusion stems from Nicolas' (1979) study of numerous heterozygous sites in *Ascobolus.* Of course, this does not imply that base-pair mismatches play no role in the conversion process.

Intrachromosomal Gene Conversion

The terms intrachromosomal recombination and intrachromosomal conversion imply either recombination between chromatids attached to the same centromere (sister-strand recombination) or recombination between repeated sequences or genes within a single chromatid (intrachromatid recombination). In both instances, it is highly probable that the underlying basic recom-

Table 10 Effect of mismatch on conversion

Locus	Genotype[a]	Asci	Aberrant segregation					Total events (%)	pms/events (%)
			6:2	2:6	5:3	3:5	other		
arg4-17	+/*o*	12,358	431	514	35	19	0	8.10	5.4
	a/o	1,118	54	37	3	1	0	8.50	4.2
his4-4	+/*o*	12,510	411	303	39	41	2	6.40	10.1
	a/o	1,118	18	20	2	3	0	3.80	11.6
trp1-1	+/*a*	17,061	48	37	10	7	0	0.59	16.7
	a/o	1,118	1	3	0	0	0	0.36	0

[a] *a* is the amber allele; *o* is the ocher allele.

binational event entails the production of an hDNA segment at the molecular level and a corresponding eventual gene conversion or pms at the cellular level.

Hinnen et al. (1978) demonstrated that transformation of yeast by a non-replicating plasmid carrying a yeast gene occurs by a homologous recombination, which results in the formation of a nontandem duplication of the yeast gene separated by the vector plasmid. This arrangement provides an excellent tool for the analysis of intrachromosomal or sister-chromatid recombinational and conversional events (see Prospectus: Recombinant DNA as a Tool in the Study of Recombination). The interaction of the heteroallelic duplicated sequences flanking the vector can lead to the excision of the insertion (a recombinational event) or the production of a homoallelic duplication (a gene-conversion event). Using this methodology, Szostak and Wu (1980) reported sister-chromatid exchange in the ribosomal DNA region. Similarly, Klein and Petes (1981) demonstrated intrachromosomal gene conversion.

ANALYSIS OF GENE CONVERSION AT *ARG4*

Recombination at the *ARG4* locus is particularly well studied in yeast and provides additional examples of the properties of gene conversion described above. Mortimer and Hawthorne (1973) and Mortimer and Schild (1980) assigned *ARG4* to a chromosome-VIII locus as follows: centromere-*PET1-ARG4-THR1-CUP1*. From our own studies, the map values of the various intervals are 4.1 centiMorgans (cM), 3.0 cM, 15.6 cM, and 30 cM. The order of the *arg4* alleles, as determined both from prototroph frequency and from outside marker relations in tetrads reciprocally recombined for the *arg4* mutant alleles, is shown in Figure 4. These analyses demonstrate polarity, the

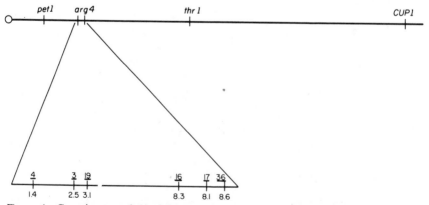

Figure 4 Genetic map of the right arm of chromosome VIII and fine-structure map of *arg4*. The value below each allele is the average conversion frequency for the site.

association of conversion with outside marker recombination, and allele-dependent pms frequencies.

Polarity and Coevents at *ARG4*

The gene-conversion frequencies for several of the alleles of *ARG4* are given on the map in Figure 4. Conversion frequencies are highest at the distal end of *ARG4*, marked by the ocher mutant *arg4-36*, and decrease toward the proximal end—an expression of polarity. The distribution of various conversion events for two- and three-point intragenic crosses is displayed in Figures 5 and 6. Included in Figure 6 are closely related hybrids comprising an isogenic subset. The latter were derived inter se by isolating specific mitotic recombinants from the appropriate progenitor strain.

In each two-point cross, the distal allele always exhibits a higher total event frequency—a reflection of the continuous polarity gradient across the locus. Continuity is apparent from the basic conversion frequency values attaching to each site in Figure 4. For two-point crosses, the events include conversion and pms for either allele, reciprocal recombination between the alleles, and events that involve both alleles in a coconversion or co-pms.

Some further aspects of events embracing two or more alleles are evident from the three-point crosses (Figs. 5 and 6). Here, each allele displays single-site events. In addition, frequent events involving two alleles were seen for the adjacent allele pairs *36-16* and *17-16*, and only infrequently were coevents encountered for the widely separated *19-17* or *36-3* combinations. The collective data from the two- and three-point crosses support the assumptions that hybrid DNA in yeast is formed in a continuous tract on a single chromatid. Also, conversion appears continuous, although the conversion or correction tract may be shorter than the hybrid DNA tract. That is, in the view that conversions reflect mismatch repair, the repair tracts can be long enough to cover all of the mismatches, and correction is confined to one strand or the other throughout the length of the event. In the same three-point hybrids homozygous for the *cor1* mutant, correction patterns are markedly altered. They are most simply accounted for by the frequent occurrence of independent correction events within a relatively long hDNA segment.

Influence of Adjacent Heterozygosity on pms Frequency

A tacit assumption inherent in virtually all recombination studies is the notion that the genetic markers utilized to study the process do not themselves modify or otherwise alter the process. Evidence to the contrary is available for closely linked markers. An illustrative marker effect reported by Gutz (1971) and Goldman (1974) concerns the M26 mutation in the *ade6* locus of *Schizosaccharomyces pombe*. This mutation converts at a rate much

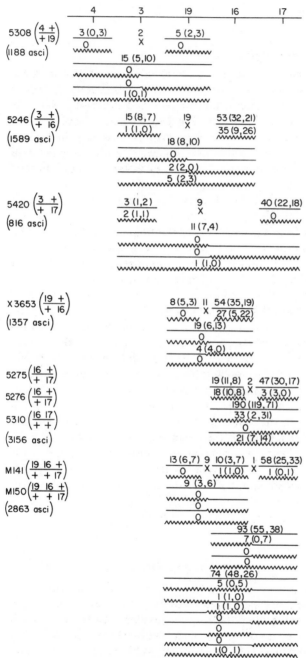

Figure 5 (See facing page for legend.)

higher than any other mutation in the gene. Moreover, unlike the other mutations that display approximate conversional parity, M26 converts almost exclusively in a single direction, i.e., from mutant to wild type, and markers nearby on either side experience coconversion. Since M26 is an amber nonsense mutant, it is a base-substitution mutation that generated a highly recombinagenic chromosomal site. *chi* sites and *cog* sites in phage λ and *Neurospora,* respectively, share similar properties.

Another class of marker effect exists for mutants in the *b2* locus of *Ascobolus.* Leblon and Rossignol (1973) demonstrated a marked depression in pms frequency for a given site when an adjacent heterozygosity was simultaneously present. We investigated the base-substitution mutant *arg4-16* with respect to this particular class of marker effect (Fogel et al. 1979). When it is present as a single heterozygous site, *arg4-16* displays a basic conversion frequency of nearly 8.5%, and about half the events are manifested as pms segregations. When a single heterozygosity is introduced to either side of *arg4-16,* the frequency of pms at *arg4-16* is reduced from about 48% to about 28%, although the basic overall conversion frequency remains effectively constant. However, if two heterozygosities are introduced within the *ARG4* locus, one immediately proximal and one closely distal to the *arg4-16* site, the overall basic conversion frequency remains unchanged and the pms value declines further to 8.5%. Thus, a fivefold shift in pms can result, depending on whether heterozygosities in close proximity are present or absent. These observations strongly support the view that $6^+:2^-$ segregations are the result of correction of a mismatched base pair.

The high pms frequency observed when *arg4-16* is present as a single heterozygosity suggests that its associated mismatches constitute poor signals for initiating excision repair. However, mismatches on either side of *arg4-16* and within the same hybrid DNA tract may provide stronger signals and provoke efficient repair. We confirm the conclusion of Leblon and Rossignol (1973) to the effect that correction may be induced by the presence of additional heterozygous sites and, once initiated, may often include other nearby sites. Clearly, these studies, as well as those concerning the further characterization of the earlier noted *cor* mutants (Williamson and Fogel 1980), open a novel experimental approach to analyzing the correction process itself.

Figure 5 Conversion events for two- and three-point crosses. (————) 6:2 or 2:6 segregations; (wwwww) 5:3 and 3:5 segregations; (×) a reciprocal crossover. The values written above each of the symbols correspond to the number of events seen for each abberant class, and the numbers in parentheses give the distribution of these classes. The number written first is the number of 6:2 or 5:3 segregations, and the second is the number of 2:6 or 3:5 segregations. For events involving two or three alleles, the numbers in parentheses always refer to the segregation of the proximal allele. Data for hybrids M141 and M150 were analyzed by F. Tavares (University of California, Berkeley) and R.K. Mortimer.

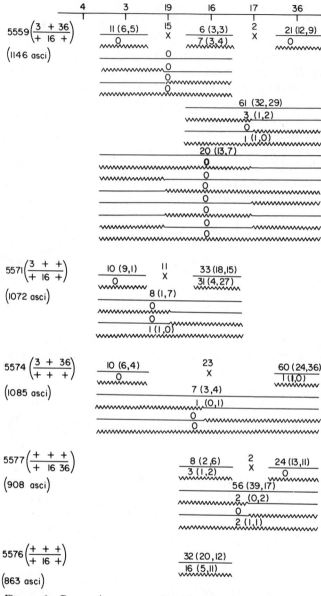

Figure 6 Conversion events for isogenic one-, two-, and three-point crosses. Symbols as in Fig. 5.

Associated Outside Marker Exchange

The early experiments of Case and Giles (1958, 1959, 1964) and Stadler and Towe (1963) in *Neurospora* indicated that approximately 50% of irregular segregations were associated with reciprocal recombination of neighboring sites. This finding was also reported by Fogel and Hurst (1967) in a study of asci selected on the basis of containing a prototrophic spore. Also, in our earlier report of conversion-associated recombination in unselected yeast tetrads, we again observed close to 50% exchange of flanking markers (Hurst et al. 1972). However, we noted that if the distance between the outside markers was greater than approximately 20 cM, the frequency rose above 50%. This suggested the occurrence of incidental exchanges as well as conversion-associated exchanges. One can ignore these incidental exchanges in shorter intervals only if all of the conversion events exert strong positive chiasma interference similar to regular meiotic exchanges. However, we have shown that only conversions that are associated with recombination cause interference; for the balance of the conversion events, additional exchanges occur at normal or above normal frequencies (Mortimer and Fogel 1974). Stadler (1973) and D. D. Perkins (pers. comm.) have pointed out, on the basis of these observations, that recombination frequencies associated with conversion events must be corrected for incidental exchanges in the flanking interval. The basic assumptions in this correction are that conversions associated with recombination interfere completely, but for the balance, incidental exchanges occur at normal frequencies. The data from 12 hybrids segregating for markers on chromosome VIII are presented in Table 11. The results show that approximately two thirds of all of the conversion events are recombined for the markers immediately flanking the converted segment. However, the observed values range from 0.27 to 0.76. When these observed values are corrected for incidental exchanges, the average fraction of conversions directly associated with crossing-over is approximately 0.35, and the corrected values range from 0.18 to 0.66. The conversion-associated outside marker exchanges permit an order-of-magnitude calculation (Fogel et al. 1971). Given the average probability of conversion and an estimate of the length of the converted segment, we can demonstrate that conversion-associated exchange alone can account for the total length of the yeast genetic map.

It is possible to test the validity of some of the assumptions made in the above analysis by an examination of the outside marker configurations in tetrads containing conversion or pms events. In such tetrads, we can classify exchanges either as incidental or as associated with the basic recombination event. Associated exchange events are defined as those that involve the converted or pms strand, whereas incidental exchange may or may not involve these same strands. Incidental exchanges of the latter type are uniquely identifiable (Fogel et al. 1979). A corrected value for the fraction of conversions that is associated with outside marker recombination may be obtained by subtracting twice the observed incidental-exchange frequency from the frequency of associated exchanges found by direct examination. For *thr1*, a total of 334 conversion events were detected in the segregations from five hybrids. Of these, 253, or 75.8%, was exchanged for the outside markers *arg4* and *CUP1*. However, 65 exchanges were identified as incidental. When the

Table 11 Association of conversion and crossing-over in *Saccharomyces*

Marker sequences	Hybrids	Total asci	Converted marker	Interval	Normal exchange frequency of interval (x)	Conversion events	Conversion events with exchange	Conversion-associated exchange frequency[a]	
								observed (F)	corrected (f)
c-m1-m2-thr-CUP	8	10,096	m1	c-m2	0.280	76	56	0.74	0.66
c-m1-m2-m3-thr-CUP	2	2,863	m2	m1-m3	0.003	11	3	0.27	0.27
c-m1-m2-thr-CUP	9	10,912	m2	m1-thr	0.278	396	219	0.553	0.39
c-m1-m2-thr-CUP	10	11,991	m1-m2	c-thr	0.527	512	327	0.639	0.24
c-m1-m2-thr-CUP	12	12,807	thr	m2-CUP	0.708	735	557	0.758	0.18

Based on data from hybrids M141, M150, X3653, 5246, 5275, 5276, 5276-3, 5308, 5310, 5313, 5420, and 5420-1.
[a]f = (F − x)/(1 − x) (D.D. Perkins, pers. comm.).

observed values are corrected for incidental exchanges, the fraction of associated exchange is 0.37. Di Caprio and Hastings (1976) also noted values significantly less than 0.50 for conversion-associated exchange in the *SUP6* region of chromosome VI in yeast.

Position of Conversion-associated Exchanges

In both the Holliday and Messelsoṅ-Radding models of recombination it is envisioned that the association of conversion and recombination reflects the resolution of a strand exchange by nicks in the uncrossed strands at the end of the exchange region (Figs. 2 and 3). Polarity is most easily explained as the result of a site or region initiating strand transfer at the high conversion end. Combined, these views lead to the expectation that conversion-associated recombinations should be on the low conversion side of the locus.

For tetrads where the central marker shows pms and the flanking markers are recombined, we may determine whether the exchange is associated or incidental and whether it is located proximally or distally (Fogel et al. 1979). These relationships, illustrated for $3^+:5^-$ asci in Table 12, also apply to $5^+:3^-$ asci. The *arg4-16* site was analyzed in this respect. In three hybrids, X3653, 5246, and 5497, 173 pms of *arg4-16* were detected. Of these, 85 were exchanged for the flanking markers, but in only 73 was the exchange associated with the chromatid that exhibited pms. Among the associated exchanges, 20 were assigned to the proximal interval and 44 to the distal region. An additional nine asci had associated exchanges in both the proximal and distal intervals, and five tetrads contained an incidental exchange in addition to a proximal- or distal-associated exchange. Finally, seven tetrads displayed an incidental exchange in the distal interval. For two crosses, X3653 and 5246, the proximal marker was another heterozygous site within the *ARG4* locus. The third cross, 5497, contained *pet1* as the proximal marker. Even though the proximal interval in the first two crosses was much

Table 12   Position of conversion—associated exchange

	Exchanged		
Not exchanged	proximal	distal	not associated
+ + +	+ + +	+ + +	+ +/m +
+ +/m +	+⌐m b	+ +/m⌐b	+ +⌐b
a m b	a⟋+/m +	a m⟋+	a m⟋+
a m b	a m b	a m b	a m b

Method used to deduce the position of associated events in three-point crosses. The assumed parental genotypes are + + + and *a m b*. The analysis applies to asci in which only the central marker exhibits pms. The presumption is that the sectored spore was originally ± , and $\times$ represents the exchange necessary to account for the observed segregation.

shorter than in 5497 (< 1 map unit compared to 3 map units), the ratio of proximal exchanges to distal exchanges was not significantly different, i.e., 7:16 versus 13:28. The data are presented in Table 13. The relatively high frequency of multiple exchanges among the asci that exhibit pms for *arg4-16* was unexpected. From the double-exchange frequency in the total sample, estimated from the nonparental ditype (four-strand double) frequency for the *pet1-thr1* interval, we expected only about two such tetrads.

Another interesting aspect of the data is that in $5^+:3^-$ tetrads the majority (13 out of 20) of the exchanges are in the proximal interval, whereas in $3^+:5^-$ tetrads, most of the associated exchanges (37 out of 44) are distal.

The finding that associated exchanges occur both proximal and distal to the *arg4-16* site is contrary to the notion that polarity reflects a fixed initiation site near the high conversion end of the *ARG4* locus with conversion tracts extending from this site into the gene for various lengths. Conceivably, it might, for different events, signal the initiation and propagation of hDNA in two different directions.

Does Gene Conversion Yield Excess Two-strand Double Exchanges?

Holliday's original model (1964) features symmetrical hybrid DNA with two dissimilar mismatched base pairs. We have argued from the paucity of the $ab4^+:4^-$ segregation class (expected if no correction occurs) that this symmetrical structure is not made. Furthermore, if the two mismatches of a symmetrical hybrid event are repaired independently, an additional prediction concerning the frequencies of other tetrad classes may be generated. This concerns the incidence of apparent two-strand double exchanges among the unconverted tetrads that have their origin in a conversion event. Such events are detectable only if the involved site is flanked by closely linked heterozygous markers. If both mismatches are corrected, but in opposite directions, half will appear as though no event had occurred, and the other half will appear as though a two-strand double exchange had occurred, with one exchange located on each side of the corrected site. Given random correction of the two mismatches and that 50% of the conversional events display

Table 13 Position of associated exchange

Hybrid	Asci	No exchange		Associated exchange				Incidental exchange				Multiple exchanges
				proximal		distal		proximal		distal		
		5:3	3:5	5:3	3:5	5:3	3:5	5:3	3:5	5:3	3:5	
X3653	1301	1	14	1	0	1	5	0	0	0	1	0
5246	1589	1	12	5	1	2	8	0	0	0	1	1
5497	2609	9	5	7	6	4	24	0	0	0	5	13
Total	5499	11	31	13	7	7	37	0	0	0	7	14

outside marker exchange, it can be shown that apparent two-strand double exchanges will be generated at one-fourth the frequency of the observed conversion events. These apparent two-strand double exchanges would be seen in the unconverted asci and would be in addition to the actual two-strand double exchanges occurring in the two intervals. Thus, instead of the expected 1:2:1 ratio of two-strand, three-strand, and four-strand double exchanges, one predicts a significant excess of the two-strand double-exchange class.

We have examined 13 hybrids with respect to the relationship between double-exchange classes and gene-conversion events. Most of the data involve genes on chromosome VIII, although results from regions on chromosomes V and VI are also included. The data used in this analysis are presented in Table 14. Data for a given cross are apportioned into the various conversion classes, because for each class a different pair of flanking intervals is involved. In the last column of Table 14, the number of apparent two-strand doubles that would arise from independent correction of both mismatches in a symmetrical heteroduplex is presented. The observed numbers of two-, three-, and four-strand doubles for each of the marked intervals associated with a class of conversions are presented in the columns under Double crossovers. In all, 1790 conversion events were observed and, from this, one would predict 447.5 excess two-strand double exchanges in the unconverted tetrads. However, only 256 two-strand doubles were observed along with 457 three-strand doubles and 200 four-strand doubles. Since these values are not significantly different from the 1:2:1 ratio expected in the absence of chromatid interference for two exchanges, we conclude that the contribution of correction-generated two-strand double exchanges is negligible. Our data favor a single heteroduplex or a constraint on symmetrical hDNA such that repair typically involves strands of opposite polarity in the two hybrid DNA segments. Stadler and Towe (1971) derived a similar conclusion from their analysis of ab5$^+$:3$^-$ and ab3$^+$:5$^-$ octads in *Ascobolus immersus* (Fig. 1d). Alternatively, Holliday (1974) has pointed out that a symmetrical structure such as that proposed by Sobell (1972) can account for low frequencies of apparent two-strand double exchanges and of ab4$^+$:4$^-$, ab5$^+$:3$^-$ and ab3$^+$:5$^-$ asci.

DISCUSSION

This paper reviews our most recent data concerning properties of gene conversion and its relationship to meiotic crossing-over in the yeast *Saccharomyces cerevisiae*. Since gene conversion has been seen in a broad range of organisms and is generally associated with recombination, we believe our results are applicable to the general problem of genetic recombination in all organisms.

As given in Table 6, it is clear that different heterozygous sites display

Table 14 Occurrence of two-strand double crossovers among nonconversion tetrads

Hybrid	Heterozygous sites	Total asci	Converted site	Conversion events					Flanking intervals		Double crossovers			Conversion[a]
				3:1	1:3	5:3	3:5	total	proximal	distal	2-strand	3-strand	4-strand	(4)
X2961	19 16 17 thr1 CUP1	2,338	arg4-19	8	11	—	—	19	pet-19	19-thr	2	1	0	4.75
			arg4-16	13	2	—	—	15	19-16	16-17	1	0	0	3.75
			arg4-17	30	10	—	—	40	16-17	17-thr	1	0	0	10.00
			arg4-19, 16	0	2	—	—	2	pet-19	16-17	0	0	0	0.50
			arg4-16, 17	42	41	—	—	83	19-16	17-thr	0	0	0	20.75
			arg4-19, 16, 17	29	20	—	—	49	pet-19	17-thr	4	3	1	9.80
			thr1	61	83	—	—	144	17-thr	thr-CUP	57	112	44	36.00
X2988	1 2 16 17 thr1	1,181	arg4-16	2	0	—	—	2	2-16	16-17	0	0	0	0.50
			arg4-17	9	4	—	—	13	16-17	17-thr	0	0	0	4.75
			arg4-2, 16	3	0	—	—	3	1-2	16-17	0	0	0	1.00
			arg4-2, 16, 17	16	11	—	—	27	1-2	17-thr	0	0	0	7.75
X3653	19 16 thr1 CUP1	1,357	arg4-16	35	19	5	22	81	19-16	16-thr	21	0	0	20.20
			thr1	42	38	1	1	82	16-thr	thr-CUP	21	43	9	20.50
3735	hom3 his1 arg6	975	his1	17	18	—	—	35	hom-his	his-arg	0	0	0	8.80
3910	hom3 his1 arg6	485	his1	11	4	—	—	15	hom-his	his-arg	0	1	0	3.80

4311-4380	his2 SUP6 met10	2,702	SUP6	153	148	22	24	347	his-SUP	SUP-met	4	4	1	86.70
5246	3 16 thr1 CUP1	1,589	arg4-16	32	21	9	26	88	3-16	16-thr	2	1	0	22.00
			thr1	41	37	0	0	78	16-thr	thr-CUP	32	56	28	19.50
5275	16 17 thr1 CUP1	1,079	arg4-17	14	5	1	0	20	16-17	17-thr	0	0	0	5.00
			thr1	37	31	1	3	72	17-thr	thr-CUP	17	36	24	18.00
5276	16 17 thr1 CUP1	1,313	arg4-17	13	9	2	0	24	16-17	17-thr	0	0	0	6.00
			thr1	41	39	1	1	82	17-thr	thr-CUP	31	44	20	20.50
5308	4 19 thr1 CUP1	1,188	thr1	43	21	0	1	65	19-thr	thr-CUP	21	48	28	16.20
5310	16 17 thr1 CUP1	763	thr1	19	22	0	0	41	17-thr	thr-CUP	8	22	12	10.20
M141	19 16 17 thr1 CUP1	1,324	arg4-17	11	15	0	1	27	16-17	17-thr	0	0	0	6.80
			arg4-16, 17	24	19	1	0	44	19-16	17-thr	0	1	0	11.00
			thr1	54	59	0	0	113	17-thr	thr-CUP	28	32	12	28.20
M150	19 16 17 thr1 CUP1	1,539	arg4-17	14	18	0	0	32	16-17	17-thr	0	0	0	8.00
			arg4-16, 17	31	19	0	0	50	19-16	17-thr	2	0	0	12.50
			thr1	60	34	2	1	97	17-thr	thr-CUP	23	53	21	24.80
Total		19,293		905	760	45	80	1790			256	457	200	447.50

[a]Excess two-strand doubles calculated on the assumption of only symmetrical hybrid DNA and independent correction of mismatches.

characteristic conversion behaviors with respect to frequency, parity or dissymmetry, and pms. It can be proposed that these differences are due to the different potential mismatches associated with the various mutant sites. Results from *Ascobolus* (Leblon 1972; Leblon and Rossignol 1973) and *Sordaria* (Yu-Sun et al. 1977) show marked differences between base-substitution mutations and single-base addition or deletion mutations. In those systems, base-substitution mutants exhibited approximate symmetry and relatively high pms frequencies, whereas the addition or deletion mutants showed marked dissymmetries (as high as 100-fold) and little pms. The yeast results presented in Table 6 are from a collection of mutants that were induced largely by UV radiation. From suppressibility, complementation, and temperature-conditional responses, it can be deduced that at least 17 of the 30 mutants represent base-substitution changes and none of the mutants are known to be of the frameshift type. All of the yeast mutants resembled the base-substitution mutants of *Sordaria* and *Ascobolus* although generally with less frequent pms and less marked dissymmetries.

With regard to the role of the specific mismatch in conversion, our results concerning the *ochre/amber* mismatch at three different sites in the genome are especially relevant. Since we found that conversion frequencies and correction patterns were characteristic of the specific site and not of the potential mismatch, we conclude that the specific region that includes the heterozygosity is of primary importance in determining the parameters of conversion.

Originally explained some years ago on the basis of a copy-choice mechanism, a group of octads provided critical evidence for the notion that gene-conversion events did not involve just a single heterozygous site but had extent (Case and Giles 1958, 1959, 1964). When two alleles at a locus were included in a cross, the conversion events frequently involved both alleles in a coordinated fashion, an event designated by us as coconversion (Fogel and Mortimer 1969; Fogel et al. 1971). In effect, coconversion results in the replacement of a segment of genetic information in one homolog with the corresponding information contained in the equivalent segment of the homolog from the opposite parent.

Figure 5 presents the results for nine hybrids. Seven hybrids are marked by two heterozygous sites within the *arg4* gene, and two hybrids carried three marked sites in this locus. The novel feature of these data concerns coevents in which one, two, or three alleles show pms. Unlike the coconversion tetrads (represented by straight lines in Fig. 5), which contain only parental strands, coevent tetrads that involve pms (wavy lines) contain three parental strands and one strand that displays sectoring for either or both parental alleles. This is illustrated by the two-point crosses (5275, 5276, and 5310) involving alleles *16* and *17*. For these two alleles, there were 190 tetrads of the coconversion type and 21 tetrads in which one spore clone was sectored for both markers. In addition, 33 tetrads exhibited sectoring for the *16* site accompanied by conversion of the *17* site. When all of the coevents are examined, they are consistent with the postulate that a tract of hybrid DNA including both sites was formed on a single chromatid. Similar conclusions apply with equal validity to the subset of isogenic hybrids represented in Figure 6.

We may pose the question, Does correction occur in discrete segments, or does it involve a continuous length of hybrid DNA? In wild-type strains, the influence of proximal or distal heterozygosity on the correction of site *arg4-16* strongly suggests that it is continuous. If two sites are included in a single stretch of hDNA (asymmetric), it can be shown that independent correction of the two mismatches will generate all of the tetrad classes that have been observed. However, single-site events should always be at least as prevalent as coevents, and their relative frequencies should be independent of the distance between the alleles. An excess of single-site events is seen for crosses involving allele pairs *3* and *16*, *3* and *36*, *3* and *17*, and *19* and *16*, all of which are widely separated. However, for the crosses involving alleles in close proximity, i.e., *36* and *16* and *16* and *17*, there is a clear excess of coevents, a result incompatible with independent correction.

Early studies on conversion (Lissouba et al. 1962; Murray 1963) demonstrated the polarized character of conversion. They found that the frequencies of conversion of different alleles decreased in a continuous fashion from one end of the locus to the other. The gradient of conversion frequencies was explained by proposing that the recombination events start at a fixed point external to the high polarity end of the gene and proceed into the locus for varying distances. The data for the *arg4* locus are, in general, consistent with this notion since allele *36* or *17* converts with the highest frequency and is involved in most of the events at the locus. One could propose a major starting point to the right of *36* or *17* with most events proceeding from this point into the locus. Many molecular recombination models propose such recombinator regions from which hybrid DNA is propagated. They further predict that if the event results in an associated outside marker exchange, this exchange will occur at the terminus of the hDNA tract. Thus, it follows that associated exchanges should all occur in the interval that includes the low polarity side of the gene. In the case of the *arg4* locus, where the high end of the polarity gradient is distal, relative to the centromere, we would expect the associated exchanges to occur proximal to the converted site. We devised a method of deducing the location of the associated exchange only in tetrads that exhibit a pms for the relevant site (see Table 12). Contrary to the expectations outlined above, we found that only about one third of the associated exchanges were in the proximal interval, whereas the remaining associated exchanges were located distally. These findings are incompatible with the above notions on polarity. At the same time, they call into question the correctness or applicability of the original Meselson-Radding model. Clearly, some modification is required. To reconcile these results we might suppose that hDNA is propagated in two directions starting from one of a number of initiation sites that are more numerous at the high polarity end of the gene. Alternatively, synapton structures, as proposed by Sobell (1974) with certain additional assumptions, can account for polarity and the occurrence of proximal and distal exchanges. P.J. Hastings (pers. comm.) has suggested that limited diffusion of the half chiasma could account for the observed data. However, why the bulk of the associated exchange should be distal is not clear at this time and, in addition, the cross connection could not migrate to the next base-pair mismatch without generating symmetrical hybrid DNA.

Studies based on the recovery of recombinants in heteroallelic crosses show that the recombinants (i.e., prototrophs) may arise from both reciprocal and nonreciprocal events (Roman 1957; Fogel and Hurst 1967). In most studies it has been found that the majority of the recombinants arise as a consequence of a nonreciprocal event. This finding is also revealed in the data presented in Figures 5 and 6. For example, in hybrid X3653, which contained alleles *19* and *16* in repulsion, 60 tetrads contained a spore clone that was wholly or partially prototrophic. Of these, 11 were reciprocally recombined for *19* and *16* as well as for the outside markers. The remaining 49 tetrads represented nonreciprocal events as follows: 6:2 for *19*, 6:2 and 5:3 for *16*, and 6:2 and 3:5 for *19* and *16*, respectively. Thus, over 80% of the prototrophic recombinants in this case have a nonreciprocal origin. A similar analysis of the other two-and three-point crosses reveals a corresponding distribution between reciprocal and nonreciprocal origins of the prototrophic recombinants.

The reciprocal events between two alleles can be accounted for in two ways. The first involves a short region of hybrid DNA between the two alleles, which is resolved by a recombination event. The second requires that hybrid DNA cover one or the other of the two alleles and that the associated exchange fall between the two alleles. An additional requirement is that correction of the mismatch restores the parental state. The contribution to the observed reciprocals from this second type of event may be estimated from the single-site event frequency if we make the following assumptions: (1) Corrections of the mismatch to yield restorations to normal $4^+ : 4^-$ (Hastings et al. 1980) or gene-conversion events are equally probable; (2) resolution of the event to yield an exchange occurs about one third of the time (see Table 11); and (3) associated exchanges may occur at either the proximal or distal end of the conversion tract with relative frequencies of 1 out of 3 and 2 out of 3 (see Table 13). With these assumptions, the number of reciprocals that arise by the second mechanism should be equal to two ninths of the number of proximal single-site conversions ($6^+:2^-$ plus $2^+:6^-$) plus one ninth of the number of conversions of the distal allele. The calculated and observed values for the various hybrids are presented in Table 15. There is agreement between the observed and calculated values when the data for all of the hybrids are pooled. However, more reciprocals are predicted to occur for the closely spaced combinations than were observed (28 vs. 9). For the remaining widely

Table 15 Origins of reciprocal crossovers between alleles by conversional restorations to normal

Cross	Genotype	Single-site conversions proximal allele (P)	Single-site conversions distal allele (D)	Predicted reciprocals (2P+D)/9	Observed reciprocals
		Closely spaced alleles			
5275	$\frac{16\ +}{+\ 17}$	7	19	3.7	1
5276	$\frac{16\ +}{+\ 17}$	8	22	4.2	0
5310	$\frac{16\ +}{+\ 17}$	5	6	1.8	1
M141, M150	$\frac{16\ +}{+\ 17}$	19	58	10.7	3
5559	$\frac{16\ +}{+\ 36}$	6	21	3.6	2
5577	$\frac{16\ 36}{+\ \ +}$	8	24	4.4	2
Subtotal		53	150	28.4	9
		Widely spaced alleles			
5240	$\frac{3\ +}{+\ 17}$	3	40	5.1	9
5246	$\frac{3\ +}{+\ 16}$	15	53	9.2	19
X3653	$\frac{19\ +}{+\ 16}$	8	54	7.8	11
M141, M150	$\frac{19\ +}{+\ 16}$	13	103	14.3	9
5308	$\frac{4\ +}{+\ 19}$	3	5	1.2	2
5559	$\frac{3\ +}{+\ 16}$	11	67	9.9	15
5571	$\frac{3\ +}{+\ 16}$	10	33	5.9	11
5559	$\frac{3[\ +\ ]36}{+[16]\ +}$	11	82	11.6	17
5574	$\frac{3\ 36}{+\ \ +}$	10	60	8.9	23
Subtotal		84	497	73.9	116
Total		137	647	102.3	125

separated allele combinations, the calculated number of reciprocals is less than the number observed (74 vs. 116). One way of explaining the discrepancy for the closely spaced pairs is to propose that a major fraction ($\sim$4 out of 5) of the observed single-site conversions actually arises as a consequence of independent mismatch corrections within a hybrid DNA tract that covers both alleles. The contribution of independent corrections within coevents to the observed single-site frequency for widely separated alleles is expected to be considerably less. Even though it may be necessary to posit some independent correction to account for the above discrepancy, it is still apparent from the large number of coconversions observed that correction for the most part is coordinated, i.e., it proceeds along a single DNA strand. Our results on the influence of adjacent heterozygosities on correction of allele *16* also indicate that correction is at least in part coordinated. An alternative way of explaining the large excess of calculated reciprocals over observed reciprocals for close alleles is to propose that the correction of the mismatches to yield a conversion tetrad is about four times more probable than to yield a $4^+:4^-$ normal tetrad.

Bearing on the question of independent mismatch corrections in a single hDNA tract is the study by Kalogeropoulos and Rossignol (1980). If two mismatches formed in the same hDNA segment can experience independent repair, then for an appropriate three-point cross, one should find instances where two repair tracts can leave a centrally located mismatch uncorrected. Convincing evidence for precisely such meiotic events was reported by these investigators: Outside sites undergo conversion, and the middle site exhibits pms as $3^+:5^-$. Fincham and Holliday (1970) also invoked independent correction as a factor contributory to map expansion. Finally, in our own studies on *mitotic* recombination with diploid 5559, where three mutant sites in *arg4* are monitored, i.e.,

+	arg4-3	+	arg4-36	+	CUP1
pet1	+	arg4-16	+	thr1	cup1

a minor but significant group of prototrophic diploids can be demonstrated by subsequent meiotic analysis to carry the genotype

+	+	+	+	+	CUP1
pet1	+	arg4-16	+	thr1	cup1

An economical hypothesis to account for such events involves the formation of a single heteroduplex covering all three sites. Correction of the *arg4-3* and *arg4-36* mismatches without correcting the mismatch at *arg4-16* could then generate the required genotype. Such an event is not explained in models invoking a single excision repair to explain coconversions.

Studies in *A. immersus* (Nicolas 1979) and *Sordaria brevicollis* suggest a correlation between the nature of the DNA sequence change in a mutation and its conversional behavior. Frameshift mutations (ICR-induced) generally show very significant disparities and very low pms frequencies, whereas base-substitution mutations (EMS-induced) may show disparity in the excess 6^+:2^- direction but exhibit high ratios of pms to total-event frequency. Comparable mutagenic specificity has not been found in yeast, although this may reflect the fact that very few yeast frameshift mutants have been analyzed from the standpoint of gene conversion. Conceivably, other determinants of conversion parameters exist (Wickramaratne and Lamb 1978), and these might include the extent of symmetrical hDNA, the nature of neighboring sites, and the region in which the mutant is localized. Physiological factors, such as temperature at critical developmental stages, e.g., during pachytene, are known to affect frequency of hybrid DNA formation at particular sites. The overall efficiency of mismatch repair and the relative frequencies of correction to wild type or mutant are described by L. S. Davidow et al. (pers. comm.) for the mutant *pac1*.

A considerable number of molecular recombination models have been published (for review, see Hotchkiss 1974) that purport to reconcile the association between gene conversion and reciprocal recombination. They may be classified into three major groups featuring (1) wholly symmetrical hybrid DNA (Whitehouse 1963; Holliday 1964; Sobell 1972); (2) varying proportions of symmetrical and asymmetrical hybrid DNA (Meselson and Radding 1975; Wagner and Radman 1975); and (3) homoduplex models, either symmetrical or asymmetrical, which carry relatively short regions of hDNA at the termini (Stahl 1969; Boon and Zinder 1971). Although nearly all models readily accommodate the available data bearing on parity, fidelity, and coconversion, they differ with respect to their prediction about the frequencies of asci yielding segregations of the aberrant 4^+:4^- type, the ab5^+:3^- type, or the ab3^+:5^- type (Stadler and Towe 1971). They also generate different predictions concerning the occurrence of co-pms tetrads, the frequency and position of conversion-associated reciprocal recombination, and the behavior of deletions and nontandem duplications during the conversion process.

The regular and frequent appearance of ab4^+:4^- octads, primarily in *Sordaria* and *Ascobolus,* comprises the principal genetic evidence for the occurrence of symmetrical hDNA (Kitani and Olive 1967; Leblon and Rossignol 1973; Kitani 1978; Paquette 1978; Rossignol and Haedens 1980). In such octads, the two spore pairs that each contain both parental alleles are taken to represent the two possible base-pair mismatches at the mutant site. The commonly observed conversion classes 6^+:2^- and 2^+:6^- are interpreted as corrections of both heteroduplexes in the same direction; the 5^+:3^- and 3^+:5^- segregations signify correction of only a single heteroduplex. Corrections of the two heteroduplexes in opposite directions would generate either an apparently normal ascus or one containing an apparent two-strand double exchange for the central site relative to the flanking markers. In yeast, however, the frequency of ab4^+:4^- tetrads is extremely low. Only 16 ab4^+:4^- tetrads were found among 9994 aberrant segregations (0.16%). If we consider the original static version of the Holliday model for symmetrical hDNA, the expected frequency of ab4$^+$:4$^-$ tetrads may be calculated on the basis of the frequencies observed for 6$^+$:2$^-$, 2$^+$:6$^-$, 5$^+$:3$^-$, and 3$^+$:5$^-$. Let a represent the probability that a site falls within symmetrical hDNA and p, the probability of correction at the mismatch. The frequencies of 6$^+$:2$^-$ plus 2$^+$:6$^-$, 5$^+$:3$^-$ plus 3$^+$:5$^-$, and ab4$^+$:4$^-$ are $\frac{1}{2}ap^2$, $2ap(1-p)$, and $a(1-p)^2$, respectively. Applying these relationships to *ade8-18* (see Table 6), we would expect to see 94 aberrant 4$^+$:4$^-$ tetrads, although only 10 were detected within a sample of 13,230 segregations tested. Similar observed deficiencies over expected deficiencies are also found for *his4-4* and *arg4-16*, the other loci to which the test is applicable.

Ghikas and Lamb (1977) have pointed out that ab4$^+$4$^-$ tetrads may arise from two independent events that involve all four chromatids at a site. In addition to the ab4$^+$:4$^-$ events, certain other aberrant classes (7$^+$:1$^-$, 1$^+$:7$^-$, ab6$^+$:2$^-$, and ab2$^+$:6$^-$) are generated by these four-chromatid events. For *ade8-18, arg4-16,* and *his4-4* there were seven, two, and one, respectively, of these aberrant tetrads. Thus, probably most of the 16 ab4$^+$:4$^-$ asci seen at these loci do not represent symmetrical hybrid DNA.

When two or more heterozygous sites are included in the same tract of hybrid DNA, a failure of correction over the region that includes the mismatches corresponding to the sites will lead to the appearance of clones that are sectored for all uncorrected sites. By analogy with coconversion events, these are designated as co-pms events. The occurrence of such co-pms events cannot be easily reconciled with recombination models that feature large segments of homoduplex DNA flanked by smaller segments of hDNA (Stahl 1969; Boon and Zinder 1971). However, their occurrence is consonant with models that feature only hDNA. In the case of symmetrical hDNA models, co-pms events require correction of only one heteroduplex, whereas in models that feature asymmetric hDNA, these events arise by a failure of correction. In Figures 5 and 6, data from several hybrids that are heterozygous for two or three sites in the *arg4* locus are summarized. In addition to single-site and coconversion events, instances of single-site pms events and co-pms events are recorded. Also presented are events that represent correction at one or two sites and failure of correction at the remaining sites. There are a total of 35 events that involve pms on two or three adjacent sites. Thirteen of these co-pms events, in hybrids involving allelic markers *4-19, 3-6, 3-17, 19-16,* and *19-17,* extend over at least half the locus. It is difficult to account for such events in terms of the homoduplex models described above, where relatively short sections of hDNA flank longer homoduplex segments.

As discussed above, long deletions show conversion with fidelity in both directions at frequencies comparable to those seen for base-substitution mutations. They also coconvert with other heterozygous sites in the same gene or adjacent genes but fail to show any pms. The behavior of deletions in the conversion process is an important consideration in assessing the merits of the various molecular models of recombination. In heteroduplex models, it seems likely that the movement of a Holliday structure (symmetrical hDNA) by rotary diffusion would be blocked by a deletion. Progression of asymmetrical hDNA by strand displacement through the deleted region would require, in addition, the uncoupling of the postulated nuclease and polymerase, and this could result in the formation of long single-stranded regions. If strand displacement can proceed through a deletion, the resultant structural mismatch is a single-strand loop whose length equals that of the deletion. Since conversion occurs in both directions, the mismatch correction process must involve precise excision or replication of the loop. Since no pms is observed, these processes must occur with high efficiency. On the basis of the above data, a recent proposal formulated by Radding (1979) satisfactorily accounts for the behavior of long deletions within the framework of an asymmetric hDNA model. In summary, it is our view that gene conversion in yeast is mediated mainly by formation of hDNA that is primarily of an asymmetric nature. Except for the data concerning the position of conversion-associated exchanges, the model proposed by Meselson and Radding (1975) accommodates our results on the paucity of $ab4^+{:}4^-$ asci and frequencies of outside marker exchange different from 50%. It is also consonant with our observations on co-pms, the absence of excess two-strand doubles, and the behavior of insertions and deletions. From a consideration of average conversion frequencies, conversion tract lengths, and frequencies of conversion-associated recombination, it is possible to show that crossing-over, associated with conversion, is sufficient to account for all meiotic recombination (Hurst et al. 1972). Thus, we view conversion as a signal of the basic recombination event.

PROSPECTUS: RECOMBINANT DNA AS A TOOL IN THE STUDY OF RECOMBINATION

We have suggested that the genetic analyses described above have revealed the form, if not the details, of the mechanism of gene conversion in yeast. What then can we expect in the future from this field of research? Happily, we are provided with several new approaches. First, a variety of mutations (*pac1, cor1, spo11, spo13*) perturb the rate and/or spectrum of gene-conversion events. These mutations should allow the genetic analysis of the synapsis, hDNA formation, and pms correction pathway proposed to account for 6:2 segregations. Secondly, the expanding technologies of recombinant DNA cloning and genetic transformation of yeast cells allow both the sequence analysis of well-characterized alleles and the introduction of new mutations of defined nature into yeast for analysis of their conversional behavior. Petes (1980a,b) and Botstein and Davis (1982) have recently reviewed the methodologies by which one can identify and isolate specific yeast genes, clone and characterize specific alleles, and reintroduce into yeast novel alleles made in vitro.

S. Fogel and D. W. Kilgore (unpubl.) recently isolated the *ADE8* gene by a screen of *Bam*HI fragments cloned into YRp16 (Fig. 7). This plasmid has the interesting and useful property that integration events can be readily recognized. *ade8 ade2* yeast colonies are white, whereas *ade2* yeast colonies are red (Roman 1957). *ade2 ade8* strains carrying YRp16 *ADE8* as an autonomous plasmid are white with red sectors in which the plasmid has stably integrated into the chromosome (Fig. 8). The conversional behavior of the nontandem duplication created by this integration at *ade8* has recently been analyzed. In haploids, these represent intrachromatid or sister-chromatid events. In diploids, conversion between homologs can be monitored. At least four kinds of questions are being addressed with these structures.

1. Is the conversion rate sensitive to the size of the insertion? A systematic study of the response of conversion rate and degree of parity to the size of the insertion is possible by varying the size of the vector. Results from this

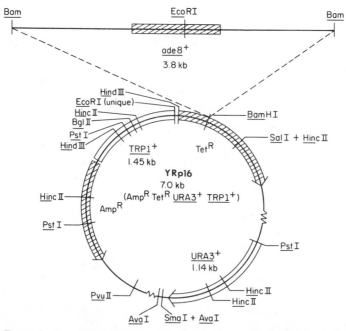

Figure 7   The YRp16 cloning vector (constructed by W.M. McDonnell and R.W. Davis, pers. comm.) is a 7.0-kb plasmid carrying pBR322 sequences and the wild-type sequences for the yeast genes *TRP1⁺* and *URA3⁺* in addition to an autonomous replication sequence, *ars*. A 3.8-kb *Bam*HI insert carrying *ADE8⁺* is shown.

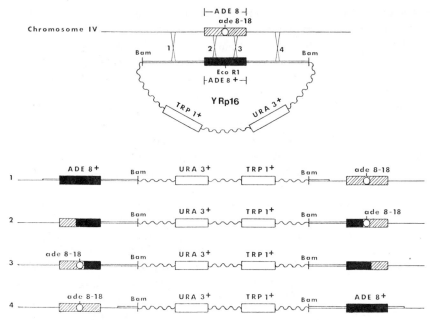

Figure 8 Integrations of the YRp16 insert-bearing cloning vehicle into the homologous chromosome-IV region in the vicinity of *ade8-18*. (*1-4*) The consequence of heteroduplex formation with associated crossing-over.

experiment should reflect both the sensitivity of prerecombinational synapsis to the insertional disruption and the size restrictions on the formation and resolution of an actual DNA strand exchange intermediate.

2. How is gene conversion in the region of the insertion affected? The conversion rate for genes near the insertion reflects the sensitivity of conversion to a local pairing disruption. A series of different size insertions at the same position in the genome could be used to determine whether pairing can be made rate-limiting for conversion. S. Fogel (unpubl.) scored 326 tetrads from a nontandem duplication heterozygote (*ADE8-vector-ade8-18/ade8-18*) via the plate-dissection method in which the vector segregated $4^+:4^-$. A total of 20 aberrant segregation events were observed with respect to *ade8*. Of these, 13 were of the $6^+:2^-$ and $2^+:6^-$ variety (11 and 2 in number, respectively), and 7 more exhibited pms in the $5^+:3^-$ and $3^+:5^-$ patterns (3 and 4 in number, respectively). Overall, we observed a basic conversion frequency of 6.1% and a ratio of pms to total events of about 35%. Comparable but conventional heterozygotes exhibit a basic conversion frequency of 9.7% and 54% pms at *ade8*. One might expect a lower-than-normal gene-conversion frequency if the mutant site in one homolog can pair equally well with the wild-type or mutant *ade8-18*

sequence carried in the nontandem duplication. Detailed studies of this sort can shed considerable light on the molecular details in chromatid pairing as well as heteroduplex formation and correction during meiosis.

3. Does the order of the duplication (i.e., *ade8-18*-vector-*ADE8* vs. *ADE8*-vector-*ade8-18;* Fig. 8) affect the conversion rate? If polarity reflects a site at which pairing and/or recombination is initiated, an insertion between the gene and that site will reduce the conversion rate. If, for example, conversion tracts initiate to the right of *ADE8* and propagate to the left, *ade8-18*-vector-*ADE8*/*ade8-18* diploids would show a higher conversion rate than *ADE8*-vector-*ade8-18*/*ade8-18* diploids. If the initiation site were to the left of *ADE8* and propagated to the right, the second diploid would show the higher conversion rate.

4. Is intrachromosomal gene conversion sensitive to the size of the insertion? Fogel (unpubl.) recently generated a nontandem duplication of the *ADE8* locus that carries two different mutant alleles: *ade8-18*-vector-*ade8-10*. Haploid strains of this type exhibit a very high rate of Ade$^+$ papilla formation, which is indicative of an intrachromosomal conversion event. We may now monitor the sensitivity of papilla formation to changes in the insert DNA (size of insertion or a screen for recombinagenic sites) and changes in the genetic background of the cell (various recombination-defective strains).

Combined, these approaches, herald the beginning of a new understanding of recombination and gene conversion at the molecular level. They bring to consciousness John Milton's closing lines of *Lycidas*:

At last he rose, and twitched his mantle blue—tomorrow to fresh woods and pastures new.

ACKNOWLEDGMENTS

This work was supported by a National Institutes of Health grant (GM-17317) to S.F. and by a grant to R.K.M. from the Division of Biomedical and Environmental Research of the Department of Energy. S.F. acknowledges with gratitude the stimulating collegiality extended to him by the Stanford University Department of Biochemistry during his sabbatical leave from 1979 to 1980. For assistance, instruction, patient criticism, and gifts of DNA, special thanks are directed to M. McDonell, T. St. John, D. Stinchcomb, M. Thomas, S. Scherer, M. Brenan, C. Mouer, M. Fasulo, J. Rine, M. Johnson, and of course to Ron W. Davis, leader of the group. S.F. also thanks J.L. Rossignol and the members of his group at the Université de Paris XI (Paris-Sud), Orsay, France, for the opportunity to learn and experience the *Ascobolus* system at first hand.

REFERENCES

Bassel, J. and R.K. Mortimer. 1971. Genetic order of the galactose structural genes in *Saccharomyces cerevisiae*. *J. Bacteriol.* **108**: 179.

Boon, T. and N.D. Zinder. 1971. A mechanism for genetic recombination generating one parent and one recombinant. *Proc. Natl. Acad. Sci.* **64**: 573.

Botstein, D. and R.W. Davis. 1982. Principles and practice of recombinant DNA research with yeast. In *The molecular biology of the yeast* Saccharomyces. II. *Metabolism and gene expression* (ed. J. Strathern et al.), Cold Spring Harbor Laboratory, Cold Spring Harbor, New York (In press.)

Brunswick, H. 1926. Die Reduktionsteilung bei den Basidiomyceten. *Z. Bot.* **18**: 481.

Case, M.E. and N.H. Giles. 1958. Evidence from tetrad analysis for both normal and abnormal recombination between allelic mutants in *Neurospora crassa*. *Proc. Natl. Acad. Sci.* **44**: 378.

―――. 1959. Recombination mechanisms at the *pan-2* locus in *Neurospora crassa*. *Cold Spring Harbor Symp. Quant. Biol.* **23**: 119.

―――. 1964. Allelic recombination in *Neurospora*: Tetrad analysis of a three-point cross within the *pan-2* locus. *Genetics* **49**: 529.

Catcheside, D.G. 1975. Occurrence in wild strains of *Neurospora crassa* of genes controlling genetic recombination. *Aust. J. Biol. Sci.* **28**: 213.

―――. 1977. *The genetics of recombination*. University Park Press, Baltimore, Maryland.

Clarke, L. and J. Carbon. 1978. Functional expression of cloned yeast DNA in *Escherichia coli*: Specific complementation of arginino succinase lyase (*arg* H) mutations. *J. Mol. Biol.* **120**: 517.

Di Caprio, L. and P.J. Hastings. 1976. Gene conversion and intragenic recombination at the *SUP6* locus and the surrounding region in *Saccharomyces cerevisiae*. *Genetics* **811**: 697.

Emerson, S. 1969. Linkage and recombination at the chromosome level. In *Genetic organization* (ed. E.W. Caspari and A.W. Ravin), p. 267. Academic Press, New York.

Esposito, M.S. 1967. X-ray and meiotic fine structure mapping of the adenine-8 locus in *Saccharomyces cerevisiae*. *Genetics* **58**: 507.

―――. 1971. Post-meiotic segregation in *Saccharomyces*. *Mol. Gen. Genet.* **111**: 297.

Esposito, M.S. and R.E. Esposito. 1977. Gene conversion, paramutations and controlling elements: A treasury of exceptions. In *Cell biology, a comparative treatise* (ed. L. Goldstein and D.M. Prescott), vol 1, p. 59. Academic Press, New York.

Esser, K. and R. Kunnen. 1967. *Genetics of fungi* (translated by E. Steiner). Springer Verlag, New York.

Fincham, J.R.S. and R. Holliday. 1970. An explanation of fine structure map expansion in terms of excision repair. *Mol. Gen. Genet.* **109**: 309.

Fincham, J.R.S., P.R. Day and A. Radford. 1979. *Fungal genetics*, 4th edition. University of California Press, Berkeley, California.

Fink, G.R. 1974. Properties of gene conversion of deletions in *Saccharomyces cerevisiae*. In *Mechanisms in recombination* (ed. R. Grell), p. 287. Plenum Press, New York.

Fink, G.R. and C. Styles. 1974. Gene conversion of deletions in the *his4* region of yeast. *Genetics* **77**: 231.

Fogel, S. and D.D. Hurst. 1967. Meiotic gene conversion in yeast tetrads and the theory of recombination. *Genetics* **57**: 455.

Fogel, S. and R.K. Mortimer. 1968. Meiotic gene conversion of nonsense mutations in yeast tetrads. *Proc. Int. Congr. Genet.* **1**: 6.

―――. 1969. Informational transfer in meiotic gene conversion. *Proc. Natl. Acad. Sci.* **62**: 96.

―――. 1970. Fidelity of gene conversion in yeast. *Mol. Gen. Genet.* **109**: 177.

―――. 1971. Recombination in yeast. *Annu. Rev. Genet.* **5**: 219.

Fogel S., D.D. Hurst, and R.K. Mortimer. 1971. Gene conversion in unselected tetrads from multipoint crosses. *Stadler Genet. Symp.* **2**: 89.

Fogel, S., R.K. Mortimer, K. Lusnak, and F. Tavares. 1979. Meiotic gene conversion: A signal of the basic recombination event in yeast. *Cold Spring Harbor Symp. Quant. Biol.* **43**: 1325.

Ghikas, A. and B.C. Lamb. 1977. The detection, in unordered octads, of 6+ :2m and 2+ :6m ratios with postmeiotic segregation, and of aberrant 4:4s, and their use in corresponding-site interference studies. *Genet. Res.* **29**: 267.

Goldman, S.L. 1974. Studies on the mechanism of the induction of the site-specific recombination in the *ade-6* locus of *Schizosaccharomyces pombe. Genetics* **132**: 347.

Grell, R.F., ed. 1974. *Mechanisms in recombination.* Plenum Press, New York.

Gutz, H. 1971. Site specific inductions of gene conversion in *Schizosaccharomyces pombe. Genetics* **69**: 317.

Hastings, P.J. 1975. Some aspects of recombination in eukaryotic organisms. *Annu. Rev. Genet.* **9**: 129.

Hastings, P.J., A. Kalogeropoulos, and J.L. Rossignol. 1980. Restoration to the parental genotype of mismatches formed in recombinant DNA heteroduplex. *Curr. Genet.* **2**: 169.

Hawthorne, D.C. 1969. Identification of nonsense codons in yeast. *J. Mol. Biol.* **43**: 71.

Hicks, J.B., A. Hinnen, and G.R. Fink. 1979. Properties of yeast transformation. *Cold Spring Harbor Symp. Quant. Biol.* **43**: 1305.

Hicks, J.B., J.N. Strathern, and I. Herskowitz. 1977. The cassette model of mating-type interconversion. In *DNA insertion elements, plasmids, and episomes* (ed. A. Bukhari et al.), p. 457. Cold Spring Harbor Laboratory, Cold Spring Harbor, New York.

Hicks, J.B., J.N. Strathern, and A.J.S. Klar. 1979. Transposable mating type genes in *Saccharomyces cerevisiae. Nature* **282**: 478.

Hinnen, A., J. Hicks, and G.R. Fink. 1978. Transformation in yeast. *Proc. Natl. Acad. Sci.* **75**: 1929.

Holliday, R. 1964. A mechanism for gene conversion in fungi. *Genet. Res.* **5**: 282.

———. 1974. Molecular aspects of genetic exchange and gene conversion. *Genetics* **78**: 273.

Hotchkiss, R.D. 1974. Models of genetic recombination. *Annu. Rev. Microbiol.* **28**: 445.

Hurst, D.D., S. Fogel, and R.K. Mortimer. 1972. Conversion associated recombination in yeast. *Proc. Natl. Acad. Sci.* **69**: 101.

Kalogeropoulos, A. and J.L. Rossignol. 1980. Evidence for independent mismatch corrections along the same hybrid DNA tract during meiotic recombination in *Ascobolus. Heredity* **45**: 263.

Kitani, Y. 1978. Aberrant 4:4 segregation. *Jpn. J. Genet.* **53**: 301.

Kitani, Y. and L.S. Olive. 1967. Genetics of *Sordaria fimicola.* VI. Gene conversion at the *g* locus in mutant × wild-type crosses. *Genetics* **57**: 767.

Kitani, Y., L.S. Olive, and A.S. El-Ani. 1962. Genetics of *Sordaria fimicola.* V. Aberrant segregation at the *g* locus. *Am. J. Bot.* **49**: 697.

Klapholz, S. and R.E. Esposito. 1980a. Isolation of *spo12-1* and *spo13-1* from a natural variant of yeast that undergoes a single meiotic division. *Genetics* **96**: 567.

———. 1980b. Recombination and chromosome segregation during the single meiosis in *spo12-1* and *spo13-1* diploids. *Genetics* **96**: 589

Klein, H.L. and T.D. Petes. 1980. Intrachromosomal gene conversion in yeast. *Nature* **289**: 144.

Kniep, H. 1928. Die Sexualitat der niederen Pflanzen. Gustav Fischer, Jena.

Korch, C.T. and R. Snow. 1973. Allelic complementation in the first gene of histidine biosynthesis in *Saccharomyces cerevisiae. Genetics* **74**: 287.

Kushev, V. 1974. *Mechanisms of genetic recombination* (translated by B. Haigh) Plenum Press, New York.

Lamb, B.C. 1972. 8:0, 0:8, 7:1 and 1:7 conversion ratios in octads from wild-type × mutant crosses of *Ascobolus immersus. Heredity* **29**: 397.

———. 1975. Cryptic mutations: Their predicted biochemical basis, frequencies and effects on gene conversion. *Mol. Gen. Genet.* **137**: 305.

Lamb, B.C. and M.R.T. Wickramaratne. 1973. Corresponding site interference, synaptinemal complex structure, and 8+ :0m and 7+ :1m octads from wild-type × mutant crosses of *Ascobolus immersus*. *Genet. Res.* **22:** 113.

Lawrence, C.W., F. Sherman, M. Jackson, and R.A. Gilmore. 1975. Mapping and gene conversion studies with the structural gene for iso-1-cytochrome *c* in yeast. *Genetics* **81:** 615.

Leblon, G. 1972. Mechanism of gene conversion in *Ascobolus immersus*. I. Existence of a correlation between the origin of mutants induced by different mutagens and their conversion spectrum. *Mol. Gen. Genet.* **115:** 36.

Leblon, G. and J.L. Rossignol. 1973. Mechanism of gene conversion in *Ascobolus immersus*. III. The interaction of heteroalleles in the conversion process. *Mol. Gen. Genet.* **122:** 165.

Lindegren, C. 1949. *The yeast cell, its genetics and cytology*. Educational Publishing, St. Louis, Missouri.

Lissouba, P., J. Mouseau, G. Rizet, and J.L. Rossignol. 1962. Fine structure of genes in the Ascomycete *Ascobolus immersus*. *Adv. Genet.* **11:** 343.

Manney, T.R. and R.K. Mortimer. 1964. Allelic mapping in yeast by X-ray induced mitotic reversion. *Science* **143:** 581.

Meselson, M.S. and C.M. Radding. 1975. A general model for genetic recombination. *Proc. Natl. Acad. Sci.* **72:** 358.

Mitchell, M.B. 1955. Aberrant recombinations of pyridoxine mutants of *Neurospora*. *Proc. Natl. Acad. Sci.* **41:** 215.

Moore, C. and F. Sherman. 1974. Lack of correspondence between genetic and physical distance in the iso-1-cytochrome *c* gene in yeast. In *Mechanisms of recombination* (ed. R. Grell), p. 295. Plenum Press, New York.

———. 1975. The role of DNA sequences in genetic recombination in the iso-1-cytochrome *c* gene in yeast. I. Discrepancies between physical distances and genetic distances by fine structure mapping procedures. *Genetics* **79:** 397.

Mortimer, R.K. and S. Fogel. 1974. Genetical interference and gene conversion. In *Mechanisms in recombination* (ed. R. Grell), p. 236. Plenum Press, New York.

Mortimer, R.K. and D.C. Hawthorne. 1969. Yeast genetics. In *The yeasts* (ed. A.H. Rose and J.S. Harrison), vol. 1, p. 385. Academic Press, New York.

———. 1973. Genetic mapping in *Saccharomyces*. IV. Mapping of temperature-sensitive genes and use of disomic strains in localizing genes and fragments. *Genetics* **74:** 33.

Mortimer, R.K. and D. Schild. 1980. Genetic map of *Saccharomyces cerevisiae*. *Microbiol. Rev.* **44:** 519.

Mundkur, B.D. 1949. Evidence excluding mutations, polysomy and polyploidy as posssible causes of non-Mendelian segregation in *Saccharomyces*. *Ann. Mo. Bot. Gard.* **36:** 259.

Murray, N.E. 1963. Polarized recombination and fine structure within the *me-2* gene of *Neurospora crassa*. *Genetics* **48:** 1163.

Nasmyth, K.A. and K. Tatchell. 1980. The structure of the transposable yeast mating type loci. *Cell* **19:** 753.

Nicolas, A. 1979. Variation of gene conversion and intragenic recombination frequencies in the genome of *Ascobolus immersus*. *Mol. Gen. Genet.* **176:** 129.

Paquette, J. 1978. Detection of aberrant 4:4 asci in *Ascobolus immersus*. *Can. J. Genet. Cytol.* **20:** 9.

Parker, J.H. and F. Sherman. 1969. Fine structure mapping and mutational studies of genes controlling yeast cytochromes. *Genetics* **62:** 9.

Paszewski, A. 1970. Gene conversion: Observations on the DNA hybrid models. *Genet. Res.* **11:** 55.

Petes, T.D. 1980a. Unequal meiotic recombination within tandem arrays of yeast ribosomal DNA genes. *Cell* **19:** 765.

———. 1980b. Molecular genetics of yeast. *Annu. Rev. Biochem.* **49:** 845.

Pukkila, P. 1977. Biochemical analysis of genetic recombination in eukaryotes. *Heredity* **39:** 193.

Radding, C.M. 1973. Molecular mechanisms in genetic recombination. *Annu. Rev. Genet.* **7:** 87.

———. 1979. The mechanism of conversion of deletions and insertions. *Cold Spring Harbor Symp. Quant. Biol.* **43:** 1315.

Resnick, M.A. 1979. The induction of molecular and genetic recombination in eukaryotic cells. *Adv. Radiat. Biol.* **8:** 175.

Roman, H. 1957. Studies of gene mutation in *Saccharomyces. Cold Spring Harbor Symp. Quant. Biol.* **21:** 175.

———. 1980. Recombination in diploid vegetative cells of *Saccharomyces cerevisiae. Carlsberg. Res. Commun.* **45:** 211.

Roman, H. and F. Jacobs. 1958. A comparison of spontaneous and ultraviolet induced allelic recombinations with reference to the recombination of outside markers. *Cold Spring Harbor Symp. Quant. Biol.* **23:** 155.

Rossignol, J.L. and V. Haedens. 1980. Relationship between asymmetrical and symmetrical hybrid DNA formations during meiotic recombination. *Curr. Genet.* **1:** 185.

Rossignol, J.L., N. Paquette, and A. Nicolas. 1979. Aberrant 4:4 asci, disparity in the direction of conversion, and the frequency of conversion in *Ascobolus immersus. Cold Spring Harbor Symp. Quant. Biol.* **43:** 1343.

Sang, H. and H.L.K. Whitehouse. 1979. Genetic recombination at the *buff* spore colour locus in *Sordaria brevicollis. Mol. Gen. Genet.* **174:** 327.

Scherer, S. and R.W. Davis. 1980. Recombination of dispersed repeated DNA sequences in yeast. *Science* **209:** 1380.

Sherman, F. and C.W. Lawrence. 1974. *Saccharomyces.* In *Handbook of genetics* (ed. R.C. King), vol. 1, p. 359. Plenum Press, New York.

Sobell, H.M. 1972. Molecular mechanisms for genetic recombination. *Proc. Natl. Acad. Sci.* **69:** 2483.

———. 1974. Concerning the stereochemistry of strand equivalence in genetic recombination. In *Mechanisms in recombination* (ed. R.F. Grell), p. 433. Plenum Press, New York.

Stadler, D.R. 1973. The mechanism of intragenic recombination. *Annu. Rev. Genet.* **7:** 113.

Stadler, D.R. and A.M. Towe. 1963. Recombination of allelic cysteine mutants in *Neurospora. Genetics* **48:** 1323.

———. 1971. Evidence for meiotic recombination in *Ascobolus* involving only one member of a tetrad. *Genetics* **68:** 401.

Stahl, F.W. 1969. One way to think about gene conversion. *Genetics* (Suppl.) **61:** 1.

———. 1978. Summary. *Cold Spring Harbor Symp. Quant. Biol.* **43:** 1353.

———. 1979a. Special sites in recombination. *Annu. Rev. Genet.* **13:** 7.

———. 1979b. *Genetic recombination. Thinking about it in phage and fungi.* Freeman, San Francisco.

Szostak, J.W. and R. Wu. 1980. Unequal crossing over in the ribosomal DNA of *Saccharomyces cerevisiae. Nature* **84:** 426.

Wagner, R.E. and M. Radman. 1975. A mechanism for initiation of genetic recombination. *Proc. Natl. Acad. Sci.* **72:** 3619.

Wettstein, F.V. 1924. Morphologie und Physiologie des Formenwechsels. *Z. Indukt. Abstammungs. Vererbungsl.* **33:** 1.

Whitehouse, H.L.K. 1963. A theory of crossing over by means of hybrid deoxyribonucleic acid. *Nature* **199:** 1034.

———. 1969. *Towards an understanding of the mechanism of heredity,* 2nd edition. St. Martin's Press, New York.

Wickramaratne, M.R.T., and B.C. Lamb. 1978. The estimation of conversion parameters and the control of conversion in *Ascobolus immersus. Mol. Gen. Genet.* **159:** 63.

Williamson, M. and S. Fogel. 1980. Correction deficient (*cor*) mutants that alter the frequency and length of correction tracts in meiotic hybrid DNA. In *Tenth International Conference on Yeast Genetics and Molecular Biology,* No. 135, p. 44. Louvain-la-Neuve, Belgium.

Winge, O. and C. Roberts. 1950. Non-Mendelian segregation from heterozygotic yeast asci. *Nature* **165:** 157.

Winkler, H. 1930. Die Konversion der Gene. Gustav Fischer, Jena.

Yu-Sun, C.C., M.R.T. Wickramaratne, and H.L.K. Whitehouse. 1977. Mutagen specificity in conversion pattern in *Sordaria brevicollis. Genet. Res.* **29:** 65.

Zimmermann, F.K. 1968. Enzyme studies on the products of mitotic gene conversion in *Saccharomyces cerevisiae. Genetics* **101:** 171.

Mechanisms of Mitotic Recombination

Michael S. Esposito
Biology and Medicine Division
Lawrence Berkeley Laboratory
University of California at Berkeley
Berkeley, California 94720

Joseph E. Wagstaff
University of Chicago
Department of Biology and Committee on Genetics
Chicago, Illinois 60637

INTRODUCTION

Chromosomal recombination in *Saccharomyces cerevisiae* occurs at three distinct life-cycle stages of diploid cells: (1) mitotic recombination, during vegetative cell division; (2) meiototic recombination, in diploid cells uncommitted to haploidization following exposure to a meiosis-inducing environment (see Esposito and Esposito 1978); and (3) meiotic recombination, in cells that complete meiosis and ascosporogenesis (see Fogel et al. 1979). Interest in the mechanisms of mitotic recombination has been stimulated by the observation that certain mutagenic carcinogens are also recombinagens in in yeast (Zimmermann et al. 1966) and data indicating that genetic recombination is sometimes involved in the control of gene expression in prokaryotes and eukaryotes (cf. McClintock 1961; Esposito and Esposito 1977; Silverman et al. 1979). Current studies of the molecular mechanisms and genic control of spontaneous and induced mitotic recombination in *S. cerevisiae* focus upon answering three basic, related questions regarding mitotic exchange: (1) At what stages of the mitotic cell cycle does mitotic exchange occur? (2) Can the properties of mitotic and meiotic recombination be accounted for by a single molecular model of DNA recombination? (3) To what extent do genes controlling mitotic exchange participate in the control of meiotic recombination, homothallic interconversion of mating-type alleles, and other cellular pathways of DNA synthesis and repair? In the following discussion we summarize the properties of intergenic and intragenic mitotic recombination of standard marker loci, aspects of mitotic exchange between sister chromatids, and the phenotypes of *rec* mutants relevant to these questions.

MITOTIC GENE CONVERSION AND RECIPROCAL EXCHANGE

Intergenic and intragenic mitotic recombination occur spontaneously during the growth of hybrid cell cultures at rates several orders of magnitude below spontaneous meiotic rates of recombination for the same genetic intervals (reviewed by Fogel and Mortimer 1971). The frequency of mitotic recombination can be stimulated by a variety of treatments, including UV light, ionizing radiation, and chemical recombinagens (James and Lee-Whiting 1955; Roman and Jacob 1957; Mortimer 1959; Zimmermann and Schwaier 1967; Esposito 1968). Genotypic analyses of recombinant diploids of spontaneous and induced origin have illustrated that mitotic cells achieve genetic recombination by both nonreciprocal events, i.e., gene conversion, and reciprocal events, as do meiotic cells. When the properties of mitotic recombination are examined in detail, however, several features emerge that distinguish mitotic recombination from its meiotic counterpart.

Intragenic Recombination

Mitotic intragenic recombination between heteroalleles (e.g., *ml-1/ml-2*) is most conveniently studied in the case of mutations conferring auxotrophy because rare prototrophic recombinants can be selected on nutrient omission media. Prototrophic heteroallelic recombinants occur spontaneously and following exposure to recombinagens at rates that are typically well in excess of those expected from reversion of either of the input heteroalleles (Roman and Jacob 1957, 1959).

The relative contributions of gene conversion and reciprocal exchange to the origin of intragenic recombinants have been assessed by determining the genotypes of diploid recombinants at the recombinant locus. Representative data are summarized in Table 1. Prototrophic recombinants of heteroallelic diploids are found to consist of $++/ml$-$1+$, $++/+ml$-2, $++/++$, and $++/ml$-1 ml-2 cells. The first two genotypes are recovered most frequently and at nearly equal frequencies, whereas the latter two genotypes are recovered least frequently but at nearly equal frequencies.

Several properties of mitotic heteroallelic recombination can be inferred from these data. First, strict origin of intragenic recombinants by reciprocal exchange can be eliminated from consideration. Such hypotheses predict that all prototrophs will be of the $++/ml$-1 ml-2 genotype in the case of reciprocal exchange between homologs before chromosomal replication, and that half will be of this genotype in the case of reciprocal exchange between nonsister chromatids following chromosomal duplication. The simplest interpretation of the genotypic arrays observed is that they result from mitotic gene conversion by heteroduplex formation and repair. Although gene conversion satisfactorily explains the apparent nonreciprocal nature of the events involved (reviewed by Roman 1963), the recovery of $++/++$ and $++/ml$-1 ml-2 mitotic recombinants at nearly equal frequencies is not anticipated if mitotic and meiotic recombination in yeast follow the same rules. Their presence, as discussed later, indicates that mitotic cells engage in the formation of symmetric Holliday structures, which appear to occur very rarely, if at all, in meiosis of wild-type yeast cells (Esposito 1978; Fogel et al. 1979).

Spontaneous mitotic intragenic recombination, like meiotic intragenic recombination, occurs in nonrandom association with reciprocal exchange of

Table 1 Genotypes at the convertant locus of mitotic prototrophic heteroallelic recombinants recovered from wild-type $MATa/MAT\alpha$ hybrids

Heteroalleles ml-$1/ml$-2[a]	Genotypes of prototrophs				Total	Origin[b]	Reference
	$++$ $1+$	$++$ $+2$	$++$ $++$	$++$ 12			
$his1$-$7/his1$-1	32	33	3	3	71	SPN	Hurst and Fogel (1964)
$his1$-$7/his1$-1	47	51	6	0	104	UV	Hurst and Fogel (1964)
$his1$-$315/his1$-1	40	66	14	9	129	X-ray	Wildenberg (1970)
$ilv1$-$2/ilv1$-1	36	31	1	0	68	SPN	Kakar (1963)
$ilv1$-$b/ilv1$-a	11	11	1	0	23	NMU	Zimmermann and Schwaier (1967)
$trp5$-$d/trp5$-c	104	102	13	20	239	SPN	(Golin 1979; M.S. Esposito, unpubl.)

[a]ml-1 indicates centromere-proximal allele; ml-2 indicates centromere-distal allele.
[b]SPN indicates spontaneous; UV indicates ultraviolet-induced; X-ray indicates X-ray-induced; and NMU indicates nitroso-methyl-urethane-induced.

heterozygous markers flanking the locus of heteroallelic conversion (Roman 1957; Roman and Jacob 1959), which suggests that mitotic gene conversion, like meiotic gene conversion (Holliday 1964), may have a molecular precursor in common with reciprocal recombination. Evidence in support of this interpretation of mitotic data is reviewed below.

Intergenic Recombination and Gene Conversion in Sectored Colonies

Mitotic intergenic recombination between heterozygous markers distributed along the length of a chromosome results in the formation of sectored colonies that can be observed visually or by the replica-plating of intact colonies to diagnostic media to detect segregation of recessive markers. The colonial color change system devised by Roman (1956, 1957) and variations of it have been widely employed in studies of spontaneous and induced intergenic recombination. Roman's technique, which illustrates the principles of sector analysis, makes use of the recessive mutations *ade1* or *ade2* that block adenine biosynthesis and result in production of a cell-limited red pigment. Other *ade* mutations block adenine biosynthesis at steps preceding the *ade1* and *ade2* blocks and thus prevent pigment accumulation in double mutant segregants. Diploids of the genotype *ade2/ade2 ADE6/ade6*, for example, are adenine-requiring and form red colonies. When populations of such cells are plated on nutrient medium, one observes rare white and red-white sectored colonies. The latter, which are recovered at a frequency of about 10^{-4}, are most often the result of either reciprocal exchange between the *ade6* locus and its centromere or mitotic conversion at the *ade6* locus.

Sectors resulting from gene conversion can be distinguished from those that result from reciprocal recombination by genotypic analysis of the red and white portions of sectored colonies (Roman 1971, 1973). Sectored colonies most simply ascribed to reciprocal exchange are homozygous *ADE6/ADE6* in the red portion, homozygous *ade6/ade6* in the white portion, and reciprocally homozygous for markers distal to the *ade6* locus that were originally present in heretozygous condition. Conversional events are detected as sectored colonies that are *ADE6/ade6* in the red portion and *ade6/ade6* in the white portion. Determination of the genotypes of both sides of sectored colonies is thus analogous to meiotic tetrad analysis and has provided several insights into the similarities and differences between mitotic and meiotic recombination.

Studies of spontaneous and induced sectoring (Esposito 1968; Nakai and Mortimer 1969; Wildenberg 1970; Johnston 1971, 1972; Roman 1971, 1973) have illustrated that joint segregation of linked markers is most often due to a single reciprocal exchange, whereas segregation of a single marker flanked by heterozygous markers is usually the result of a conversional event. The latter is signaled by retention of heterozygosity (+ /m) in one portion of the sectored colony and recessive homozygosis (m/m) in the other portion. Conversion of one of a set of linked markers and coincident reciprocal exchange at an adjacent site has also been seen and provides added evidence of a mechanistic association between mitotic conversion and reciprocal recombination.

A residuum of genotypes have been observed that are not anticipated from meiotic data. These include sectors in which widely separated chromosomal markers exhibit coincident conversion (Nakai and Mortimer 1969; Wildenberg 1970; Johnston 1971, 1972; Esposito 1978) and instances in which both sides of a sectored colony are homozygous for the same allele of a site heterozygous in the original hybrid, with the latter homozygosity occurring in the vicinity of a reciprocal exchange. These exceptions are not easily explained by four-strand, i.e., postreplicative, models of mitotic recombination and are now recognized as evidence that mitotic recombination of homologs is initiated at the two-strand stage (Esposito 1978; Golin 1979).

MITOTIC RECOMBINATION BY REPAIR AND REPLICATIVE RESOLUTION OF HOLLIDAY STRUCTURES FORMED AT THE TWO-STRAND STAGE

It has been generally accepted that mitotic recombination occurs between nonsister chromatids at the four-strand stage, as originally proposed by Stern (1936). Stern's interpretation has gained acceptance because it has not been obvious how reciprocal recombination between unreplicated homologs can result in sectored colonies exhibiting reciprocal segregation of markers distal to the site of exchange (Fig. 1). This can occur, however, as illustrated in Figure 2, by formation of a Holliday structure at the two-strand stage, which is not cleaved and persists through duplication of the chromosomes in the S phase of mitosis. This mechanism, like exchange between nonsister chromatids at the four-strand stage, results in a diploid cell containing two parental and two recombinant chromatids (Fig. 2). Given the appropriate segregation of chromosomes, a pair of daughter cells reciprocally homozygous for markers distal to the site of exchange gives rise to a sectored colony.

We recently proposed a molecular model of two-strand intragenic and intergenic mitotic recombination (Esposito 1978) incorporating features of the Meselson-Radding general mechanism of genetic exchange (Meselson and Radding 1975). The two-strand model illustrated in Figure 3 accounts satisfactorily for the exceptional properties of spontaneous and induced mitotic recombination mentioned in the previous section and the differences between mitotic and meiotic recombination to be discussed below.

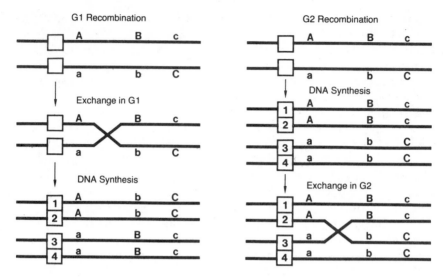

Figure 1 Reciprocal exchange between unreplicated chromosomes in G_1 and exchange between nonsister chromatids during G_2. (*Left*) After replication of the recombinant G_1 chromosomes, two modes of chromosomal disjunction can follow. Chromatids 1 and 3 separate from 2 and 4, or chromatids 1 and 4 separate from 2 and 3. Neither disjunction pattern can result in a sectored colony or twin spot for the recessive markers *b* and *c*. (*Right*) In the case of recombination during G_2, a twin spot or sectored colony exhibiting segregation for *b* and *c* results when chromatids 1 and 3 separate from 2 and 4. When chromatids 1 and 4 separate from 2 and 3, a twin spot or sectored colony does not result and the exchange event escapes detection.

Evidence Supporting the Two-strand Model

Genetic data indicating that mitotic recombination occurs at the two-strand stage come from studies of both spontaneous and induced mitotic exchange in wild-type strains and diploids homozygous for the hyper-recombination mutation *rem1-1*. Current evidence supporting the two-strand model with respect to spontaneous mitotic recombination has been obtained by genotypic analysis of prototrophic red-white sectored colonies in which an event of heteroallelic recombination at either the *leu1* locus or the *trp5* locus yielded a pair of Leu$^+$ or Trp$^+$ prototrophic daughter cells reciprocally recombined for markers distal to the site of conversion, including *ade5* (Esposito 1978; Golin 1979; J. Wagstaff and M. S. Esposito, unpubl.). Such prototrophic colonies are recovered at rates indicating that they are the result of a single conversion-associated exchange event (Esposito 1978; Golin 1979).

The two-strand model of mitotic recombination predicts recovery of nine different genotypes at the locus of conversion, resulting in a prototrophic colony sectored for markers distal to the site of intragenic recombination (Fig. 3 and Table 2). The four-strand model of recombination of Meselson

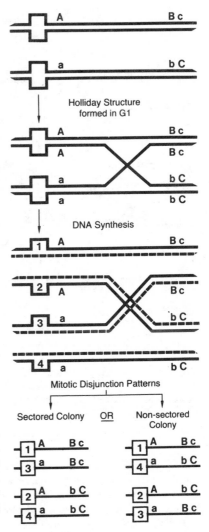

Figure 2 Holliday structure formation during G_1, followed by chromosome replication, resulting in sectored colonies or twin spots. Prior to the S phase, a Holliday structure is formed by exchange of single strands of DNA of the same chemical polarity. When the crossing strands are not cleaved prior to duplication of the chromosomes, the resulting chromatids have the same genotypes as those generated by crossing-over in G_2 between nonsister chromatids (compare with Fig. 1). When chromatids 1 and 3 separate from 2 and 4, a sectored colony results, exhibiting segregation of the recessive markers *b* and *c*. (----) DNA synthesized during the S phase of mitosis.

and Radding (1975), which satisfactorily explains the salient properties of meiotic recombination, predicts recovery of only one genotypic class of sectored prototrophic colonies. Such sectored prototrophic colonies must contain both input heteroalleles, i.e., they must be ++/*m1-1* + on one side and ++/+ *m1-2* on the other, with the *m1-1* and *m1-2* alleles being contributed by the pair of nonsister chromatids that were not involved in the conversion-associated event (Fig. 4).

Genotypic analysis of prototrophic sectored colonies has resulted in the recovery of genotypes diagnostic of two-strand recombination (Table 3). The single genotypic class expected from four-strand-stage events is rarely found

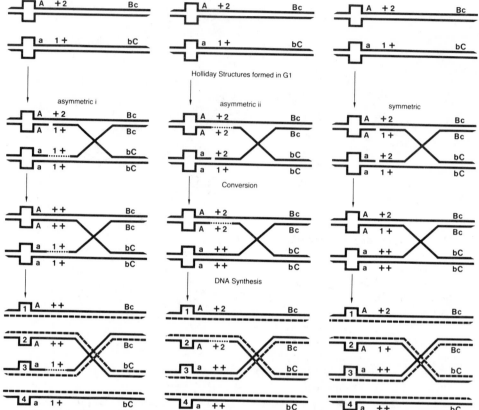

Figure 3 Heteroallelic recombination and associated exchange at the two-strand (e.g., G_1) stage. Conversion of the heteroalleles with concomitant exchange in the distal interval can result by formation of the three Holliday structures shown, as described by Meselson and Radding (1975). Two types of asymmetric structures (*i* and *ii*) can be formed in the region of the heteroallelic locus. (....) Regions of DNA synthesis during asymmetric strand transfer. The Holliday structure that is symmetric in the region of the heteroalleles contains heteroduplex regions in both arms. Each of the three structures is shown as symmetric distal to the heteroallelic locus to facilitate visualization of conversion and the consequence of chromosome duplication without cleavage of the crossing strands of the Holliday structure. In each case, the sectored prototrophic colony results from repair (conversion) of *1+/+2* to *++/++*, followed by the S phase and segregation of chromatids 1 and 3 from 2 and 4. The asymmetric structures yield sectored clones heterozygous for the same allele in the *Bc/Bc* and *bC/bC* portions. Table 2 lists the various genotypes expected from the symmetric structure depending upon the conversion (repair) pattern of the remaining *1+/+2* heteroduplex. (---) The DNA chains synthesized during the S phase.

348

Table 2 Chromosomal genotypes at the convertant locus of prototrophic daughter cells

Chromosomes	Genotypes at the locus of conversion[a]								
	I	II	III	IV	V	VI	VII	VIII	IX
1 and 3	+ +	+ +	+ +	+ +	+ +	+ +	+ +	+ +	+ +
	1 +	+ *2*	+ +	*1 2*	+ *2*	*1* +	+ *2*	*1* +	*1* +
---	---	---	---	---	---	---	---	---	---
2 and 4	+ +	+ +	+ +	+ +	+ +	+ +	+ +	+ +	+ +
	1 +	+ *2*	+ +	*1 2*	+ +	+ +	*1 2*	*1 2*	+ *2*

Daughter cells originate from the Holliday structures of Fig. 3.

[a]Classes I through VIII are diagnostic of two-strand-stage exchange. Class IX may result from either two-strand-stage or four-strand-stage exchange. Classes I and II may result from either asymmetric or symmetric Holliday structures. Classes III–IX are diagnostic of symmetric Holliday structures.

and can be generated by the two-strand model as well (Table 2). The two-strand model we have proposed also accounts for the data of Wildenberg (1970), who conducted an extensive pedigree analysis of prototrophic hetero-allelic recombinants induced by exposure of cells in G_1 of the mitotic cell cycle to X-rays. Wildenberg reported 129 independently arising pedigrees containing at least one His$^+$ prototrophic cell following exposure of *his1-315/his1-1* cells to 3.3 kr or 6.6 kr of X-rays. In 77 of 129 pedigrees, the mother cell and first bud were both His$^+$ and therefore can be analyzed in the same manner as sectored prototrophic colonies. The results of this analysis

Table 3   Genotypes at the convertant locus of prototrophic sectored colonies and pedigrees

Strain	Heteroalleles *m1-1/m1-2*	I + + *1* + --- + + *1* +	II + + + *2* --- + + + *2*	III + + + + --- + + + +	IV + + *1 2* --- + + *1 2*	V + + + *2* --- + + + +	VI + + *1* + --- + + + +	VII + + + *2* --- + + *1 2*	VIII + + *1* + --- + + *1 2*	IX + + *1* + --- + + + *2*	Total
MATa/MATα	*trp5-d/trp5-c*	32	20	2	5	2	7	0	1	1	70
	leu1-c/leu1-12	7	7	1	0	2	3	0	0	0	20
MATa/MATα; *rem1-1/rem1-1*	*trp5-d/trp5-c*	22	11	40	3	0	5	0	0	0	81
	leu1-c/leu1-12	11	7	43	0	5	5	1	0	2	74
MATa/MATa	*trp5-d/trp5-c*	10	19	2	0	0	1	0	0	0	32
MATa/MATα[a]	*his1-315/his1-1*	27	33	2	4	1	3	2	2	3	77

[a]Pedigree data of Wildenberg (1970).

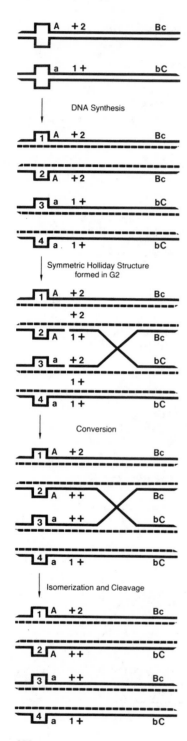

Figure 4 Heteroallelic recombination and associated exchange at the four-strand (e.g., G₂) stage. After the S phase, two nonsister chromatids (2 and 3) participate in the formation of a symmetric Holliday structure. To obtain sectored prototrophic clones, conversion on both chromatids 2 and 3 must yield a ++ recombinant. Isomerization and cleavage of the crossing strands of the Holliday structure in the recombinant configuration with respect to the distal markers and segregation of chromatids 1 and 3 from 2 and 4 at mitosis results in a ++ colony exhibiting segregation of the distal marker. The *Bc/Bc* portion of the sector is of genotype ++/+2 and the *bC/bC* portion has the genotype ++/1+. The genotype of sectored colonies is the same even when the recombinant chromatids (2 and 3) are heteroduplex at one of the mutant sites. The generation of such sectors requires only that the ++ recombinant configuration of each chromatid be borne on the transcribed DNA strand. (---) DNA chains synthesized during the S phase.

are tabulated together with our data in Table 3. Wildenberg's data illustrate that X-ray-induced heteroallelic recombination, whether associated with reciprocal exchange or not, occurs almost exclusively prior to chromosomal duplication in the treated G_1 cells.

Heteroduplexes in Mitotic Recombination

Analysis of prototrophic sectors and pedigrees further shows that DNA heteroduplexes involved in mitotic recombination differ from their meiotic counterparts with respect to two properties. First, mitotic heteroduplex regions are often symmetric (Esposito 1978), whereas meiotic heteroduplexes are almost exclusively asymmetric (Fogel et al. 1979). Second, heteroduplexes appear to be more extensive in mitosis than in meiosis.

Asymmetric heteroduplexes at the two-strand stage can generate only the two prototrophic sector genotypes designated I and II in Table 2. Symmetric heteroduplexes can generate all nine sector genotypes. Therefore, the fraction of sectors of genotypes III–IX represents a minimum estimate of the fraction of mitotic gene conversion that involves symmetric heteroduplexes. As Table 3 shows, at least about 25% of spontaneous conversion events at *trp5* and *leu1* and X-ray-induced events at *his1* in wild-type strains must involve a symmetric intermediate; in diploids homozygous for the hyper-recombination mutation *rem1-1*, the minimum fraction of symmetric heteroduplexes is even greater than that of wild-type strains.

Other evidence illustrates that mitotic heteroduplexes are longer on average than meiotic heteroduplexes. One observes nearly equal recovery of the centromere-proximal and the centromere-distal allele in sectored prototrophic colonies. In such a situation a single-site heteroduplex incorporating only the centromere-proximal site cannot result in a sectored prototrophic colony. In contrast, single-site heteroduplexes incorporating the distal site as well as two-site heteroduplexes can result in sectored prototrophic colonies. The nearly equal recovery of centromere-proximal and centromere-distal alleles in sectored prototrophic colonies thus indicates that two-site heteroduplexes are common. Input heteroalleles are also recovered at nearly equal frequencies among random prototrophic colonies and pedigrees in which there is no bias against recovery of one or the other. The very same alleles, however, can exhibit strong polarity in meiotic conversion; the *leu1* and *trp5* heteroalleles for which we find no mitotic polarity exhibit strong polarity in meiotic conversion. Among 55 single-site *leu1* meiotic conversions, for example, 54 were conversions of *leu1-c* to + or of + to *leu1-c*, whereas only 1 was a conversion at *leu1-12* (Plotkin 1978). Thus, the extent and/or repair of heteroduplex regions differs depending upon whether mitotic or meiotic structures are involved.

The second manifestation of long mitotic heteroduplexes is the high rate of conversion at linked heteroallelic sites. In the case of the *leu1* and *trp5* loci, which are about 16 centiMorgans (cM) from one another on chromosome VII, we have observed that cells doubly prototrophic at *leu1* and *trp5* due to conversion arise at a rate at least 500 times greater than expected for independent events. The frequency of coincident mitotic gene conversion between markers on chromosome VII declines as the map distance between the loci in question increases. This is compatible with the hypothesis that these events arise from long heteroduplexes (Golin 1979). Coincident gene-conversion frequencies, moreover, are subject to experimental modification: Arrest of *MATa/MATa* diploid cells in G_1 with the mating pheromone α-factor for 3–6 hours results both in an increase in the frequency of prototrophic sectors and in an approximately tenfold increase in the fraction of sectors showing coincident gene conversion at *leu1* and *trp5* (J. Wagstaff, unpubl.). We interpret this as indicating that arrest of cells at the two-strand stage permits extension of heteroduplex regions beyond their normal lengths.

It would be misleading to insist on a strict dichotomy between mitotic cells and meiotic cells with respect to the lengths of their conversion tracts and the heteroduplex lengths inferred from them. Although the analysis of unselected tetrads by Mortimer and Fogel (1974) showed no "negative interference" between conversion events at closely linked loci, Di Caprio and Hastings (1976) have shown a high frequency of coincident conversion events for markers linked to the *SUP6* gene. These results emphasize the importance of comparing mitotic and meiotic recombination parameters for the same markers in closely related strains.

Mismatch Repair

The genotypes at the convertant locus of sectored prototrophic colonies and pedigrees summarized in Table 3 demonstrate that mismatched base pairs unselected for correction are repaired with fairly high efficiency in mitotic cells. A lower limit of the efficiency of repair can be calculated by considering only those sectored prototrophic colonies and pedigrees that definitely originated from symmetric Holliday structures. These are designated classes III–IX in Table 2. Since repair of only one of the putative *m1-1 +/+ m1-2* heteroduplexes to + +/+ + is sufficient to generate a sectored colony, the genotype of the remaining heteroduplex is *unselected* with respect to mismatch repair (Table 4). The data for *MATa/MATα* hybrids given in Table 3 may be analyzed as follows: In wild-type strains at *trp5*, classes III–IX include 18 sectors (i.e., 36 mismatches), and 24 of 36 (67%) at a minimum were repaired. At *leu1* in wild-type strains, 7 of 12 (58%) mismatches at a minimum were repaired. At *his1* in wild-type strains, 20 of 34 (59%) were repaired.

In contrast to the above, the analysis of sectored colonies exhibiting single-site gene conversion for a marker originally present in heterozygous condition (e.g., +/*m* on one side and *m/m* on the other side) cannot be used to estimate mismatch repair efficiency, because a sector of this genotype can

Table 4   Interpretation of genotypic classes III–IX on the assumption that the unselected heteroduplex includes both *ml-1* and *ml-2*

Genotypic class[a]	Mismatch repair of unselected heteroduplex	Direction of repair
	Repair of one mismatch	
V	$\dfrac{ml\text{-}1 \quad +}{+ \quad ml\text{-}2} \longrightarrow \dfrac{+ \quad +}{+ \quad ml\text{-}2}$	$ml\text{-}1 \longrightarrow +$
VIII	$\dfrac{ml\text{-}1 \quad +}{+ \quad ml\text{-}2} \longrightarrow \dfrac{ml\text{-}1 \quad +}{ml\text{-}1 \ ml\text{-}2}$	$+ \longrightarrow ml\text{-}1$
VI	$\dfrac{ml\text{-}1 \quad +}{+ \quad ml\text{-}2} \longrightarrow \dfrac{ml\text{-}1 \quad +}{+ \quad +}$	$ml\text{-}2 \longrightarrow +$
VII	$\dfrac{ml\text{-}1 \quad +}{+ \quad ml\text{-}2} \longrightarrow \dfrac{ml\text{-}1 \ ml\text{-}2}{+ \quad ml\text{-}2}$	$+ \longrightarrow ml\text{-}2$
	Repair of both mismatches	
III	$\dfrac{ml\text{-}1 \quad +}{+ \quad ml\text{-}2} \longrightarrow \dfrac{+ \quad +}{+ \quad +}$	$ml\text{-}1 \longrightarrow +;\ ml\text{-}2 \longrightarrow +$
IV	$\dfrac{ml\text{-}1 \quad +}{+ \quad ml\text{-}2} \longrightarrow \dfrac{ml\text{-}1 \ ml\text{-}2}{ml\text{-}1 \ ml\text{-}2}$	$+ \longrightarrow ml\text{-}1;\ + \longrightarrow ml\text{-}2$
	No repair	
IX	$\dfrac{ml\text{-}1 \quad +}{+ \quad ml\text{-}2} \longrightarrow \dfrac{ml\text{-}1 \quad +}{+ \quad ml\text{-}2}$	

[a]Classes I and II are not shown since they may result from either asymmetric or symmetric heteroduplexes. When they are of symmetric origin, classes I and II result from repair of both mismatches of the unselected heteroduplex.

result either from a symmetric or an asymmetric heteroduplex. For a symmetric heteroduplex at the two-strand stage, correction of one of the two mismatches yields a 1:3 conversion sector; for an asymmetric heteroduplex, failure to correct the resultant mismatch yields the 1:3 conversion sector. A consequence of the two-strand model, therefore, is that a 1:3 conversion sector requires failure to correct a mismatch.

The genotypes of sectored prototrophic colonies (classes III–IX; Tables 2 and 3) also provide information with regard to the relative frequencies with which an unselected mismatch ($+/m$) is repaired to m/m versus $+/+$ (Table 4). In wild-type strains, the sample sizes are small but show no tendency toward extreme dissymmetry in the directionality of repair.

The mismatch repair parameters obtained from a similar analysis of the genotypes of sectored prototrophic colonies in *rem1-1/rem1-1* hybrids are strikingly different from those in the wild-type. These differences arise from the fact that at least 50% of *LEU1* and *TRP5* colonies in *rem1-1/rem1-1* diploids are homozygous + +/+ + on both sides of the sectored colony. This genotype is found in only about 5% of sectored colonies in wild-type strains. Table 3 shows that, at *leu1*, 97 of 112 (87%) unselected mismatches were repaired in a *rem1-1* homozygous diploid; of 48 *leu1-c/+* mismatches that were repaired, all were repaired to +/+, whereas of 49 *leu1-12/+* mismatches repaired, 48 were repaired to +/+ and 1 was repaired to *leu1-12/leu1-12*. At *trp5*, 91 of 96 unselected mismatches (95%) were repaired. Of 48 *trp5-c/+* mismatches repaired, 45 were repaired to +/+, and 3 to *trp5-c/trp5-c;* 43 of 48 *trp5-d/+* mismatches were repaired, 40 to +/+ and 3 to *trp5-d/trp5-d.*

These data, taken at face value, suggest that mismatch repair in *rem1-1/rem1-1* diploids occurs more efficiently than in the wild-type and that for four different mismatches (*leu1-c/+*, *leu1-12/+*, *trp5-c/+*, and *trp5-d/+*) there is an extreme dissymmetry of repair, with preferential repair to wild-type in each of the four cases. It is unclear whether the apparent difference in repair patterns in *rem1-1* strains actually reflects a molecular similarity of the four mismatches in question or whether the apparent repair phenotypes of *rem1-1* reflect some difference from the wild-type in the mechanism of heteroduplex formation or resolution caused by this mutation. (For example, branch migration in one direction followed by branch migration in the opposite direction, in the presence of efficient mismatch repair, would result in an enrichment of + +/+ + : + +/+ + sectors at the expense of other sector genotypes in *rem1-1*.) Resolution of this question will require analysis of recombination between mutant alleles other than those used in these studies and analysis of recombinants unselected for prototrophy.

Association of Gene Conversion and Crossing-over

Spontaneous intragenic mitotic recombination resulting in prototrophic heteroallelic recombinants occurs in nonrandom association with reciprocal exchange in the vicinity of the site of gene conversion (Roman 1957; Roman and Jacob 1959). Extraction of precise association values from the literature on spontaneous recombination is difficult since such values can only be calculated from data in which the prototrophs analyzed are of independent origin or in instances where the mitotic rates have been measured for the relevant genotypic classes of convertants and nonconvertants. The association of mitotic gene conversion and exchange can be defined as follows: association = $([x/z] - y) \times 100$, where x is the rate of conversion to prototrophy accompanied by exchange, y is the rate of flanking marker exchange in the general population, and z is the total rate of conversion to prototrophy. Since spontaneous rates of mitotic recombination are low and on the order of 10^{-4}, it is possible to obtain a fairly accurate measure of association even if the y term is unknown. The association values observed in yeast range from about 10% to 55% (Fogel and Hurst 1963; Kakar 1963; Hurst and Fogel 1964; Zimmermann and Schwaier 1967; Esposito 1978; Golin 1979). These values are similar to the distribution of association values of approximately 17–64% reported in the case of spontaneous meiotic gene conversion and reciprocal exchange of flanking markers (Hurst et al. 1972). Given this concordance, it is useful to note that the mitotic association observed reflects replicative resolution of Holliday structures on the two-strand model of Esposito (1978) and that the meiotic association reflects resolution by cleavage of four-strand-stage Holliday structures according to the model of Holliday (1964) and Meselson and Radding (1975).

Heteroallelic recombination induced by UV light (Roman and Jacob 1959; Fogel and Hurst 1963; Hurst and Fogel 1964) and chemical recombinagens (Zimmerman and Schwaier 1967) also exhibits a variable association with adjacent reciprocal exchange.

Exceptions Explained by the Two-strand Model

Several investigators, as noted previously, have recorded instances of apparent coincident conversion of widely spaced chromosomal markers as well as homozygosity for the same allele in both portions of sectored colonies in studies that would have revealed their origin by conventional mechanisms, including deletion and chromosomal loss (Nakai and Mortimer 1969; Wildenberg 1970; Johnston 1971, 1972). Coincident conversion of widely spaced chromosomal markers may reflect branch migration of a Holliday structure driven by chromosomal replication in the S phase following its establishment. Markers ahead of replication forks may thereby be transiently brought into heteroduplex formation affording an opportunity for gene conversion (Esposito 1978). This is unlikely to be the sole basis of coincident conversion since coincident conversion frequencies are stimulated by arrest of cells at the two-strand stage as noted earlier. Homozygosis of sectored colonies for the same allele in both portions of a sectored colony can also be explained since this is a direct outcome of gene conversion at the two-strand stage.

MECHANISTIC INTERPRETATIONS OF INDUCTION DATA

The detailed phenomenology of induced mitotic recombination provides additional insights into the mechanisms of mitotic recombination. Mechanistic interpretations of induced mitotic recombination can be based on several types of genetic evidence, e.g., dose-response curves, differential responses of conversion and reciprocal exchange to specific recombinagens, and variations in inducibility of recombination during the mitotic cell cycle. In this section, we summarize the relevant evidence from each of these experimental approaches and then attempt to decide what mechanistic conclusions are justified by the available evidence.

Dose-response Curves

Mitotic recombination, like mutation (Haynes and Kunz, this volume), is induced by different agents with characteristic dose-response kinetics. Analysis of the apparent multiplicity of "hits" necessary for a specific agent to induce a specific genetic endpoint has been the basis of numerous molecular models for the induction of mutation and mitotic exchange.

A distinction that was noted soon after the first reports of mitotic recombination in yeast was the difference between dose-response curves for X-ray- and UV-induced gene conversion. X-rays in sublethal doses induced heteroallelic conversion with linear kinetics (Manney and Mortimer 1964), whereas UV-induced conversion showed approximately dose-squared kinetics (Roman and Jacob 1957). The linear response of gene conversion to X-irradiation has been confirmed in a number of studies (for a recent example, see Moore and Sherman 1975), as has the nonlinearity of the UV response (although Moore and Sherman [1975] report that some combinations of *cyc1* alleles show UV kinetics as linear as those obtained with X-rays). The linear response of heteroallelic recombination to X-rays led Manney and Mortimer (1964) to suggest that a single X-ray lesion anywhere in the region separating two alleles could lead to gene conversion. The higher-order kinetics of the UV-induced event would thus indicate that more than one lesion is required.

This simple view is challenged by the recent work of Ito and Kobayashi (1975), who have shown that the apparent number of hits necessary to produce UV-induced heteroallelic recombination is a function of the cell-cycle stage of the irradiated cells. Their data can best be fitted to equations of the form $f_i = (a_i t)^{\alpha_i}$, where a_i and α_i are cell-cycle stage-specific parameters, f_i is the recombination rate, and t is the UV dose. α_i ranges from 1 (single-hit kinetics) for cells with large buds (presumably G_2 cells) to 2 (two-hit kinetics) for unbudded cells. a_i, a stage-specific "scaling factor," varies approximately 30-fold during the cell cycle in their experiments, with a maximum in G_1 cells and a minimum in G_2 cells. We suspect that many recombinagenic agents, if examined carefully, would show a similar cell-cycle dependence of recombination induction kinetics. This represents a potential source of nonreproducibility of mitotic recombination results but, more importantly, it underscores the inadequacy of dose-response curves obtained by exposure of asynchronous cells to describe the recombinational responses of mitotic cells.

Gene Conversion versus Reciprocal Exchange

A number of studies have also been devoted to the examination of the extent to which gene conversion and reciprocal exchange respond in a correlated fashion to agents that induce mitotic recombination. The intent of these studies has been to determine whether the two phenomena are the manifestations of a single molecular process or are mechanistically distinct.

On the most basic level, agents that induce mitotic gene conversion almost invariably induce mitotic crossing-over. An examination of numerous chemical mutagens[1] by Fahrig (1979) revealed that, with only one exception, each of the agents induces both conversion and exchange in yeast. The exception was acridine orange, which induces conversion but has not been demonstrated to induce crossing-over.

Although most recombinagenic agents induce both gene conversion and crossing-over, the quantitative responses of the two processes may be partially uncoupled from each other. When compared at identical survival frequencies, four inducing agents (UV, nitrous acid, EMS, and gamma rays) produce identical levels of mitotic crossing-over over a wide range of survival values (Davies et al. 1975); for a given survival value, however, frequencies of heteroallelic gene conversion induced by these agents vary over a tenfold range. In the above study, the efficiency of induction of mitotic conversion was greatest for UV, less for nitrous acid and gamma rays, and lowest for EMS.

Differences in the relative rates of induction of conversion and crossing-over by specific agents have also been detected by Roman (1973), who observed that most *ade6* sectoring induced by UV is due to crossing-over, whereas that induced by EMS is mostly due to gene conversion. Although this result seems to be in conflict with that cited above, Parry (1969) also reported in an earlier paper, using a different yeast strain, that EMS induced gene conversion much more efficiently than did UV.

Parry and Cox (1965) have further demonstrated partial dissociation of conversion from reciprocal exchange by the effects of photoreactivation after UV irradiation. They found that, whereas gene conversion and survival were partially photoreactivable, no effect of photoreactivating light on intergenic exchange could be demonstrated.

The data summarized above are compatible with conversion and reciprocal exchange either as two separate processes (induced by the same lesions) or two outcomes of a single process. The evidence favoring the second hypothesis is as follows: (1) Spontaneous gene conversion is positively associated with crossing-over in adjacent intervals. (2) Mitotic gene conversion and exchange are maximally inducible during the same portions of the cell cycle. (3) Mitotic gene conversion and exchange respond similarly to a number of *rec* mutations that either elevate or depress recombination rates, as will be discussed later. These observations are most easily reconciled with the hypothesis that gene conversion arises from the establishment of hetero-

---

[1](EMS) Ethylmethanesulfonate, (MMS) methylmethanesulfonate, (MNNG) *N*-methyl-*N'*-nitro-*N*-nitrosoguanidine, (TEM) triethylenemelamine, (4-NQO) 4-nitroquinoline 1-oxide, (2,4-DNFB) 1-fluoro-2,4-dinitrobenzene, and (AO) acridine orange.

duplex DNA and the action of mismatch repair functions on the same molecular structure that generates reciprocal recombinants.

A hypothesis of conversion and crossing-over as outcomes of a single molecular process must account for those instances where the two events respond differently to recombinagenic agents. Our best guess as to the cause of differences in the relative efficiencies of agents in inducing conversion versus crossing-over is an involvement of mismatch repair systems. Mutagens that induce genetic damage that acts as a substrate for mismatch repair systems might show low conversion frequencies because mismatches induced directly by the mutagen compete for repair with mismatches generated by recombination. Alternatively, mismatch repair might be induced separately from recombination, and different agents might induce mismatch repair with different efficiencies. No evidence exists in support of either of these possibilities in yeast.

Induction of Mitotic Recombination in Synchronized Cells

Two major conclusions have emerged from studies of induced mitotic recombination in synchronized yeast cells. Conversion and crossing-over are each induced maximally at the same point in the cell cycle (Esposito 1968; Davies et al. 1978), and the point at which maximum induction occurs is dependent upon the agent used to induce recombination (Davies et al. 1978).

Esposito (1968) found that, for nonlethal doses of both UV and X-rays administered to synchronized diploids, frequencies of conversion and reciprocal exchange rise shortly before the initiation of DNA synthesis, reach a peak, and then decline as replication begins. In her study, the difference between minimal and maximal recombination frequencies was approximately fivefold. Correlated with the cyclic variation in recombination inducibility was a cyclic fluctuation in the sensitivity of the cells to lethal effects of irradiation (at higher doses than those used for the recombination studies). Late G_1 cells, which were maximally inducible for recombination, also showed maximal sensitivity to UV killing, whereas G_2 cells showed minimal inducibility and maximal radioresistance.

The more recent results of Davies et al. (1978) confirm those described above in the essential details. They used synchronous cells obtained by zonal rotor centrifugation (rather than by feeding and starving), and they found even larger differences in UV inducibility of mitotic recombination between prereplicative cells and cells in S or G_2 phases. Nitrous acid, by contrast, showed a very pronounced maximum of induced recombination (again, both conversion and crossing-over) in S-phase cells with a corresponding minimum in survival.

The cyclic fluctuations in radiation sensitivity observed by the authors mentioned above and by Brunborg and Williamson (1978) suggest the operation of a cell-cycle stage-specific mode of repair, which is dependent upon the presence of replicated chromosomes. We propose that this type of repair occurs by sister-chromatid exchange, i.e., at the four-strand stage, and that the apparent fluctuation in recombination inducibility during the cell cycle reflects a cyclic variation in the partitioning of recombination between two-strand homologous chromosome exchange and sister-chromatid exchange at the four-strand stage. Induced recombination in S and G_2 phases on this model involves primarily sister chromatids. Survival data suggest that if recombination makes a contribution to DNA repair in yeast (as Resnick's [1975] results suggest for X-ray-induced mitotic gene conversion), this recombinational repair in S and G_2 phases is more efficient than recombinational repair in G_1 diploids involving homologous chromosomes exclusively. The recent availability of genetic systems allowing measurement of sister chromatid exchange (Szostak and Wu 1980) makes this proposal amenable to direct experimental test.

Studies of induced recombination in synchronized cells suffer from the fact that the actual recombination events detected need not have taken place during the cell-cycle interval when the cells were treated. This weakness has been cleverly avoided by Fabre (1978), who used diploid cells heteroallelic for thermosensitive *cdc* mutations with arrest points in G_1 (*cdc4* or *cdc25*) to show that UV- and gamma-ray-induced gene conversion to Cdc$^+$ (and therefore to temperature-resistance) can take place in G_1 phase. Since the treated cells were unable to proceed in the cell cycle past their arrest points in the absence of heteroallelic recombination, the actual events induced must have been consummated during G_1, i.e., at the two-strand stage.

Induction of Mitotic Recombination by Treatment of One Haploid Parent

Campbell (1973) demonstrated that mating of X-irradiated haploid yeast cells with unirradiated haploids results in induced mitotic recombination within the diploids formed. As in the case of X-irradiation of diploid cells, recombination frequency increased as a linear function of dose; however, segregation of recessive markers resulted primarily from gene conversion rather than from crossing-over, which is the major cause of segregation after irradiation of diploid cells.

The fact that irradiation of haploid cells before mating is recombinagenic in the diploid could be either a direct effect of lesions induced in the DNA of the irradiated parent, or an indirect effect of the induction of recombinational repair in the irradiated parent, or a combination of the two. An elegant series of experiments by Fabre and Roman (1977) indicate that an inducible system is responsible, at least in part. Their experimental system involved an irradiated haploid that is doubly mutant in the *ade6* gene (*ade6-21, 45*) and a mating-capable (*MATa/MATa* or *MATα/MATα*) diploid heteroallelic at *ade6* (*ade6-21/ade6-45*). Irradiation of the haploid parent followed by mating with the heteroallelic diploid resulted in a substantial induction of mitotic gene conversion, despite the fact that the irradiated double-mutant chromosome could not have been involved in the conversion event. This inducibility in *trans* was demonstrated to be independent of nuclear fusion in crosses where the irradiated haploid parent carried the *kar1-1* mutation. Fabre and Roman (1977) ascribed these results to an inducible recombination system involving a diffusible and non-nucleus-limited gene product.

Spontaneous versus Induced Rates of Intragenic and Intergenic Recombination

Spontaneous mitotic recombination, like gene mutation, occurs during the cell divisions that establish mitotic cultures. The statistical methods developed for estimation of mutation rates have been employed to measure rates of mitotic intragenic and intergenic recombination. These techniques include the method of Luria and Delbruck (1943), by which the rate of recombination may be calculated from the number of cultures in a series that contained no recombinants; the method of Lea and Coulson (1949), by which the rate of recombination may be calculated from the median number of recombinants per culture; and the method of Drake (1970), by which the rate of recombination may be calculated for each culture of a series and the arithmetic and geometric mean rates determined (Malone et al. 1980).

The spontaneous rates of intragenic recombination of a series of heteroallelic diploids involving four heteroalleles of the *cyc1* gene were determined by Moore and Sherman (1975). The order of the four mutations and the DNA sequences surrounding them were previously inferred from amino acid sequence analysis of revertants of the four heteroalleles. Moore and Sherman (1975) compared the spontaneous rates of intragenic recombination with the rates of X-ray-, UV-, and near-UV-induced mitotic intragenic recombination and spontaneous meiotic rates as well. None of the five sets of recombination data correctly ordered the mutant sites and thus none routinely reflects the true physical distances involved. It was further shown that the relationships of the recombination rates determined for UV light, near-UV light, and X-rays were very similar but different from spontaneous mitotic and meiotic values. The latter also differed inter se.

Malone et al. (1980) measured the spontaneous rates of intergenic recombination for three intervals of chromosome VII employing the median method of Lea and Coulson (1949) and the method of Drake (1970). The results of this study demonstrated that spontaneous mitotic intergenic exchange occurs relatively more frequently in centromere-proximal intervals than in distal intervals when compared with the meiotic map. In earlier experiments, Nakai and Mortimer (1969) and Campbell (1973) studied X-ray-induced mitotic sectoring of the same genetic markers and observed a closer correspondence of the relative rates with the meiotic map.

These results taken together suggest that different gene products are involved in the various pathways leading to recombination and that they may respond differently to the same DNA sequence with regard to the probability of recombination.

SISTER-CHROMATID EXCHANGE

The genetic systems that we have described thus far permit detection only of recombination events involving homologous chromosomes or nonsister chromatids. Crossing-over or conversion involving sister chromatids is, in general, genetically not detectable because the sister chromatids have identical DNA sequences. The only exceptions to this generalization are those cases where unequal sister-chromatid exchange produces deletions and duplications of chromosomal regions.

Detection of such unequal sister-chromatid exchanges in yeast has recently become feasible through the construction of two types of yeast strains: strains having yeast genes, such as *LEU2*, inserted into the tandemly repeated rDNA gene cluster (Szostak and Wu 1979) and strains having tandem duplications of normally nonrepeated genes, produced by transformation with cloned yeast DNA (Hinnen et al. 1978). The mechanism by which unequal sister-chromatid crossing-over can generate new genotypes in each of these two systems is diagramed in Figure 5. Two other mechanisms not shown in Figure 5 may also account for the generation of new genotypes from these systems: intrachromatid exchange resulting in loss of DNA, perhaps as a circular molecule, from the chromosome and gene conversion.

Szostak and Wu (1980), using a haploid yeast strain with *LEU2* inserted into the tandem array of rDNA repeats, found that Leu⁻ segregants were produced at a rate of approximately 5×10^{-4}/generation. They were able to distinguish sister-chromatid exchange from either intrastrand exchange or gene conversion by the analysis of sectored Leu⁺/Leu⁻ colonies, which presumably arose from recombination events near the time of plating. Unequal sister-chromatid crossing-over will produce one chromatid with 2 copies of *LEU2* and one with 0 copies. Intrastrand exchange or gene conversion yields one chromatid with 1 copy of *LEU2* and one with 0 copies. Quantitative filter hybridization and restriction enzyme analysis showed that approximately half of their sectored colonies (4/10 UV-induced sectors and 6/7 spontaneous sectors) contained the *LEU2* duplication expected from unequal sister-chromatid crossing-over (Fig. 5). The distance between the two *LEU2* insertions in the duplication sides of their sectors varied from 1 to 8 rDNA repeat units. This range reflects the degree of mispairing preceding crossing-over, although the failure to detect greater mispairing may be due to the inviability of larger deletions.

On the basis of the average displacement observed during spontaneous

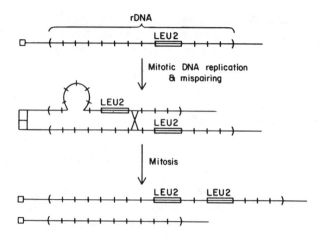

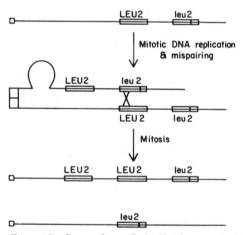

Figure 5 Generation of duplications and deletions by unequal sister-chromatid exchange. (*Top*) The parental haploid strain carries a single copy of the *LEU2* gene inserted in the tandem array of rDNA repeats. Mispairing of sister chromatids followed by crossing-over results in one chromatid with 2 copies of the *LEU2* insert and 1 with none. (*Bottom*) The parental haploid strain contains a copy of the *LEU2* gene inserted in tandem wilth a *leu2* mutant allele. Unequal sister-chromatid exchange involving the *LEU2* and *leu2* alleles can yield one chromatid with 2 copies of *LEU2* and 1 copy of *leu2* and one chromatid bearing a single *leu2* allele.

unequal crossing-over, Szostak and Wu (1980) estimated that the rate of unequal crossing-over between sister chromatids within the entire rDNA cluster is on the order of 10^{-2} events per generation. Assessment of the rate of equal sister-chromatid exchange remains a difficult problem; analysis of a large enough number of sectors to permit extrapolation to a displacement of 0 repeats may provide an indication of this rate.

GENETIC CONTROL OF MITOTIC RECOMBINATION

Gene mutations affecting mitotic recombination, which have been recognized by a variety of techniques (Catcheside 1974; Baker et al. 1976), provide additional insights into the relationship between mitotic and meiotic recombination. In addition to *rec* mutants isolated directly by their hypo-recombination or hyper-recombination phenotypes, several have been detected among *cdc* (cell-division cycle) mutants, *rad* (radiation-sensitive) mutants, and other mutations affecting repair of damaged DNA. Some have extensive pleiotropic effects including recombination defects in meiosis and diminished ascospore formation and ascospore survival. Although a full characterization of the available variants with respect to mitotic and meiotic intragenic and intergenic recombination has not been completed, their phenotypes provide unambiguous evidence for the existence of mitosis-specific and meiosis-specific *rec* functions and *rec* functions of a more generalized nature.

Control of Mitotic Recombination by the Mating-type Alleles

The genotype at the mating-type locus of diploid cells has been demonstrated to be involved in the expression of mating ability, sporulation, mitotic recombination proficiency, and resistance to ionizing radiation (see Baker et al. 1976; Crandall et al. 1977). The effect of the mating-type alleles on mitotic recombination was observed by Friis and Roman (1968), who determined that $MATa/MATa$ and $MAT\alpha/MAT\alpha$ diploids yield three to six times fewer UV-induced prototrophic intragenic recombinants than closely related $MATa/MAT\alpha$ hybrids. J. Wagstaff and M. S. Esposito (unpubl.) subsequently measured the spontaneous rates of intragenic recombination at four heteroallelic loci and intergenic recombination on chromosome VII. They also found the spontaneous rates of recombination to be three to five times lower in $MATa/MATa$ and $MAT\alpha/MAT\alpha$ hybrids than in a congenic $MATa/MAT\alpha$ hybrid. Analysis of the genotypes of Trp$^+$ red-white sectored colonies of the $MATa/MATa$ hybrid, in which conversion at *trp5* was accompanied by a reciprocal exchange of distal markers, indicated that the recombinational events observed involved unreplicated homologs and thus occurred at the two-strand stage as in $MATa/MAT\alpha$ hybrids (Table 3).

Thus, it appears that $MATa/MAT\alpha$ diploids, which are typically non-mating and capable of meiosis, engage in mitotic recombination more frequently than diploids homoallelic for the mating-type alleles, which are typically incapable of meiosis but capable of mating. The residual recombination that occurs in diploids homoallelic at the mating-type locus may reflect an alternate minor pathway of mitotic recombination and/or leakiness of the control exerted by the mating-type genotype.

Mitotic Hypo-recombination Mutants

Since heteroallelic recombination results primarily from the establishment of heteroduplex regions and mismatch repair, and intergenic recombination reflects the manner of resolution of Holliday structures, one may anticipate that some *rec* mutants might be defective in intragenic recombination but not intergenic recombination and vice versa. *rec1* and *rec3* (Rodarte-Ramon and Mortimer 1972) behave as mutations of the former type. Both are defective in spontaneous and X-ray-induced heteroallelic recombination but are apparently normal with respect to X-ray-induced intergenic recombination. The *rec1* mutant does not affect sporulation, whereas the *rec3* mutant depresses it severely, suggesting a role in meiosis of the *REC3*-gene product.

Several mutants depress both spontaneous and induced intragenic and intergenic mitotic recombination and exhibit meiotic defects as well. These include *rec2* (allelic to *rad52*), *rad51* (allelic to *mut5*), and *mms1* (Reno 1969; Rodarte-Ramon and Mortimer 1972; Game and Mortimer 1974; Sanfilippo 1977; Morrison and Hastings 1979). Among these, *rad51* and *rad52* have been demonstrated to be defective in meiotic recombination (Morrison and Hastings 1979; Game et al. 1980; Prakash et al. 1980). The *rad52* mutant also blocks homothallic interconversion of mating-type alleles, a further manifestation of its role in recombination (Malone and Esposito 1980).

Apparently meiosis-specific *rec* mutants have also been observed. These include the *con1*, *con2*, and *con3* mutants isolated by Roth and Fogel (1971), Fogel and Roth (1974), and Roth (1976). These mutants exhibit UV-induced mitotic heteroallelic recombination leading to prototrophy but are defective in meiotic heteroallelic recombination resulting in prototrophy.

The *rec4* mutant (Rodarte-Ramon and Mortimer 1972), whose mitotic hypo-recombination effect is limited to heteroallelic recombination in the *arg4* region, also reduces the frequency of meiotic single-site conversions that lead to prototrophy by a relative increase in the frequency of meiotic coconversion of linked sites (Sanfilippo 1977). Coconversion does not result in recombination of alleles. The *REC4*-gene product thus affects the manner in which heteroduplex regions are repaired or formed at *arg4* in both mitosis and meiosis.

Mitotic Hyper-recombination Mutants

Several *rad* mutants defective in repair of UV damage, including both excision-repair-defective and error-prone repair-defective mutants (see Baker et al. 1976; Haynes and Kunz, this volume), exhibit a hyper-recombination phenotype following UV irradiation. Kern and Zimmermann (1978) demonstrated that *rad3* (excision repair-defective) and *rad6* (error-prone repair-defective) exhibit a hyper-recombination phenotype for both spontaneous intragenic recombination (i.e., gene conversion) and spontaneous intergenic recombination. These results and those of Boram and Roman (1976) involving *rad18* provide evidence that some *rad* genes specify functions that repair spontaneously occurring lesions that act as substrates for the onset of mitotic recombination. The function specified by *RAD6* is also involved in meiosis since *rad6* mutants fail to sporulate and are defective in meiototic intragenic recombination (Game et al. 1980). Two temperature-sensitive *cdc* mutants, *cdc9* (DNA ligase defective) and *cdc21* (thymidylate synthetase defective) have been demonstrated to exhibit spontaneous mitotic hyper-recombination activity following brief exposure to high temperature or growth at a semipermissive temperature (Game et al. 1979; J. Game, pers. comm.). Given their enzymatic defects, strains incorporating these mutations are likely to accumulate recombinagenic DNA damage at restrictive or semipermissive temperatures (Game et al. 1979).

Dominant hyper-recombination mutants affecting spontaneous gene conversion, referred to as *MIC* mutants, have been described by Maloney (1977). Some affect both spontaneous and UV-induced intragenic recombination, but none apparently affects meiosis. A semidominant mutation, *rem1-1*, isolated as a spontaneous mutator by Golin and Esposito (1977), stimulates both spontaneous intragenic and intergenic recombination (Golin 1979).

Conversion-associated exchange in *rem1-1* diploids was studied at the *leu1* and *trp5* loci on chromosome VII, as mentioned previously, to determine whether sectored colonies have the genotypes expected from two-strand or four-strand exchange (Golin 1979). As in wild-type strains, the sectored colonies exhibited genotypes reflecting two-strand-stage recombination (Table 3).

PERSPECTIVES AND ALTERNATIVE HYPOTHESES

The results described above demonstrate that mitotic and meiotic recombination exhibit certain basic mechanistic features in common but differ from one another in several respects. Both mitotic and meiotic recombination derive from conversion events as well as reciprocal exchanges. Unlike meiotic conversions, mitotic conversions frequently reflect the establishment of symmetric heteroduplexes. Mitotic heteroduplexes also appear longer than meiotic heteroduplexes since coincident conversion of widely spaced chromosomal markers is relatively more frequent in mitosis than in meiosis. Both mitotic and meiotic reciprocal exchange can be explained by the formation and resolution of Holliday structures. In mitosis, in which chromosomal recombination appears to occur prereplicatively, reciprocal exchange is detected when Holliday structures are not cleaved endonucleolytically but are instead resolved by DNA replication (Esposito 1978). In meiosis, in which exchange of homologs appears to occur postreplicatively, recombinant chromosomes can be generated by endonucleolytic cleavage of Holliday structures in the recombinant configuration (Holliday 1964).

The present evidence for mitotic recombination at the two-strand stage consists of the interpretation of the genotypes of sectored conversion-associated colonies and pedigrees and the induction experiments described above. Although the data for two-strand exchange are compelling, one can formulate alternative four-strand mitotic models by invoking multiple conversion and exchange events involving both nonsister and sister chromatids. Additional tests of the validity of the two-strand model are therefore desirable. The two-strand model makes two predictions that can be tested. Mitotic double exchanges should include only two-strand and four-strand doubles since three-strand doubles cannot result from the interaction of two DNA double helices. In addition, the model predicts that conversion-associated sectored colonies obtained by exposure of G_2 cells to recombinagens should have the genotype (class IX; Table 2) diagnostic of four-strand-stage events in the instance of recombination events consummated before cell division.

The two-strand model also predicts the phenotype of mutants specifically defective in both mitotic and meiotic endonucleolytic cleavage of Holliday structures. They would be expected to be hyper-recombinant with respect to detectable mitotic intergenic recombination and hypo-recombinant in meiotic intergenic recombination. Such mutants are likely to sporulate poorly and/or form inviable ascospores owing to the ensuing aneuploidy of the meiotic products.

The availability of genetic systems allowing direct detection of sister-chromatid exchange provides an opportunity to inquire whether sister-chromatid exchange and exchange between homologs are dependent upon the same gene functions. Mutations that selectively block one or the other process may prove useful in describing how exchange is partitioned between unreplicated homologs, sister chromatids, and nonsister chromatids over the yeast life cycle and the roles played by each type of recombination in DNA repair and generation of novel genotypes. The recent demonstration that *rad52-1* strains are capable of mitotic sister-chromatid exchange (T. Petes and T. Zamb, pers. comm.), though they are defective in mitotic and meiotic recombination between homologous chromosomes, is a harbinger of future insights.

Our interest in the mechanisms of mitotic recombination derives in part from the implication of two-strand-stage gene conversion for mutagenesis in somatic cells exposed to recombinagenic mutagen-carcinogens. Coupled mutation and gene conversion in G_1 phase may result in genotypic changes from homozygous wild-type (A/A) to heterozygosity for an induced recessive mutation (A/a) and, finally, homozygosity (a/a) without an intervening cell division. The operation of such a pathway, testable in yeast, would provide a simple explanation for the expression and recovery of recessive mutations in diploid cells and the origin of recessive oncogenic genotypes in particular.

ACKNOWLEDGMENTS

We thank J. Golin, E. Jones, S. Klapholz, R. Malone, H. Roman, and J. Strathern for their valuable comments. Research and preparation of this manuscript was supported by grants from the National Institutes of Health (GM-23277 and GM-29002) and Lawrence Berkeley Laboratory, Biology and Medicine Division, Director's Program Fund (3669-33). J. E. W. was supported by a National Science Foundation predoctoral fellowship and a U. S. Public Health Service training grant (GM-07197).

REFERENCES

Baker, B.S., A.T.C. Carpenter, M.S. Esposito, R.E. Esposito, and L. Sandler. 1976. The genetic control of meiosis. *Annu. Rev. Genet.* **10**: 53.

Boram, W.R. and H. Roman. 1976. Recombination in *Saccharomyces cerevisiae*: A DNA repair mutation associated with elevated mitotic gene conversion. *Proc. Natl. Acad. Sci.* **73**: 2828.

Brunborg, G. and D.H. Williamson. 1978. The relevance of the nuclear division cycle to radiosensitivity in yeast. *Mol. Gen. Genet.* **162**: 277.

Campbell, D.A. 1973. The induction of mitotic gene conversion by X-irradiation of haploid *Saccharomyces cerevisiae*. *Genetics* **74**: 243.

Catcheside, D.G. 1974. Fungal genetics. *Annu. Rev. Genet.* **8**: 279.

Crandall, M., R. Egel, and V. MacKay. 1977. Physiology of mating in three yeasts. *Adv. Microb. Physiol.* **20**: 307.

Davies, P.J., W.E. Evans, and J.M. Parry. 1975. Mitotic recombination induced by chemical and physical agents in the yeast *Saccharomyces cerevisiae*. *Mutat. Res.* **29**: 301.

Davies, P.J., R.S. Tippins, and J.M. Parry. 1978. Cell-cycle variation in the inducibility of lethality and mitotic recombination after treatment with UV and nitrous acid in the yeast *Saccharomyces cerevisiae*. *Mutat. Res.* **51**: 327.

Di Caprio, L. and P.J. Hastings. 1976. Gene conversion and intragenic recombination at the *SUP6* locus and the surrounding region in *Saccharomyces cerevisiae*. *Genetics* **84**: 697.

Drake, J. 1970. *The molecular basis of mutation.* Holden-Day, San Francisco.

Esposito, M.S. 1978. Evidence that spontaneous mitotic recombination occurs at the two-strand stage. *Proc. Natl. Acad. Sci.* **75**: 4436.

Esposito, M.S. and R.E. Esposito. 1977. Gene conversion, paramutation and controlling elements: A treasure of exceptions. *Cell Biol.* **1**: 59.

———. 1978. Aspects of the genetic control of meiosis and ascospore development inferred from the study of *spo* (sporulation-deficient) mutants of *Saccharomyces cerevisiae*. *Biol. Cell.* **33**: 93.

Esposito, R.E. 1968. Genetic recombination in synchronized cultures of *Saccharomyces cerevisiae*. *Genetics* **59**: 191.

Fabre, F. 1978. Induced intragenic recombination in yeast can occur during the G1 mitotic phase. *Nature* **272**: 795.

Fabre, F. and H. Roman. 1977. Genetic evidence for inducibility of recombination competence in yeast. *Proc. Natl. Acad. Sci.* **74**: 1667.

Fahrig, R. 1979. Evidence that induction and suppression of mutations and recombinations by chemical mutagens in *S. cerevisiae* during mitosis are jointly correlated. *Mol. Gen. Genet.* **168**: 125.

Fogel, S. and D.D. Hurst. 1963. Coincidence relations between gene conversion and mitotic recombination in *Saccharomyces*. *Genetics* **48**: 321.

Fogel, S. and R.K. Mortimer. 1971. Recombination in yeast. *Annu. Rev. Genet.* **5**: 219.

Fogel, S. and R. Roth. 1974. Mutations affecting meiotic gene conversion in yeast. *Mol. Gen. Genet.* **130**: 189.

Fogel, S., R. Mortimer, K. Lusnak, and F. Tavares. 1979. Meiotic gene conversion: A signal of the basic recombination event in yeast. *Cold Spring Harbor Symp. Quant. Biol.* **43**: 1325.

Friis, J. and H. Roman. 1968. The effect of the mating type alleles on intragenic recombination in yeast. *Genetics* **59**: 33.

Game, J.C. and R.K. Mortimer. 1974. A genetic study of X-ray sensitive mutants in yeast. *Mutat. Res.* **24**: 281.

Game, J.C., L.H. Johnston, and R.C. von Borstel. 1979. Enhanced mitotic recombination in a ligase-defective mutant of the yeast *Saccharomyces cerevisiae*. *Proc. Natl. Acad. Sci.* **76**: 4589.

Game, J., T. Zamb, R. Braun, M. Resnick, and R. Roth. 1980. The role of radiation (*rad*) genes in meiotic recombination in yeast. *Genetics* **94**: 51.

Golin, J.E. 1979. "The properties of spontaneous mitotic recombination in *Saccharomyces cerevisiae*." Ph.D. thesis, University of Chicago, Illinois.

Golin, J.E. and M.S. Esposito. 1977. Evidence for joint genic control of spontaneous mutation and genetic recombination during mitosis in *Saccharomyces*. *Mol. Gen. Genet.* **135**: 91.

Hinnen, A., J.B. Hicks, and G.R. Fink. 1978. Transformation in yeast. *Proc. Natl. Acad. Sci.* **75**: 1929.

Holliday, R. 1964. A mechanism for gene conversion in fungi. *Genet. Res.* **5**: 282.

Hurst, D.D. and S. Fogel. 1964. Mitotic recombination and heteroallelic repair in *Saccharomyces cerevisiae*. *Genetics* **50**: 435.

Hurst, D.D., S. Fogel, and R.K. Mortimer. 1972. Conversion-associated recombination in yeast. *Proc. Natl. Acad. Sci.* **69**: 101.

Ito, T. and K. Kobayashi. 1975. Studies on the induction of mitotic gene conversion by ultraviolet irradiation. *Mutat. Res.* **30**: 33.

James, A.P. and B. Lee-Whiting. 1955. Radiation-induced genetic segregations in vegetative cells of yeast. *Genetics* **40**: 826.

Johnston, J.R. 1971. Genetic analysis of spontaneous half-sectored colonies of *Saccharomyces cerevisiae*. *Genet. Res.* **18**: 179.

―――. 1972. Genetic analysis of X-ray induced half-sectored colonies of *Saccharomyces cerevisiae*. *Radiat. Res.* **49**: 558.

Kakar, S.N. 1963. Allelic recombination and its relation to recombination of outside markers in yeast. *Genetics* **48**: 957.

Kern, R. and F.K. Zimmermann. 1978. The influence of defects in excision and error prone repair on spontaneous and induced mitotic recombination and mutation in *Saccharomyces cerevisiae*. *Mol. Gen. Genet.* **161**: 81.

Lea, D.E. and C.A. Coulson. 1949. The distribution of numbers of mutants in bacterial populations. *J. Genet.* **49**: 264.

Luria, S.E. and M. Delbruck. 1943. Mutations of bacteria from virus sensitivity to virus resistance. *Genetics* **28**: 491.

Malone, R.E. and R. Easton Esposito. 1980. The *RAD52* gene is required for homothallic interconversion of mating types and spontaneous mitotic recombination in yeast. *Proc. Natl. Acad. Sci.* **77**: 503.

Malone, R.E., J.E. Golin, and M.S. Esposito. 1980. Mitotic versus meiotic recombination in *Saccharomyces cerevisiae*. *Curr. Genet.* **1**: 241.

Maloney, D.H. 1979. "Mutational analysis of mitotic recombination in yeast." Ph.D. thesis, University of California, Berkeley.

Manney, T.R. and R.K. Mortimer. 1964. Allelic mapping in yeast by X-ray induced mitotic reversion. *Science* **143**: 581.

McClintock, B. 1961. Some parallels between gene control systems in maize and bacteria. *Am. Nat.* **95**: 265.

Meselson, M.S. and C.M. Radding. 1975. A general model for genetic recombination. *Proc. Natl. Acad. Sci.* **72**: 358.

Moore, C.W. and F. Sherman. 1975. Role of DNA sequences in genetic recombination in the iso-1-cytochrome *c* gene of yeast. I. Discrepancies between physical distances and genetic distances determined by five mapping procedures. *Genetics* **79**: 397.

Morrison, D.P. and P.J. Hastings. 1979. Characterization of the mutator *mut5-1*. *Mol. Gen. Genet.* **175**: 57.

Mortimer, R.K. 1959. Invited discussion. *Radiat. Res.* (Suppl.) **1**: 394.

Mortimer, R.K. and S. Fogel. 1974. Genetical interference and gene conversion. In *Mechanisms in recombination* (ed. R.F. Grell), p. 236. Plenum Press, New York.

Nakai, S., and R.K. Mortimer. 1969. Studies of the genetic mechanism of radiation-induced mitotic segregation in yeast. *Mol. Gen. Genet.* **103**: 329.

Parry, J.M. 1969. Comparison of the effects of ultraviolet light and ethylmethanesulphonate upon the frequency of mitotic recombination in yeast. *Mol. Gen. Genet.* **106**: 66.

Parry, J.M. and B.S. Cox. 1965. Photoreactivation of ultraviolet induced reciprocal recombination, gene conversion and mutation to prototrophy in *Saccharomyces cerevisiae*. *J. Gen. Microbiol.* **402**: 235.

Plotkin, D. 1978. "Commitment to meiotic recombination: A temporal analysis." Ph.D. thesis, University of Chicago, Illinois.

Prakash, S., L. Prakash, W. Burke, and B.A. Montelone. 1980. Effects of the *RAD52* gene on recombination in *Saccharomyces cerevisiae*. *Genetics* **94**: 31.

Reno, B. 1969. "Genetic studies of radiation sensitive mutants of *Saccharomyces cerevisiae*." M.S. thesis, University of California, Davis.

Resnick, M.A. 1975. The repair of double-strand breaks in chromosomal DNA of yeast. In *Molecular mechanisms for repair of DNA* (ed. P. Hanawalt and R.B. Setlow), p. 549. Plenum Press, New York.

Rodarte-Ramon, U.S. and R.K. Mortimer. 1972. Radiation-induced recombination in *Saccharomyces*: Isolation and genetic study of recombination deficient mutants. *Radiat. Res.* **49**: 133.

Roman, H. 1956. A system selective for mutations affecting the synthesis of adenine in yeast. *C.R. Trav. Lab. Carlsberg* **26**: 299.

———. 1957. Studies of gene mutation in *Saccharomyces*. *Cold Spring Harbor Symp. Quant. Biol.* **21**: 175.

———. 1963. Genic conversion in fungi. In *Methodology in basic genetics* (ed. W.J. Burdette), p. 209. Holden-Day, San Francisco.

———. 1971. Induced recombination in mitotic diploid cells of *Saccharomyces*. *Genet. Lect.* **2**: 43.

———. 1973. Studies of recombination in yeast. *Stadler Genet. Symp.* **5**: 35.

Roman, H. and F. Jacob. 1957. Effet de la lumiere ultraviolette sur la recombinaison genetique entre alleles chez la levure. *C.R. Acad. Sci.* **245**: 1032.

———. 1959. A comparison of spontaneous and ultraviolet induced allelic recombination with reference to the recombination of outside markers. *Cold Spring Harbor Symp. Quant. Biol.* **23**: 155.

Roth, R. 1976. Temperature sensitive yeast mutants defective in meiotic recombination and replication. *Genetics* **83**: 675.

Roth, R. and S. Fogel. 1971. A system selective for yeast mutants deficient in meiotic recombination. *Mol. Gen. Genet.* **112**: 295.

Sanfilippo, D. 1977. "A mutational analysis of gene conversion and reciprocal recombination in the yeast, *Saccharomyces cerevisiae*." Ph.D. thesis, University of California, Berkeley.

Silverman, M., J. Zieg, M. Hilmen, and M. Simon. 1979. Phase variation in *Salmonella*: Genetic analysis of a recombinational switch. *Proc. Natl. Acad. Sci.* **76**:391.

Stern, C. 1936. Somatic crossing over and segregation in *Drosophila melanogaster. Genetics* **21**:625.

Szostak, J.W. and R.Wu. 1979. Insertion of a genetic marker into the ribosomal DNA of yeast. *Plasmid* **2**:536.

———. 1980. Unequal crossing over in the ribosomal DNA of *Saccharomyces cerevisiae. Nature* **284**:426.

Wildenberg, J. 1970. The relation of mitotic recombination to DNA replication in yeast pedigrees. *Genetics* **66**:291.

Zimmermann, F.K. and R. Schwaier. 1967. Induction of mitotic gene conversion with nitrous acid, 1-methyl-3-nitro-1-nitrosoguanidine and other alkylating agents in *Saccharomyces cerevisiae. Mol. Gen. Genet.* **100**:63.

Zimmermann, F.K., R. Schwaier, and U.V. Laer. 1966. Mitotic recombination induced in *Saccharomyces cerevisiae* with nitrous acid, diethyl sulfate, and carcinogenic alkylating nitrosamides. *Z. Vererbungsl.* **98**:230.

DNA Repair and Mutagenesis in Yeast

Robert H. Haynes and Bernard A. Kunz*
Department of Biology
York University
Toronto, Canada M3J 1P3

INTRODUCTION

Physical and chemical mutagens can produce various kinds of structural defects in yeast DNA. Under appropriate vegetative growth conditions, such changes in DNA may lead to killing (measured as inhibition of macrocolony formation), mutation, or mitotic recombination. Sensitivity to these effects depends on the nature of the mutagen, on the genetic constitution of the cells, and on their physiological state before, during, and after exposure. For reviews of early work in this area see Mortimer (1961), Haynes (1964a, 1975a), James and Werner (1965), and Kilbey (1975).

Our knowledge of the mechanisms of DNA repair and its involvement in mutagenesis has come primarily from genetic and biochemical studies on the effects of ultraviolet light (254 nm UV) on microorganisms, including yeast (Haynes et al. 1966; Hanawalt and Setlow 1975; Hanawalt et al. 1978, 1979). Therefore, in this chapter, our discussion of general concepts is based

---

*Present address: Laboratory of Molecular Genetics, National Institute of Environmental Health Sciences, Research Triangle Park, North Carolina 27709

largely on experimental results from the UV photobiology of *Saccharomyces cerevisiae*. However, most of the ideas are applicable in broad outline, if not in precise genetic or biochemical detail, to the corresponding cellular responses to ionizing radiation and chemical mutagens. For information concerning these latter mutagens, as well as UV, see the recent articles by Kircher et al. (1979), Resnick (1979), Ruhland and Brendel (1979), Lemontt (1980), Prakash and Prakash (1980) and Lawrence (1981).

At the observational level in large populations, the genetic effects of UV are randomly distributed, generally independent of one another, and the dose-response relations can be described formally by single-event[1] Poisson statistics (Haynes and Eckardt 1979a). Whether a given cell type under a given set of physiological conditions survives and whether it is also an induced mutant or recombinant, depends not only on the average number of DNA lesions initially produced but also on the presence and efficiency of various enzymic mechanisms for DNA repair. The lesions that produce these effects can be classified in a number of ways, but from a biological standpoint, it is convenient to recognize three categories: (1) those that are potentially lethal, (2) those that are potentially mutagenic, and (3) those that are potentially recombinagenic. These lesions are not necessarily different chemical entities and, for example, UV-induced pyrimidine dimers can produce all three biological effects.

DNA DAMAGE-REPAIR CONCEPT

UV-induced killing, mutation, and mitotic recombination (both gene conversion and mitotic crossing-over) are all enhanced over wild-type responses in mutant strains defective in certain modes of DNA repair. This is taken to indicate that many of the lesions capable of stimulating these effects may be removed by repair in wild-type strains before they have had an opportunity to provoke genetic effects.

On the other hand, mutants also exist that are sensitive to the lethal effects of mutagens but in which induced mutagenesis is suppressed or absent.

---

[1]At the observational level, in the assays used, each treated cell is ultimately scored either as a survivor or a nonsurvivor; survivors are scored either as mutants or nonmutants or as recombinants or nonrecombinants. The all-or-none character of the biological endpoints therefore allows the application of single-event Poisson statistics. Ambiguities of interpretation are common throughout the literature because of the uncritical or confused use of the word "hit" in both biological and physical contexts. For our purpose a physical hit is a potentially lethal, mutagenic, or recombinagenic lesion in a relevant macromolecular target. Complex physicochemical mechanisms are involved in the formation of physical hits, and different but equally complex biochemical and physiological processes are involved in the conversion of physical hits to biological hits. We refer to the initial physical hits as lesions and use the word hit only in the biological context of lethal, mutational, or recombinational hits. In a population of irradiated cells, one lethal hit corresponds to a dose that leaves a fraction e^{-1} (the reciprocal of the base of the natural logarithms) of the cells as survivors. For further details, see Haynes and Eckardt (1979a or 1980) and Eckardt and Haynes (1980).

These results have led to the view, first enunciated for UV mutagenesis in *Escherichia coli* (Witkin 1967, 1969) and subsequently applied to yeast (Lemontt 1971a,b, 1972; Lawrence et al. 1974; Prakash 1974; 1976a,b; Lawrence and Christensen 1976), that DNA repair systems are involved as causative as well as ameliorative factors in mutagenesis. From an operational standpoint, repair systems that are involved in the production of mutations are said to be error-prone or mutagenic. All other modes of repair are said to be error-free. These observations form the basis of the so-called DNA damage-repair concept, which has been used to account qualitatively for the existence of strains both sensitive and refractory to the effects of mutagens and, quantitatively, for the dose-response relations for cell killing and mutagenesis by UV (Fig. 1). However, it is important to remember that, unlike the biochemically defined process of error-free excision repair (Haseltine et al. 1980), error-prone repair is still a hypothesis for which no direct biochemical evidence exists. In view of recent studies on the genetic control of deoxythymidine-5'-monophosphate (dTMP)-induced mutagenesis in yeast (Barclay and Little 1981), it is altogether possible that misrepair and misreplication mutagenesis are not as different biochemically as one tends to assume on the basis of current terminology (Haynes et al. 1981; Lawrence 1981).

In its simplest mathematical form, as applied to cell killing, the DNA damage-repair hypothesis asserts that the surviving fraction of cells in an irradiated or chemically treated population is related statistically to the number of DNA structural defects that remain unrepaired. On average, the number of such unrepaired defects is jointly proportional to the number of potentially lethal lesions formed initially and to the probability that they are not repaired by any repair mechanism (Haynes 1964b, 1966; Haynes and Eckardt 1979b, 1980). Pyrimidine dimers are the DNA defects primarily responsible for cell killing in UV-irradiated yeast. Measurements of the number of dimers initially formed and of the number removed by repair show that the quantitative difference in survival between repair-proficient wild-type and UV-sensitive strains can be accounted for entirely on the basis of the deficiency in repair. The magnitude of the resistant "shoulder" and the increasing slope of wild-type survival curves are quantitatively consistent with the decline in repair efficiency that occurs as dose increases (Haynes 1975b; Wheatcroft et al. 1975).

The repair of potentially lethal damage in wild-type yeast is surprisingly effective, and such cells have the capacity to eliminate large numbers of chemically diverse lesions from their DNA. For example, in a mutant strain in which all three known modes of repair are blocked, the UV dose necessary to kill 63% of a haploid population (LD_{37}) produces only one or two pyrimidine dimers per nuclear genome (Cox and Game 1974). On the other hand, the LD_{37} for a wild-type population generates about 18,000 dimers per genome, essentially all of which must be removed or bypassed for a cell to

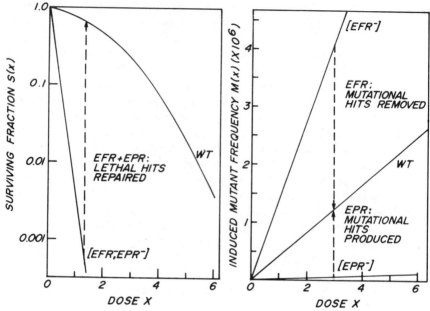

Figure 1   Schematic illustration of the DNA damage-repair concept. (*Left*) Typical UV survival curves for a haploid *RAD* wild-type strain of yeast together with that for a mutant deficient both in error-prone and error-free modes of repair. The shoulder and increasing (negative) slope of the wild-type curve reflect the decline in repair efficiency with increasing dose; because of the absence of repair, no such shoulder appears on the mutant curve. The difference in log survival levels between these two curves for any given dose gives the number of lethal hits removed by repair. (*Right*) Typical UV-induced mutation frequency curves for a haploid *RAD* wild-type strain (WT) together with those for two UV-sensitive mutants, one deficient in error-free repair (EFR⁻) and the other deficient in error-prone repair (EPR⁻). For stochastically independent mutation and killing events, the difference between EFR⁻ and WT for any given dose is equal to the number of mutational hits removed by error-free repair, whereas the difference between WT and EPR⁻ is equal to the number of mutational hits produced by error-prone repair. It is important to note that we use the word hit in the biological context of lethal or mutational hits. We use the word lesion to refer to the alterations produced in DNA by the radiation. Note also that in this diagram, the dose-mutation curves are plotted on rectilinear scales; in practice, such curves are often plotted on double-log scales (cf. Fig. 6).

survive, since one or two unrepaired dimers is sufficient to kill. However, this remarkable efficiency of dimer removal should not be taken to imply that all strains sensitive to killing by mutagens are necessarily defective in some mode of repair. A number of mutations that confer sensitivity to radiation and chemical mutagens are highly pleiotropic, and so it could be misleading to assume that DNA repair is the primary biological process controlled by such loci.

The DNA damage-repair hypothesis has also been applied to the analysis of UV dose-response relations for mutagenesis (Haynes and Eckardt 1979a). Here, the basic mathematical assumption is that the frequency of induced mutants is proportional to the number of premutational lesions that are not removed by error-free repair but are processed by error-prone repair. Again, this does not imply that all mutant strains with enhanced or depressed mutabilities are necessarily deficient in error-free or error-prone repair, respectively. Neither does it imply, at least in the case of UV, that the mutant base sequence necessarily occurs at, or even near, the site of the dimer (Lawrence and Christensen 1979).

Preliminary studies indicate that it is possible to extend the DNA damage-repair concept to the interpretation of the UV dose-response relations for induced mitotic recombination (Kunz and Haynes 1981a). In analogy with the case for UV mutagenesis, the assumption here would be that the frequency of induced recombinants is proportional to the number of potentially recombinagenic lesions that are not removed by nonrecombinational modes of repair but are processed by repair mechanisms involving recombination.[2] Again, this should not be taken to imply that all mutant strains with depressed recombinagenicities are necessarily deficient in some mode of repair (Rodarte-Ramon and Mortimer 1972; Cox 1978) nor that induced recombination necessarily occurs in the vicinity of the lesions (Fabre and Roman 1977). For more details regarding this latter point, see Esposito and Wagstaff (this volume).

BIOCHEMICAL MECHANISMS FOR DNA REPAIR

Several different mechanisms for DNA repair have been described in various organisms, although the relations among these processes and, in turn, their relations with normal replication and recombination are far from clear (Radding 1978; Hanawalt et al. 1979; Kornberg 1980). In the case of repair of UV-induced pyrimidine dimers, it is useful to distinguish broadly between

---

[2]It is possible to distinguish a priori among four categories of repair: (1) repair unaccompanied by any genetically detectable change (error-free repair); (2) repair that generates point mutations, a process appropriately called mutagenic repair (Hastings et al. 1976); (3) repair that involves mitotic recombination or DNA strand exchange (recombinagenic repair); and (4) repair that may be accompanied by both point mutations and mitotic recombination. Under conventional terminology, 1 and 3 together would be classed as error-free repair, whereas 2 and 4 together would constitute error-prone repair. It is unfortunate, however, that the word "repair" is used to describe mutagenic processes, since the initial damage may not actually be removed but rather "bypassed" or "tolerated" in the cell. There is no direct evidence that any of the excision processes illustrated in Fig. 2 are mutagenic. Thus, it would be preferable to restrict the word "repair" to error-free processes and to use the more neutral word "process" or, possibly, "recovery" for the error-prone case.

photoreactivation, which involves exposure to visible light, and dark reactivation or repair, which does not.

Photoreactivation

Photoreactivation is an apparently ubiquitous and highly specific process
that consists of the direct enzymic cleavage of cyclobutane-type pyrimidine
dimers in DNA by the so-called photoreactivating (PR) enzyme. The DNA-
PR enzyme complex absorbs visible light, which is required as an energy
source for the reaction (Harm 1976). A yeast mutant deficient in photoreactivation has been isolated (Resnick 1969a), and large-scale extraction and
purification of the PR enzyme has been carried out with yeast (Muhammed
1966).

Photoreactivation and certain modes of dark repair in yeast attack the
same damaged substrate. This is seen in the gradual decline in photo-
reactivability and concomitant loss of dimers from DNA, which occur when
UV-irradiated cells are held in buffer or distilled water prior to exposure to
visible light and plating for viability (Parry and Cox 1968; Ferguson and Cox
1980). Because of its great specificity, photoreactivation provides a valuable
technique for determining whether pyrimidine dimers are the lesions
responsible for any given UV-stimulated biological effect. Indeed, it is on this
basis that the common involvement of pyrimidine dimers in UV mutagenesis
and recombinagenesis has been established; however, such studies also
indicate that other nonphotoreactivable lesions, which may not be dimers, are
involved in these effects (Parry and Cox 1968; Hunnable and Cox 1971).

Dark Repair

Dark-repair mechanisms, described primarily in *E. coli* and its phages,
include the various modes of excision repair (Fig. 2; Hanawalt et al. 1979);
heteroduplex repair of mismatched bases (Nevers and Spatz 1975; Wildenberg and Meselson 1975; Glickman and Radman 1980); repair of single- and
double-strand breaks (Hutchinson 1978); "proofreading repair," which is
effected by the $3' \rightarrow 5'$ exonuclease activity of DNA polymerase I (Brutlag
and Kornberg 1972); nicked-strand ligation (Lehman 1974); purine-base
insertion (Deutsch and Linn 1979); and daughter-strand gap repair (Rupp
and Howard-Flanders 1968). The proposed inducible mode of error-prone
"SOS" repair (Witkin 1976) may be based on the suppression of the proofreading exonuclease and concomitant modification of the polymerase
activity in replication forks blocked at pyrimidine dimer sites; such modified
replication complexes are thought to be capable of error-prone transdimer
synthesis (Villani et al. 1978). The enzymic basis of these processes has been
reviewed recently by Kornberg (1980).

There is evidence, though often indirect, for the existence in yeast of most
of these processes. However, it is not known whether error-prone transdimer

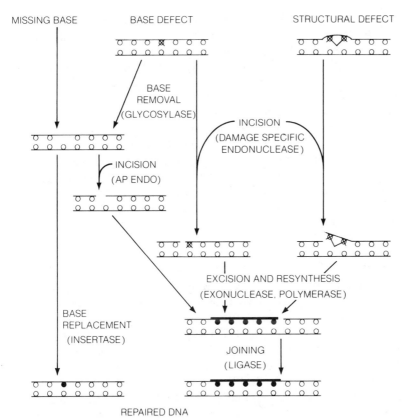

MISSING BASE   BASE DEFECT   STRUCTURAL DEFECT

BASE
REMOVAL
(GLYCOSYLASE)

INCISION
(DAMAGE SPECIFIC
ENDONUCLEASE)

INCISION
(AP ENDO)

EXCISION AND RESYNTHESIS
(EXONUCLEASE, POLYMERASE)

BASE
REPLACEMENT
(INSERTASE)

JOINING
(LIGASE)

REPAIRED DNA

Figure 2 Schematic overview of excision repair. Three types of lesions are illustrated: missing bases, base defects unaccompanied by any significant change in DNA secondary structure, and structural defects that are accompanied by distortions in the phosphodiester backbone. In the classical scheme, the damage is recognized and a phosphodiester strand scission is made; the lesion, along with adjacent nucleotides, is excised, and the deleted stretch is resynthesized (repair replication) using the intact complementary strand as template. Repair is completed by ligating the 3' end of the repair patch to the adjacent original strand. In some instances, a glycosylase initiates the whole process by removing an altered or incorrect base (e.g., uracil) from DNA. The resulting apyrimidinic or apurinic (AP) site may then be recognized by a specific AP endonuclease, or it may become the substrate for direct replacement of the missing base by "insertase" activity. Recently, it has been shown that the incision reactions catalyzed by the *Micrococcus luteus* and T4 UV endonucleases are initiated by cutting the glycolytic bond of the 5' thymine in the dimer. (Diagram courtesy of Professor Philip C. Hanawalt, Stanford University; for further details see Hanawalt et al. 1979).

synthesis occurs in yeast or whether any UV-induced mutations arise during excision repair. There are two main reasons for the relative paucity of information in this area: (1) Yeast lacks thymidine kinase (Grivell and Jackson 1968); thus, the bacterial type of thymine-requiring auxotrophs, which would allow the specific radioactive labeling of yeast DNA, are not available. (2) Very few studies have been carried out on the enzymes of DNA synthesis and degradation in yeast. In recent years, however, several mutants permeable to, and/or auxotrophic for, the nucleotide dTMP have been isolated, and these can be used in experiments on DNA repair and related phenomena (Brendel and Haynes 1972; Brendel and Fäth 1974; Wickner 1974; Little and Haynes 1979). Also, two nuclear DNA polymerases, designated I and II, a topoisomerase, and four endonucleases have been isolated from *S. cerevisiae*. DNA polymerase II has an associated $3' \rightarrow 5'$ exonuclease activity capable of removing replication errors; one of the endonucleases, designated α, shows increased activity against UV-irradiated DNA, and another endonuclease appears to be specific for apurinic sites (Piñon and Leney 1975; Chang 1977; Chlebowicz and Jachymczyk 1977; Bryant and Haynes 1978a,b; Durnford and Champoux 1978).

Excision Repair

The two repair processes that have received greatest attention at the molecular level are the excision repair of pyrimidine dimers and the repair of X-ray-induced double-strand breaks (DSB repair). Dimer excision has been demonstrated in yeast (Unrau et al. 1971). Highly sensitive techniques for the detection of the loss of sites sensitive to dimer-specific endonucleases have been used to show that the gene products of at least nine loci are involved in dimer excision repair (Prakash 1977a,b; Reynolds 1978; Prakash and Prakash 1979). Also, Goldberg and Little (1978) have measured repair replication in a Pettijohn-Hanawalt assay (Pettijohn and Hanawalt 1964) after UV irradiation of a ρ^0 haploid strain auxotrophic for dTMP. The DNA of such strains can be density-labeled with 5-bromo-2'-deoxyuridine-5'-monophosphate (BrdUMP). In unirradiated cultures, a large shift in the buoyant density of DNA occurs during replication and, after one generation, the DNA bands entirely at the hybrid-density position. Following various UV doses, the labeled BrdUMP is incorporated into DNA banding at the parental density. A rough estimate of the average patch size in these experiments indicates a length of 60 nucleotides (see also Haynes et al. 1981).

DSB Repair

The repair of DSBs produced by X-rays (Ho 1975; Resnick and Martin 1976; Luchnik et al. 1979; Brunborg et al. 1980; Frankenberg-Schwager et al. 1980) has been demonstrated using sucrose gradients, which separate DNA molecules on the basis of size. It would appear that DSB repair requires the simultaneous presence of two DNA duplexes (Luchnik et al. 1977; Resnick 1978). This fact makes it likely that DSB repair is responsible for (1) the resistant "tail" observed on X-ray survival curves of haploid log-phase populations that contain budding cells (Bird and Manney 1974; Benathen and Beam 1977); (2) the increase in X-ray resistance that occurs in haploids during G_2 phase (Brunborg and Williamson 1978; Brunborg et al. 1980); and (3) in part at least, the occurrence of liquid-holding recovery in X-irradiated stationary-phase diploids, a process that does not occur in G_1 haploids (Patrick et al. 1964; Luchnik et al. 1977; Frankenberg-Schwager et al. 1980). Whether these DSBs are directly responsible for cell killing has not yet been established (Frankenberg-Schwager et al. 1979).

PHENOMENA INDICATING THE REPAIR OF GENETIC DAMAGE IN YEAST

Repair of Potentially Lethal Lesions

Indirect evidence for the existence of enzymic, energy-requiring, dark-repair processes in yeast came first from studies on the synergistic killing that occurs in cells exposed to both radiation and chemical mutagens (Haynes and Inch 1963) and, more directly, from studies of liquid-holding recovery (LHR). This latter phenomenon is manifest in the enhanced survival of cells held in nonnutritive suspension for some hours prior to plating; liquid holding of unirradiated cells has no effect on their radiation sensitivity (Patrick et al. 1962, 1964; Patrick and Haynes 1964, 1968; Korogodin et al. 1966). The energy requirements and other physiological conditions that affect LHR in X-irradiated cells have been studied extensively by various workers, especially Pohlit and his collaborators (see, e.g., Jain and Pohlit 1972 and Rao and Murthy 1978).

Figure 3 illustrates the LHR effect seen in UV-irradiated, stationary-phase diploid cells held for varying times at 30°C in distilled water prior to plating on a complete growth medium. Notice that recovery of viability occurs with a dose-dependent lag but generally is complete after 18 hours. Similar effects also occur after treatment with the bifunctional alkylating agent nitrogen mustard (Patrick et al. 1964). It is abundantly clear from the increases in viability shown in Figure 3 that potentially lethal lesions are repaired under liquid-holding conditions. The problem is, what is actually going on inside the cells to effect this increase in survival? Originally, it was simply assumed that liquid holding modifies the extent of excision repair. If the amount that normally occurs in cells plated immediately after irradiation is relatively low, liquid holding tends to increase it, but if repair is very extensive upon immediate plating, liquid holding has no effect or may even decrease recovery.

Subsequent studies have shown that LHR after UV is under genetic control (Parry and Parry 1976) and involves (but oddly enough is not wholly dependent upon) dimer excision (Ferguson and Cox 1980). Three main interpretations have been offered for this phenomenon: (1) Liquid holding under nonnutritive conditions allows more time for repair to take place

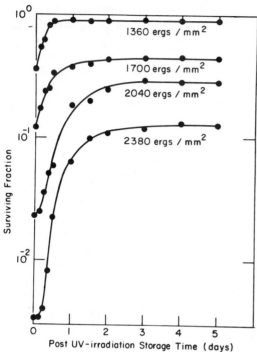

Figure 3   Recovery of UV-irradiated diploid cells as a function of storage time in distilled water in the dark (details in text).

before the cells are stimulated to replicate their DNA; (2) for some unknown reason, repair is more efficient in cells that have been subjected to liquid holding prior to plating; and (3) during LHR, the protein synthesis that is induced (Eckardt et al. 1978) includes the synthesis of products that improve the efficiency of excision repair (Ferguson and Cox 1980). The latter authors concluded that none of the above explanations is solely sufficient to account for the behavior of UV-irradiated haploid yeast. Their work showed that the major process involved is excision, even though certain excision-deficient strains show LHR, whereas other excision-proficient strains do not; however, in addition, synthesis of proteins that improve the efficiency of excision certainly occurs.

This latter process can be invoked to account for the phenomenon of radiation-induced UV resistance that is seen in irradiated yeast after LHR (Patrick and Haynes 1968; Parry and Parry 1972). Cells irradiated with either UV or X-rays, and then allowed LHR, are found to have acquired a greatly enhanced resistance to subsequent UV killing but not to X-ray inactivation. The fact that this enhancement of resistance can be inhibited by cycloheximide (Eckardt et al. 1978), and that no enhancement of X-ray resistance is seen, is fully consistent with the induction during LHR of systems (both error-free and error-prone) that serve to enhance the efficiency of pyrimidine dimer repair.

The most extensive biological evidence for the repair of potentially lethal damage in yeast has been provided by the discovery of a large number of mutant strains that are more sensitive than the wild-type parent to the lethal effects of various mutagens. The loci involved can be classified into three epistasis groups on the basis of the sensitivity of double mutants formed from various pairwise combinations of the available alleles (see below, Repair-deficient Mutants: Characterization and Epistasis Groups). In Figure 4 we show UV survival curves for a diploid repair-proficient *RAD* wild-type strain together with curves for three other diploids that are homozygous for loci belonging to each of the three recognized epistasis groups. It is clear from the relative sensitivities of the three mutants shown in Figure 4 that the process controlled by *RAD3* is capable of repairing a larger fraction of the potentially

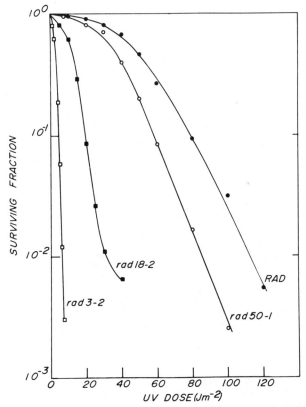

Figure 4   UV survival curves for mutants in the three epistasis groups. Stationary-phase diploid cells were irradiated in suspension and plated on synthetic complete medium to determine survival. (Jm⁻², joules per square meter.)

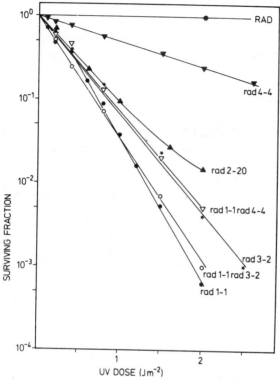

Figure 5 UV survival of single- and double-mutant (excision-defective) haploid strains. Stationary-phase cells were irradiated in suspension and plated on complete medium to determine survival.

lethal lesions produced by UV than is the process controlled by *RAD50*. In Figure 5, we show, on a greatly expanded scale, UV survival curves for a repair-proficient haploid *RAD* wild-type strain together with curves for various repair-deficient haploid single and double mutants. Note here that the double mutants *rad1-1 rad3-2* and *rad1-1 rad4-4* are no more sensitive than the most sensitive single-mutant parent. On this basis, the loci *RAD1, RAD3,* and *RAD4* all belong to the same epistasis group with respect to UV killing (Game and Cox 1972; Brendel and Haynes 1973).[3]

---

[3]The classification of the mutants *rad1-1* and *rad4-4* as members of the same epistasis group is unaffected by the fact that the double-mutant (*rad1-1 rad4-4*) is actually more resistant than the most sensitive single-mutant parent (cf. Fig. 5). The reason for this resistance is unknown, although it is possible that partial complementation occurs between the alleles involved.

Repair of Premutational Lesions

UV-induced reversions to prototrophy for a number of nutritional markers have been observed to decline under liquid-holding conditions; this has been taken to indicate that the repair systems associated with LHR are capable of repairing premutational damage (Parry and Parry 1972). However, it would appear that one of the new systems induced by liquid-holding conditions is itself error-prone (Eckardt et al. 1978; Ferguson and Cox 1980). Presumably, therefore, whether reversion frequencies are observed to decline or increase after LHR will depend on the balance of the effects of error-free dimer excision and the extent of induction of error-prone repair activity.

UV-induced mutation frequencies for reversion to prototrophy for various nutritional markers are greater in certain UV-sensitive mutants than in their wild-type counterparts. An example of this is shown in Figure 6 for induced

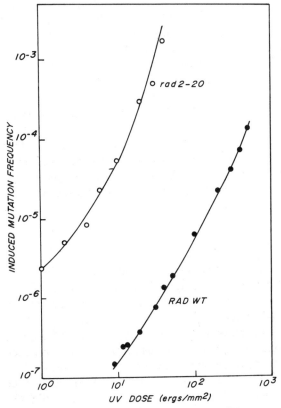

Figure 6   UV-induced reversion in *RAD* wild-type and *rad2-20* (excision-defective) haploids. Stationary-phase cells were UV-irradiated and plated on selective medium to score reversion of *lys2-1*.

lys2-1 locus revertants in the excision-deficient strain *rad2-20*. (These curves look different from the schematic diagram shown in Fig. 1, because the latter was drawn on rectilinear paper for clarity, whereas the data in Fig. 6 are presented in the customary way on double logarithmic scales.) The fact that the points for *rad2-20* lie above those for *RAD* wild type reflects the activity of error-free repair processes in removing premutational lesions in the wild-type strains (see Fig. 1).

A number of UV-sensitive mutants exist in which various categories of induced mutagenesis are markedly depressed or absent. Obviously, if there is no induced mutagenesis, the data cannot be plotted meaningfully on a graph such as that shown in Figure 6. Any residual level of induced mutagenesis would give points falling much below the curve for *RAD* wild type. Thus, the level of induced mutation frequencies in wild-type strains is considered to represent the net result of the opposing effects of error-free and error-prone repair processes that are stimulated to act by the presence of premutational lesions in DNA (see Fig. 1).

Strictly interpreted, the DNA damage-repair hypothesis implies that error-prone repair is the source of all induced mutations. On this basis, one would expect some mutagen-sensitive strains to be deficient in induced mutagenesis and all strains defective in induced mutagenesis to be mutagen sensitive. Whereas the former expectation has been met (e.g., *rad6* as shown by Lawrence and Christensen [1976]), Lemontt (1977) has isolated mutants (*umr*) with depressed levels of UV-induced mutation but that are only slightly, if at all, sensitive to UV; indeed, *umr4* and *umr5* have wild-type resistance. This suggests that steps of mutagenesis exist that may be independent of repair; however, if some error-prone repair process is controlled by these genes, then its contribution to survival is so small, it cannot be detected using standard assay techniques.

Repair of Prerecombinational Lesions

Mitotic recombination is a term used to refer collectively to the phenomena of gene coversion and mitotic crossing-over (see Esposito and Wagstaff, this volume). Gene conversion is generally measured by the appearance of revertants to prototrophy at heteroallelic loci in diploid strains, whereas mitotic crossing-over can be measured by the segregation of recessive homozygotes from heterozygous diploids. Various special strains have been set up to measure induced mitotic recombination, but one of the most useful is the strain D7, constructed by Zimmermann et al. (1975), in which gene conversion is measured at *TRP5* and mitotic crossing-over at *ADE2*. Figure 7 shows curves for UV-induced gene conversion and mitotic crossing-over in D7 *RAD* wild type and in a derivative of D7, which is also homozygous for the excision deficiency *rad3-2*. If the curves for *RAD* wild type are extrapolated to low doses, it is evident that they lie below the corresponding curves

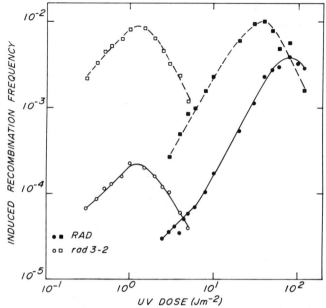

Figure 7 UV-induced mitotic recombination in *RAD* wild-type and *rad3-2* (excision-defective) diploids. Crossing-over and gene conversion were determined as described by Zimmermann et al. (1975). Stationary-phase diploid cells were irradiated in suspension and plated on minimal media selective for the genetic endpoints scored. (□, ■) Mitotic crossing-over; (○, ●) mitotic gene conversion.

for *rad3-2*. In analogy with the situation for mutation, the relative positions of the curves in Figure 7 for *RAD* wild-type and *rad3-2* may be taken as evidence for the repair of prerecombinational lesions in wild-type yeast. The decline in recombination frequencies at high doses can be interpreted on the basis that killing and recombination are stochastically dependent processes, i.e., at high doses, the probability of recombinant clones surviving is less than that of nonrecombinants (see Eckardt and Haynes 1977b).

The effects of liquid holding on UV-induced mitotic recombination are somewhat more complex than the effects of LHR on killing and mutagenesis (Parry and Cox 1968). Here it has been observed that although induced mitotic crossing-over frequencies are reduced by LHR, gene-conversion frequencies increase. The reasons for this are not understood. However, there is evidence for a net induction of repair during liquid holding (Eckardt et al. 1978; Ferguson and Cox 1980). It may be that the increased production of convertants, as a result of enhanced repair processes, more than compensates for the loss of prerecombinagenic lesions. Unlike gene conversion, resolution of crossovers requires progression through S and G_2 phases. Prolonged holding in the G_1 phase may not only reduce prerecombinagenic lesions that could lead to crossing-over but also might allow additional time for dissociation of any Holliday structures that may have formed (see Esposito and Wagstaff, this volume).

According to a strict interpretation of the DNA damage-repair hypothesis, all induced mitotic recombination would be the result of recombinagenic repair. Thus, one would expect that certain radiation- and/or chemical-sensitive mutants would be defective for induced recombination and that all recombination-deficient mutants would be more sensitive than wild-type strains to these agents. Although mutants of the former type have been found, Rodarte-Ramon and Mortimer (1972) described mutants with reduced frequencies of UV- and X-ray-induced recombination that show wild-type sensitivities to these types of radiation. This suggests that some steps of recombination may be independent of any mode of repair. At the molecular level, a transient accumulation of strand breaks leads to recombination (Game et al. 1979), but it has not been determined whether the breaks themselves provoke recombination independently of repair, or whether they are dealt with by a recombinational repair process. Paradoxically, mutants defective in strand-break repair can exhibit either decreased or increased frequencies of spontaneous and/or induced recombination (Hunnable and Cox 1971; Jachmyczyk et al. 1977; Kern and Zimmermann 1978; Strike 1978; Chlebowicz and Jachmyczyk 1979; Fabre and Roman 1979; Game et al. 1979; Mowat and Hastings 1979; Prakash et al. 1980; Saeki et al. 1980). Clearly, we cannot say for certain whether recombination is a repair process, or whether some modes of repair are merely recombinagenic (Cox 1978).

REPAIR-DEFICIENT MUTANTS:
CHARACTERIZATION AND EPISTASIS GROUPS

The first radiation-sensitive mutants of *S. cerevisiae* were reported by Nakai and Matsumoto (1967) and Snow (1967). Subsequently, Cox and Parry (1968) deliberately attempted to saturate with mutations the dark repair genes that might exist. They isolated 96 recessive UV-sensitive mutants that occupied 22 distinct complementation groups. A number of these mutants were also shown to be sensitive to methylmethanesulfonate (MMS) and/or nitrous acid (Zimmermann 1968). Despite the large number of loci conferring radiosensitivity, Cox and Parry (1968) argued, on statistical grounds, that there probably existed additional, independent genes controlling radiosensitivity for which no mutant alleles had been found. To date, at least 95 mutations are known that confer sensitivity to radiation and/or chemicals (Table 1); however, the allelic relations of all of these mutations have not yet been established.

A useful approach for sorting out the relations among the loci is to study the interactions, with respect to radiosensitivity, of various mutant genes when combined in multiply mutant strains (Nakai and Matsumoto 1967;

Khan et al. 1970; Lemontt 1971b; Game and Cox 1972, 1973; Brendel and Haynes 1973; Cox and Game 1974; Prakash and Prakash 1979; McKee and Lawrence 1980). Double mutants have been used most frequently in these studies. In principle, three different situations may arise (Fig. 8): (1) The two genes are epistatic, i.e., the double mutant is no more sensitive than its most sensitive single-mutant parent; (2) the two genes have an additive effect on each other, i.e., the killing produced by a given dose in the double mutant is equal to the sum of the killing in the wild type plus the incremental killing of the two parents over wild type; and (3) the two genes have a synergistic effect on each other, i.e., the interaction is more than additive. Epistasis implies that

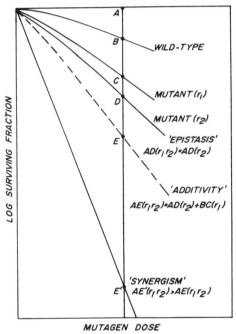

Figure 8 Schematic diagram showing the three types of gene interactions that might occur in yeast strains containing two mutant alleles at different loci controlling sensitivity to a mutagen. Points *B, C,* and *D* lie on the survival curves for *RAD* wild type and two single mutants, r_1 and r_2, respectively. If the sensitivity of the double mutant (r_1r_2) is no greater than that of the most sensitive single-mutant parent (assuming r_1 and r_2 are nonleaky), then the loci are said to be epistatic. If the two repair pathways controlled by r_1 and r_2 act independently on different types of lesions, the interaction is said to be additive, and the survival curve of the double mutant r_1r_2 will pass through point *E*, where AE = AD + BC. If the repair pathways compete for a single class of lesions, the interaction is said to be synergistic, and the survival curve will pass through some point *E'*, where AE' > AE. (For mathematical details, see Brendel and Haynes [1973].)

Table 1 Properties of mutant genes that confer sensitivity to radiation and/or chemicals

Mutant	Sensitivity to[a]					Dimer excision[b]	LHR[c]	Sporulation[d]	Mutation[e]				Recombination[e]				
	UV	X/γ	MMS	HN2	8MOP+UV				spont.	UV	X/γ	chem.	spont.	UV	X/γ	chem.	meiotic
α/α, **a/a**	0[16]	+32	+/−37				−26	−67					−74	−25			−67
ant1	+/−61	0[81]							−61	0[81]							
cdc2			+8					−79									−78
cdc8	+57							−79		−57							−78
cdc9[80]	+/−28	0[13]	+28					−79	+61								
cdc31		+69							+51				+17			−13	
gam1		+/−15							+15								
gam3		+15							+15		+15						
gam5		++15							+15		+15						
MIC1	+/−40	0[40]	0[40]					−40	+40	−40			+40	0[40]			
MIC5	+/−40	0[40]	0[40]					+40	+40				+40	+40	0[40]		
MIC8	+/−40	0[40]	0[40]					−40	+40				+40	+40			
MIC9	0[40]	0[40]	+/−40					+40	+40				+40	0[40]	0[40]		
MIC15	+40	0[40]	+40					−40	+40				+40	0[40]	0[40]		
MIC19	+40	0[40]	+/−40					−40	+40				+40	0[40]	0[40]		
MIC23	+/−40	0[40]	+40					−40	0[40]				+40	0[40]	0[40]		
mms1	0[58]	0[58]	+58					−58									
mms2	0[58]	0[58]	+58					+58				+56					
mms3	+58	+58	+58					+58	−82	−41							
mms4	0[58]	0[58]	+58					−58		−41	−41		0[82]	−56			
mms5	0[58]	0[58]	+58					−58									
mms6	+58	0[58]	+58					+58									
mms7	+58	+58	+58					−58				−56					

mms8[80]	0[58]	+[58]	+[58]					−[58]	0[80]				+[59]				
mms9	0[58]	+[58]	+[58]					−[58]					+[59]				
mms10	+[58]	0[58]	+[58]					−[58]									
mms11	+[58]	+[58]	+[58]					+[58]									
mms12	+[58]	+[58]	+[58]					−[58]				−[56]					
mms13	+[58]	0[58]	+[58]					−[58]					+[59]				
mms14	0[58]	+[58]	+[58]					−[58]									
mms15	+[58]	+[58]	+[58]					−[58]									
mms16	0[58]	+[58]	+[58]					−[58]									
mms17	+[58]	+[58]	+[58]					−[58]									
mms18	+[58]	0[58]	+[58]					−[58]									
mms19	+[58]	0[58]	+[58]			−[60]		−[58]		+[60]							
mms20	0[58]	+[58]	+[58]					−[58]				−[56]					
mms21	+[58]	+[58]	+[58]					−[58]	+[44]	0[44]		−[45]	+[59]				
mms22	0[58]	0[58]	+[58]					−[58]									
mut2	0[21]	0[21]	+[21]						+[21]								
mut3	+/−[21]	0[21]	+[21]						+[21]								
mut4	+/−[21]	0[21]	+[21]						+[21]								
mut9	+/−[21]	+/−[21]	+[21]						+[21]								
mut10	0[21]	0[21]	+[21]						+[21]								
pso1	+[24]	+[24]		++[24]	+[24]			−[24]	−[7]	−[7]	−[7]	−[7]					
pso2	0[24]	0[24]		+/−[24]	++[24]			+[24]	0[7]	−[7]	−[7]	−[7]					
pso3	0[24]	0[24]		+/−[24]	+/−[24]			−[24]	0[7]	−[7]	+[7]	+[7]					
rad1	++[10]	+/−[42]	+/−[58]	+[5]	+/−[2]	−[72]	−[50]	+[10]	+[61]	+[33]	0[42]	+[53]	0[70]	+[70]			0[70]
rad2	++[10]	0[47]	+/−[76]	+[5]	+/−[2]	−[63]	−[50]	+[10]	0[6]	+[12]	0[42]	+[53]	0[70]	+[70]			0[70]
rad3	++[10]	0[76]	+[76]		+/−[27]	−[54]	−[50]	+[10]	+[6]	+[33]		+[53]	+[29]	+[70]	0[11]	+[29]	0[70]
rad4	++[10]	0[76]	+/−[76]			−[54]	+[50]	+[10]	0[6]	+[33]	0[42]	+[53]	0[70]	+[70]			0[70]

Table 1 (Continued)

Mutant	Sensitivity to[a]					Dimer excision[b]	LHR[c]	Sporu-lation[d]	Mutation[e]				Recombination[e]				
	UV	X/γ	MMS	HN2	8MOP+UV				spont.	UV	X/γ	chem.	spont.	UV	X/γ	chem.	meiotic
rad5	+10	+/−34	+/−76			+14	+50	+10	+35	−23	−42	−53	0[70]	+70	+35		0[70]
(rev2)[18]																	
rad6	+10	+76	+76	+52	+/−22	+54	+50	−10	+21	−33	−42	−53	+29	+29	−9	+29	*[f]
rad7	+10	0[76]	0[76]			−60	+50	+10		+33							
rad8	+10	+18	+/−76				+50	+10		−33	−42						
rad9	+10	+18	+76	+/−52	+/−2	+54	+50	+10	0[61]	−12	0[42]	+53	0[11]	0[11]			
rad10	+/−10	+/−76	+/−76			−55	+50	+10		+33		+53					
rad11	+/−10	+18	+76				+50	+10									
rad12	+10	+18	+76				+50	+10		+33		+53					
rad13	+10	0[76]	+76					+10		−33							
rad14	+10	0[76]	+76			−60	+50	+10		+33		+53	0[70]	+70			0[70]
rad15	+10	+18	0[76]			+56	−50	+10		−33	−42	−53					
rad16	+10	0[76]	+/−76			−55	+50	+10		+33		+53					
rad17	+10	+18	+76			+56	+50	+10		+33		−53					
rad18	+62	+62	+5	+5		+64	+50	+18	+73	−33	−42	−53	+4	+31	+77		0[4]
rad19	+/−10	0[76]	+/−76				+50	−10									
rad20	+10	0[76]	+/−76				+50	−10									
rad21	+/−10	0[76]	+76				+50	−10									
rad22	+10	0[76]	+76				+50	+10									
rad23	+43	0[43]							+43								
rad50	+/−10	++71	++71		+23	+14	−71	−18	0[61]	−33	0[42]	−71	+38	−71	−71	−71	−19
rad51	+/−47	++47	+71		+23	+14	+71	−18	+21	−71	0[42]	−71	−77	−71	−71	−71	−46
(mut5)[46]																	

390

	1	2	3	4	5	6	7	8	9	10	11	12	13	14	15	16	17
rad52	+/−[62]	++[62]	+[71]	0[5]	+[23]		+[71]	−[18]	+[73]	0[33]	−[42]	−[53]	−[39]	−[71]	−[71]	−[71]	−[19]
(rec2)[66]																	
rad53	+/−[47]	++[47]	+[71]		0[23]		+[71]	−[18]	0[61]	0[71]	−[42]	−[71]	0[77]	−[71]	−[71]	0[71]	−[31]
rad54	+/−[71]	++[71]	+[71]		+[23]		+[71]	−[18]	+[61]	−[71]	0[42]	−[71]	−[77]	−[71]	−[71]	−[71]	
rad55	+/−[71]	++[71]	+[71]		+[23]		+[71]	−[18]	+[61]	−[71]	0[42]	−[71]	−[77]	0[71]	+[71]	+[71]	
rad56	+/−[71]	+/−[71]	+/−[71]		+[23]		+[71]	+[18]	+[61]	−[71]	0[42]	−[71]	−[77]	−[71]	0[71]	−[71]	−[19]
rad57	+/−[71]	+[71]	+[71]		+[23]		+[71]	−[18]	+[61]	−[71]	−[42]	−[71]	−[77]	−[71]	−[71]	−[71]	
r_1^s	+[1]	+[11]						+[11]		0[12]			+[30]	+[30]			
rec5	0[66]	+[66]												−[66]	−[66]		
(2C4)[66]	+[66]																
rec7	+[66]	+[66]												−[66]	−[66]		
(2C16)[66]	+[66]																
rev1	+[34]	+/−[34]	+[71]			+[14]	+[14]	+[34]	0[35]	−[34]	−[42]	−[53]	0[35]	+[35]	+[35]		0[35]
rev3	+[34]	+/−[34]	+[81]					+[34]	−[61]	−[34]	−[42]	−[53]	0[35]	+[35]	+[35]		0[35]
hn2/1	+/−[68]			+[68]													
hn2/2	+[68]			+/−[68]													
hn2/3	+/−[68]			+[68]													
hn2/6	+/−[68]			++[68]				+[68]		0[68]		0[68]		0[68]		0[68]	
hn2/9	+[68]			+[68]													
hn2/103	+[68]			+/−[68]													
hn2/106	+/−[68]			+[68]													
spo8	+/−[3]							−[3]					+[3]	−[3]			−[3]
umr1	+/−[36]							+[36]		−[36]							
umr2	+/−[36]							−[36]		−[36]							
umr3	+/−[36]							−[36]		−[36]							
umr6	+/−[36]							−[36]		−[36]							
umr7	+/−[36]							−[36]		−[36]							

Table 1 (Continued)

Mutant	Sensitivity to[a]					Dimer excision[b]	LHR[c]	Sporulation[d]	Mutation[c]				Recombination[e]				
	UV	X/γ	MMS	HN2	8MOP+UV				spont.	UV	X/γ	chem.	spont.	UV	X/γ	chem.	meiotic
cor1[g]	0[75]	0[75]						+[75]	+[75]				+[75]				+[75]
cor2[g]	0[75]	0[75]						+[75]	+[75]				+[75]				+[75]
cor3[g]	0[75]	0[75]						+[75]	+[75]				+[75]				+[75]
cor4[g]	0[75]	0[75]						+[75]	+[75]				+[75]				+[75]
gam2[g]		0[15]							+[15]								
gam4[g]		0[15]							+[15]								
MIC12[g]	0[40]	0[40]	0[40]					+[40]	+[40]				+[40]	−[40]			
mut1[g]	0[21]	0[21]	0[21]						+[21]								
MUT6[g]	0[21]	0[21]	0[21]						+[21]								
mut7[g]	0[48]	0[48]	0[48]						+[21]				+[49]				
mut8[g]	0[48]	0[48]	0[48]						+[21]								
rec1[g]	0[66]	0[66]												−[66]	−[66]		
rec3[g]	0[66]	0[66]						−[65]					−[65]	−[66]	−[66]		−[65]
rec4[g]	0[66]	0[66]											−[65]	−[66]	−[66]		−[65]
rec6[g]	0[66]	0[66]												−[66]	−[66]		
(2D11)[66 g]																	
rem1[g]								−20	+20				+20				0[20]
spo7[g]	0[3]							−3	−3					−3			−3
umr4[g]	0[36]							−36		−36							
umr5[g]	0[36]							+36		−36							

A minimum number of symbols have been used to indicate the responses of the mutants to various treatments and their influence on certain processes. It should be noted that although merely an increase or decrease is shown, the actual degree of response depends on genetic background, the particular allele at the defective locus and, for induced effects, the dose range of interest. Mutants may affect forward or reverse mutation or both. Similarly, they may influence mitotic gene conversion or crossing-over or both. An indicated response (+, −, or 0) for mutation or recombination does not necessarily imply that both forward and reverse mutation or that both gene conversion and crossing-over are affected equally. For this information, the particular reference given (superscript numbers) should be consulted. Not all of the allelic relationships among these mutants have been determined.

Abbreviations: UV, ultraviolet light; X/γ, X-ray or gamma radiation; MMS, methylmethanesulfonate; HN2, nitrogen mustard; 8MOP + UV, 8-methoxy-psoralen plus 365 nm light; spont., spontaneous; chem., chemicals.

References: [1]Averbeck et al. (1970); [2]Averbeck et al. (1978); [3]Baker et al. (1976); [4]Boram and Roman (1976); [5]Brendel et al. (1970); [6]Brychcy and von Borstel (1977); [7]Cassier et al. (1980); [8]M.N. Conrad and C.S. Newlon (pers. comm.); [9]Cox and Game (1974); [10]Cox and Parry (1968); [11]F. Eckardt (pers. comm.); [12]Eckardt et al. (1975); [13]Fabre and Roman (1979); [14]Ferguson and Cox (1980); [15]Foury and Goffeau (1979); [16]Friis and Roman (1968); [17]Game et al. (1979); [18]Game and Mortimer (1974); [19]Game et al. (1980); [20]Golin and Esposito (1977); [21]Hastings et al. (1976); [22]J.A.P. Henriques (pers. comm.); [23]Henriques and Moustacchi (1980a); [24]Henriques and Moustacchi (1980b); [25]Hopper et al. (1975); [26]Hunnable (1972); [27]W. Jachmczyk and R.C. von Borstel (pers. comm.); [28]Johnston and Nasmyth (1978); [29]Kern and Zimmermann (1978); [30]Kowalski and Laskowski (1975); [31]B.A. Kunz (unpubl.); [32]Laskowski (1960); [33]Lawrence and Christensen (1976); [34]Lemontt (1971a); [35]Lemontt (1973); [36]Lemontt (1977); [37]Livi and Mackay (1980); [38]R.E. Malone and R. Easton Esposito (pers. comm.); [39]Malone and Easton Esposito (1980); [40]Maloney and Fogel (1980); [41]Martin et al. (1978); [42]McKee and Lawrence (1979); [43]G. McKnight (pers. comm.); [44]Monteleone et al. (1978); [45]B. Monteleone et al. (pers. comm.); [46]Morrison and Hastings (1979); [47]Nakai and Matsumoto (1967); [48]Nasim and Brychcy (1979); [49]Ord et al. (1978); [50]Parry and Parry (1969); [51]M. Pastorcic et al. (pers. comm.); [52]Prakash (1974); [53]Prakash (1976b); [54]Prakash (1977a); [55]Prakash (1977b); [56]L. Prakash (pers. comm.); [57]Prakash et al. (1979); [58]Prakash and Prakash (1977a); [59]Prakash and Prakash (1977b); [60]Prakash and Prakash (1977b); [61]S.-K. Quah and R.C. von Borstel (pers. comm.); [62]Resnick (1969b); [63]Resnick and Setlow (1972); [64]Reynolds and Friedberg (1981); [65]Rodarte-Ramon (1972); [66]Rodarte-Ramon and Mortimer (1972); [67]Roman and Sands (1953); [68]A. Ruhland et al. (pers. comm.); [69]D. Schild et al. (pers. comm.); [70]Snow (1968); [71]Strike (1978); [72]Unrau et al. (1971); [73]von Borstel et al. (1971); [74]J.E. Wagstaff et al. (pers. comm.); [75]M. Williamson and S. Fogel (pers. comm.); [76]Zimmermann (1968); [77]Saeki et al. (1980); [78]Schild and Byers (1978); [79]Simchen (1974); [80]Monteleone et al. (1981); [81]Quah et al. (1980); [82]Martin et al. (1981).

[a]++, very sensitive; +, moderately sensitive; +/−, slightly sensitive; 0, wild-type sensitivity.

[b]+, possesses ability to excise pyrimidine dimers; −, lacks ability.

[c]Liquid-holding recovery (see references for type of radiation). +, liquid holding increases survival; −, liquid holding decreases survival.

[d]Sporulation and/or spore viability. +, normal; −, reduced.

[e]+, increased response; −, decreased response; 0, wild-type response.

[f]rad6 may be defective in meiotic recombination, but since it fails to complete sporulation, this interpretation is open to question.

[g]These mutants have been included because they appear to affect repair-related processes although they have not yet been reported to be sensitive to radiation or chemicals.

Table 2   Epistasis groups

RAD3	RAD6	RAD52
rad1[2]	rad5[8a]	rad50[8]
rad2[2]	rad6[5]	rad51[8]
rad3[2]	rad8[8a]	rad52[8]
rad4[2]	rad9[5]	rad53[8]
rad7[11]	rad15[8a]	rad54[8]
rad10[9]	rad18[5]	rad55[8]
rad14[11a]	rev1[5]	rad56[8]
rad16[9]	rev3[5]	rad57[8]
$r_1^{s\,3\,b}$	umr1[6a]	$r_1^{s\,3}$
mms19[11a]	umr2[6a]	cdc9[4]
cdc8[10b]	umr3[6a]	
	mms3[7]	
	cdc8[10]	
	pso1[1]	

In the *RAD3* and *RAD6* groups, epistasis is based on UV sensitivity. In the *RAD52* group, epistasis is based on sensitivity to ionizing radiation.

References (superscript numbers): [1]Cassier et al. (1981); [2]Game and Cox (1972); [3]F. Eckardt (pers. comm.); [4]Game et al. (1978); [5]Lawrence and Christensen (1976); [6]Lemontt (1977); [7]Martin et al. (1978); [8]McKee and Lawrence (1980); [9]Prakash (1977b); [10]Prakash et al. (1979); [11]Prakash and Prakash (1979).

[a]Included on a phenotypic basis. Where multiple-mutant studies have been performed, we do not consider the data to be conclusive.

[b]Dimer excision not investigated.

the gene products mediate steps in the same repair pathway or that they are part of a multimeric repair complex; additivity implies that the gene products act independently on different substrates, and synergism implies that the genes control steps in two repair systems that compete for the same lesions.[4] Three epistasis groups have been identified, and loci in these groups affect at least three modes of DNA repair as well as other phenotypic characteristics of the cells. As a matter of convenience the three groups are named by a prominent locus in each.

---

[4]In other words, synergism will be observed, if, when two competing repair pathways are present, it is not necessary for either to operate at maximum capacity to remove all the lesions that can be repaired. If one pathway is eliminated, the other can then work at a higher capacity so that survival of the single mutant is greater than if the two repair processes were totally independent. Thus, the double mutant will appear to be more sensitive than the sum of the sensitivities of the two single mutants. (For a more precise explanation, see the mathematical appendix of Brendel and Haynes [1973].)

RAD3 Group

Loci of the *RAD3* group (Table 2) are sensitive to UV, in general show enhanced UV mutagenesis and, with the exception of *mms19*, are sporulation-proficient (Table 3). At least nine of these loci are known to control error-free excision of pyrimidine dimers (Table 1). Reynolds and Friedberg (1980, 1981) have found, in whole cell studies, that *rad1–rad4* and *rad14*, but not *rad7*, are wholly or partially defective in the production of single-strand DNA breaks during post-UV incubation. However, if UV-irradiated DNA is preincised with a dimer-specific endonuclease, then cell extracts from *rad1–rad4, rad7, rad10, rad14,* and *rad16* can excise the dimers (Reynolds et al. 1981). These results suggest that it is the initial incision step of repair that is defective in these mutants. Bekker et al. (1980) have reported that cell extracts of *rad1–rad4, rad10,* and *rad16* possess a dimer-specific endonuclease activity. In addition to defects in pyrimidine dimer excision, *rad3* is unable to remove psoralen plus 360-nm light-induced DNA interstrand cross-links or monoadducts (W. J. Jachmczyk and R. C. von Borstel, pers. comm.).

RAD52 Group

The *RAD52* group (Table 2) is sensitive primarily to X-rays (Game and Mortimer 1974), although loci in this group also control a so-called minor pathway for repair of UV damage (Cox and Game 1974). Most mutants in this group affect sporulation and recombination, and the majority have been shown to depress UV-induced and chemically induced mutation (Table 3). However, the effects of these mutants on recombination are much greater than their effects on mutation. *rad51, rad52, rad54,* and *rad57* are defective in the repair of DNA double-strand breaks (Ho 1975; Resnick and Martin 1976; Mowat and Hastings 1979; W. J. Jachymczyk and R. C. von Borstel; D. Schild and R. K. Mortimer; both pers. comm.), but *rad51* is capable of repairing induced single-strand breaks (Mowat and Hastings 1979); *rad51* can also remove DNA interstrand cross-links induced by psoralen plus 360-nm light (W. J. Jachymczyk and R. C. von Borstel, pers. comm.). As might be expected, these strains are defective in induced recombination (Table 1). Moreover, other mutants incapable of repairing strand breaks (e.g., *rad6, cdc9*) also show decreased levels of induced recombination. Thus, defects in recombination may reflect, among other things, an inability to repair strand breaks. On this basis, other members of the *RAD52* group, which reduce or eliminate meiotic recombination and/or spontaneous and induced mitotic recombination (e.g., *rad50, rad53, rad56*), may also be unable to repair breaks in DNA strands.

Homozygosity for mating type confers sensitivity to X-rays but not to UV (Mortimer 1958; Laskowski 1960; Friis and Roman 1968). Diploids homozygous for mating type, and also for *rad51* or *rad52*, are no more

Table 3 Properties of epistasis groups

Group	Sensitivity to[a]					Dimer excision[b]	LHR[c]	Sporulation[d] meiotic
	UV	X/γ	MMS	HN2	8MOP+UV			
RAD3	+(11/11)	0(7/10)	+(8/9)		+(3/3)	-(9/9)	+(5/8)	+(8/10)
RAD6	+(14/14)	+(9/10)	+(6/7)	+(4/4)		+(6/6)	+(6/7)	+(8/12)
RAD52	+(10/10)	+(9/10)	+(9/9)		+(7/8)		+(7/8)	-(8/9)

Group	Mutation[e]				Recombination[e]				
	spont.	UV	X/γ	chem.	spont.	UV	X/γ	chem.	meiotic
RAD3	+(3/5)	+(9/11)	0(3/3)	+(7/7)	0(3/5)	+(6/6)			0(5/5)
RAD6	+(4/8) 0(2/8) -(2/8)	-(14/14)	+(8/9)	-(8/8)	+(2/6) 0(3/6) -(1/6)	+(5/7)	+(4/5)		0(4/5)
RAD52	+(7/9)	-(6/9)	0(5/8)	-(8/8)	-(6/9)	-(7/9)	-(6/9)	-(5/8)	-(4/5)

If less than three members of a group have been tested for a given property, the results are not included. The numbers in parentheses give the fraction of group members, tested for the particular property, that exhibit the indicated response. To date, the *RAD3*, *RAD6*, and *RAD52* groups contain 11, 14, and 10 members, respectively. Thus, it is obvious that, in some cases, <50% of the members of a given group have been tested for a specific property. As more data are reported, certain group characteristics indicated here may change. For definitions of abbreviations, see footnote to Table 1.

[a] Sensitivities vary substantially, both between and within the epistasis groups. See Table 1 for more details on relative sensitivities of particular mutants. +, any measurable degree of sensitivity; 0, wild-type response.

[b] +, possesses ability to excise pyrimidine dimers; -, lacks ability.

[c] +, liquid holding increases survival.

[d] +, normal; -, reduced.

[e] +, increased response; -, decreased response; 0, wild-type response.

sensitive to X-rays than are similar strains heterozygous for mating type (Ho and Mortimer 1973; Morrison and Hastings 1979). However, we have not included the mating-type locus in the *RAD52* group because (1) the epistatic effect obviously can be observed only in diploids, and (2) no actual mutation is involved in the change from *MATa/MATα* to *MATa/MATa* or *MATα/MATα*, since *MATa* and *MATα* are codominant wild-type alleles. Cells homozygous for mating type are also sensitive to MMS, display reduced frequencies of spontaneous and UV-induced recombination, and are unable to sporulate (Table 1). Morrison et al. (1980) have suggested that an error-free repair process is present in **a**/α diploids but is absent from strains homozygous for mating type.

RAD6 Group

Loci of the *RAD6* group (Table 2) influence sensitivity to both UV and X-rays and control error-prone repair (Tables 1 and 3); some of these mutants may also affect error-free repair. In addition, they are remarkably pleiotropic in that some of them are required for meiotic and mitotic recombination, for sporulation (Table 1), and even for resistance to the antifolate drug trimethoprim (Game et al. 1975). McKee and Lawrence (1979) have suggested that loci of the *RAD6* epistasis group can be subdivided into four categories. The first contains only the *RAD6* locus itself; the function of this gene is required for error-prone repair of lesions produced by most mutagens so far examined, as well as for the component of error-free repair associated with this epistasis group. The mutant *rad6* can excise thymine dimers (Prakash 1977a), although dimer excision may be inefficient (Haladus and Zuk 1979), and it is unable to repair DNA strand breaks induced by MMS (Jachymczyk et al. 1977; Chlebowicz and Jachymczyk 1979). The second category, which includes the *RAD18* locus, appears to be concerned only with *RAD6*-dependent, error-free repair. *rad18* accumulates single- and double-strand DNA breaks during post-X-irradiation incubation (Mowat and Hastings 1979) and may be partially defective in dimer excision (Boram and Roman 1976). However, this latter deficiency does not appear to be in the endonucleolytic or exonucleolytic steps of excision repair (Reynolds and Friedberg 1981). The third category, which includes *RAD9* and *RAD15* (both capable of thymine dimer excision [Table 1]), contains genes that function in an error-free or error-prone mode, depending on the nature of the lesions attacked. The fourth category, which contains the *REV* loci, consists of genes concerned exclusively with error-prone repair and conveys only a moderate degree of radiation sensitivity.

Finally, it should be noted that no mutation that is fully analogous with *recA* in *E. coli* has been discovered in yeast, i.e., no mutation depresses the induction of mutation and recombination by all agents. However, the situation is complicated by the fact that there are 30 genes (see Table 1) that exhibit some degree of coordinate control over mutation and recombination (either spontaneous or induced by specific agents.)

Significance of Epistasis Groups

Despite the increasing number of loci that have been found to affect sensitivity to radiation and/or chemicals in yeast, it is doubtful that any additional UV-epistasis groups exist. This surmise is based on the observation that in a haploid triple mutant containing one allele from each group, the UV dose for one lethal hit is sufficient to produce only one or two pyrimidine dimers per genome (Cox and Game 1974). However, as can be seen in Table 2, the epistasis groups are not mutually exclusive. For example, it has been found that some alleles (r_1^s, $cdc8$) appear to belong simultaneously to two epistasis groups. In addition, $rad6$ and $rad18$, both members of the same UV-epistasis group, are not epistatic with regard to X-ray sensitivity (Game and Mortimer 1974). Finally, members of two or more groups have various properties in common. Thus, all three groups contain mutants that are mutators, that reduce or eliminate sporulation, and that decrease the yield of induced recombinants.

We are left with a picture of a large number of loci that fall into three slightly overlapping epistasis groups. These groups contain genes that are pleiotropic and effect a variety of functions associated with nucleic acid metabolism. This suggests that we should enlarge our view of the macromolecular significance of these groups. It was argued initially that each group corresponds to a distinct "biochemical pathway" for repair and, certainly, genes coding for enzymes involved in sequential repair reactions would be expected to be epistatic to one another. However, the fact that the *RAD3* group contains nine loci controlling dimer excision (and probably more are involved) suggests that a larger number of gene products are required for the process than might be expected on strict enzymological grounds. Thus, it is possible that the epistasis groups may control the formation of macromolecular complexes that are required to mediate the three major manifestations of DNA metabolism, namely, excision repair, normal replication (possibly including some steps of precursor synthesis), and recombination. If such "metabolic complexes" exist in yeast, then it is not surprising that there should be a substantial number of loci epistatic to one another. If these complexes share any common proteins, or in some way interact with or overlap one another, then it is also not surprising that some loci might exist that appear to belong to two epistasis groups. Although at present we have no biochemical evidence for the existence of such complexes in yeast, enzyme complexes associated with DNA precursor synthesis have been isolated from bacteriophage-T4-infected bacteria (Matthews et al. 1979) and Ehrlich ascites carcinoma cells (Shoaf and Jones 1973). Similarly, DNA replication complexes have been detected in adenovirus-infected HeLa nuclei (Abboud and Horowitz 1979) and in Chinese hamster embryo fibroblasts (Reddy and Pardee 1980). Finally, several enzymes involved in excision repair in *E. coli* are believed to function as a unit (Seeberg 1978). Thus, it would not be surprising to find such complexes in yeast.

REPAIR AND THE ORIGIN OF GENETIC CHANGE

It is clear that yeast possesses several enzymic systems for the repair of induced DNA defects that differ in their chemical structures and biological consequences. The selective advantage arising from the potential for repair inherent in double-stranded DNA could account for the ubiquity of this molecule as the genetic material of cells. Error-free repair can be viewed as a homeostatic device that maintains the structural and informational integrity of the genome. It exists in organisms as simple as mycoplasms (Smith and Hanawalt 1969) and as complex as man (Friedberg et al. 1979). Paradoxically, the two principal mechanisms of genetic change also involve DNA repair. Despite the caveats mentioned (see Repair of Premutational Lesions and Repair of Prerecombinational Lesions), current evidence suggests that induced mutation and recombination may arise in large part from processes of error-prone and recombinagenic repair. Indeed, with respect to error-prone repair, Drake (1977) has suggested that there might be a phylogenetic misrepair trend by which mutagenicity by direct base mispairing has been reduced progressively so as to leave error-prone repair as the principal mutagenic mechanism in higher organisms.

Spontaneous mutagenesis in yeast has been studied extensively by von Borstel and co-workers (von Borstel et al. 1971, 1973; Hastings et al. 1976; von Borstel 1978). In their analysis of various mutants with altered spontaneous mutabilities, they have made extensive use of the repair channeling hypothesis (von Borstel and Hastings 1977), which asserts that when a repair pathway is blocked, the lesions that ordinarily would have been repaired by that pathway may be channeled along other, normally competitive, pathways (Brendel et al. 1970; Game and Cox 1973). Within our present theoretical framework, it is natural to invoke this idea because of the frequent association between increased spontaneous mutability and UV sensitivity. Thus, in analogy with the mechanism of induced mutagenesis, Hastings et al. (1976) have argued that much spontaneous mutation arises by mutagenic repair of spontaneous lesions. On this basis, the reason *rad3* is a mutator is that if the error-free process controlled by *RAD3* is blocked, spontaneous lesions are channeled into one or more mutagenic repair pathways (Brychcy and von Borstel 1977).

The general observations that emerge from the work on spontaneous mutagenesis and that transcend any particular mechanistic theory are, first, that the maximum increase in mutation rates in mutator strains is small compared with those often found in prokaryotes and, second, that the mutator effect varies in different test systems, with suppressor loci usually showing the greatest responses. Finally, it should be noted that the only mutant with an abnormal enzyme involved in DNA replication that also is a mutator is *cdc9* (Johnston and Nasmyth 1978).

Enhanced frequencies of spontaneous recombination have been observed in 17 repair-deficient mutants, as well as in 7 other mutants for which no repair defect has been reported (Tables 1 and 3). Eleven of the former mutants and all seven of the latter strains also exhibit enhanced levels of spontaneous mutagenesis. Thus, it would appear that spontaneous recombination might arise, at least in part, from the repair of spontaneous lesions in DNA. Clearly then, DNA repair may be of great evolutionary importance both in preserving the integrity of the genome and in mediating the appearance of genetic variation.

INTERPRETATION OF DOSE-RESPONSE RELATIONS
FOR INDUCED MUTAGENESIS

Perhaps because of the endemic *odium mathematica* among biologists, it is almost fashionable to claim that nothing can be learned from the analysis of dose-response curves. However, such data are the basic product of many experiments in mutation research, and, at the very least, mathematical analysis compels one to make explicit the assumptions used in their interpretation. Furthermore, much interesting information may be gathered from careful, quantitative examination of curve shapes. In such studies there are two basic quantities to be considered: yield, $Y(x)$, and frequency, $M(x)$, where x is UV dose. Yield is the number of induced mutant clones per cell treated, whereas frequency is the number of induced mutants per survivor. Thus, by definition, frequency is the ratio of yield to surviving fraction; and so to report frequency data alone is to suppress information, since yield and survival are independent observables (Haynes and Eckardt 1979a, 1980). Frequency curves are important theoretically in studies on the mechanisms of mutagenesis, whereas yield data are of great practical value in quantifying the relative mutagenicities of different agents or the relative mutabilities of different cellular systems (Eckardt and Haynes 1980).

Frequency curves for both mutation (Fig. 6) and mitotic recombination (Fig. 7) induced by UV are often linear at low doses but become nonlinear at higher doses. Such nonlinear responses can reflect the existence of multihit processes in mutation (or recombination) fixation. However, they also arise if the lethal and genetic effects of mutagens are not statistically independent of one another, i.e., if the probability of clone formation is different for mutant and nonmutant cells in the assay system employed. In this latter case, there are said to be "δ effects" in the systems (Eckardt and Haynes 1977b), which may be associated with phenomena occurring subsequent to mutation fixation (Auerbach 1976).

The high-dose decline in recombination frequencies shown in Figure 7 and similar features seen in the frequency curves for the induction of forward mutations in the *ADE2* system (Eckardt and Haynes 1977a) can be accounted for in terms of δ effects in which the mutants (or recombinants, as the case may be) have a lower probability of clone formation than the rest of the cells in the population. Conversely, the upward curvature of the frequency curve for reversions at *LYS2* in *rad2-20* (Fig. 6) can be attributed to an increased probability of clone formation by the induced prototrophs (Eckardt and Haynes 1977b). On the other hand, the superficially similar curvature in the *RAD* wild-type case (Fig. 6) apparently is not due to a δ effect, but rather to the induction of an inducible, cycloheximide-sensitive component of error-prone repair, accompanied by some net inactivation of error-free repair (Eckardt et al. 1978). These latter conclusions were derived by using the DNA damage-repair hypothesis to construct a mechanistic model for the interpretation of mutation-frequency curves (Haynes and Eckardt 1979b). On this model also, the existence of a linear frequency response at low doses (Fig. 6) implies that there must exist in yeast a constitutive component of error-prone repair and, further, that the efficiency of error-free repair cannot be 100% at vanishingly small doses. However, it is not known why this inducible component of error-prone repair is not manifest in the *rad2* strain.

Analysis of yield data is very important for practical reasons, primarily because yield data reveal the net effect of the lethal and mutagenic actions of radiation and chemicals in cells. Although the induced mutation frequency in excision-deficient strains is higher than that in the corresponding *RAD* wild type (Fig. 6), the actual mutant yield is very much reduced. It has been shown quantitatively that the mutational sensitivity of the excision-deficient strain *rad1-1* is 6-fold greater than that of *RAD* wild type; however, this increase in sensitivity is accompanied by an 11-fold decline in mutant yield (Fig. 9). This means that for the identification of weak mutagens, it may not be wise to use tester strains carrying repair deficiencies. Finally, plots of yield versus lethal hits can be used to quantify the relative mutagenic efficiencies of agents whose exposure doses would otherwise be incommensurable (Eckardt and Haynes 1980).

MUTATIONAL MOSAICISM AND HETERODUPLEX REPAIR

Prior to 1953, when genes were considered analogous to beads on a single chromosomal string, a vexing problem in mutation research was the occasional appearance of sectored mutant clones, since only pure mutant clones were to be expected in experiments with G_1 haploid cells. With the discovery of the double-strand nature of the genetic material, the problem was reversed and became much more serious. Since pyrimidine dimers are

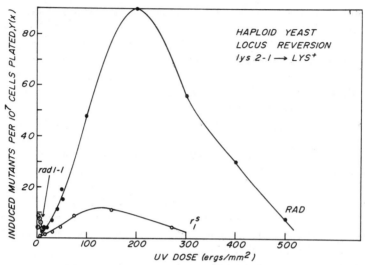

Figure 9   Mutant yield curves for UV-induced locus reversions of *lys2-1 RAD* wild type and UV-sensitive strains *rad1-1* and r_1^s. The UV sensitivity of r_1^s is roughly intermediate between that of *RAD* and *rad1-1*. It is evident that the *RAD* wild-type strain gives a higher yield of mutants than either of the UV-sensitive strains, except at very low doses where *rad1-1* yields more mutants than *RAD* or r_1^s. The initial slopes of the yield curves are proportional to the mutational sensitivities of the strains, and these were found to be in the ratio 33: 5.5: 1 for *rad1-1: RAD: r_1^s*. The yield maxima are in the ratio 11: 1.3: 1 for *RAD: r_1^s: rad1-1*. Thus, the 6-fold increase in mutational sensitivity associated with the *rad1-1* excision deficiency is accompanied by an 11-fold decline in maximum mutant yield.

intrastrand structures and since UV mutagenesis at low doses generally is a single-hit phenomenon, one would expect that the initial mutational change would occur in one strand of the DNA and, therefore, that primarily sectored (mutant/parental) clones would be formed. However, depending on the yeast species, the strain, and the mutational system used, up to ten times more pure than sectored mutant clones are produced in repair-competent cells. In contrast, sectors outnumber pure mutant clones, often substantially, in excision-deficient mutants (Abbandandolo and Simi 1971; Hannan et al. 1976; Eckardt and Haynes 1977a; Eckardt et al. 1980). Since lethal sectoring cannot account quantitatively for the loss of the parental information after a mutation has taken place in one strand, it is generally believed that some mode of heteroduplex repair effects the transfer of the mutational information to the otherwise unaltered strand (Nasim and Auerbach 1967; Nasim 1968). Heteroduplex repair is a well-delineated process in bacteria (Wildenberg and Meselson 1975; Glickman and Radman 1980), but it has not as yet been demonstrated biochemically in yeast.

In bacteria it would appear that the major mutagenic repair pathway for UV acts postreplicatively and that prereplicative error-prone repair is a rare and perhaps locus-specific event (Bridges 1977); in yeast this seems not to be the case (Kilbey et al. 1978). To account for the differing ratios of pure to sectored mutant clones in repair-proficient as compared with repair-deficient haploid G_1 cells of yeast, two hypotheses are required: (1) that error-prone repair generally occurs prior to DNA replication in repair-competent strains but after replication in excision-deficient strains (James and Kilbey 1977; James et al. 1978) and (2) that heteroduplex repair is an integral part of the mutagenic process (Eckardt et al. 1980). If error-prone repair should act postreplicatively in *RAD* wild type, a rather complicated series of events must be postulated to account for the transfer of the mutant base sequence from one mutated to the three nonmutated DNA strands in the absence of lethal sectoring; if error-prone repair acts prereplicatively, the mutation has to be transferred only from one single duplex strand to the other (Eckardt and Haynes 1977a).

It has been suggested that certain enzymes that take part in excision repair could also be involved in heteroduplex repair (Wildenberg and Meselson 1975). The enhanced frequency of sectored mutant clones among total UV-induced mutants in excision-deficient yeast therefore could be interpreted as an effect on heteroduplex repair of the genes controlling excision. However, in pedigree studies, James et al. (1978) and Kilbey et al. (1978) have shown that in *rad1*, mismatches must be generated by a postreplicative process presumed to be some mode of error-prone repair. The only way half-sectored mutant clones could arise in such a situation would be if heteroduplex repair copied the mutant information from one DNA strand into its complementary strand prior to the first postirradiation cell division; otherwise only quarter sectors would appear. Thus, the fact that a high percentage ($\sim 40\%$) of half-mutant/half-parental clones occurs after low UV doses in excision-deficient strains implies that heteroduplex repair is not blocked in these strains (Eckardt et al. 1980).

It is important to note that what are called prereplicative and postreplicative error-prone-repair events in yeast need not necessarily occur in regions of DNA remote from any putative replication complex. Depending on its size and structural organization, it is conceivable that prereplicative error-prone repair could occur within such a complex in advance of the replication fork (defined as the 3′-OH priming terminus of the leading strand), whereas postreplicative error-prone-repair events could occur in the immediate wake of the fork.

Finally, it should be noted that the very existence of heteroduplex repair poses questions of evolutionary interest: Why should a process exist that, in principle, can both eliminate and duplicate base-pairing mistakes, and is there any discrimination in the process either for or against the parental base sequence? Unfortunately, at present, we cannot answer these important questions, and so we do not know whether heteroduplex repair should be regarded primarily as a system for the avoidance, or rather the efficient fixation, of genetic change.

GENETIC EFFECTS OF THYMIDYLATE STRESS

In yeast, the only biosynthetic pathway for thymidylate (dTMP) is via methylation of deoxyuridine-5'-monophosphate (dUMP). This reaction is catalyzed by thymidylate synthetase and is tetrahydrofolate-dependent. Thus, starvation for thymine nucleotides can be effected in three distinct ways: (1) incubation of strains auxotrophic for dTMP in the absence of thymidylate (Brendel and Langjahr 1974; Barclay and Little 1978; Little and Haynes 1979); (2) inhibition of thymidylate synthetase by 5-fluoro-2'-deoxy-uridine-5'-monophosphate (FdUMP); and (3) inhibition of folate metabolism, either by sulfonamides, which block de novo folate synthesis, or by aminopterin or methotrexate, which inhibit dihydrofolate reductase (Kunz et al. 1980a,b). Thymine nucleotide excess can be produced by incubating permeable cells in high concentrations of dTMP (Barclay and Little 1981).

Depletion of thymine nucleotide pools results in cell death. It also induces cytoplasmic petites, mitochondrial point mutations, and mitotic recombination (both gene conversion and crossing-over) but does not mutagenize nuclear genes. All of these genetic effects are very large (induction frequencies can be several orders of magnitude over spontaneous levels), and they can be prevented by concomitant provision of exogenous dTMP. Although thymidylate starvation is recombinagenic (but not mutagenic) for nuclear genes, dTMP excess is highly mutagenic; preliminary tests indicate that dTMP is probably not recombinagenic. rad6-1 strains are neither induced to recombine by dTMP starvation nor mutagenized by dTMP excess (Barclay and Little 1978, 1981; Little and Haynes 1979; Kunz et al. 1980b; Kunz and Haynes 1981b).

To account for the recombinagenic effect of thymidylate starvation, it has been suggested that dTMP deprivation increases the incorporation of uracil into DNA. This uracil could be excised by uracil-DNA glycosylase, an enzyme known to be present in yeast. (W. Crosby et al., pers. comm.). Under thymineless conditions, incomplete uracil excision repair could generate single-strand breaks or gaps that then could act as substrates for recombination (Kunz et al. 1980b). High concentrations of dTMP could induce mutagenesis either by pyrimidine pool imbalances resulting in erroneous insertion of 2'-deoxythymidine-triphosphate (dTTP) in place of deoxycytidine triphosphate (dCTP) or by signaling an alteration in the DNA replication complex, which leads to decreased fidelity during DNA synthesis (Barclay and Little 1981).

The fact that the effects of thymidylate stress are mediated by the RAD6-gene product tends to blur the distinction between misrepair and misreplication as mechanisms for the induction of genetic change. Furthermore, it is evident that genetic effects can be produced not only by radiation and alien chemical attack on DNA but more broadly by disturbances in the precursor pools of DNA nucleotides and related metabolities. A challenge for future research will be to determine whether the DNA damage-repair concept can be elaborated at the biochemical level to accommodate these findings (Barclay et al. 1982).

SUMMARY

Mechanisms for the repair of various kinds of structural alterations or defects that can arise in DNA are widespread in nature. The discovery of repair and its involvement in mutagenesis and recombinagenesis, as well as in the protection against the lethal effects of radiation and chemical mutagens, has led to the formulation of the DNA damage-repair concept, which is described in this paper. Ninety-five mutations have been described in yeast that affect sensitivity to radiation and chemical mutagens, as well as other cellular characteristics. However, the actual number of distinct loci involved in repair is not known nor are the allelic relations of all the putative repair-related loci established. Studies on the mutagen sensitivity of various double mutants have led to the recognition of three epistasis groups among these loci. One epistasis group (*RAD3*) appears to control the error-free excision repair of pyrimidine dimers and certain other bulky lesions in DNA. A second group (*RAD6*) appears to control, among other things, a mode of error-prone repair, which is responsible for most, if not all, forms of induced mutagenesis. The third group (*RAD52*) is required for the repair of DNA strand breaks and appears to play an important, though not fully elucidated, role in genetic recombination. The DNA damage-repair hypothesis has also been invoked to account for spontaneous mutagenesis and to interpret the shapes of the UV dose-response curves for killing, mutation, and recombination. The induction of high frequencies of pure mutant clones by UV light in repair-proficient strains provides evidence for the existence of heteroduplex repair in yeast. Finally, the recent discovery that starvation for thymine nucleotides is highly recombinagenic, whereas excess thymidylate is highly mutagenic in yeast, promises to open the way to a deeper understanding than we have at present of the biochemical relations among repair, replication, and recombination in eukaryotes.

ACKNOWLEDGMENTS

We thank Drs. B. J. Barclay, M. Brendel, B. S. Cox, F. Eckardt, J. C. Game, N. A. Kahn, J. G. Little, and M. H. Patrick for many helpful discussions over the years on the subject of this paper. In particular, we wish to mention that

Figure 3 is based on experiments carried out by M. H. Patrick and Figures 5, 6, and 9, on measurements made by Dr. F. Eckardt. We thank Dr. F. Zimmermann for the gift of strain D7 used in the experiments shown in Figures 4 and 7. Mrs. Ingeborg Vranesic typed the manuscript, and we thank Kathryn Pepper, Anna Stih, and Gordon Temple for expert technical assistance.

This work was supported by grants from the National Research Council and the Natural Sciences and Engineering Research Council of Canada.

REFERENCES

Abbandandolo, A. and S. Simi. 1971. Mosaicism and lethal sectoring in G1 cells of *Schizosaccharomyces pombe*. *Mutat. Res.* **12**: 143.

Abboud, M.M. and M.S. Horowitz. 1979. The DNA polymerases associated with the adenovirus type 2 replication complex: Effect of 2'-3' dideoxythymidine-5'-triphosphate on viral DNA synthesis. *Nucleic Acids Res.* **6**: 1025.

Auerbach, C. 1976. *Mutation research*. Chapman and Hall, London.

Averbeck, D., E. Moustacchi, and E. Bisagni. 1978. Biological effects and repair of damage photoinduced by a derivative of psoralen substituted at the 3,4 reaction site. Photoreactivity of this compound and lethal effect in yeast. *Biochim. Biophys. Acta* **518**: 464.

Averbeck, D., W. Laskowski, F. Eckardt, and E. Lehmann-Brauns. 1970. Four radiation sensitive mutants of *Saccharomyces*. *Mol. Gen. Genet.* **107**: 117.

Baker, B.S., A.T.C. Carpenter, M.S. Esposito, R. Esposito, and L. Sandler. 1976. The genetic control of meiosis. *Annu. Rev. Genet.* **10**: 53.

Barclay, B.J. and J.G. Little. 1978. Genetic damage during thymidylate starvation in *Saccharomyces cerevisiae*. *Mol. Gen. Genet.* **160**: 33.

———. 1981. Mutations in yeast induced by thymidine monophosphate. *Mol. Gen. Genet.* **181**: 279.

Barclay, B.J., B.A. Kunz, J.G. Little, and R.H. Haynes. 1982. Genetic and biochemical consequences of thymidylate stress. *Can. J. Biochem.* (in press).

Bekker, M.L., O.K. Kaboev, A.T. Akhmedov, and L.A. Luchinska. 1980. Ultraviolet-endonuclease activity in cell extracts of *Saccharomyces cerevisiae* defective in excision of pyrimidine dimers. *J. Bacteriol.* **142**: 322.

Benathen, I.A. and C.A. Beam. 1977. Genetic control of X-ray resistance in budding yeast cells. *Radiat. Res.* **69**: 99.

Bird, R.P. and T.R. Manney. 1974. Radioresistance in yeast (*S. cerevisiae*) is associated with DNA synthesis but not budding. *Radiat. Res.* **59**: 287.

Boram, W.R. and H. Roman. 1976. Recombination in *Saccharomyces cerevisiae*: A DNA repair mutation associated with elevated mitotic gene conversion. *Proc. Natl. Acad. Sci.* **73**: 2828.

Brendel, M. and W.W. Fäth. 1974. Isolation and characterization of mutants of *Saccharomyces cerevisiae* auxotrophic and conditionally auxotrophic for 5'-dTMP. *Z. Naturforsch.* **29c**: 733.

Brendel, M. and R.H. Haynes. 1972. Kinetics and genetic control of the incorporation of thymidine monophosphate in yeast DNA. *Mol. Gen. Genet.* **117**: 39.

———. 1973. Interactions among genes controlling sensitivity to radiation and alkylation in yeast. *Mol. Gen. Genet.* **125**: 197.

Brendel, M. and U.G. Langjahr. 1974. "Thymineless death" in a strain of *Saccharomyces cerevisiae* auxotrophic for deoxythymidine-5'-monophosphate. *Mol. Gen. Genet.* **131**: 351.

Brendel, M., N.A. Khan, and R.H. Haynes. 1970. Common steps in the repair of radiation and alkylation damage in yeast. *Mol. Gen. Genet.* **106**: 289.

Bridges, B.A. 1977. Recent advances in basic mutation research. *Mutat. Res.* **44**: 149.

Brunborg, G. and D.H. Williamson. 1978. The relevance of the nuclear division cycle to radiosensitivity in yeast. *Mol. Gen. Genet.* **162**: 277.

Brunborg, G., M.A. Resnick, and D.H. Williamson. 1980. Cell cycle specific repair of DNA double strand breaks in *Saccharomyces cerevisiae*. *Radiat. Res.* **82**: 547.

Brutlag, D. and A. Kornberg. 1972. Enzymatic synthesis of deoxyribonucleic acid. XXXVI. A proofreading function for the 3′ → 5′ exonuclease activity in deoxyribonucleic acid polymerases. *J. Biol. Chem.* **247**: 241.

Bryant, D.W. and R.H. Haynes. 1978a. A DNA endonuclease isolated from yeast nuclear extract. *Can. J. Biochem.* **56**: 181.

──────. 1978b. Endonuclease α from *Saccharomyces cerevisiae* shows increased activity on ultraviolet irradiated native DNA. *Mol. Gen. Genet.* **167**: 139.

Brychcy, T. and R.C. von Borstel. 1977. Spontaneous mutability in UV sensitive excision deficient strains of *Saccharomyces*. *Mutat. Res.* **45**: 185.

Cassier, C., R. Chanet, J.A.P. Henriques, and E. Moustacchi. 1981. The effects of three *PSO* genes on induced mutagenesis: A novel class of mutationally defective yeast. *Genetics* **96**: 841.

Chang, L.M.S. 1977. DNA polymerases from baker's yeast. *J. Biol. Chem.* **252**: 1873.

Chlebowicz, E. and W.J. Jachymczyk. 1977. Endonuclease for apurinic sites in yeast. Comparison of the enzyme activity in the wild-type and *rad* mutants of *Saccharomyces cerevisiae* to mms. *Mol. Gen. Genet.* **154**: 221.

──────. 1979. Repair of mms-induced double-strand breaks in haploid cells of *Saccharomyces cerevisiae*, which requires the presence of a duplicate genome. *Mol. Gen. Genet.* **167**: 279.

Cox, B.S. 1978. Recombination and repair in simple eucaryotes. In *DNA repair mechanisms* (ed. P.C. Hanawalt et al.), p. 429. Academic Press, New York.

Cox, B.S. and J.C. Game. 1974. Repair systems in *Saccharomyces*. *Mutat. Res.* **20**: 257.

Cox, B.S. and J.M. Parry. 1968. The isolation, genetics and survival characteristics of ultraviolet light sensitive mutants in yeast. *Mutat. Res.* **6**: 37.

Deutsch, W.A. and S. Linn. 1979. DNA binding activity from cultured human fibroblasts that is specific for partially depurinated DNA and that inserts purines into apurinic sites. *Proc. Natl. Acad. Sci.* **76**: 141.

Drake, J.W. 1977. Fundamental mutagenic mechanisms and their significance for environmental mutagenesis. In *Progress in genetic toxicology* (ed. D. Scott et al.), p. 43. Elsevier/North Holland, New York.

Durnford, J.M. and J.J. Champoux. 1978. The DNA untwisting enzyme from *Saccharomyces cerevisiae*. *J. Biol. Chem.* **253**: 1086.

Eckardt, F. and R.H. Haynes. 1977a. Induction of pure and sectored mutant clones in excision proficient and deficient strains of yeast. *Mutat. Res.* **43**: 327.

──────. 1977b. Kinetics of mutation induction by ultraviolet light in excision-deficient yeast. *Genetics* **85**: 225.

──────. 1980. Quantitative measures of mutagenicity and mutability based on mutant yield data. *Mutat. Res.* **74**: 439.

Eckardt, F., S. Kowalski, and W. Laskowski. 1975. The effect of three *rad* genes on UV-induced mutation rates in haploid and diploid *Saccharomyces* cells. *Mol. Gen. Genet.* **136**: 261.

Eckardt, F., E. Moustacchi, and R.H. Haynes. 1978. On the inducibility of error prone repair in yeast. In *DNA repair mechanisms*, (ed. P.C. Hanawalt et al.), p. 421. Academic Press, New York.

Eckardt, F., S.J. Teh, and R.H. Haynes. 1980. Heteroduplex repair as an intermediate step of mutagenesis in yeast. *Genetics* **95**: 63.

Fabre, F. and H. Roman. 1977. Genetic evidence for inducibility of recombination competence in yeast. *Proc. Natl. Acad. Sci.* **74**: 1667.

──────. 1979. Evidence that a single DNA ligase is involved in replication and recombination in yeast. *Proc. Natl. Acad. Sci.* **76**: 4586.

Ferguson, L.R. and B.S. Cox. 1980. The role of dimer excision in liquid holding recovery of UV-irradiated yeast. *Mutat. Res.* **69**: 19.

Foury, F. and A. Goffeau. 1979. Genetic control of enhanced mutability of mitochondrial DNA and X-ray sensitivity in *Saccharomyces cerevisiae*. *Proc. Natl. Acad. Sci.* **76**: 6529.

Frankenberg-Schwager, M., D. Frankenberg, D. Blöcher, and C. Adamczyk. 1979. The influence of oxygen in the survival and yield of DNA double-strand breaks in irradiated yeast cells. *Int. J. Radiat. Biol.* **36**: 261.

———. 1980. Repair of DNA double strand breaks in irradiated yeast cells under nongrowth conditions. *Radiat. Res.* **82**: 498.

Friedberg, E.C., U.K. Ehmann, and J.I. Williams. 1979. Human diseases associated with defective DNA repair. *Adv. Radiat. Biol.* **8**: 86.

Friis, J. and H. Roman. 1968. The effect of the mating type alleles on intragenic recombination in yeast. *Genetics* **59**: 33.

Game, J.C. and B.S. Cox. 1972. Epistatic interactions between four *rad* loci in yeast. *Mutat. Res.* **16**: 353.

———. 1973. Synergistic interactions between *RAD* mutations in yeast. *Mutat. Res.* **20**: 35.

Game, J.C. and R.K. Mortimer 1974. A genetic study of X-ray sensitive mutants in yeast. *Mutat. Res.* **24**: 281.

Game, J.C., L.H. Johnston, and R.C. von Borstel. 1978. Studies on repair and recombination in the ligase mutant *cdc9*. *9th International Conference on Yeast Genetics and Molecular Biology*, Rochester, New York. p. 41. (Abstr.)

———. 1979. Enhanced mitotic recombination in a ligase deficient mutant of the yeast *Saccharomyces cerevisiae*. *Proc. Natl. Acad. Sci.* **76**: 4589.

Game, J.C., J.G. Little, and R.H. Haynes. 1975. Yeast mutants sensitive to trimethoprim. *Mutat. Res.* **28**: 175.

Game, J.C., T.J. Zamb, R.J. Braun, M.A. Resnick, and R.M. Roth. 1980. The role of radiation (*rad*) genes in meiotic recombination in yeast. *Genetics* **94**: 51.

Glickman, B.W. and M. Radman. 1980. *Escherichia coli* mutator mutants deficient in methylation instructed DNA mismatch correction. *Proc. Natl. Acad. Sci.* **77**: 1063.

Goldberg, O. and J.G. Little. 1978. DNA repair replication in the yeast *Saccharomyces cerevisiae*. *9th International Conference on Yeast Genetics and Molecular Biology*, Rochester, New York. p. 40 (Abstr.)

Golin, J.E. and M.S. Esposito. 1977. Evidence for joint genic control of spontaneous mutation and genetic recombination during mitosis in *Saccharomyces*. *Mol. Gen. Genet.* **150**: 127.

Grivell, A.R. and J.F. Jackson. 1968. Thymidine kinase: Its absence from *Neurospora crassa* and some other microorganisms, and the relevance of this to the specific labelling of DNA. *J. Gen. Microbiol.* **54**: 307.

Haladus, E. and J. Zuk. 1979. Mitotic recombination in *rad*6-1 mutant of *Saccharomyces cerevisiae*. *Stud. Biophys.* **76**: 61.

Hanawalt, P.C. and R.B. Setlow, eds. 1975. *Molecular mechanisms for repair of DNA*. Plenum Press, New York.

Hanawalt, P.C., E.C. Friedberg, and C.F. Fox, eds. 1978. *DNA repair mechanisms*. Academic Press, New York.

Hanawalt, P.C., P.K. Cooper, A.K. Ganesan, and C.A. Smith. 1979. DNA repair in bacteria and mammalian cells. *Annu. Rev. Biochem.* **48**: 783.

Hannan, M.A., P. Duck, and A. Nasim. 1976. UV-induced lethal sectoring and pure mutant clones in yeast. *Mutat. Res.* **36**: 171.

Harm, H. 1976. Repair of UV-irradiated biological systems: Photoreactivation. In *Photochemistry and photobiology of nucleic acids* (ed. S.Y. Wang), vol. 2, p. 219, Academic Press, New York.

Haseltine, W.A., L.K. Gordon, C.P. Lindan, R.H. Grafstrom, N.L. Shaper, and L. Grossman. 1980. Cleavage of pyrimidine dimers in specific DNA sequences by a pyrimidine dimer DNA-glycosylase of *M. luteus*. *Nature* **285**: 634.

Hastings, P.J., S.-K. Quah, and R.C. von Borstel. 1976. Spontaneous mutation by mutagenic repair of spontaneous lesions in DNA. *Nature* **264**: 719.

Haynes, R.H. 1964a. Role of DNA repair mechanisms in microbial inactivation and recovery phenomena. *Photochem. Photobiol.* **3**: 429.

———. 1964b. Molecular localization of radiation damage relevant to bacterial inactivation. In *Physical processes in radiation biology* (ed. L. Augenstein et al.), p. 51. Academic Press, New York.

———. 1966. The interpretation of microbial inactivation and recovery phenomena. *Radiat. Res.* (Suppl.) **6**: 1.

———. 1975a. DNA repair and the genetic control of radiosensitivity in yeast. In *Molecular mechanisms for repair of DNA* (ed. P.C. Hanawalt and R.B. Setlow), p. 529. Plenum Press, New York.

———. 1975b. The influence of repair processes on radiobiological survival curves. In *Proceedings of the 6th L.H. Gray Memorial Conference* (ed. T. Alper), p. 197. Wiley, London.

Haynes, R.H. and F. Eckardt. 1979a. Analysis of dose response patterns in mutation research. *Can. J. Genet. Cytol.* **21**: 277.

———. 1979b. Complexity of DNA repair in a simple eukaryote. In *Proceedings of the 6th International Congress on Radiation Research* (ed. S. Okada et al.), p. 454. Japanese Association for Radiation Research, University of Tokyo.

———. 1980. Mathematical analysis of mutation induction kinetics. *Chem. Mutagens* **6**: 271.

Haynes, R.H. and W.R. Inch. 1963. Synergistic action of nitrogen mustard and radiation in microorganisms. *Proc. Natl. Acad. Sci.* **50**: 839.

Haynes, R.H., S. Wolff, and J. Till, eds. 1966. Structural defects in DNA and their repair in microorganisms. *Radiat. Res.* (Suppl.) **6**: 1.

Haynes, R.H., B.J. Barclay, F. Eckardt, O. Landman, B. Kunz, and J.G. Little. 1981. Genetic control of DNA repair in yeast. In *Proceedings of the XIV International Congress of Genetics*, Moscow, 1978. (ed. B.D.K. Beliayev), NAUKA, Moscow. (In press.)

Henriques, J.A.P. and E. Moustacchi. 1980a. Sensitivity to photoaddition of mono- and bifunctional furocoumarins of x-ray sensitive mutants of *Saccharomyces cerevisiae*. *Photochem. Photobiol.* **31**: 557.

———. 1980b. Isolation and characterization of *pso* mutants sensitive to photoaddition of psoralen derivatives in *Saccharomyces cerevisiae*. *Genetics* **95**: 273.

Ho, K.S.Y. 1975. Induction of DNA double strand breaks by X-rays in a radiosensitive strain of the yeast *Saccharomyces cerevisiae*. *Mutat. Res.* **30**: 327.

Ho, K.S.Y. and R.K. Mortimer. 1973. Induction of dominant lethality by x-rays in a radiosensitive strain of yeast. *Mutat. Res.* **20**: 45.

Hopper, A.K., J. Kirsch, and B.D. Hall. 1975. Mating type and sporulation in yeast. II. Meiosis, recombination and radiation sensitivity in an αα diploid with altered sporulation control. *Genetics* **80**: 61.

Hunnable, E.G. 1972. "Genes controlling genetic variation in yeast." Ph.D. thesis, Oxford University, Great Britain.

Hunnable, E.G. and B.S. Cox. 1971. The genetic control of dark recombination in yeast. *Mutat. Res.* **13**: 297.

Hutchison, F. 1978. Workshop summary: DNA strand break repair in eukaryotes. In *DNA repair mechanisms* (ed. P.C. Hanawalt et al.), p. 457. Academic Press, New York.

Jachymczyk, W.J., E. Chlebowicz, Z. Swietlinska, and J. Zuk. 1977. Alkaline sucrose sedimentation studies of mms-induced DNA single strand breakage and rejoining in the wild-type and UV-sensitive mutants of *Saccharomyces cerevisiae*. *Mutat. Res.* **43**: 1.

Jain, V.K. and W. Pohlit. 1972. Influence of energy metabolism on the repair of X-ray damage in living cells. *Biophysik* **8**: 254.

James, A.P. and B.J. Kilbey. 1977. The timing of UV-mutagenesis in yeast: A pedigree analysis of induced recessive mutation. *Genetics* **87**: 237.

James, A.P. and M.M. Werner. 1965. The radiobiology of yeast. *Radiat. Bot.* **5:** 359.

James, A.P., B.J. Kilbey, and G.J. Prefontaine. 1978. The timing of UV-mutagenesis in yeast: Continuing mutation in an excision defective (*rad*1-1) strain. *Mol. Gen. Genet.* **165:** 207.

Johnston, L.H. and K.A. Nasmyth. 1978. *Saccharomyces cerevisiae* cell cycle mutant *cdc9* is defective in DNA ligase. *Nature* **274:** 891.

Kern, R. and F.K. Zimmermann. 1978. The influence of defects in excision and error prone repair on spontaneous and induced mitotic recombination and mutation in *Saccharomyces cerevisiae. Mol. Gen. Genet.* **161:** 81.

Khan, N.A., M. Brendel, and R.H. Haynes. 1970. Supersensitive double mutants in yeast. *Mol. Gen. Genet.* **107:** 376.

Kilbey, B.J. 1975. Mutagenesis in yeast. *Methods Cell Biol.* **12:** 209.

Kilbey, B.J., T. Brychcy, and A. Nasim. 1978. Initiation of UV mutagenesis in *Saccharomyces cerevisiae. Nature* **274:** 889.

Kircher, M., R. Fleer, A. Ruhland, and M. Brendel. 1979. Biological and chemical effects of mustard gas in yeast. *Mutat. Res.* **63:** 273.

Kornberg, A. 1980. *DNA replication.* Freeman, San Francisco.

Korogodin, V.I., M.N. Meissel, and T.S. Remesova. 1966. Post-irradiation recovery of yeast. In *Radiation research* (ed. G. Silini), p. 538. North Holland, Amsterdam.

Kowalski, S. and W. Laskowski. 1975. The effects of three *rad* genes on survival, inter- and intragenic mitotic recombination in *Saccharomyces.* I. UV irradiation without photoreactivation or liquid holding post-treatment. *Mol. Gen. Genet.* **136:** 75.

Kunz, B.A. and R.H. Haynes. 1981a. Phenomenology and genetic control of mitotic recombination in yeast. *Annu. Rev. Genet.* **15:** 57.

———. 1981b. DNA repair and the genetic effects of thymidylate stress in yeast. *Mutat. Res.* (in press).

Kunz, B.A., B.J. Barclay, and R.H. Haynes. 1980a. A simple, rapid assay for mitotic recombination. *Mutat. Res.* **73:** 215.

Kunz, B.A., B.J. Barclay, J.C. Game, J.G. Little, and R.H. Haynes. 1980b. Induction of mitotic recombination in yeast by starvation for thymine nucleotides. *Proc. Natl. Acad. Sci.* **77:** 6057.

Laskowski, W. 1960. Inactivierungsversuche mit homozygoten Hefstammen verschieden ploidiegrades. *Z. Naturforsch.* **15b:** 495.

Lawrence, C.W. 1981. Mutagenesis in *Saccharomyces cerevisiae. Adv. Genet.* **21:** (in press).

Lawrence, C.W. and R. Christensen. 1976. UV mutagenesis in radiation sensitive strains of yeast. *Genetics* **82:** 207.

———. 1979. Absence of relationship between UV-induced reversion frequency and nucleotide sequence at the *cyc*1 locus of yeast. *Mol. Gen. Genet.* **177:** 31.

Lawrence, C.W., J.W. Stewart, F. Sherman, and R.B. Christensen. 1974. Specificity and frequency of ultraviolet-induced reversion of an iso-1-cytochrome *c* ochre mutant in radiation-sensitive strains of yeast. *J. Mol. Biol.* **85:** 137.

Lehman, I.R. 1974. DNA ligase: Structure, mechanism and function. *Science* **186:** 790.

Lemontt, J.F. 1971a. Mutants of yeast defective in mutation induced by ultraviolet light. *Genetics* **68:** 21.

———. 1971b. Pathways of ultraviolet mutability in *Saccharomyces cerevisiae.* I. Some properties of double mutants involving *uvs*9 and *rev. Mutat. Res.* **13:** 311.

———. 1972. Induction of forward mutations in mutationally defective yeast. *Mol. Gen. Genet.* **119:** 27.

———. 1973. Genes controlling ultraviolet mutability in yeast. *Genetics* (Suppl.) **73:** 153.

———. 1977. Pathways of ultraviolet mutability in *Saccharomyces cerevisiae.* III. Genetic analysis and properties of mutants resistant to ultraviolet-induced forward mutation. *Mutat. Res.* **43:** 179.

———. 1980. Genetic and physiological factors affecting repair and mutagenesis in yeast. In *DNA repair and mutagenesis in eukaryotes* (ed. F.J. de Serres et al.), p. 85. Plenum Press, New York.

Little, J.G. and R.H. Haynes. 1979. Isolation and characterization of yeast mutants auxotrophic for 2'-deoxythymidine 5'-monophosphate. *Mol. Gen. Genet.* **168:** 141.

Livi, G.P. and V.L. Mackay. 1980. Mating-type regulation of methyl methane-sulfonate sensitivity in *Saccharomyces cerevisiae. Genetics* **95:** 259.

Luchnik, A.N., V.M. Glaser, and S.V. Shestakov. 1977. Repair of DNA double strand breaks requires two homologous DNA duplexes. *Mol. Biol. Rep.* **3:** 43.

Luchnik, A.N., V.M. Glaser, S.P. Soldatov, and S.V. Shestakov. 1979. On the mechanism of DNA double-strand break repair in the yeast *Saccharomyces cerevisiae. Dokl. Akad. Nauk. S.S.R.* **244:** 213.

Malone, R.E. and R. Easton Esposito. 1979. The *RAD52* gene product is required for homothallic interconversion of mating types and spontaneous mitotic recombination. *Proc. Natl. Acad. Sci.* **77:** 503.

Maloney, D.H. and S. Fogel. 1980. Mitotic recombination in yeast: Isolation and characterization of mutants with enhanced spontaneous mitotic gene conversion rates. *Genetics* **94:** 825.

Martin, P., L. Prakash, and S. Prakash. 1978. A gene involved in error prone repair of radiation damage in diploid but not haploid yeast. *9th International Conference on Yeast Genetics and Molecular Biology*, Rochester, New York. p. 44 (Abstr.).

––––––. 1981. **a**/α-specific effect of the *mms3* mutation on ultraviolet mutagenesis in *Saccharomyces cerevisiae. J. Bacteriol.* **146:** 684.

Matthews, C.K., T.W. North, and G.P.V. Reddy. 1979. Multienzyme complexes in DNA precursor biosynthesis. *Adv. Enzyme Regul.* **17:** 133.

McKee, R. and C.W. Lawrence. 1979. Genetic analysis of gamma-ray mutagenesis in yeast. I. Reversion in radiation sensitive strains. *Genetics* **93:** 361.

––––––. 1980. Genetic analysis of gamma-ray mutagenesis in yeast. III. Double mutant strains. *Mutat. Res.* **70:** 37.

Monteleone, B., L. Prakash, and S. Prakash. 1978. Genetic studies on the *mms21* mutant of *Saccharomyces cerevisiae. 9th International Conference on Yeast Genetics and Molecular Biology*, Rochester, New York. p. 44 (Abstr.).

Monteleone, B.A., S. Prakash, and L. Prakash. 1981. Spontaneous mutation recombination in *mms8-1*, an allele of the *CDC9* gene of *Saccharomyces cerevisiae. J. Bacteriol.* **147:** 517.

Morrison, D.P. and P.J. Hastings. 1979. Characterization of the mutator mutation *mut5-1. Mol. Gen. Genet.* **175:** 57.

Morrison, D.P., S.-K. Quah, and P.J. Hastings. 1980. Expression in diploids of the mutator phenotype of some mutator mutants of *Saccharomyces cerevisiae. Can. J. Genet. Cytol.* **22:** 51.

Mortimer, R.K. 1958. Radiobiological and genetic studies on a polyploid series (haploid to hexaploid) of *Saccharomyces cerevisiae. Radiat. Res.* **9:** 312.

––––––. 1961. Factors controlling the radiosensitivity of yeast cells. *Brookhaven Symp. Biol.* **14:** 62.

Mowat M. and P.J. Hastings. 1979. Repair of γ-ray induced DNA strand breaks in radiation sensitive (*rad*) mutants of *Saccharomyces cerevisiae. Can. J. Genet. Cytol.* **21:** 574.

Muhammed, A. 1966. Studies on yeast photoreactivating enzyme. *J. Biol. Chem.* **241:** 516.

Nakai, S. and S. Matsumoto. 1967. Two types of radiation sensitive mutants in yeast. *Mutat. Res.* **4:** 129.

Nasim, A. 1968. Repair mechanisms and radiation-induced mutations in fission yeast. *Genetics* **59:** 327.

Nasim, A. and C. Auerbach. 1967. The origin of complete and mosaic mutants from mutagenic treatment of single cells. *Mutat. Res.* **4:** 1.

Nasim, A. and T. Brychcy. 1979. Cross sensitivity of mutator strains to physical and chemical mutagens. *Can. J. Genet. Cytol.* **21:** 129.

Nevers, P. and H.C. Spatz. 1975. *Escherichia coli* mutants *uvr*D and *uvr*E are deficient in gene conversion of heteroduplexes. *Mol. Gen. Genet.* **139:** 233.

Ord, W.R., S.-K. Quah, P.J. Hastings, and R.C. von Borstel. 1978. Characterization of the *mut7* mutator mutant of *Saccharomyces cerevisiae*. *9th International Conference on Yeast Genetics and Molecular Biology*, Rochester, New York. p.56. (Abstr.).

Parry, J.M. and B.S. Cox. 1968. The effects of dark holding and photoreactivation on ultraviolet light-induced mitotic recombination and survival in yeast. *Genet. Res.* **12**: 187.

Parry, E.M. and J.M. Parry. 1976. The genetic control of liquid-holding recovery and U.V.-induced repair resistance in the yeast, *Saccharomyces cerevisiae*. *Int. J. Radiat. Biol.* **30**: 13.

Parry, J.M. and E.M. Parry. 1969. The effect of UV light post-treatments on the survival characteristics of 21 UV sensitive mutants of *Saccharomyces cerevisiae*. *Mutat. Res.* **8**: 545.

―――. 1972. The genetic implications of UV light exposure and liquid holding treatment in the yeast *Saccharomyces cerevisiae*. *Genet. Res.* **19**: 1.

Patrick, M.H. and R.H. Haynes. 1964. Dark recovery phenomena in yeast. II. Conditions that modify the recovery process. *Radiat. Res.* **23**: 564.

―――. 1968. Repair induced changes in yeast radiosensitivity. *J. Bacteriol.* **95**: 1350.

Patrick, M.H., R.H. Haynes, and R.B. Uretz. 1962. The possibility of repair of primary radiation damage. *Radiat. Res.* **16**: 610.

―――. 1964. Dark recovery phenomena in yeast. I. Comparative effects with various inactivating agents. *Radiat. Res.* **21**: 144.

Pettijohn, D. and P.C. Hanawalt. 1964. Evidence for repair replication of ultraviolet damaged DNA in bacteria. *J. Mol. Biol.* **9**: 395.

Piñon, R. and E. Leney. 1975. Studies on deoxyribonucleases from *Saccharomyces cerevisiae*. Characterization of two endonucleases with a preference for double-stranded DNA. *Nucleic Acids Res.* **2**: 1023.

Prakash, L. 1974. Lack of chemically induced mutation in repair deficient mutants of yeast. *Genetics* **78**: 1101.

―――. 1976a. The relation between repair of DNA and radiation and chemical mutagenesis in *Saccharomyces cerevisiae*. *Mutat. Res.* **41**: 241.

―――. 1976b. Effects of genes controlling radiation sensitivity on chemically induced mutations in *Saccharomyces cerevisiae*. *Genetics* **83**: 285.

―――. 1977a. Repair of pyrimidine dimers in radiation sensitive mutants *rad3*, *rad4*, *rad6* and *rad9* of *Saccharomyces cerevisiae*. *Mutat. Res.* **45**: 13.

―――. 1977b. Defective thymine dimer excision in radiation sensitive mutants *rad10* and *rad16* of *Saccharomyces cerevisiae*. *Mol. Gen. Genet.* **152**: 125.

Prakash, L. and S. Prakash. 1977a. Isolation and characterization of mms-sensitive mutants of *Saccharomyces cerevisiae*. *Genetics* **86**: 83.

―――. 1977b. Increased spontaneous mitotic segregation in mms-sensitive mutants of *Saccharomyces cerevisiae*. *Genetics* **87**: 229.

―――. 1979. Three additional genes involved in pyrimidine dimer removal in *Saccharomyces cerevisiae*: *RAD7*, *RAD14* and *MMS19*. *Mol. Gen. Genet.* **176**: 351.

―――. 1980. Genetic analysis of error prone repair systems in *Saccharomyces cerevisiae*. In *DNA repair and mutagenesis in eukaryotes* (ed. F.J. de. Serres et al.), p. 141. Plenum Press, New York.

Prakash, L., D. Hinkle, and S. Prakash. 1979. Decreased UV mutagenesis in *cdc8*, a DNA replication mutant of *Saccharomyces cerevisiae*. *Mol. Gen. Genet.* **172**: 249.

Prakash, S., L. Prakash, W. Burke, and B.A. Monteleone. 1980. Effects of the *RAD52* gene on recombination in *Saccharomyces cerevisiae*. *Genetics* **94**: 31.

Quah, S.K., R.C. von Borstel, and P.J. Hastings. 1980. The origin of spontaneous mutation in *Saccharomyces cerevisiae*. *Genetics* **96**: 819.

Radding, C.M. 1978. Genetic recombination: Strand transfer and mismatch repair. *Annu. Rev. Biochem.* **47**: 847.

Rao, B.S. and M.S.S. Murthy. 1978. On the nature of damage involved in liquid-holding recovery in diploid yeast after gamma- and alpha-irradiation. *Int. J. Radiat. Biol.* **34**: 17.

Reddy, G.P.V. and A.B. Pardee. 1980. Multienzyme complex for metabolic channeling in mammalian DNA replication. *Proc. Natl. Acad. Sci.* **77**: 3312.

Resnick, M.A. 1969a. A photoreactivationless mutant of *Saccharomyces cerevisiae*. *Photochem. Photobiol.* **9**: 307.

―――. 1969b. Genetic control of radiation sensitivity in *Saccharomyces cerevisiae*. *Genetics* **62**: 519.

―――. 1978. The importance of DNA double strand break repair in yeast. In *DNA repair mechanisms* (ed. P.C. Hanawalt et al.), p. 417. Academic Press, New York.

―――. 1979. The induction of molecular and genetic recombination in eukaryotic cells. *Adv. Radiat. Biol.* **8**: 175.

Resnick, M.A. and P. Martin. 1976. The repair of double strand breaks in the nuclear DNA of *Saccharomyces cerevisiae* and its genetic control. *Mol. Gen. Genet.* **143**: 119.

Resnick, M.A. and J.K. Setlow. 1972. Repair of pyrimidine dimer damage induced in yeast by ultraviolet light. *J. Bacteriol.* **109**: 976.

Reynolds, R. 1978. Removal of pyrimidine dimers from *Saccharomyces cerevisiae* nuclear DNA under nongrowth conditions as detected by a sensitive enzymatic assay. *Mutat. Res.* **50**: 43.

Reynolds, R.J. and E.C. Friedberg. 1980. The molecular mechanism of pyrimidine dimer excision in *Saccharomyces cerevisiae*. I. Studies with intact cells and cell free systems. In *DNA repair and mutagenesis in eukaryotes.* (ed. A. Hollaender), p. 121. Plenum Press, New York.

―――. 1981. The molecular mechanism of pyrimidine dimer excision in *Saccharomyces cerevisiae*. II. The incision of UV-irradiated DNA *in vivo*. *J. Bacteriol.* **146**: 692.

Reynolds, R.J., J.D. Love, and E.C. Friedberg. 1981. Molecular mechanisms of pyrimidine dimer excision in *Saccharomyces cerevisiae:* Excision of dimers in cell extracts. *J. Bacteriol.* **147**: 705.

Rodarte-Ramon, U.S. 1972. The genetic control of recombination in mitosis and meiosis. *Radiat. Res.* **49**: 148.

Rodarte-Ramon, U.S. and R.K. Mortimer. 1972. Radiation induced recombination in *Saccharomyces*: Isolation and genetic study of recombination deficient mutants. *Radiat. Res.* **49**: 133.

Roman, H. and S.M. Sands. 1953. Heterogeneity of clones of *Saccharomyces* derived from haploid ascospores. *Proc. Natl. Acad. Sci.* **39**: 171.

Ruhland, A. and M. Brendel. 1979. Mutagenesis by cytostatic alkylating agents in yeast strains of differing repair capacities. *Genetics* **92**: 83.

Rupp, W.D. and P. Howard-Flanders. 1968. Discontinuities in the DNA of an excision-deficient strain of *Escherichia coli* following ultraviolet irradiation. *J. Mol. Biol.* **31**: 291.

Saeki, T., I. Machida, and S. Nakai. 1980. Genetic control of diploid recovery after γ-irradiation in the yeast *Saccharomyces cerevisiae*. *Mutat. Res.* **73**: 251.

Seeberg, E. 1978. Reconstitution of an *Escherichia coli* repair endonuclease activity from the separated $uvrA^+$, $uvrB^+$, $uvrC^+$ gene products. *Proc. Natl. Acad. Sci.* **75**: 2569.

Shoaf, T. and M.E. Jones. 1973. Uridylic acid synthesis in Ehrlich ascites carcinoma cells. Properties, subcellular distribution and nature of enzyme complexes of the six biosynthetic enzymes. *Biochemistry* **12**: 4039.

Smith, D.W. and P.C. Hanawalt. 1969. Repair replication of DNA in ultraviolet irradiated *Mycoplasma laidlawii* B. *J. Mol. Biol.* **46**: 57.

Snow, R. 1967. Mutants of yeast sensitive to ultraviolet light. *J. Bacteriol.* **94**: 571.

―――. 1968. Recombination in ultraviolet sensitive strains of *Saccharomyces cerevisiae*. *Mutat. Res.* **6**: 409.

Strike, T.L. 1978. "Characterization of mutants sensitive to X-rays." Ph.D. thesis, University of California, Davis.

Unrau, P., R. Wheatcroft, and B.S. Cox. 1971. The excision of pyrimidine dimers from DNA of ultraviolet irradiated yeast. *Mol. Gen. Genet.* **113**: 359.

Villani, G.S., S. Boiteux, and M. Radman. 1978. Mechanism of ultraviolet-induced mutagenesis. Extent and fidelity of *in vitro* DNA synthesis on irradiated templates. *Proc. Natl. Acad. Sci.* **75**: 3037.

von Borstel, R.C. 1978. Measuring spontaneous mutation rates in yeast. *Methods Cell Biol.* **20**: 1.

von Borstel, R.C. and P.J. Hastings. 1977. Mutagenic repair pathways in yeast. In *Research in photobiology* (ed. A. Castellani), p. 683. Plenum Press, New York.

von Borstel, R.C., K.T. Cain, and C.M. Steinberg. 1971. Inheritance of spontaneous mutability in yeast. *Genetics* **69**: 17.

von Borstel, R.C., S.-K. Quah, C.M. Steinberg, F. Flury, and D.J.C. Gottlieb. 1973. Mutants of yeast with enhanced spontaneous mutation rates. *Genetics* (Suppl.) **73**: 141.

Wheatcroft, R., B.S. Cox, and R.H. Haynes. 1975. Repair of UV-induced DNA damage and survival in yeast. I. Dimer excision. *Mutat. Res.* **30**: 209.

Wickner, R.B. 1974. Mutants of *Saccharomyces cerevisiae* that incorporate dTMP into DNA *in vivo*. *J. Bacteriol.* **117**: 252.

Wildenberg, J. and M. Meselson. 1975. Mismatch repair in heteroduplex DNA. *Proc. Natl. Acad. Sci.* **72**: 2202.

Witkin, E.M. 1967. Mutation proof and mutation prone modes of survival in derivatives of *Escherichia coli* B differing in sensitivity to ultraviolet light. *Brookhaven Symp. Biol.* **20**: 17.

———. 1969. Ultraviolet-induced mutation and DNA repair. *Annu. Rev. Genet.* **3**: 525.

———. 1976. Ultraviolet mutagenesis and inducible DNA repair in *Escherchia coli*. *Bacteriol. Rev.* **40**: 869.

Zimmermann, F.K. 1968. Sensitivity to methylmethane sulfonate and nitrous acid of ultraviolet light sensitive mutants of *Saccharomyces cerevisiae*. *Mol. Gen. Genet.* **102**: 247.

Zimmermann, F.K., R. Kern, and H. Rasenberger. 1975. A yeast strain for simultaneous detection of induced mitotic crossing over, mitotic gene conversion and reverse mutation. *Mutat. Res.* **28**: 381.

Killer Systems
in *Saccharomyces cerevisiae*

Reed B. Wickner
Laboratory of Biochemical Pharmacology
National Institute of Arthritis, Metabolism, and Digestive Diseases
National Institutes of Health
Bethesda, Maryland 20205

INTRODUCTION

Killer strains of *Saccharomyces cerevisiae* are those secreting a protein toxin that is lethal to most nonkiller *S. cerevisiae* strains (Makower and Bevan 1963; Woods and Bevan 1968) (Fig. 1). The two distinct killer traits (K_1 and K_2) are carried by double-stranded (DS) RNA plasmids in intracellular particles. These particles are not infectious, but they have an RNA polymerase activity that produces a single-stranded (SS) product. The killer plasmids interact with each other and with other non-Mendelian genetic elements. These DS RNA genomes depend heavily on the host genome for their replication, expression, and regulation. The K_1 killer toxin is a small protein that binds to sensitive cells and kills them, apparently by inducing the release of cellular ATP and potassium into the medium.

The killer systems provide useful models for the study of the control of viral or plasmid replication, especially the role of the host, plasmid or viral interference, defective-interfering (DI) particles, protein processing and secretion, and toxin action and receptors. The facts, models, and problems of

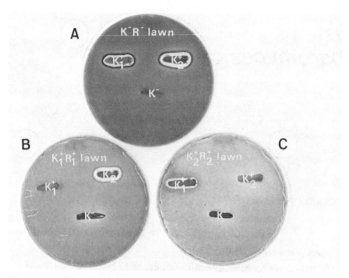

Figure 1   The killer phenomenon. (*A*) Lawn of [KIL-o] strain. (*B*) Lawn of [KIL-k₁] strain. (*C*) Lawn of [KIL-k₂] strain. A K_1^+, K_2^+, and K^- strain are streaked on each lawn.

various aspects of these systems will be discussed in this paper. Several recent reviews have also appeared (Pietras and Bruenn 1976; Wickner 1976a, 1979a; Toh-e 1979; Vodkin and Alianell 1979; Bruenn 1980; Bussey 1981).

Killer strains have been identified in many fungal genera, including *Saccharomyces, Ustilago, Torulopsis, Debaromyces, Hansenula, Kluyveromyces, Candida, Pichia,* and *Cryptococcus* (Makower and Bevan 1963; Puhalla 1968; Hankin and Puhalla 1971; Maule and Thomas 1973; Naumov and Naumova 1973; Bussey and Skipper 1975; Philliskirk and Young 1975; Stumm et al. 1977; Kandel and Stern 1979). In addition to the *Saccharomyces* killers, those in *Ustilago* have also been studied extensively (Kotlin and Day 1975, 1976a,b; Koltin 1977; Koltin and Kandel 1978; Koltin et al. 1978; Levine et al. 1979). Killer strains (K^+ phenotype) are generally resistant to the toxin they produce (R^+ phenotype). Most laboratory strains of *S. cerevisiae* are either killer (K^+R^+) or sensitive (K^-R^-), with an occasional neutral (K^-R^+) strain. Killer strains of *S. cerevisiae* may have one of two killer-resistance specificities, $K_1^+R_1^+$ or $K_2^+R_2^+$. The K_1 strains are those first discovered by Makower and Bevan (1963) and widely distributed in laboratory yeast (Fink and Styles 1972). K_1 and K_2 strains are sensitive to each other's toxin (Fig. 1). K_2 killers have been found among wine yeasts (Naumova and Naumov 1973; Naumov et al. 1973), as brewery contaminants (Maule and Thomas 1973; Ouchi and Akiyama 1976; Rogers and Bevan 1978), and in the National Collection of Yeast cultures (Young and Yagiu 1978). The K_3 strain kills different strains than the K_2 killer does (Young and Yagiu 1978), but K_2 and

K_3, when transferred into laboratory yeast, are cross-immune (R. B. Wickner, unpubl.).

Crosses of $K_1^+R_1^+$ with K^-R^- strains produced 4 $K_1^+R_1^+$:0 meiotic segregation (Somers and Bevan 1969), and crosses of $K_1^+R_1^+$ with $K_1^-R_1^+$ haploid strains showed mitotic segregation of $K_1^+R_1^+$ and $K_1^-R_1^+$ in the diploids (Bevan and Somers 1969), which indicates that a non-Mendelian genetic element (the killer plasmid) controls the killer trait. The K_1 killer trait can also be transferred by cytoplasmic mixing using the *kar1* (karyogamy-defective) mutant (Conde and Fink 1976).

Nomenclature is summarized in Table 1. Plasmid genes are in brackets; chromosomal genes are italicized (Table 3).

DS RNA AND VIRUSLIKE PARTICLES

L DS RNA

Almost all *S. cerevisiae* carry *L* DS RNA, a linear molecule composed of two equal-size strands. The size of *L* varies little from strain to strain, and its

Table 1 Notation

Phenotype[a]	
K_1^+ or K_1^-	ability or inability to secrete active K_1 toxin
K_1^{++}	superkiller phenotype
R_1^+ or R_1^-	resistance or sensitivity to killer toxin
Killer plasmid genotypes[b]	
[KIL-k_1] and [KIL-k_2]	wild-type K_1 and K_2 killer plasmids
[KIL-o]	no killer plasmid
[KIL-n_1]	plasmid conferring resistance to K_1 toxin but not toxin production
[KIL-ts]	killing is temperature-sensitive
[KIL-i]	confers toxin production but not resistance
[KIL-s]	defective-interfering plasmid (suppressive); prevents replication of [KIL-k]
[KIL-d]	defective maintenance and expression in haploid strains; normal in a/α diploid strains
[KIL-b]	bypasses needed for some *mak* genes; also confers superkiller phenotype
[KIL-sk]	confers superkiller phenotype
[KIL-sd]	plasmid that is dependent on a *ski* mutation for replication
[KIL-kd]	deletion mutant of a nonessential region of [KIL-k_1]

[a]The killer or resistance phenotype of a particular specificity is denoted by a subscript, e.g., K_2 or R_1. If no subscript is shown, the K_1 or R_1 specificity is intended.

[b]Specific mutant alleles may be indicated by a number following the lower-case letter inside the brackets, e.g., [KIL-s3]. To indicate that a particular mutant killer plasmid is derived from [KIL-k_1] or [KIL-k_2], a subscript 1 or 2 is used, e.g., [KIL-n_2]. In the absence of a subscript, derivation from [KIL-k_1] is implied.

molecular weight is between 2.5×10^6 and 3.4×10^6 as measured by gel electrophoresis, electron microscopy, sedimentation rate, and diffusion rate, end-group analysis, and denatured single-strand molecular weight (Bevan et al. 1973; Vodkin et al. 1974; Sweeney et al. 1976; Wickner and Leibowitz 1976a; Herring and Bevan 1977; Bruenn and Kane 1978; Holm et al. 1978). The disagreements are due in part to different molecular-weight standards used in gel electrophoresis and different values used for the internucleotide distance for DS RNA. L has pppGp at both of its 5' ends (Bruenn and Keitz 1976), and its 3' ends show some sequence heterogeneity, with the major 3' sequences being GUGUGCGAGUGGAAAAAUGCA-OH and GAAUUUAAAAAUUUUU-CA-OH (Bruenn and Brennan 1980). Since the 3' ends are both A, and the 5' ends are both G, the ends are either not flush or not paired.

L DS RNA is found in intracellular 150S–160S viruslike particles of 35–40 nm (Fig. 2) with little or no free DS RNA detected (Bevan et al. 1973; Buck et al. 1973; Herring and Bevan 1974; Weber and Lindner 1975; Adler et al. 1976; Hopper et al. 1977; Oliver et al. 1977; Harris 1978; Bostian et al. 1980a; Toh-e and Wickner 1980). These particules are isometric with a density in CsCl of 1.40. Oliver et al. (1977) reported that the major protein of these particles has a molecular weight of 75,000 with minor components of 53,000 and 37,000 m.w. present at about one-tenth the amount of the 75,000-m.w. species. Hopper et al. (1977) also found a single major particle protein of 88,000 m.w. and minor components of 140,000 m.w., 82,000 m.w., and 78,000 m.w.

The major product of translation of the denatured L DS RNA in a wheat germ system coelectrophoresed with the major particle protein, had an indistinguishable peptide map, and was precipitated by antibody prepared against the particles (Hopper et al. 1977). Thus, L codes for the major protein of the particles in which it is encapsulated.

RNA polymerase activity has been detected in viruslike particles containing L DS RNA (Herring and Bevan 1977). The product is a full-length SS copy of L that does not self-anneal (Herring and Bevan 1977; Welsh and Leibowitz 1980) and is released from the viruslike particles (Welsh et al.

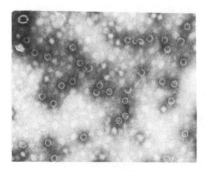

Figure 2 Viruslike particles from a killer strain of *S. cerevisiae*. (Photo from Welsh et al. 1980.)

1980). Small amounts of DS RNA synthesis by particles have also been reported (Bevan and Herring 1976; Herring and Bevan 1977). The role of these activities in the replication and expression of L DS RNA is not yet known.

Strains lacking L DS RNA have been described previously (Mitchell et al. 1976; Livingston 1977; Vodkin 1977b). Defects in growth, mating, or other cellular functions have so far not been identified in these strains, which suggests that L DS RNA may be essentially parasitic.

Two reports have claimed that L and M DS RNAs (see M DS RNA) hybridize with cellular DNA (Fischer and Shalitin 1977; Vodkin 1977a), but since others have been unable to confirm this claim (Wickner and Leibowitz 1976a; Hastie et al. 1978) and the original data are questionable, it seems unlikely that either M or L is present as a DNA copy.

M DS RNA

In addition to L DS RNA, killer strains have another smaller species: M_1 (1.1×10^6 to 1.7×10^6) in K_1 killers, M_2 (1.0×10^6) in K_2 killers, and M_3 (0.87×10^6) in K_3 killers (Bevan et al. 1973; Vodkin et al. 1974; Adler et al. 1976; Nesterova et al. 1976; Sweeney et al. 1976; Wickner and Leibowitz 1976a; Young and Yagiu 1978) (Fig. 3). In this case, as well as with L DS RNA, part of the disagreement about the size of M DS RNA can be traced to the use of different size standards.

The first evidence that M_1 DS RNA is the K_1 killer plasmid was its presence and absence in killer and sensitive strains, respectively (Bevan et al. 1973; Vodkin et al. 1974). M DS RNA (not L) is eliminated by the treatments that cure the killer plasmid, namely, cycloheximide (Fink and Styles 1972; Vodkin et al. 1974) and high temperature (Wickner 1974b). Chromosomal genes needed for maintenance of the killer plasmid are needed for maintenance of M, whereas chromosomal mutations affecting only expression of plasmid information do not result in loss of M (see detailed discussion below). Other evidence comes from the study of suppressive mutants of the K_1 killer plasmid (see below). A strain with a killer plasmid mutation causing the superkiller phenotype (Vodkin et al. 1974) produces a more stable toxin than wild-type killers (Palfree and Bussey 1979), which suggests that the toxin is encoded by a plasmid.

Recently, Bostian et al. (1980b) showed directly that the M_1 DS RNA codes for the K_1 killer toxin. Translation of the denatured M_1 DS RNA in a wheat germ system produced a major product of about 32,000 m.w., which could be precipitated by antibody to the purified K_1 killer toxin (11,000 m.w.; see below). Moreover, most of the tryptic peptides of the toxin could be detected in the tryptic digest of the 32,000-m.w. in vitro product.

M_1 DS RNA is linear (Sweeney et al. 1976) and consists of two single strands, each half the molecular weight of the double-stranded molecule

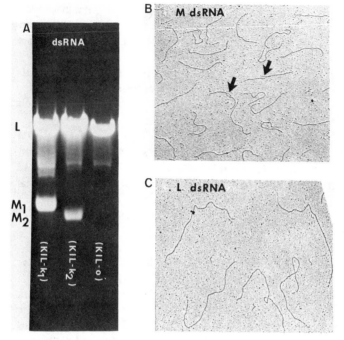

Figure 3   DS RNA from various yeast strains. (*A*) Gel electrophoresis of DS RNA isolated from [KIL-k₁], [KIL-k₂], and [KIL-o] strains. (*B*) Electron micrograph of M_1 DS RNA. (*C*) Electron micrograph of L DS RNA.

(Bruenn and Kane 1978; Fried and Fink 1978). It has an easily denatured (and thus AU-rich) region about 200 nucleotides long near the center of the molecule (Fried and Fink 1978).

In most killer strains of *S. cerevisiae*, about 0.1% of total nucleic acid is DS RNA. About 6% of this is *M* DS RNA (12 copies/cell) and 90% is *L* DS RNA (100 copies/cell) (Vodkin et al. 1974; Wickner and Leibowitz 1976a). However, the [KIL-o] strain S7 has up to several percent of its total nucleic acids as *L* DS RNA (Oliver et al. 1977). Chromosomal *ski⁻* (super*k*iller) mutants and some *mak10-1 ρ⁻* [KIL-k₁] mutants with increased *M* DS RNA have also been described (Wickner 1977; Toh-e et al. 1978; see below).

M DS RNA and *L* DS RNA do not have substantial sequence homology as judged by heteroduplex studies (Fried and Fink 1978) and RNase T1 fingerprinting (Bruenn and Kane 1978). *M*, like *L*, has pppGp at its 5′ ends (Bruenn and Keitz 1976) and sequence heterogeneity at its 3′ ends with the major 3′ sequences being GAAACACCCAUCA-OH and GUCAUUUC-UUUAUUUUCA-OH (Bruenn and Brennan 1980). The last eight nucleotides of one 3′ end of *M* (underlined) are identical to the last eight nucleotides of one 3′ end of *L*.

M_1 DS RNA, like L DS RNA, is found in intracellular viruslike particles (Herring and Bevan 1974; Adler et al. 1976; Harris 1978). M-containing particles can be separated from those containing L DS RNA by velocity sedimentation (Herring and Bevan 1974; Harris 1978; Bostian et al. 1980a). Antibody made to L-containing particles precipitates both L- and M-containing particles but not the naked DS RNAs (Harris 1978). Furthermore, rapid peptide maps of proteins from separated L- and M-containing particles were the same. Thus, L and M are separately encapsidated, but the coat proteins are similar, if not identical.

Viruslike particles containing M DS RNA also have an RNA polymerase activity that produces a full-length copy of one strand and releases it from the viruslike particles (Welsh and Leibowitz 1980; Welsh et al. 1980).

In spite of efforts in several laboratories, infection by K_1 killer viruslike particles has not been reported, although a report of yeast infection by K_2 particles has appeared (Nesterova 1974; Nesterova et al. 1976).

M DS RNA, the killer plasmid, is distinct from other cytoplasmic entities as shown by different behavior in curing experiments and different dependence on chromosomal genes. Curing killer strains of their killer plasmid by growth in low concentrations of cycloheximide (Fink and Styles 1972; Young and Yagiu 1978) or at elevated temperatures (Wickner 1974b; Young and Yagiu 1978) results in loss of M DS RNA but retention of L DS RNA (Table 2). Both treatments increase the frequency of loss of ρ (the mitochondrial genome) (Sherman 1959), but both $\rho^- K^+$ and $\rho^+ K^-$ clones are observed. Also, elimination of the mitochondrial genome with ethidium bromide (Goldring et al. 1970) does not eliminate the killer plasmid (Fink and Styles 1972; Al-Aidroos et al. 1973). The ψ plasmid increases the level of ocher and frameshift suppression (Cox 1965, 1971; Young and Cox 1971; Leibman et al. 1975; Culbertson et al. 1977); neither heat (Singh et al. 1979) nor cycloheximide (Wickner 1976a) induces loss of ψ under conditions where the same strain is losing [KIL-k_1]. Exposure of ψ^+ strains to hypertonic media induces their conversion to ψ^- (Singh et al. 1979), but this treatment does not eliminate [KIL-k_1], [KIL-k_2], [EXL], [NEX], or [HOK] (Wickner 1980 and unpubl.; see below).

The [URE3] plasmid allows cells growing on an ammonia or glutamate nitrogen source to utilize exogenous ureidosuccinate to bypass *ura2* (aspartate transcarbamylase) mutations (Lacroute 1971; Drillien et al. 1973; Aigle and Lacroute 1975). This plasmid, unlike other non-Mendelian elements in *S. cerevisiae*, is not distributed to all progeny in meiosis. It can be transferred by cytoduction (heterokaryon formation). Although the chromosomal *PET18* gene (see below) is required for maintenance of [KIL-k_1], [KIL-k_2], and ρ, it is not needed for either [URE3] or ψ (Leibowitz and Wickner 1978).

The 2-micron DNA plasmid is present in most laboratory yeast strains (Sinclair et al. 1967; Hollenberg et al. 1970; Stevens and Moustacchi 1971; Clark-Walker 1972; Leth et al. 1972). Its replication control resembles that of

Table 2 Plasmids of yeast

Plasmid	Trait	Nucleic acid	Chromosomal maintenance genes		Curing agents
			needed	not needed	
[KIL-k₁]	K₁⁺R₁⁺ viruslike particle protein	1.5 × 10⁶-m.w. DS RNA	26 *MAK* genes, *PET18, SPE2*	*MKT1*	cycloheximide, high temperature
[KIL-k₂]	K₂⁺R₂⁺	1.0 × 10⁶-m.w. DS RNA	*MAK16, MAK10, MKT1*		cycloheximide, high temperature
L DS RNA	viruslike particle protein	3 × 10⁶-m.w. DS RNA	*MAK3* (partial)	all other *MAK* genes *PET18, SPE2*	
[EXL]	excludes [KIL-k₂]	?	*MAK10, PET18, MAK1, MAK3*		
[NEX]	prevents [EXL] actions; helps *mkt1* exclude [KIL-k₂]	?	*MAK3, MAK10*	*MAK4, MAK6 MAK27, MKT1*	
[HOK]	helper function for [KIL-sd]	?	*PET18, MAK10 MAK3*	*MAK1, MAK4, MAK5, MAK6, MAK14, MAK15, MAK16, MAK19, MAK26, MAK27, MKT1*	

	Phenotype	Physical nature		Nuclear genes	Agents causing loss
ρ	respiration	50×10^6-m.w. circular DNA	PET18, CDC21	all other MAK genes, SPE2	ethidium, etc., high temperature
ψ	increased ocher and frameshift suppression	?	PNM	PET18	hypertonicity, guanidine
[cir]	?	2-micron DNA		PET18	
[URE3]	ureidosuccinate utilization	?		PET18	
Maltose fermentation if ρ^0	maltose fermentation if ρ^0				
20S RNA	20S RNA synthesis in acetate medium	20S RNA		PET18, MAK1, MAK3, MAK7, MAK9	

nuclear DNA (Livingston and Kupfer 1977) and, unlike [KIL-k$_1$], it is not lost from a *pet18* mutant (Livingston 1977).

Certain maltose-fermenting grande strains become maltose nonfermenters when made ρ^- by acriflavine or ethidium bromide, whereas other maltose-fermenting grande strains remain maltose fermenters when converted to petite strains. This difference shows non-Mendelian inheritance (Schamhart et al. 1975) but has not been studied in relation to the killer character.

Several other DS RNA species have been detected by agarose or acrylamide gel electrophoresis. One of these, called *XL*, migrates slightly more slowly than *L* DS RNA (Wickner and Leibowitz 1976a). A peak of DS RNA located similarly has been observed by Bevan and Herring (1976), whose data suggest that this band contains replicative intermediates of *L*. Several minor bands located between *M* and *L* are present in DS RNA preparations purified by CF11 cellulose chromatography (Wickner and Leibowitz 1976a). These minor bands vary in amount from one strain to another and occur in strains lacking *M* DS RNA, including *mak* mutants. They may be related to *L* DS RNA.

KILLER PLASMID REPLICATION

Genes Required for Plasmid Maintenance (*mak*)

Mutations in any of 29 chromosomal genes result in loss of [KIL-k$_1$] (Table 3). These include the *mak* (*ma*intenance of *k*iller) genes: *mak1* and *mak3–mak28* (Somers and Bevan 1969; Wickner 1974a, 1978, 1979b; Wickner and Leibowitz 1976a, 1979; Guerry-Kopecko and Wickner 1980), *pet18* (Fink and Styles 1972; Wickner and Leibowitz 1976a; Leibowitz and Wickner 1978), and *spe2* (Cohn et al. 1978b). In each case, these mutants lose *M*$_1$ DS RNA and retain *L* DS RNA. In *mak3-1* strains, the amount of *L* DS RNA per cell is reduced tenfold or more, but it is not lost (Wickner and Leibowitz 1979). *mak10* (Naumov and Naumova 1973), *mak3, mak16,* and *pet18* (Wickner 1980) are also needed for maintenance of [KIL-k$_2$], but dependence on the others has not been tested. There is also at least one gene, *MKT1* (*ma*intenance of *K$_2$*), that is needed for maintenance of [KIL-k$_2$] but not for [KIL-k$_1$] (Wickner 1980).

Most of the *mak* genes have been located on the genetic map (Fig. 4) and, like other groups of related genes, seem to be scattered at random with at least one *mak* gene on 15 of the 17 known chromosomes. More *mak* genes remain to be defined, since only a single mutation has been found for 23 of the 29 known genes.

In view of precedents in bacterial phage and plasmid systems, each *mak* gene will probably have a host-specific function independent of the presence or absence of *M*$_1$ DS RNA. In fact, *mak1, mak16,* and *pet18* mutants are temperature-sensitive for growth, although the killer plasmid is lost at temperatures permissive for growth (Wickner and Leibowitz 1976a, 1979;

Table 3 Chromosomal genes affecting the killer character of *S. cerevisiae*

I. Expression

    kex1, kex2; $K_1^-R_1^+$ [KIL-k$_1$]

        kex2 gene also needed for mating by α strains; for meiotic sporulation; changes in many extracellular proteins

    rex1 $K_1^+R_1^-$ [KIL-k$_1$]

II. Killer plasmid maintenance

    mak1, mak3–mak28; $K_1^-R_1^-$ [KIL-o]

        mak1 and *mak16* also temperature-sensitive for growth; *mak3* and *mak10* mutants lose [NEX], [HOK], and [EXL]

    pet18 also needed for growth and for ρ maintenance

    spe2 adenosylmethionine decarboxylase (spermidine and spermine biosynthesis) also needed for meiotic sporulation and optimal growth

    mkt1 needed for [KIL-k$_2$] maintenance if [NEX] is present

III. Regulation

    ski1–ski4; $K_1^{++}R_1^+$ [KIL-k$_1$]

        certain *ski mak* double mutants are K_1^+ or K_1^{++}; *ski$^-$* mutants can maintain [KIL-sd$_1$]

    KRB1 $K_1^+R_1^+$ [KIL-k$_1$]

        dominant; bypasses needed for *mak7* or *pet18*; located on a new centromere

Leibowitz and Wickner 1978). Mutations in *mak13, mak15, mak17, mak20, mak22,* and *mak27* result in slow growth at any temperature (Wickner and Leibowitz 1979). *pet18* strains also lose the mitochondrial genome, but not ψ, [URE3] (Leibowitz and Wickner 1978), or 2-micron DNA (Livingston 1977).

    SPE2 codes for adenosylmethionine decarboxylase (Cohn et al. 1978a), an enzyme in the pathway for biosynthesis of the polyamines spermidine and spermine. *spe2* mutants are completely deficient in spermidine and spermine, grow at one-sixth the normal rate, cannot undergo meiotic sporulation, and lose the *M* DS RNA when starved of polyamines (Cohn et al. 1978a,b). All of these defects are prevented by supplying either spermidine or spermine.

    Among mutants not losing the killer plasmid are auxotrophic mutants starved for required adenine and histidine; *ade2-1* (ocher) *SUQ5* [PSI$^+$] strains during adenine-limited growth; *cdc* temperature-sensitive mutants, and *rna1* and *rna2* temperature-sensitive strains grown at semipermissive temperatures; *ski1* strains, which grow slowly but are superkillers (see below); *pet18 KRB1* and various *mak$^-$ ski$^-$* [KIL-k$_1$] strains, where the *mak$^-$* defect is suppressed (see below) but the slow growth defect remains; ρ^0 strains and nuclear *pet* mutants, except *pet18*; normal strains growing slowly on glycerol or minimal medium; and most (> 99%) of the slow-growing colonies of a mutagenized stock.

    Thus, *mak* mutants define a very specific class of genes. This specificity does not, however, imply that every *MAK*-gene product is involved in

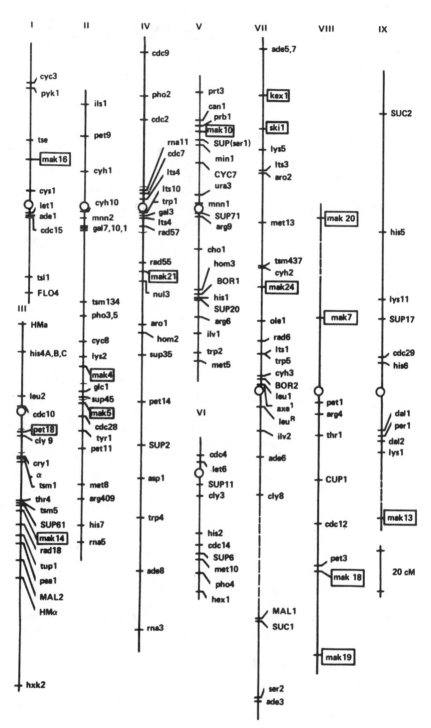

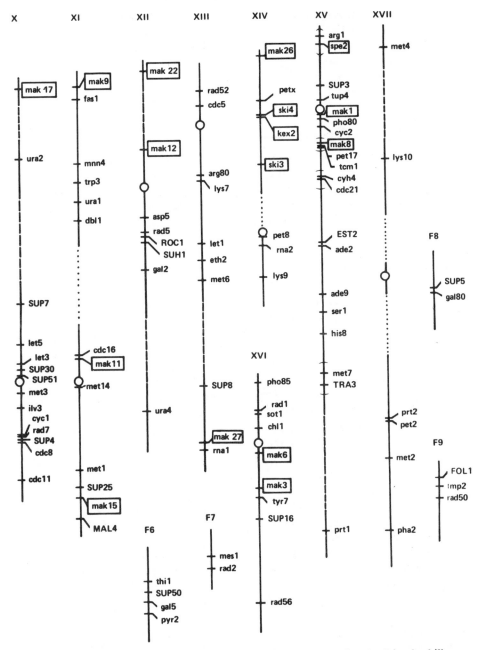

Figure 4 Genetic map of *S. cerevisiae* emphasizing genes involved in the killer systems. Parentheses indicate sequences not established with genes outside parentheses; (———) mitotic linkage; (. . .) trisomic linkage. References for most of the other markers are given in Mortimer and Schild (1980).

nucleotide polymerization of the M DS RNA. The maintenance of the killer plasmid may involve initiation, elongation, and termination steps of the DS RNA synthesis, functions coordinating this synthesis with cell growth, packaging of the new DS RNA molecules, segregation of molecules into daughter cells, protection of DS RNA from unfriendly nucleases, and regulation of each of these processes.

Regulatory Genes Affecting Killer Plasmid Replication

Recessive mutations in any of four chromosomal genes (*ski1–ski4*) increase production of killer toxin actitivy (Toh-e et al. 1978). The *ski* mutants still depend on the killer plasmid for toxin production, and the toxin they produce has the same specificity and stability as that produced by the parent strain (K. Ouchi and R. B. Wickner, unpubl.). Three of the *SKI* genes have been mapped (Fig. 4) (Toh-e et al. 1978). The four *SKI*-gene products must have a negative influence on some aspect of killer-plasmid maintenance, replication, or expression. Three lines of evidence indicate that the *SKI* genes are involved in maintenance or replication.

The *ski2, ski3,* and *ski4* mutants appear to have an increased amount of M_1 DS RNA, which suggests a role for *SKI* genes in plasmid maintenance control, but the variability of the data presented requires confirmation by other methods (Toh-e et al. 1978). The *ski1-1* mutation results in decreased cellular growth rate, as well as the superkiller phenotype, but apparently does not result in an increased cellular content of M_1 DS RNA.

When various *ski⁻ mak⁻* double mutants were constructed, many were stable killers (Table 4; Cohn et al. 1978b; Toh-e and Wickner 1980), i.e., *ski⁻* mutations bypass *mak⁻* mutations, which again suggests that *SKI*-gene products are involved in plasmid maintenance or replication. In contrast, both *kex1* and *kex2* mutations (involved in killer *ex*pression, see below) are epistatic to each *ski* mutation; that is, *kex⁻ ski⁻* double mutants are K⁻R⁺ in phenotype. The *ski1-1* mutation, which has no apparent effect on M DS RNA copy number, is the most effective in suppressing *mak* mutations, bypassing all of those tested except *mak16-1*. This result suggests that the *SKI*-gene products do not simply regulate copy number.

The third line of evidence linking *SKI* genes and plasmid maintenance and replication is the [KIL-sd₁] mutation (Toh-e and Wickner 1979; see below). This mutant M DS RNA cannot be maintained in wild-type strains (clearly a maintenance or replication defect), but it can be maintained stably in any *ski⁻* mutant.

One possible model to explain the effects of the *ski⁻* mutants is shown in Figure 5 (Toh-e and Wickner 1980). We propose that two enzymatic pathways can replicate the killer plasmid. Only one of these pathways is utilized in wild-type killer strains (the normal pathway), and it is this pathway

Table 4 Bypass of *MAK* genes for killer plasmid replication

	mak-gene mutation														*kex*-gene mutation	
	16	3	10	pet18	12	21	26	1	4	6	7	11	spe2	[KIL-sd]	1	2
ρ⁰	–	–	+	–				–	–	–	–		–	–	–	–
KRB1	–	–	–	+	–			–	–	–	+	+	–		–	–
[KIL-b]	–	–	–	–	–	–	–	+	±	+	+	+	+	+	–	–
ski2	–	–	–	–	–	–	+	+	+	+	+	+	+	+	–	–
ski3	–	–	–	–		+	+	+	+	+	+	+	+	+	–	–
ski4	–	–	–	–	+	+	+	+	+	+	+	+	+	+	–	–
ski1	–	+	+	+	+			+	+	+	+	+	+	+	–	–

Minus (–) denotes double mutant is K⁻. Plus (+) denotes double mutant is K⁺R⁺ or K⁺⁺R⁺.

429

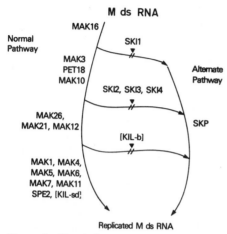

Figure 5   Heuristic model of *M* DS RNA replication to explain interaction of *mak,* *ski,* [KIL-b], and [KIL-sd] mutations. An alternate pathway for killer-plasmid replication is supposed to be normally inaccessible to *M* DS RNA because of the *ski* proteins and the [KIL-b] site or function. It is assumed that plasmid replication by the alternate pathway results in the superkiller phenotype. This model predicts the existence of *skp* genes, whose products are needed for the alternate pathway but not for the normal pathway.

that is blocked in *mak* mutants and in [KIL-sd₁] mutants. Access to an alternate pathway is controlled negatively at various points by the products of the *SKI* genes. In this model, the *ski⁻ mak⁻* double-mutant combinations that allow replication of the killer plasmid are determined by the sequence of action of *MAK* genes in the normal pathway and the points at which the various *SKI*-gene products block transfer from the normal to the alternate pathway. This model predicts the existence of new genes, called *skp* (second killer pathway or *ski* pathway), needed for maintenance of the killer plasmid only when the alternate pathway is required (e.g., in a *mak6 ski2* strain). It is assumed that replication by the alternate pathway results in the superkiller phenotype. This model will be referred to again when killer-plasmid mutants are discussed below.

Other genes involved in killer-plasmid replication and relationships among such genes may be defined by direct isolation of suppressors of *mak* mutations. These should include *ski* mutants as one class, which has been observed (Toh-e and Wickner 1980), but it also could include other mutations not producing a superkiller phenotype. *KRB1* (*killer replication bypass*) mutants are dominant chromosomal mutations that make the *MAK7*- and *PET18*-gene products dispensable for killer-plasmid replication (Wickner and Leibowitz 1977). *KRB1* bypasses the killer-plasmid replication defect of *pet18* mutants but not their defect in maintenance of mitochondrial DNA or their temperature sensitivity for growth. Of the *mak* genes tested, only *mak7* and *pet18* are suppressed. *KRB1* strains are not superkillers, and *KRB1* is not a translational suppressor. *KRB1* mutants may be constitutive for an enzyme that is needed by the killer plasmid for replication and that is normally induced by *MAK7* and *PET18*.

The *KRB1* mutation arises with very high frequency and has an unusually high reversion frequency. It would be of interest to study the structure of the *KRB1* locus.

A mutation of the mitochondrial genome specifically bypasses the *mak10* gene (Wickner 1977). This effect is not due simply to respiratory deficiency of the mutant. In fact, blocking mitochondrial protein synthesis with high concentrations of chloramphenicol or erythromycin does not result in the bypass effect. This suggests that the mitochondrial bypass mutation affects a gene whose product need not be translated to act. Since some spontaneous ρ^- strains bypass *mak10* and others do not, it should be possible to map the mitochondrial locus involved. The relationship of the mitochondrial genome (ρ) and the killer plasmid [KIL-k$_1$] may be summarized as follows: (1) Neither ρ nor [KIL-k$_1$] is needed for maintenance of the other; (2) *PET18* is needed for maintenance of both ρ and [KIL-k$_1$]; and (3) ρ^0 strains do not need *MAK10* for replication of [KIL-k$_1$].

KILLER PLASMID MUTATIONS AFFECTING MAINTENANCE

[KIL-sd]

The *M* DS RNA carries a gene (site or product) that, in many wild-type strains, is essential for the maintenance or replication of *M*$_1$ DS RNA (Toh-e and Wickner 1979). A mutant killer plasmid was isolated that could be maintained in *ski*$^-$ strains but not in wild-type strains. This mutation is called the *ski*-dependent plasmid [KIL-sd]. Its *M*$_1$ DS RNA is unaltered in size. Since this killer plasmid is lost in wild-type strains, it is clearly defective in some function required for its maintenance (Toh-e and Wickner 1979). The bypass of this defect by *ski*$^-$ mutations is similar to the bypass of chromosomal *mak*$^-$ mutations by *ski*$^-$ mutations (see above), which suggests that *MAK*-gene products act with the site or product defined by [KIL-sd] to replicate the killer plasmid. Whether or not the [KIL-sd] mutation is complementable by a wild–type killer plasmid is not known. We propose (Toh-e and Wickner 1980) that [KIL-sd] is mutant in a site or product of *M*$_1$ DS RNA that is involved in the action of the group of *MAK* genes bypassed by all of the *ski*-gene mutations (*mak1, mak4, mak5, mak6, mak7, mak11, spe2,* and, possibly, *mak12, mak21,* and *mak26*) (see Table 4 and Fig. 5).

The [KIL-sd] mutant plasmid was first isolated from the original *ski2-2* mutant strain. The plasmid in this strain was heat-cured and replaced with a wild-type killer plasmid. When the replaced plasmid in the *ski2-2* strain was then tested, it too had become a [KIL-sd] mutant plasmid. It is not yet clear whether this represents selection in the *ski2-2* strain of a frequent "variant" or an increased frequency of this particular type of plasmid mutation.

Further studies of [KIL-sd] mutants have shown that many SKI^+ strains are unexpectedly able to maintain this mutant killer plasmid. Crosses between these strains and SKI^+ strains unable to maintain [KIL-sd] show that this property exhibits non-Mendelian inheritance. The presence of a plasmid, which we call [HOK] (*h*elper *o*f *k*iller), allows [KIL-sd] replication and maintenance even in a SKI^+ background (R. B. Wickner, in prep.). [HOK] requires *MAK3, MAK10,* and *PET18* for its maintenance and replication, but it does not require the products of 11 other *MAK* genes or *MKT1*. [HOK] is distinct from the previously described plasmids ψ, ρ, [URE3], 2-micron DNA, and 20S RNA.

[KIL-b]

Another plasmid site or product has been identified by the [KIL-b] (*b*ypass) plasmid (Toh-e and Wickner 1980). This killer plasmid was found in a laboratory stock and has the same K_1 killing and R_1 immunity specificity as the normal killer plasmid. However, [KIL-b] does not require the products of the *MAK4, MAK7, MAK11,* or *MAK17* genes for its maintenance. These are some, but not all, of the *MAK* genes bypassed by all of the *ski⁻* mutations. Strains carrying this plasmid are also superkillers. Thus, the [KIL-b] plasmid can be viewed as carrying a *ski⁻* mutation of the killer plasmid.

Suppressive Mutants

This class of mutants is interesting because of what it tells us about the killer system and because it closely parallels the defective-interfering virus mutants found in most animal virus systems. Suppressive plasmids, designated [KIL-s_1], are deletion mutants of M_1 DS RNA that confer the K^-R^- phenotype on their host and interfere with the replication of the normal killer plasmid (Somers 1973; Vodkin et al. 1974; Sweeney et al. 1976; Bruenn and Kane 1978; Fried and Fink 1978). The term suppressive is perhaps unfortunate because of frequent confusion with nonsense suppressors, but it is derived historically from analogous mutations of the mitochondrial genome (Ephrussi et al. 1955).

The suppressive plasmids remain dependent on the *MAK10*-gene product for their maintenance (Somers 1973) and can be cured by cycloheximide (Sweeney et al. 1976), as in the case of the normal plasmid. Although the [KIL-s_1] mutants in some way cause elimination of the wild-type killer plasmid, they do not interfere with L DS RNA, nor, surprisingly, do they interfere with each other. Strains stably maintaining two or even three different size suppressive plasmids have been constructed (Sweeney et al. 1976).

Several [KIL-s₁] strains have been examined by heteroduplex analysis (Fried and Fink 1978), T1 ribonuclease fingerprinting (Bruenn and Kane 1978; Kane et al. 1979), and 3' end sequencing (Bruenn and Brennan 1980). The ends of the [KIL-s₁] mutants are the same as the ends of [KIL-k₁], but they have a large internal deletion or a deletion followed by a direct repeat of the remaining sequences. These studies also provide evidence that, once formed, suppressive plasmids may be unstable, accumulating secondary deletions and duplications. M_1 and L DS RNA showed no homology by these methods, except for the identical eight bases at one 3' end mentioned above.

The deleted sequences of suppressive plasmids must be necessary for killing and resistance. Also, the smallest suppressive plasmids (only about 250,000 m.w.) must have sufficient information for their replication.

The small S DS RNAs of suppressive mutants, like L and M, are found in intracellular particles of the same density in CsCl as those containing M (Harris 1978; Kane et al. 1979).

[KIL-d]

The [KIL-d] (*d*iploid-dependent) mutants show a defective phenotype in haploid strains (e.g., K⁻R⁺), and the mutant plasmid is easily lost; but diploid strains carrying [KIL-d] are stably K⁺R⁺ (Wickner 1976b). A potentially analogous phenomenon has been described by Rothstein and Sherman (1980): **a** or α haploids with the *cyc7-2* mutation overproduce iso-2-cytochrome *c*, but **a**/α diploids homozygous for the same mutation do not. Their discovery that **a**/**a** or α/α diploids overproduce the enzyme was the clue that iso-2-cytochrome *c* had come under control of the mating type. [KIL-d] mutants possibly represent a similar phenomenon, but further work will be necessary; in particular **a**/**a** and α/α diploids should be examined.

A Mutation Affecting L DS RNA

Mitchell et al. (1976) have reported a cross of a killer, K7, carrying both L and M, with a *mak10* strain, S3 (not a suppressive DS RNA), lacking both L and M DS RNA, in which the nonkiller segregants lacked both L and M DS RNA. Crosses of K7 with *mak10* strains carrying L segregated two killers with L and M: two *mak10* nonkillers with L. Likewise, crosses of S3 and other killer strains yielded two killers with L and M: two *mak10* nonkillers with L. In this paper evidence is presented that the difference between K7 and other killer strains is a cytoplasmic difference. These results suggest that K7 has a mutant L DS RNA that has become dependent on the *mak10* gene for its replication. Does L normally use either the *mak* replication pathway or its own pathway? Is the normal L pathway identical to the postulated alternate pathway for M replication in *ski* mutants? These and other questions could be approached by isolation of chromosomal mutants unable to maintain L.

Although several different killer plasmid mutations have been described, it is at present impossible to do complementation tests with plasmid mutants because two killer-plasmid genomes cannot be maintained stably in the same cell. Also, there are no reports of genetic recombination involving DS RNA genomes in yeast or other systems (not including the segment reassortment seen in reovirus) so that genetic mapping of these mutations is not yet possible.

MOLECULAR BIOLOGY OF DS RNA REPLICATION

Little is known about the molecular mechanism of L or M DS RNA replication or transcription. Whether replication is conservative (as in reovirus) or semiconservative is not known. The synthesis and degradation of L DS RNA is affected by the physiological state of the cell. Growth in ethanol results in severalfold more L DS RNA per cell than growth in glucose (Oliver et al. 1977). Preventing protein synthesis either by starvation for a required amino acid or by cycloheximide inhibits L DS RNA synthesis, but degradation of parental L DS RNA is not observed. In contrast, nitrogen starvation has little effect on the rate of synthesis of L DS RNA, but degradation of 85% of parental L DS RNA is observed. This degradation can be prevented by the addition of cycloheximide to the nitrogen-starved culture (Clare and Oliver 1979).

Studies on the physiology and molecular biology of L and especially M DS RNA are difficult because they cannot be labeled specifically and they represent only a very small proportion of the total cellular RNA. Oliver et al. (1977) and Clare and Oliver (1979) were able to study L metabolism in strain $S7$ because it carries 10–20 times more L than most strains.

PLASMID EXCLUSION (INTERFERENCE) PHENOMENA

Mating [KIL-k_1] and [KIL-n_1] haploids results in killer diploids that show mitotic segregation of pure [KIL-n_1] and pure [KIL-k_1] clones with comparable frequencies (Bevan and Somers 1969). This is incompatibility between plasmids and can be explained by fluctuations in the random selection of genomes for replication and segregation (Ishii et al. 1978; Novick and Hoppensteadt 1978). Mating [KIL-s_1] mutants with [KIL-k_1] strains produces diploid clones most or all of which are [KIL-s_1]. Vodkin et al. (1974) found that two suppressive plasmids were maintained at about twice the copy number at which M DS RNA is maintained. This may contribute to their preferential inheritance.

When [KIL-k_1] and [KIL-k_2] haploids were mated, the diploids formed rapidly and completely lost [KIL-k_2] (Naumov and Naumova 1973; R. B. Wickner, unpubl.), although their copy numbers are comparable. This type of phenomenon is referred to as plasmid exclusion. [KIL-s_1] also excludes [KIL-k_2] even more efficiently than it excludes [KIL-k_1] (Wickner 1980).

Another non-Mendelian element called [EXL] efficiently excludes [KIL-k_2] but has no effect on [KIL-k_1] (Wickner 1980). [EXL] depends on at least *PET18, MAK1, MAK3,* and *MAK10* and is unaffected by curing mitochondrial DNA with ethidium bromide. [EXL] has the properties expected of a suppressive mutant of [KIL-k_2], but a corresponding DS RNA has not been detected (R. B. Wickner, unpubl.).

Some [KIL-k_2] strains are nonexcludable by [EXL]; that is, mating these [KIL-k_2] haploids with haploids carrying [EXL] produced only $K_2^+R_2^+$ diploids. These strains carry another non-Mendelian trait, called [NEX] (*nonex*cludable). [NEX] is distinct from [KIL-k_2] since [NEX] could be recovered from strains where [KIL-k_2] had been eliminated. [NEX] also was not eliminated by ethidium bromide. It is not yet known whether [EXL] or [NEX] is cured by heat or cycloheximide. Although [NEX] protects [KIL-k_2] from [EXL], it does not prevent [KIL-k_1] or [KIL-s_1] from excluding [KIL-k_2], nor does it prevent [KIL-s_1] from excluding [KIL-k_1]. [NEX] is independent of the *MAK4-, MAK6-,* and *MAK27*-gene products, but it depends upon *MAK3* and *MAK10* for its maintenance (Wickner 1980 and unpubl.).

[NEX] has another effect on [KIL-k_2] maintenance. Strains carrying the recessive allele of the chromosomal gene *mkt1* (*m*aintenance of K_2) can maintain [KIL-k_2] only if [NEX] is absent. Introduction of [NEX] by either crosses or cytoduction results in loss of [KIL-k_2] from these strains (Wickner 1980). This is analogous to previous findings with *MAK10*, ρ, and [KIL-k_1]. The *MAK10* gene is not needed for [KIL-k_1] maintenance in the absence of ρ (Wickner 1977; see above). Similarly, the *MKT1* gene is not needed for [KIL-k_2] in the absence of [NEX]. *MKT1* is not linked to *MAK10*, and [NEX] is not mitochondrial. *MKT1* is not needed to maintain [KIL-k_1], even if [NEX] is present.

The nucleic acid corresponding to [NEX] has not yet been identified.

20S RNA PLASMID

As an outgrowth of studies of RNA synthesis in sporulation, another new non-Mendelian genetic element has been discovered with a possible relationship to the K_1 killer trait. When most haploid or diploid yeast strains are placed in acetate medium without glucose or a nitrogen source (the conditions used to induce meiotic sporulation in normal diploids), the synthesis of a 20S (about 0.8×10^6 m.w.) SS RNA is induced, accounting for as much as 15% of new RNA synthesis (Kadowaki and Halvorson 1971a,b; Garvik and Haber 1978). This RNA species is not detectable in vegetative cells (Kadowaki and Halvorson 1971a,b), it is not methylated (Sogin et al. 1972), and it is not a form of ribosomal RNA (Wejksnora and Haber 1978).

The ability to synthesize the 20S RNA is controlled by a cytoplasmic genetic element (Garvik and Haber 1978). The 20S RNA plasmid is distinct from *L* or *M* DS RNA. However, a possible relationship was suggested to the killer phenomenon (Garvik and Haber 1978). All killer strains examined produced 20S RNA as did many nonkillers, but all nonproducers of 20S RNA were also nonkillers. Garvik and Haber (1978) suggest that the 20S RNA plasmid may be necessary for the killer phenotype or for *M* DS RNA maintenance, but, as is indicated in this paper, the evidence is still entirely circumstantial.

KILLER TOXIN

The K_1 killer toxin has recently been purified to homogeneity by Palfree and Bussey (1979). The active species is an 11,470-m.w. monomer that comprises about 5-10% of secreted protein. Toxin activity is produced during the log phase of growth of killer strains, and little accumulates during the stationary phase (Palfree and Bussey 1979).

Direct translation of denatured *M* DS RNA in vitro produced a 32,000-m.w. major product that is precipitated by antibody to the purified toxin and contains tryptic peptides of the purified toxin (Bostian et al. 1980; see above). This might represent a toxin precursor molecule, but attempts to find this precursor in wild-type or *kex2* mutant intracellular or extracellular material have not yet succeeded (Rogers et al. 1979).

In addition to *M* DS RNA, the two chromosomal genes *KEX1* and *KEX2* are needed for production or excretion of an active killer toxin (Fig.4; Wickner 1974a; Wickner and Leibowitz 1976b). Mutants defective in these genes are K^-R^+ and still maintain the killer plasmid. In addition to their inability to secrete the killer toxin, *kex2* strains of the α mating type show a marked defect in mating ability, do not secrete the α pheromone, and do not respond fully to the **a** pheromone (Leibowitz and Wickner 1976). The mating defect is not corrected by adding normal α pheromone. Cells of **a** *kex2* strains mate normally and secrete **a** pheromone. Homozygous *kex2/kex2* diploids also failed to complete the final stages of meiotic spore formation. Although they undergo DNA replication, meiotic recombination, and nuclear division, the meiotic spores are not formed. None of these defects is seen in *kex1* mutants (Leibowitz and Wickner 1976).

kex2 strains show alterations of size and isoelectric pH of several of their major extracellular proteins (Rogers et al. 1979). In contrast, *kex1* strains showed no such alterations. No precursor of killer toxin was observed in extracts of *kex2* strains or their extracellular materials.

MECHANISM OF TOXIN ACTION

Three chromosomal genes, called *kre1, kre2,* and *kre3* (*k*iller *re*sistant), are necessary for the K_1 killer toxin to be able to kill an otherwise sensitive strain (Al-Aidroos and Bussey 1978). Mutants defective in *kre1* are also resistant to the K_2 killer toxin. Mutants defective in *kre1* and *kre2* showed decreased binding of crude K_1 toxin activity to whole cells compared with wild type, but the *kre3* mutant was unaffected.

By using these mutants and in-vivo-labeled ^{35}S-toxin purified to homogeneity, Bussey et al. (1979) have defined the cell wall receptor for the killer toxin. Wild-type cells bind the toxin with an association constant of 2×10^6 M^{-1} and a binding site number of 1.1×10^7 per cell, whereas *kre1* and *kre2* mutants show only a very low level of binding. Unlabeled toxin competes with labeled toxin for binding to wild-type sensitive cells but does not eliminate the low level of binding to *kre1* cells. This correlation of binding with the *kre* mutants proves that the binding observed is involved in the action of the killer toxin and thus may be called a receptor. Clearly, the low level of binding to *kre1* strains is nonspecific binding. About 2.8×10^4 toxin molecules must bind to a cell to kill it.

Digestion of sensitive cells with zymolyase released a soluble, nondialyzable component with the properties of the receptor (Bussey et al. 1979). This material competed with sensitive cells for binding of radioactive killer toxin and protected these sensitive cells from being killed by the toxin. A similar fraction prepared from a *kre1* strain had little or no activity of this kind, which again indicates that the binding is related to toxin action.

Treatment of sensitive cells with the K_1 killer toxin produces—after a delay of 40–50 minutes at 22–24° C—a selective leakage of intracellular ATP and potassium with a concomitant coordinate inhibition of all macromolecular synthesis (Bussey 1972, 1974; Bussey and Sherman 1973; Bussey et al. 1973). The evidence clearly indicates that this is an effect of the killer toxin, but the long delay suggests that intermediate steps may be discernible. The effects of the K_2 killer toxin are similar (Rogers 1976).

By studying a killer strain from sake yeast (Imamura et al. 1974) whose toxin has the K_1 killer specificity (M. J. Leibowitz, pers. comm.), Kotani et al. (1977) showed that after binding of the K_1 toxin to sensitive cells had occurred, when the toxin had been diluted out and cells were plated on rich growth medium, cell death was almost completely prevented by inclusion of 0.1 M calcium in the medium. If the calcium was removed from the medium at a later time, toxin action resumed as judged by ATP leakage. In contrast, if the toxin-treated cells were plated on minimal medium, little killing occurred. Inclusion of 1 mM ADP in the minimal medium promoted killing substantially. The authors propose that killing occurs in two stages: (1) reversible and (2) irreversible, blocked by Ca^{++}, promoted by ADP, and accompanied by ATP leakage from the cell.

In another approach to dissecting the killing process, Skipper and Bussey (1977) used potassium release as a marker of toxin action. The energy poisons carbonyl cyanide M-chlorophenylhydrazone and 2,4-dinitrophenol prevented potassium release from toxin-treated cells but did not affect toxin binding to the cells. This led Skipper and Bussey (1977) to also propose a two-stage action of the killer toxin, the first stage being the toxin binding and the second stage being the drug–sensitive potassium leakage.

Thus, studies to date have defined clearly the binding of toxin to the receptor. Several approaches are available for discerning subsequent events. The *kre3* mutant allows toxin binding to proceed normally and so must be affected at a later stage in the process. Similarly, the killer plasmid R^+ function, with the product of the *REX1* (resistance *ex*pression) chromosomal gene (Wickner 1974a), prevents toxin action at some point after toxin binding (H. Bussey, pers. comm.). The toxin action requires energy at some stage after toxin binding. Finally, Ca^{++} stops membrane leaks produced by toxin action.

The *KRE1*-gene product is necessary for K_2 toxin action, but *KRE2* and *KRE3* are dispensible. This suggests that the K_2 toxin receptor overlaps with the K_1 toxin receptor but that the K_2 toxin has a distinct mechanism of killing sensitive cells (Rogers and Bevan 1978).

PROSPECTS AND PROBLEMS

The killer system may provide a useful model for several phenomena that are widespread among eukaryotes. Control of replication of viruses and plasmids is of general interest; however, in higher eukaryotes, research of the host role in this process is severely limited because of the difficulties in studying the genetics of these systems. The sort of complex interactions outlined above would be difficult to approach in those systems. Moreover, the role of the host seems to be far from trivial. The same is true of the phenomena of defective-interfering viruses and viral interactions such as interference.

The mechanism of protein processing and secretion has broad interest for higher systems as does the mechanism of toxin action; they are analogous to hormone production and action as well. For all of these problems, the possibility of a combined genetic and biochemical approach promises to make the killer system very fruitful.

Many host genes and other plasmids affect the killer systems: *MAK, KEX, SKI, MKT,* and *REX* genes and ρ, [EXL], [NEX], and [HOK] plasmids. Almost all of the host genes must have host-specific functions. In many cases, there is evidence for this since defects in growth, mating, mitochondrial DNA maintenance, recombination, or meiotic spore formation accompany alterations in maintenance or expression of the killer plasmid. Future work will be directed toward finding the enzymatic functions of these gene products and how they carry out their functions for the host and for the killer plasmids. Also, it is not yet known what nucleic acids correspond to [EXL], [NEX], and [HOK] and how they interact with the killer systems.

Although the number of *MAK* genes is large, this is a very specific class of genes. The role of some *MAK*-gene products in killer plasmid replication may be indirect; however, each *mak* mutation is specific in that the effect on plasmid replication is greater than any effect on cell growth. If cell growth rate and plasmid replication rate were each halved by a particular mutation, the plasmid would not be lost from the cell and the mutation would not be scored as a *mak* mutation. At least 24 proteins (16 of them host proteins) are involved in $\phi\chi 174$ phage DNA synthesis in *Escherichia coli*. T4 phage finds it necessary to bring into its host well over 100 genes, but T4 growth is independent of most of the host genes on which $\phi\chi 174$ depends. In this light, it is not surprising that many host genes are needed to replicate and maintain the small killer plasmid genome.

Dissecting this system requires a combination of genetic and biochemical approaches. The mapping of mutant genes facilitates their identification with genes previously mapped. The physiology of temperature-sensitive mutants may yield valuable clues about gene function. Cloning of the *MAK* genes will facilitate studies of their mechanism of expression, the effects of their overproduction, and their gene structure. It will also allow localized mutagenesis of specific *MAK* genes. Examination of the viruslike particle RNA polymerase activities in mutants that are temperature sensitive for maintenance of *L* and/or *M* may be revealing. Another approach, similar to that devised by Bonhoeffer and co-workers for *E. coli* DNA replication and later widely used in that field, is to devise an in vitro RNA synthesis reaction using thermolability in a *mak* temperature-sensitive mutant as the main criterion that one is detecting the desired reaction. Then in vitro complementation between extracts of mutants that complement each other in vivo could be used as an assay for purification of the normal *MAK*-gene products. This approach is especially good for a multicomponent system because it allows one to work on one component at a time.

REFERENCES

Adler, J., H.A. Wood, and R.F. Bozarth. 1976. Virus-like particles from killer, neutral, and sensitive strains of *Saccharomyces cerevisiae*. *J. Virol.* **17**: 472.

Aigle, M. and F. Lacroute. 1975. Genetic aspects of [URE3], a nonmitochondrial, cyto-plasmically inherited mutation in yeast. *Mol. Gen. Genet.* **136**: 327.

Al-Aidroos, K. and H. Bussey. 1978. Chromosomal mutants of *Saccharomyces cerevisiae* affecting the cell wall binding site for killer factor. *Can. J. Microbiol.* **24**: 228.

Al-Aidroos, K., J.M. Somers, and H. Bussey. 1973. Retention of cytoplasmic killer determinants in yeast cells after removal of mitochondrial DNA by ethidium bromide. *Mol. Gen. Genet.* **122**: 323.

Bevan, E.A. and A.J. Herring. 1976. The killer character in yeast: Preliminary studies of virus-like particle replication. In *Genetics, biogenesis and bioenergetics of mitochondria* (ed. W. Bandlow et al.), p. 153. de Gruyter, Berlin.

Bevan, E.A. and J.M. Somers. 1969. Somatic segregation of the killer (k) and neutral (n) cytoplasmic genetic determinants in yeast. *Genet. Res.* **14**: 71.

Bevan, E.A., A.J. Herring, and D.J. Mitchell. 1973. Preliminary characterization of two species of ds RNA in yeast and their relationship to the "killer" character. *Nature* **245**: 81.

Bostian, K.A., J.A. Sturgeon, and D.J. Tipper. 1980a. Encapsidation of yeast killer ds RNAs: Dependence of *M* on *L*. *J. Bacteriol.* **143**: 463.

Bostian, K.A., J.E. Hopper, D.T. Rogers, and D.J. Tipper. 1980b. Translational analysis of the killer-associated virus-like particle ds RNA genome of *Saccharomyces cerevisiae: M*-ds RNA encodes toxin. *Cell* **19:** 403.

Bruenn, J.A. 1980. Virus-like particles of yeast. *Annu. Rev. Microbiol.* **34:** 49.

Bruenn, J.A. and V.R. Brennan. 1980. Yeast viral double-stranded RNAs have heterogeneous 3′ termini. *Cell* **19:** 923.

Bruenn, J. and W. Kane. 1978. Relatedness of the double-stranded RNAs present in yeast virus-like particles. *J. Virol* **26:** 762.

Bruenn, J. and B. Keitz. 1976. The 5′ ends of yeast killer factor RNAs are pppGp. *Nucleic Acids Res.* **3:** 2427.

Buck, K.W., P. Lhoas, and B.K. Street. 1973. Virus particles in yeast. *Biochem. Soc. Trans.* **1:** 1141.

Bussey, H. 1972. Effects of yeast killer factor on sensitive cells. *Nat. New Biol.* **325:** 73.

———. 1974. Yeast killer factor-induced turbidity changes in cells and spheroplasts of a sensitive strain. *J. Gen. Microbiol.* **82:** 171.

———. 1981. Physiology of killer factor in yeast. *Adv. Microb. Physiol.* **22:** (in press).

Bussey, H. and D. Sherman. 1973. Yeast killer factor: ATP leakage and coordinate inhibition of macromolecular synthesis in sensitive cells. *Biochim. Biophys. Acta* **298:** 868.

Bussey, H. and N. Skipper. 1975. Membrane-mediated killing of *Saccharomyces cerevisiae* by glycoproteins from *Torulopsis glabrata. J. Bacteriol.* **124:** 476.

———. 1976. Killing of *Torulopsis glabrata* by *Saccharomyces cerevisiae* killer factor. *Antimicrob. Agents Chemother.* **9:** 352.

Bussey, H., D. Sherman, and J.M. Somers. 1973. Action of yeast killer factor: A resistant mutant with sensitive spheroplasts. *J. Bacteriol.* **113:** 1193.

Bussey, H., D. Saville, K. Hutchins, and R.G.E.Palfree. 1979. Binding of yeast killer toxin to a cell wall receptor on sensitive *Saccharomyces cerevisiae. J. Bacteriol.* **140:** 888.

Clare, J.J. and S.G. Oliver. 1979. The regulation of RNA synthesis in yeast. IV. Synthesis of double-stranded RNA. *Mol. Gen. Genet.* **171:** 161.

Clark-Walker, G.D. 1972. Isolation of circular DNA from a mitochondrial fraction of yeast. *Proc. Natl. Acad. Sci.* **69:** 388.

Cohn, M.S., C.W. Tabor, and H. Tabor. 1978a. Isolation and characterization of *Saccharomyces cerevisiae* mutants deficient in *S*-adenosylmethionine decarboxylase, spermidine, and spermine. *J. Bacteriol.* **134:** 208.

Cohn, M.S., C.W. Tabor, H. Tabor, and R.B. Wickner. 1978b. Spermidine or spermine requirement for killer double-stranded RNA plasmid replication in yeast. *J. Biol. Chem.* **253:** 5225.

Conde, J. and G.R. Fink. 1976. A mutant of *Saccharomyces cerevisiae* defective for nuclear fusion. *Proc. Natl. Acad. Sci.* **73:** 3651.

Cox, B.S. 1965. ψ, a cytoplasmic suppressor of super-suppressor in yeast. *Heredity* **20:** 505.

——— ———. 1971. A recessive lethal super-suppressor mutation in yeast and other ψ phenomena. *Heredity* **26:** 211.

Culbertson, M.R., L. Charnas, M.T. Johnson, and G.R. Fink. 1977. Frameshifts and frameshift suppressors in *Saccharomyces cerevisiae. Genetics* **86:** 745.

Drillien, R., M. Aigle, and F. Lacroute. 1973. Yeast mutants pleiotropically impaired in the regulation of the two glutamate dehydrogenases. *Biochem. Biophys. Res. Commun.* **53:** 367.

Ephrussi, B., H. de Margerie-Hottinguer, and H. Roman. 1955. Suppressiveness: A new factor in the genetic determinism of the synthesis of respiratory enzymes in yeast. *Proc. Natl. Acad. Sci.* **41:** 1065.

Fink, G.R. and C.A. Styles. 1972. Curing of a killer factor in *Saccharomyces cerevisiae. Proc. Natl. Acad. Sci.* **69:** 2846.

Fischer, I. and C. Shalitin. 1977. Increased synthesis of abundant poly(a)-containing RNA in a DNA defective mutant of *Saccharomyces cerevisiae* containing the "killer character." *Biochim. Biophys. Acta* **475:** 64.

Fried, H.M. and G.R. Fink. 1978. Electron microscopic heteroduplex analysis of "killer" double-stranded RNA species from yeast. *Proc. Natl. Acad. Sci.* **75:** 4224.

Garvik, B. and J.E. Haber. 1978. New cytoplasmic genetic element that controls 20S RNA synthesis during sporulation in yeast. *J. Bacteriol.* **134:** 261.

Goldring, E.S., L.I. Grossman, D. Krupnick, D.R. Cryer, and J. Marmur. 1970. The petite mutation in yeast. Loss of mitochondrial deoxyribonucleic acid during induction of petites with ethidium bromide. *J. Mol. Biol.* **52:** 323.

Guerry-Kopecko, P. and R.B. Wickner. 1980. Isolation and characterization of temperature-sensitive *mak* mutants of *Saccharomyces cerevisiae. J. Bacteriol.* **144:** 1113.

Hankin, L. and J.E. Puhalla. 1971. Nature of a factor causing interstrain lethality in *Ustilago maydis. Phytopathology* **61:** 50.

Harris, M.S. 1978. Virus-like particles and double-stranded RNA from killer and nonkiller strains of *Saccharomyces cerevisiae. Microbios* **21:** 161.

Hastie, N.D., V. Brennan, and J.A. Bruenn. 1978. No homology between double-stranded RNA and nuclear DNA of yeast. *J. Virol.* **28:** 1002.

Herring, A.J. and E.A. Bevan. 1974. Virus-like particles associated with the double-stranded RNA species found in killer and sensitive strains of the yeast *Saccharomyces cerevisiae. J. Gen. Virol.* **22:** 387.

———. 1977. Yeast virus-like particles possess a capsid-associated single-stranded RNA polymerase. *Nature* **268:** 464.

Hollenberg, C.P., P. Borst, and E.F.J. van Bruggen. 1970. Mitochondrial DNA. V. A 25μ closed circular duplex DNA molecule in wild-type yeast mitochondria. Structure and genetic complexity. *Biochim. Biophys. Acta* **209:** 1.

Hollings, M. 1978. Mycoviruses: Viruses that infect fungi. *Adv. Virus Res.* **22:** 1.

Holm, C.A., S.G. Oliver, A.M. Newman, L.E. Holland, C.S. McLaughlin, E.K. Wagner, and R.C. Warner. 1978. The molecular weight of yeast P1 double-stranded RNA. *J. Biol. Chem.* **253:** 8332.

Hopper, J.E., K.A. Bostian, L.B. Rowe, and D.J. Tipper. 1977. Translation of the *L*-species ds RNA genome of the killer-associated virus-like particles of *Saccharomyces cerevisiae. J. Biol. Chem.* **252:** 9010.

Imamura, T., M. Kawamoto, and Y. Takaoka. 1974. Characteristics of main mash infected by killer yeast in *Saké* brewing and the nature of its killer factor. *J. Ferment. Technol.* **52:** 293.

Ishii, K., T. Hashimoto-Gotoh, and K. Matsubara. 1978. Random replication and random assortment model for plasmid incompatibility in bacteria. *Plasmid* **1:** 435.

Kadowaki, K. and H.O. Halvorson. 1971a. Appearance of a new species of ribonucleic acid during sporulation in *Saccharomyces cerevisiae. J. Bacteriol.* **105:** 826.

———. 1971b. Isolation and properties of a new species of ribonucleic acid synthesized in sporulating cells of *Saccharomyces cerevisiae. J. Bacteriol.* **105:** 831.

Kandel, J.S. and T.A. Stern. 1979. Killer phenomenon in pathogenic yeast. *Antimicrob. Agents Chemother.* **15:** 568.

Kane, W.P., D.F. Pietras, and J.A. Bruenn. 1979. Evolution of defective-interfering double-stranded RNAs of the yeast killer virus. *J. Virol.* **32:** 692.

Koltin, Y. 1977. Virus-like particles in *Ustilago maydis:* Mutants with partial genomes. *Genetics* **86:** 527.

Koltin, Y. and P.R. Day. 1975. Specificity of *Ustilago maydis* killer proteins. *Appl. Microbiol.* **30:** 694.

———. 1976a. Inheritance of killer phenotypes and double-stranded RNA in *Ustilago maydis. Proc. Natl. Acad. Sci.* **73:** 594.

———. 1976b. Suppression of the killer phenotype in *Ustilago maydis. Genetics* **82:** 629.

Koltin, Y. and J.S. Kandel. 1978. Killer phenomenon in *Ustilago maydis*: The organization of the viral genome. *Genetics* **88:** 267.

Koltin, Y., I. Mayer, and R. Steinlauf. 1978. Killer phenomenon in *Ustilago maydis*: Mapping viral functions. *Mol. Gen. Genet.* **166:** 181.

Kotani, H., A. Shinmyo, and T. Enatsu. 1977. Killer toxin for sake yeast: Properties and effects of adenosine-5′-diphosphate and calcium ion on killing action. *J. Bacteriol.* **129**: 640.

Lacroute, F. 1971. Non-Mendelian mutation allowing ureidosuccinic acid uptake in yeast. *J. Bacteriol.* **106**: 519.

Leibman, S.W., J.W. Stewart, and F. Sherman. 1975. Serine substitutions caused by an ochre suppressor in yeast. *J. Mol. Biol.* **94**: 595.

Leibowitz, M.J. and R.B. Wickner. 1976. A chromosomal gene required for killer plasmid expression, mating, and sporulation in *Saccharomyces cerevisiae*. *Proc. Natl. Acad. Sci.* **73**: 2061.

————. 1978. *pet18*: A chromosomal gene required for cell growth and the maintenance of mitochondrial DNA and the killer plasmid of yeast. *Mol. Gen. Genet.* **165**: 115.

Leth, B., C. Christiansen, and G. Christiansen. 1972. Circular, repetitive DNA in yeast. *Biochim. Biophys. Acta* **269**: 527.

Levine, R., Y. Koltin, and J. Kandel. 1979. Nuclease activity associated with the *Ustilago maydis* virus induced killer proteins. *Nucleic Acids Res.* **6**: 3717.

Livingston, D.M. 1977. Inheritance of the 2 μm DNA plasmid from *Saccharomyces*. *Genetics* **86**: 73.

Livingston, D.M. and D.M. Kupfer. 1977. Control of *Saccharomyces cerevisiae* 2 μ DNA replication by cell division cycle genes that control nuclear DNA replication. *J. Mol. Biol.* **116**: 249.

Makower, M. and E.A. Bevan. 1963. The inheritance of a killer character in yeast (*Saccharomyces cerevisiae*). *Proc. Int. Congr. Genet. XI* **1**: 202.

Maule, A.P. and P.D. Thomas. 1973. Strains of yeast lethal to brewery yeasts. *J. Inst. Brew.* **79**: 137.

Mitchell, D.J., A.J. Herring, and E.A. Bevan. 1976. The genetic control of ds RNA virus-like particles associated with *Saccharomyces cerevisiae* killer yeast. *Heredity* **37**: 129.

Mortimer, R.K. and D. Schild. 1980. Genetic map of *Saccharomyces cerevisiae*. *Microbiol. Rev.* **44**: 519.

Naumov, G.I. and T.I. Naumova. 1973. Comparative genetics of yeast. XIII. Comparative study of killer strains of *Saccharomyces* from different collections. *Genetika* **9**: 140.

Naumov, G.I., L.V. Tyurina, N.I. Bur'Yan, and T.I. Naumova. 1973. Wine-making, an ecological niche of type k_2 killer *Saccharomyces*. *Biol. Nauki* **16**: 103.

Naumova, T.I. and G.I. Naumov. 1973. Comparative genetics of yeast. XII. Study of antagonistic interrelations in *Saccharomyces* yeast. *Genetika* **9**: 85.

Nesterova, G.F. 1974. Infection of germinating spores of *Saccharomyces cerevisiae* by cytoplasmic factor k. *Genetika* **10**: 78.

Nesterova, G.F., Y.O. Soom, and A.P. Perevoshikov. 1976. Double-stranded RNA of homothallic *Saccharomycetes* with various cytoplasmic determinants of an antagonistic activity. *Dokl. Akad. Nauk SSSR Ser. Biol.* **226**: 951.

Novick, R.P. and F.C. Hoppensteadt. 1978. On plasmid incompatibility. *Plasmid* **1**: 421.

Oliver, S.G., S.J. McCready, C. Holm, P.A. Sutherland, C.S. McLaughlin, and B.S. Cox. 1977. Biochemical and physiological studies of the yeast virus-like particle. *J. Bacteriol.* **130**: 1303.

Ouchi, K. and H. Akiyama. 1976. Breeding of useful killer *Saké* yeast by repeated back-crossing. *J. Ferment. Technol.* **54**: 615.

Palfree, R. and H. Bussey. 1979. Yeast killer toxin: Purification and characterization of the protein toxin from *Saccharomyces cerevisiae*. *Eur. J. Biochem.* **93**: 487.

Philliskirk, G. and T.W. Young. 1975. The occurrence of killer character in yeasts of various genera. *Antonie van Leeuwenhoek J. Microbiol. Serol.* **41**: 147.

Pietras, D.F. and J.A. Bruenn. 1976. The molecular biology of yeast killer factor. *Int. J. Biochem.* **7**: 173.

Puhalla, J.E. 1968. Compatibility reactions on solid medium and interstrain inhibition in *Ustilago maydis*. *Genetics* **60**: 461.

Rogers, D.T. 1976. "The genetic and phenotypic characterization of killer strains of yeast isolated from different sources." Ph. D. thesis, Queen Mary College, University of London.

Rogers, D.T. and E.A. Bevan. 1978. Group classification of killer yeasts based on cross-reactions between strains of different species and origin. *J. Gen. Microbiol.* **105**: 199.

Rogers, D.T., D. Saville, and H. Bussey. 1979. *Saccharomyces cerevisiae* killer expression mutant *KEX2* has altered secretory proteins and glycoproteins. *Biochem. Biophys. Res. Commun.* **90**: 187.

Rothstein, R.J. and F.Sherman. 1980. Dependence on mating type for the overproduction of iso-2-cytochrome *c* in the yeast mutant *CYC7-H2*. *Genetics* **94**: 891.

Schamhart, D.H.J., A.M.A. Ten Berge, and K.W. Van De Poll. 1975. Isolation of a catabolite repression mutant of yeast as a revertant of a strain that is maltose negative in the respiratory-deficient state. *J. Bacteriol.* **121**: 747.

Sherman, F. 1959. The effects of elevated temperatures on yeast. II. Induction of respiratory-deficient mutants. *J. Cell. Comp. Physiol.* **54**: 37.

Sinclair, J.H., R.J. Stevens, P. Sanghavi, and M. Rabinowitz. 1967. Mitochondrial satellite and circular DNA filaments in yeast. *Science* **156**: 1234.

Singh, A., C. Helms, and F. Sherman. 1979. Mutation of the non-Mendelian suppressor, ψ^+, in yeast by hypertonic media. *Proc. Natl. Acad. Sci.* **76**: 1952.

Skipper, N. and H. Bussey. 1977. Mode of action of yeast toxins: Energy requirements for *Saccharomyces cerevisiae* killer toxin. *J. Bacteriol.* **129**: 668.

Sogin, S.J., J.E. Haber, and H.O. Halvorson. 1972. Relationship between sporulation-specific 20S ribonucleic acid and ribosomal ribonucleic acid processing in *Saccharomyces cerevisiae*. *J. Bacteriol.* **112**: 806.

Somers, J.M. 1973. Isolation of suppressive mutants from killer and neutral strains of *Saccharomyces cerevisiae*. *Genetics* **74**: 571.

Somers, J.M. and E.A. Bevan. 1969. The inheritance of the killer character in yeast. *Genet. Res.* **13**: 71.

Stevens, B.J. and E. Moustacchi. 1971. ADN satellite γ et molécules circulaires torsadées de petite taille chez la levure *Saccharomyces cerevisiae*. *Exp. Cell. Res.* **64**: 259.

Stumm, C., J.M.H. Hermans, E.J. Middelbeek, A.F. Croes, and G.J.M.L. De Vries. 1977. Killer-sensitive relationships in yeasts from natural habitats. *Antonie van Leeuwenhoek J. Microbiol. Serol.* **43**: 125.

Sweeney, T.K., A. Tate, and G.R. Fink. 1976. A study of the transmission and structure of double-stranded RNAs associated with the killer phenomenon in *Saccharomyces cerevisiae*. *Genetics* **84**: 27.

Toh-e, A. 1979. Genetics of the killer character of yeast. *Tampaku-shitsu Kakusan Koso (Japan)* **24**: 1169.

Toh-e, A. and R.B. Wickner. 1979. A mutant killer plasmid whose replication depends on a chromosomal "superkiller" mutation. *Genetics* **91**: 673.

――――. 1980. "Superkiller" mutations suppress chromosomal mutations affecting double-stranded RNA killer plasmid replication in *Saccharomyces cerevisiae*. *Proc. Natl. Acad. Sci.* **77**: 527.

Toh-e, A., P. Guerry, and R.B. Wickner. 1978. Chromosomal superkiller mutants of *Saccharomyces cerevisiae*. *J. Bacteriol.* **136**: 1002.

Vodkin, M. 1977a. Homology between double-stranded RNA and nuclear DNA of yeast. *J. Virol.* **21**: 516.

――――. 1977b. Induction of yeast killer factor mutations. *J. Bacteriol.* **132**: 346.

Vodkin, M.H. and G.A. Alianell. 1979. Fungal viruses and killer factors (*Saccharomyces cerevisiae*). In *Fungal viruses* (ed. H.P. Molitoris et al.), p. 108. Springer-Verlag, New York.

Vodkin, M., F. Katterman, and G.R. Fink. 1974. Yeast killer mutants with altered double-stranded ribonucleic acid. *J. Bacteriol.* **117**: 681.

Weber, H. and R. Lindner. 1975. Virus-like particles in yeast protoplasts. I. Demonstration by means of electron microscopy. *Z. Allg. Mikrobiol.* **15**: 631.

Wejksnora, P.J. and J.E. Haber. 1978. Ribonucleoprotein particle appearing during sporulation in yeast. *J. Bacteriol.* **134:** 246.

Welsh, J.D. and M.J. Leibowitz. 1980. Transcription of killer virion double-stranded RNA *in vitro. Nucleic Acids Res.* **8:** 2365.

Welsh, J.D., M.J. Leibowitz, and R.B. Wickner. 1980. Virion DNA-dependent RNA polymerase from *Saccharomyces cerevisiae. Nucleic Acids Res.* **8:** 2349.

Wickner, R.B. 1974a. Chromosomal and nonchromosomal mutations affecting the "killer character" of *Saccharomyces cerevisiae. Genetics* **76:** 423.

―――. 1974b. "Killer character" of *Saccharomyces cerevisiae*: Curing by growth at elevated temperature. *J. Bacteriol.* **117:** 1356.

―――. 1976a. Killer of *Saccharomyces cerevisiae*: A double-stranded RNA plasmid. *Bacteriol. Rev.* **40:** 757.

―――. 1976b. Mutants of the killer plasmid of *Saccharomyces cerevisiae* dependent on chromosomal diploidy for expression and maintenance. *Genetics* **82:** 273.

―――. 1977. Deletion of mitochondrial DNA bypassing a chromosomal gene needed for maintenance of the killer plasmid of yeast. *Genetics* **87:** 441.

―――. 1978. Twenty-six chromosomal genes needed to maintain the killer double-stranded RNA plasmid of *Saccharomyces cerevisiae. Genetics* **88:** 419.

―――. 1979a. The killer double-stranded RNA plasmids of yeast. *Plasmid* **2:** 303.

―――. 1979b. Mapping chromosomal genes of *Saccharomyces cerevisiae* using an improved genetic mapping method. *Genetics* **92:** 803.

―――. 1980. Plasmids controlling exclusion of the k_2 killer double-stranded RNA plasmid of yeast. *Cell* **21:** 217.

Wickner, R.B. and M.J. Leibowitz. 1976a. Chromosomal genes essential for replication of a double-stranded RNA plasmid of *Saccharomyces cerevisiae*: The killer character of yeast. *J. Mol. Biol.* **105:** 427.

―――. 1976b. Two chromosomal genes required for killing expression in killer strains of *Saccharomyces cerevisiae. Genetics* **82:** 429.

―――. 1977. Dominant chromosomal mutation bypassing chromosomal genes needed for killer RNA plasmid replication in yeast. *Genetics* **87:** 453.

―――. 1979. *mak* mutants of yeast: Mapping and characterization. *J. Bacteriol.* **140:** 154.

Woods, D.R. and E.A. Bevan. 1968. Studies on the nature of the killer factor produced by *Saccharomyces cerevisiae. J. Gen. Microbiol.* **51:** 115.

Young, C.S.H. and B.S. Cox. 1971. Extrachromosomal elements in a super-suppression system of yeast. I. A nuclear gene controlling the inheritance of the extrachromosomal elements. *Heredity* **26:** 413.

Young, T.W. and M. Yagiu. 1978. A comparison of the killer character in different yeasts and its classification. *Antonie van Leeuwenhoek J. Microbiol. Serol.* **44:** 59.

The Yeast Plasmid 2μ Circle

James R. Broach
Department of Microbiology
State University of New York
Stony Brook, New York 11794

INTRODUCTION

Investigation of the yeast plasmid 2μ circle has been pursued in a number of laboratories as a result of the recognition of its potential both for the study of eukaryotic DNA replication and as a cloning vector for yeast. In addition, its ability to interconvert between two separate structures by intramolecular recombination has fostered speculation that expression of the molecule may be modulated by DNA rearrangement. Thus, 2μ circle can also serve as a model system for specialized recombination in a eukaryotic cell with the possibility that this system will provide insights into an unusual form of gene regulation.

In this article I plan to review current knowledge on the cellular location, structure, expression, and functions of 2μ circle with an emphasis on those results that concern the above three topics. By necessity, certain aspects of these topics are covered in other chapters in this volume. The similarities of 2μ circle and chromosomal DNA replication—specifically the dependence of 2μ circle DNA replication on the cell cycle and on chromosomally encoded replication functions—are reviewed by Fangman and Zakian (this volume)

445

and, as a consequence, will not be recapitulated here. However, I will discuss those aspects of 2μ circle replication that distinguish it from chromosomal DNA replication. Similarly, although the use of 2μ circle as a cloning vector is described by Botstein and Davis (1982), the distinctive characteristics of 2μ-circle-mediated transformation, as well as the use of yeast transformation as a means of analyzing 2μ circle functions, is covered in this article.

CELLULAR LOCATION

2μ circle is a 6300-bp circular, double-stranded-DNA species present in most *Saccharomyces cerevisiae* strains at 60–100 copies per diploid cell (Clark-Walker and Miklos 1974). By genetic criteria, it has been functionally classified as a cytoplasmic element, although circumstantial evidence suggests that its actual location within the cell is the nucleoplasm.

Extrachromosomal Inheritance

The genetic experiments establishing non-Mendelian inheritance of 2μ circles were performed by D. Livingston, who monitored the segregation of 2μ circles following mating and sporulation of appropriate strains (Livingston 1977). Since, at the time, no phenotype had been consistently associated with the presence of 2μ circles in a cell, segregational analysis was accomplished by physical means, made possible by the identification of strains lacking 2μ circles and strains containing physical variants of 2μ circle (see below). Livingston found that after mating a strain containing 2μ circles with one lacking 2μ circles and after sporulating the resultant diploid, all four spores in each tetrad contained 2μ circles. Similarly, in crosses between a strain containing a normal 2μ circle and one containing a 2μ circle variant, all spores contained a mixture of the two 2μ circle species. Thus, 2μ circles are inherited cytoplasmically. A corollary of this experiment, as well as other subsequent experiments of this type, is that all *S. cerevisiae* or *S. carlsbergensis* strains that do not have endogenous 2μ circles do not possess any obvious barrier to their propagation once they are introduced into the strain.

Nucleoplasmic Location

There are several lines of evidence suggesting that 2μ circles reside in the nucleoplasm. First, 2μ circle plasmids have been found to interact with genomic DNA both by genetic and by biochemical criteria. Autonomously replicating hybrid plasmids, consisting of 2μ circle DNA and DNA containing a specific yeast gene, can recombine with the homologous chromosomal DNA, which indicates that 2μ circle plasmids can come within close proximity to chromosomes in the cell (Kielland-Brandt et al. 1981; J. Hicks, pers. comm.). Biochemical evidence for this association comes from observations by Taketo et al. (1980), who demonstrated that 2μ circle DNA cosediments with the nuclear DNA complex, called folded chromosome, obtained by lysis of yeast spheroplasts in low salt buffer containing a nonionic detergent. (The sedimentation value for this complex [4000S] is distinct from that of super-coiled 2μ circle DNA [25S] or nucleosome-packaged 2μ circle DNA [75S].)

A second line of evidence suggesting a nuclear location for 2μ circles is its interaction with proteins that are presumed to be nuclearly limited. For example, 2μ circle DNA replication requires the same gene products that are used to replicate chromosomal DNA (Petes and Williamson 1975; Livingston and Kupfer 1977). In addition, unlike mitochondrial DNA but like chromo-somal DNA, 2μ circle DNA is packaged into nucleosomes containing the normal composition of core histones (Livingston and Hahne 1979; Nelson and Fangman 1979). Finally, 2μ circle is extensively transcribed, presumably by RNA polymerase II (Broach et al. 1979). Futhermore, at the restrictive temperature in strains containing a temperature-sensitive mutation in the nuclear RNA transport gene *rna1*, the steady-state level of 2μ circle mRNA species is substantially reduced (J.R. Broach, unpubl.). Thus, since many of these proteins are limited to the nucleus, 2μ circles must spend at least part of the time in the nucleoplasm.

A third line of evidence suggesting a nuclear location for 2μ circles has been obtained from cytoduction experiments. Livingston crossed a [*cir*^{0}, *rho*^{0}] *cyh*^{R} strain[1] with a [*cir*$^+$, *rho*$^+$] *CYH*^{S} *kar1* strain to produce hetero-karyons. (Nuclear fusion is substantially reduced after normal zygote forma-tion following mating if one of the parent strains carries a *kar1* mutation.) From the heterokaryons, he selected haploid segregants that were [*rho*$^+$] *cyh*^{R}, i.e., segregants that had the nucleus of the [*cir*^{0}] parent and the cyto-plasm of the [*cir*$^+$] parent. Only 50% of these [*rho*$^+$] *cyh*^{R} segregants contained 2μ circles, even though random assortment of freely diffusible 2μ circle molecules would predict that nearly 100% of the segregants should have acquired 2μ circles (Livingston 1977). Thus, although 2μ circles are transmit-ted efficiently to [*cir*^{0}] strains if nuclear fusion occurs (i.e., following normal mating and sporulation), they are not transmitted if only cytoplasmic fusion occurs. Consequently, one can conclude that 2μ circles are sequestered in the cell and, since they must spend some time in the nucleus, that they are sequestered in the nucleus.

In the cytoduction experiments described above, the fact that 2μ circle DNA is transmitted to the [*cir*^{0}] strain does not constitute a counterargument for a nuclear location of 2μ circles. It has been demonstrated that in similar cytoduction experiments, there is a low-frequency internuclear transfer of individual chromosomes (S.K. Dutcher, in prep.). In addition, the frequen-cy with which an individual chromosome is transferred is inversely propor-tional to its size. Given this information, the observed transfer of 2μ circles

---

[1]Strains lacking 2μ circles are indicated by [*cir*^{0}], whereas those having a normal complement of wild-type 2μ circles are indicated by [*cir*$^+$].

is actually lower than that predicted by extrapolation of the percentage of transfer versus chromosomal DNA size.

The only other evidence against a nuclear location of 2μ circles is a reported failure to observe copurification of 2μ circles and nuclei in a subcellular fractionation experiment (Clark-Walker and Miklos 1974). However, yeast nuclei are unusually fragile, and many components, such as DNA polymerase, that are nucleoplasmic in the cell are not enriched in nuclei preparation. Also, as described above, Taketo et al. (1980), using different lysis procedures, have shown that 90% of 2μ circles are associated with the nuclear matrix of the cell. Thus, the reported localization of 2μ circles in the particulate fraction of the cell can be explained by leakage of circles from nuclei during preparation and by the similarities in sedimentation behavior of nucleosomal 2μ circle DNA and particulate material of the cell.

Absence of Integrated Copies

Although 2μ circles appear to reside in the nucleus, there does not appear to be any 2μ circle DNA covalently associated with chromosomal DNA. It is clear that in the few [cir^0] *S. carlsbergensis* and *S. cerevisiae* strains examined, there are no DNA sequences homologous to 2μ circles in the genome, as determined both by Southern hybridization and by genomic DNA driver hybridizations (< 0.1 molecule per genome equivalent; J.R. Broach, unpubl.; Sigurdson et al. 1981). In addition, there are no RNA species in [cir^0] strains homologous to 2μ circle DNA (Broach et al. 1979). The presence of integrated copies of 2μ circle DNA in [cir^+] strains cannot be rigorously excluded due to the high copy number of free plasmids and to the existence in the cell of multimeric forms of the plasmid (see below). However, by Southern analysis of genomic DNA digested with different restriction enzymes, there are no unaccountable restriction fragments homologous to 2μ circle DNA that would indicate the presence of a copy of 2μ circle integrated at a unique site in chromosomal DNA (Cameron et al. 1977). However, the possibility of short-lived, randomly occurring integration events cannot be excluded. In addition, there is no specific exclusion of 2μ circle DNA from chromosomal integration, since hybrid plasmids containing a portion of 2μ circle DNA can integrate by homologous recombination through a chromosomal DNA segment carried on the plasmid (Kielland-Brandt et al. 1981; J.R. Broach, unpubl.).[2]

---

[2]The frequency with which such events occur is very low. The extra restriction fragments found in yeast transformants obtained with hybrid 2μ circle plasmids, which have been proffered as evidence for high-frequency integration of such plasmids into a chromosome (Struhl et al. 1979), are most likely due to recombination of the transforming plasmid with endogenous 2μ circle molecules (Broach and Hicks 1980). Genetic evidence suggests that chromosomal integration of such plasmids is relatively rare.

STRUCTURE OF 2μ CIRCLE

The primary structure of 2μ circle DNA has been probed by a variety of techniques, including heteroduplex analysis of native and cloned 2μ circle sequences, restriction analysis of native and cloned DNA, and, most recently, direct sequence determination of the entire 2μ circle molecule (Beggs et al. 1976; Guerineau et al. 1976; Hollenberg et al. 1976b; Gubbins et al. 1977; Livingston and Klein 1977; Hartley and Donelson 1980). The location of restriction sites for a selected set of enzymes is schematically represented in Figure 1. These sites are determined from the primary sequence of the molecule and are consistent with data obtained by restriction analysis from a number of laboratories. A more detailed listing of restriction sites is given in Tables 1 and 2 (see Appendix to this chapter).

Inverted Repeat Sequences

The most striking structural feature of 2μ circles is the presence of two regions, of 599 bp each, which are precise inverted repeats of each other. The two inverted repeats separate the molecule into two regions of unique sequence, the larger of which is 2774 bp long and the smaller, 2346 bp (Hartley and Donelson 1980). In yeast, recombination occurs readily between the two inverted repeat regions, the result of which is an inversion of one unique region with respect to the other (Beggs 1978). As a consequence of this recombination, 2μ circle DNA isolated from yeast is a mixed population of equal amounts of two plasmids that differ in the orientation of one unique region with respect to the other. These two forms of the molecule have been variously designated form A, 23, XY′, or R, and form B, 14, XY, or L. Preparations of 2μ circle DNA isolated from yeast contain, in addition to the two forms of the molecule, multimeric plasmids (i.e., 4μ circle, 6μ circle, etc.), constituting approximately 20% of the total population of plasmid molecules (Petes and Williamson 1975; Guerineau et al. 1976). Recent experiments using hybrid plasmids containing 2μ circle sequences suggest that these multimeric forms arise by intermolecular recombination through the inverted repeat sequences. The purpose of the inverted repeats and the function served by intramolecular and intermolecular recombination is discussed below.

Within both inverted repeat regions there are several symmetrical sequences (Hartley and Donelson 1980; see Fig. 4). Centered near the *Xba*I restriction site is a near perfect dyad symmetry (abc-c′b′a′), which could give rise to a stem-loop structure with a 50-bp stem and a 22-base loop. In addition, within each 50-bp region that constitutes the stem portion of this structure is a secondary dyad symmetry of 16 bp in length, the two halves of which are separated by 5 bp. In one region, the 16-bp match is perfect, and in the other, it is near perfect. Thus, an alternative secondary structure in the form of a cloverleaf can be drawn. Finally, there is a 12-bp sequence within the

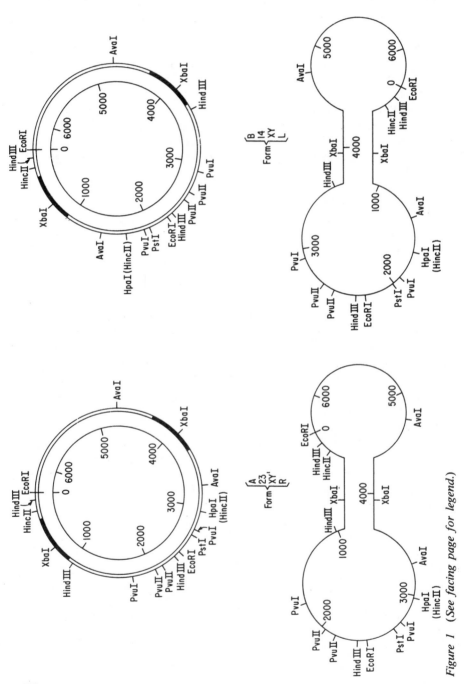

Figure 1 (*See facing page for legend.*)

large-loop region that is precisely repeated, after a 1-base gap, in the stem region. The function served, if any, by these elaborate symmetrical features is unknown, although some speculation is offered below, in Functions of 2μ Circle, Origin of Replication.

Other Structural Features

In the region spanning the *Hpa*I and *Ava*I restriction sites in the large unique region are 5.5 copies of a sequence 62–63 bases in length arranged as direct tandem repeats (Hartley and Donelson 1980). These repeats share 78–97% homology with a consensus sequence of 5′-TTTTTPuPy-AGAACAAAAA-TGCAACGCGAGCGCTAATTTTCAAACAAAGAATCTGAGCTTCA-3′. The functions of these sequences are unknown, but they do lie near the 5′ end of one of the 2μ circle transcripts.

Variants

At least two physical variants of 2μ circle have been observed as natural isolates from various strains of *S. cerevisiae* (Cameron et al. 1977; Livingston 1977). In strains in which these variants are obtained, each variant constitutes the sole 2μ circle species in that strain. In addition, no consistent or atypical phenotype can be ascribed to strains that harbor these variants.

As determined by restriction analysis, one variant designated Scp2 contains a 125-bp deletion, which removes the *Hpa*I restriction site in the large unique region. The second variant, designated Scp3, contains a 220-bp deletion at approximately the same site as that in Scp2 and removes both the *Hpa*I and the *Ava*I restriction sites. In addition, Scp3 is missing the *Eco*RI restriction site in the large unique region, an alteration formally unrelated to the deletion. Scp3 is apparently identical to variant *cir⁺-E2*, described by Livingston (1977). *cir⁺-E2* lacks the *Eco*RI site in the large unique region and contains a 200-bp deletion that removes the *Hpa*I and *Ava*I restriction sites. Since there is no detailed genealogy available for the two strains in which Scp3 has been found, it is not clear whether Scp3 arose independently on at

Figure 1   Restriction map of 2μ circle. The locations of various restriction sites are indicated on two schematic representations of the two forms of 2μ circle. In the upper representation, the inverted repeat sequences are indicated by the filled lines. In the lower representation, the inverted repeats are shown as parallel lines. The various designations for each of the two forms are given between the two representations of each form. The base-pair numbering system is shown inside each diagram and is consistent with that of Hartley and Donelson (1980), which uses the *Eco*RI site in the small unique region as an arbitrary zero reference point. The positions of the restriction sites are derived from the sequence of 2μ circle (Hartley and Donelson 1980) and were compiled by J. Hartley.

least two occasions or whether both strains had a common ancestor in which Scp3 was resident. Since the deletions in Scp2 and Scp3 are located at a site in 2μ circle at which there are a series of tandem repeats, it seems likely that these variants arose by intramolecular or unequal intermolecular recombination at these repeats. The sizes of the deletions are consistent with the removal of 2 and 4 copies of these repeated sequences, respectively, in Scp2 and Scp3. If this is the correct interpretation, it should be noted, however, that at least 1.5 copies of the repeated sequence would be retained even in Scp3.

Higher-order Structure

As mentioned in an earlier section, 2μ circle DNA is packaged by a normal complement of core histones into chromatinlike nucleosomal DNA (Livingston and Hahne 1979; Nelson and Fangman 1979). Thus, 2μ circle chromatin isolated from cells under nondenaturing conditions sediments at a substantially faster rate than purified 2μ circle DNA, has a beaded structure when visualized by electron microscopy, and is cut by micrococcal nuclease into discrete fragments whose lengths are integral multiples of 165 bp. Recent analysis (Seligy et al. 1980) has shown that this discrete banding pattern is maintained up to fragment sizes corresponding to 35 repeat units (38–39 core particles can be accommodated on a molecule the size of 2μ circle). These results suggest, although they do not conclusively prove, that most 2μ circle molecules in the cell do not contain unpackaged regions but, rather, are completely covered by nucleosomes. It is not known, however, whether the nucleosomes are uniquely phased along the molecule or whether 2μ circle molecules are randomly packaged.

EXPRESSION OF 2μ CIRCLE

Coding Regions

There are several lines of evidence indicating that 2μ circle contains several genes that are expressed in yeast. First, there are at least three open coding regions in the molecule that could give rise to proteins of substantial size. These regions are labeled A, B, and C in Figure 2 and would, if transcribed in vivo and translated from the first available AUG codon, produce peptides of 49,000 daltons, 43,000 daltons, and 33,000 daltons, respectively (Hartley and Donelson 1980). In addition, two smaller open coding regions (labeled D and E in Fig. 2) could produce peptides of approximately 18,000 daltons and 12,000 daltons. Prior confirmation of the existence of these open reading frames has been provided by examination of the expression of cloned 2μ circle DNA in *Escherichia coli* minicells. Hollenberg and his colleagues found that 2μ circle DNA could program the synthesis of at least three polypeptides in *E. coli* (Hollenberg et al. 1976a; Hollenberg 1978). The sizes of these polypeptides and the regions and orientations from which they are read are

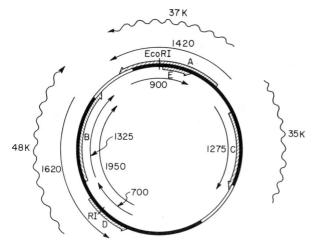

Figure 2 Expression of 2μ circle. Superimposed on a schematic representation of the A form of 2μ circle (as shown in Fig. 1) are the locations of the open coding frames (hatched, abutting lines), the positions of some of the polyadenylated 2μ circle transcripts present in [*cir*⁺] yeast strains (→), and the approximate positions of inferred coding frames determined from polypeptide synthesis in *E. coli* minicells programmed with 2μ circle DNA (wavy arrow). The 5′- to-3′ orientation of the open coding frames is indicated by the half arrowheads. The regions, as drawn, extend from the first AUG triplet to the first nonsense codon and are designated *A–E* on the basis of length. The positions are determined from the 2μ circle sequence of Hartley and Donelson (1980). The locations of the polyadenylated 2μ circle transcripts are obtained from Broach et al. (1979). The arrowheads lie at the 3′ end of the transcripts, each of which is designated by its length in bases. The locations of coding regions for synthesis of polypeptides in *E. coli* minicells are taken from Hollenberg (1978). The arrowheads indicate the inferred carboxyl ends of the coding sequences, where known. Each region is designated by the size polypeptide synthesized in *E. coli* minicell programmed by the corresponding DNA.

indicated in Figure 2. Although the sizes of the polypeptides obtained in *E. coli* are not precisely those predicted by the sequence of 2μ circle DNA—a discrepancy perhaps due to translational initiation at other than the initial AUG and/or to posttranslational modifications—the locations and orientations of the coding sequences of 2μ circle that function in *E. coli* indicate that there are indeed open coding frames in those regions predicted by sequence analysis.

Evidence indicating that these open reading frames constitute functional 2μ circle genes has been obtained by analysis of 2μ circle transcription in vivo and by studies of 2μ circle mutants (Broach et al. 1979). Those regions of the 2μ circle genome transcribed into stable polyadenylated RNA have been identified by hybridization of labeled 2μ circle DNA restriction fragments to total yeast polyadenylated RNA fractionated on denaturing gels and immobilized on diazotized paper. The two major transcripts of 2μ circle, which are also found on polysomes in vivo, are transcribed in the appropriate directions from regions spanning open coding frames B and C (Fig. 2). In addition, a less prevalent polyadenylated RNA is transcribed in the appropriate orientation spanning coding-region A. There do not appear to be stable polyadenylated RNAs strictly correlated with regions D and E, although both regions are transcribed in the appropriate direction in vivo and could be represented by mRNA species that are below the level of detection by the methods used. Thus, at least the three largest open coding regions of 2μ circle are transcribed into polyadenylated RNAs in vivo.

The other line of evidence indicating that the open coding frames are in fact 2μ circle genes has been obtained from mutational analysis of 2μ circle, which is described in detail in subsequent sections. Basically, these results demonstrate that lesions introduced in vitro into regions A, B, or C cause specific defects in 2μ circle replication or recombination when the altered molecules are reintroduced into yeast.

Transcription

A map of the approximate positions on the 2μ circle genome from which each of the detectable polyadenylated RNA species is transcribed in vivo is shown in Figure 3. In addition to those transcripts indicated, there are also two transcripts in the cell (500 bases and 350 bases in length) that are com-

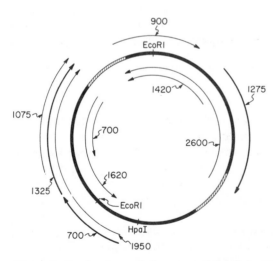

Figure 3 In vivo transcription map of 2μ circle. On a schematic diagram of the A form of 2μ circle (as shown in Fig. 1) are indicated the positions from which polyadenylated RNA species found in [*cir*+] yeast strains are transcribed. (Adapted from Broach et al. 1979.)

plementary to the inverted repeat regions. The particular inverted repeat from which each is transcribed and the direction of transcription has not been determined. The techniques used to map 2μ circle transcripts would only have detected polyadenylated RNA species larger than 300 bases. Thus, whether minor, nonpolyadenylated 2μ circle transcripts or transcripts smaller than 300 bases are present in the cell has not been determined. The 1325-base and 1275-base RNAs are the 2μ circle transcripts present in highest abundance in the cell and constitute approximately 0.1% of the total polyadenylated RNA of the cell. The remaining transcripts are approximately tenfold less abundant. The precise locations of the 5' ends of the transcripts have not been determined. As a consequence, no useful sequence data on potential promoter sites or processing sites are available.

In common with other small genomes, such as SV40 and human mitochondrial DNA, almost the entire 2μ circle genome is represented by polyadenylated RNA species in the cell. Since only a limited number of these transcripts can function as mRNAs, we previously proposed that many of the transcripts represent various stages of synthesis and decay of the actual mRNA species (Broach et al. 1979). That is, the mature mRNAs are obtained by specific cleavage of larger precursor molecules and then decay in part by specific endonucleolytic cleavage. We further hypothesized that transcription of the small unique region is initiated at a promoter in the large unique region, thus producing two distinct sets of transcripts from the two forms of the molecule. Analysis of in vivo transcription from hybrid 2μ circle plasmids frozen in one orientation or the other has not provided confirmation of these hypotheses, although these experiments have not excluded them either (J. R. Broach, unpubl.). Thus, further analysis is required. It is intriguing that the proposed location of the primary promoter in this model is near the tandem repeat sequences in the large unique region.

FUNCTIONS OF 2μ CIRCLE

Attempts to identify a cellular phenotype associated with 2μ circles have met with only limited success. Several *S. carlsbergensis* and *S. cerevisiae* strains completely lacking 2μ circles have been identified (Guerineau et al. 1974; Livingston 1977; D.C. Sigurdson et al., in prep.). These have been tested for growth rates in various media, sporulation ability, conjugation, and heterothallic mating-type switching and have been found to be indistinguishable from isogenic [*cir*+] strains (D.M. Livingston, pers. comm.; J.N. Strathern, pers. comm.; J.R. Broach, unpubl.). Only two phenotypes have been reported to be associated with 2μ circles: oligomycin resistance (Guerineau et al. 1974), which, as discussed below, is as yet only tenuously associated with 2μ circles, and lethal sectoring, which is manifested only in specific mutant strains (C. Holm, pers. comm.). Nonetheless, through cloning technology in conjunction with yeast transformation, the functions of several 2μ circle

456

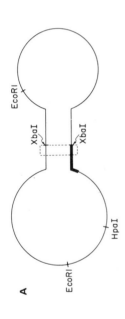

A

EcoRI
XbaI
XbaI
HpaI
EcoRI

B

inverted
repeat

3551
ATTCTTCATT GGTCAGAAAA TTATGAACGG TTTCTTCTAT TTTGTCTCTA TATACTACGT ATAGGAAATG TTTACATTTT CGTATTGTTT TCGATTCACT
TAAGAAGTAA CCAGTCTTTT AATACTTGCC AAAGAAGATA AAACAGAGAT ATATGATGCA TATCCTTTAC AAATGTAAAA GCATAACAAA AGCTAAGTGA

3651
CTATGAATAG TTCTTACTAC AATTTTTTTG TCTAAAGAGT AATACTAGAG ATAAACATAA AAAATGTAGA GGTCGAGTTT AGATGCAAGT TCAAGGAGCG
GATACTTATC AAGAATGATG TTAAAAAAAC AGATTTCTCA TTATGATCTC TATTTGTATT TTTTACATCT CCAGCTCAAA TCTACGTTCA AGTTCCTCGC

3751
AAAGGTGGAT GGGTAGGTTA TATAGGGATA TAGCACAGAG ATATATAGCA AAGAGATACT TTTGAGCAAT GTTTGTGGAA GCGGTATTCG CAATATTTTA
TTTCCACCTA CCCATCCAAT ATATCCCTAT ATCGTGTCTC TATATATCGT TTCTCTATGA CAAACACTTA CAAACACCTT CGCCATAAGC GTTATAAAAT
③ ② ③ XbaI

3851
GTAGCTCGTT ACAGTCCGGT GCGTTTTTGG TTTTTTGAAA GTGGCGTTTC AGAGCGCTTT TGGTTTTCAA AAGGCGCTCTG AAGTTCCTAT ACTTTCTAGA
CATCGAGCAA TGTCAGGCCA CGCAAAAACC AAAAAACTTT CACGCGAAAG TCTCGCGAAA ACCAAAAGTT TTCGCGAGAC TTCAAGGATA TGAAAGATCT
④ ② ① ②

3951
GAATAGGAAC TTCGAATAG GAACTTCAAA GCGTTTCCGA AAACGAGCGC TTCCGAAAAT GCAACGCGAG CTGCGCCACAT ACAGCTCACT GTTCACGTCG
CTTATCCTTG AAGCCTTATC CTTGAAGTTT CGCAAAGGCT TTTGCTCGCG AAGGCTTTTA CGTTGCGCTC GACGGCGTGTA TGTCGAGTGA CAAGTGCAGC
④

Figure 4 (See facing page for legend.)

genes have been identified. These appear to be involved solely in 2μ circle maintenance, i.e., in 2μ circle replication and in 2μ circle interconversion.

Replication

Origin of Replication

Certainly, one of the obvious properties of 2μ circles is their ability to replicate autonomously in the yeast cell. Thus, each molecule should contain a region required for this process, i.e., an origin of replication. However, attempts to determine, with electron microscopy, either the site of 2μ circle replication initiation in vivo or the direction of replication have so far been thwarted by the low percentage of 2μ circle replication intermediates and by the presence in the cell of two forms of the molecule (C.S. Newlon, pers. comm.). However, by transformation of yeast with plasmids containing various fragments of 2μ circle, a small unique region of the molecule has been shown to promote autonomous replication in yeast of plasmids that contain it (Broach and Hicks 1980).

The identification of this site was accomplished by determining which restriction fragments of 2μ circle, when incorporated into a hybrid plasmid containing the *LEU2* gene of yeast, could promote autonomous replication of the hybrid plasmid in yeast following transformation of a [*cir*⁰] *leu2* strain. By subsequent analysis of the effect of deletions introduced into one such hybrid plasmid on its replication potential, the replication site could be localized to a 350-bp sequence lying predominantly within one inverted repeat region but extending approximately 100 bp into the contiguous large unique region (see Fig. 4).

Several lines of evidence suggest that this site is indeed the 2μ circle origin of replication. First, this site has the properties of an origin of replication: It is *cis*-acting, and it is the only region of 2μ circle that will promote autonomous replication in vivo of plasmids in which it is contained. Second, preliminary electron microscopy studies on the locations of replication bubbles on 2μ circle molecules isolated from the cell, although not adequate to unambiguously localize the origin, are consistent with this assignment (Newlon et al. 1981). Third, analysis of 2μ circle replication in vitro indicates that under conditions of inhibition of DNA elongation, a tenfold higher incorpo-

Figure 4   2μ circle origin of replication. (*A*) The approximate location of the 2μ circle origin of replication is indicated by the heavy line on a schematic diagram of the A form of 2μ circle. (Adapted from Broach and Hicks 1980.) The region to which site-specific recombination has been delimited is indicated by the dashed rectangle. (*B*) Nucleotide sequence of the region spanning the 2μ circle origin of replication. Regions of dyad symmetry or direct repeats are indicated by the enumerated arrows above the sequence. The region in which site-specific recombination occurs is indicated by the dashed line below the sequence. (Adapted from Hartley and Donelson 1980.)

ration of label occurs at this site than at any other site on the molecule (Broach et al. 1981).

The sequence of the 2μ circle origin of replication is shown in Figure 4. As can be seen, the replication origin encompasses most of the symmetrical features found within the inverted repeat region. In addition, the 100-bp region adjacent to the inverted repeat has a very high A + T content (80%). It is this latter feature, rather than the symmetrical sequences within the inverted repeat, that the 2μ circle origin of replication shares with at least one putative yeast chromosomal replication site (G. Tschumper and J. Carbon, in prep.; D.T. Stinchcomb, pers. comm.). It is possible that this A + T-rich region could either function as a transcriptional initiation site for primer RNA synthesis or promote local melting of the DNA helix as a prelude to replication initiation. Elaborate symmetrical features, similar to those of the 2μ circle origin, have been observed in the sequence of replication origins from a variety of viruses, including phage λ, SV40, and polyoma virus (Hobom et al. 1979; Soeda et al. 1979; Subramanian et al. 1979). In these organisms, in addition to the replication origin and the host replication apparatus, successful replication requires *trans*-acting replication functions encoded by the organism itself, which apparently interact with its origin of replication (Furth et al. 1979; Shortle et al. 1979; Tjian 1979). As described in the next section, 2μ circle also encodes specific, *trans*-acting replication functions. Thus, the dyad symmetries within the 2μ circle origin of replication, as well as in that of these other organisms, may represent repeated binding domains for specific replication proteins, or, less likely, may form unusual secondary structures that are recognized and stabilized by these proteins.

REP *Genes*

Efficient in vivo replication of hybrid 2μ circle plasmids—and, by inference, of 2μ circle itself—requires, in addition to the replication origin, *trans*-acting functions that are encoded by 2μ circle. Hybrid plasmids containing a restriction fragment of 2μ circle spanning only the origin of replication can replicate autonomously in [cir^0] yeast strains but are present in low copy number (~ 2 copies/cell) and are rapidly lost if grown in the absence of selective pressure. However, the same plasmids in [cir^+] strains are maintained at high copy number (~ 30 copies/cell) and are quite stable mitotically and meiotically. Similarly, a hybrid plasmid containing the entire 2μ circle genome is maintained stably and at high copy number in [cir^0] strains. Thus, some region of 2μ circle, other than the origin, is required for efficient replication (Broach and Hicks 1980).

The sites at which these replication functions are encoded have been determined by analysis of the effects of insertions within various regions of the 2μ circle moiety of a hybrid plasmid (pCV20) containing the entire 2μ circle genome. Random insertions of the bacterial transposon Tn5 into pCV20 have been obtained in *E. coli*, and the locations of the various insertions have been determined by restriction analysis. Those pCV20::Tn5 plasmids in which the transposon is inserted in coding-region B or C (see Fig. 2) display impaired replication in yeast; they are present at low copy number and are lost rapidly in the absence of selective pressure. On the other hand, both pCV20 and pCV20::Tn5, in which the transposon has inserted at other regions of the 2μ circle moiety, replicate stably and at high copy number in yeast. Thus, regions B and C, designated *REP1* and *REP2*, respectively, encode proteins that promote high copy levels of 2μ circle plasmids (Broach et al. 1981).

Under normal growth conditions, replication of 2μ circle is strictly under cell-cycle control; the plasmid population is doubled during a cell division by a single round of replication of each molecule during the early portion of the S phase (Zakian et al. 1979). *REP* proteins apparently act to override cell-cycle control of 2μ circle replication—i.e., they can promote multiple rounds of replication during a single cell cycle—when the plasmid copy number is reduced. This conclusion is derived from several observations. A Leu$^+$ transformant of a [*cir^0*] *leu2* strain obtained with pCV20 *rep1*::Tn5 has a plasmid copy number of 1–2/cell. If the *REP1* gene is introduced into this strain through conjugation with a [*cir$^+$*] strain, then by the time the diploid has grown sufficiently to allow analysis, the level of pCV20 *rep1*::Tn5 has increased to approximately 30 copies/cell. That is, during the first 22 cell divisions of the diploid, the population of pCV20 *rep1*::Tn5 has doubled 27 times. It is unlikely that the high copy number arises from asymmetric segregation of plasmid molecules during the first few cell divisions, followed by clonal selection of those cells that have acquired large numbers of plasmid molecules: *leu2* [pCV20 *rep1*::Tn5, *cir^0*] and *leu2* [pCV20 *rep1*::Tn5, *cir$^+$*] strains have the same growth rate on selective media. Thus, there is no selective advantage of cells with higher plasmid levels.

A similar demonstration of 2μ circle copy control has been obtained by recent cytoduction experiments reported by Sigurdson et al. (1981). By performing cytoduction experiments with a donor strain in which two distinguishable 2μ circle variants were resident and then analyzing the 2μ circle content of the resultant cytoductants, they concluded that those recipient cells that had acquired 2μ circles had, in most cases, received only one molecule during the transfer. Nonetheless, by the time they had grown sufficiently to allow analysis, all [*cir$^+$*] segregants from these crosses had a normal complement of 2μ circle molecules. Thus, in cells growing under normal conditions and without any obvious selective pressure, 2μ circle copy number can increase from 1 to 30.

Thus, 2μ circle contains an origin of replication that is recognized by the replication apparatus of the cell and, conseqently, in steady-state conditions conforms to cellular control of DNA replication. However, in addition, 2μ circle contains two genes, *REP1* and *REP2*, that have the potential to bypass normal cell-cycle control of 2μ circle replication and to induce multiple rounds of replication during a single cell-division cycle. These two aspects of 2μ circle replication may be reflected in the sequence of the replication origin:

an A+T-rich region, which resembles a chromosomal replicon, and a region of high symmetry, defining the site of action of *REP* proteins. Finally, by analogy to copy control system of *E. coli* plasmids, it is likely that the activity and/or synthesis of one or both of the *REP* proteins is regulated in some way in response to the copy level of 2μ circle.

Interconversion

One of the more remarkable aspects of the biology of 2μ circle is its ability to undergo intramolecular recombination and, thus, interconvert between two distinct forms. As mentioned previously, no function has been ascribed to this interconversion, although we have speculated that transcription from the two forms is different. Nonetheless, through the use of hybrid 2μ circle plasmids, the mechanism of interconversion has been explored. The results of this analysis indicate that 2μ circle constitutes a specialized recombination system, containing a specific site at which recombination occurs and encoding an enzyme required for the recombination event.

FLP *Gene*

Several laboratories have demonstrated that 2μ circle encodes a protein required for 2μ circle interconversion (Gerbaud et al. 1979; Broach and Hicks 1980). Experiments establishing this fact were performed basically as follows.

Various yeast strains were transformed with a hybrid plasmid consisting of the entire 2μ circle plasmid, which contains an *E. coli* plasmid cloned into the *Eco*RI site within the small unique region of 2μ circle (pCV20 is an example of such a plasmid). Since the *rec* functions of *E. coli* do not promote recombination between the inverted repeat sequences of 2μ circle, such hybrid plasmids can be obtained from *E. coli* as uniform populations of molecules consisting of the 2μ circle moiety in only one orientation. Following transformation of yeast, the hybrid plasmid was examined to determine whether it was present in one orientation or in both orientations, i.e., whether interconversion of 2μ circle took place in the yeast strain. The results were that in transformants of strains containing endogenous 2μ circles, both orientations of the hybrid plasmid were present in equal amounts; however, in transformants of strains lacking endogenous 2μ circles, the hybrid plasmid was present only in the orientation identical to that of the purified DNA used to obtain the transformant. This was true regardless of the form of the plasmid used for transformation. Thus, inactivation of the gene designated *FLP*, spanning the *Eco*RI site in the small unique region (coding-region A, Fig. 2), prevents interconversion of 2μ circle during mitotic growth. Furthermore, the *flp* defect can be complemented in *trans* by the presence of wild-type 2μ circles. Thus, *FLP* appears to correspond to coding-region A and to encode a protein required for recombination with the inverted repeat region of 2μ circle.

Recombination catalyzed by *FLP* is quite efficient. In those cases in which hybrid 2μ circle plasmid DNA containing 2μ circle moiety in only one orientation is used to transform [*cir*⁺] yeast strains (i.e., strains that are Flp⁺), the resultant transformant, by the time it has grown sufficiently to allow analysis, contains equal proportions of both forms of the hybrid plasmid. Experiments of this type place a lower limit on the recombination frequency at 0.1 recombination events per cell per generation.

By analogy to other specialized recombination systems, such as phage λ integration, it is likely that *FLP*-gene product is not solely responsible for 2μ circle interconversion but that chromosomally encoded functions are required as well. However, at least one nuclear gene product, that from *RAD52*, has been demonstrated not to be involved in the 2μ-circle, *FLP*-mediated recombination process (D.M. Livingston, pers. comm.).

Recombination can occur at a low level within the inverted repeat sequences in the absence of *FLP* function, apparently mediated by the general recombination system of the cell. Using a hybrid plasmid constructed in such a fashion that the inverted repeat sequences are present as direct repeats, an extremely sensitive assay for the recombination event has been developed. In this plasmid, the direct repeats bracket in one half of the molecule all of the genes required for selection and propagation of the plasmid in yeast and *E. coli*. In the other half of the molecule, the repeats bracket the *lacZ* gene of *E. coli*. Thus, recombination through the repeats results in the deletion of *lacZ*. Consequently, recombination that occurs in an in vivo or in vitro system can be assessed by extracting plasmid DNA and using it to transform a *lacZ⁻ E. coli* strain. The proportion of Lac⁻ to Lac⁺ transformants is a measure of the extent of recombination. Through the use of this assay, it has been shown that after mitotic propagation of the plasmid in Flp⁻ yeast cells for 30 generations, approximately 1% of the molecules have undergone a recombination event between the two repeat sequences. On the other hand, after propagation in Flp⁺ yeast cells for a similar length of time, more than 99% of the molecules have recombinants within the inverted repeat region. Thus, general mitotic recombination can occur within the inverted repeat regions but at a frequency at least 100-fold less than that mediated by the *FLP*-gene product. Using the same assay, it has been demonstrated that recombination within the inverted repeat region also occurs in *E. coli* at approximately the same frequency that it does in Flp⁻ yeast strains. That this recombination is mediated by the general recombination system of *E. coli* is demonstrated by the fact that this low-level recombination is abolished in *recA* strains of *E. coli*.

Site-specific Recombination

We have recently demonstrated that *FLP*-mediated recombination does not occur throughout the inverted repeat region but that the recombination event occurs at a specific site within the inverted repeat. These results were obtained by examining the *FLP*-generated recombination products of hybrid plasmids in which the bacterial transposon Tn5 had been inserted at various points within the inverted repeat region. If recombination occurs randomly throughout the inverted repeat region, then the orientation of the Tn5 insertion would be randomized with respect to both unique regions. The results, however, demonstrated that *FLP*-mediated recombination randomized the orientation of the Tn5 with respect to only one of the unique regions. The particular unique region with regard to which the Tn5 was randomized was dependent on the location of the Tn5 within the inverted repeat. Thus, by using several different insertions, we could localize the site for *FLP*-mediated recombination to within the 120 bp spanning the *Xba*I site in the inverted repeat region (see Fig. 4).

As described previously in the section on the origin of replication, this region in which the site for *FLP*-mediated recombination is located is replete with symmetrical sequences. The precise sequences required for site-specific recombination and the role in recombination, if any, of the symmetrical regions within the inverted repeats must await further analysis. The fact that the site for recombination lies within the origin of replication circumstantially suggests a connection between recombination and replication. Both the A and B forms of hybrid *flp* plasmids replicate perfectly well in yeast in the absence of recombination. Thus, there does not appear to be an obligatory role for recombination in replication, although a subtle, as yet undetected, enhancement of replication potential may accrue from recombination. On the other hand, it is entirely possible that recombination is dependent upon replication, a hypothesis that awaits further analysis.

As indicated, the function of interconversion is unknown. However, that interconversion serves some useful role in maintenance of 2μ circles is suggested from the fact that 40% of the molecule is devoted to this process. In several prokaryotic systems, analogous site-specific recombination events function as genetic switches, allowing either alternative expression of two sets of genes—as in the *H2* locus of *Salmonella* (Zeig et al. 1977) or the G loop of phage Mu (Bukhari and Ambrosio 1978)—or alternative pursuit of two different growth cycles—as in phage λ integration (Gottesman and Weisberg 1971). Thus, it is compelling to envision that interconversion of 2μ circle reflects alternation between two phenotypically distinct states. However, it is also possible that this recombination may serve solely as a replication function, allowing resolution of catenated molecules arising from replication or promoting dimer formation as a mechanism for bypass of cell-cycle control of replication.

2μ-circle-associated Phenotypes

Oligomycin Resistance

The evidence for the association of oligomycin resistance with 2μ circles has recently been reviewed by Guerineau (1979) and is summarized as follows. A [*cir*⁺] strain of yeast exhibiting multiple antibiotic resistance was isolated and, from this strain, revertants that had lost resistance to one or more of the antibiotics were obtained. In every case in which resistance to oligomycin was lost, the cells had also lost detectable 2μ circle sequences (Guerineau et al. 1974). Clearly, although these results do establish a correlation between the loss of oligomycin resistance and the absence of 2μ circles, they do not establish a causal relationship. It is possible, for example, that loss of oligomycin resistance results from a membrane alteration that also impairs the maintenance of 2μ circles. Thus, unless it can be shown that the reintroduction of 2μ circles into the oligomycin-sensitive strain reestablishes resistance or that 2μ circles isolated from the oligomycin-resistant strain can transform some other strain to oligomycin resistance, one cannot conclude that oligomycin resistance is a phenotype ascribable to 2μ circles.

Lethal Sectoring

On the basis of observations by Holm, a particular morphological colony change in a particular strain of *S. carlsbergensis* has been correlated with the acquisition of 2μ circles (C. Holm, pers. comm.). From crosses between a *karl* [*cir⁺*, *rho⁺*] strain and a particular *cyh*ᴿ [*cir⁰*, *rho⁰*] strain, Livingston (1977) isolated *cyh*ᴷ [*rho⁺*] cytoductants. Holm found that these conductants exhibited two distinct colony morphologies. One class had a normal morphology similar to that of the parental recipient strain. The other class had a rough or nibbled appearance (Fig. 5). All cytoductants in the first class had not acquired 2μ circles, whereas all those of the second class had acquired 2μ circles. Thus, the presence of 2μ circles in this strain of *S. carlsbergensis* induces a nibbled colony morphology.

Subsequent work on this problem by Holm (pers. comm.) has provided information on the phenomenology and genetics of 2μ-circle-induced nibbled-colony morphology. In summary, her results demonstrated that, in addition to the presence of 2μ circles, nibbled colony morphology is correlated with a single, recessive chromosomal allele, which has been designated *nib1*. That is, *nib1* [*cir⁺*] strains have a nibbled morphology, whereas *NIB1* [*cir⁺*], *nib1* [*cir⁰*], and *NIB1/nib1* [*cir⁺*] strains all have smooth colony morphologies. The nibbled morphology apparently results from the existence of cells in the population that have become mortal. In a culture of a *nib1* [*cir⁺*] strain, there are large and small cells. Any single small cell will give rise to a nibbled colony. However, most single large cells give rise to micro colonies, all the cells of which are nonviable. Thus, the nibbled colony morphology results from a continual sectoring of mortal cells. These results can be explained by assuming that there is a certain probability that a *nib1* [*cir⁺*] cell will undergo some event that causes, after a delay of several generations, cell death among all of its progeny. Thus, the concurrence of three elements in a yeast cell produces mortality: the chromosomal *nib1* allele, the nonchromosomal 2μ circle plasmid, and an as yet unspecified, randomly occurring change in the cell. Holm (pers. comm.) has obtained evidence that once this random event

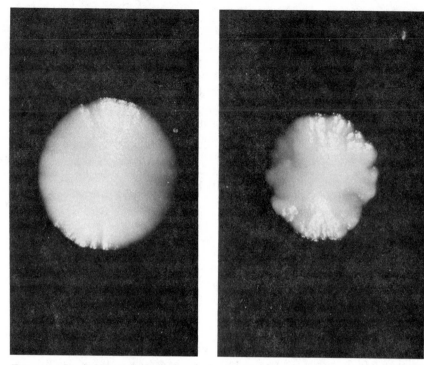

Figure 5 2μ-circle-induced lethal sectoring in *nib* yeast strain. Photomicrographs of two cytoductants of a *nib⁻* yeast strain are shown. (*Left*) Colony derived from a cell that did not acquire 2μ circles during cytoduction; (*right*) colony derived from a cell that did acquire 2μ circles.

has occurred, it is heritable, although in a non-Mendelian fashion, and it is recessive.

The precise role of 2μ circle in lethal sectoring is not known. It is not even clear at the moment whether the role of 2μ circle is a passive one—i.e., that any extrachromosomal, circular plasmid would induce lethal sectoring—or whether lethal sectoring results from the action of one or more of the 2μ circle gene products. Further investigation of the molecular mechanism of lethal sectoring is certainly warranted.

CONCLUSIONS AND SPECULATIONS

2μ circle appears to be an unusual molecular species. Except for a few viruses, such as Epstein-Barr virus and bovine papilloma virus, which exist in transformed cells in part, if not exclusively, as free plasmid DNA (Amtmann et al. 1980; Zur Hausen 1980), autonomously replicating double-stranded circular DNA molecules are not found in higher eukaryotic cells. (Free circular DNA molecules have been detected in a number of eukaryotic cells but, as in the case of those from *Drosophila*, they appear to be heterogeneous in size and composition and thus do not constitute a discrete molecular species [Stanfield and Helinski 1976].) Nor do prokaryotic systems provide an analogous species. Although superficially similar to prokaryotic plasmids, 2μ circle is distinguishable from these elements by its adherence to the cell cycle for its replication. It is not known whether other fungi possess similar plasmid elements. It is clear, however, that among a wide variety of yeast, 2μ circle sequences are found only in *S. cerevisiae* and the closely related species *S. carlsbergensis* and *S. italicus*.

2μ circles do not appear to provide any advantage for cells that harbor them. As mentioned, isogenic [*cir⁰*] strains are phenotypically indistinguishable by a variety of criteria. In addition, three of the 2μ-circle gene products appear to be devoted solely to the biology of 2μ circle itself, i.e., to its replication and interconversion. It is certainly possible that 2μ circles do confer some, as yet unobserved, selective advantage to strains in which they are resident. It is also possible, however, that 2μ circles constitute an excellent example of what has been termed selfish DNA (Doolittle and Sapienza 1980; Orgel and Crick 1980). In this latter context, it is noteworthy that 2μ circles constitute only a small (2%) percentage of the total genomic DNA. In addition, the molecules use the cellular apparatus for DNA replication efficiently, while at the same time incorporating a mechanism that can bypass cellular control and promote self amplification.

The strict adherence of 2μ circles to cell-cycle control for DNA replication suggests a possible origin of the species. This property, in addition to the absence of any encoded virionlike proteins, argues against a viral origin for 2μ circle. Rather, it seems more plausible that its progenitor was a chromosomal replicon that acquired the potential for bypassing cell-cycle control to promote more than one round of DNA replication during a single S phase. This amplification potential would have allowed both its excision from the chromosome—by a mechanism similar to the induction of virus from cells containing integrated copies of SV40 DNA (Botchan et al. 1979)—and its maintenance and propagation in the absence of attachment to the mitotic and meiotic apparatuses of the cell.

It is clear from this review that there are certainly several puzzling aspects of the biology of 2μ circle. What is the function of interconversion? How can 2μ circle bypass cell-cycle control of DNA replication? What is the role of 2μ circle in cell mortality in *nib* strains? Certainly the availability of cloning technology and site-specific mutagenesis will allow us to further define sequences of 2μ circle involved in many of these processes. In addition, in vitro replication and recombination systems from yeast, both of which appear to be imminent, should allow us to probe the detailed mechanisms of these two 2μ circle properties. Results from these analyses will undoubtedly yield insights into the control of eukaryotic DNA replication, as well as provide a framework for examining the mechanisms of developmentally relevant genetic arrangements in yeast and other eukaryotes.

APPENDIX: A COMPREHENSIVE RESTRICTION ANALYSIS OF 2μ CIRCLE

Table 1 Restriction sites of 2μ circle

AluI

106	
417	
617	
631	
797	
1018	1199
1350	1483
2030	1608
2134	1994
2285	2074
2309	2128
2332	2255
2395	2321
2522	2341
2576	2365
2656	2516
3042	2620
3167	3300
3451	3632
3853	
4019	
4033	
4233	
4614	
4666	
4692	

HaeII

652	
726	
745	
2050	1386
2830	1448
3014	1511
3077	1573
3139	1636
3202	1820
3264	2600
3903	
3922	
3996	
5560	

BalI

264

AvaII

33	
1333	1295
2470	2180
3355	3317

HaeIII

152	
206	
265	
1459	2161
1620	2859
1691	2959
1791	3030
2489	3191
4388	
5002	
5029	
5326	
5425	

MboII

44	
112	
171	
558	
753	
957	1067
1005	1097
1206	1252
1338	1280
1438	1556

HhaI

122	1136
627	1336
653	1385
727	1447
746	1510
2051	1572
2831	1635
3015	1819
3078	2599
3140	3904
3203	3923
3265	3997
3314	4023
3514	4981
	5421
	5561
	6270

BclI

304

HinfI

1032	1007
1239	1100
1590	1233
1751	1428
2548	1490
3035	1553
3097	1615
3160	2102
3222	2899
3417	3060
3550	3411
3643	3618
4535	
4582	
4900	
4954	
4972	
4987	
5020	
5171	
5857	
5951	
6141	

Sau3A

93	1920
305	2109
344	2133
1204	2864
1357	2894
1756	3293
1786	3446
2517	
2541	
2730	
4306	
4622	
5270	
5556	
6037	
6164	
6283	
6287	

EcoRI

1	2243
2407	

HpaII

36	
784	
1816	1734
2916	2834
3866	
4586	
5008	
5427	

ThaI

635	1221
1054	1454
2085	1517
2171	1579
2559	1642
3008	2091
3071	2479
3133	2565
3196	3596
3429	
4015	
5422	
5448	

TaqI

927	1009
1030	1978
1544	2440
1754	2531
1758	2733
1917	2892
2119	2896
2210	3106
2672	3620
3641	
3723	
4898	
5439	
5481	
5778	
5949	
6074	
6285	

RsaI					AvaI
4746	1546	1809	97	1916	3258 1392
4825	1556	2095	331	2200	4764
5056	2301	2349	482	2710	
5148	2555	3094	524	2813	
5744	2841	3104	1128	2878	
6118	3094	3212	1309	3341	
	3398	3444	1772	3522	
	3553	3645	1837		
	3583	3693	1940		
	3896		2450		
	4091		2734		
	4468		4126		
	4678		4168		
	4681		4714		
	4751		5152		
	4773		5305		
	4876		5392		
	4922		5410		
	4960				
	4975				
	5380				
	5960				
	6111				
	6138				

ClaI	HincII	EcoRII		PvuI	HindIII	HphI	
6284	217	267		1755 1921	105 2319	1248 2523	
	2964 1686	2771 1785		2729 2895	1017 3633	1741 2909	
		2865 1879			2331	2127 3402	
		4455				4600	
		5044				5651	
						6020	

PvuII	HpaI	AccI	
2029 2517	2964 1686	277	
2133 2621		977 2207	
		2351 2216	
		2434 2299	
		2443 3673	
		6067	

PstI	HgiA	XbaI
2652 1998	513	703
	756	3945
	2648 2002	
	3893	
	4136	
	4656	
	5139	
	5815	

No sites for

BamHI
BglI
BglII
KpnI
SalI
SmaI
SstI
SstII
XhoI
BstEII

The location of cleavage sites for each of the listed restriction enzymes is indicated using the numbering system presented in Fig. 1. The central and left-hand columns under each restriction enzyme listing give the restriction sites for the A form of the molecule. The central and right-hand columns give the restriction sites for the B form of the molecule. (Adapted from Hartley and Donelson 1980.)

Table 2 Restriction Fragments of 2μ Circle

AluI: 680, 596, 402, 386, 381, 374, 332, 311, 306, 284, 231, 221, 200, 200, 166, 166, 151, 127, 125, 104, 92, 80, 79, 63, 54, 54, 52, 26, 24, 23, 14, 14

HaeII: 1564, 1410, 1305, 780, 639, 184, 74, 74, 63, 63, 62, 62, 19, 19

HaeIII: 1899, 1194, 1045, 698, 614, 297, 161, 100, 99, 71, 59, 54, 27

HhaI: 1305, 958, 784, 709, 505, 440, 390, 200, 184, 170, 140, 74, 74, 63, 63, 62, 62, 49, 26, 26, 19, 19

HinfI: (1209), [1184], [917], (892), 797, 686, 487, 351, 318, 207, 195, 190, 161, 151, 133, 94, 93, 63, 62, 62, 54, 47, 33, 18, 15

HpaII: 1100, 1032, 950, 927, 748, 720, 422, 419

ThaI: 1505, 1407, 1031, 586, 449, 419, 388, 233, 86, 63, 63, 62, 26

Sau3A: 1576, 860, 731, 648, 481, 399, 316, 286, 212, 189, 153, 127, 124, 119, 39, 30, 24, 4

TaqI: 1175, 969, 960, 541, 514, 462, 297, 211, 210, 202, 171, 159, 125, 103, 91, 82, 42, 4

RsaI: 1392, 1005, 604, 546, 510, 463, 438, 284, 234, 181, 153, 151, 103, 87, 65, 42, 42, 18

MboII: 745, 580, 405, 387, 377, 313, 286, 276, 254, 253, 224, 210, 204, 201, 195, 195, 155, 151, 132, 108, 103, 100, 70, 68, 59, 48, 46, 38, 30, 28, 27, 22, 15, 10, 3

EcoRII: [2576], (2504), (1590), 1541, [1518], 589, 94

HphI: 2473, 1546, 1051, 493, 386, 369

AccI: (3624), [2394], [1930], 1374, (700), 528, 83, 9

HgiA: 1892, 1245, 1016, 676, 520, 483, 243, 243

AvaI: (4812), [3372], [2946], (1506)

AvaII: [3034], (2996), (1300), [1262], 1137, 885

EcoRI: [4076], (3912), (2406), [2242]

HincII: [4849], (3571), (2747), [1469]

HindIII: [4092], [2790], [2214], 1314, (912)

HpaI: 6318

ClaI: 6318

PstI: 6318

XbaI: 3242, 3076

PvuI: 5344, 974

PvuII: 6214, 104

BalI: 6318

BclI: 6318

The lengths (in bp) of fragments generated by complete digestion of 2μ circle by the indicated restriction enzyme are listed in decreasing size. Numbers in parentheses denote fragments derived from the A form of the molecule. Numbers in brackets denote fragments derived from the B form of the molecule. (Adapted from Hartley and Donelson 1980.)

REFERENCES

Amtmann, E., H. Muller, and G. Sauer. 1980. Equine connective tissue tumors contain unintegrated bovine papilloma virus DNA. *J. Virol.* **35**: 962.

Beggs, J.D. 1978. Transformation of yeast by a replicating hybrid plasmid. *Nature* **275**: 104.

Beggs, J.D., M. Guerineau, and J.F. Atkins. 1976. A map of the restriction targets in yeast 2 micron plasmid DNA cloned on bacteriophage lambda. *Mol. Gen. Genet.* **148**: 287.

Botchan, M., W. Topp, and J. Sambrook. 1979. Studies on simian virus 40 excision from cellular chromosomes. *Cold Spring Harbor Symp. Quant. Biol.* **43**: 709.

Botstein, D. and R.W. Davis. 1982. Principles and practice of recombinant DNA research with yeast. In *The molecular biology of the yeast* Saccharomyces. II. *Metabolism and gene expression* (ed. J. Strathern et al.), Cold Spring Harbor Laboratory, Cold Spring Harbor, New York (In press.)

Broach, J.R. and J.B. Hicks. 1980. Replication and recombination functions associated with the yeast plasmid, 2μ circle. *Cell* **21**: 501.

Broach, J.R., J.F. Atkins, C. McGill, and L. Chow. 1979. Identification and mapping of the transcriptional and translational products of the yeast plasmid 2μ circle. *Cell* **16**: 827.

Broach, J.R., V.R. Guaracio, M.H. Misiewicz, and J.L. Campbell. 1981. Replication of the yeast plasmid, 2μ circle. In *Molecular genetics in yeast, Alfred Benzon Symposium 16* (ed. D. von Wettstein et al.), p. 31. Munksgaard, Copenhagen.

Bukhari, A. and L. Ambrosio. 1978. The invertible segment of bacteriophage Mu DNA determines the absorption properties of Mu particles. *Nature* **271**: 575.

Cameron, J.R., P. Philippson, and R.W. Davis. 1977. Analysis of chromosomal integration and deletions of yeast plasmids. *Nucleic Acids Res.* **4**: 1429.

Clark-Walker, G.D., and G.L.G. Miklos. 1974. Localization and quantification of circular DNA in yeast. *Eur. J. Biochem.* **41**: 359.

Doolittle, W.F. and C. Sapienza. 1980. Selfish genes, the phenotype paradigm and genome evolution. *Nature* **284**: 601.

Furth, M.E., J.L. Yates, and W.F. Dove. 1979. Positive and negative control of bacteriophage lambda DNA replication. *Cold Spring Harbor Symp. Quant. Biol.* **43**: 147.

Gerbaud, C., P. Fournier, H. Blanc, M. Aigle, H. Heslot, and M. Guerineau. 1979. High frequency of yeast transformation by plasmids carrying part or entire 2-μm yeast plasmid. *Gene* **5**: 233.

Gottesman, M.E. and R. Weisberg. 1971. Prophage insertion and excision. In *The bacteriophage lambda* (ed. A.D. Hershey), p. 113. Cold Spring Harbor Laboratory, Cold Spring Harbor, New York.

Gubbins, E.J., C.S. Newlon, M.D. Kann, and J.E. Donelson. 1977. Sequence organization and expression of a yeast plasmid gene. **1**: 185.

Guerineau, M. 1979. Plasmid DNA in yeast. In *Viruses and plasmids in fungi*, p. 539. Marcel Dekker, New York.

Guerineau, M., C. Grandchamp, and P. Slonimski. 1976. Circular DNA of a yeast episome with two inverted repeats: Structural analysis by a restriction enzyme and electron microscopy. *Proc. Natl. Acad. Sci.* **73**: 3030.

Guerineau, M., P.P. Slonimski, and P.R. Avner. 1974. Yeast episone: Oligomycin resistance associated with a small covalently closed non-mitochondrial circular DNA. *Biochem. Biophys. Res. Commun.* **61**: 462.

Hartley, J.L. and J.E. Donelson. 1980. Nucleotide sequence of the yeast plasmid. *Nature* **286**: 860.

Hobom, G., R. Grosschadl, M. Lusky, G. Scherer, E. Schwartz, and H. Kossel. 1979. Functional analysis of the replication structure of lambdoid bacteriophage DNAs. *Cold Spring Harbor Symp. Quant. Biol.* **43**: 165.

Hollenberg, C.P. 1978. Mapping of regions on cloned *Saccharomyces cerevisie* 2-μm DNA coding for polypeptides synthesized in *E. coli* minicells. *Mol. Gen. Genet.* **162**: 23.

Hollenberg, C.P., B. Kusterman-Kuhn, and H.D. Royer. 1976a. Synthesis of high molecular weight polypeptides in *E. coli* minicells directed by cloned *Saccharomices cerevisiae* 2μm DNA. *Gene* **1**: 33.

Hollenberg, C.P., A. Degelman, B. Kusterman-Kuhn, and H.D. Royer. 1976b. Characterization of 2-μm DNA of *Saccharomyces cerevisiae* by restriction fragment analysis and integration in an *Escherichia coli* plasmid. *Proc. Natl. Acad. Sci.* **73**: 2072.

Kielland-Brandt, M.C., T. Nilsson-Tillgren, J.R. Litske Peterson, and S. Holmberg. 1981. Transformation in yeast without the involvement of bacterial plasmids. In *Molecular genetics in yeast, Alfred Benzon Symposium 16.* Munksgaard, Copenhagen. (In press.)

Livingston, D.M. 1977. Inheritance of 2 μm DNA plasmid from *Saccharomyces*. *Genetics* **86**: 73.

Livingston, D.M. and S. Hahne. 1979. Isolation of condensed intracellular form of the 2μm DNA plasmid of *Saccharomyces cerevisiae*. *Proc. Natl. Acad. Sci.* **76**: 3727.

Livingston, D.M. and H.L. Klein. 1977. Deoxyribonucleic acid sequence organization of a yeast plasmid. *J. Bacteriol.* **129**: 472.

Livingston, D.M. and D.M. Kupfer. 1977. Control of *Saccharomyces cerevisiae* 2μm DNA replication by cell division cycle genes that control nuclear DNA replication. *J. Mol. Biol.* **116**: 249.

Nelson, R.G. and W.L. Fangman. 1979. Nucleosome organization of the yeast 2μm DNA plasmid: A eucaryotic minichromosome. *Proc. Natl. Acad. Sci.* **76**: 6515.

Newlon C.S., R.J. Devenish, P.A. Suci, and C.J. Roffis. 1981. Replication origins used *in vivo* in yeast. *ICN-UCLA Symp. Mol. Cell. Biol.* **26**: (In press).

Orgel, L.E. and F.H.C. Crick. 1980. Selfish DNA: The ultimate parasite. *Nature* **284**: 604.

Petes, T.D. and D.H. Williamson. 1975. Replicating circular DNA molecules in yeast. *Cell* **4**: 249.

Seligy, V.L., D.Y. Thomas, and B.L.A. Miki. 1980. *Saccharomyces cerevisiae* plasmid, Scp or 2μm: Intracellular distribution, stability and nucleosomal-like packaging. *Nucleic Acids Res.* **8**: 3371.

Shortle, D.R., R.F. Margolskee, and D. Nathans. 1979. Mutational analysis of the simian virus 40 replicon: Pseudorevertants of mutants with a defective replication origin. *Proc. Natl. Acad. Sci.* **76**: 6128.

Sigurdson, D.C., M.E. Gaarder, and D.M. Livingston. 1981. Characterization of the transmission during cytoductant formation of the 2μ DNA plasmid from *Saccharomyces*. *Mol. Gen. Genet.* (in press).

Soeda, E., J.R. Arrand, N. Smolar, and B.E. Griffin. 1979. Sequence from early region of polyoma virus DNA containing viral replication origin and encoding small, middle, and (part of) large T antigens. *Cell* **17**: 357.

Stanfield, S. and D.R. Helinski. 1976. Small circular DNA in *Drosophila melanogaster*. *Cell* **9**: 333.

Struhl, K., D.T. Stinchcomb, S. Scherer, and R.W. Davis. 1979. High-frequency transformation of yeast: Autonomous replication of hybrid DNA molecules. *Proc. Natl. Acad. Sci.* **76**: 1035.

Subramanian, K.N., R. Dahr, and S.M. Weissman. 1979. Nucleotide sequence of a fragment of SV40 DNA that contains the origin of DNA replication and specifies the 5' ends of "early" and "late" viral RNA. *J. Biol. Chem.* **252**: 355.

Taketo, M., S.M. Jazwinski, and G.M. Edelman. 1980. Association of the 2μm DNA plasmid with yeast folded chromosome. *Proc. Natl. Acad. Sci.* **77**: 3144.

Tjian, R. 1979. Protein-DNA interactions at the origin of simian virus 40 DNA replication. *Cold Spring Harbor Symp. Quart. Biol.* **43**: 655.

Zakian, V.A., B.J. Brewer, and W.L. Fangman. 1979. Replication of each copy of the yeast 2 micron DNA plasmid occurs during the S phase. *Cell* **17**: 923.

Zeig, J., M. Silverman, M. Hilmen, and M.I. Simon. 1977. Recombinational switch for gene expression. *Science* **196**: 170.

Zur Hausen, H. 1980. Oncogenic herpesviruses. In *DNA tumor viruses* (ed. J. Tooze), p. 747. Cold Spring Harbor Laboratory, Cold Spring Harbor, New York.

Mitochondrial Structure

Barbara Stevens
Centre de Recherche de Biochimie et de Génétique Cellulaires
du Centre National de la Recherches et Moyens d'Essais
118 route de Narbonne
31062 Toulouse Cedex, France

INTRODUCTION

The study of the structure and organization of mitochondria in *Saccharomyces cerevisiae* offers a unique opportunity to relate the considerable metabolic, genetic, and biochemical data on this organism to visible morphological entities. The ability of yeast to thrive under a variety of growth conditions permits their mitochondria to exist in a wide range of functional states and to maintain genetic deficiencies that would otherwise be lethal to an organism. In many cases, these functional states and mutations have their morphological counterpart in defined fine structural modifications of the mitochondrial population. In addition, the complex interaction of the nuclear and mitochondrial genomes in the formation of mitochondria can be dissociated in yeast by making use of ρ^- strains, which lack a functional mitochondrial genome and thus allow the contribution of the nuclear genome to be visualized directly.

TECHNIQUES OF PREPARATION

Although there are favorable aspects of yeast as an organism for the study of mitochondrial structure, an inherent and major difficulty lies in its rigid and impermeable cell wall. This wall, which enables yeast to survive drastic growth conditions, is a definite impediment to the standard preparation procedures for electron microscopy. The penetration of fixatives, notably osmium tetroxide, and the infiltration of embedding media are hindered, if not completely blocked, by the wall. To overcome this handicap, one may

471

resort to the use of chemical fixatives that penetrate more readily, such as potassium permanganate. Alternatively, one may weaken or completely remove the cell wall by using digestive enzymes to allow the penetration and oxidizing action of osmium and to permit a thorough infiltration and polymerization of the embedding medium.

Potassium-permanganate fixation has been the most widely employed procedure for demonstrating mitochondria in thin sections of yeast (Yotsuyanagi 1962a,b; Marchant and Smith 1968). It readily penetrates the cell wall under all growth conditions, and its powerful oxidizing action removes many cell components, leaving the mitochondrial profiles clearly visible against a uniform cytoplasmic background (Plate 1).[1] In general, prefixation with a buffered mixture of paraformaldehyde and glutaraldehyde, followed by a 4% aqueous permanganate fixation and uranyl-acetate staining en bloc gives the best results. The outer and inner mitochondrial membranes and cristae are well demonstrated by this procedure, but the matrix is emptied of content, and only traces of DNA can be found (Plate 6a). There is no evidence available that this procedure introduces artifactual distortion or rearrangement of mitochondria.

It was first shown by Eddy and Williamson (1957) that the digestive juice of the snail, *Helix pommatia*, could be used to attack the cell wall and to form protoplasts. Since then, the procedure has been considerably refined and adapted for the formation of yeast spheroplasts, which are amenable for preparation by standard aldehyde and osmium fixatives and a variety of embedding media (Peterson et al. 1972; Zickler and Olson 1975; Stevens 1977). Enzymes of high purity (Glusulase [Endo]) and specificity (Zymolyase, from *Arthrobacter luteus* [Kirin Brewery]) act mainly to digest the cell wall and produce little cytoplasmic damage. In the presence of sulfhydryl reagents, the wall can be attacked even after preliminary aldehyde fixation.

The sensitivity of the cell wall to enzymatic digestion varies greatly during the life cycle, and slight differences in cell permeability are immediately reflected in the preservation of mitochondria. Although slightly hypertonic conditions produce compact and well-fixed cells, mitochondrial profiles tend to be obscured by the density of surrounding cytoplasmic ribosomes (Plate 8b). In contrast, when cells are slightly expanded after partial removal of the cell wall, mitochondria are more clearly visible but are often swollen and distorted (Plate 6b-d). Mitochondrial swelling can be avoided to some extent by preliminary glutaraldehyde fixation prior to enzymatic attack of the cell wall, but overall cell preservation still suffers from even a slight overdigestion (Plate 10b). A further problem lies in the binding of osmium to cell membranes; even when cell-wall digestion and fixation appear optimum, mitochondrial membranes can remain unstained and display a negative contrast (Plate 8b). Postfixation with Dalton's chrome osmium and staining with uranyl acetate en bloc can achieve positively stained membranes in this case (Zickler and Olson 1975) (Plates 2, 3, and 10a).

[1]For Plates 1–16, see pp. 489–504.

Other additional fixing and staining techniques are useful in the study of mitochondrial structure. Aldehyde fixation alone reveals the ribosomal and DNA components, although cristae are less distinct following aldehyde fixation than following osmium fixation (Plate 15a–d). The electron-dense product of the diaminobenzidine (DAB) reaction for the demonstration of cytochrome oxidase and cytochrome c peroxidase (Todd and Vigil 1972) clearly defines cristae and inner membranes, even in respiratory-deficient mutants (Plates 4a and 14b). A staining procedure for the demonstration of polysaccharides, PATAg (Thièry 1967), is effective in exposing mitochondrial profiles against a pale cytoplasm after standard aldehyde and osmium fixation (Plates 4b and 14c). Selective staining of mitochondrial DNA by the fluorescent dye DAPI provides a valuable tool for tracing mitochondrial behavior at the light microscope level (Plates 5 and 12; Williamson and Fennell 1975).

STRUCTURE OF ACTIVELY RESPIRING MITOCHONDRIA

General Features

The mitochondria of yeast resemble those found in other unicellular and multicellular organisms. In thin sections, profiles of actively respiring mitochondria are rounded or elongated and are regularly distributed in the peripheral cytoplasm (Plates 1 and 2). The average cross-sectional diameter of the profiles is on the order of 0.3–0.4 μm. Long filamentous forms, branching profiles, and figure-eight shapes are typical of certain growth conditions or life-cycle stages (Plates 10 and 11). Details are given later in the text.

In well-fixed preparations, the inner and outer membranes are closely apposed. The inner membrane is evaginated into the matrix at irregular intervals to form short, platelike cristae. These usually lie perpendicular to the surface and extend only partially across the mitochondrion, terminating freely in the matrix (Plates 1, 3, 4, and 8a).

Mitochondria never appear to make direct contact with the plasma membrane. They lie just centripetal to the peripheral cisterna of endoplasmic reticulum and the occasional, associated peroxisomes (Plate 6d). Continuity of the outer mitochondrial membrane with other cellular membranes is also not observed. Although the remainder of the cytoplasm is generally packed full of ribosomes, a genuine association of these cytoplasmic ribosomes with the mitochondrial surface, as reported by Kellems et al. (1975), is not usually encountered. The observations of these investigators, showing cytoplasmic ribosomes attached in situ to the outer mitochondrial membrane, were made on cells converted to spheroplasts and allowed to grow in rich media for an additional 2½ hours. The phenomenon was not observed when spheroplasts were fixed immediately after the hour-long treatment with Glusulase. The significance of this difference is not clear, but it can be stated that when intact cells are fixed directly in glutaraldehyde, followed or not by spheroplast formation and osmium postfixation, no specific association of cytoplasmic ribosomes to the outer membrane can be observed.

Number and Volume

Data such as the position of mitochondria in the cell or the arrangement of their cristae may be obtained from random thin sections. However, knowledge of the entire mitochondrial population of a cell, the chondriome, must be obtained by other means. To date, this information has been gained primarily from serial thin sections of whole cells (Hoffmann and Avers 1973; Grimes et al. 1974; Stevens 1977). Another technique consists in the examination of serial thick sections by high-voltage electron microscopy (Davidson and Garland 1975).

Until recently, it was generally accepted that many mitochondria were present in a yeast cell. In the light microscope, using selective stains, numerous units were reported (Ephrussi et al. 1956; Avers et al. 1965). However, phase microscopy of living cells shows a dynamic system of long filamentous forms, evoking a reticular structure of the chondriome. The presence of a single highly branched, "giant" mitochondrion per cell was first proposed by Hoffmann and Avers (1973). Subsequent studies (Grimes et al. 1974; Stevens 1977) have demonstrated a more complex situation. In general, it has been shown that the chondriome is not a static entity, but its form and volume undergo rapid and extensive modifications in accord with changes in the life cycle and physiological state.

Examination and measurement of mitochondrial profiles in serial thin sections of 34 entire cells from several different strains growing vegetatively under a variety of conditions allow the following basic conclusions (Table 1) (Stevens 1977).

1. Rapidly growing cells contain few mitochondria, usually less than 10 (Fig. 1).
2. Stationary-phase cells contain numerous mitochondria, up to 50/cell (Fig. 2).
3. The chondriome of all but two cells was composed of more than one mitochondrion. The exceptions were cells in exponential phase under glucose repression.
4. The three-dimensional form varies, with small oval shapes predominant in stationary-phase cells, and some elongated, curved, and branching forms in all states. However, one mitochondrion or, rarely, two is always significantly larger and more complex than the others in the cell (Fig. 1).
5. The volume of the chondriome is clearly a function of the physiological state of the cell: In glucose-repressed cells, it represents only about 3% of the cell volume, whereas in fully respiring cells, it represents 10–12%

Table 1 Number of mitochondria and relation of giant mitochondrion to chondriome

Strain	Genotype	Growth conditions	Number of cells	Number of mitochondria (mean ±S.E.M.)	Percentage volume of chondriome in giant mitochondrion (mean ±S.E.M.)
IL46	diploid ρ^+	exponential, glucose-repressed	6	2 ± 1	79 ± 15%
IL46		exponential, glycerol	8	8 ± 3	58 ± 16%
R$_1$	haploid ρ^+	exponential, glucose-derepressed in chemostat	4	5 ± 1	49 ± 12%
N123	haploid ρ^+	exponential, glycerol	2	8 ± 3	49 ± 9%
IL46		stationary, glucose-derepressed	9	34 ± 11	19 ± 12%
IL46		stationary, glycerol	3	21 ± 3	29 ± 3%
DV147	haploid ρ^+	stationary, glucose-derepressed	2	44 ± 1	10 ± 1%

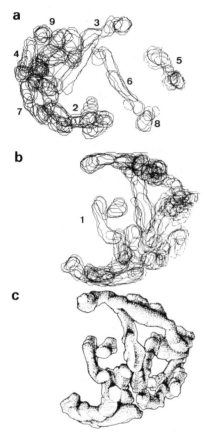

a
9 3
4 5
6
7 2
8

b
1

Figure 1 Computer reconstruction of the chondriome in an exponential-phase cell. Strain IL46. Glycerol medium. Tracings of the nine individual mitochondria of the cell are displayed in relation to their position in the cell. (*a*) Mitochondria *2-9* are displayed together. (*b*) The giant mitochondrion (*1*) forms a highly branched reticulum. (*c*) Reconstruction of mitochondrion *1* after viewing the tracings in stereo. (Drawing by A. Charrier.)

c

(Table 2). These values are in agreement with those given by Grimes et al. (1974) and Damsky (1976).

The giant mitochondrion observed in all cells represents a different proportion of the chondriome according to growth conditions: In exponential-phase cells, more than half of the volume of the chondriome was in this large mitochondrion, whereas in stationary-phase cells, it represented 10–30% (Table 1). Its position in the cell varied from branching into all parts to being confined to one end of the cell. Out of 15 budding cells, the giant mitochondrion was continuous into the bud in 5 of them, but there was no continuity in the others, or another smaller mitochondrion was continuous into the bud.

Visualization of the form, number, and position of mitochondria from serial sections was greatly enhanced by computer-aided, three-dimensional reconstruction procedures (Stevens and White 1979). Tracings made from the micrographs were digitized and entered into a computer. The data on

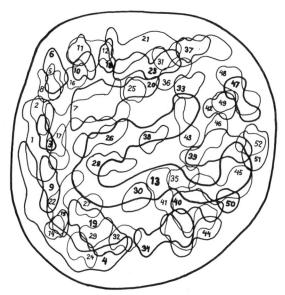

Figure 2   Schematic representation of the chondriome in a stationary-phase cell. Strain IL46. Glucose medium. Outlines of computer reconstructions of the tracings of the 52 individual mitochondria are drawn in their relative position in the cell. (Drawing by M. Klapholz.)

individual mitochondria, or the whole population, could be retrieved, displayed, rotated in space, and printed out. The third dimension is achieved by plotting stereo pairs, which differ by a 6-degree rotation about the vertical axis, and viewing them with a stereo viewer (Fig. 1 and Plate 11b).

In light of these observations and those on sporulating cells (see below, Meiosis and Sporulation), it appears that the chondriome is in a mobile and flexible state in the living yeast cell. It forms a branching network that ramifies throughout the peripheral cytoplasm; its branches constantly fuse together, pinch apart, and change location. The membranes, enzymes, and nucleic acid components are thus being continually redistributed within a "fluid" mitochondrial system (Williamson 1976).

Visualization of Mitochondrial DNA and Ribosomes

The size of the yeast mitochondrial DNA molecule (50×10^6 daltons) and its AT-rich composition fortuitously allow its visualization in whole cells at the light microscope level using the DNA-complexing fluorescent stain DAPI (4′, 6-diamidino-2-phenylindole) (Williamson and Fennell 1975, 1979). In addition to the brightly staining nucleus, small, discrete fluorescent spots are discerned in the cytoplasm at the periphery of vegetative wild-type cells (Plate 5).

Table 2 Volumetric data: Strains IL46 and R1

Strain	Growth conditions	Number of cells	Cell volume mean value (μm^3)	Chondriome volume	
				mean value (μm^3)	Percentage of cell volume
IL46	exponential, glucose	3	64.1	2.0	3.2
IL46	exponential, glycerol	2	45.0	5.7	12.6
IL46	stationary, glucose	4	56.6	6.4	11.3
IL46	stationary, glycerol	2	76.4	7.2	9.5
R1	exponential, glucose-derepressed in chemostat	1	32.9	2.2	6.6

These have been shown to represent one or more mitochondrial DNA molecules and have been termed chondriolites (Williamson 1976). When the wet-mount preparation is more "squashed," these spots are observed in a regular distribution over the cell surface. Staining is completely absent in the cytoplasm in ρ^0 strains, making this procedure a rapid and reliable diagnostic test for these mutant strains (Plate 5).

The number of fluorescent chondriolites per cell was observed to vary with the physiological state of the culture (Williamson 1976). In exponential-phase cells of several haploid and diploid strains, under glucose repression or growing on nonfermentable catabolites, about 10–20 spots could be counted. The same strains in stationary phase exhibited many more (in the range of 20–40). Simple calculations indicate that each chondriolite contains from two to eight mitochondrial DNA molecules. The larger mitochondria would contain several chondriolites; the small mitochondria in stationary-phase cells would contain only one. A rapid decrease in numbers was observed when stationary-phase cells were changed to exponential-phase conditions by inoculation into fresh media. Since there was no net loss of mitochondrial DNA, a logical explanation would be that aggregation of some chondriolites had occurred, perhaps as a result of mitochondrial fusion.

Although these fluorescent spots do represent mitochondrial DNA, their number, size, and distribution vary. They cannot therefore be considered as permanent genetic structures nor as an indication of the number of mitochondria. As pointed out for the chondriome, the highly mobile nature of the mitochondrial system also involves the pool of mitochondrial DNA molecules. These are no doubt undergoing constant reassociation and re-distribution in the same way as the mitochondria themselves.

Visualization of mitochondrial DNA in situ in thin sections is greatly dependent on the fixation employed. In permanganate-fixed cells, tiny electron-dense spots lying in an electron-transparent area of the matrix can sometimes be observed (Plate 6a). Following aldehyde or aldehyde-osmium fixation, mitochondrial DNA is visible as filaments or bundles of filaments in prominent electron-translucent matrix areas (Plate 6b,c). These areas are usually devoid of cristae and of mitochondrial ribosomes (mitoribosomes) as well. The DNA filaments do not show any definite attachment to the membranes. The clumped appearance of mitochondrial DNA is similar to that of bacterial DNA in situ. No doubt, the almost complete absence of associated protein in both cases is responsible for the clumping of DNA. In fact, when uranyl-acetate staining is carried out before dehydration, instead of distinct bundles of filaments, fine strands are discernible in the clear matrix areas (Swift et al. 1968; Plate 6d).

The mitochondrial DNA areas do not appear to form a continuous reticulum inside the organelle. From serial sections, they were observed to be discrete. Their numbers could be estimated: In stationary-phase cells, all mitochondria, with the possible exception of the very small ones, contained at least one DNA area. In the large branching mitochondria of rapidly growing cells, up to 20 DNA areas could be counted (Stevens 1977). The distribution along the branches was suggestive of a succession of elementary units, each with a DNA area surrounded by a finite volume of the mitochondrion. In fortuitous single sections, a similar distribution can be directly observed (Plate 7). The demonstration of a cyclic fusion and fragmentation within the chondriome during growth and the accompanying redistribution of the DNA areas thus provide strong morphological support for the considerable genetic data concerning recombination and segregation of mitochondrial genomes in yeast (see Dujon, this volume).

The presence of ribosomelike particles within yeast mitochondria was first observed by Swift et al. (1968). Subsequent work in a number of laboratories has confirmed that these particles are indeed mitoribosomes, and considerable biochemical data on the mitochondrial protein synthetic machinery in yeast have been amassed (e.g., Stegeman et al. 1970).

Following aldehyde or aldehyde-osmium fixation, mitoribosomes appear as small, 120–150 Å, electron-dense particles situated mainly along the inner membrane at the periphery of the organelle. They also occur within the matrix area where they are frequently associated with cristae membranes (Plate 8). Their association with the mitochondrial membrane is not as intimate as that between cytoribosomes and endoplasmic reticulum: Mitoribosomes often do not appear to "sit" directly on the membrane. Although yeast mitoribosomes fall into the 70S–74S class of organelle ribosomes, they cannot be distinguished in thin sections from cytoribosomes on the basis of their size.

Regularly spaced alignments of these particles along the inner membrane or cristae evoke the notion of polysomes. However, sections tangential to the membranes rarely if ever show typical polysome whorls or rosettes. Striking polysome arrays have been observed in thin sections of isolated mitochondria from *Candida* (Vignais et al. 1972), but there is little morphological evidence for mitochondrial polysomes in *S. cerevisiae*. The presence of mitoribosomes in certain ρ^- mutants is discussed in a later section.

MODIFICATIONS OF THE CHONDRIOME

Behavior during the Cell Cycle

One of the earliest events following the formation of a new bud is the entry of mitochondria into the bud. Between this moment and the final separation of the daughter cell from the mother, the mitochondrial population of the bud is mobile and unstable. Phase microscopy reveals a constant movement of mitochondria in and out through the neck between mother and daughter. In thin sections, mitochondrial profiles are observed in the neck before, during, and after migration of the nucleus into the bud (Plates 7 and 9). Often, these mitochondria appear to be long and filamentous. Indeed, from serial sections, the giant mitochondrion of the mother cell was observed to extend into the bud in some but not all of the dividing cells studied. A continuity might be expected, since the giant mitochondrion represents more than half of the chondriome in rapidly dividing cells and is present throughout most of the cell.

By using the DAPI strain, mitochondrial DNA areas can be visualized within even the smallest bud (Plate 5), but since the population does not remain fixed, it is difficult to estimate the number of genomes that a bud receives. In addition, during the period of bud formation and growth, the synthesis of new mitochondrial DNA occurs both in the mother and the bud, and the newly synthesized molecules are no doubt distributed throughout the dividing cell.

The ultimate distribution of mitochondria and mitochondrial DNA between mother and bud appears to be a completely random process. No morphological evidence for any other mechanism has been recorded. The movement of mitochondria in and out of the bud is simply stopped by the formation of the septum. It is interesting to note that in one completed bud whose cytoplasm was separated from the mother cell, the volume of the chondriome represented almost 20% of the bud (Stevens 1977). This is a much higher proportion than in fully respiring adult cells and may indicate an overprovision of mitochondria in young cells.

Behavior during the Life Cycle

Meiosis and Sporulation

In diploid cells induced to sporulate by inoculation into sporulation medium lacking a nitrogen source and containing potassium acetate, there is a single and highly branched mitochondrion per cell. The chondriome is unitary throughout all stages of meiosis and up to the formation of the four ascospores. Each ascospore also contains a single, branched mitochondrion that closely encircles the nucleus.

The foregoing statements are based on data derived from complete or nearly complete serial sections of 12 entire cells of three different strains in various stages of meiosis and sporulation (Stevens 1978 and unpubl.; Table 3). In each cell, the single mitochondrion was highly reticulated and extended in a complex system of branching and fusing arms throughout the cytoplasm. Even in the earliest meiotic prophase stage represented, where the spindle plaque was single but the electron-dense material in the vacuoles indicated entry into sporulation, the chondriome was composed of a single mitochondrion.

During the meiotic prophase stages, the majority of mitochondrial profiles are small cross-sections about 0.4 μm in diameter. These profiles are situated in the peripheral cytoplasm as in vegetative cells. In tangential sections of this

Table 3 Number of mitochondria during sporulation

Strain	Cell	Stage in cycle	Number of mitochondria[a]
112	1	MI	1
	2	AF	1 in each ascospore
	3	AF	1 in each ascospore
Z193	1	SP	1
	2	SP, DB	1
	3	LE, DB	1
	4	SPD, LE	1
	5	AF	1 in each ascospore
S41	1	SC, DB	1
	2	SPD, LE	1 plus 1 small
	3	SPD, SC, DB	1
	4	MII	1

[a]Growth in acetate-sporulation medium.

Abbreviations: AF, ascospore formation; DB, dense body in nucleolus; LE, lateral elements of complexes; MI, meiosis-I division; MII, meiosis-II division; SC, synaptonemal complexes; SP, single spindle plaque; SPD, spindle-plaque duplication.

region, mitochondrial profiles are long and sinuous, indicating a reticulum (Plate 10a). The location of mitochondrial DNA by DAPI staining of cells in these stages coincides with this peripheral distribution (Plate 12).

At the time of the first and second meiotic divisions, mitochondrial profiles are observed close to the nucleus and are generally absent in the peripheral cytoplasm. The branches of the mitochondrial network closely encircle the dividing nucleus and lie in depressions at the nuclear surface (Plate 10b). When the prospore membranes are forming around the nuclear products, there is a striking and close association of mitochondrial profiles with the central nuclear region. In DAPI-stained cells, a network of fluorescent mitochondrial DNA areas in the central region points out this association (Plate 12).

Branches of the complex mitochondrion are enclosed within each spore as the prospore membrane closes around the spore nucleus. These branches form the single organelle of each spore (Plate 11a,b). In DAPI-stained asci, a small fluorescent network is visible around each spore nucleus (Plate 12). A significant portion of the chondriome is not incorporated into the four ascospores; numerous, normal-looking profiles are excluded from the ascospores and remain behind in the residual cytoplasm of the ascus (Plate 11a).

Our data indicating a multiple chondriome in vegetative cells and a unitary one in sporulating cells appear to be at variance. We tend to believe in the accuracy of these two phenomena because of the number of cells and strains examined in each situation. Also, in one cell from the preparation of sporulating cells of strain 112, none of the morphological indications of sporulation were observed and it contained 23 individual mitochondria. Although we cannot rule out strain differences at present, the most appealing and logical interpretation is that the chondriome responds to a signal at the start of sporulation, and individual mitochondria fuse to form the single, highly branched mitochondrion observed in all sporulating cells.

Segregation of mitochondria into the four ascospores is apparently a random process that involves the loss of a considerable mitochondrial mass. The intimate association of the branches of the organelle with the four daughter nuclei can be interpreted as a mechanism to ensure that each spore receives at least a certain proportion of the chondriome.

Zygote Formation

The behavior of mitochondria during zygote formation and budding has not been well studied. In view of the detailed and fascinating genetic data being accumulated on the transmission of mitochondrial genes to the progeny of a cross (e.g., Birky et al. 1978; Strausberg and Perlman 1978; Dujon, this volume), morphological data on the form and distribution of the mitochondria from the two parent cells in the zygote and in the successive zygotic buds would be extremely valuable.

Some early observations on the fine structure of zygotes (Smith et al. 1972; Osumi et al. 1974) indicate a disorganization of the mitochondria at the time of nuclear fusion. However, in a preliminary study of yeast zygotes, we do not confirm this; no unusual forms of structures could be discerned. It is clear though from the data on segregation of mitochondrial genomes in zygotic clones that some mechanism(s) is operating to produce nonrandom patterns. Whether there is a morphological basis for the patterns cannot be answered at present, although slow mixing of parental mitochondrial genomes and the position of the first bud have primary roles in biparental inheritance patterns (Strausberg and Perlman 1978). Uniparental inheritance may involve more complex mechanisms (Birky et al. 1978), and information on mitochondrial behavior in these crosses would add insight into the problem.

Structure according to Nutritional Conditions

The mitochondria of cells growing aerobically under nonrepressive conditions are well developed and have prominent cristae (Plates 1, 3, and 4). In general, whether the carbon source is a nonfermentable respiratory substrate such as glycerol, acetate, or ethanol or a fermentable nonrepressing substrate such as galactose or melibiose, mitochondrial structure is the same. Although a careful comparative study has not been done, no specific effects of any of these substrates have been distinguished (Marchant and Smith 1968).

When yeast is grown on a high concentration of glucose (> 0.1%) (which is the standard fermentable substrate) in the presence of oxygen, catabolite repression reduces the synthesis of certain respiratory enzymes and the substrate is partially fermented (Ephrussi et al. 1956). Mitochondrial structure becomes drastically modified: Profiles are simple spheres or long, sinuous, and branching shapes; the cristae are rare and irregularly positioned. Some sections may lack mitochondrial profiles altogether (Plate 13) (Yotsuyangi 1962a). Mitochondrial DNA and mitoribosomes are still observed under these conditions. More important than the structural changes, however, is the reduction in volume of the chondriome. The mitochondria of glucose-repressed cells represent only 3–4% of the cell volume (Stevens 1977). When cultures reach stationary phase and glucose becomes limiting, respiratory activity recommences. Mitochondrial structure returns to normal, and the chondriome volume increases to the value for respiring cells (Table 2).

Under continuous aerobic culture conditions in a chemostat under glucose limitation and in the absence of catabolite repression, mitochondria were large and had closely packed cristae (Stevens 1977). The relative volume of the chondriome measured in one cell was 6.6% (Table 2).

There is now general agreement that a mitochondrial system is present in anaerobically growing cells using glucose as the major carbon source (Marchant and Smith 1968; Swift et al. 1968; Damsky et al. 1969; Plattner et al. 1970). Mitochondrial membranes in anaerobically grown, lipid-limited cells are difficult to demonstrate, particularly with permanganate fixation (Damsky et al. 1969). With this fixation, membranes become visible only when unsaturated fatty acids and sterol are supplied or when cells are exposed to air. Following aldehyde-osmium fixation, anaerobically growing, lipid-limited cells contain recognizable mitochondrial profiles with a double membrane and few cristae. Mitochondrial DNA and mitoribosomes are clearly present in these mitochondria (Rabinowitz and Swift 1970). The volume density of mitochondria in anaerobically grown *S. carlsbergensis* is about 4%. Upon aerobic adaptation, the volume increases to 6–7% by 6 hours, and the surface density of the inner membrane doubles (Damsky 1976). Comparable data for *S. cerevisiae* are lacking, but the inner membrane appears to be less well developed under these conditions.

Modifications of Structure in Mutants

Respiratory-deficient Mutants

The well-known and well-studied petite mutation is characterized by defective mitochondria and the inability of strains to grow on nonfermentable substrates. Respiratory-deficient mutants demonstrate either cytoplasmic

(ρ^-) or nuclear (*pet*) inheritance. Although mitochondrial respiratory functions are lacking, all of these mutants contain recognizable mitochondria that are the expression of the nuclear or cytoplasmic genes still functioning. The mitochondrial structure in nuclear *pet* strains is essentially normal (Yotsuyanagi 1962b), but some aberrant aspects can be demonstrated (e.g., the inability to produce the DAB reaction product in one *pet* mutant [M. Waxman, unpubl.]). Conversely, the mitochondria in ρ^- strains invariably show distinct and unmistakable morphological alterations (Yotsuyanagi 1962b; Smith et al. 1969; Stevens and Moustacchi 1976).

Mitochondrial profiles in cytoplasmic petite mutants are most often small and spherical or slightly elongated. Occasional filamentous forms may be present. Outer and inner membranes of the mitochondrial envelope can always be demonstrated. The major alteration in these mitochondria is the loss of normal cristae, concomitant with the absence of the inner-membrane protein subunits specified by the mitochondrial genome. In sections, many of these mitochondria have no discernible internal organization; others may contain paired membranes, or lamellae, that lie parallel to the mitochondrial surface or that extend across one end to form a curved partition (Plate 14a–c). More aberrant forms of membranes as concentric lamellae can also be found. The electron-dense product of the DAB reaction is present in lamellae of ρ^- mitochondria (Plate 14b). In the absence of cytochrome oxidase, it is likely that in ρ^- mitochondria, DAB is reacting with the nuclear-coded mitochondrial enzyme, cytochrome c peroxidase (Todd and Vigil 1972). These abnormal lamellar forms are unambiguous features of ρ^- mitochondria; they are not observed in healthy ρ^+ cells under any growth condition.

In contrast with wild-type cells, the distribution of mitochondria in ρ^- cells is irregular; they tend to aggregate in some cytoplasmic regions and are absent in others (Plate 16a). The number of organelles is variable: In serial sections of two ρ^0 cells, one had 6 mitochondria and the other had 22.

Mitochondrial DNA can be visualized in many ρ^- strains, according to the amount retained in these mutants. Exceptional amounts can even be demonstrated in some mutants (Plate 15c), perhaps in relation to amplification of the retained sequences.

Although ρ^- cells are unable to carry out mitochondrial protein synthesis, mitoribosomes may be present. In an unpublished study of the fine structure of several petite mutants that retain the 16S or the 23S ribosomal RNA genes or both (Faye 1977), we have observed mitoribosomal particles in all mutants. Particles were abundant in the CEP2 strain (containing 16S and 23S genes) and in the D21 strain (containing only the 23S gene); they were more difficult to demonstrate in the O_1P2 strain, which retains only the 16S gene (Plate 15a–d). In thin sections, it was not possible to distinguish subunits from complete ribosomes by the particle sizes. These observations support the findings of Faye (1977), who was able to isolate 37S and 50S subunits from the O_1P2 and D21 strains, respectively, and they confirm the interpretation that apparently normal but nonfunctional subunits are produced and accumulated when one or both rRNAs is transcribed.

ρ^0 strains, which lack detectable mitochondrial DNA, contain similar aberrant forms. In these, the matrix is uniformly electron-opaque, and internal lamellae are extremely rare. Most organelles are only double-membrane-enclosed spheres or ellipsoids (Plates 15d and 16a).

Other Mutants

In spite of the genetic characterization of numerous nuclear mutations affecting mitochondrial function, there are few reports on the morphological aspects of these mutations (see Lloyd 1974). No doubt, if well chosen and carefully studied under proper growth conditions, some of these mutants should reveal structural abnormalities in their chondriomes. It is less sure, however, whether any modifications noted would be a direct effect of the mutation.

In one such study of a mutant showing enhanced suceptibility to petite induction by UV irradiation and other agents, we found morphological evidence of an unstable chondriome and defective mitochondrial biogenesis (Stevens and Moustacchi 1976). In cells growing on nonfermentable media, doublet mitochondria, in which a transverse septum produces figure-eight shapes, were common, suggesting a rearrangement of cristae and a partial transformation to the petite state. When grown in glucose to permit the survival of ρ^- cells, a small but significant population of cells appeared to have a mixed mitochondrial population: some abnormal petite-type mitochondria in the same cytoplasm with normal ones (Plate 16b).

Other descriptions of the transformation of a wild-type cell to a ρ^- cell have not been reported. The process may be a rapid one, and cells with a mixed phenotype may be rare. More likely, though, the lack of examples is due to the difficulty in recognizing intermediate morphological forms in random thin sections.

CONCLUDING REMARKS

The main emphasis in this description of yeast mitochondria has been on the flexibility and mobility of the system. Data on the number of individual mitochondria and their distribution lose importance when it is realized that these data only represent moments in time. Of greater interest is the demonstration of continual fusion and fragmentation of organelles, no doubt accompanied by continual recombination and segregation of mitochondrial genomes. Some conditions, such as exponential growth or sporulation, appear to favor these genetic exchanges.

Some of the most interesting phenomena have not been well studied. For example, little information is available on the behavior of mitochondria during zygote formation and budding. Transformation of the chondriome to the petite phenotype has not been studied as a dynamic process. Whether all mitochondria are transformed synchronously or whether each responds independently to the mitochondrial DNA lesion is not known. Information such as this can be gained by studies on the fine structure of wild-type and mutant cells.

ACKNOWLEDGMENTS

I am grateful to my many friends and colleagues who have given me support, counsel, and assistance during the long course of this work. I am particularly indebted to D. Williamson, E. Moustacchi, P. Slonimski, J. André, J. White, S. Klapholz, R. Esposito, and H. Swift.

Much of this work was carried out in the Laboratoire de Biologie Cellulaire 4, Université de Paris-Sud, Orsay, France. Support was from the Centre National de la Recherches et Moyens d'Essais (Equipe de Recherche Associée 174) and the Direction des Recherches et Moyens d'Essais (contract 73/844). Some of the work was done in the Department of Biology, University of Chicago, and was supported by grants from the U.S. Public Health Service (GM-23277-02 and CA-19265-03).

REFERENCES

Avers, C.J., M.W. Rancourt, and F.H. Lin. 1965. Intracellular mitochondrial diversity in various strains of *Saccharomyces cerevisiae*. *Proc. Natl. Acad. Sci.* **54**: 527.

Birky, C.W., Jr., C.A. Demko, P.S. Perlman, and R. Strausberg. 1978. Uniparental inheritance of mitochondrial genes in yeast: Dependence on input bias of mitochondrial DNA and preliminary investigations of the mechanism. *Genetics* **89**: 615.

Damsky, C.H. 1976. Environmentally induced changes in mitochondria and endoplasmic reticulum of *Saccharomyces carlsbergensis* yeast. *J. Cell Biol.* **71**: 123.

Damsky, C.H., W.M. Nelson, and A. Claude. 1969. Mitochondria in anaerobically grown, lipid-limited brewer's yeast. *J. Cell Biol.* **43**: 174.

Davidson, M.T. and P.B. Garland. 1975. Mitochondrial structure studied by high voltage electron microscopy of thick sections of *Candida utilis*. *J. Gen. Microbiol.* **91**: 127.

Eddy, A.A. and D.H. Williamson. 1957. A method of isolating protoplasts from yeast. *Nature* **179**: 1252.

Ephrussi, B., P.P. Slonimski, Y. Yotsuyanagi, and J. Tavlitzki. 1956. Variations physiologiques et cytologiques de la levure au cours du cycle de la croissance aerobie. *C. R. Trav. Lab. Carlsberg* **26**: 87.

Faye, G. 1977. RNA mitochondriaux chez la levure *Saccharomyces cerevisiae*. *Biol. Cell.* **28**: 93.

Grimes, G.W., H.R. Mahler, and P.S. Perlman. 1974. Nuclear gene dosage effects on mitochondrial mass and DNA. *J. Cell Biol.* **61**: 565.

Hoffman, H.P. and C.J. Avers. 1973. Mitochondrion of yeast: Ultrastructural evidence for one giant, branched organelle per cell. *Science* **181**: 749.

Kellems, R.E., V.F. Allison, and R.A. Butow. 1975. Cytoplasmic type 80S ribosomes associated with yeast mitochondria. IV. Attachment of ribosomes to the outer membrane of isolated mitochondria. *J. Cell Biol.* **65**: 1.

Lloyd, D. 1974. *The mitochondria of microorganisms*. Academic Press, New York.

Marchant, R. and D.G. Smith. 1968. Membranous structures in yeast. *Biol. Rev.* **43**: 459.

Osumi, M., C. Shimoda, and N. Yanagishima. 1974. Mating reaction in *Saccharomyces cerevisiae*. V. Changes in the fine structure during the mating reaction. *Arch. Microbiol.* **97**: 27.

Peterson, J.B., R.H. Gray, and H. Ris. 1972. Meiotic spindle plaques in *Saccharomyces cerevisiae*. *J. Cell Biol.* **53**: 837.

Plattner, H., M.M. Salpeter, J. Saltzgaber, and G. Schatz. 1970. Promitochondria of anaerobically grown yeast. IV. Conversion into respiring mitochondria. *Proc. Natl. Acad. Sci.* **66**: 1252.

Rabinowitz, M. and H. Swift. 1970. Mitochondrial nucleic acids and their relation to the biogenesis of mitochondria. *Physiol. Rev.* **50**: 376.

Smith, D.G., D. Wilkie, and K.C. Srivastava. 1972. Ultrastructural changes in mitochondria of zygotes in *Saccharomyces cerevisiae*. *Microbios* **6**: 231.

Smith, D.G., R. Marchant, N.G. Maroudas, and D. Wilkie. 1969. A comparative study of the mitochondrial structure of petite strains of *Saccharomyces cerevisiae*. *J. Gen. Microbiol.* **56**: 47.

Stegeman, W.J., C.S. Cooper, and C.J. Avers. 1970. Physical characterization of ribosomes from purified mitochondria of yeast. *Biochem. Biophys. Res. Commun.* **39**: 69.

Stevens, B.J. 1977. Variation in number and volume of the mitochondria in yeast according to growth conditions. A study based on serial sectioning and computer graphics reconstruction. *Biol. Cell.* **28**: 37.

———. 1978. Behavior of mitochondria during sporulation in yeast. In *Electron microscopy 1978* (ed. J.M. Sturgess), vol. 2, p. 406. Microscopical Society of Canada, Toronto.

Stevens, B.J. and E. Moustacchi. 1976. Ultrastructural characterization of mitochondria from a yeast mutant sensitive to *petite* induction (uvs ρ72). In *Genetics, biogenesis and bioenergetics of mitochondria* (ed. W. Bandlow et al.), p. 137. de Gruyter, Berlin.

Stevens, B.J. and J.G. White. 1979. Computer reconstruction of mitochondria from yeast. *Methods Enzymol.* **56**: 718.

Strausberg, R.L. and P.S. Perlman. 1978. Effect of zygote bud position on transmission of mitochondrial genes in *Saccharomyces cerevisiae*. *Mol. Gen. Genet.* **163**: 131.

Swift, H., M. Rabinowitz, and G. Getz. 1968. Cytochemical studies on mitochondrial nucleic acids. In *Biochemical aspects of the biogenesis of mitochondria* (ed. E.C. Slater et al.), p.3. Adriatica Editrice, Bari.

Thiéry, J.P. 1967. Mise en évidence des polysaccharides sur coupes fines en microscopie électronique. *J. Microsc.* **6**: 987.

Todd, M.M. and E.L. Vigil. 1972. Cytochemical localization of peroxidase activity in *Saccharomyces cerevisiae*. *J. Histochem. Cytochem.* **20**: 344.

Vignais, P.V., B.J. Stevens, J. Huet, and J. André. 1972. Mitoribosomes from *Candida utilis*. Morphological, physical and chemical characterization of the monomer form and of its subunits. *J. Cell Biol.* **54**: 468.

Williamson, D.H. 1976. Packaging and recombination of mitochondrial DNA in vegetatively growing yeast cells. In *Genetics, biogenesis and bioenergetics of mitochondria* (ed. W. Bandlow et al.), p. 117. de Gruyter, Berlin.

Williamson, D.H. and D.J. Fennell. 1975. The use of fluorescent DNA-binding agent for detecting and separating yeast mitochondrial DNA. *Methods Cell Biol.* **12**: 351.

———. 1979. Visualization of yeast mitochondrial DNA with the fluorescent stain "DAPI." *Methods Enzymol.* **56**: 728.

Yotsuyanagi, Y. 1962a. Etudes sur le chondriome de la levure. I. Variation de l'ultrastructure du chondriome au cours du cycle de la croissance aérobie. *J. Ultrastruct. Res.* **7**: 121.

———. 1962b. Etudes sur le chondriome de la levure. II. Chondriomes des mutants à déficience respiratoire. *J. Ultrastruct. Res.* **7**: 141.

Zickler, D. and L.W. Olson. 1975. The synaptonemal complex and the spindle plaque during meiosis in yeast. *Chromosoma* **50**: 1.

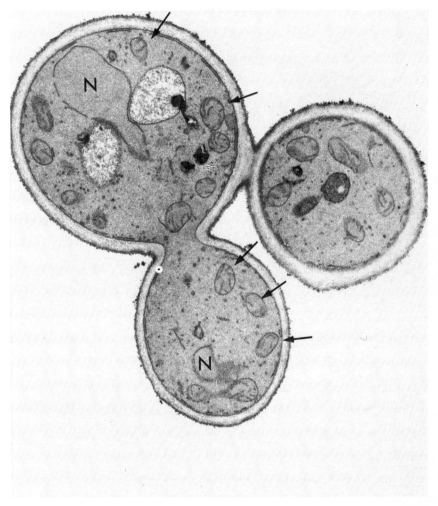

Plate 1 Typical example of potassium-permanganate fixation. Mitochondria (→) are regularly distributed near the cell periphery. (N) Nucleus. Strain N123. Early stationary phase in glycerol. Paraformaldehyde and glutaraldehyde prefixation, $KMnO_4$ fixation, uranyl-acetate stain en bloc; lead-citrate stain. Magnification, 11,160 × .

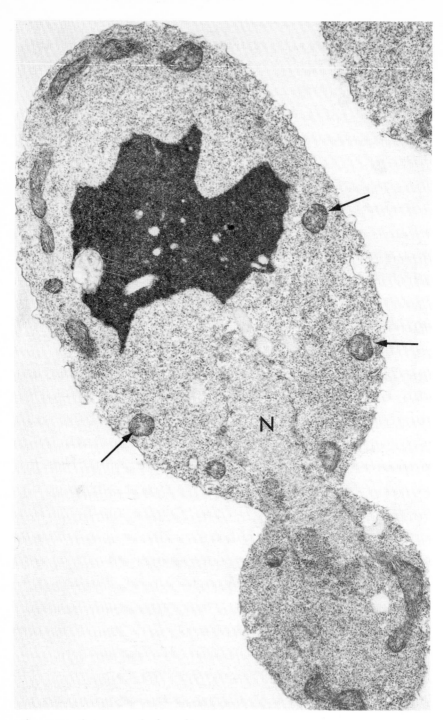

Plate 2 (See facing page for legend.)

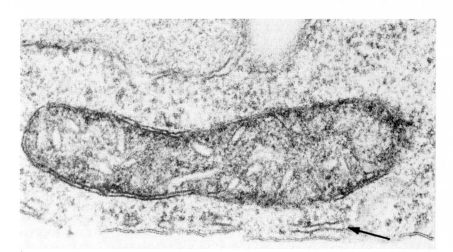

Plate 3 Thin section of an elongated mitochondrion. The outer and inner membranes are closely apposed. Short cristae extend partway into the matrix. A peripheral cisternum of endoplasmic reticulum lies between the plasma membrane and the mitochondrion (→). Strain S41. The preparation is as described in the legend to Plate 2. Magnification, 70,000 × .

Plate 2 Typical example of osmium fixation after spheroplasting. Mitochondria (→) are distributed in the peripheral cytoplasm. The nucleus (N) is elongating into the bud. Strain K65-3D. 1-hr sporulation. Glutaraldehyde prefixation, Zymolyase digestion, osmium postfixation, uranyl-acetate stain en bloc; lead-citrate stain. Magnification, 16,000 × .

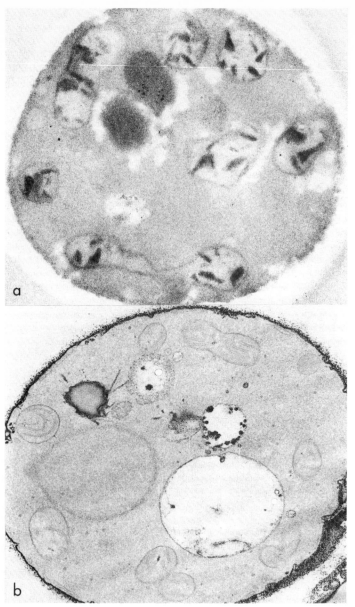

Plate 4 Examples of staining techniques to demonstrate mitochondria. (*a*) Cytochemical demonstration of cytochrome oxidase and cytochrome *c* peroxidase by the DAB reaction. Cristae and inner membranes show electron-dense reaction product. Strain N123. Stationary phase in glucose. Magnification, 41,000 ×. (*b*) Cytochemical demonstration of polysaccharides in thin section by the PATAg technique (Thiéry 1967). Following Glusulase digestion, cell-wall remnants stain strongly; some vacuolar contents also stain. Mitochondrial profiles are clearly visible against a homogeneous background. Strain N123. Early stationary phase in glucose. Magnification, 26,240 ×.

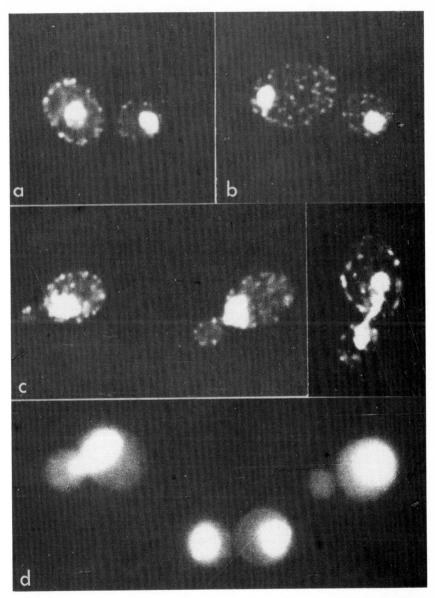

Plate 5 Examples of DNA staining by the fluorescent stain DAPI. Nuclei are brightly stained. Small fluorescent dots in the cytoplasm represent mitochondrial DNA areas. Strain K65-3D. Magnification, 3300 ×. (*a*) In a slightly flattened preparation, mitochondrial DNA is distributed in the periphery. (*b*) In a more flattened preparation, mitochondrial DNA dots are scattered over the cell surface. (*c*) Mitochondrial DNA is visible in very small buds, before and after nuclear migration. (*d*) A ρ^0 mutant shows only nuclear staining.

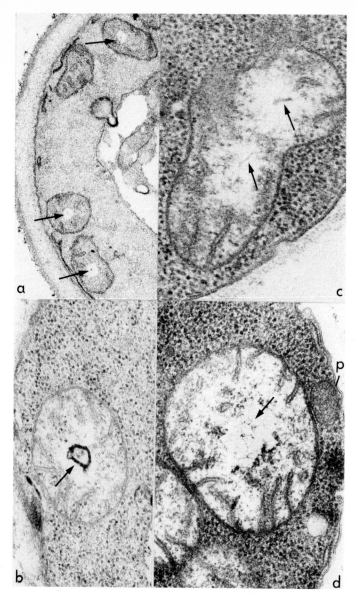

Plate 6 Visualization of mitochondrial DNA in situ in thin sections. Strain N123. (*a*)
Permanganate fixation. Mitochondrial DNA is visible as tiny spots (→) in clear areas
of the matrix. Magnification, 11,200 × . (*b*) Aldehyde-osmium fixation after Glusulase
digestion. Mitochondrial DNA appears as a prominent electron-dense clump (→).
Magnification, 48,000 × . (*c*) Same preparation technique as in *b*. Improved preservation of mitochondrial DNA, but clumped aspect remains (→). Magnification,
56,000 × . (*d*) Same preparation technique as in *b*, followed by uranyl-acetate staining
en bloc. Mitochondrial DNA is visible as fine filaments (→). Definition of
membranes is improved by the uranyl-acetate stain. (p) Peroxisome. Magnification,
56,000.× .

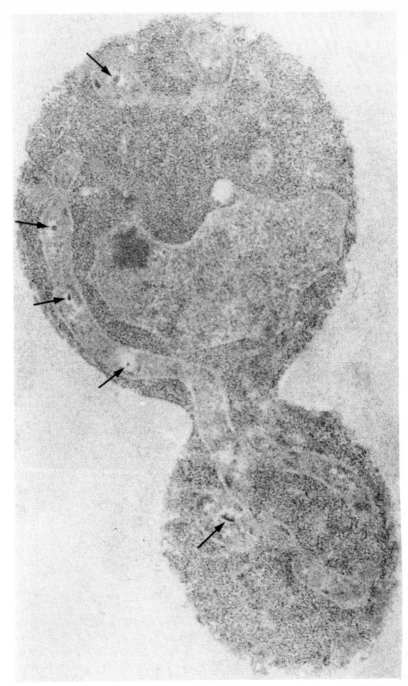

Plate 7 Thin section of a dividing cell showing several mitochondrial DNA areas regularly spaced within one long, filamentous mitochondrion (→). Strain S41. Exponential phase in acetate medium. Paraformaldehyde fixation and Zymolyase digestion. Magnification 27,900 × .

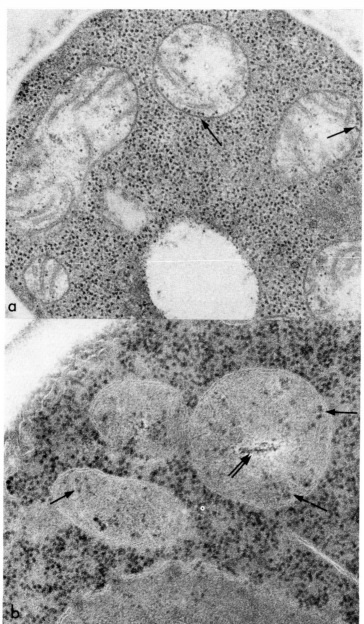

Plate 8 Visualization of mitoribosomes in situ. (*a*) Mitoribosomal particles are distributed along the inner membrane (→) and occasionally along the cristae. Strain N123. Glusulase digestion followed by paraformaldehyde and glutaraldehyde pre-fixation and osmium postfixation. Magnification 51,600 × . (*b*) In well-fixed and compact mitochondria whose membranes are in negative contrast, mitoribosomes (→) and mitochondrial DNA (⇉) are clearly visible. Sensitive strain SX. Glutaralydehyde and osmium fixation of intact cells; uranyl-acetate and lead-citrate stain. Magnification, 60,200 × .

496

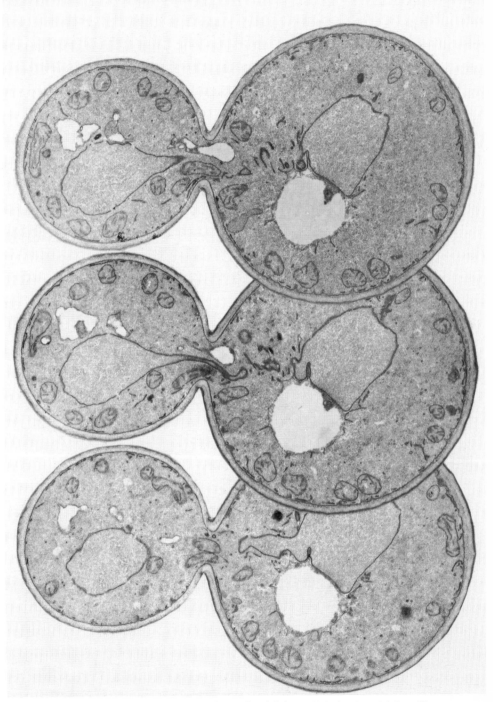

Plate 9   Three consecutive thin sections of a dividing cell. Mitochondrial profiles are observed in the neck during nuclear elongation and migration. Strain IL46. Early stationary phase in glucose. Preparation is as described in the legend to Plate 1. Magnification, 9600 × .

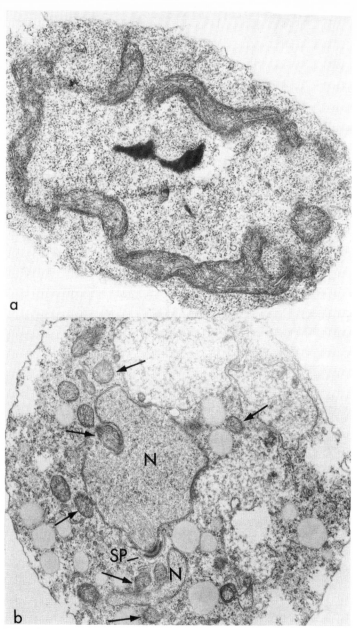

Plate 10   Behavior of mitochondria during sporulation. (*a*) Tangential section at the cell periphery showing a continuous and branching profile of the unitary chondriome. Strain K65-3D. 1-hr sporulation. Preparation is as described in the legend to Plate 2. Magnification, 17,200 ×. (*b*) Meiosis-II division. Mitochondrial profiles (→) are located close to the dividing nucleus (N). (SP) Meiotic spindle plaque. Cell preservation is poor due to slight overdigestion with Zymolyase. Strain S41. 20-hr sporulation. Preparation is as described in the legend to Plate 2. Magnification, 15,480 ×.

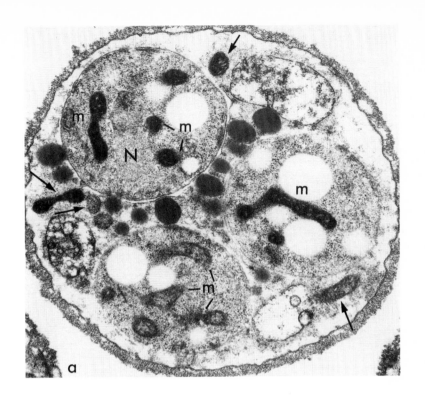

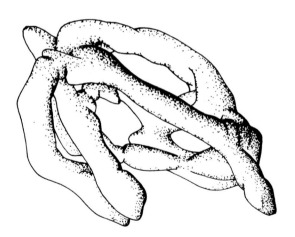

b

Plate 11 Mitochondria in ascospores. (*a*) Thin section of an ascus showing three of the four ascospores. Profiles of the single mitochondrion (m) in each ascospore closely encircle the nuclei (N). Normal profiles are also observed in the residual cytoplasm of the ascus (→). Strain 112, prepared by Zickler and Olson (1975). Magnification, 21,600 × . (*b*) Computer reconstruction of the single mitochondrion in one ascospore from the ascus in *a*. (Drawing by A. Charrier.)

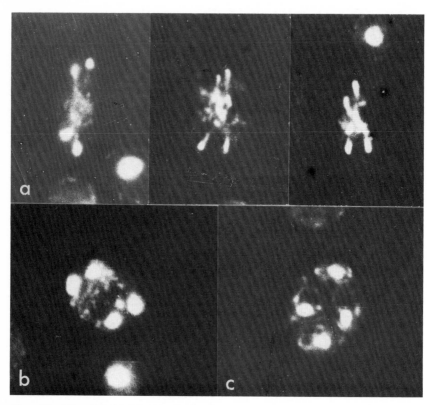

Plate 12   Mitochondrial DNA distribution during sporulation by DAPI staining. Strains K65-3D and S41. Magnification, 3300 ×. (*a*) Examples of meiosis-II division. Mitochondrial DNA is visible in the central nuclear region. (*b, c*) Mitochondrial DNA forms a small network associated with each spore nucleus.

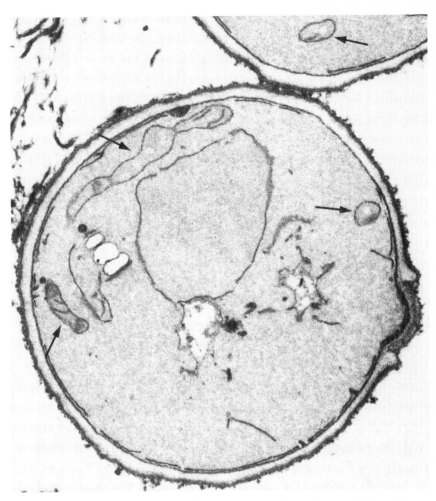

Plate 13 Mitochondrial structure in glucose repression. A reduced chondriome shows rare profiles, long and sinuous, or simple spheres (→). Large cytoplasmic areas lack profiles. Strain IL46. Exponential phase in glucose. Preparation is as described in the legend to Plate 1. Magnification, 22,000 × .

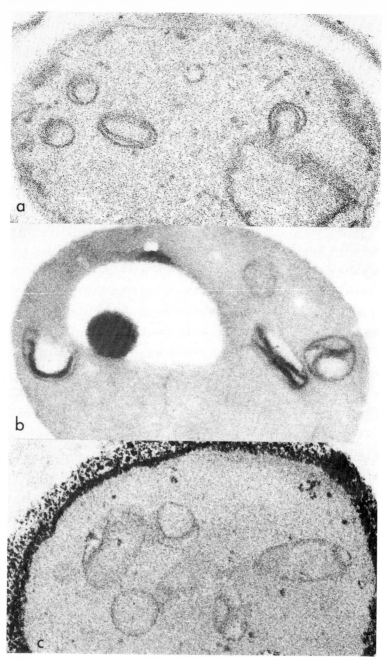

Plate 14 Mitochondrial structure in ρ⁻ strains. Profiles are round or slightly elongated vesicles. Some contain internal lamellae, which form curved partitions or lie parallel to the mitochondrial surface. Strain ρ⁻24 from N123. Magnification, 45,000 × . (*a*) Permanganate fixation is as described in the legend to Plate 1. (*b*) DAB reaction. Lamellae contain the electron-dense reaction product. (*c*) PATAg staining. Mitochondrial profiles are simple vesicles.

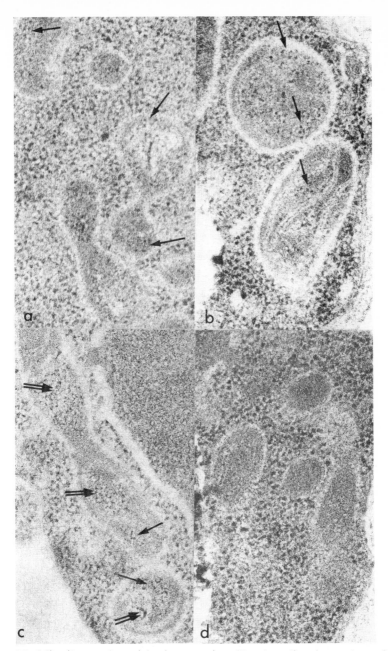

Plate 15 Mitoribosomal particles in ρ^- strains. Glusulase digestion and paraformaldehyde and glutaraldehyde fixation. Magnification 69,600 × . (*a*) Strain CEP2 (16S + 23S genes present). Mitoribosomal particles are visible (→). (*b*) Strain D21 (23S gene present). Particles appear abundant (→). (*c*) Strain O_1P2 (16S gene present). Some indication of particles (→). Mitochondrial DNA is particularly abundant (⇉). (*d*) Strain H71 (ρ^0). Mitochondria show a uniformly electron-opaque matrix.

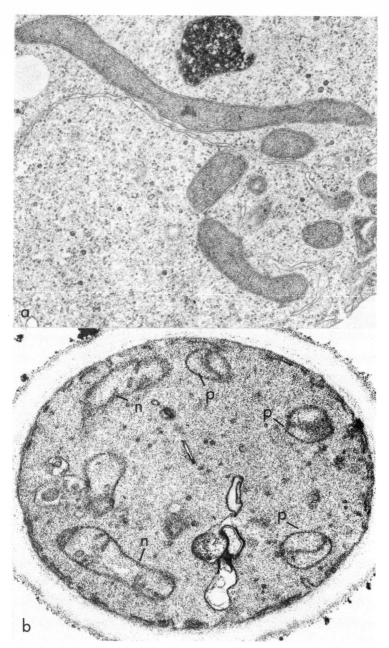

Plate 16 (a) ρ^0 mitochondria. Mitochondrial profiles show clear, double membranes and no internal organization. Some profiles can be long and filamentous. ρ^0 mutant of strain 308/6C (*tmp1-6*). Preparation is as described in the legend to Plate 2. Magnification, 26,100 × . (b) Example of a "mixed" mitochondrial phenotype. Normal profiles (n) and petite-type profiles (p) appear in the same cytoplasm. uvs ρ72 (ρ^+) mutant derived from N123. Stationary phase in glucose. Preparation is as described in the legend to Plate 1. Magnification, 33,060 × .

Mitochondrial Genetics and Functions

Bernard Dujon
Centre de Génétique Moléculaire
du Centre National de la Recherche Scientifique
91190 Gif sur Yvette, France

INTRODUCTION AND BRIEF HISTORY

Mitochondria are complex organelles specialized in respiration and oxidative phosphorylation, which possess their own genetic system and their own protein synthetic machinery. The mitochondrial genome carries the genetic information for only a few, essential mitochondrial components. The formation of mitochondria is a complex process that requires the intricate assembly of components from two different origins: the nucleus and the mitochondria. The majority of the mitochondrial proteins, in mass as well as in number of species, are encoded by the nuclear genome and synthesized on cytoplasmic ribosomes. The minority (~ 5% of all mitochondrial proteins) are encoded and synthesized within the mitochondria. The dual origin is even more apparent if one examines individually the different mitochondrial components, because each of the mitochondrially encoded proteins is eventually assembled into functional complexes comprising nuclear-encoded proteins as well. Despite their numerical importance, thus far very little is known of the mitochondrial components of nuclear origin and of their genes. However, a considerable effort during the past several years has resulted in a very detailed description of the mitochondrial genes and their products to the point that the mitochondrial genome of yeast is, at present, one of the best characterized eukaryotic genomes. This paper is therefore devoted to a detailed description of the mitochondrial genes, their genetic properties, and their functions, with only brief mention of nuclear genes. This apparently limited topic (the mitochondrial genome is, after all, very small) has recently undergone such a complex evolution that, during the past 10 years, more than a thousand original articles, numerous review articles (Borst 1972; Linnane et al. 1972; Mahler 1973b; Tzagoloff et al. 1973; Dujon and Michaelis 1974; Kovac 1974; Schatz and Mason 1974; Michaelis and Somlo 1976; Schatz 1976; Borst et al. 1977; Nagley et al. 1977; Birky 1978; Borst and Grivell 1978; Butow and Strausberg 1979; Locker et al. 1979; Tzagoloff et al. 1979a; Perlman et al. 1980), and several colloquia and symposia (Miller 1970; Boardman et al. 1971; Kroon and Saccone 1974, 1980; Birky et al. 1975; Puiseux-Dao 1975; Bandlow et al. 1976, 1977; Bücher et al. 1976; Saccone and Kroon 1976; Bacila et al. 1978; Cummings et al. 1979), or books (Sager 1972; Gillham 1977) have been devoted to the various aspects of the mitochondrial genome of yeast.

Because this paper will focus essentially on the most recent developments (during the last 10 years and, more particularly, during the last 5 years) and because its different sections are articulated in a logical rather than a historical presentation, it may be appropriate to briefly mention a few steps that have marked major advances in the development of mitochondrial genetics and have framed the present aspect of this field.

Evidence of organelle heredity was already presented more than 70 years ago when Correns (1909) and Baur (1909) discovered that some genetic characters, affecting the plastid phenotype of higher plants, show a non-Mendelian type of inheritance. The genetics of mitochondria in yeast was initiated about 30 years ago when Ephrussi and collaborators showed that some respiratory-deficient mutants (called petite colony or vegetative petites; ρ^-), obtained after induction by acriflavine, affect several respiratory enzymes at the same time (Slonimski and Ephrussi 1949) and do not follow the Mendelian rules of mitotic and meiotic segregations. Instead, they show a mode of inheritance indicative of cytoplasmic particles (Ephrussi et al. 1949a,b). It became clear that some mitochondrial functions, particularly the formation of respiratory enzymes, were under the control of self-reproducible cytoplasmic factor (Slonimski 1953; Ephrussi and Slonimski 1955), called ρ. The existence of a DNA species, specific to mitochondria, was recognized several years later in vertebrates and then in yeast (Schatz et al. 1964; Tewari et al. 1965). The observation that mitochondrial DNA (mtDNA) is grossly altered in ρ^- mutants, but not in nuclear respiratory-deficient mutants (*pet*-), demonstrated conclusively that the mtDNA is the hereditary material defined by ρ^- mutants (Mounolou et al. 1966, 1968). But, because they are pleiotropic and all show the same phenotype, ρ^- mutants alone were of limited help in further unraveling the details of the genetic functions of mtDNA. About this time, it was shown (Clark-Walker and Linnane 1966; Lamb et al. 1968) that some antibiotics, active on the bacterial ribosomes, inhibit mitochondrial protein synthesis without affecting cytoplasmic protein synthesis, producing phenocopies of ρ^- mutants that cannot grow on a nonfermentable carbon source. This differential effect offered the possibility of rapidly selecting the first mitochondrial mutants different from ρ^- mutants (Linnane et al. 1968b; Thomas and Wilkie 1968b; Coen et al. 1970). These mutants, conferring resistance to the yeast cell against the antibiotic erythromycin or chloramphenicol, opened a new dimension in the field of mitochondrial genetics, because point mutations could now be analyzed. From this time, several hundred mitochondrially inherited mutants, conferring resistance to various antibiotics or drugs, were discovered and analyzed (reviewed in Dujon et al. 1977b). The next important step in the isolation of mitochondrial mutations was achieved when it was recognized that some respiratory-deficient mutants (called *mit*-), isolated after a different type of mutagenesis, are not ρ^- mutants but are point mutations deficient for one (or only a few) specific functions (Flury et al. 1974; Tzagoloff et al. 1975a,b; Slonimski and Tzagoloff 1976). Again, hundreds of these mutants have now been isolated and analyzed (reviewed in Dujon et. al. 1977b). This brief history would not be complete without mentioning the elucidation of the molecular nature of the ρ^- mutation. Numerous studies (see below, Organization of mtDNA in ρ^- Mutants) have shown that ρ^- mutations result from very extensive deletions of wild-type mtDNA and regular repetitions of the conserved sequence. The faithful replication of the conserved sequence creates a natural in vivo purification of various and precisely defined mtDNA fragments into different ρ^- clones. It is not at all certain whether this property is of any interest to the cell or represents a by-product of more fundamental mechanisms, but it is a most valuable tool to the molecular geneticist.

GENETIC PROPERTIES OF THE MITOCHONDRIAL GENOME*

Mutants of the Mitochondrial Genome and General Background

ρ^- Mutants

The wide variety of mitochondrial mutants is unique to the mitochondrial genetics of *Saccharomyces cerevisiae*. The different classes of mitochondrial mutations are summarized in Table 1. The ρ^- or "petite" mutants are nonreverting pleiotropic mutations with extensively deleted mtDNA or no mtDNA at all (in the latter case, they are called ρ^0). *S. cerevisiae*, a facultative anaerobe, is one of the very few eukaryotic organisms that (provided a fermentable carbon source is available) can survive the complete loss or the gross alterations of its mtDNA and grow indefinitely as a ρ^- mutant (yeast species of this type are called petite-positive as opposed to the petite-negative species in which no stable ρ^- mutants have been found). ρ^- mutants exhibit a number of unusual genetic properties that will be discussed at length later in the text. Due to the extensive deletions of mtDNA, they cannot carry out mitochondrial protein synthesis (because components of the mitochondrial protein synthesizing machinery are coded for by the mtDNA) and, consequently, they are deficient for all polypeptides synthesized on mitochondrial ribosomes. These mitochondrial polypeptides are elements of the respiratory chain (e.g., cytochrome b and cytochrome oxidase) and of the ATPase complex. Any mitochondrial protein present in a ρ^- mutant is therefore not dependent upon mitochondrial protein synthesis and must be synthesized on cytoplasmic ribosomes and imported into the mitochondria. These include the proteins of the outer mitochondrial membrane, the numerous enzymes of the matrix (e.g., the enzymes of the citric

---

*Several nomenclatures have successively been proposed for the genetics of mitochondria (Von Borstel, 1969; Sherman and Lawrence 1974; Plischke et al. 1976), but each, inspired from nuclear genetics, proved to be poorly adapted to mitochondrial genetics and, in fact, almost half a dozen different nomenclatures can be found in the literature to designate mitochondrial genes, mutants, and genetic loci in *Saccharomyces cerevisiae*.

At the Interdisciplinary Conference on the Genetics and Biogenesis of Chloroplasts and Mitochondria at Munich (August 1976), a serious effort was made to agree upon a unique nomenclature for organelle genetics in general. The results of this meeting were summarized in a proposal, drafted by C.W. Birky, A.M. Colson, and P.S. Perlman, which was widely circulated but never published, since difficulties were encountered in finding a nomenclature perfectly adapted to the specificity and the degree of sophistication of mitochondrial genetics and in the need to respect some common usages and to easily allow subsequent developments.

One of the major nomenclatural problems in yeast mitochondrial genetics is the distinction between the genes and the genetic loci. This derives from the fact that recombination is extremely high, so that genetic defects altering the same product can be readily separable or even genetically unlinked. This problem is particularly evident in cases where defects at separable sites within a single gene lead to different phenotypes. For example, the gene for the large rRNA includes sites that control sensitivity to chloramphenicol and others that control sensitivity to erythromycin, which are not closely linked. This had led to the designation of three genetic loci (*rib1, rib2, rib3*) within what turns out to be a single gene. At this time it is still convenient to designate mutations in terms of these recombinationally defined clusters (a locus being defined only as a cluster of mutational sites closely linked to one another but unlinked, or loosely linked, to mutational sites at other loci). Similarly, the gene for the apocytochrome b, *cob-box*, extends over a region so long as to contain ten different regions within it, not linked to

acid cycle and many others), almost all of the proteins of the mitochondrial ribosomes, many of the proteins of the inner mitochondrial membrane and, finally, several subunits of the ATPase complex and of the respiratory-chain complexes (e.g., 4 subunits of cytochrome oxidase, cytochrome c_1, cytochrome c, etc.).

Dogma

It is important to stress that the site of synthesis of the mitochondrial proteins (i.e., the cytoplasmic ribosomes or the mitochondrial ribosomes) obviously does not indicate the origin of the genetic message used to direct their synthesis (nuclear or mitochondrial genes). There is, however, a general dogma in mitochondrial genetics at present specifying that no exchange of RNA molecules occurs between the cytoplasm and the mitochondria. According to this rule, all RNA species present in mitochondria are transcripts of the mitochondrial genome, and mitochondrial ribosomes only synthesize proteins encoded by mitochondrial genes, whereas cytoplasmic ribosomes only synthesize proteins encoded by nuclear genes. All of the available experimental evidence has so far never invalidated this dogma. Any mitochondrial protein present in a ρ^- mutant must therefore be encoded by nuclear genes, and a mitochondrial gene whose message is translated can never be expressed in a ρ^- mutant. It follows that all ρ^- mutants (assuming they have identical nuclear genetic backgrounds) always have identical phenotypes irrespective of the fragment(s) of wild-type mtDNA they retain. If the presence of a component in a ρ^- mutant constitutes the definitive demonstration of its nuclear genetic origin, the absence of a component in a ρ^- mutant does not necessarily demonstrate its mitochondrial

---

one another (hence, the various *box* loci designations). The fact that multiple proteins are produced from some regions of the mitochondrial genome by differential mRNA splicing patterns poses another important nomenclatural problem. For example, several distinct products can be synthesized from the *cob-box* gene after different pre-mRNA splicing events. Other problems may also arise from the pleiotropic phenotype resulting from a number of mitochondrial mutations and from the numerous polymorphic variations of the same mitochondrial gene.

The nomenclature used in this review follows the main recommendations of the Munich Conference, with the adjustments necessary to account for the developments since this Conference. Genes and loci designations consist of three italicized lowercase letters and a number. Because classical dominance tests cannot be done in mitochondria, the use of uppercase letters is avoided. The wild-type allele of a locus is designated by a superscript plus (e.g., $oxi1^+$), the mutant allele leading to a deficiency by a minus (e.g., $oxi1^-$). The use of superscript letters denoting sensitivity (e.g., $oli1^S$) or resistance (e.g., $oli1^R$) is conserved in cases in which both alleles give active products. Allele numbers are also italicized and are separated from the gene or locus designation by a hyphen (e.g., *par1-454*). The distinction between nuclear mutations and mitochondrial mutations is generally clear from the context, and no confusion is possible if care is taken to avoid the same symbols in naming new nuclear or mitochondrial genes. Greek symbols ρ and ω have been conserved as recommended at the Conference. As far as possible the designations of genes and loci used here follow the most common usage in the recent literature. In many cases, genes are designated from the most conspicuous genetic cluster they contain (e.g., *oli1*). To facilitate comparisons, the compilation of published mutations, genes, and loci with their original nomenclature and map assignments, made available at the Schliersee Meeting in 1977, should be consulted (Dujon et al. 1977b).

Table 1 General classes of mitochondrial mutations and polymorphic variations

General symbol	Functions concerned	Allelic symbol	Description and specific properties
ρ	all	ρ^+	wild type, respiratory competent
		ρ^-	respiratory-deficient mutant due to very large deletions in the mitochondrial genome and regular repetitions of the conserved segment
		ρ^0	respiratory-deficient mutant due to complete loss of mtDNA
ant	resistance or sensitivity to specific drugs or inhibitors (mainly antibiotics but also others)	ant^S	sensitivity to the inhibitor (wild type)
		ant^R	resistance to the inhibitor
		ant^0	gene lost from mtDNA in a ρ^- mutant
mit	respiratory chain and oxidative phosphorylation	mit^+	wild type
		mit^-	mutant deficient in one or a few components of the respiratory chain or of the ATPase complex
		mit^0	gene lost from mtDNA in a ρ^- mutant
syn	mitochondrial protein synthesis	syn^+	wild type
		syn^-	mutant deficient for mitochondrial protein synthesis due to specific and limited lesion of mtDNA (different from ρ^-)
		syn^0	gene lost from mtDNA in a ρ^- mutant

mim	interactions between mitochondrial genes (suppressors)	
	mim^+	wild type, inactive suppressor
	mim^-	mutant, active suppressor
	mim^0	gene lost from mtDNA in a ρ^- mutant
var	mitochondrial ribosome	
	$var1$[a]	a series of variant forms of a mitochondrially synthesized polypeptide resulting from a natural polymorphism at a specific mitochondrial locus called $var1$ determinant
ω	polarity of recombination of flanking genetic markers	
	ω^+	one of the two natural polymorphic forms at the ω locus
	ω^-	the other natural polymorphic form at the ω locus
	ω^n	mutant of ω^-, deficient for the polarity of recombination

This table gives the definitions and symbols of the different classes of mitochondrial mutations as they appear in this paper. The nomenclature follows the recommendations of the 1976 Interdisciplinary Conference held in Munich. Other designations currently used are indicated as follows: ρ^- and ρ^0 mutants are often referred to as petite (colony), cytoplasmic petite, or vegetative petite. ρ^0 mutants, which are all nonsuppressive, are sometimes called neutral petite in the literature. However, ρ^- mutants that are nonsuppressive also exist. The corresponding wild type (ρ^+) is often designated as grande. *ant* mutants are sometimes called drug-resistant or drug-sensitive. The *gin* class, as defined by Mahler et al. (1976), comprises both the *mit* and the *syn* classes.

[a]Several natural polymorphic forms of the *var1* determinant, as well as recombinants between these forms, have been described above (see Elusive Mitochondrial Genes: The *var1* determinant).

genetic origin, because this absence may reflect indirect regulatory effects of the ρ^- mutation on the expression of nuclear-encoded products or their assembly into mitochondria.

Genetic Rules

Historically, ρ^- mutants allowed the definition of the first rules of mitochondrial inheritance so that when mutants of the next classes became available, clearcut criteria to define mitochondrial mutations already existed. Briefly, these rules are as follows: (1) After a cross between a ρ^+ and a ρ^- mutant, the ρ^+ and ρ^- characters segregate during the mitotic cell divisions so as to generate a mixture of ρ^+ cells and ρ^- cells. (2) When a ρ^+ cell issued from such a cross undergoes meiosis, segregation of the ρ^+ and ρ^- characters does not occur (only tetrads issued from ρ^+ cells can be tested because the respiratory-deficient mutants do not sporulate). Following these two rules is a prerequisite for any mitochondrial mutation. In addition, because different ρ^- mutants are large deletions in different parts of the mitochondrial genome, they offer reference backgrounds against which the exact nature and precise map localization of other mitochondrial mutations can be defined.

Point Mutations

The second class of mitochondrial mutations is referred to in this paper by the general symbol *ant* to indicate that the first mutants of this class conferred resistance to antibiotics but, at present, this class also includes mutations conferring resistance to other types of drugs or inhibitors. *ant*^{R} mutations concern various types of functions depending upon the inhibitor against which they confer resistance. Some directly inhibit oxidative phosphorylation (e.g., oligomycin, venturicidin, or ossamycin) or the respiratory chain at the level of cytochrome b (e.g., antimycin A, mucidin, diuron, or funiculosin); others inhibit the mitochondrial protein synthesis (e.g., chloramphenicol, erythromycin, spiromycin, or paromomycin) and that, in turn, affects the formation of the elements of the respiratory chain and of the ATPase. In all cases, therefore, the final result is to prevent normal respiratory functions and, at the appropriate concentration, each inhibitor will prevent the wild-type cell (*ant*^{S}) from growing on a nonfermentable carbon source. Mutants that can grow under such conditions can be easily selected. Mitochondrial *ant*^{R} mutants follow the same genetic rules as ρ^- mutants: mitotic segregation after crosses and the absence of meiotic segregation. In addition, their mitochondrial determinism is demonstrated by the fact that they are eliminated in all ρ^0 mutants and in some, but not all, ρ^- mutants. Because they behave genetically as point mutations, often with little or no deleterious effects to the cell, and offer a large diversity of phenotypes, the *ant*^{R} mutations have been extremely useful in determining the precise quantitative rules for the transmission and recombination of mitochondrial genes and in constructing the first mitochondrial genetic map.

Mitochondrial mutations that affect a respiratory component but—unlike ρ^- mutants—not mitochondrial protein synthesis, constitute a third class of mitochondrial mutants, referred to as mit^-. Many mit^- mutants, either conditional (e.g., temperature-sensitive) or not, have now been isolated and characterized. They define the different mitochondrial genes that will be examined in detail below (see Genetic Functions of the Mitochondrial Genome). Like the ant^R mutants, the mit^- mutants behave as point mutations and obey the same rules of mitochondrial genetics. Particularly, they are eliminated in all ρ^0 mutants and in some, but not all, ρ^- mutants. Another class of point mutations, called syn^-, behave genetically as the mit^- mutant but specifically affect mitochondrial protein synthesis and, hence, are respiratory-deficient. Unlike ρ^- mutants, mit^- or syn^- mutants can recombine with one another and complement one another to restore a wild-type phenotype. Similarly, they can recombine with or complement ρ^- mutants that retain the corresponding wild-type allele, but not with mit^0 (or syn^0) ρ^- mutants.

The last class of mitochondrial mutants, recently designated mim (or MIM) refers to mitochondrial mutations that are able to suppress mit^- mutants and restore a wild-type (or pseudo-wild-type) phenotype. This class contains allele-specific and locus nonspecific suppressors (probably informational suppressors) as well as locus-specific but allele nonspecific suppressors (suggesting direct interactions between mitochondrial genes or loci).

Finally, comparison of the mitochondrial genomes between different *S. cerevisiae* strains shows an important natural polymorphism due to the presence or the absence, at specific locations in the mitochondrial genome, of inserts varying in size from a few dozen base pairs to more than a thousand base pairs. The ω^+ and ω^- alleles, the different $var1$ forms, and the short and long forms of the mosaic genes are examples of this polymorphism.

Homoplasmic vs. Heteroplasmic Cells

Both biochemical evidence and genetic evidence (see later in the text) indicate that mitochondria are a multiple-copy genetic system. It follows that a given yeast cell may be either homoplasmic, when all mtDNA copies are genetically identical (this definition applies either to all loci over the whole mitochondrial genome or to the particular locus considered), or heteroplasmic, in the opposite case. The determination of these two types of cells relies on a clonal analysis (Fig. 1). Whenever a clone issued from a single cell is entirely composed of cells having the same mitochondrial genotypes (ex-

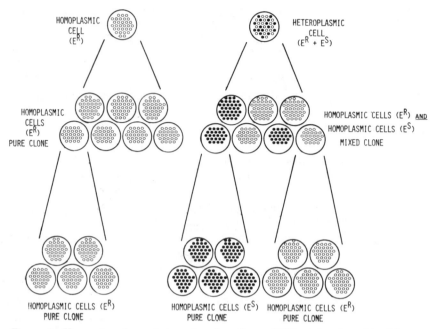

Figure 1 Homoplasmic vs. heteroplasmic cells. mtDNA molecules with different genetic markers: (●) *ery*^{S} and (○) *ery*^{R}.

cluding spontaneous mutants appearing during the growth of the clone), the original cell is said to be homoplasmic. In contrast, a heteroplasmic cell will always give rise to a mixed clone composed of cells of different mitochondrial genotypes with frequencies easily recognizable from those of spontaneous mutants. This operational definition is limited by the degree of significance with which one determines the mixed or pure character of the clone. It is, however, unambiguous in practice for the following reasons. First, clones issued from typical heteroplasmic cells (e.g., zygotes) usually do not segregate cells, even those of the less frequent genotypic class, with frequencies lower than 10^{-4}, a figure much higher than that of spontaneous mutants (Coen et al. 1970). Second, the interclonal variance of the frequency of spontaneous mutants (the classical Luria-Delbrück fluctuation test) is extremely low (Dujon et al. 1976). Therefore, the clones actually issued from homoplasmic cells, but that would show, by chance, a high frequency of mutants and be undistinguishable from clones issued from heteroplasmic cells, either do not exist or are extremely rare.

Abundant experimental evidence from several laboratories demonstrates that all of the cells of a clone issued from a homoplasmic cell are themselves homoplasmic (Fig. 1, left) and will later give rise to clones entirely composed of homoplasmic cells of the same genotype. Thus, stable homoplasmic cell lines are easily formed and maintained. In contrast, the cells of a clone issued from a heteroplasmic cell (Fig. 1, right) can be either heteroplasmic or homoplasmic, but the kinetics of segregation are such that the frequency of the first type diminishes very rapidly with the age of the clone and, after a limited number of cell divisions, the large majority of the cells are homoplasmic. Thus, the heteroplasmic state is always transient and, so far, there exists no method to maintain stable heteroplasmic cell lines.

Mitochondrial Crosses

Mitochondria are part of the cell, not distinct organisms, and one cannot score the phenotype of mitochondria but only the phenotype of the cell in which they reside. A particular methodology was therefore needed for mitochondrial crosses. A typical mitochondrial cross involves the formation of zygotes from the fusion of homoplasmic yeast cells (of opposite cellular mating types) with appropriate mitochondrial genotypes. Zygotes issued from the cross of two nonisomitochondrial cells are obviously heteroplasmic as they initially contain the two parental mtDNA populations and, later, the recombinants formed between them. The vegetative multiplication of these zygotes is accompanied by a rapid segregation of mtDNA molecules such that cells issued after some 20–25 generations from the zygote (the size of a visible colony on a petri dish) are virtually all homoplasmic. The results of mitochondrial crosses can, therefore, be analyzed at two different levels: the final progeny of the cross after segregation has formed homoplasmic cells or the transient heteroplasmic phase that immediately follows zygote formation. Analyses at the final level are the basis of all studies on the transmission and recombination of mitochondrial genes and of the construction of various homoplasmic recombinant cell lines. It is the construction of such recombinants that allows the studies of the interactions between mitochondrial genes or mutations associated in a *cis* array. In contrast, analyses at the initial zygote level, the phenotype of the heteroplasmic phase, provide information on the interactions of the mitochondrial genes or mutations associated in a *trans* array, i.e., mitochondrial genes on separate mtDNA molecules of the same cell (see below, Complementation Tests, Complementation Map, Mitochondrial Genes, and Loci; Mosaic Mitochondrial Genes and their Interactions).

Three different methods have been used in the experiments reported in the next section to genetically analyze the progeny of mitochondrial crosses after segregation has produced homoplasmic cells (Fig. 2). In the "standard cross," a large population of zygotes is formed by random mass mating between the two parents and allowed to grow on a nonselective medium. The entire progeny, issued from all zygotic clones mixed together, is then quantitatively analyzed using a representative sample of cells taken at random. This method gives quantitative information on the genotypic composi-

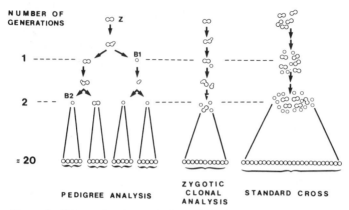

NUMBER OF GENERATIONS

PEDIGREE ANALYSIS   **ZYGOTIC CLONAL ANALYSIS**   **STANDARD CROSS**

Figure 2   Schematic representation of the three methods for mitochondrial crosses. In each case, the genotypic composition of the progeny after ~ 20 cell divisions from the zygote(s) can be determined from the quantitative analysis of a representative sample of homoplasmic cells taken at random. For detailed experimental procedures, see Coen et al. (1970). (Z) zygote; (B1) first zygotic bud; (B2) second zygotic bud.

tion of the progeny as a whole, a characteristic and reproducible figure for a given mitochondrial cross (i.e., between two given parental strains defined by their nuclear and mitochondrial genetic backgrounds). In the "zygotic clonal analysis," on the other hand, the genetic composition of each zygotic clone is analyzed individually (again using representative samples of cells taken at random). This method gives quantitative information on the genotypic composition of the progeny of each individual zygote, a variable figure with a broad interclonal distribution within a single mitochondrial cross (see below, Segregation of Mitochondrial Genomes). Finally, in the "pedigree analysis," the cells issued from the first few divisions of the zygote are separated from one another by micromanipulation, and each is allowed to form a colony. The genetic composition of each colony is again determined using representative samples of cells taken at random. This method gives detailed descriptions of the patterns of segregation and recombination during the early cell divisions after the zygote, but since it is very laborious, its use has been limited only to some specific cases. Obviously, the sum of frequencies of each genotype for all clones issued from a pedigree analysis (pondered for the rank of each clone) is equivalent to a zygotic clone in the clonal analysis. Similarly, the sum of a great number of zygotic clones in the clonal analysis is equivalent to a standard cross.

Mechanisms and Rules for Mitochondrial Inheritance: Multifactorial Crosses Involving Antibiotic-resistant Mutations

The genetic properties of mitochondrial genes have now been precisely described from the results of numerous crosses performed by different laboratories and involving various combinations of ant^R mitochondrial markers (Bolotin et al. 1971; Kleese et al. 1972a; Rank and Bech-Hansen 1972a; Avner et al. 1973; Howell et al. 1973; Rank 1973; Wolf et al. 1973; Callen 1974a,b; Netter et al. 1974; Suda and Uchida 1974; Wakabayashi 1974a,b; Dujon et al. 1975, 1976; Gunge 1975; Kaldma 1975a,b; Schweyen and Kaudewitz 1976; Backhaus et al. 1978). Mitochondrial crosses reveal all the properties of a multiple-copy genetic system and can best be treated in terms of population genetics at the intracellular level, bearing some similarities to bacteriophage crosses. The first explicit statement of a model analogous to bacteriophage genetics was developed several years ago (Dujon 1974, 1976; Dujon et al. 1974, 1976; Dujon and Slonimski 1976). This model, although deliberately simplified in its formalism, is still the only attempt to provide an integrated view of the genetic properties of mitochondrial crosses and to give rise to specific predictions amenable to a quantitative experimental verification. Its premises, which will frame the discussions of this paragraph, are (1) random pairing and recombination between the different mtDNA molecules of a cell (a panmictic pool), (2) multiple rounds of pairing and recombination, and (3) random segregation. Furthermore, the input ratio, i.e., the genetic equivalent of the relative proportion of the two parental mtDNA molecules that enters the zygote and contributes to the panmictic genetic pool, is variable from one cross to the next in a relatively broad range. The input ratio is obviously not directly measurable, but the consequences of its variations are reflected in the output of the cross in a way that is predictable by the model and verifiable by the experiment. Summarized below are the major features of mitochondrial crosses and their implications for the mechanisms involved. For the sake of clarity, all phenomenon relevant to the polarity of recombination, although important in the original literature, will be referred to later in this section.

The Coordinated Variation of the Output Allele Ratio and Its Significance

In a mitochondrial cross, the respective frequencies with which the two alleles of a locus are represented in the progeny (called the output or transmission of each allele) are the same for all loci considered (this is obviously under experimental conditions that do not favor cells of a particular mitochondrial genotype over the others). The output allele ratio is a characteristic figure of a given cross. It varies largely from one cross to the next (if nonisonuclear crosses are compared or after specific treatments of the parents) but always in a strictly coordinated fashion for all mitochondrial loci (Fig. 3a). This strongly implies that all mitochondrial loci are carried by the same mtDNA molecule and that the cause of the variation concerns the entire mtDNA molecule. The coordinated output is so characteristic of mitochondrial genes that it can be used as a test to determine whether or not a newly isolated mutation is carried by the mtDNA.

The variation observed is a direct argument for a multiple-copy genetic system with random pairing and variable input, because if a limited number

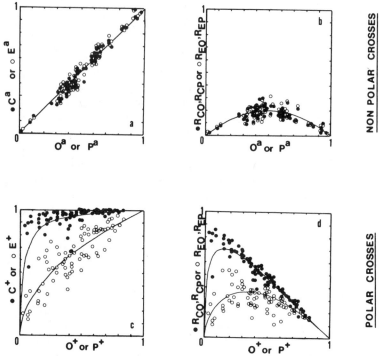

Figure 3 Output allele ratio and frequency of recombinants in mitochondrial crosses. Each point represents the result of a single multifactorial mitochondrial cross. All crosses involve the same mitochondrial genetic markers (cap^R-321, ery^R-221 [or ery^R-514], oli^R-1, and par^R-454, respectively, at $rib1$, $rib3$, $oli1$, and $par1$ loci) but differ from one another by their input ratio (see text). The input ratio is calculated from the output of alleles at the $oli1$ and/or $par1$ loci, unlinked to the polar region. The abscissas are the input frequencies of alleles from one of the two parents at these loci (either the $MATa$ parent O^a or P^a or the ω^+ parent O^+ or P^+). (*a,c*) The output allele ratios at the $rib1$ loci (●) or $rib3$ loci (○), respectively, in nonpolar and in polar crosses (ordinates are the output frequencies of alleles from the same parent as defines the input ratio). (*b,d*) The frequencies of recombinants (ordinates) between genetically unlinked markers, respectively, in nonpolar and in polar crosses. (●) Recombinants between $rib1$ and either $oli1$ or $par1$ loci; (○) recombinants between $rib3$ and either $oli1$ or $par1$ loci. The curves are drawn using the formalism of Dujon et al. (1974) with the best estimates of each parameter for the set of experimental data. (Adapted from Dujon et al. (1974, 1975, 1976) and Dujon (1976).

of copies were present in each cell, variation of this number by one unit would result in a discrete distribution of the output allele ratio. In contrast, the continuous variation observed reflects the sum of discrete variations among a relatively large number of copies. Furthermore, a proportional relationship between the input ratio and the output ratio is expected of a random mating population in population genetics. (The Hardy-Weinberg equation specifies that in a random mating population, as long as there is no selective pressure in favor of one of the alleles, the allele ratio will remain constant from one generation to the next.)

A number of experimental treatments of the parental cells prior to crossing are able to modify the output to a broad range. These include UV-irradiation (Dujon et al. 1975), glucose repression (Goldthwaite et al. 1974; Birky 1975b; Boker et al. 1976), hydroxyurea (Dujardin et al. 1978), and the growth at nonpermissive temperature of temperature-sensitive nuclear or mitochondrial mutants (Backhaus et al. 1978). All of these treatments always modify the output allele ratios in a coordinated fashion and are best interpreted as variations of the input ratio. Similar effects are due to the ploidy number of the parents (Gunge 1976).

The input ratio is defined until now only in formal terms as the genetic equivalent of the relative number of mtDNA molecules that each parent contributes to the genetic pool. The biological bases for its variation may be several: (1) the variations of the actual number of mtDNA molecules from each parent, and (2) the possible subsequent modifications of this initial number due to the mode of replication and/or segregation of mtDNA molecules. It is now clear that the amount of mtDNA varies to some extent among different strains and according to the physiological conditions of the cells, but these variations usually do not extend to a range as wide as the variations of input (from almost 0 to almost 1; see Fig. 3). It should be remembered, however, that no valid correlation can be established between the input ratio and these amounts of mtDNA because the amounts of mtDNA in the cells that form the zygotes (which are in a different physiological state from the rest of the cells) may not be identical to the average amount of mtDNA measured on the whole cell population (see, e.g., Lee and Johnson [1977] for the variations of mtDNA quantity according to cell volume). Nevertheless, it seems unlikely that a range of variations identical to the range of input variations can be found, and the second mechanisms must also be considered.

The Frequency of the Recombinant and Its Significance

Cells carrying nonparental mitochondrial genotypes issued from mitochondrial crosses were first reported by Thomas and Wilkie (1968a). That such novel cell types arise from the molecular recombination between different mtDNA molecules and not from the reassortment of independent genetic units requires the demonstration of the genetic linkage between the markers considered, the independent determination that they are carried on the same DNA molecule or, finally, the demonstration of the existence of physically recombined mtDNA molecules. In the case of yeast mitochondria, all three demonstrations have been given successively.

To properly interpret the significance of the recombinant frequencies in crosses, several issues must be examined: the degree of randomness of pairing between mtDNA molecules of the same cell, the average number of pairing events that a given mtDNA molecule can undergo before being segregated into a homoplasmic cell and, finally, the molecular mechanism of recombination itself. The latter two issues are discussed later in this section because they are linked to the polarity of recombination; the randomness of pairing is discussed here.

The frequencies of recombinants between mitochondrial genetic markers show a number of properties that indicate random pairing of mtDNA molecules of an intracellular pool with variable input. First, the maximal frequency of recombinants observed between any two pairs of alleles does not exceed 20–25% of the total progeny (Fig. 3b). This recombinant frequency, significantly lower than the 50% classically observed for unlinked nuclear genes, does, however, represent the absence of genetic linkage between distant mitochondrial markers and is never exceeded. This low upper limit is expected from a random pairing population because some pairing events will occur between genetically identical mtDNA molecules and lead to undetectable recombinants, hence, reducing the observed frequency of recombinants. Second, the random pairing predicts that the frequency of recombinants between two given markers must vary with the input ratio, because the probability of a pairing between two mtDNA molecules of given genotypes depends upon the frequencies of each genotype in the intracellular population, a prediction also met by the experimental result (Fig. 3b). Because the overall frequency of recombinants between two markers is a function of the input ratio, the frequency of recombination (which measures the genetic linkage) cannot be directly deduced from the frequency of recombinants unless an independent estimate of the input ratio is available in the same cross. Finally, in a random pairing system, the frequency of double recombinants observed should be at least two times higher than predicted from the single recombinant frequencies (and even more for highly biased input), a prediction also verified by previous experiments (Dujon et al. 1975; Dujon and Slonimski 1976).

In yeast mitochondrial crosses, markers distant from each other by only short physical distances are genetically unlinked (remember that as discussed above, this corresponds to 10–12% of wild-type recombinants). A series of crosses indicate that genetic linkage is only apparent between mitochondrial markers separated from each other by less than about 1000 bp (B. Dujon, unpubl.). This shows the very high efficiency of mitochondrial recombination and is easily interpretable with multiple mating rounds. Assuming recombination is homogenous over the entire genome, more than 50 linkage groups could exist in the 75,000-bp-long genome. This makes difficult the general mapping of mitochondrial genes on the basis of recombination. For the linked markers, figures of 0.06–0.14% of wild-type recombinants for distances of 30 bp have been reported (Sebald et al. 1979b;

Lazowska et al. 1980), but due to the limited number of examples available and the absence of control for the other parameters, no precise relationship between frequencies of recombination and physical distances is available as yet.

The Polarity of Recombination and Its Significance

A particular phenomenon, called polarity of recombination, occurs in a specific region of the mitochondrial genome (the polar region) in the immediate vicinity of the locus ω and, since the first ant^R mutants available (cap^R and ery^R) are localized in this region, this phenomenon has played a great importance in elucidating the mechanisms involved in mitochondrial crosses (Coen et al. 1970; Bolotin et al. 1971; Dujon et al. 1976). We now know that the polar region is the gene for the large rRNA, with cap^R and ery^R being point mutations in this gene (see below, Ribosomal RNA Genes). On the basis of the polarity of recombination in crosses, all yeast strains can be unambiguously classified as ω^+ or ω^-. When ω^+ strains or ω^- strains are crossed respectively between themselves, no polarity occurs. The two reciprocal recombinants between the markers cap^R and ery^R appear with similar frequencies, and the crosses follow all of the rules described previously. On the other hand, when an ω^+ strain is crossed with an ω^- strain, the two reciprocal recombinants between the same two markers occur with extremely unequal frequencies, and the allele ratios at loci of the polar region are no longer proportional to the input ratios (Fig. 3c). Furthermore, the ω locus itself is part of the nonreciprocity, since the large majority of the recombinants are themselves ω^+, and the ω^- allele is virtually eliminated from the progeny. The intensity of the polar effect can be measured from the extent of the deviation from the coordinated output (Fig. 3c) and from the frequencies of the two reciprocal recombinants (Fig. 3d). Both indicate that the polar effect is stronger on the *rib1* locus than on the *rib3* locus. A third locus of the polar region (*rib2*) is affected with an intermediate intensity (Netter et al. 1974). This defines the *cis* effect of polarity: The closer a locus is to ω, the greater the polar effect is on this locus. The *cis* effect demonstrates that it is the allelic difference at the ω locus itself that is the point of initiation of the polarity.

The ω^+ / ω^- polymorphism is correlated with the presence or absence of an intron in the large rRNA gene (see below, Ribosomal RNA Genes). This correlation suggests a possible role for the intron in the mechanism that determines the polarity. More recently, a second polymorphic insert (a 66-bp mini-insert), separable from the intron by recombination, has been detected in the large rRNA gene (Dujon 1980). Due to the polarity mechanism, when one parental strain possesses the intron in its large rRNA gene and the other does not, most of the progeny inherits the intron. This unique property offers an example of a mechanism by which new introns can be inserted into preexisting simple genes.

Mutants of polarity, derived from ω^- and called ω^n (neutral) because they show no polarity in crosses to ω^+ and in crosses to ω^-, can be isolated associated with mutations at the *rib1* locus (Dujon et al. 1976). The ω^n phenotype results from a base substitution very close to the point of insertion of the intron in ω^+ strains, in the segment of the rRNA gene that determines the chloramphenicol-resistant or -sensitive phenotype (Dujon 1980).

The molecular mechanism that transfers the intron and its flanking regions from the rRNA gene of the ω^+ strain to the rRNA gene of the ω^- strain is not completely understood. A mechanism of dissymetrical gene conversion has initially been proposed (Dujon et al. 1974). In this mechanism, the pairing of the polar regions of mtDNA molecules from the ω^+ parent and the ω^- parent, i.e., the formation of a heteroduplex between the ω^+ and ω^- DNA strands, creates a specific and dissymetrical structure that triggers specific enzymes to excise the short strand (without intron) and, after a limited degradation and resynthesis of the flanking regions, to transform the previous ω^- strand into a copy of the ω^+ strand. It suffices to assume that the extent of the degradation and resynthesis has a greater probability of reaching the loci in the immediate vicinity of the excision point (such as *rib1*) than the more distant loci (such as *rib3*) to account for the *cis* effect. Reiteration of such a mechanism during the multiple rounds of pairing accounts for the frequencies of recombinants observed. In this scheme, the ω^n mutants represent mutations at the site recognized by the specific excision enzyme(s).

Nothing is known about the nature of the enzyme(s) responsible for the polarity. Results of $\rho^- \times \rho^-$ crosses or inhibition of mitochondrial protein synthesis by erythromycin (Strausberg and Birky 1979) suggest that the polarity takes place in the absence of mitochondrial protein synthesis and that the enzyme(s) required must therefore be imported from the cytoplasm. The existence of a long open reading frame within the intron in ω^+ strains leads us to suspect, however, that it could code for a polypeptide involved in the polarity (Dujon 1980).

With the exception of the *var1* determinant, which shows a polarity phenomenon comparable to ω, but much less pronounced (Strausberg and Butow 1981), no other phenomenon of polarity has been discovered in the mitochondrial genome. Particularly, the other optional introns (e.g., in the *cob-box* gene or in the *oxi3* gene) do not introduce a polarity of recombination between flanking markers. Thus, the presence alone of an intron is not sufficient to determine the polarity, and more specific mechanisms must take place. Perhaps, being constrained to form a functional rRNA molecule in which secondary structures obviously play an important role, an rRNA gene is more prone to contain sequences capable of forming specific secondary structures at the DNA level as well. Such structures might also play a role in the polarity.

Multiple Mating Rounds and the Mechanism of Genetic Recombination

All of the properties of mitochondrial crosses indicate multiple rounds of pairing and recombination of mtDNA molecules. Such a system makes difficult the genetic analysis of the mechanism of recombination because the only observable recombinants are the final result of successive recombination events, and there is no direct access to the product of each individual recombination event. Indirect information on the mechanism of recombination can, nevertheless, be gained from a closer examination of the number of mating rounds, because the final frequency of a recombinant is the function of both the probability of recombination per mating round and the number of mating rounds. One possible method of estimating the number of mating rounds comes from the polarity of recombination. Assuming an absolute efficiency of dissymetrical gene conversion at each pairing event (i.e., assuming symmetrical heteroduplex formation and an obligatory correction of the two heteroduplexes, despite the opposite orientations of the DNA strands), a minimum number of three to four mating rounds is needed to account for the observed frequencies of recombinants. This number must obviously be greater if the efficiency of gene conversion is not absolute. Another estimate of the number of mating rounds, based on classical hypotheses on the probabilities of double recombinations versus single ones in nonpolar crosses, leads to a maximum number of one to two mating rounds. This paradox can be partially solved if one further assumes that the major molecular mechanism of recombination of all mitochondrial genes, even in the absence of polarity, includes frequent gene-conversion events (for a discussion of this problem, see Dujon 1976; Dujon et al. 1974). Another possibility is to consider that the mechanism of polarity of recombination is not initiated by the random pairing and heteroduplex formation of the mechanism of generalized recombination, but itself initiates recombination at a specific point and with greater efficiency.

In any case, the average number of mating rounds calculated as above has no direct physical significance. It is only the genetic equivalent of the actual average number of pairings and recombination and of other phenomena resulting from the mode of replication and segregation of the mtDNA molecules that may tend to modify the intracellular genetic pool. Considering that most cells are already homoplasmic after a few divisions from the zygote, the numbers of mating rounds calculated above (from one to four) are compatible with about one actual mating round per cell generation of a heteroplasmic cell.

In conclusion, a better understanding of the mechanism of mtDNA genetic recombination probably awaits molecular studies during the heteroplasmic phase in which recombinations take place and not just the examination of the homoplasmic progeny several generations later. Furthermore, the formalism used so far incorporates only a minimal number of parameters. A more complete formalism should take into account the modes of replication and segregation of mtDNA molecules and the interclonal distributions of the output ratios (and possibly of input ratios as well), and it should integrate varying degrees of randomness for pairing or segregation. But it is not certain whether this approach, which needs complex mathematical formulations, will lead to the discovery of new biological properties of mitochondrial crosses, since the introduction of many new parameters will result in a more complex model that will tend to become unfalsifiable and, hence, of little value.

Segregation of Mitochondrial Genomes

It is a characteristic feature of all mitochondrial alleles that they continuously segregate during mitotic cell divisions. The mechanism of segregation, however, remains poorly understood. It is commonly observed that, during the first few cell divisions after the zygote, an important fraction of the cells is already homoplasmic for one allele or another (Coen et al. 1970; Lukins et al. 1973; Waxman et al. 1973; Wilkie and Thomas 1973; Callen 1974c; Dujon et al. 1974; Aufderheide 1975; Forster and Kleese 1975a,b; Uchida and Suda 1976, 1978; Birky 1978; Strausberg and Perlman 1978). Such a rapid purification is remarkable because, with about 100 copies of mtDNA initially present in the zygote, a random partitioning during the cell divisions would very seldom produce a homoplasmic cell. Therefore, either the segregation is not random, i.e., there is a selection based on the genotype of each copy of mtDNA, or the segregation is random but the number of copies that enter the bud is very small.

Experiments in which the first zygotic buds were separated from the zygotes by micromanipulation showed that the frequencies of first buds homoplasmic for a given locus are not only relatively high but are also a function of the input ratio of the particular cross (Dujon et al. 1974; Dujon and Slonimski 1976; Dujon 1976). This is predicted from a random segregation, because the greater the number of copies carrying one allele in the zygote, the greater is the probability of finding this allele in the bud. Using an intentionally oversimplified hypothesis of random segregation with nonexhaustive sampling, it was calculated that the number of "segregating units" entering the first bud must be very small (around three on the average), a figure that is of little physical significance since the average quantity of mtDNA in a first bud, just after its formation, is about one-third that of the zygote. (Sena et al. [1976] found more than 40 mtDNA molecules in the bud.) Reconsidering this question later with a hypothesis of random segregation but with exhaustive sampling, Birky et al. (1978b) obtained similarly small numbers of segregating units entering the bud and in the zygote. To try to reconcile the genetic and physical figures, several modes of "packaging" of mtDNA molecules have been proposed in such a way that several mtDNA molecules with identical genotypes would tend to segregate in groups instead of independently, thereby reducing the randomness of pairing and segregation. However, no stable physical entity corresponding to a segregating unit composed of several mtDNA molecules resists a closer examination, and one is forced to conclude that such segregating units are missing entities (see Williamson et al. 1977 and references therein). This conclusion is relatively obvious, and all of the previous reasoning suffers from two points: (1) The genetic composition of the intracellular mtDNA population in each zygote at the time of bud formation needs not necessarily be identical to the average input ratio and, thus, is not directly measurable; (2) The genetic composition of the intracellular mtDNA population in the bud is not directly measurable either, since it can only be examined genetically after the bud has formed a clone including subsequent segregation events. Actually, the rapid purification of heteroplasmic cells is compatible with a random segregation of a relatively large number of mtDNA molecules if one assumes that some molecular mechanisms tend to amplify biases in the intracellular genetic pool and, thus, accelerate the purification. This may result from gene conversion between mtDNA molecules and/or from a random choice of mtDNA molecules for replication. That such mechanisms play an important role in segregation is consistent with the fact that, although zygotic buds homoplasmic at one locus are relatively frequent, those that are homoplasmic at all loci at the same time are rarer (Lukins et al. 1973; Forster and Kleese 1975b; Dujon 1976). It is also consistent with the high variance of the frequencies of parental or recombinant classes in interclonal distributions (Coen et al. 1970; Lukins et al. 1973). Some zygotic clones, with frequencies dependent upon the input, are even entirely composed of one parental genotype (uniparental zygotes; Birky 1975a; Birky et al. 1978a).

Finally, an effect of the bud position (central or apical) on the frequency of recombinants and parental classes has been reported (Callen 1974c; Strausberg and Perlman 1978) as an indication of incomplete panmixia. But, in other crosses, such an effect is not found, which indicates that a complete panmixia is rapidly established soon after zygote formation (Aufderheide 1975).

Origin of Mitochondrial Mutants

An important question in a multiple-copy genetic system concerns the origin of mutants. Assuming that one mutation has occurred in one mtDNA molecule, which mechanism will sort out this particular mtDNA molecule from the population of nonmutated mtDNA molecules in order to give rise to a homoplasmic mutant cell line? In practice, how many cell divisions and/or what selective pressures are needed to obtain a mitochondrial mutant?

Two types of experiments, inspired from classic bacterial genetics, have provided some insight into this question. Using the Newcombe respreading test, Birky (1973) concluded that an intracellular selection of spontaneous mitochondrial mutations takes place. In the absence of the selective agent (in this case, erythromycin), the resistant mitochondria arising from a spontaneous mutation must be at a selective disadvantage as compared with the wild-type mitochondria and cannot accumulate to yield homoplasmic cells resistant to the selected agent. However, in the presence of the erythromycin, the mutant genome can multiply until a sufficient number is reached to allow the cell to grow. Similar conclusions have been reached from the quantitative analysis of the formation of cap^R and ery^R mutants using the Luria-Delbruck fluctuation test (Dujon et al. 1976). In each case, the interclonal variance of the number of mutants is slightly greater than the mean but not as high as is observed in cases of single-copy genetic systems. Again, the distributions found can be easily interpreted in a multiple-copy model by an intracellular selection of spontaneous mutation that preexists the action of the antibiotic. Also, consistent with an intracellular selection is the fact that the growth of the resistant colonies corresponding to new spontaneous mitochondrial mutants on the selective medium always shows a lag as compared with the growth of colonies from homoplasmic cells of a resistant mutant. But, in addition to the intracellular selection, several other mechanisms can also contribute to the observed phenomena. First, in all of these experiments, there is some residual growth of the sensitive cells on the selective medium, and the segregation during this residual growth can certainly contribute to the formation of homoplasmic mutant cells. Second, an intracellular genetic drift due to nonreciprocal recombination may mimic an intracellular selection by eliminating the rare molecules. Third, the mode of replication of mtDNA molecules is a critical parameter for the distribution of the mutated copies. If replication is not "geometric" (i.e., if each mtDNA molecule does not replicate once per cell cycle), then a mutation appearing in one molecule of the cell will have a high probability of being eliminated (because it has not replicated) and a low probability of being replicated several times. It is probably the combination of all of these factors that finally accounts for the formation of mitochondrial mutants in yeast. That these phenomena may not be as efficient in other organisms (e.g., *Paramecium, Neurospora,* mammalian cells, etc.) may explain the difficulty encountered in isolating mitochondrial mutants from these organisms.

All that has been said so far concerns spontaneous mutations conferring a phenotype that allows the direct selection of the mutants. But the frequency of spontaneous mitochondrial mutants is low enough to make difficult the isolation of mutants for which no selective pressure can be applied. Thus, a mutagen that would specifically induce mitochondrial mutations is needed. Most classical mutagens induce ρ^- mutants with extremely high efficiencies (see below, Formation of ρ^- Mutants) and are therefore inefficient for the isolation of point mutations. Others, such as nitrosoguanidine or 2-amino-purine, induce mitochondrial mutations but also induce nuclear mutations with much greater efficiency (Handwerker et al. 1973; Dawes and Carter 1974; Flury et al. 1974; Storm and Marmur 1975; Tzagoloff et al. 1975b). Putrament et al. (1973) argued that an efficient induction of mitochondrial mutations would likely result due to errors during replication rather than repair, because mutagens interacting with DNA itself are unable to induce mitochondrial mutations (except ρ^- mutants), and repair of UV-induced lesions in mtDNA is very (if not totally) inefficient. Using this rationale, they discovered that the addition of manganese in the culture medium during a few cell divisions increases the number of cap^R and ery^R mitochondrial mutations up to 100-fold. From different considerations about the properties of the mtDNA polymerase, the competition with other cations, and the action of inhibitors of replication, they concluded that Mn^{++} acts as an error-producing factor during the replication of mtDNA molecules because of its primary interaction with the mtDNA polymerase (Putrament et al. 1975b, 1977). Mn^{++} also efficiently induces ρ^- mutants, and this may help the sorting out of homoplasmic cells carrying point mutations in the mitochondrial genome by reducing the number of ρ^+ genomes in the mutagenized cells (see above). Furthermore, if the mutagenized culture is allowed a few cell divisions after Mn^{++} is removed from the medium, the total number of mutant cells does not increase (or increases very little) as if the previously mutated mtDNA molecules were distributed to only one of the two daughter cells during division (Putrament et al. 1975a, 1976). This observation, consistent with our previous discussion on the origin of spontaneous mutants, is of great practical importance because, after a short mutagenic treatment with Mn^{++}, the large majority of mitochondrial mutants isolated are issued from independent mutagenic events.

A few mitochondrially inherited respiratory-deficient mutants different from ρ^- had just been discovered at this time (Handwerker et al. 1973; Flury et al. 1974; Storm and Marmur 1975). The introduction of the Mn^{++} mutagenesis promoted large-scale screenings of such new mutants. Tzagoloff et al. (1975a,b) reported the first extensive isolation of *mit⁻* mutants. After mutagenesis, they picked up all respiratory-deficient mutants (characterized by their inability to grow on a nonfermentable carbon source) and examined each one for its mitochondrial protein synthesis ability in order to distinguish *mit⁻* from ρ^- mutants. The mitochondrial inheritance of *mit⁻* mutants isolated was subsequently determined by crosses to ρ^0 mutants (see Table 2).

Kotylak and Slonimski (1977a) introduced a new isolation procedure, considerably less laborious and more efficient, on the basis of the previous observation by Kovacova et al. (1968) that the nuclear mutant *op1*, deficient in the adenyl nucleotide translocase, is lethal if in combination with a ρ^- mutation. They observed however, that the *op1 (pet9)* mutant is not lethal in combination with *mit⁻* mutations. The reason for this difference is still unclear, but it offers a screening procedure whose rationale is as follows.

Table 2   Genetic tests to differentiate between the possible types of respiratory-deficient mutants

Type of Mutant	Complete genotype	Respiratory ability of diploids issued from crosses to	
		a ρ^0 PET^+ tester	different ρ^- PET^+ testers
Mitochondrial ρ^- or ρ^{0a}	ρ^- mit^+ PET^+ or ρ^- mit^- PET^+ or ρ^0 mit^0 PET^+	always $\ominus$	always $\ominus$
Mitochondrial mit^- (or syn^-)	ρ^+ mit^- PET^+	always $\ominus$	$\oplus$ with some testers (ρ^- mit^+) $\ominus$ with other testers (ρ^- mit^0)
Recessive nuclear pet^-	ρ^+ mit^+ pet^-	always $\oplus$	always $\oplus$

Crosses are made between the respiratory-deficient mutants to be tested and the tester strains as indicated. The respiratory ability of the resulting diploids is determined from their ability to grow on a nonfermentable carbon source: $\oplus$ indicates growth; $\ominus$ indicates absence of growth. Only the mit^- (or syn^-) mutants can give different results with different ρ^- PET^+ tester strains.
[a]In cases where the *op1* method of selection is used, this type does not exist.

op1 cells are unable to grow (or, more precisely, grow extremely slowly) on a nonfermentable carbon source but, if mated to a wild-type strain (OP^+), the resulting diploids ($op1/OP^+$) recover a normal capacity to grow on this medium. Therefore, after mutagenesis, if a viable *op1* clone is crossed to an OP^+ ρ^0 tester strain and the resulting diploid is found not to grow on glycerol, then this clone must carry a mutation in the mtDNA. This mutation cannot be ρ^- or ρ^0 nor can it be a recessive nuclear mutation. Theoretically, it could be a dominant nuclear mutation (although no such case has yet been reported), but crosses with appropriate ρ^- tester strains easily distinguish it from a *mit*⁻ mutant (see Table 2). Another advantage to this method is that some *mit*⁻ mutants are relatively unstable and accumulate ρ^- or ρ^0 mutants that must be eliminated by subcloning and appropriate tests. The use of *op1* eliminates this problem. Furthermore, *mit*⁺ revertants do not accumulate because of the epistatic effect of *op1* with respect to growth, whereas they easily overgrow *mit*⁻ parental cells in OP⁺ nuclear background. The only disadvantage to this method is that it is not certain that particular types of mitochondrial mutants will not be lethal with *op1*. This might be the case for some *syn*⁻ mutants, for example. Other methods to isolate *mit*⁻ mutants have been reported previously (Mahler et al. 1976; Rytka et al. 1976), but they do not combine all the advantages of the *op1* method and have not been used as extensively.

The distinction between *mit*⁻ mutants and ρ^- mutants (in the event that the *op1* selection method is not used), or nuclear *pet*⁻ mutants, relies on a simple genetic test based on the crosses between the respiratory-deficient mutant to be tested and a series of ρ^- mutants carrying different parts of the mitochondrial genome. This test (Table 2) combines all of the advantages. First, it unambiguously discriminates between a *mit*⁻ (or a *syn*⁻) mutant and any other type of mutation as the only one able to yield opposite results with different isonuclear ρ^- tester strains. Second, it allows the simultaneous screening of all types of *mit⁻* or *syn⁻* mutants (point mutation, small deletions, etc.). Third, assuming one uses a collection of ρ^- mutants as testers (in practice, a limited number of well-defined ρ^- mutants is sufficient), it indicates at the same time the map location of new *mit⁻* or *syn⁻* mutants on the mitochondrial genome (Fig. 4; also see Mitochondrial Genetic Map).

Complementation Tests, Complementation Map, Mitochondrial Genes, and Loci

The efficient complementation to restore the respiratory activity of zygotes made between nuclear *pet⁻* mutants and mitochondrial ρ^- mutants was demonstrated many years ago (Sels and Jakob 1967). If one now considers the functional complementation among mitochondrial mutations, several difficulties can be foreseen: the absence of stable heteroplasmic cell lines, the

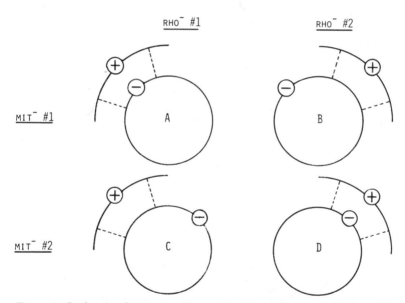

Figure 4 Rationale of the restoration test in ρ^- by *mit⁻* crosses. The test is based on the recombination between the mitochondrial genome of the ρ^+ *mit⁻* parent (complete circles) and the fragment conserved in the ρ^- parent (arcs). (−) A *mit⁻* allele at a given locus; (+) its corresponding wild-type allele. The fragment conserved in *rho⁻ #1* is allelic to the region where *mit⁻ #1* is localized but not to the region of *mit⁻ #2*. The opposite is true for *rho⁻ #2*. Therefore, in crosses *A* and *D*, the *mit⁺* allele carried by the ρ^- parent can be integrated by recombination in place of the *mit⁻* allele of the ρ^+ parent, leading to the formation of ρ^+ *mit⁺* recombinants, able to grow on a nonfermentable carbon source. In crosses *B* and *C*, no ρ^+ *mit⁺* recombinants can be formed. (Adapted from Slonimski and Tzagoloff 1976).

extensive genetic recombination between mtDNA molecules prone to rapidly create wild-type recombinant molecules within heteroplasmic cells, and, finally, some intracellular compartmentalization if the two mitochondrial populations of the parents are not rapidly intermingled with each other. However, the time competition between these different phenomena is such that a functional complementation test can nevertheless be devised if one follows the time course of the appearance of respiratory activity of synchronously formed zygotes in crosses between various combinations of *mit⁻* mutants. Two characteristic situations can be observed: (1) For some combinations of *mit⁻* mutants, full restoration of the respiration (similar oxygen consumption per cell per time unit as in the wild type) is usually obtained about 5–8 hours after the mixing of the parents. Considering that the zygotes appear, in most cases, 2–4 hours after the mixing of the parents, the respiration is restored in these combinations very rapidly after zygote formation. (2) For other combinations of *mit⁻* mutants, no significant amount of respiration is obtained before 15–20 hours, the time in which wild-type recombinants are formed.

Using this test, Foury and Tzagoloff (1978), Slonimski et al. (1978b), and Tzagoloff et al. (1978) have examined numerous pairwise combinations of *mit⁻* mutants in the *oxi1*, *oxi2*, *oxi3*, and *cob-box* genes. The complete absence of correlation between the final frequency of wild-type recombinant formed and the early respiration of zygotes demonstrates that this respiration results from functional complementation and not from recombination. Furthermore, at different time intervals, the contribution of the recombinants to the total respiratory activity of the zygotes can be estimated from the number of zygotes able to form a colony when plated onto a nonfermentable carbon source (to form such a colony, a zygote must contain recombined mtDNA molecules at the time of plating or be able to produce such molecules by genetic recombination from its heteroplasmic pool). All of these experiments demonstrated that mutations representing heteroalleles located at the same genetic locus from recombination tests (see below, Mitochondrial Genetic Map) are unable to complement one another. A locus, as defined by recombination, therefore constitutes a single cistron in the functional sense. In contrast, mutations located at different genetic loci in the recombination test may or may not complement one another, depending on the functions affected. Two *mit⁻* mutants, affecting different gene products (e.g., a subunit of cytochrome oxidase and cytochrome *b*), complement each other, but two *mit⁻* mutants, belonging to genetically unlinked loci affecting a unique gene product (e.g., different unlinked mutants of *oxi3*), do not complement each other. Therefore, some mutations, genetically unlinked by recombination, may nevertheless belong to the same cistron. The special case of the different *box* loci and several cistrons intermingled with one another is described below, in Mosaic Mitochondrial Genes and their Interactions.

The above results allow the construction of a simple complementation map (Table 3) with three different cistrons for the three subunits of cytochrome oxidase and one cistron for the exons of the cytochrome *b* gene. In addition, there are two genes (*oli1* and *oli2*) in the mitochondrial genome, specifying two subunits of the ATPase complex (see below, Simple Mitochondrial Genes: The *oli1* and *oli2* Genes for Two ATPase Subunits). A few *mit*⁻ mutants (*pho1* and *pho2*) are known in these genes, but they cannot be analyzed by complementation test because they are themselves able to respire.

A particular type of complementation test, called "zygotic gene rescue," was originally developed for the analysis of the *var1* determinant (Strausberg and Butow 1977) but has a much wider potential applicability. In a zygotic gene rescue experiment, one of the two parents is a ρ^- mutant, and consequently, is unable to express its own genetic message into proteins (transcription is still active). One asks for the mitochondria of the ρ^+ parent to translate the messages of the ρ^- parental genome into proteins prior to the recombination of the mtDNA sequences retained by the ρ^- parent into the ρ^+ genome of the other parent. The fact that the mtDNA sequences carried by the ρ^- parent are amplified probably helps them to successfully compete with those of ρ^+ for transcription and translation. Theoretically, zygotic gene rescue should be the method of choice in studying the expression of specific mtDNA segments, precisely delimited by ρ^- deletions in a completely homologous genetic system. But it is not always easy to precisely control the actual occurrence of the expression of ρ^- messages prior to the genetic recombination of ρ^+ and ρ^- mtDNA molecules.

Table 3 Complementation matrix between mutants in different mitochondrial genes

	oxi1	*oxi2*	*oxi3*	*box*[a]
oxi1	–	+	+	+
oxi2		–	+	+
oxi3			–	+
box				+ or – [a]

Complementation between various *mit*⁻ mutants mapping in three genetic loci, *oxi1*, *oxi2*, and *oxi3*, and in the different *box* loci has been tested by the ability of the heteroplasmic zygotes to respire shortly after mating. + indicates respiration; – indicates absence of respiration. (Data compiled from Foury and Tzagoloff 1978; Slonimski et al. 1978b; Tzagoloff et al. 1978.)

[a]The complementation matrix between *box*⁻ mutants is detailed in Table 7.

In all cases of complementation between *mit⁻* mutants, fusion or reshuffling of mitochondria organelles must occur in the zygote in order to account for the reconstitution of inner-membrane-bound active complexes of the respiratory chain. There are basically two classes of mechanisms by which complementation can take place: (1) either the exchange or reassembly of previously synthesized polypeptides to reform a functional respiratory chain or (2) a de novo protein synthesis. A priori, the reassembly of preexisting elements could account for the complementation between mutants affecting different enzymatic complexes of the respiratory chain (especially because cytochrome *c* is soluble) or even between mutants affecting different subunits of the same complex; but it is not at all certain whether such a mechanism can take place. However, there is evidence arguing that de novo synthesis must take place, i.e., at least a new translation of preexisting mRNAs and possibly also a new transcription. For example, erythromycin, which inhibits mitochondrial protein synthesis, seems to inhibit the complementation (Foury and Tzagoloff 1978). Furthermore, there are a number of cases, such as the zygotic gene rescue (Strausberg and Butow 1977), the complementation between a *syn⁻* mutant and a *mit⁻* mutant (Foury and Tzagoloff 1978), and the complementation between intron mutants and exon mutants of the *cox-box* gene, in which a synthesis of new polypeptides is required since no material preexists in the parental cells.

Formation of ρ⁻ Mutants

Induction of ρ⁻ Mutants

Although ρ⁻ mutants were discovered more than 30 years ago and form the basis of mitochondrial genetics, the mechanism of their formation is not well understood. It is well known that the mutation from ρ⁺ to ρ⁻ occurs spontaneously at a high rate (often ~1% or 2% of mutants per cell per generation). It can be induced with extremely high efficiency by acriflavine (Ephrussi et al. 1949a,b) or, more precisely, by its euflavine component (Marcovich 1951) or ethidium bromide (Slonimski et al. 1968) and, with varying efficiencies, by an impressive list of agents or special treatments, including classic mutagens, e.g., nitrosoguanidine (Nordstrom 1967; Mayer and Legator 1970; Dawes and Carter 1974); UV-irradiation (Wilkie 1963; Allen and McQuillan 1969; Mayer and Legator 1970; Moustacchi 1971; Deutsch et al. 1974; Dujon et al. 1975); starvation or unsaturated fatty acid depletion (Nagai 1969; Marzuki et al. 1974; Barclay and Little 1978); intra-mitochondrial ATP depletion (Subik et al. 1978); high or low temperatures (Weislogel and Butow 1970; Butow et al. 1973); heavy metals (Lindergreen et al. 1958); protein synthesis inhibitors (Carnevali et al. 1971; Williamson et al. 1971a); and, in brief, almost any condition under which yeast grows poorly.

Kinetics studies of the process of induction from ρ⁺ to ρ⁻ by ethidium bromide or UV, interpreted by the classic target theory, leads to a paradox since the inactivation curves extrapolate to a number of targets that is small (from 2–20 and most often ~2) and of no significance with respect to the number of mtDNA molecules (Allen and McQuillan 1969; Deutsch et al. 1974). Although a number of hypotheses have been proposed to explain this phenomenon (see references in Deutsch et al. 1974), the paradox is only apparent because the target theory is not meant to determine the absolute number of genetic units. In these experiments, one particularly observes mutant cells after several cell divisions and not the original mutational event at the level of mtDNA. Thus, in addition to mechanisms that specifically induce ρ^- mutations, those phenomena that play a role in the origin of mitochondrial point mutants and in the segregation of mitochondrial genetic markers also operate in the formation of a ρ^- mutant. From elegant pedigree experiments, James et al. (1975) showed that the spontaneous mutability of a ρ^+ cell to ρ^- decreases as it produces consecutive buds, in a manner indicating that the increasing cell volume is accompanied by an increasing number of mtDNA molecules and, hence, a change in targets.

Ethidium bromide is most commonly used to induce ρ^- mutants. Several authors have examined in detail the molecular mechanisms of this induction (Goldring et al. 1970, 1971; Nagley and Linnane 1970, 1972b; Perlman and Mahler 1971b; Mahler and Perlman 1972a; Mahler 1973b; Nagley et al. 1973; Bastos and Mahler 1974; Mahler and Bastos 1974a; Mahler et al. 1975; Perlman 1975; Criddle et al. 1976a; Hall et al. 1977). Mahler and collaborators distinguished several steps. First, as a result of ethidium bromide binding to mtDNA (possibly a covalent binding [Mahler and Bastos 1974b]), a degradation of the mtDNA occurs, probably corresponding to the fragmentation of preexisting ρ^+ mtDNA molecules to create ρ^- mutants. At this step the cells are in a reversible condition. During subsequent energy-dependent steps, a further fragmentation occurs, with an eventual release of ethidium bromide and free nucleotides. This process, if permitted to go to completion, results in the formation of ρ^0 mutants. If ethidium bromide is eliminated before an energy source is provided, ρ^- mutants can be formed. Many different treatments can modulate the induction of ρ^- mutants by ethidium bromide (Hollenberg and Borst 1971; Perlman and Mahler 1971a; Bech-Hansen and Rank 1972; Mahler and Perlman 1972b; Vidova and Kovac 1972; Whittaker and Wright 1972; Mahler 1973a; Pinto et al. 1975; Hall et al. 1976b; Juliani et al. 1976; Wolf and Kaudewitz 1976; Nagley and Mattick 1977; Fukunaga and Yielding 1978; Hixon et al. 1979). Among these treatments, the action of a number of inhibitors suggests some role of the mitochondrial ATPase in the damage-repair system of mtDNA. With respect to the mechanism of ρ^- induction by ethidium bromide, two lines of research may prove to be promising. First, Paoletti et al. (1972) have found a DNase activity from yeast mitochondria that is stimulated in the presence of ethidium bromide. Second, a number of mutations have been selected that either specifically increase or decrease the mutability of the cell after ethidium bromide treatment (Dujon et al. 1977a; Dujardin and Dujon 1979). Their mode of action is still unknown.

The mode of action of other efficient inducers of the ρ^- mutation, such as berenil and various acridines, has also been investigated (Mahler 1973a; Perlman and Mahler 1973; Nagley et al. 1975; Mattick and Nagley 1977; Nagley and Mattick 1977). Unlike ethidium bromide, berenil does not intercalate into DNA. All acridines tested (except one) only induce ρ^- mutants under growing conditions, whereas berenil also induces ρ^- mutants in nongrowing cells, provided an energy source is available. Finally, different mutations, both nuclear and mitochondrial, point out functions needed for the maintenance of the ρ^+ state. For example, some *pet*- mutants, as well as a number of nuclear temperature-dependent mutations (*cdc, tsp, tpi*), yield the rapid accumulation of ρ^- mutants at the restrictive temperature (Rubin and Blamire 1977, 1979; Schweizer et al. 1977; Backhaus et al. 1978; Johnston 1979; Newlon et al. 1979; Marmiroli et al. 1980b). One mitochondrial mutant (*tsm8*) also accumulates ρ^- mutants when grown at high temperature (Handwerker et al. 1973; Bandlow and Schweyen 1975; Bandlow 1976, 1977). *tsm8* maps within the main tRNA cluster in the vicinity of the gene for tRNA^{Ile} (Schweyen and Kaudewitz 1976; Monnerot et al. 1977). The mechanism by which it produces ρ^- mutants is unclear.

Primary and Secondary Clones

We have previously examined the sorting out of a single mutated mtDNA molecule from the intracellular population of nonmutated ones. The very high frequency of mutation from ρ^+ to ρ^- further complicates this situation. The induction of ρ^- mutations results in a heterogeneous population of differently mutated mtDNA molecules within the mutagenized cell. Subsequent recombination between these molecules may even increase the heterogeneity. Primary clones, issued from the mutagenized cell, will therefore be a highly heterogeneous population of cells resulting from the segregation of the heterogeneous population of mtDNA molecules. Heterogeneous clones are obviously of little use for genetic or molecular studies. Subcloning this cell population yields mostly pure secondary clones, because each of them is issued from a homoplasmic ρ^- cell. Secondary clones are therefore the first level at which a defined ρ^- mutant can be isolated. The purity of further subclones is then only dependent upon the intrinsic stability of such a mutant (see below, Genetic Properties of ρ^- Mutants).

Loss and Retention of Genetic Markers

Because ρ^- mutants do not express their mitochondrial genetic information, the determination of the markers retained in a ρ^- mutant requires the rescue of these markers by recombination into a ρ^+ genome after crosses with appropriate ρ^+ strains. Using this method, Deutsch et al. (1974) examined the kinetics of loss of the markers cap^R, ery^R, and oli^R as a function of the time of treatment by ethidium bromide and found that the cap^R and ery^R markers are very frequently retained or lost together, whereas the oli^R marker is less frequently associated with the other two. This provided the groundwork for a mapping method based on the frequencies of coretention of markers during mutation from ρ^+ to ρ^-. Molloy et al. (1975) and Schweyen et al. (1976) extended this methodology, using primary, secondary, or tertiary clones, to a larger number of mitochondrial markers. In such experiments, it is often found that markers in different regions of the map are not equally lost or retained. For example, the markers in the *oli1* and *cob-box* regions or in the *oxi1* region are frequently retained, whereas markers in the *oxi3* and *oli2* regions are less frequently retained (Fukuhara and Wesolowski 1977; Mathews et al. 1977; Heude et al. 1979). This may result from the proximity of each marker to a *rep* region (see below, Replication of mtDNA) and/or may reflect specific sequences preferred for the formation of ρ^- mutants or for their stability.

Organization of mtDNA in ρ^+

For the sake of clarity, wild-type mtDNA will be examined first and, later, the mtDNA of ρ^- mutants will be reviewed. This distinction is, however, largely artificial because, as we shall see, the peculiarities of ρ^+ mtDNA are integrally associated with the nature and properties of the ρ^- mutation. Furthermore, our present picture of ρ^+ mtDNA has actually been elaborated largely from the comparison with the mtDNA of ρ^- mutants.

It is now clear that the wild-type mtDNA of yeast has an overall complexity of 50×10^6 daltons (corresponding to about 75,000 bp or 25 μm) and it is admitted that, in vivo, the mtDNA molecules are circular. However, it is only after very mild conditions of extraction that a few circular molecules with a perimeter of 25 μm have been observed (Hollenberg et al. 1970; Christiansen and Christiansen 1976). Most methods of mtDNA preparation usually yield linear molecules, resulting from the random fragmentation of the native circles, with average sizes of about one third or, at best, one half of the complete circles (Goldring et al. 1970; Bernardi et al. 1972; Locker et al. 1974b; Guerineau et al. 1975; Morimoto et al. 1975; Sanders et al. 1977b).

With an average of only 18% of G + C, yeast mtDNA is one of the most biased of functional DNAs in composition. Furthermore, its extreme compositional heterogeneity, with GC and AT pairs being highly clustered, makes the mtDNA of yeast even more unusual. This heterogeneity, which was directly demonstrated by "nearest-neighbor" analysis (Grossman et al. 1971), results in a number of apparently "abnormal" physicochemical properties that have been the subject of many detailed investigations (Bernardi et al. 1968, 1970; Leth-Bak et al. 1969; Bernardi and Timasheff 1970; Christiansen et al. 1971, 1974, 1975; Blamire et al. 1972; Carnevali and Leoni 1972;

Ehrlich et al. 1972; Hollenberg et al. 1972a; Faye et al. 1973; Fauman and Rabinowitz 1974; Fukuhara et al. 1974; Locker and Rabinowitz 1974; Rabinowitz et al. 1974; Guerineau and Paoletti 1975; Borst and Flavell 1976; Christiansen and Christiansen 1976; Sanders and Borst 1977). In particular, the thermal denaturation profiles are unusual, with about half of the DNA melting at a very low temperature and the rest melting over a broad temperature range in a characteristic multimodal fashion that can be dissected to a high degree of resolution using various ρ^- mutants (Bernardi et al. 1970; Casey et al. 1974a; Michel 1974; Michel et al. 1974). From partial digestions of ρ^+ mtDNA with composition-sensitive nucleases (spleen acid DNase and micrococcal nuclease), Bernardi and collaborators proposed the existence of four structural elements (Bernardi et al. 1972; Piperno et al. 1972; Prunell and Bernardi 1974, 1977; Bernardi 1976; Prunell et al. 1977; Fonty et al. 1978). About 50% of all ρ^+ mtDNA sequences are made of A + T-rich segments containing less than 5% G + C and ranging in size from 150 bp to 1500 bp. (They were called "spacers," but this nomenclature is to be used with caution because the terms genes and spacers normally refer to function and not to structure, as it is meant here.) A number of short sequences (each ~50–80 bp long) with a very high G + C content (> 50% G + C) constitute another 2–3% of the total mtDNA. Roughly half of them contain clusters of sites for the two restriction endonucleases HaeIII (5'-GGCC-3') and HpaII (5'-CCGG-3'), and the other half do not. (They were called, respectively, "site clusters" and "GC clusters.") The rest of all ρ^+ mtDNA sequences are composed of segments (that were called genes) with a size range similar to the A + T-rich segments and with variable but moderate G + C content (an average of 26% G + C). Because the number of each type of structural element is about equal in the whole ρ^+ genome (~30–40), Prunell and Bernardi (1977) and Prunell et al. (1977) proposed a regular arrangement, -- "GC-cluster" - "site - cluster" - "gene" - "spacer" --, as the basic structure of the mitochondrial genome, with the G + C-rich clusters serving hypothetical regulatory functions.

A much more precise description of the nature and arrangement of these structural elements is now available from the recent determination of the DNA sequence of large portions of the mitochondrial genome, in which actual genes are clearly defined by the existence of mutations (see below, Genetic Functions of the Mitochondrial Genome). All sequences known to specify a polypeptide (i.e., the exon sequences and the intron-encoded reading frames) have a G + C content of 18–28%, with relatively homogeneous distributions of GC pairs. They correspond reasonably well in size and composition with the "genes" of Bernardi and collaborators (see above), except that they are not genes but, most often, pieces of genes. Long A + T-rich stretches, with very rare and usually clustered GC pairs, exist both between and within known genes (especially in the introns or parts of introns without reading frames). Finally, the short G + C-rich segments are not at the beginning of the genes but within the long intergenic or intragenic

A + T-rich stretches. They are usually not found within known genes, with the exception of the mini-insert in the 21S rRNA gene (Dujon 1980). Interestingly, some of the short G + C-rich segments share sequences homologies and/or are palindromic (Cosson and Tzagoloff 1979; Tzagoloff et al. 1979b; Dujon 1980). Their role is not clear but, at least in some cases, they are polymorphic variants (e.g., the mini-insert and the *var1* determinant).

The reason for the high A + T content of yeast mtDNA is not known. mtDNAs of several other species also have high A + T contents but usually not as extreme as yeast (see Borst 1972; Borst and Flavell 1976). A fashionable hypothesis is that yeast mitochondria may be devoid of a uracil-excision mechanism. Uracil, however, is continuously created in every DNA from the deamination of cytosine. If it is not excised from mtDNA, a constant mutagenic force creating GC-to-AT transitions should result. Possibly, the mtDNA tolerates the gradual accumulation of AT pairs up to a point that is still compatible with normal gene function, and the deficient mtDNA molecules (or cells) created by this mutagenic force are simply eliminated by selection instead of being repaired.

Organization of mtDNA in ρ^- Mutants

Numerous and detailed studies have focused on the mtDNA of ρ^- mutants. These include buoyant density determinations (Mounolou et al. 1966; Mehrotra and Mahler 1968; Carnevali et al. 1969; Bernardi et al. 1970; Carnevali and Leoni 1972; Hollenberg et al. 1972a; Faye et al. 1973; Lazowska et al. 1974); DNA/DNA and RNA/DNA hybridizations (Fukuhara et al. 1969, 1974; Cohen et al. 1972; Fauman et al. 1973; Faye et al. 1973; Gordon and Rabinowitz 1973; Casey et al. 1974a; Fauman and Rabinowitz 1974; Gordon et al. 1974; Lazowska et al. 1974); thermal denaturation and reassociation analyses and renaturation kinetics (Bernardi et al. 1970; Hollenberg et al. 1972a; Faye et al. 1973; Casey et al. 1974a; Locker et al. 1974b; Michel 1974; Michel et al. 1974, 1979); electron microscopy of native or partially denatured molecules or heteroduplex mapping (Faye et al. 1973; Locker et al. 1974a,b; Lazowska et al. 1976; Lazowska and Slonimski 1976; Locker and Rabinowitz 1976; Wakabayashi 1978); and, finally, restriction mapping analysis (Morimoto et al. 1975; DiFranco et al. 1976; Lewin et al. 1978; Shepherd et al. 1978; Faugeron-Fonty et al. 1979; Bos et al. 1980a) and DNA sequencing (see below, Genetic Functions of the Mitochondrial Genome, and references in Fig. 7). By the combination of these methods, the nature and organization of the mtDNA in ρ^- mutants are now well understood.

Deletion and Amplification

ρ^- mutants result from very large deletions of the ρ^+ genome, the fragment of mtDNA sequences conserved always being short (usually less than one third of the genome) and sometimes very short (the extent of the deletion is easily estimated from the loss of different genetic markers and the physical analysis of the conserved mtDNA). ρ^- mutants retaining only 1% of the wild-type genome are common, and ρ^- mutants with a conserved fragment as small as 66 bp have been reported (Mol et al. 1974; Van Kreijl and Bos 1977; Bos et al. 1978b). Very rare examples of ρ^- mutants retaining more than 50% of the wild-type genome have been reported, but it is not always certain whether the fragment retained is colinear with the wild-type genome over its entire length (in some cases, it is not even certain whether the complete sequence is actually carried by a unique type of mtDNA molecule within homoplasmic ρ^- cells).

Since the total amount of mtDNA in a ρ^- mutant is similar to the ρ^+ mutant from which it derives, irrespective of the size of the deletion, the number of copies of the conserved fragment per cell must be correspondingly amplified (Fukuhara 1969; Hollenberg et al. 1972a; Nagley and Linnane 1972b; Borst et al. 1976a). If only a very short sequence is conserved, the amplification factor may be considerable (e.g., 100 times or more).

Arrangements of Repeats

To achieve the amplification, the conserved sequence is regularly repeated along the mtDNA molecules of ρ^- mutants. The arrangements of the repeats have been elucidated from numerous studies by electron microscopy (Locker et al. 1974a,b; Lazowska et al. 1976), denaturation/reassociation (Michel 1974; Michel et al. 1974, 1979; Michaelis et al. 1976b), and restriction enzyme digests (Morimoto et al. 1975, 1976; DiFranco et al. 1976; Heyting and Sanders 1976; Jacq et al. 1977; Heyting et al. 1979a,b). The arrangements are of two major types (Fig. 5). In the simplest type, straight head to tail (ρ^-[A]), the repeat unit (efghi) is directly identical with the conserved sequence of ρ^+ mtDNA. In restriction endonuclease digests, only one new junction fragment (ie), which is not present in the wild type, is found; and all fragments, including the junction, are in stoichiometric amounts. This type of arrangement has been found for ρ^- mutants with conserved sequences of various sizes issued from various regions of the map. ρ^- mutants with relatively short conserved sequences (i.e., < 1000 bp) are almost always arranged in this way. In the other type of arrangement, inverted or palindromic (ρ^-[B]), the repeat unit (edcc′d′e′f′) is an inverted duplication of the conserved sequence (cdef), creating two different junction fragments in substoichiometric amounts of the other fragments. In many cases, ρ^- mutants of this type are not perfect repeats, and a short sequence (f′) may be incompletely repeated at the junctions. After denaturation, the mtDNA of such ρ^- mutants renatures with first-order kinetics and can form hairpin structures observable by electron microscopy. It is not clear whether palindromic ρ^- mutants can be formed with similar frequencies from any part of the genome or whether they are particularly frequent in some regions (e.g., the gene for the 21S rRNA). Finally, other ρ^- mutants contain mixed

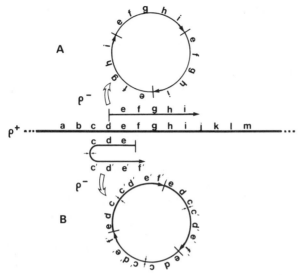

Figure 5 Schematic representation of the two major types of arrangement of mtDNA in ρ^- mutants. The middle line represents a segment of ρ^+ mtDNA with the unique sequence abcdefghijklm. The repeat unit of ρ^- (*A*) is composed of the conserved sequence efghi, which is regularly repeated with head-to-tail junctions (*top*). The repetition creates a single new sequence, ie, at the junctions of the repeat units. In ρ^- (*B*), the repeat unit edcc'd'e'f' is composed of an inverted duplication of the entire (or almost entire) conserved sequence cdef, creating a palindromic junction cc' (*small arrows*) and a second junction sequence e'f'e. The segment f', which is not duplicated, may vary in size. If negligible or absent, a second palindromic junction also exists. Recombination between the homologous segments (either intramolecular or intermolecular) results in the inversion of the f' segment, so that the two mirror-image junction sequences e'f'e and ef'e' can be found in contiguous repeats. mtDNA of ρ^- mutants of type *B* contain self-renaturing fraction able to form typical hairpin figures. In both types of arrangement, ρ^- mtDNA can form multimeric series of circles (here only trimers have been shown). Other types of arrangements, including mixed repeats, can also be found in some ρ^- mutants.

repeats in which the repetitions of the conserved sequence may be direct or inverted along the same molecule (Lazowska and Slonimski 1976; Heyting et al. 1979a). In this case, a local inverted duplication is normally found at the site where the repeat units can insert in two orientations (Heyting et al. 1979a).

Until now, we have only considered "simple" ρ^- mutants, in which the conserved sequence is colinear with the corresponding sequence of the ρ^+ genome. Other ρ^- mutants show internal rearrangement of the conserved sequence or even "scrambled" sequences, in which different segments of the wild-type mitochondrial genome that were originally separated have been made continuous in the ρ^- mutant. This may result from internal deletions within the conserved sequence or from the illegitimate recombination between different ρ^- mtDNA molecules after mutagenesis (see Genetic Properties of ρ^- Mutants). The rearrangements create many new junction sequences recognizable by restriction pattern analysis. These types of complex rearrangements have to be carefully distinguished from the complex population of mtDNA molecules issued from heteroplasmic cells. Rearranged ρ^- mutants show the same types of regular repeats as simple ρ^- mutants.

Circles

The size of the native mtDNA molecules of ρ^- mutants has been a controversial question (Carnevali et al. 1969; Goldring et al. 1971; Hollenberg et al. 1972b; Weth and Michaelis 1974). The bulk of mtDNA purified from ρ^- mutants generally consists of linear fragments, varying in length in the same range as the fragments purified from ρ^+ mutants. This indicates that an important fraction of the mtDNA in ρ^- mutants must be composed of relatively long molecules, possibly circular molecules similar in size to the ρ^+ genome. But, in addition, some ρ^- mutants contain a certain fraction of relatively small circular mtDNA molecules. The fraction of circles depends on the ρ^- mutant, and some mutants do not show an appreciable amount of circles   (Hollenberg et al. 1972a; Faye et al. 1973; Borst et al. 1976b; Lazowska and Slonimski 1976; Shepherd et al. 1978; Heyting et al. 1979a; Locker et al. 1979). All evidence indicates that the small circles and the linear fragments are, however, of the same nature and contain the same basic sequence.

From the contour lengths of the small circular molecules, Locker et al. (1974b) and Lazowska and Slonimski (1976) concluded that the circles are of the repeat unit size, as measured from their DNA complexity in renaturation kinetics or restriction mapping, or they are oligomers of this unit. The oligomeric series contains odd and even numbers of the conserved sequence if the repeats are straight but contains mainly even numbers if the repeats are inverted (Lazowska and Slonimski 1976). Thus, in the last case, the unit of circularization is made of two times the conserved sequence. Finally, in some ρ^- mutants, two (or more) multimeric series of circles coexist in the cell and, by a mechanism not very well understood, do not seem to segregate after subcloning (Lazowska and Slonimski 1976).

In each series, the monomers are more frequent than the dimers, the dimers are more frequent than the trimers, etc., in such a way that an empirical rule for the frequency of circles could be proposed in which the frequency of a given oligomer (an n-mer) is inversely proportional to the number (n) of repeat units it contains (Lazowska and Slonimski 1976). Although this rule fits the observed distributions for the lower oligomers well (up to about tetramer size), it cannot be extrapolated to higher oligomers because the random breakage of DNA seriously interferes with the frequencies of large observable circles and, fundamentally, because it needs an arbitrary limit for the total mass of mtDNA ($\Sigma \frac{1}{n}$) to be defined. Despite these limitations, the frequency distributions of circles point out that, at least in some ρ^- mutants, a dynamic equilibrium between the different oligomers must exist, probably mediated by recombination between mtDNA molecules and/or their replication. The monomers, dimers, trimers, etc., are only moments in time in a pool with a constant exchange of genetic material.

The Problem of New Sequences

Because the mtDNA of ρ^- mutants is not expressed into a functional product, any mutation should be absolutely neutral to the cell, and the mtDNA sequences should be entirely free to mutate as long as they can be replicated, quite a unique situation in nature. Conceivably then, new sequences could be created from extensive reshuffling of the sequences initially inherited from the ρ^+ parent. This idea was popular a few years ago because some filter-bound DNA-DNA hybridizations suggested the presence of sequences in some ρ^- mutants that were unable to exhaust the DNA sequences of ρ^+ (Gordon and Rabinowitz 1973; Fauman and Rabinowitz 1974; Gordon et al. 1974; Lazowska et al. 1974). However, further studies have shown that this phenomenon was due to the rapidly renaturing fraction of ρ^- mtDNA, so that some sequences hybridize to themselves much more rapidly than in heterologous reactions. More recently, precise restriction mapping and determination of the complete DNA sequence of numerous ρ^- mutants have also failed to show any evidence for new sequences (except, of course, the junction sequences). However, all observations indicate a remarkable fidelity of replication of the sequences conserved from the wild type, even in the absence of expression.

Mechanisms of ρ^- Mutant Formation

If the precise arrangement of repeats in the established ρ^- mutants are well described, the mechanisms that create these repeats are not yet understood, and several hypotheses have been proposed (Clark-Walker and Gabor Miklos 1974; Perlman and Birky 1974; Perlman 1975; Slonimski and Lazowska 1977; Baldacci et al. 1980). The formation of direct repeats is the easiest to imagine by analogy with the integration/excision mechanism of bacteriophage λ. The excision could create a circular monomer from which the multimers are easy to obtain either from replication (e.g., by a rolling circle mechanism) or from the legitimate recombination between two monomers to form a dimer. The determination of the nucleotide sequence at the junctions of the direct repeats from a sufficient number of ρ^- mutants should rapidly inform us about the nature of the excision sequences. At present, short direct repeats of 9 bp (Gaillard et al. 1980) or 6 bp (B. Dujon, unpubl.) have been found in the wild-type genome at the point of excision of several

ρ^- mutants arranged in straight head-to-tail repeats. Recombination between these short direct repeats results in the excision of the segment in between them, which constitutes the conserved sequence of the ρ^- mutant. It is not certain whether all ρ^- mutants with straight arrangement are formed by this mechanism.

The formation of a palindromic ρ^- is much more difficult to imagine. The analogy with bacteriophage λ cannot be carried further because λ normally integrates into the bacterial chromosome in only one orientation. Because the point of inverted duplication coincides or almost coincides with the ends of the conserved sequence, it is possible that such ρ^- mutants are created by an abnormal scission of the replication fork or of a recombination intermediate. The structural analogy with λdv plasmids may reflect a similarity in their mechanism of formation. Another attractive hypothesis is that ρ^- mutants with inverted repeats are not created unless a short inverted repeat preexisted in the ρ^+ mtDNA. Numerous short palindromic sequences are known in the genome (e.g., in the rRNA and tRNA genes or in the short G + C-rich segments). Furthermore, short inverted repeats have been found in the A + T-rich segments (Blanc and Dujon 1980). These two hypotheses are not mutually exclusive and, once again, the determination of the nucleotide sequence at the junctions of palindromic ρ^- mutants should be informative.

Genetic Properties of ρ^- Mutants

Intrinsic Stability of Genetically Defined ρ^- Clones

Indication of the genetic stability of ρ^- mutants was first reported by Ephrussi and Grandchamp (1965), who showed that upon subcloning a ρ^- mutant, the degree of suppressiveness is transmitted unchanged to the majority of the subclones, but not to all, with some subclones having a degree of suppressiveness different from the original mutant. The new suppressiveness is a mutation since it is, itself, a heritable character. Similar observations on the heritability of suppressiveness as well as on the retention of genetic markers in subclones of ρ^- mutants were later reported by Saunders et al. (1970), indicating that in the progeny of a ρ^- mutant, heritable alterations of the mitochondrial genotype may be found with relatively high frequencies.

Due to their instability, in every culture of ρ^- mutants a fraction of cells are ρ^0 or have mitochondrial genotypes different from the original ρ^- strain (eventually the instability leads to the transformation of the whole culture into ρ^0 cells). To determine this fraction is of great practical importance if one wants to correlate, in a meaningful way, the genotype and genetic properties of a ρ^- mutant with the sequences of mtDNA conserved or their arrangement. (This is the basis of many experiments described below, in Genetic Functions of Mitochondrial Genome.) However, in a given culture at a given stage, the fraction of mutant cells has no particular significance, being a function of both the intrinsic stability of the ρ^- strain and the history of the culture.

That the stability itself is a hereditary property of each ρ^- mutant can be demonstrated if one determines the proportion of ρ^- cells retaining the original mitochondrial genotype as a function of the number of cell generations in the culture. If the probability of mutation per cell per generation ([1-S] if S is stability) is constant and the mutation is irreversible, then the proportion of cells of the original genotype in the culture (P) should decrease as the number of generations (n) increases, according to the equation

$$P/P_0 = e^{-(1-S)n} \tag{1}$$

This exponential decrease has been directly verified by serial transfer experiments for several hundred cell generations (Faye et al. 1973; Michel et al. 1979). From such curves one can easily calculate the intrinsic stability of a ρ^- mutant as follows:

$$S = 1 + \frac{1}{n} \ln [^P/_{P_0}] \tag{2}$$

The stability as determined above is a characteristic hereditary property of a given ρ^- mutant and is transmitted, unchanged, to its subclones with the exception of rare stability mutants. The stability mutants, themselves, show exponential decreases of the proportion of the original genotype in the cultures, but the slopes differ from the parental ρ^- strain. Stability mutants may retain the same genetic markers as the original ρ^- parent, but their conserved sequence and/or the arrangements of the repeat units is altered (Michel et al. 1979).

Cultures of ρ^- strains with more than one genetic marker undergo even more complex evolutions since the kinetics of loss of each individual marker may not be the same. This may result from the unequal probabilities of losing different segments of the mtDNA sequence conserved in the original ρ^- strain or from the fact that the different mutants formed by the loss of some markers have, themselves, different intrinsic stabilities. It must be remembered that, since the ρ^- mutants do not express their mitochondrial genotypes, there is absolutely no selection at the cellular level, and each exponential slope directly reflects the stability of the mtDNA itself.

The intrinsic stability of a ρ^- clone is primarily determined by its mitochondrial genome, i.e., the conserved sequences and their arrangements. This stability varies widely among different isonuclear ρ^- mutants. The stability is also influenced by the nuclear genetic background of the strain. Although we do not know which genes are responsible for this, a certain correlation with the degree of spontaneous mutability of the strain from ρ^+ to ρ^- is possible. A priori, the instability could result from partially defective replication, illegitimate recombination, or internal deletion of the conserved sequence of mtDNA. So far, it has not been possible to correlate specific degrees of stability with specific sequences or arrangements of mtDNA in the ρ^- strain. Thus, the molecular basis of the instability remains obscure. This situation results mainly from the fact that only ρ^- mutants with the highest stability can be analyzed by classic genetics or molecular methods. After 20 cell divisions (a colony size), only 98% of cells remain of the original genotype if the ρ^- mutant has a stability of 0.999, and only 81% remain if its stability is 0.99. After 40 cell divisions, a minimum number of generations for most usual preparations of mtDNA, the figures are 96% and 67%, respectively.

Crosses with ρ^+ Strains: Suppressiveness and Recombination

Abundant information on the properties of crosses between ρ^+ and ρ^- strains has been published (Coen et al. 1970; Bolotin et al. 1971; Uchida and Suda 1973; Deutsch et al. 1974; Forster and Kleese 1975a; Bolotin-Fukuhara and Fukuhara 1976; Perlman 1976; Devenish et al. 1979; Michel et al. 1979; Strausberg and Butow 1981), and even more information is in unpublished form from the Gif Laboratory. In such crosses a certain fraction of the diploid progeny is composed of ρ^- cells, which usually are not analyzed genetically since this requires triploid crosses, and the rest are composed of ρ^+ cells, some of which have inherited the markers of the ρ^- parent. In the latter case, the markers of the ρ^- mutants are integrated by recombination into the ρ^+ genome in place of the corresponding markers of the ρ^+ parent.

In view of the organization of their mtDNA, ρ^- mutants should, a priori, yield different results in crosses from their parental ρ^+ strain. The direct quantitative comparison of ρ^+ subpopulations in the progenies of these two types of crosses (which has sometimes been used in the literature to classify ρ^- mutants) can only be invalid. A closer examination of the phenomena encountered is adequate at this point. Particularly, one must first examine how the rules for mitochondrial crosses, previously established from $\rho^+ \times \rho^+$ crosses, can be interpreted in the case of ρ^- mutants to discover whether specific mechanisms exist in these crosses and, if so, what is their nature. The predictions are as follows. (1) In $\rho^+ \times \rho^+$ crosses, a certain fraction of the zygotic clones is entirely composed of only one parental genotype. This fraction is, however, generally limited. Extrapolation of this observation to $\rho^+ \times \rho^-$ crosses (even considering the amplification factor) predicts that only a relatively small proportion of the zygotic clones should be entirely composed of ρ^- cells. Therefore, each ρ^- mutant should be of low or moderate zygotic suppressiveness. (2) Because the ρ^- mutants contain as much mtDNA as do the ρ^+ mutants and many more copies of the conserved sequence, the random segregation of mtDNA molecules should result in at least half of the final progeny of $\rho^+ \times \rho^-$ standard cross being composed of ρ^- cells. The final proportion of ρ^- cells cannot, however, be measured accurately, because there is no growth condition that does not select slightly against ρ^- cells; hence, the ρ^- cells are always underrepresented. (3) For the same reasons as in $\rho^+ \times \rho^+$ crosses and because the random pairing between the conserved sequence of ρ^- and the corresponding sequence of the ρ^+ mtDNA must obviously be limited by the availability of ρ^+ sequences, the frequency of ρ^+ recombinants carrying genetic markers of ρ^- should also be limited. Furthermore, it should increase proportionally to the amplification factor of the conserved sequence in ρ^-, i.e., in a first approximation, in the inverse proportion to the size of the conserved sequence. (4) From the rule of the coordinate output, when several genetic markers are carried by ρ^-, the frequency of integration of each marker in the ρ^+ genome should be identical, and the frequencies of all types of recombinants should be strictly predictable. This also holds true for the polar region, which has been one of the most intensively studied in crosses involving ρ^- mutants. A typical ρ^- mutant should behave as described above if there is no mechanism specific to the ρ^- mutation.

In fact, a number of ρ^- mutants exist that fulfill the above predictions. They do not need particular attention here, since all of their properties are predictable from previously known mechanisms. In particular, they respond as predicted to the variations of the input of the ρ^+ partner (Dujon 1976). However, a number of ρ^- mutants also exist that differ from this scheme and reveal new mechanisms that deserve mention here, such as hypersuppressiveness, "effacement," high frequency of integration and, finally, boundary effects.

Hypersuppressiveness and Effacement.   The degree of suppressiveness, i.e., the proportion of zygotic clones entirely composed of ρ^- cells, is characteristic of a given ρ^- mutant and varies widely from one ρ^- mutant to the next. Several hypotheses have been advanced to explain suppressiveness, assuming there is a preferential replication of ρ^- mtDNA molecules (Slonimski 1968; Carnevali et al. 1969; Rank 1970a,b; Rank and Bech-Hansen 1972b) or a particular recombination between ρ^+ and ρ^- mtDNA molecules leading to the genetic, or even the physical, destruction of the ρ^+ mtDNA (Coen et al. 1970; Michaelis et al. 1973; Deutsch et al. 1974; Perlman and Birky 1974; Blamire et al. 1976; Slonimski and Lazowska 1977). As has been mentioned, however, no special mechanism need be invoked for low or moderate suppressiveness, since it is expected from the random segregation of mtDNA molecules. The same is true for ρ^- mutants that are nonsuppressive at the level of zygotic clones but produce a sizable fraction of ρ^- cells in the final progeny of the cross. All such ρ^- mutants are true "neutral" in terms of the mechanisms. But two other classes of ρ^- mutants should be mentioned. The first consists of ρ^- mutants that produce virtually no ρ^- cells in the final progeny (ρ^0 mutants, which are obviously of this class [see Michaelis et al. 1971], are disregarded here). In this case, although ρ^- genomes are originally present in the zygotes, they are not transmitted to their progeny without recombination with ρ^+ mtDNA molecules. To explain this phenomenon, a mechanism of effacement must be postulated. The nature of this mechanism is not yet understood. In contrast, hypersuppressive ρ^- mutants, i.e., those that produce almost 100% zygotic clones entirely composed of ρ^- cells, have recently been elucidated. In this case, although ρ^+ genomes are originally present in the zygotes, they are not transmitted to their progeny. The genetic and molecular analysis of a collection of hypersuppressive ρ^- mutants has revealed that the hypersuppressiveness results from a preferential replication of ρ^- mtDNA molecules (Blanc and Dujon 1980; Dujon and Blanc 1980). All hypersuppressive ρ^- mutants share a 300-bp-long common sequence, which represents one of the few replication origins of mtDNA (see below, Replication of mtDNA). It is the increased density of *rep* sequences per unit length of mtDNA—by virtue of the deletion of 98–99% of the rest of the genome (all hypersuppressive ρ^- mutants have short conserved sequences)—that accounts for the preferential replication of these mtDNA molecules during the transient heteroplasmic phase following zygote formation. This can also be directly verified in crosses to other ρ^- mutants devoid of any *rep* sequence because ρ^- mutants formed in the progeny are indistinguishable from the hypersuppressive ρ^- parent. It is conceivable that the presence of a *rep* sequence within a larger conserved sequence (hence, a lower amplification factor) accounts for the high, but not extreme, suppressiveness of other ρ^- mutants.

High Integration Frequency.   Another phenomenon specific to ρ^- mutants, high integration frequency (HIF), has been revealed from studies of some ρ^- mutants of the polar region (Michel et al. 1979). Namely, the frequencies of ρ^+ recombinants in crosses with some ρ^- mutants largely exceed those predicted from a random pairing and recombination of mtDNA molecules (in some cases, almost the entire ρ^+ progeny is composed of recombinants). The mechanism responsible for HIF is not well understood. Since it is often the same ρ^- mutants that exhibit a HIF activity and efface themselves in crosses, the two phenomena may reflect a common mechanism. Possibly, these mtDNA molecules, devoid of any *rep* sequence, are not replicated if in competition with ρ^+ mtDNA molecules but contain some sites or particular structures responsible for an efficient recombination without long sequence homology (HIF sites). Recombination at these sites would result in the integration of the ρ^- conserved sequence into ρ^+ mtDNA and, hence, a tandem duplicated structure that should eventually be resolved to yield ρ^+ recombinant molecules. The nature of these sites remains to be determined, but the very efficient recombination found between ρ^- mutants without significant homology (below) supports this idea.

Boundary Effects. Boundary effects could be detected by a systematic study of crosses between precisely delimited ρ^- mutants and ρ^+ strains with ant^R

point mutations and various ω alleles (Michel et al. 1979). For markers located near the ends of the conserved sequence in the ρ^- mutant or near the point of divergence of sequences (e.g., ω-allele-specific introns), the frequency of marker rescue into the ρ^+ genome is reduced. A similar effect at the edge of the conserved sequence has been observed in crosses of ρ^- mutants to the various *box*⁻ mutants (Slonimski et al. 1978b). This phenomenon is readily explainable by the low probability at which a genetic exchange occurs between the marker considered and the end of the region of homology, but it points out that, in some ρ^- mutants (that do not show HIF), legitimate recombination is the major mechanism by which the integration of markers into the ρ^- genome takes place.

Crosses between ρ^- Mutants

Homoplasmic ρ^- mutants do not complement one another and do not recombine with one another to produce a wild-type phenotype. Some years ago, in the search for physical evidence of mtDNA recombination, Michaelis et al. (1973) examined the results of crosses between ρ^- mutants retaining different (but adjacent) segments of the mitochondrial genome (one parent is $cap^R\ ery^0$, and the other is $cap^0\ ery^R$. Diploid cells resulting from such crosses were analyzed genetically by marker rescue in triploid crosses with a ρ^+ strain. In the progeny of such crosses, diploid ρ^- clones carrying the two markers cap^R and ery^R, associated in a stable fashion, are very frequent. Each clone contains a single new species of mtDNA with buoyant density intermediate between that of the parents, indicating that a fusion has occurred between the two parental mtDNA molecules to form the recombinants. The investigators also noted that, in some crosses, $cap^R\ ery^R\ \rho^-$ recombinants were of different degrees of suppressiveness. Two such recombinants, one with low suppressiveness (like the two parents) and the other with high suppressiveness (but not hypersuppressive) were studied in more detail by DNA-DNA hydridization, high-resolution melting, and reassociation kinetics. It was found that recombination takes place with high efficiency between two ρ^- mtDNAs with very little sequence homology and that the resulting recombinant molecules contain the two parental sequences in a 1:1 ratio (Michaelis et al. 1976b). Electron microscopy (Lazowska and Slonimski 1977) showed that, like the two parents, the recombinant mtDNAs form oligomeric series of circles and display a very regular arrangement of inverted repeats, each repeat unit being formed from one repeat unit of the $cap^R\ ery^0$ parent (which is composed of the inverted duplication of the conserved sequence) and one repeat unit from the $cap^0\ ery^R$ parent (which is also composed of the inverted duplication of the conserved sequence). Recombination, therefore, must occur by the joining of one repeat unit from each parent followed by the regular amplification of the recombinant sequence formed. In the case of low suppressive recombinants, the wild-type sequence has been restored by the recombination, whereas, in the case of suppressive recombinants, a small gap seems to exist in the sequence, but its location and origin have not been precisely determined.

To examine the recombination between ρ^- mutants without any possible natural sequence homology, Michaelis et al. (1976a) performed crosses between ρ^- mutants retaining markers widely separated on the map, instead of contiguous as above. From all pairwise combinations between six different $cap^0\,par^R$ mutants and nine different $cap^R\,par^0$ mutants, they found that, in many cases, very high frequencies of $cap^R\,par^R$ recombinants were formed, whereas, in other cases, no such recombinants could be detected at all. The very high efficiency of recombination occurs even between ρ^- mutants that do not share natural sequence overlap. A similar recombination matrix, but performed between more precisely defined (physically) ρ^- mutants retaining segments of the polar region, also indicates the existence of very efficient recombination between some ρ^- mutants and the complete absence of recombination between others and shows that a large spectrum of different recombinants can be formed (Michel 1981).

Recombination between ρ^- mtDNA molecules demonstrates that the enzyme(s) required are nuclearly encoded, but the precise mechanism for recombination between segments of mtDNA without sequence homology is not completely understood. Dujon et al. (1977a) distinguished two different mechanisms of mtDNA recombination: a legitimate recombination involving heteroduplex formation on long stretches of homology and an "illegitimate" recombination without long stretches of homology. Slonimski and Lazowska (1977) proposed that the highly efficient recombination between ρ^- mutants results from the second mechanism. It is clear that all of the previous studies are not resolutive enough to determine the nature of the sites used for the fusion of mtDNA segments. If a mechanism similar to the formation of ρ^- operates, then sequences as short as 6 bp might be sufficient. In principle, the ρ^- mutants, offering a wide diversity of precisely delimited sequences, a different arrangement of repeats, and a complete absence of expression of the sequences involved, should be choice material for the study of the various mechanisms of genetic recombination using the rapid and powerful modern methods to analyze mtDNA.

GENETIC FUNCTIONS OF THE MITOCHONDRIAL GENOME

Mitochondrial Genetic Map

Complete and detailed maps of the yeast mitochondrial genome from different strains are now available. They have been elaborated on from a combination of genetic and physical methods, some of which, specific to mitochondria, are detailed below.

Recombination Mapping

Although it is the most classic method of genetic mapping in other systems, recombination mapping has been of limited help in constructing the general map of the mitochondrial genome because, due to efficient recombination and multiple mating rounds, mitochondrial loci distant from one another by only about 1500 bp are genetically unlinked (B. Dujon, unpubl.). Recombination frequencies can, however, be used for fine-structure mapping of mitochondrial genes (i.e., on segments shorter than ~1000 bp), assuming that, for each cross, the other parameters such as the number of mating rounds and the input ratio are carefully controlled. Finally, recombination mapping of very closely linked markers is further complicated by the fact that the classic phenomena, such as the expansion or contraction of the map, positive or negative interference, recombination hotspots, etc., which may interfere with fine-structure mapping, have not yet been studied in detail in the case of mitochondria.

Frequency of Marker Separation in ρ^-

A method specific to yeast mitochondria, the quantitative analysis of the loss or retention of genetic markers during the formation of primary, secondary, or tertiary ρ^- clones, offers the possibility of constructing a general map of the mitochondrial genome (Molloy et al. 1975; Schweyen et al. 1976). According to this method, the genetic markers retained by each ρ^- clone are determined after their rescue by recombination into a ρ^+ genome, and the order and map distances of the genetic markers are obtained using the two following criteria (see Schweyen et al. 1976): (1) Considering the markers two by two, the observed frequency of separation between them (i.e., the frequency of ρ^- clones with genotypes A^+B^0 and A^0B^+), as compared with the frequency of separation expected from the retention of each individual marker, determines their genetic distance. Such distances are additive over large sectors of the mitochondrial map; (2) considering the genetic markers four by four, three possible orders exist on a circular map. The "double loss/double retention" criteria determine the correct order, because the probability of codeletion of two contiguous markers is greater than the probability of double deletion of two noncontiguous markers and retention of the other two.

Application of these two criteria, using appropriate genetic markers, has allowed the construction of a general circular mitochondrial map (Schweyen et al. 1978).

Petite Deletion Mapping

Like the previous method, petite deletion mapping is based on the rescue of the genetic markers carried by the ρ^- genome by recombination into a ρ^+

genome (Fig. 5). In this case, a series of point mutations are mapped from a matrix of crosses to a series of ρ^- mutants that can discriminate between the point mutations. From this matrix, a unique order of point mutations can be deduced, assuming that the sequence retained in the ρ^- mutants does not contain internal deletion(s) or rearrangements (Fig. 6). The matrix is self-sufficient in ordering all mutations in a number of intervals defined by the ends of the ρ^- deletions. The major interest in this method lies in the ability to correlate the genetic intervals with the physical intervals of the mtDNA, defined by the physical analysis of the ρ^- sequences (Fig. 6). Each physical interval can be located very precisely on the restriction map of the mitochondrial genome by a variety of classic methods (e.g., restriction endonuclease cleavage, hybridization).

Physical Maps

Restriction mapping and DNA sequencing are obviously not methods specific to yeast mitochondrial DNA and therefore need not be discussed here. Largely, this also holds true for the mapping of the tRNA or rRNA genes by RNA-DNA hybridization.

General Map of the Mitochondrial Genome

By combining all of the above methods, the detailed maps of the mitochondrial genomes of several yeast strains have been established (Fig. 7). The comparison of the physical maps of mtDNA from different yeast strains reveals the same overall gene organization but several major and minor variations in sequences. These variations represent an interesting type of polymorphism due to specific inserts at specific positions. We now know that all of the large inserts (from ~200 bp to >1000 bp in length) represent optional introns of the three mosaic genes. It is probably very significant that these large inserts have only been found within genetically active regions and not between the known genes. The minor variations, however, are insertions or deletions of short segments (<100 bp each) and are widespread on the genome. It is possible that most of them correspond to the short G + C-rich segments.

Two types of functions are coded for by the mitochondrial genome (Table 4). A first set of genes, identified by the *mit⁻* and/or the *ant*^R mutations, code for hydrophobic mitochondrial polypeptides (three subunits of cytochrome oxidase, two subunits of ATPase and apocytochrome *b*). These are translated on the mitochondrial ribosomes and assembled into functional complexes bound to the inner mitochondrial membrane. The other genes, identified by *syn⁻* and/or *ant*^R mutations, code for some elements of the mitochondrial protein synthetic machinery: the two rRNAs, a complete set of tRNAs, and the product of the *var1* determinant, a protein associated with the mitochondrial ribosome.

As will be shown in this section, almost all known genetic loci have now

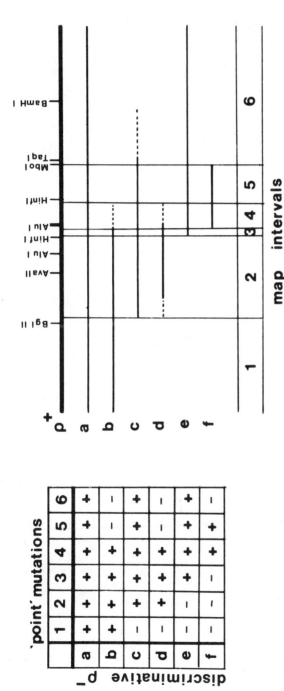

Figure 6 Rationale of petite deletion mapping. In the first part of the method, a restoration matrix is built up (*left*) by crossing, in all pairwise combinations, the discriminative ρ⁻ mutants (*a, b, c, d, e, f*) and the point mutations to be mapped (*1, 2, 3, 4, 5, 6*). A unique ordering of the point mutations can be obtained, assuming the sequences conserved in the discriminative ρ⁻ mutants are not rearranged. In the second part of the method (*right*), the restriction maps of the mtDNAs from the discriminative ρ⁻ mutants are aligned with respect to the ρ⁺ map and/or with respect to one another. (——) Conserved sequence; (------) the uncertainties of its limits. The interval in which each point mutation is localized is then delimited from the overlaps of the conserved sequences in the discriminative ρ⁻ mutant. The example shown here is a fragment of the *cob-box* gene (adapted from Lazowska et al. 1980). The same rationale applies for the mapping of any locus on the mitochondrial genome using an appropriate collection of discriminative ρ⁻ mutants.

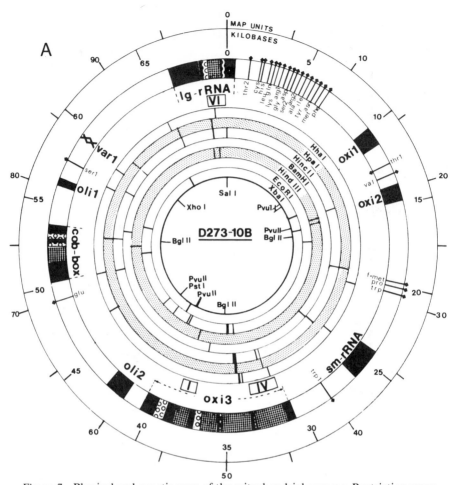

Figure 7 Physical and genetic map of the mitochondrial genome. Restriction maps (inner circles) of the short strain (D273-10B; 70 kb) (*A*) and the long strain (KL14-4A; 75.8 kb) (*B*) have been redrawn to scale from the original maps published by Borst et al. (1977), Sanders et al. (1977a,b), and Morimoto and Rabinowitz (1979a,b). For each strain, the genes have been placed on the map (outer circle) on scale with respect to the restriction sites (except for the main tRNA cluster, which is extended) using, for their alignment, the most precise physical maps available and/or the complete nucleotide sequences. References are Heyting et al. (1979), Tabak et al. (1979), and Dujon (1980) for the rRNA genes; Hensgens et al. (1979), Macino et al. (1979), and Macino and Tzagoloff (1980) for the *oli* genes; Grivell and Moorman (1977), Coruzzi and Tzagoloff (1979), Fox (1979a), Bonitz et al. (1980a,b), and Thalenfeld and Tzagoloff (1980) for the *oxi* genes; Lazowska et al. (1980) and Nobrega and Tzagoloff (1980a,b) for the *cob-box* gene; Vincent et al. (1980) for the *var1* determinant and Bos et al. (1979), Li and Tzagoloff (1979), Martin et al. (1979), Miller et al. (1979), Wesolowski and Fukuhara (1979), Berlani et al. (1980), Bonitz et al. (1980c), Martin et al. (1980), and Nobrega and Tzagoloff (1980a) for the tRNA genes. Gene designations refer to Table 4, which should be consulted for all genetic loci. (■) Exon sequences; (▦) an open reading frame; and (▩) a sequence blocked in all reading frames. To eliminate the background of random sequences, only those

552

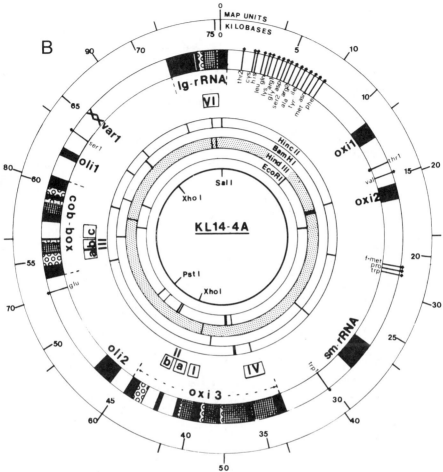

reading frames are considered as open that (1) either start with AUG or are linked, in register, with the preceding exon and (2) are significantly longer than the average distance between two stop codons in a random sequence. In cases where an open reading frame is found, all other frames of the same sequence are blocked. Blanks are either intron sequences not yet determined or intergenic sequences. (ⅹ) Localizes the *var1* determinant. The structural sequences for the var 1 polypeptide are not known. Most tRNA genes (referred to here by the corresponding amino acid; see Table 5) have been mapped in strain D273-10B, and it has been assumed that the same map of tRNA genes applies to strain KL14-4A in view of their similarities of 4S hybridization. Roman numerals in small boxes refer to the optional introns of the three mosaic genes, originally defined as large inserts by comparison between the two maps shown and with the map of the strain with the shortest mitochondrial genome known (NCYC74, 68 kb; Sanders et al. 1977a). Transcription and translation are clockwise for all mitochondrial genes (including the intron-encoded open reading frames), with the exception of the tRNA$_1^{Thr}$ gene. Other information on mitochondrial maps can be found in Groot-Obbink et al. (1976); Nagley et al. (1976); Sriprakash et al. (1976a,b); Bolotin-Fukuhara et al. (1977); Choo et al. (1977); Schweyen et al. (1977, 1978); Linnane and Nagley (1978a,b); Borst et al. (1979).

Table 4 Mitochondrial genes, loci, and mutants in *S. cerevisiae*

Gene[a]	Final gene product	Genetic loci	Class of mutants[b]	Phenotype of mutants	Current mutant designation[c]
oxi1	subunit II of cytochrome oxidase	*oxi1*	*mit⁻*	respiratory-deficient	*oxi1⁻*
oxi2	subunit III of cytochrome oxidase	*oxi2*	*mit⁻*	respiratory-deficient	*oxi2⁻*
oxi3	subunit I of cytochrome oxidase	*oxi3*	*mit⁻*	respiratory-deficient	*oxi3⁻*
		mim2	*mim*	suppress *box7* mutants locus-specific, allele-nonspecific	*mim2* or *MIM2*
cob-box	cytochrome *b*	*box1–box10*	*mit⁻*	respiratory-deficient	*box⁻*
		diu1, diu2	*ant^R*	diuron-resistant	*diu^R* or *D^R*
		muc1–muc3	*ant^R*	mucidin-resistant	*muc^R* or *M^R*
		ana1, ana2	*ant^R*	antimycin-resistant	*ana^R* or *A^R*
		fun1	*ant^R*	funiculosin-resistant	*fun^R* or *F^R*
oli1	subunit 9 of ATPase	*oli1, oli3*	*ant^R*	oligomycin-resistant	*oli^R* or *O^R*
		oli5?[d]	*ant^R*	oligomycin-resistant	
		ven1	*ant^R*	venturicidin-resistant	*ven^R* or *V^R*
		oss2	*ant^R*	ossamycin-resistant	*oss^R*
		pho2[e]	*mit⁻*	phosphorylation-deficient	*pho⁻*
oli2	subunit 6(?) of ATPase	*oli2, oli4*	*ant^R*	oligomycin-resistant	*oli^R* or *O^R*
		oss1	*ant^R*	ossamycin-resistant	*oss^R*
		pho1[e]	*mit⁻*	phosphorylation-deficient	*pho⁻*
var1	mitoribosomal polypeptide	*var1*			

"lg-rRNA"	21S rRNA	chloramphenicol-resistant	rib1–rib3	ant^R	cap^R or C^R
		erthromycin-resistant			ery^R or E^R
		spiramycin-resistant			spi^R or S^R
		suppresses various mit$^-$ mutants, allele-specific, locus-nonspecific	mim1	mim	mim1 or MIM1
		polarity of recombination between flanking markers (loci rib1, rib2, and rib3)	ω	ω	ω$^+$, ω$^-$, ω^n
"sm-rRNA"	15S rRNA	paromomycin-resistant	par1?^{d}	ant^R	par^R or P^R
"tRNA$^{(a.a)}$"	transfer RNAs	mitochondrial protein, synthesis-deficient or altered	syn$^-$	syn$^-$	
rep^f	replication origins		rep1–rep3		

The nomenclature of genetic loci and mutants follows the recommendations of the 1976 Interdisciplinary Conference held in Munich (Bücher et al. 1976).

^{a}Genes have been designated here (as in Fig. 7 and text) from the symbol of their most conspicuous genetic locus (or loci). lg-rRNA and sm-rRNA designate, respectively, the genes for the large and the small ribosomal RNAs. tRNA$^{(a.a)}$ symbolizes all genes for transfer RNAs (see Table 5).

^{b}The class of mutants refers to Table 1.

^{c}The current designation of mutants used in this paper is that used at present by the majority of authors. The original designations and map locations of most mutants described in the original literature have been compiled previously (Dujon et al. 1977b).

^{d}The genetic loci oli5 and par1 are linked to the corresponding genes oli1 and sm-rRNA but have not yet been conclusively demonstrated as elements of these genes.

^{e}Note that two nuclear genes in the phosphatase pathway are also currently designated PHO1 and PHO2.

^{f}rep stands for several similar sequences (each ~ 300 bp long) that act as replication origins and, when cloned into a ρ$^-$ mutant, confer extreme suppressiveness.

been correlated with specific gene products, which raises the issue of the saturation of the mitochondrial map. That all known *mit⁻* mutants fall into the limited number of genetic loci listed in Table 4 suggests that the map is saturated for the genes directly involved in the respiratory functions. In contrast, only a small number of *syn⁻* mutants have been isolated, and the saturation of the map for this class of mutants is much more questionable. Nevertheless, all genes involved in mitochondrial protein synthesis are probably also known from physical methods (except, possibly, a few tRNA genes). The map therefore seems saturated for the functions mentioned above. However, the recent discoveries of intron-encoded reading frames in the three mitochondrial mosaic genes suggest that entirely new types of functions may be coded for by the mitochondrial genome. The existence of mutants in some of these reading frames confirms this idea (see below, Mosaic Mitochondrial Genes and Their Interactions). If an appropriate selection is applied, new types of mitochondrial mutations, which would affect the mitochondrial functions or regulations in a subtle way, can probably be expected. Remarkably, the grand total of all sequences whose functions are known or suspected (including the introns of the mosaic genes, the *rep* sequences [see below, Replication of mtDNA], and a hypothetical structural gene for the var 1 polypeptide) amounts to only 45% of the total genome. The roles of the other sequences are not known.

Transcription and Transcripts

All evidence presently available clearly indicates that all RNA molecules present in mitochondria represent the transcripts of the mitochondrial genome (see, e.g., Reijnders et al. 1972; Mahler and Dawidowicz 1973). Despite the relative simplicity of the mitochondrial genome, relatively little is known about the mode of transcription, the nature, the number and localization of promoters, and the nature of the primary transcripts and their processing mechanisms (the special case of splicing intron sequences will be considered separately).

Maps of about 25 discrete RNA species, separated by their sedimentation coefficients and representing the most prominent transcripts of ρ^+ mitochondria (except the rRNAs and tRNAs), have been constructed by hybridization to the fragments of ρ^+ mtDNA from a "long" strain (KL14-4A) and a "short" strain (NCYC74) (Borst and Grivell 1978; Grivell et al. 1979; Van Ommen et al. 1979). By examining the transcripts from a series of different ρ^- mutants covering the various parts of the genome, Morimoto et al. (1979b) arrived at an essentially similar map. These maps reveal at least one transcript and often multiple overlapping transcripts for all known mitochondrial genes. Overlapping transcripts probably represent various steps of complex processing schemes, although the possibility of multiple alternative promoters and/or terminators has not been conclusively eliminated in every case. "Intergenic" regions do not usually show prominent transcripts. There are, however, no real gaps in the transcription maps, and high-molecular-weight transcripts can be found that hybridize weakly to DNA fragments between known genes. It is not always easy, in this system, to precisely define the mature mRNA among several overlapping transcripts. Fractions of RNAs with messenger activity for cytochrome oxidase or ATPase subunits (12S and 6S, respectively) have, nevertheless, been successfully characterized in cell-free translation systems (Padmadaban et al. 1975; Moorman et al. 1978; Grivell et al. 1979). In many cases, the mature transcripts are much longer than necessary to specify the final gene product, and long untranslated leader sequences have been described (Grivell et al. 1979).

By analogy to the mitochondria of HeLa cells, the possibility of a complete and symmetrical transcription in yeast mitochondria has been discussed. A few years ago, Hendler et al. (1976) and Rabinowitz et al. (1976) reported a certain degree of self-annealing of mitochondrial RNAs. However, the self-complementary components are present at very low concentrations (Lewin et al. 1977; Borst and Grivell 1978), and a complete symmetrical transcription is now regarded as a very unlikely possibility. A priori, a complete transcription of only one strand from a single promoter, with all RNA species being the products of processing of the unique precursor, would be possible because all mitochondrial genes are encoded in the same DNA strand (except one tRNA gene). This possibility now also seems unlikely because recent in vitro transcription experiments suggest the presence of at least two major promotors in ρ^+ mtDNA (Levens et al. 1981a). Furthermore, in vitro capping experiments suggest several sites for initiation of transcription (Levens et al. 1981b).

All ρ^- mutants seem able to transcribe their conserved sequences (Faye et al. 1974; Morimoto et al. 1979b), and the fact that transcripts with (apparently) normal characteristics can be found in many ρ^- mutants containing only small parts of the genome also argues that the initiation of transcription is either poorly specific or that the number of initiation sites in the wild-type genome is large enough so that most ρ^- mutants have a chance to retain at least one of them. Some ρ^- mutants also accumulate transcripts not normally found in ρ^+. In this respect, a striking observation, suggesting long-distance interactions for the processing of some mitochondrial transcripts, deserves mention here (Morimoto et al. 1979b). In ρ^- mutants retaining only the main cluster of tRNA genes, a general transcript of this region is observed but no mature tRNAs are formed. However, if the conserved sequence also extends over the *par1* locus, then mature tRNAs are formed. This suggests the existence of a region on the mitochondrial genome, around the *par1* locus, that plays a role in the processing of the tRNAs transcribed from distant genes. Since this phenomenon occurs in ρ^- mutants, no mitochondrially encoded polypeptide can be involved. It is possible that a particular transcript of the *par1* region (possibly the small rRNA gene itself) is required for the processing of the tRNAs.

The posttranscriptional modifications of the mitochondrial RNAs are not completely known. The existence of poly(A) tails has been disputed. Hendler et al. (1975) reported that a small fraction of the total mtRNA must contain poly(A) stretches since it binds to poly(U)-Sepharose. However, Groot et al. (1974) and Moorman et al. (1978) showed that the bulk of mtRNA with messenger activity (as well as some partially purified mRNAs) do not contain poly(A) tails. They concluded that the binding of mtRNAs to poly(U)-Sepharose is merely fortuitous, due probably to the presence of long internal A-A or A-U tracts. Certainly, the most remarkable posttranscriptional modification is the circularization of some mtRNAs. For example, the 11S and 18S fractions of yeast mtRNA contain more than 50% single-stranded circular molecules (Grivell et al. 1979; Arnberg et al. 1980). The resistance of such circles to a variety of denaturing agents and enzymatic digestions indicate that they are covalently closed. We now know that these circles correspond to splicing products of some intron sequences in the *oxi3* and *cob-box* mosaic genes (see below, Mosaic Mitochondrial Genes and Their Interactions).

A DNA-dependent RNA polymerase from yeast mitochondria was originally reported by Tsai et al. (1971), Wintersberger (1972), and Scragg (1976), but these preparations showed relatively low specific activities and differed in properties and composition. In a more recent study, a DNA-dependent RNA polymerase preparation with a high specific activity was described (Levens et al. 1981a,b). The activity cosediments with a 45,000-dalton polypeptide and is inhibited by antibodies against this polypeptide. It shows a number of properties that distinguish it from the nuclear RNA polymerases (insensitivity to α-amanitin or rifampicin, requirements for Mg^{++}, inhibition by Mn^{++}, and high ionic strength). It prefers poly d(A-T) or mtDNA as template, but the template preference is not as pronounced as with other RNA polymerases. Further studies are needed to determine whether, in vivo, the mitochondrial RNA polymerase is a multimer of the 45,000-dalton polypeptide or whether it contains several nonidentical subunits of the same size. In view of the apparent simplicity of the mtRNA polymerase (by comparison to the RNA polymerases of bacteria or eukaryotic nucleus), it is possible that little specificity exists at the level of transcription (possibly with rather imprecise start and stop signals) and that all regulatory controls take place at the level of posttranscriptional processing.

In Vivo and In Vitro Mitochondrial Protein Synthesis

The mitochondrial translation system differs from the cytoplasmic and the prokaryotic systems by a number of properties (see O'Brien and Matthews 1976). The mitochondrial ribosomes differ from their cytoplasmic counterparts by their structure (size and number of ribosomal proteins and RNAs) and by their properties (Reijnders et al. 1973; Groot 1974; Faye and Sor 1977; Faye et al. 1979b). They are of dual genetic origin. The two rRNAs are transcribed from mitochondrial genes (no 5S RNA equivalent has been found in mitochondria), whereas all of the ribosomal proteins (~ 70 in total) are synthesized on cytoplasmic ribosomes and transported across the mitochondrial membranes, with the exception of the var 1 polypeptide, a part of the small ribosomal subunit, which is synthesized on mitochondrial ribosomes (see below, Elusive Mitochondrial Gene: The *var1* Determinant). Mitochondrial ribosomes are sensitive to inhibitors of the large subunit of the prokaryotic ribosome (such as chloramphenicol, erythromycin, spiramycin, carbomycin, oleandomycin, lincomycin, and thiostrepton) but not to the inhibitors of the small subunit of the prokaryotic ribosome (such as kanamycin and streptomycin), with the exception of paromomycin and neomycin (Lamb et al. 1968; Linnane et al. 1968a; Davey et al. 1970; Mahler 1973b; Molloy et al. 1973; Schatz and Mason 1974). In contrast to cytoplasmic ribosomes, they are insensitive to cycloheximide. This last property, which allows for the distinction in vivo among the products of the mitochondrial protein synthesis, has largely contributed to our present understanding of the genetic functions of the mitochondrial genome. The use of inorganic $^{35}SO_4^-$ as label in the presence of cycloheximide (Douglas and Butow 1976; Douglas et al. 1979a) provided a rapid and particularly useful method for analyzing mitochondrially synthesized polypeptides from a large number of mutants. Labeled polypeptides can be simply identified after separation of the total mitochondrial polypeptides by gel electrophoresis and autoradiography (see also Claisse et al. 1977; Groot et al. 1978; Finkelstein and Butow 1979). Under such conditions, only a few bands with high radioactivity are found from a wild-type yeast: the three mitochondrially encoded polypeptides of cytochrome oxidase, at least one of the two of ATPase, the apocytochrome *b*, and the var 1 polypeptide.

Attempts to translate mitochondrial mRNAs using standard cell-free systems, such as *E. coli*, wheat germ, or reticulocytes, have yielded relatively few results. In all cases, only low-molecular-weight products could be detected (Moorman et al. 1977, 1978). Some of these products could, nevertheless, be identified as fragments of cytochrome oxidase or ATPase subunits by their immunoreactivity. The reason for this apparent untranslatability of mitochondrial mRNAs in heterologous systems is now understandable considering the presence of UGA codons in mitochondrial messages (see below, Transfer RNA Genes and the Genetic Code) and in vitro systems able to overcome this difficulty are now being studied (De Ronde et al. 1980).

Ribosomal RNA Genes

mtDNAs of all species studied contain one gene for each rRNA. But the mtDNA of yeast is unique in the sense that the two rRNA genes are far apart on the mitochondrial map and are separated by other genes (Fig. 7), making the transcript of common precursor highly improbable (Faye et al. 1975, 1976b; Sanders et al. 1975, 1976; Sriprakash et al. 1976a). Mature rRNAs differ in length and base composition from their cytoplasmic counterparts (Reijnders et al. 1973; Tabak et al. 1979; Sriprakash and Clark-Walker 1980).

Because it is the site of the ω locus and of the first mitochondrial ant^R mutations isolated, the large rRNA gene has been the object of numerous genetic and molecular studies during the last few years. Three genetic loci (*rib1, rib2,* and *rib3*), linked to one another, have been described previously (Bolotin et al. 1971; Netter et al. 1974). Mutations conferring resistance to chloramphenicol map at the *rib1* locus, whereas mutations conferring resistance to either erythromycin or spiramycin or both define the *rib2* and *rib3* loci (see Table 4 and Fig. 8). All cap^R, ery^R, or spi^R mutations are highly clustered into these three genetic loci (Coen et al. 1970; Trembath et al. 1973; Netter et al. 1974). Very rare cap^R or ery^R mutants, not exactly allelic to those at the *rib1, rib2,* or *rib3* loci, though very closely linked to them, have been described previously (Kleese et al. 1972b; Knight 1980). In vitro resistance of the mitochondrial ribosome originally indicated that the *rib* loci are part of the large rRNA gene (Grivell et al. 1973). Cross-resistance and other phenotypic interactions between the ant^R mutations at the three *rib* loci are often observed. They probably point out interactions between different regions of the 21S rRNA molecules after assembly in the mitochondrial ribosomes.

Mutations in the large rRNA gene that would affect the mitochondrial protein synthesis are much more difficult to obtain than ant^R mutants, and only a few have been characterized so far. Mutants with cold-sensitive defects in mitochondrial protein synthesis have been mapped in the *rib* loci (Kaldma 1975c; Mason et al. 1976, 1979; Devenish et al. 1978, 1979; Singh et al. 1978; Spithill et al. 1978). From a more extensive search for temperature-sensitive mutants specific to mitochondrial functions, Bolotin-Fukuhara (1979) found a few mutants that map within the large rRNA gene and affect the large rRNA at the restrictive temperature (Sor and Faye 1979). A new locus, *mim1* (or *MIM1*) has recently been described in the large rRNA gene in the vicinity of the *rib3* locus (Dujardin et al. 1980a). The *mim1* mutant was originally selected as a suppressor of a *box⁻* mutant, but its large spectrum of action on various *mit⁻* mutations suggests that it is able to restore a pseudo-wild-type phenotype to *mit⁻* mutants by decreasing the fidelity of translation.

Detailed molecular mapping of the large rRNA gene using numerous genetically defined ρ^- mutants has allowed the precise localization of the *rib* loci (Dujon and Michel 1976; Heyting and Sanders 1976; Jacq et al. 1977; Atchinson et al. 1979; Heyting and Menke 1979; Heyting et al. 1979a,b; Michel et al. 1979) and has revealed the existence of two different forms of the large rRNA gene in different yeast strains (Fig. 8). In ω^+ strains, the gene

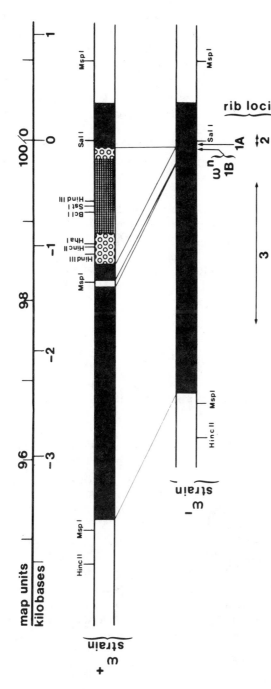

Figure 8 Physical and genetic map of the large rRNA gene from the ω^+ and ω^- strains. The map has been established from a combination of genetic and physical methods as described in the text. The location of the three *rib* genetic loci is indicated. The *rib1* locus is composed of two sites (1A and 1B), one of which is also the site for the ω^n mutation (Dujon et al. 1980). The location of *rib3* is imprecise. ω^+ and ω^- strains differ by the presence of the intron and the mini-insert containing an *MspI* site. (■) Exon sequences (the position of the 5' end of the gene is only approximative); (▦) intron-encoded open reading frame; and (▨) intron sequences blocked in all reading frames. Transcription is from left to right. (Adapted from Heyting et al. 1979 and Dujon 1980.)

561

is split by a 1.1-kb intron corresponding to insert VI (see Fig. 7), whereas in ω^- strains the same gene is not split (Bos et al. 1978a; Faye et al. 1979a). The correlation between the presence or absence of the intron and the ω^+ or ω^- allele has now been established on a large number of wild-type strains from different laboratories as well as in the progeny of crosses (B. Dujon, unpubl.). This correlation indicates that the intron itself plays an important role in the polarity of recombination (B. Dujon, unpubl.). In addition to the intron, other polymorphic variations occur in the large rRNA gene. A mini-insert with a palindromic $G + C$-rich structure is found, upstream from the intron, in a region of the molecule that is very variable among different species (Dujon 1980). The mini-insert is separable from the intron by recombination, and its structure is reminiscent of a transposable element. Other polymorphic variations also exist at the 3' end of the molecule (B. Dujon, unpubl.).

The determination of the DNA sequence of the large RNA gene first revealed the existence of long open reading frames within mitochondrial introns (Dujon 1980). The intron of the large rRNA gene contains genetic information for a polypeptide of 235 amino acids. The corresponding poly-peptide has not yet been found. If it is actually synthesized, this polypeptide cannot, however, play an essential role to the mitochondria because ω^- strains do not have this DNA sequence. Two dispensable functions can be easily imagined: the splicing of the intron sequence to form the mature 21S rRNA and the polarity of recombination. A function in splicing is improbable because ρ^- mutants that retain the entire large rRNA gene synthesize a mature-size rRNA (Faye et al. 1974) and accumulate an RNA species corresponding to the excised intron sequence, as do ρ^+ strains. A function in the polarity of recombination has not yet been conclusively eliminated but also seems unlikely. If this reading frame is not functional, its occurrence raises the question of the origin of the DNA sequence forming the introns of the mitochondrial mosaic genes, because the probability of the random occurrence of such a reading frame is extremely low. Furthermore, a long reading frame exists in only one of the six possible frames (the other five have densities of stop codons as predicted for a random sequence). The finding of long reading frames in most other mitochondrial introns (see below, Mosaic Mitochondrial Genes and Their Interactions) possibly underlines a fundamental mechanism for generating and maintaining these sequences.

The primary sequence of the large rRNA gene of yeast mitochondria, as compared with the large rRNA genes of other species, indicates regions that are highly evolutionarily conserved, whereas others are very variable (Dujon 1980). This points out essential regions versus less essential regions for the structure and/or function of rRNA molecules. The determination of the nucleotide sequences of appropriately constructed strains has identified the nucleotide monosubstitutions in the 3' region of the large rRNA gene that correspond to cap^R mutations (Dujon 1980). Interestingly, two sites (*rib1A* and $rib1B$), relatively close to each other in ω^- strains but separated by the intron in the ω^+ strains, contain cap^R mutations. The two sites interact with each other to determine the final resistance or sensitivity to chloramphenicol. Mutations at one site ($rib1B$) can suppress resistance due to mutations at the other site ($rib1A$). The fact that cap^R mutations are monosubstitutions within two stretches (each one 10 nucleotides long) entirely conserved in very different organisms (e.g., mammals, insects, bacteria, or plant chloroplasts) points out severe functional constraints on these parts of the rRNA molecule. Appropriate folding of this part of the 21S rRNA molecule into secondary structure (F. Michel, unpubl.) shows that $rib1A$ and $rib1B$ segments constitute a unique active site in the mitochondrial ribosome (probably the peptidyl transferase site). It is precisely within these conserved sequences, corresponding to a region of the rRNA molecule believed to be at the interface of the two ribosomal subunits, that the intron is inserted in ω^+ strains. It is conceivable then that the splicing takes place after the assembly of the large ribosomal subunit, which would provide the proper alignment of the sequences to be spliced. The $rib1B$ site, located very near the point of insertion of the intron in ω^+ strains, also contains ω^n mutations. This strengthens the idea that the sequences at the intron-exon junctions are important to initiate the polarity of recombination.

Fewer studies have been devoted to the small rRNA gene. Its precise physical localization has only recently been determined from S1 mapping experiments (Tabak et al. 1979). Since paromomycin is known to act on the small subunit of the bacterial ribosome, it is tempting to assume a relationship between the $par1$ locus and the small ribosomal subunit. Recently, Spithill et al. (1979) showed that paromomycin and neomycin at low concentrations increase two to three times the misreading of poly(U) by isolated mitochondrial ribosomes. Mitochondrial ribosomes prepared from par^R mutants are more resistant to the misreading effect. The $par1$ locus maps genetically in the region of the small rRNA gene (Wolf et al. 1973; Schweyen et al. 1978), and it has been suggested that par^R mutations are within the small rRNA gene (Linnane et al. 1976; Linnane and Nagley 1978a). However, ρ^- mutants can retain the $par1$ locus and lose all sequences that hybridize to the 15S rRNA (Faye et al. 1975, 1976b; Sanders et al. 1976). Furthermore, these ρ^- mutants hybridize to a specific segment of the mitochondrial genome that is more than 4000 nucleotides upstream from the small rRNA gene (Faye et al. 1976b; Sanders et al. 1976). The exact nature of the mutations at the $par1$ locus therefore remains unresolved. The $par1$ locus could be involved in the formation of a ribosomal protein of the small subunit, but Faye and Sor (1977) found no difference in electrophoretic mobility of mitoribosomal proteins on two-dimensional gels between two different $par1$ mutants and the wild type.

The transcription and the processing of the two rRNA genes are not fully understood, and the separation between the two genes raises a problem for their coordinate expression. From R-loop mapping and RNA blotting experiments, Merten et al. (1980) have observed a large transcript (> 5 kb) containing the sequence of the mature 21S rRNA (~ 3.1 kb), the sequence of the intron (1.1 kb), and additional sequences at the 3′ end of the molecule. Using the intron as a probe, a number of other transcripts of intermediate sizes have also been observed, but the exact processing scheme is not yet known (Bos et al. 1980b; R. Morimoto and B. Dujon, unpubl.). Interestingly, the excised intron sequences accumulate as a relatively stable RNA (possibly circular), more abundant than the precursors. This RNA molecule may be the mature mRNA for the intron-encoded polypeptide. No post-transcriptional processing occurs at the 5′ end of the 21S rRNA molecule (Levens et al. 1981c). However, the 15S rRNA is probably transcribed as a precursor extending approximately 80 nucleotides at the 5′ end of the molecule (Osinga et al. 1981).

Other posttranscriptional modifications of the rRNAs are extremely limited. Only two methylated residues have been found in the 21S rRNA and none in the 15S rRNA, and each rRNA contains only one pseudo-uridine residue (Klootwijk et al. 1975), indicating that such modified groups are not essential for the assembly of the mitochondrial ribosome.

Transfer RNA Genes and the Genetic Code

Mitochondrial tRNAs and Their Genes

Mitochondria contain a set of tRNA species that hybridize with the mtDNA and differ from the corresponding cell-sap tRNAs by various physico-chemical properties, such as their behavior in chromatographic systems, electrophoretic mobilities, acylating properties, base compositions, degree of methylation, rare nucleosides and, finally, by the presence of a $tRNA^{fMet}$ species (Accoceberry and Stahl 1971; Halbreich and Rabinowitz 1971; Schneller et al. 1975c,d; R. P. Martin et al. 1976, 1977). Earlier, it was suspected that the number of tRNA genes in the mitochondrial genome (Reinjders and Borst 1972; Schneller et al. 1975c), as well as the number of major tRNA species in the mitochondria (R. P. Martin et al. 1977), were fewer than the 32 minimally required by the "wobble hypothesis" for a completely autonomous translation system. Hence, mitochondria must either import specific tRNAs translated from nuclear genes or be able to decipher all codons of the genetic code with an unusually low number of tRNAs using rules different from the classical wobble hypothesis. To examine this question the nature and number of the mitochondrial tRNA species and of the mitochondrial tRNA genes have been determined in a series of recent studies (Table 5). All 20 natural amino acids can be charged by mitochondrially encoded tRNAs and, in many cases, two or more major and/or minor tRNA species are observed for each amino acid (the number of isoacceptor species and/or their relative importance sometimes differ between different determinations). In some cases, the number and relative abundance of different isoacceptors may also depend upon physiological conditions (e.g., $tRNA_2^{Ser}$, and $tRNA^{Ala}$; Baldacci et al. 1977, 1978; Colletti et al. 1979).

The map location of the tRNA genes has been determined from different methods. Hybridizations of iodinated 4S RNA to restriction fragments of ρ^+ or ρ^- mutants showed that about 18 tRNA genes span a segment of less than 10 kbp, which constitutes the main cluster of tRNA genes, whereas 5–8 other genes are dispersed in the genome (Van Ommen et al. 1977). Petite deletion mapping, using various ρ^- mutants previously characterized with respect to the major mitochondrial loci and hybridizations of individually labeled tRNAs to specific restriction fragments of ρ^+ or ρ^- mtDNA, have resulted in a precise map of all tRNA genes (Morimoto et al. 1978; Wesolowski and Fukuhara 1979). The determination of the nucleotide sequence from segments of the mtDNA characterized by hybridization with individual tRNA species has confirmed and extended this map (Bos et al. 1979; Li and Tzagoloff 1979; N. C. Martin et al. 1979; Miller et al. 1979; Berlani et al. 1980; Martin et al. 1980; Nobrega and Tzagoloff 1980a). Within the DNA sequences of these fragments, each tRNA gene is defined as a sequence able to generate a typical cloverleaf structure. The nature of each gene found is then determined from the anticodon sequence. The tRNA genes discovered by this method are in agreement with the previous map of tRNA genes. The general map of all tRNA genes presently known is shown in Figure 7.

The primary sequences of some mitochondrial tRNAs have been determined (Martin et al. 1978, 1980; Sibler et al. 1979; Canaday et al. 1980). Their colinearity with the gene sequences definitively determines their relationships. In other cases, however, the assignment of specific isoacceptors (when several are found) to the tRNA genes is not always straightforward. On the one hand, different isoacceptors may not always be resolved by RPC-5 chromatography. On the other hand, tRNA species, separable by chromatography, may not necessarily correspond to different sequences but may result from posttranscriptional modifications. Furthermore, some tRNA genes, characterized from their DNA sequence, may not be transcribed or may be transcribed only under some conditions and/or with low efficiency. Finally, the possibility of gene duplications recently suggested for the two $tRNA^{Trp}$ genes (Martin et al. 1980) may even further complicate the situation.

Overall, it is now clear that mitochondria code for a full set of tRNA species able to charge all 20 natural amino acids and able to interpret all codons with only a limited number of tRNAs (24 or 25). The deficit is not compensated by the import of specific tRNA species but rather by wobble rules slightly different from the "universal" code (see below). The import of one tRNA species from the cytoplasm ($tRNA^{Lys}$) has, nevertheless, been recently reported (R. P. Martin et al. 1979). But this tRNA does not extend

Table 5 Mitochondrial tRNAs and their genes

Charged amino acid	Isoacceptor species (no.)	Designation and anticodon	Map location of gene	Remarks	References
Ala	1 (+2 minor)	$tRNA^{Ala}$ (CGU)	main cluster	variable number of species	1,2,3,4
Arg	1 (+1 minor)	$tRNA_1^{Arg}$ (UCU) $tRNA_2^{Arg}$ (GCA)	main cluster main cluster		1,3,5
Asp	1	$tRNA^{Asp}$ (CUG)	main cluster	mutants 170 and M7-37 do not charge asp	1,3,6,7
Asn	1	$tRNA^{Asn}$ (UUG)	main cluster		3
Cys	2 or 1 (+1 minor)	$tRNA^{Cys}$ (ACG)	main cluster	mutant G76-26 does not charge cys on major species	1,3,8,9
Gln	1	$tRNA^{Gln}$ (GUU)	main cluster	previously called gluII	1,3,6,10,30
Glu	1	$tRNA^{Glu}$ (CUU)	oli2-box	previously called gluI	1,3,6,10,11
Gly	1 (+1 minor)	$tRNA^{Gly}$ (CCU)	main cluster		1,3,4,5
His	1 or 3-4	$tRNA^{His}$ (GUG)	main cluster	mutant G76-35 abolishes all species	1,3,6,8,9, 12
Ile	1 (+1 minor)	$tRNA^{Ile}$ (UAG)	main cluster		1,3,4,17
Leu	1	$tRNA^{Leu}$ (AAU)	main cluster		1,3,13,14, 15,16
Lys	1 or 2	$tRNA^{Lys}$ (UUU)	main cluster		1,3,6,17
Met	1	$tRNA^{Met}$ (UAC)	main cluster	previously called metI	1,3

		tRNA^fMet (UAC)	oxi2-par1	previously called metII	1,3,18,19,20
fMet	1				
Phe	1 or 2	tRNA^Phe (AAG)	main cluster		1,3,4,5,21, 22,23,24
Pro	1	tRNA^Pro (GGU)	oxi2-par1		1,3,6
Ser	1, 2, or 3	tRNA$_1^{Ser}$ (AGU)	oli1-var1	variable number of species	1,3,5,6,17, 22,25,26,31
		tRNA$_2^{Ser}$ (UCG)	main cluster		
Thr	2	tRNA$_1^{Thr}$ (GAU)	oxi1-oxi2	reads Leu codons	1,3,8,27, 28,29
		tRNA$_2^{Thr}$ (UGU)	main cluster		
Trp	1 (+1 or 2 minor)	tRNA$_1^{Trp}$ (ACU)	par1-oxi3		1,3,20
		tRNA^{Trp} (?)	oxi2-par1		
Tyr	1 (or up to 4)	tRNA^Tyr (AUG)	main cluster		1,3,6,13,14
Val	2 (+1 minor)	tRNA^Val (CAU)	oxi1-oxi2		1,3,5,13, 15,16,28

The number of major isoacceptor tRNA species as determined from different laboratories (by RPC-5 chromatography of mitochondrial tRNA charged with individual amino acids) is indicated along with estimates of some minor species. The mitochondrial genes for tRNAs have been characterized by a variety of methods as described in the text. The anticodons are given in the orientation 3'NNN5'. The map location of each gene is indicated with respect to the flanking genes (see Fig. 7). The main cluster of tRNA genes is between the large rRNA gene and the *oxi1* gene.

References: (1) Martin and Rabinowitz (1978); (2) Baldacci et al. (1978); (3) Wesolowski and Fukuhara (1979); (4) Cohen and Rabinowitz (1972); (5) N.C. Martin et al. (1979); (6) Casey et al. (1974b); (7) Faye et al. (1976a); (8) Berlani et al. (1980); (9) Bos et al. (1979); (10) N. Martin et al. (1976); (11) Nobrega and Tzagoloff (1980a); (12) Sibler et al. (1979); (13) Schneller et al. (1975b); (14) Schneller et al. (1975a); (15) Casey et al. (1972); (16) Cohen et al. (1972); (17) Tzagoloff et al. (1978); (18) Halbreich and Rabinowitz (1971); (19) Canaday et al. (1980); (20) Martin et al. (1980); (21) Schneller et al. (1975d); (22) Miller et al. (1979); (23) Martin et al. (1978); (24) Baldacci et al. (1975); (25) Coletti et al. (1979); (26) Baldacci et al. (1977); (27) Macino and Tzagoloff (1979b); (28) Li and Tzagoloff (1979); (29) Trembath et al. (1977); (30) N. Martin et al. (1977); (31) Baldacci et al. (1976).

the reading capacity of the mitochondria (a mitochondrial tRNA^{Lys} that is able to read the two lysine codons already exists), and it is not used for mitochondrial protein synthesis (it cannot be charged by mitochondrial aminoacyl-tRNA synthetase preparations). The specificity of this import (only one tRNA has been found) suggests an active transport across the mitochondrial membranes and might reflect important communication mechanisms between the cytoplasm and the mitochondria that we do not yet properly understand.

All tRNA genes known so far (with the unique exception of tRNA$_1^{Thr}$; Li and Tzagoloff 1979) are encoded in the same DNA strand, with transcription being clockwise on the map of Figure 7 (Bos et al. 1979; Miller et al. 1979; Bonitz et al. 1980c). In some cases, two successive tRNA genes of the main cluster were found extremely close to each other with no reasonable space between them for a possible promoter (e.g., tRNA$_2^{Ser}$ and tRNA$_2^{Arg}$; N. C. Martin et al. 1979). From this and from the observation of the transcripts in ρ^- mutants (Morimoto et al. 1979b), it seems probable that all tRNA genes of the main cluster are transcribed in a unique precursor, which is subsequently processed into individual tRNAs. The mode of transcription of the other tRNA genes dispersed in the genome is unknown. At least some ρ^- clones that retain tRNA genes can transcribe these genes and form normal tRNAs that can be acylated in vitro (Casey et al. 1974c). The terminal -CCA-3′ sequence is never encoded in the mtDNA but is a posttranscriptional addition.

Only a limited number of syn^- mutants, deficient for mitochondrial protein synthesis, are known at present. Some appear as good candidates for mutants in tRNA genes on the basis of their map location (Faye et al. 1976a; Trembath et al. 1977; Wesolowski and Fukuhara 1979; Berlani et al. 1980) or their deficiency for the charging of the corresponding amino acid (see Table 5).

In contrast to tRNAs, our knowledge of mitochondrial aminoacyl-tRNA synthetases is sketchy. It is clear that mitochondria contain at least some synthetases distinct from their cytoplasmic counterparts (e.g., mitochondrial synthetases for phenylalanine, leucine, and methionine have been purified; Baldacci et al. 1975; Schneller et al. 1976a,b; Diatewa and Stahl 1980). These enzymes, present in ρ^- mutants, are coded for by nuclear genes synthesized on cytoplasmic ribosomes and imported into mitochondria (Schneller et al. 1976b). It is an attractive hypothesis to believe that the same is true for all mitochondrial synthetases. Finally, an activity able to formylate tRNA^{fMet} after it has been charged with methionyl has been described previously (Mahler 1973b).

Codon Recognition Rules and Codon Usage

Bonitz et al. (1980c) have recently compiled the nucleotide sequences of 24 tRNA genes and, assuming that this represents all tRNA genes, have proposed the following codon recognition rules. In every case of an unmixed codon family (all four codons specifying the same amino acid independent of the wobble position), only one tRNA is found and it always contains a U at the wobble position of the anticodon. These investigators proposed that such a tRNA is able to recognize all four bases of the wobble position of the codon. The CGN family is an exception to this rule since the corresponding tRNA has an A at the wobble position of the anticodon, but this family is not found in genes and only rarely in intron-encoded reading frames (see Table 6). In every case of mixed family (except the AUN family) two tRNAs are found, one with a G in the wobble position that reads codons terminating with U or C, and the other with a U in the wobble position that reads codons terminating with A or G. It is not absolutely clear what prevents the latter tRNAs from reading the other two codons of the family if the previous rule for unmixed families is correct. There will be a better understanding of these problems when more sequencing information becomes available on the tRNAs themselves to determine the nature of the modified bases. In the case of the AUN family, the tRNA^{Ile} has a G at the wobble position of the anticodon. Bonitz et al. (1980c) proposed that this G is modified in order to recognize U, A, and C (note, however, that the codon usage for genes is strongly biased in favor of AUU instead of AUA although both are equally probable; Table 6). Both tRNA^{Met} and tRNA^{fMet} have a C at the wobble position so that only AUG can be read.

Departures of the mitochondrial genetic code from the universal genetic code were surprising, though they are, after all, limited. The major difference is certainly the fact that yeast mitochondria translate the UGA codon as tryptophan (Fox 1979a; Macino et al. 1979). The same is true for human mitochondria, and this may then be universal to mitochondria of all organisms. The UGA codon has been found in almost all mitochondrial genes sequenced so far as well as in all intron-encoded reading frames (Table 6). Another surprising difference was found in the *oli1* gene, where one CUA codon is translated into threonine instead of leucine in the subunit 9 of ATPase (Sebald and Wachter 1979). Several other CUA codons exist in other mitochondrial genes but, unfortunately, no information on the protein sequences is available. However, Li and Tzagoloff (1979) have found that the mtDNA fragment that hybridizes to one tRNA^{Thr} contains a single tRNA gene (tRNA$_1^{\text{Thr}}$) that differs from normal tRNAs by showing eight nucleotides in the anticodon loop instead of seven. They concluded that this abnormal tRNA is charged by threonine but possesses the anticodon 3'-GAU-5'. No other tRNA able to charge leucine and to translate the CUN codon family has been discovered. As a consequence of their rules, Bonitz et al. (1980c) proposed that all four codons of the CUN family are also read as threonine by the abnormal tRNA$_1^{\text{Thr}}$.

The complete nucleotide sequences of all known mitochondrial genes and of many intron-encoded reading frames provide enough statistical signifi-

Table 6 Codon usage for mitochondrial genes and reading frames

	Codon	Genes	IRF		Codon	Genes	IRF		Codon	Genes	IRF		Codon	Genes	IRF
Phe	UUU	41	36	Ser	UCU	18	20	Tyr	UAU	42	54	Cys	UGU	7	15
	UUC	36	5		UCC	1	4		UAC	7	5		UGC	1	1
Leu	UUA	131	92		UCA	41	18	Stop	UAA	3	2	Trp	UGA	21	12
	UUG	1	4		UCG	0	1		UAG	0	0		UGG	0	1
Leu	CUU	1	5	Pro	CCU	21	18	His	CAU	27	18	Arg	CGU	0	3
	CUC	1	0		CCC	1	1		CAC	2	2		CGC	0	1
	CUA	9	6		CCA	20	5	Gln	CAA	18	21		CGA	0	0
(Thr)	CUG	1	0		CCG	0	2		CAG	2	2		CGG	0	0
Ile	AUU	84	71	Thr	ACU	20	20	Asn	AAU	36	96	Ser	AGU	9	19
	AUC	15	6		ACC	1	4		AAC	5	8		AGC	0	1
	AUA	4	37		ACA	29	21	Lys	AAA	18	99	Arg	AGA	21	35
Met	AUG	38	12		ACG	0	4		AAG	1	8		AGG	0	2
Val	GUU	25	17	Ala	GCU	38	16	Asp	GAU	23	37	Gly	GGU	51	33
	GUC	2	1		GCC	3	3		GAC	1	3		GGC	0	2
	GUA	44	25		GCA	30	8	Glu	GAA	22	30		GGA	16	15
	GUG	3	2		GCG	1	2		GAG	2	5		GGG	3	4

The utilization of codons is determined from the DNA sequence of four simple genes (*oli1*, *oli2*, *oxi1*, and *oxi2*) and the exons of two mosaic genes (*oxi3* and *cob-box*) for the genes, and of seven intron-encoded reading frames (IRF) in the three mosaic genes for the reading frames. The figures have been normalized for 1000 codons. Total codons for genes: 1742. Total codons for reading frames: 3120. (Data compiled from Coruzzi and Tzagoloff 1979; Fox 1979a; Hensgens et al. 1979; Macino and Tzagoloff 1979a; Bonitz et al. 1980b; Dujon 1980; Lazowska et al. 1980; Macino and Tzagoloff 1980; Nobrega and Tzagoloff 1980c; Thalenfeld and Tzagoloff 1980.) Note that the exact limits between introns and exons are not always defined with precision. In cases where several options are possible, the compilation has been done using the limits presented in the original articles. These uncertainties do not significantly modify the frequencies of codon utilization. Departures of the yeast mitochondrial genetic code from the "universal" code are the use of UGA for Trp (Fox 1979a; Macino et al. 1979) and the use of the CUN family for Thr (Li and Tzagoloff 1979).

cance to examine the codon preferences (Table 6). As a general rule, yeast mitochondria preferentially select codons that terminate in A or U among synonymous codons, in agreement with its low average G + C content. Despite this bias, almost all codons have been found at least once. The most striking exception is probably the complete absence of codons of the CGN family from the genes and their rare occurrence in intron-encoded reading frames. This is possibly in correlation with the fact that the $tRNA_2^{Arg}$ is the only one with an A at the wobble position of the anticodon (Bonitz et al. 1980c). Similarly, the rare occurrence of the codons of the CUN family, both in genes and in reading frames, should probably be correlated with the abnormal $tRNA_1^{Thr}$. All mitochondrial genes are terminated by UAA and, in many cases, several UAA codons are found at the end of the coding sequence as if repeated stops were favorable.

Bias in codon usage is the rule for most organisms, and it is not surprising to also find some bias in mitochondria. Probably more interesting, however, is the fact that a number of significant differences can be found if one compares the codon usages for the genes with those for the intron-encoded reading frames (Table 6). Some of these differences only reflect the different amino acid compositions between the products of the genes that are highly hydrophobic and the products of the intron reading frames that are hydrophylic. For example, all intron reading frames show a significant excess of Asp and Lys codons, a fact that cannot be explained by the high A + T content alone, because other A + T-rich codons (e.g., Phe, Leu, Ile, or Tyr) are not in excess. But other differences must be relevant to the genetic code itself because they do not affect the amino acid composition. For example, a striking difference is found in the use of the two Phe codons. In the case of the intron reading frames, an excess of UUU versus UUC is observed, as expected from the average base composition but, in the case of genes, this ratio is close to 1. This may be related to pecularities of $tRNA^{Phe}$ (Martin et al. 1978) so that arrays of 5 or 6 U's in the message tend to create frameshifts during translation, a situation that the genes avoid by using, preferentially, the UUC codon. Another significant difference is found for the AUA codon. In human mitochondria, the AUA codon is believed to specify methionine instead of isoleucine (Barrell et al. 1979). Further studies of $tRNA^{Ile}$ will probably reveal whether the AUA codon can actually specify isoleucine in yeast.

Simple Mitochondrial Genes:
The *oli1* and *oli2* Genes for Two ATPase Subunits

*Two Genetically Unlinked Clusters of Mutational Sites Carry
a Variety of* ant^R *and* mit⁻ *Mutants*

Mitochondrially inherited oligomycin-resistant mutants were first reported more than 10 years ago (Avner and Griffiths 1970; Stuart 1970; Wakabayashi and Gunge 1970). They must be carefully distinguished from other oligomycin-resistant mutants of yeast (class I; Avner and Griffiths 1973a), which are cross-resistant to a variety of unrelated inhibitors (e.g., uncoupling agents, chloramphenicol, spiramycin, antimycin A) and exhibit a papillated phenotype on the selective medium and an apparent mitotic segregation that suggest a cytoplasmic determinant. Such mutants are, however, not mitochondrially inherited but are related to a general class of yeast mutants, resulting from nuclear mutations and increasing the overall resistance of the cells (Rank and Bech-Hansen 1973; Guerineau et al. 1974; Lancashire and Griffiths 1975a). The mitochondrial mutants (class II; Avner and Griffiths 1973a) are specifically resistant to oligomycin (or rutamycin) and obviously follow all rules of mitochondrial inheritance.

Allelism tests between mitochondrial oli^R mutants (Avner and Griffiths 1973b) and multifactorial crosses using other ant^R mutants as genetic markers (Avner et al. 1973) initially revealed the existence of two loci on the mitochondrial genome (*oli1* and *oli2*), genetically unlinked from each other (Table 4 and Fig. 7), with the majority of the mutants falling at the *oli1* locus. The subsequent isolation and genetic characterization of venturicidin-resistant mutants (Griffiths et al. 1975; Lancashire and Griffiths 1975b) allowed the definition of a third genetic locus, called *oli3*, closely linked to *oli1* (~1% of recombinants in most crosses). A fourth locus, *oli4*, closely linked to *oli2*, is represented by a single allele conferring a weak resistance to oligomycin in vivo (but not to ATPase in vitro) and isolated from a naturally resistant laboratory strain (Clavilier 1976). Finally, from a systematic analysis of a large number of mutants, induced by Mn^{++} and selected as resistant to oligomycin, venturicidin, or ossamycin (an oligomycinlike inhibitor with a different structure), Lancashire and Mattoon (1979) have recently completed this map with two new genetic loci for the mutants conferring resistance of ossamycin (*oss1* and *oss2*) and, possibly, a new oligomycin locus (*oli5*), comprising rare oli^R mutants distantly linked to *oli1*. The *oss1* and *oss2* loci are closely linked to *oli2* and *oli1*, respectively. It is probable that these loci saturate the map for this type of mutant. All mutants resistant to inhibitors of the ATPase function are thus clustered in only two segments of the mitochondrial genome, each containing three closely linked genetic loci (*oli1*, *oli3*, and *oss2*, respectively, for one segment and *oli2*, *oli4*, and *oss1* for the other). Unfortunately, the fine-structure genetic maps of these two segments, including the whole variety of mutants, have not yet been established.

The mutants at the *oss1* and *oss2* loci are specifically resistant to ossamycin, but a variety of cross-resistances between the three drugs exist at the other loci. For example, venturicidin-resistant mutants at the *oli3* locus are also resistant to oligomycin. No venturicidin-resistant mutant, however, has been found in the *oli2* region. Furthermore, many oli^R mutants are cold-sensitive (Avner and Griffiths 1973a; Lancashire 1976). One of them, mapping at the *oli3* locus, was found to affect the assembly of mitochondrial ATPase at low temperature (Trembath et al. 1975a, 1976). The extent of the cold sensitivity results from interactions with a series of nuclear genes (Trembath et al. 1975b).

The map is completed by a few *mit⁻* mutants. These mutants, deficient for mitochondrial oligomycin-sensitive ATPase activity, belong to two mitochondrial genetic loci: *pho1* (see nomenclature footnote to Table 4), closely linked to *oli2,* and *pho2,* closely linked to *oli1* (Foury and Tzagoloff 1976a; Corruzzi et al. 1978; Darlison and Lancashire 1980). The revertants of *pho2* mutants are frequently oligomycin-resistant, suggesting that *oli1* and *pho2* affect the same gene product. Because the ATPase activity of mitochondria or of submitochondrial particles from the oli^R mutants is more resistant to oligomycin than the wild type (although not completely insensitive, as in the case of ρ^- mutants in which the membrane factor is missing), the components of the membrane factor must be modified in the oli^R mutants, and the two genetic segments mentioned above are the obvious candidates as the genes to specify these polypeptides (Shannon et al. 1973; Griffiths and Houghton 1974; Somlo et al. 1974; Somlo and Cosson 1976; Somlo 1977; Murphy et al. 1978; Somlo and Krupa 1979). Several properties of the mutants at the *oli1, oli3,* and *pho2* loci suggest that they are parts of the structural gene for the subunit 9 (8 kD) of ATPase (Tzagoloff et al. 1975c, 1976a). For example, in *pho2* mutants and *oli1* mutants, the hexameric form of the subunit 9 is not found, indicating a structural alteration of this subunit. Similarly the *pho1⁻* mutants affect the 21.9-kD subunit of ATPase, which suggests that the second segment encodes this subunit (Roberts et al. 1979).

Each Cluster of Mutational Sites Is a Structural Gene for One Mitochondrially Synthesized Subunit of the ATPase Complex

The complete nucleotide sequences of the *oli1* and *oli2* regions have recently been determined, providing direct evidence that the *oli1* segment is the structural gene for subunit 9 of the ATPase complex and very strong suggestions that the *oli2* segment is the structural gene for the 21.9-kD subunit of the same complex. The *oli1* region contains a coding sequence (231 nucleotides long) (Hensgens et al. 1979; Macino and Tzagoloff 1979a) that is colinear with the amino acid sequence of subunit 9 of the ATPase complex, demonstrating that it is the structural gene for this polypeptide (Sebald and Wachter 1979). Furthermore, the few mutants that have been sequenced (some oli^R mutants at the *oli1* and *oli3* loci and a revertant of the *pho2* mutant) are nucleotide monosubstitutions leading to amino acid substitutions in a region of the subunit-9 polypeptide flanking the dicyclohexylcarbodiimide (DCCD) binding site (Glu at position 59; Sebald et al. 1979b).

The nucleotide sequence of the *oli2* region also reveals (Macino and Tzagoloff 1980) a unique coding sequence, able to specify a polypeptide that is 259 amino acids long, a size compatible with the 21.9-kD subunit of ATPase. So far, no information on the sequence or composition of the 21.9-kD subunit of yeast ATPase is available because this polypeptide has not yet been purified. However, the predicted amino acid composition of the polypeptide encoded by the *oli2* gene closely resembles the amino acid composition of the subunit of *Neurospora* ATPase equivalent to the 21.9-kD subunit of yeast ATPase, and Roberts et al. (1979) have found that a mit⁻ mutant of the *oli2* region affects this subunit. Furthermore, the two mutants that have been sequenced (at the *oli2* and *oli4* loci) are nucleotide monosubstitutions within this unique gene. Assuming that the genetic order - *oli2* - *pho1* - *oli4* - (Somlo et al. 1977) is correct, then the *pho1* mutants should also occur within the same gene.

Simple Mitochondrial Genes: The *oxi1* and *oxi2* Genes for Two Cytochrome-oxidase Subunits

The *oxi⁻* mutants, deficient for cytochrome-*c* oxidase activity and spectrum (cytochrome *a* plus cytochrome *a₃*), represent the most frequent class of *mit⁻* mutants. All *oxi⁻* mutants isolated so far (several hundred in total) fall into three distinct genetic loci (*oxi1, oxi2,* and *oxi3*), distant from one another on the mitochondrial genetic map (Fig. 7) and constituting three separate complementation groups (Table 3). Two, *oxi1* and *oxi2*, are simple genes and are discussed here; *oxi3* is mosaic and is discussed below.

From their study of mitochondrial polypeptides synthesized by *oxi1⁻* mutants, Cabral et al. (1978) concluded that the *oxi1* locus is the structural gene for the subunit II of cytochrome oxidase. This conclusion is based on the fact that most *oxi1⁻* mutants are deficient for the 33.5-kD polypeptide, identified as the subunit II of cytochrome oxidase, but synthesize shorter polypeptides (in the range of 28–31 kD) and some "extra" bands (in the range of 10–20 kD). The 28–31-kD polypeptides are fragments of the subunit II, as shown by peptide mapping experiments and cross-reactivity with antisera prepared against the subunit II. These results suggest that *oxi1⁻* mutants are either small deletions or frameshift mutations leading to premature chain termination. The colinearity between the map location of *oxi1⁻* mutants and the sizes of the polypeptides they synthesize support the second hypothesis (Fox 1979b; Weiss-Brummer et al. 1979). Furthermore, Cabral et al. (1978) have shown that pseudo-wild-type revertants of *oxi1⁻* frameshift mutants, resulting from a second mutation within the *oxi1* locus, synthesize active polypeptides slightly shorter than the wild-type subunit II but with strong immunological cross-reactivity with this subunit.

More recently, the complete nucleotide sequence of the *oxi1* region has been determined (Corruzzi and Tzagoloff 1979; Fox 1979a). A unique coding sequence of 251 codons is found in only one of all possible reading frames (or 238, since two AUG codons separated by 13 codons are found at the aminoterminal end of the gene, and it is not known which one is normally used). To directly confirm that this sequence is the structural gene for the subunit II, Fox (1979a) has shown that the isoelectric differences predicted from the DNA sequence of three pseudo-wild-type revertants, restoring the reading frame of a frameshift mutant within 15 nucleotides of the original mutation, are actually observed in the polypeptides synthesized by the revertants.

Less detailed studies have been performed on *oxi2⁻* mutants. The absence of the subunit III of cytochrome oxidase and its replacement by new polypeptides in some mutants suggest that the *oxi2* locus is the structural gene for the subunit III (Cabral et al. 1978). This is confirmed by the nucleotide sequence of the *oxi2* region (Thalenfeld and Tzagoloff 1980), in which a unique coding sequence is found. It is able to specify a polypeptide that is 269 amino acids long, whose predicted amino acid composition is in agreement with the composition of the subunit III.

Mosaic Mitochondrial Genes and Their Interactions

A first series of *mit⁻* mutants (called *cob⁻*), deficient for the coenzyme Q-H$_2$ cytochrome-*c* reductase activity and often also lacking spectroscopically detectable cytochrome *b*, has first been isolated from the "short" strain (D273-10B; Tzagoloff et al. 1975a,b). *cob⁻* mutants allowed the localization of a segment of the mitochondrial map carrying the gene for cytochrome *b* (Slonimski and Tzagoloff 1976). But more attention has been given to the cytochrome-*b* gene, after the isolation of a second series of *mit⁻* mutants from a "long" strain revealed that many of them (called *box⁻*) also affect the synthesis of subunit I of the cytochrome-oxidase complex, coded for by the *oxi3* gene and distant from the cytochrome-*b* gene on the map (Cobon et al. 1976; Kotylak and Slonimski 1976, 1977b; Pajot et al. 1976, 1977).

Mosaic Genetic and Physical Map of the cob-box Gene

All *box⁻* mutants are localized in a unique region of the mtDNA, flanked by the *oli1* and *oli2* loci (Figs. 7 and 9; Foury and Tzagoloff 1976b; Kotylak and Slonimski 1976, 1977b; Slonimski and Tzagoloff 1976; Tzagoloff et al. 1976c,d; Schweyen et al. 1977; Slonimski et al. 1978b; for review, see Dujon 1979). They fall into ten genetic loci (*box1–box10* in historical order), defined on the basis of recombination frequencies in crosses between *box⁻* mutants (interloci crosses give frequencies of recombinants close or equal to the maximal frequency of recombinants corresponding to unlinked genetic markers) and on the basis of the complementation map (see below). The *box* region has been further dissected into 39 intervals by petite deletion mapping, each mutant being unambiguously located in a given interval (Jacq et al. 1980; Lazowska et al. 1980; P. Slonimski, unpubl.).

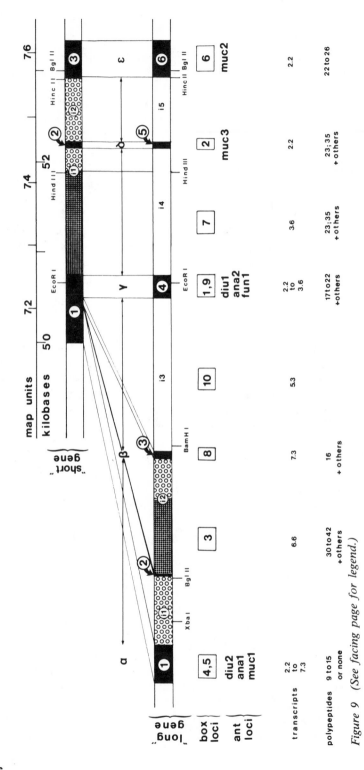

Figure 9 (See facing page for legend.)

The map is completed by a different set of mutants, conferring resistance against mucidin, antimycin A, diuron, or funiculosin (Subik 1975; Burger et al. 1976; Michaelis 1976; Colson et al. 1977; Groot-Obbink et al. 1977; Subik et al. 1977), which define a few ant^R loci (Pratje and Michaelis 1976, 1977; Subik et al. 1977; Subik and Takacsova 1978; Subik and Goffeau 1980). Petite deletion mapping and crosses between the ant^R mutants and the box^- mutants show that each ant^R locus is closely linked (or even allelic) to one of the box loci (Michaelis and Pratje 1977; Colson and Slonimski 1979; Colson et al. 1979; Takacsova et al. 1980).

Overall, the box^- mutations span a segment of about 7500 bp (roughly eight times longer than needed to specify the apocytochrome b sequence) and are highly clustered in a limited number of loci separated by relatively long, genetically "silent" regions. The ant^R mutants are even more clustered, being allelic to only a subset of the box loci ($box\ 5,4$; $box1$; $box2$; and $box6$; see Fig. 9). The map of the box^- mutants and other properties discussed later originally suggested a mosaic organization for the cob-box gene (Slonimski et al. 1978a). Since, contrary to the box^- mutants, the ant^R mutants must affect the sensitivity of cytochrome b to inhibitors but not affect its synthesis, the ant^R mutants must define the structural segments within the mosaic gene (the exons) and indicate which of the box loci fall in these segments.

The mosaic organization has been fully confirmed by the complete nucleotide sequences of the cob-box gene from the short strain and of fragments of the gene from the long strain (Lazowska et al. 1980; Nobrega

Figure 9 Physical and genetic map of the cytochrome b gene from the long and short strains. The map has been established by a combination of genetic and molecular methods as described in the text. The nomenclature of the *box* and *ant* loci is historical and thus fully extendable to new developments. The exons (■) and the introns (i1–i5 in the long strain and i1 and i2 in the short strain) are numbered using a logical nonextendable nomenclature. Equivalence with a previous nonextendable logical nomenclature of exons (α, β, γ, δ, and ϵ) is given for references with original publications. (▦) Intron-encoded open reading frames; (▨) intron sequences blocked in all reading frames. Intron sequences not yet determined are blank. All available evidence indicates that introns i4 and i5 of the long strain are very similar, if not identical, to introns i1 and i2 of the short strain. The gene contains a long ($\sim$ 1000 nucleotides) leader sequence and a short ($\sim$ 50 nucleotides) trailer sequence not shown in the figure. The mature transcript for the apocytochrome b is only 2.2 kb long, but splicing deficient *box*⁻ mutants accumulate higher molecular weight transcripts up to 7.3 kb in length (see text). The size (in kb) of the smallest transcript found in representative *box*⁻ mutants of each cluster is indicated. Similarly, chain termination mutants or splicing defective mutants accumulate polypeptides not seen in wild-type strains. The apparent molecular weight (in kD) of the major polypeptides found in representative *box*⁻ mutants of each cluster is indicated. Transcription and translation is from left to right. (Adapted from Slonimski et al. 1978b; Colson et al. 1979; Grivell et al. 1979; Haid et al. 1979; Perlman et al. 1979; Jacq et al. 1980b; Lazowska et al. 1980; Nobrega and Tzagoloff 1980; Takascova et al. 1980.)

and Tzagoloff 1980a,b), as well as by R-loop mapping using the 2.2-kb RNA that is believed to represent the mature mRNA (Grivell et al. 1979). This latter method has initially localized five exons (numbered α to ϵ; see Fig. 9) in the long gene, but comparison of the DNA sequences of the long and short genes (Lazowska et al. 1980) has recently revealed an additional mini-exon that constitutes the second exon of the long gene (Fig. 9). The short gene is missing the first three introns that correspond to the polymorphic insert III found previously (see Fig. 7). Recombination studies between the long and short forms of the gene to construct genes with different combinations of introns are now in progress (Perlman et al. 1980).

Although mosaic genes are obviously not unique to mitochondria, the *cob-box* gene presents an original aspect in that some mutants occur within the introns (*box3, box10,* and *box7* loci), which indicates that the introns perform genetic functions essential to the expression of the mosaic gene. Furthermore, all intron mutants (as well as some exon mutants) not only prevent the expression of cytochrome *b* but also that of the *oxi3* gene.

Truncated and Readthrough Polypeptides.   It is now well established that the final product of the *cob-box* gene is apocytochrome *b*, a polypeptide of complex III of the respiratory chain, synthesized on mitochondrial ribosomes with an apparent molecular weight of 30 kD (see below, Biogenesis of Mitochondria). In vivo labeling of the mitochondrially synthesized polypeptides in the presence of cycloheximide previously demonstrated that most of the exon mutants at the *box4, box1,* and *box6* loci lack apocytochrome *b* and synthesize a new and shorter polypeptide whose apparent molecular weight increases in correlation with the map position of the mutation from the *box4* locus to the *box6* locus (Fig. 9; Claisse et al. 1977, 1978; Spyridakis and Claisse 1978). This correlation strongly suggests that the new polypeptides are fragments of apocytochrome *b*, resulting from premature chain termination with sizes increasing in the direction of translation. This idea has been confirmed by the similarity of the tryptic fingerprints (Haid et al. 1979; Hanson et al. 1979; Solioz and Schatz 1979) and by the fact that all polypeptides, even the shorter ones found in the *box4* mutants, carry cytochrome-*b* antigenic determinants (Kreike et al. 1979). Many *box⁻* mutants in exons are, therefore, either nonsense mutations or frameshift mutations leading to a stop codon in the segment immediately downstream. Other *box⁻* mutants synthesize a complete but inactive apocytochrome *b* (they show a cytochrome-*b* spectrum and a 30-kD polypeptide but no reductase activity) as expected of missense mutations (Claisse et al. 1977, 1978). Finally, the *box5⁻* mutants, which do not seem to synthesize any polypeptides related to cytochrome *b* (although short polypeptides might have escaped the analysis), may be mutants in the promoter or in the leader sequence (Claisse et al. 1978; Haid et al. 1979; Kreike et al. 1979; Alexander et al. 1980).

The above results are simply expected from mutants in a single cistron. But the most interesting feature of the *cob-box* gene is that other *box⁻* mutants, produce—instead of the 30-kD apocytochrome *b* or in addition to it—multiple new polypeptides of apparent molecular weight ranging from about 13 kD to about 56 kD (Claisse et al. 1977, 1978, 1980). These include the intron mutants (*box3⁻* and *box7⁻*), as well as some exon mutants flanking the introns (e.g., *box9⁻* and many *box2⁻* and *box8⁻* mutants). From peptide mapping studies of a series of *box7⁻*, *box2⁻*, and *box9⁻* mutants, Claisse et al. (1980) demonstrated that the multiple new polypeptides synthesized share a common sequence with apocytochrome *b* in their aminoterminal ends but diverge in their carboxyterminal ends. Similarily, *box3⁻* mutants synthesize, in addition to bands of lower intensity, a 42-kD polypeptide (Claisse et al. 1977, 1978) that cross-reacts with anti-apocytochrome-*b* antiserum (Kreike et al. 1979; Solioz and Schatz 1979). The multiplicity of the new polypeptides and the sizes of the larger ones argue against them being the result of frameshift mutations and suggest that they must be translational hybrids from both exon and intron sequences as the result of a readthrough of intron sequences that are normally spliced prior to translation. This conclusion is strengthened by the nature of the polypeptides observed in double mutants constructed by recombination between single *box⁻* mutants (Lamouroux 1979; Alexander et al. 1980; Perlman et al. 1980). From all four possible combinations of double mutants examined (exon-exon, intron-exon, exon-intron, and intron-intron), it has been found that a chain termination mutant in an upstream exon is always epistatic over the downstream mutation (either in intron or in exon), which indicates that the hybrid polypeptides made by the intron mutants require the correct translation of the preceding exon(s). Conversely, an intron mutant is always epistatic over a downstream mutation (either in intron or in exon), which indicates that the translation of the exons cannot proceed beyond the mutated intron. Note that the pleiotropic effect of the intron mutants on the synthesis of the subunit I of cytochrome oxidase is always epistatic regardless of the position, which indicates the need for specific functions encoded by the *cob-box* introns for the expression of the *oxi3* gene (see below).

Transcripts, Processing Schemes, and Splicing Deficiencies.   In wild-type mitochondria, the major transcript of the *cob-box* region is an 18S RNA (2200 nucleotides), which is likely to be the mature mRNA for apocytochrome *b*, because it is able to direct the synthesis, in vitro, of antigenic determinants of cytochrome *b*, and it hybridizes with exon probes but not with intron probes (Grivell et al. 1979). The lengths of a series of larger transcripts and their hybridization with different probes suggest that they are intermediates in a splicing process that generates the 18S RNA from a primary transcript of at least 7500 nucleotides, covering the entire *cob-box* region and including a long (~1000 nucleotides) untranslated leader and probably a short (~ 50 nucleotides) untranslated trailer remaining in the mature mRNA (Church et al. 1979; Grivell et al. 1980; Haid et al. 1980; Halbreich et al. 1980; Van Ommen et al. 1980). The 18S RNA must therefore arise by a series of cut-and-splice events as for other mosaic genes. The *cob-box* gene is of further interest because intron mutants are blocked in the processing pathway at different steps, each specific to its own intron, the deficient splicing of an upstream intron leading to the deficient splicing of a downstream intron (e.g., *box3⁻* mutants lead to the accumulation of long precursors, whereas *box10⁻* and *box7⁻* mutants accumulate smaller precursors). This has been demonstrated by hybridization of the transcripts from a variety of *box⁻* mutants with different intron or exon mtDNA fragments (Church et al. 1979; Haid et al. 1980; Halbreich et al. 1980; Van Ommen et al. 1980). Examination of the precursors accumulated by the different *box⁻* mutants suggests the following splicing scheme for the long gene (Van Ommen et al. 1980). The first intron is excised first, resulting in the formation of a stable 10S RNA (Van Ommen et al. 1979) that is circular (Halbreich et al. 1980; see also Transcription and Transcripts, above), and a 28S RNA in which the *box4,5* exon, the mini-exon, and the reading frame of the *box3* intron are linked together in register (see Fig. 9). Then, the 28S transcript undergoes further processing of the other introns, but there does not appear to be an obligate order for splicing of these intermediates, although preferred pathways are possible. The excision of the *box3* intron frequently seems to be a late event and, possibly, a holding point in the processing.

The 10S RNA must be excised by nuclearly encoded enzymes, because ρ⁻ mutants can produce an apparently normal 10S RNA (Halbreich et al. 1980). However, two *box⁻* mutants affect the excision of the 10S RNA. One of them may be altered at the junction of the first intron with the *box4,5* exon, making the splicing impossible (Haid et al. 1980), but the other one maps in the more distant *box8* locus and its mode of action is more mysterious (Halbreich et al. 1980). Whether the abundant circular RNA plays a role or represents a by-product of a general splicing mechanism is still unknown, but other circular RNA molecules transcribed and spliced from the *oxi3* gene have also been found (Grivell et al. 1979; Arnberg et al. 1980).

Mosaic Complementation Map of the cob-box *Gene*

Complementation tests (Table 7) between numerous pairwise combinations of *box⁻* mutants, representative of each genetic locus, indicate that, as expected, all exon mutants (*box5,4; box1; box2;* and *box6* loci) constitute a single cistron. However, interspersed among the elements of this cistron, the intron mutants (*box3, box10,* and *box7* loci) belong to other cistrons since they complement the exon mutants. Furthermore, the mutants in each intron constitute a different cistron since they complement one another (Slonimski et al. 1978b; Kochko 1979; Lamouroux et al. 1980).

Table 7 Complementation matrix between representative
mutants at the various *box* loci

	box4,5	box3	box10	box1	box7	box2	box6
box4,5	0.2	0.4		0.1	7.1	0.1	0.2
box3		0.1	8.9	4.9	5.1	4.3	4.2
box10			0.8	10.2	10.3		6.1
box1				0.3	3.1	0.1	0.4
box7					0.1	4.6	5.5
box2						0.3	0.2
box6							0.4

The complementation between *box⁻* mutants is tested by the ability of the
heteroplasmic zygotes to respire shortly after mating. Respiration is expressed
as μmoles of O_2/hr/10^8 zygotes. In most cases, the figures given are averages
from several crosses involving different mutants of a given locus. (Data
compiled from Slonimski et al. 1978b; Kochko 1979; Lamouroux et al. 1980).
Figures lower than 1 indicate the absence of complementation; figures of ca.
3–10 indicate complementation.

The complementation map demonstrates that the *box3, box10,* and *box7*
loci must be responsible for the synthesis of three *trans*-acting elements,
different from one another and necessary for the synthesis of apocytochrome *b* by the exons (*box5,4; box1; box2;* and *box6* loci). Since intron
mutants are deficient for splicing their own introns, it must be postulated
that during complementation, the *trans*-acting element synthesized by the
wild-type intron of the exon mutant allows the excision of the intron
sequence from the precursor RNA of the intron mutant. After this splicing,
the wild-type exons of the intron mutant can be used to direct the synthesis
of apocytochrome *b*. The use of *ant*^R mutations as markers of the exons
demonstrated that all exons of one parent are always expressed in *cis* during
complementation to direct the synthesis of apocytochrome *b* (Kochko et al.
1979). The complementation of the splicing deficiency occurs in *trans*, but
the result of the splicing only ligates exons in *cis*. The complementation map
shows a polar effect of chain termination mutants in an upstream exon
(Lamouroux et al. 1980). For example, the chain termination mutants at the
box4 locus are unable to complement the *box3⁻* mutants. Similarly, some
box1⁻ mutants do not complement the *box7⁻* mutants (downstream), but
they complement the *box3⁻* mutants (upstream). This polar effect indicates
that an upstream exon needs to be properly translated for the *trans*-acting
element from the downstream intron to be expressed. The polar effect,
however, falls off with distance, and some chain termination mutants in the
box4 locus efficiently complement the *box7⁻* mutants.

Functions for Introns

All properties of intron mutants indicate that some of the introns of the
cob-box gene encode separate functions required for the splicing of the

cytochrome-*b* pre-mRNA and are able to act in *trans*. Two classes of hypotheses have been proposed for the nature of these functions. Initially, the idea that the introns code for a "guide RNA" was favored (Church et al. 1979; Halbreich et al. 1980). In this model, a sequence present in the intron RNA serves as a guide to align the intron-exon boundaries in the pre-mRNA for correct splicing of the intron. The intron mutants were supposed to alter the guide-RNA sequence. More recently, however, several observations have contributed to reject the guide-RNA hypothesis and to propose that the introns code for proteins required for RNA splicing. These observations were, first, the absence of splicing of the *box3, box7,* and *box10* introns in ρ^- mutants, whereas other introns, such as the one from the large rRNA gene, are spliced (Church and Gilbert 1980). This suggests that the splicing of some mitochondrial introns requires mitochondrial protein synthesis. Second, the synthesis by the splicing-deficient mutants of hybrid polypeptides resulting from long readthrough within the introns (Claisse et al. 1978; Haid et al. 1979; Kreike et al. 1979; Solioz and Schatz 1979) indicated that, as for the intron of the large rRNA gene discovered previously (Dujon 1980), long open reading frames should also exist within the *cob-box* introns, a fact entirely confirmed later by the DNA sequence (Lazowska et al. 1980; Nobrega and Tzagoloff 1980a,b). Finally, the existence of wild-type revertants from splicing deficient intron mutants indicated that the translation of the intron reading frames is required prior to the splicing of the intron and not in an already excised intron since, in the latter case, no revertants could ever emerge (Kochko 1979). These observations are integrated in a unique mechanism if one assumes that the hybrid polypeptides resulting from the readthrough of the long intron-encoded reading frames before their excision help the splicing of their own introns and/or the ligation of the adjacent exons. Two explicit forms of this model have now been proposed.

Church and Gilbert (1980) postulated that the three introns (*box3, box10,* and *box7*) are successively used to synthesize three hybrid polypeptides and that each polypeptide ("spligase") is required for the splicing of its own intron. They further proposed that the translation of the *box4,5* exon of cytochrome *b* continues in phase within the *box3* intron and produces the first spligase (*box3* spligase) which, after excision of the 10S RNA by cell-sap-splicing enzymes, in turn, excises the *box3* intron and ligates the *box4,5* and *box8* exons. Another translation can now proceed through the ligated *box4,5* and *box8* cytochrome-*b* exons and then enter the *box10* intron, making the second spligase which, in turn, excises the *box10* intron and ligates the *box8* and *box1,9* exons. The same mechanism allows the synthesis of the third spligase (*box7*) and, at last, the synthesis of cytochrome *b*. The *cob-box* mosaic gene appears as a nested set of modular units: The introns for one polypeptide are the exons for another and, as they are sequentially removed, a cascade of new "genes" is formed.

Jacq et al. (1980a), who reported the occurrence of an open reading frame in a segment of the *box3* intron and showed that two *box3⁻* mutants must either arrest the reading or modify the translation product, also postulated that an intron-encoded polypeptide (that they called "maturase") is required for the splicing of the *box3* intron.

A more precise formulation of this model was proposed after Jacq et al. (1980b) and Lazowska et al. (1980) determined the nucleotide sequence of the first two introns of the long gene and showed that a long reading frame in the *box3* intron was indeed fused in phase with the preceding exon (mini-exon); the mini-exon was discovered after comparison of the DNA sequence of the short gene. In contrast, in the first intron, which is spliced by cell-sap enzymes to produce the 10S circular RNA and contains no mutant, all reading frames are blocked. Furthermore, they showed that three *box3⁻* mutants, deficient for splicing the *box3* intron, result from a sequence alteration within the intron-encoded reading frame (two nonsense mutants and one double missense mutant, respectively), altering the hybrid polypeptide in its carboxyterminal part. The length of the truncated polypeptides predicted from the positions of the two nonsense mutants is in agreement with the apparent molecular weight determinations of Claisse et al. (1978). On the basis of these observations, Lazowska et al. (1980) proposed the following mechanism for the early steps of the expression of the cytochrome-*b* gene. After transcription of the entire *cob-box* gene to form the large pre-mRNA molecule (7.5 kb), the first splicing event, catalyzed by cell-sap enzymes, removes the first intron, giving rise to the 10S circular RNA that accumulates. This first splicing event links, in phase, the first exon (*box4,5*) with the second exon (mini-exon) and, consequently, with the open reading frame of the *box3* intron fused to it. The resulting RNA molecule can now be translated (in mitochondria, transcription and translation are in the same compartment) to produce a hybrid protein, with an apparent molecular weight of 42 kD, called the "maturase." Maturase activity is required for the second splice, which ligates the third exon (*box8*) in phase with the two preceding ones and, at the same time, eliminates the *box3* reading frame so that no new synthesis of maturase molecules can take place. It is assumed that the *box3⁻* mutants alter the maturase in a way that is essential for its activity so that the splicing of the *box3* intron does not occur, leading to an overproduction of defective maturase (the 34–42-kD polypeptides observed in *box3⁻* mutants).

Although the splicing activity of the maturase remains to be directly demonstrated, all the properties of the *box⁻* mutants (complementation, epistatic upstream effect, splicing deficiency of *box3⁻* and *ρ⁻* mutants) strongly support the idea that the synthesis of the *box3* maturase is essential for the splicing of the *box3* intron. Furthermore, the observation (Dujardin et al. 1980b) that some *box3⁻* mutants can be suppressed by the *mim1* mutation of the large rRNA gene (which probably decreases the translational fidelity) or by paromomycin (which causes misreading of mitochondrial ribosomes) suggests that a minute amount of misreading is sufficient to overcome the *box3⁻* deficiency but not enough to alter the expression of the exons of the *cob-box* gene. This supports the idea that only catalytic amounts of the maturase are sufficient to ensure splicing of the *box3* intron. Whether the *box3* maturase is itself the splicing enzyme or only a protein that influences the specificity of splicing is not known, but the second possibility seems more plausible.

That the 42-kD *box3* maturase is a hybrid protein containing the first 143 amino acids of cytochrome *b*, fused with 280 amino acids of the intron reading frame, is now demonstrated by the fact that the defective maturase overproduced in *box3⁻* mutants reacts with the antibodies against cytochrome *b* (Kreike et al. 1979; Solioz and Schatz 1979) and the fact that the chain termination mutants in the *box4,5* exon abolish the synthesis of the maturase (Alexander et al. 1980).

Whether the maturase model also applies to the *box10* and *box7* introns has not yet been directly verified. A long open reading frame fused in phase with the preceding exon has been found in the *box7* intron of the short strain (Nobrega and Tzagoloff 1980b,c) and could specify a *box7* maturase. However, the intron-encoded parts of the *box3* and *box7* maturases do not bear obvious amino-acid-sequence homologies.

oxi3 *Mosaic Gene*

That the *oxi3* locus represents the structural gene for the subunit I of cytochrome oxidase is suggested from the absence of this subunit in the *oxi3⁻* mutants (Tzagoloff et al. 1975a,b; Slonimski and Tzagoloff 1976; Ecclesall et al. 1978; Keyhani 1979) and by the fact that the transcripts of the *oxi3* region are able to direct the synthesis in vitro of fragments of polypeptides with antigenic determinants of the subunit I (Grivell and Moorman 1977). However, the first attempt to map *oxi3⁻* mutants already indicated that they span a sector of the mitochondrial map, flanked by the genetic loci *par1* and *oli2*, several times larger than needed to synthesize the subunit I of cytochrome oxidase (Slonimski and Tzagoloff 1976), as is now expected of a mosaic gene. Subsequent petite deletion mappings of *oxi3⁻* mutants (Carignani et al. 1979; Weiss-Brummer et al. 1979) and restriction mapping of the *oxi3* region showed that the mutations span a segment about 10 kb long (Grivell and Moorman 1977; Morimoto et al. 1979a).

As in the case of the other mitochondrial mosaic genes, several polymorphic forms of the *oxi3* gene are known (Grivell et al. 1980), differing by the long inserts I, IIA, IIB, and IV, previously characterized (see Fig. 7). The complete nucleotide sequence of the *oxi3* gene from the short strain (D273-10B) has recently been determined (Bonitz et al. 1980a,b). In the absence of information on the amino acid sequence of the subunit I, the high homology between the sequence predicted from the DNA sequence and the amino acid sequence of the cytochrome oxidase subunit of human mitochondria corresponding to the subunit I of yeast has allowed the identification of the exons of the *oxi3* gene and the proposition of relatively precise locations for some intron-exon junctions. At least six exons and five introns are found in the short strain (and possibly up to eight exons, since it is conceivable that two very short introns exist in the last exon). The long strain contains two long additional introns within the equivalent of the fifth exon (see Fig. 7), and an even shorter strain (NCYC74) is missing the first and fourth introns of D273-10B.

But the major interest in the DNA sequence of the *oxi3* genes is undoubtedly that it also reveals very long open reading frames in the first four introns, each fused in phase with the preceding exon. The fifth large intron is blocked in all reading frames. Despite the large number of *oxi3*⁻ mutants and the existence of very long intron-encoded reading frames, no *oxi3*⁻ mutant seems to occur within the introns, contrary to the situation in the *cob-box* gene. The reason for this striking difference is unknown but, interestingly, the reading frames of the first two introns (779 and 786 codons long, respectively) share about 50% sequence homology, whereas the reading frame of the fourth intron (317 codons) is about 70% homologous to the *box7* intron of the *cob-box* gene.

Splicing Interactions between Two Mosaic Genes

As mentioned earlier, intron mutants of the *cob-box* gene (as well as some chain termination mutants in the *box4,5* exon) fail to express the *oxi3* gene. Hybridizations of mtRNAs with mtDNA fragments of the *oxi3* region have shown that the interaction occurs at the level of splicing, i.e., *box3*⁻ and *box7*⁻ mutants fail to synthesize the mature mRNA (2300 nucleotides long) of the *oxi3* gene but accumulate a slightly larger precursor (Church et al. 1979). The size difference between the precursor and the mature mRNA suggests that the pleiotropic effect prevents the splicing of only one intron of the *oxi3* gene. A single mutation in the *cob-box* gene is therefore able to alter the splicing of two different mosaic genes, offering an interesting mechanism of coordinated gene regulation. The interaction between the two mosaic genes is not reciprocal because complete (or very large) deletions of the *oxi3* gene (Slonimski and Tzagoloff 1976; Mahler et al. 1977; Morimoto et al. 1979a; Grivell et al. 1980) do not alter the expression of the *cob-box* gene. Thus, the expression of the *oxi3* gene is not required for that of the *cob-box* gene. However, two *oxi3*⁻ mutants exist that affect the splicing of the *cob-box* pre-mRNA at steps that mimic the action of *box3*⁻ or *box7*⁻ mutants. Possibly, these mutants synthesize a product that interferes with specific steps of the *cob-box* pre-mRNA processing.

The mechanism of interaction between the two mosaic genes is not completely understood. It is probable that only the *box7* intron is directly involved in the interaction and that the pleiotropic effect of the other *box⁻* mutants is a secondary consequence of their primary effect on the expression of the *box7* intron. This is suggested because a *box7⁻* mutant, introduced by recombination into the short *cob-box* gene, still exerts the same pleiotropic effect on *oxi3* gene expression (Perlman et al. 1980). The important sequence homology between the *box7* intron and the fourth intron of the *oxi3* gene tends to suggest a direct interaction between these two introns. It is conceivable that the splicing requirements for these two introns is similar and ensured only by a *box7* maturase, because the fourth intron of *oxi3* would encode a deficient maturase. In this respect, the fact that a mitochondrial mutation (*mim2*), which supresses all *box7⁻* mutants but no other *box⁻* mutants, maps in the *oxi3* gene (Dujardin et al. 1980b) may be in favor of such a hypothesis. The mechanism of action of *mim2*, the nature of the mutation, and even its precise locations are not known as yet. But it is conceivable that the *mim2* mutation restores the activity of the corresponding *oxi3* maturase so that it can now splice its own transcript and also the *box7* intron.

Elusive Mitochondrial Genes: The *var1* Determinant

Douglas and Butow (1976) discovered a few years ago that one of the polypeptides translated on mitochondrial ribosomes (var 1) shows a remarkable strain-dependent variation of mobility on SDS-polyacrylamide gel electrophoresis. Since then, several detailed studies have been devoted to this polypeptide and its genetic determinant. At present, more than 15 different forms of the var 1 polypeptide have been identified among different laboratory wild-type yeast strains or the recombinants constructed from them, varying in apparent molecular weight from 40 kD to 44 kD (Butow et al. 1980; Strausberg and Butow 1981).

All genetic criteria (coordinated output, recombinant frequencies, petite deletion mapping, and zygotic gene rescue, with ρ^- mutants retaining only a short segment of the mitochondrial genome) indicate that the genetic determinant of the polymorphic variation is located at a unique locus ("*var1* determinant") in the segment of the mitochondrial genome flanked by the *oli1* and *rib3* loci (Douglas et al. 1976; Perlman et al. 1977) (see Fig. 7). Formally, all polymorphic forms of the var 1 polypeptide can be interpreted by the combinations of two short insertions (< 100 bp each) within this segment of the mitochondrial genome, called *a* and *b*, respectively. Because the presence or absence of two inserts can produce only four possible types, it is further assumed that portions of the *b* insert exist (called b_p) so that numerous var 1 forms can be generated (e.g., a^+b^+ [44 kD], a^+b^- [41.8 kD], a^-b^+ [42.2 kD], a^-b^- [40 kD], or $a^-b^+_p$). Different sizes of b_p inserts account for all intermediate sizes of the polypeptide. Test crosses with the standard forms a^-b^-, a^+b^-, and a^-b^+ offer the possibility to unambiguously determine the genotype of each strain from the different forms of the var 1 polypeptide generated in the progeny (Strausberg and Butow 1981).

Crosses between strains with different var 1 polypeptides often yield nonparental forms of the var 1 polypeptide in the progeny (Strausberg et al. 1978). In crosses of the type $a^+b^+ \times a^-b^-$, the two reciprocal recombinants (a^+b^- and a^-b^+) appear with significantly unequal frequencies. By analogy to the polarity of recombination controlled by the ω locus (see above, Mechanisms and Rules for Mitochondrial Inheritance: Multifactorial Crosses Involving Antibiotic-resistant Mutations), this phenomenon has been interpreted by dissymmetrical gene conversions of the a and b inserts after the pairing of two mtDNA molecules. The gene conversion of a^- into a copy of a^+ is more efficient than the gene conversion of b^- into a copy of b^+ and, in the latter case, the conversion takes only part of the b segment about one fifth of the time, creating the various b_p's. In fact, this phenomenon is similar to the polarity of recombination at the ω locus only in the sense that it is always the inserts that are preferentially recovered in the progeny, but the efficiency of the dissymmetrical gene conversion at the *var1* determinant is much lower than that at the ω locus.

Petite deletion mapping, using a large collection of discriminative ρ^- mutants, has allowed the precise localization of the a and b inserts inside a short fragment (*Hinc*II no. 10, see Fig. 7) (Butow et al. 1980; Vincent et al. 1980). Because this DNA fragment itself appears variable in size in a manner coordinated with the size variations of the var 1 polypeptide, it is likely that the a and b inserts are directly responsible for the polymorphic variation of the var 1 polypeptide (Strausberg et al. 1978). However, the determination of the DNA sequence of this fragment from a ρ^- mutant derived from the short strain (D273-10B; see Fig. 7) showed that at least 90% of the *var1* determinant region is made of A + T-rich stretches with only a few, highly clustered GC base pairs (Tzagoloff et al. 1979b). Thus, it is improbable that this region can specify a polypeptide of the size of var 1. If these results are correct, the origin of the structural information for the var 1 polypeptide is mysterious. Because the genetic location of the *var1* determinant does not seem questionable, it is postulated that it is responsible for the final size of the var 1 polypeptide but that the bulk of the structural information is at a distance from the genetic determinant. The region of the *var1* determinant is not, however, genetically silent, and a few mutants conferring *mit⁻* or *syn⁻* phenotypes have been mapped in this region (Butow et al. 1980); but no mutant specifically affected in the var 1 polypeptide has been isolated so far.

Zygotic gene rescue experiments with ρ^- mutants retaining only the *var1* determinant have recently been improved by the utilization of a double nuclear mutant (*met⁻ cys⁻*) as the ρ^+ parent, so that only the mitochondrial translation products made in the zygote can be labeled by $^{35}SO_4^-$ in the presence of cycloheximide (Butow et al. 1980). Under such conditions, it is observed that most of the var 1 polypeptides synthesized in the zygote are specified by the ρ^- parent. Experiments of this type demonstrate that the 1-kb fragment carrying the *a* and *b* inserts is sufficient to act in *trans* to specify the var 1 polypeptide; this suggests an intermolecular splicing between the transcripts of the *var1* determinant and the transcript of the structural gene (Butow et al. 1980). Van Ommen et al. (1979) and Grivell et al. (1979) originally identified two transcripts (17.5S and 19S, respectively) that hybridize with the mtDNA fragments overlapping the *var1* determinant. More recently, Butow et al. (1980) have reported that three major transcripts (16S, 15S, and 10S) hybridize to the region of the *var1* determinant, but they believe that the 15S species is indistinguishable from the small rRNA and that the 10S species might be a derivative of the 15S species. Particularly remarkable is the fact that the 16S species (1900–2000 nucleotides long and probably identical to the 17.5S species of Van Ommen et al. [1979]) shows sequence homology to a short DNA segment (only 250 nucleotides long) in the region of the *var1* determinant but not to the flanking segments. Furthermore, the 16S species varies in length (1900–2000 nucleotides long) in a strain-dependent manner as does the var 1 polypeptide, as if this transcript contained the inserts *a* and *b*. This transcript is present in ρ^- deleted for all of the mitochondrial genome except the region of the *var1* determinant. Butow et al. (1980) speculated that the 16S variable RNA may be the result of an intermolecular splicing between an RNA species carrying the bulk of the structural information for the var 1 polypeptide and another RNA species transcribed from the *var1* determinant region. The origin of the structural information for the var 1 polypeptide is still unknown, but Butow et al. (1980) envisaged the possibility that it could be of nuclear origin.

Unlike the other mitochondrially synthesized polypeptides, which are highly hydrophobic, the var 1 polypeptide is a water-soluble basic protein. That the var 1 polypeptide is found associated with the small subunit of the mitochondrial ribosome and is in roughly equimolar ratio with the other proteins of this subunit suggests that it is an integral component of the mitochondrial ribosome (Terpstra et al. 1979). The comigration, on SDS-polyacrylamide gel electrophoresis, of the different forms of the var 1 polypeptide with the unique mitochondrially synthesized protein of the small ribosomal subunit previously identified (Groot 1974) conclusively identifies var 1 with this ribosomal protein (Groot et al. 1977, 1979; Butow and Strausberg 1979). Pulse-chase experiments indicate that the var 1 polypeptide is first assembled into a small RNase-sensitive particle (15S–20S), which represents a precursor in the formation of the small ribosomal subunit (38S). Furthermore, erythromycin inhibition of mitochondrial protein synthesis shows that the presence of the var 1 polypeptide is required for the assembly of this ribosomal subunit (Terpstra and Butow 1979).

Replication of mtDNA

mtDNA Amount, Variation, and Control

mtDNA represents an average of about 15% (5–25% range) of the total DNA of yeast (Williamson et al. 1971b; Nagley and Linnane 1972b; Grimes et al. 1974; Hall et al. 1976a). Considering that the total DNA content of a haploid cell is 2.3×10^{-14} g (Ogur et al. 1952), a haploid cell contains about 50 molecules of mtDNA of 75,000 bp each and a diploid cell contains roughly twice as many copies (Grimes et al. 1974). The total amount of mtDNA, however, varies to some extent between different nonisogenic strains under the control of multiple nuclear genes (Hall et al. 1976a) and can also vary in the same strain in response to different physiological conditions. For example, under glucose repression, the amount of mtDNA is generally lower than in fully derepressed cells (Williamson 1970; Bleeg et al. 1972; Cryer et al. 1974; Goldthwaite et al. 1974). In contrast, anaerobiosis does not lower the amount of mtDNA as long as ergosterol and unsaturated fatty acids are present, but it does if the cells are partially depleted for ergosterol and unsaturated fatty acids (Fukuhara 1969; Nagley and Linnane 1972a). In the latter case, the aeration of the cells results in a rapid synthesis of mtDNA (Mounolou et al. 1968; Rabinowitz et al. 1969; Nagley and Linnane 1972a). Finally, Lee and Johnson (1977) showed that the amount of mtDNA increases significantly (up to 4 times) during conjugation proportional to the cell volume.

The question of how mtDNA molecules are distributed among the different mitochondria is a traditional one among people who learn mitochondrial genetics. However, this question loses all significance when it is realized that the different mitochondria of a cell, observed as individual entities at a given time, actually undergo continual fusion and fragmentation (see Stevens, this volume), forming a single dynamic syncytium in which mtDNA molecules can undergo genetic exchanges. Considering the total volume of mitochondria per cell ($\sim 5 \ \mu m^3$) and the total amount of mtDNA, the packaging of the mtDNA into mitochondria does not appear as acute a problem as it is for the nucleus. However, the mtDNA is associated in vivo with a protein (HM) of 20 kD that is rich in lysines and arginines (Caron et al. 1979). The HM protein represents, in mass, about the same amount as the mtDNA itself, i.e., there must be an average of 2500 HM proteins per ρ^+ mtDNA molecule or 1 HM protein/30 bp of mtDNA. It shows a high affinity for all double-stranded DNAs and resembles the histones in the sense that it induces superhelical turns in relaxed closed circular DNA molecules. The HM protein, present in ρ^- mutants, is nuclearly encoded.

Timing and Mode of mtDNA Synthesis

By all criteria, mtDNA replication shows a high degree of independence from nuclear DNA replication. It takes place continuously throughout the cell cycle, whereas nuclear DNA shows a stepwise increase at the S phase (Williamson and Moustacchi 1971; Sena et al. 1975). A similar uncoupling is observed during meiosis (Kuenzi and Roth 1974; Piñon et al. 1974). Furthermore, mtDNA replication continues in α-factor-arrested cells (Petes and Fangman 1973; Cryer et al. 1974) or in the presence of cytoplasmic protein synthesis inhibitors that block the reinitiation of nuclear DNA synthesis. Prolonged incubation of the cells under such conditions results in the accumulation of mtDNA to about two to three times the normal value (Grossman et al. 1969). This phenomenon is reminiscent of the accumulation of mtDNA during prolonged incubation of cells in the stationary phase. Independence between nuclear and mtDNA replications is also attested by the *cdc* mutants. mtDNA replication is active in the mutants deficient for the initiation of nuclear DNA synthesis (*cdc4, cdc7,* and *cdc28*) but not in the mutants deficient for continued replication during the S phase (*cdc8, cdc21*) (Cottrell et al. 1973; Wintersberger et al. 1974; Newlon and Fangman 1975). *cdc8* and *cdc21* mutants accumulate ρ^- mutants at elevated temperature (Newlon et al. 1979).

Classic density-shift experiments, in which cells grown in the presence of ^{15}N are transferred to ^{14}N medium, show a single peak of mtDNA with a gradual shift in density as if mtDNA replication were dispersive (Williamson and Fennell 1974; Sena et al. 1975). The same is observed during zygote formation and maturation (Sena et al. 1976). This apparent dispersive replication has been attributed to the high rate of recombination among the different mtDNA molecules of the cells that are nonsynchronized for replication, the replication itself being semiconservative. On the contrary, when BrdU is used as the density label, discrete classes of intermediate density are observed (Leff and Eccleshall 1978). The reason for this difference between the two methods of labeling is not clear.

Replication Machinery

Since every stable ρ^- mutant must obviously replicate its mtDNA faithfully, the essential components of the mtDNA replication machinery and controls must be nuclearly encoded and synthesized on cytoplasmic ribosomes. This simple fact also demonstrates that mtDNA replication can occur faithfully even in the absence of mitochondrial protein synthesis. That treatment by chloramphenicol or erythromycin results in the instability of the mitochondrial genome and the production of the ρ^- mutation in some strains (Carnevali et al. 1971; Williamson et al. 1971a; Strausberg and Birky 1979; Marmiroli et al. 1980a) is then probably not directly relevant to mtDNA replication but rather to some other mechanisms specific to the ρ^- formation.

A mtDNA polymerase, different from the two nuclear DNA polymerases, was reported previously (Iwashima and Rabinowitz 1969; Wintersberger and Wintersberger 1970). As expected, this enzyme is present in ρ^- mutants. More recently, Wintersberger and Blutsch (1976) reexamined in detail the properties of a more highly purified preparation of mtDNA polymerase. The enzyme is very efficient in catalyzing gap-filling reaction. No exonuclease activity has been detected with the purified enzyme, and this may be in agreement with its relatively low fidelity during replication. The mtDNA polymerase activity is higher in mitochondrial preparations from derepressed cells than from glucose-repressed cells.

Isolated mitochondria, in the presence of an appropriate source of energy, are able to incorporate exogeneous deoxyribonucleotides into mtDNA molecules. Using BrdU for density labeling, Mattick and Hall (1977) observed the synthesis of mtDNA molecules of about 10% the size of the wild-type genome and of a density compatible with these molecules being a hybrid DNA made by semiconservative replication. A similar synthesis was also found in mitochondria isolated from a ρ^- mutant. Due to the low rate and extent of such a synthesis, it is difficult to distinguish whether it represents a replicative process or a repair synthesis.

Initiation of mtDNA Replication and Its Control

The problem of the initiation of mtDNA replication has remained obscure until very recently because of the difficulty in isolating intact mtDNA molecules and the high rate of genetic recombination. Furthermore, the fact that different ρ^- mutants can faithfully replicate different short segments of the mitochondrial genome without sequence homology among them even shed some doubt on the existence of specific sequences initiating mtDNA replication (see Borst et al. 1976a). But major developments on these questions have recently resulted from the genetic and molecular characterization of hypersuppressive ρ^- mutants (see above, Genetic Properties of ρ^- Mutants). The comparison of the mtDNA from these mutants with that of their progeny (in crosses to either ρ^+ or other nonhypersuppressive ρ^- mutants) shows that hypersuppressive ρ^- mutants preferentially replicate their mtDNA during the transient heteroplasmic phase following zygote formation (Blanc and Dujon 1980). The fact that all hypersuppressive ρ^- mutants share a common sequence of 300 bp strongly suggests that this sequence (*rep*) represents the origin of replication of mtDNA (Blanc 1979; Blanc and Dujon 1980; Dujon and Blanc 1980). The molecular analysis of a complete collection of hypersuppressive ρ^- mutants revealed the existence of three different *rep* sequences, sharing extensive, but not completely identical, sequence homology between them (Blanc and Dujon 1980; Dujon and Blanc 1980). They probably result from duplications of a unique ancestral sequence followed by independent mutations in the nonessential parts. The three *rep* sequences are localized in three different regions of the wild-type genome (*rep1* is between the *var1* determinant and the large rRNA gene, *rep2* is between the *oxi2* locus and the *par1* locus, and *rep3* is probably between the *oli2* gene and the *cob-box* gene). De Zamaroczy et al. (1979) and Goursot et al. (1980) have also reported the sequence of a fragment of mtDNA from ρ^- mutants, which they initially claimed was the origin of replication of mtDNA, but this sequence is only 83 bp long. Subsequent studies have shown that their 83-bp sequence was, in fact, a fragment of the 300-bp *rep* sequences (Bernardi et al. 1980).

The idea that *rep* sequences control the initiation of mtDNA replication easily accounts for the replication of ρ^+ mtDNA molecules, using different *rep* sequences, and of ρ^- mutants that retain a *rep* sequence. The fact that other ρ^- mutants do not retain any *rep* sequences (*rep⁰*) indicates that, in the absence of *rep*, mtDNA replication can nevertheless be initiated either at different sequences or by a different mechanism that is not sequence-specific. However, mtDNA molecules devoid of *rep* can only replicate as long as they are in a homoplasmic cell but fail to replicate if they are in competition with other mtDNA molecules carrying a *rep* sequence (e.g., in crosses between a hypersuppressive ρ^- mutant and a *rep⁰* ρ^- mutant, only hypersuppressive molecules replicate). From this we can conclude that the *rep* sequences have a high affinity for a factor of initiation of mtDNA replication, the initiation being the limiting step in the replication of mtDNA (Blanc and Dujon 1980). That an increased density of *rep* sequences per unit length of mtDNA is necessary for a ρ^- mutant to be hypersuppressive (all hypersuppressive ρ^- mutants have a short conserved sequence and, hence, a relative amplification of the *rep* sequence) indicates that the probability of a mtDNA molecule being replicated depends upon the number of *rep* sequences it carries. The amount of mtDNA of a cell could be controlled by the limiting quantity (or activity) of the initiation factor (whose nature is still unknown) present at a given time. The cell may simply "count" the number of initiation events per cell cycle and/or per time unit, disregarding which molecule is replicated but ensuring that a correct number of replication cycles have taken place to conserve a regulated amount of mtDNA.

The in vivo cloning of *rep* sequences in hypersuppressive ρ^- mutants offers a means of directly investigating these questions in the near future. It has already been demonstrated that *rep* sequences inserted into a chimeric plasmid by in vitro recombination are able to allow the replication of these plasmids when introduced into yeast by transformation (Blanc and Dujon 1981). This result directly demonstrates that the *rep* sequences are replication origins.

BIOGENESIS OF MITOCHONDRIA

This section is intentionally limited to the aspects of the biogenesis of mito-chondria directly relevant to the functions of the mitochondrial genome and their interactions with the nuclear genome. We have not attempted to dis-cuss in detail the structure and function of the enzymatic complexes of the inner mitochondrial membrane or their relationship with the membrane. For these topics, the reader is referred to recent review articles (Mahler 1973b, 1976; Tzagoloff et al. 1973; Kovac 1974; Schatz and Mason 1974; Butow et al. 1975; Linnane and Crowfoot 1975; Mahler et al. 1975; Sebald 1977; Senior 1977) and symposia (Tzagoloff 1975; Packer and Gomez-Puyou 1976; Lee et al. 1979).

Nuclear Accommodation of Mitochondria

Because the number of genes on the mitochondrial genome is limited, it is obvious that a large number of nuclear genes must be involved in the synthe-sis of mitochondrial components (e.g., enzymes of the respiratory chain or of oxidative phosphorylation). In addition, a number of nuclear genes must be implicated directly with the expression of the mitochondrial genome if only because many elements of the mitochondrial protein synthesis machin-ery or of the replication and transcription of mtDNA are nuclearly encoded. In total, many nuclear genes must be devoted to the accommodation of mitochondria. However, in contrast to the detailed information gathered on mitochondrial genes and although the first *pet* genes were found many years ago (Sherman 1963; Sherman and Slonimski 1964), our knowledge of *pet* genes is still very limited. One of the reasons for this may be that the majority of *pet⁻* mutants are affected in several respiratory functions at the same time and often accumulate secondary ρ^- mutations. At present, three approaches to the nuclear accommodation of mitochondria have been attempted.

In the first approach, a limited number of *pet⁻* mutants are examined in some detail to determine the biochemical defect due to the mutation. For example, from their isolation of 18 new *pet⁻* mutants that retain functional mitochondrial protein synthesis and do not accumulate secondary ρ^- mutants, Ebner et al. (1973b) found seven new complementation groups. One mutant, lacking F1 ATPase, and a few others, affecting cytochrome oxidase, have been the object of more detailed studies (Ebner and Schatz 1973; Ebner et al. 1973a; Todd et al. 1979). The fact that some *pet⁻* mutants are deficient for mitochondrially synthesized polypeptides and cytoplasmi-cally synthesized subunits are in almost normal amount is probably more interesting to consider with respect to the expression of the mitochondrial genome or the assembly of its products. For example, two *pet⁻* mutants synthesize the subunit III of cytochrome oxidase but not the subunit II and only traces of the subunit I (Cabral and Schatz 1978), whereas another mutant (*pet494-1*) specifically lacks the subunit III (Ebner et al. 1973a). The exact molecular mechanism by which these nuclear genes specifically affect the synthesis of single mitochondrially encoded polypeptides is not yet known, but the phenotypic suppressibility of the *pet494-1* mutation by a nuclear amber suppressor demonstrates that this gene must be translated into a protein (Ono et al. 1975). Tzagoloff et al. (1975c) also isolated a new collection of *pet⁻* mutants retaining functional mitochondrial protein synthesis and affecting cytochrome-*c* reductase (16 complementation groups for 32 mutants), cytochrome-*c* oxidase (10 complementation groups for 20 mutants), or ATPase (3 complementation groups for 3 mutants). They also observed that some of the *pet⁻* mutants specifically lack one of the mitochondrially encoded polypeptides.

In the second approach, a large number of mutations are isolated with the eventual goal of defining all possible nuclear complementation groups involved in the mitochondrial functions. Using this approach, Burkl et al. (1976) isolated a large collection of temperature-sensitive respiratory-deficient nuclear mutants. From all possible pairwise combinations among 270 mutants, 106 complementation groups have been found with only one or a few mutants in most groups and up to 10–25 mutants in the 10 larger groups. Therefore, even more complementation groups are likely to exist. Mutants in roughly one quarter of the complementation groups still exhibit respiratory activity at the nonpermissive temperature and are possibly deficient for oxidative phosphorylation or energy coupling. About the same proportion is converted into ρ^- mutants when grown for a few cell divisions at the nonpermissive temperature (Schweizer et al. 1977). *pet⁻* mutants affected in mitochondrial protein synthesis tend to accumulate ρ^- mutants when grown at the restrictive temperature.

The last approach is limited to the nuclear genes directly involved in the expression of the mitochondrial genome. In this respect, Dujardin et al. (1980a,b) recently found, from a first isolation of suppressors of specific *mit⁻* mutants, three nuclear suppressors (called *NAM1* [nuclear accomodation of mitochondria], *NAM2*, and *nam3*. *NAM1* (dominant) and *nam3* (recessive), selected to suppress an *oxi3⁻* and a *box⁻* mutant, respectively, have a large spectrum of action on *mit⁻* mutants in different genes (*cob-box, oxi1, oxi2,* and *oxi3*). They are allele-specific but locus-nonspecific and are probably informational suppressors. In contrast, four *NAM2* (dominant) mutants (heteroalleles of the same nuclear gene) originally selected to suppress a *box7⁻* mutant have a spectrum of action limited to *box7⁻* mutants. It is probable that such suppressors restore the deficiency of splicing of *box7⁻* mutants, hence, offering a unique control of mitochondrial gene expression. Further analysis of such mutations should probably reveal intricate mechanisms of genetic expression between the mitochondrial and nuclear genomes.

ATPase Complex

The Subunits and Their Sites of Synthesis

The energy-transducing, oligomycin-sensitive ATPase complex of yeast mitochondria is a major constituent of the mitochondrial inner membrane, coupling the proton translocation across the inner membrane and the synthesis of ATP. Classically, the ATPase complex, extracted with mild detergents and purified by density-gradient centrifugation, has been described as being composed of 10 nonidentical subunits, corresponding to an overall molecular weight of ca. 630,000 daltons (Tzagoloff and Meagher 1971; Todd et al. 1980). The complex can be resolved in three functional parts: the F1, the oligomycin-sensitive conferring protein (OSCP), and the membrane factor (Table 8). The purified F1 is able to catalyze the hydrolysis of ATP (but in an oligomycin-insensitive fashion) but neither ATP synthesis nor the exchange reaction with inorganic phosphate. Since ρ^- mutants are able to synthesize the F1 ATPase, all subunits of the F1 must be translated on cytoplasmic ribosomes (Kovac and Weissova 1968; Schatz 1968). The extraction of the F1 from submitochondrial particles (using NaBr) leaves behind depleted membranes, incapable of hydrolyzing ATP but capable of binding a purified F1 fraction to reconstitute the oligomycin-sensitive activity. Further treatment of these depleted membranes (with alkali) removes the second functional part, OSCP, and these membranes are now unable to bind a purified F1 unless the OSCP is added. Thus, the OSCP is believed to function as a link between the soluble F1, protruding in the matrix and carrying the catalytic activity, and the membrane factor, carrying the oligomycin-sensitive sites. The OSCP is a single polypeptide that is assigned to the subunit 7 by analogy to the ATPase of beef-heart mitochondria. Its synthesis also takes place on cytoplasmic ribosomes (Tzagoloff 1970). The membrane factor has originally been described as composed of four nonidentical subunits, which are highly hydrophobic polypeptides, tightly associated with the inner membrane (Tzagoloff and Meagher 1971, 1972). The subunits of the membrane factor are not found in ρ^- mutants and, from a series of in vivo labeling experiments of yeast cells undergoing glucose derepression, Tzagoloff and Meagher (1972) and Tzagoloff and Akai (1972) originally concluded that all four polypeptides of the membrane factor are synthesized on mitochondrial ribosomes. However, only two mitochondrial genes have been found, coding, respectively, for the 8-kD and 21.9-kD subunits, and it is not very probable that other mitochondrial genes of this type can be found (see above, Simple Mitochondrial Genes: The *oli1* and *oli2* Genes for Two ATPase Subunits). The exact nature of the other subunits of the membrane factor therefore remains unclear. Although the two mitochondrial genes contain oli^R mutants, it is the subunit 9 (8kD) that is directly responsible for the oligomycin sensitivity, as can be shown by

Table 8 Oligomeric enzyme complexes containing mitochondrially encoded subunits

Polypeptide		Site of synthesis	Apparent molecular weight (kD)[a]	Additional mass of precursors (kD)	Final destination	Stoichiometry in complex	Genes	References
a. Oligomycin-sensitive ATP synthetase complex								
F1 ATPase	subunit 1 (or α)	cytoplasm	56	6	protruding in	3		1,2,3,4,5,6,7,8
	subunit 2 (or β)	cytoplasm	52	2	matrix bound	3	AUR	1,2,3,4,5,6,7,8
	subunit 3[b] (or γ)	cytoplasm	32	6	to OSCP	1		1,2,3,4,5,6,7,8
	subunit 4	cytoplasm	25.3			2		1,2,3,4,5,6
	subunit 8a[b]	cytoplasm?	9.6			1[b]		1,2,3,4,5,6
OSCP	subunit 7	cytoplasm?	13.7		bound to membrane factor	1		1,9
Membrane factor	subunit 5	?	23.7		inner	1	?	1,10,11
	subunit 6[b]	mitochondria	21.9			4[b]	oli2	1,10,11
	subunit 8b[b]	?	9.6		membrane	1[b]	?	1,10,11
	Subunit 9	mitochondria	8			probably 6	oli1	1,10,11
b. Cytochrome-c oxidase complex								
	subunit I	mitochondria	42		inner membrane		oxi3	12,13,14,15,20,21
	subunit II	mitochondria	34.5	1.5	inner membrane (outer side)		oxi1	12,13,14,15,16, 20,21,27
	subunit III	mitochondria	23		inner membrane (outer side)		oxi2	12,13,14,15,20,21
	subunit IV	cytoplasm	14	3	inner membrane (inner side)			12,13,14,15,17, 18,20,21

						References
subunit V	cytoplasm	12.5	2		inner membrane	12,13,14,15,19,20,21
subunit VI	cytoplasm	12.5	6		inner membrane	12,13,14,15,19,20,21
subunit VII	cytoplasm	4.5	0		inner membrane	12,13,14,15,19,20,21

c. Cytochrome b-c₁ complex

							References
subunit I	cytoplasm	44			inner membrane		22
subunit II	cytoplasm	40			inner membrane		22
cytochrome c_1	cytoplasm	32	6		inner membrane	1	19,22,22,23,24
cytochrome b	mitochondria	30	2	cob-box	inner membrane	2	22,25
subunit V	cytoplasm	17	2		inner membrane		22,26
subunit VI	cytoplasm	14			inner membrane		22
subunit VII	cytoplasm	11			inner membrane		22

[a] The apparent molecular weights may differ among estimations. Refer to original articles.

[b] Subunits 6 and 8 have a total stoichiometry of 4 and 2, respectively. Each can be separated into two bands, probably isostoichiometric. For the two bands of subunit 6, only the 21.9-kD polypeptide is mitochondrially synthesized (Somlo et al. 1981). Subunit 3 may be a proteolytic product of Subunit 1 (Ryrie and Gallagher 1979).

References: (1) Todd et al. (1980); (2) Kovac and Weisskova (1968); (3) Schatz (1968); (4) Tzagoloff et al. (1972); (5) Tzagoloff (1969); (6) Tzagoloff et al. (1973); (7) Schatz (1979); (8) Maccacchini et al. (1979b); (9) Tzagoloff (1970); (10) Tzagoloff and Meagher (1971); (11) Tzagoloff and Meagher (1972); (12) Dockter et al. (1978); (13) Eytan and Schatz (1975); (14) Birchmeier et al. (1976); (15) Cerletti and Schatz (1979); (16) Woodrow and Schatz (1979); (17) Lewin et al. (1980); (18) Mihara and Blobel (1980); (19) Nelson and Schatz (1979); (20) Ross et al. (1974); (21) Rubin and Tzagoloff (1973a,b); (22) Katan et al. (1976a,b); (23) Claisse and Pajot (1974); (24) Nelson and Schatz (1979); (25) Kreike et al. (1979); (26) Cote et al. (1979); (27) Sevarino and Poynton (1980).

affinity-labeling experiments (Criddle et al. 1976b) and by irreversible labeling with radioactive DCCD (a drug with an inhibitory effect similar to oligomycin; Sebald et al. 1979a). Although the subunit 9 is the most hydrophobic subunit (polarity $< 25\%$), the fact that it can easily be solubilized from submitochondrial particles by chloroform/methanol extractions has permitted its purification (Tzagoloff and Akai 1972; Sierra and Tzagoloff 1973; Tzagoloff et al. 1973). At present, it is the only polypeptide encoded by the yeast mitochondrial genome whose amino acid sequence has been directly determined (Wachter et al. 1977). From studies of the number of inhibitor binding sites per ATPase complex equivalent, Sebald et al. (1979a) determined the presence of six subunits 9 per complex. The high-molecular-weight form of the subunit 9 (45,000 daltons), which can be extracted from the ATPase complex or from submitochondrial particles under some conditions (Tzagoloff and Akai 1972; Sierra and Tzagoloff 1973; Tzagoloff et al. 1976a) probably corresponds to the native hexameric form of this subunit. Finally, a protein that is a strong inhibitor of the activity of the soluble F1 ATPase has recently been discovered in *Saccharomyces cerevisiae* (Ebner and Maier 1977), although it was previously detected only in beef heart mitochondria (see Senior 1977). As in beef-heart mitochondria, the protein inhibitor is believed to play a regulatory role in the energy conservation functions by controlling the backflow of energy from ATP to the mitochondrial inner membrane. It is most probably nuclearly encoded because its synthesis is inhibited by cycloheximide. However, it is absent in ρ^- mutants, suggesting a mitochondrial control over its synthesis or transport into mitochondria (Ebner and Maier 1977).

Cross-linking experiments and immunoprecipitation of uniformly labeled submitochondrial preparations indicate a minimum of 20 polypeptide chains per ATPase complex. The arrangement and the stoichiometry of the different subunits are not yet completely elucidated, because additional subunits of low molecular weight have been reported and because some of the subunits originally described may be the products of proteolytic degradations of other subunits (Enns and Criddle 1977; Ryrie and Gallagher 1979; Todd et al. 1980).

Nuclear Genes for Cytoplasmically Synthesized Subunits

Little is known about the genes for the cytoplasmically synthesized subunits of ATPase. Many mutants resistant to various inhibitors of ATPase functions have been isolated and characterized (Griffiths et al. 1974). A number of such mutants confer resistance to several unrelated inhibitors, which suggests that they may affect general membrane functions instead of specific subunits of ATPase. At present, only the gene coding for the β subunit has been characterized. The β subunit carries the aurovertin binding site, and unassembled β subunits are able to bind aurovertin with roughly the same affinity as the F1 ATPase (Douglas et al. 1977). Four aurovertin-resistant mutants (all resulting from single mutations in a unique nuclear gene, *aur1*) have been found to confer resistance to the isolated F1 ATPase in vitro (Douglas et al. 1979b). The fact that aurovertin binding does not occur with β subunits isolated from *aur1* mutants suggests that the *aur1* gene is the structural gene for this polypeptide.

Enzymes of the Respiratory Chain

Complex IV or Cytochrome-c Oxidase Complex

The purified cytochrome-*c* oxidase complex contains seven nonidentical subunits (Table 8) for an overall mass of 190–226 kD (Mason and Schatz 1973; Mason et al. 1973; Rubin and Tzagoloff 1973a). All seven subunits are integral parts of the complex necessary for the oxidation of cytochrome *c*, as can be shown by inhibition and immunoprecipitation by antisera prepared against purified subunits (Poynton and Schatz 1975a). It is now clear that the three largest subunits of the complex are synthesized on mitochondrial ribosomes (their synthesis is sensitive to erythromycin or chloramphenicol but not to cycloheximide; Mason and Schatz 1973; Rubin and Tzagoloff 1973b), and they are encoded by the three mitochondrial genes *oxi1, oxi2,* and *oxi3*. On the other hand, the four small subunits are synthesized on cytoplasmic ribosomes. However, the subunit VII is not detectable in ρ^- mutants (Ebner et al. 1973a), pointing out an independent regulation of the synthesis or the assembly of the different subunits. As in the case of the ATPase complex, the mitochondrially made subunits have a very low polarity (34.7% and 42.1%, respectively, for the subunits I and II) corresponding to membrane polypeptides, whereas the cytoplasmically made subunits are more water soluble (polarity is 48–49% for the subunits IV and VI) (Poynton and Schatz 1975b).

To investigate the assembly mechanism of the different subunits and the regulation of their synthesis, Poynton and Groot (1975) initially developed an in vitro system, using isolated mitochondria, that is able to synthesize the three mitochondrially made subunits of cytochrome oxidase. Furthermore, Groot and Poynton (1975) showed that, in their system, the synthesis of these three subunits is subject to regulation by oxygen similar to that found in vivo by Mason and Schatz (1973). The subunit III is synthesized in the absence of oxygen, whereas the subunits I and II are not. Poynton and Kavanagh (1976) observed that the protein synthesis by isolated mitochondria is dependent upon an endogenous pool of cytoplasmically synthesized proteins present within mitochondria at the time of isolation and that mitochondria depleted for this pool can be stimulated by a fraction of the postribosomal supernatant. Poynton and McKemmie (1976) argued that this stimulating factor is proteic in nature because if the supernatant is treated with antisera against the subunits IV or VI of cytochrome oxidase, no stimulation of the mitochondrial synthesis occurs. More recently, however, Everett et al. (1980) and Ohashi and Schatz (1980) pointed out that the cytoplasmic stimulating factor is not species-specific, and they identified it as guanyl nucleotides.

In these experiments, Poynton and McKemmie (1976, 1979a,b) observed that the immunoprecipitate from the cytoplasmic supernatant contains a single labeled polypeptide of molecular weight 55 kD. Because this polypeptide can be immunoprecipitated by antisera against each cytoplasmically made subunit and contains nearly all of the tryptic peptides found in these four subunits (plus a few additional ones), they concluded that it represents a polyprotein precursor to all four cytoplasmically synthesized subunits, which is processed posttranslationally during the transport or the assembly. More recently, however, Mihara and Blobel (1980) and Lewin et al. (1980) reported new results that do not support these conclusions. Using subunit-specific antisera, they found that the subunits IV, V, VI, and VII (made either in vitro in a cell-free system programmed with total yeast mRNAs or in vivo in intact cells or in spheroplasts) are synthesized as individual proteins. Aminoterminal labeling (with N-formyl [^{35}S]tRNA^{Met}) confirms that each subunit is synthesized as a separate entity and shows that most of them (except the subunit VII) are synthesized as a precursor, 2–6 kD larger than the mature products (Table 8). The additional protein sequence of the larger precursor is considered to serve as a signal sequence for the unidirectional posttranslational translocation from the cytoplasm into the mitochondria (see Mihara and Blobel 1980). The signal sequence is cleaved either during or shortly after import into mitochondria. It is possible that the subunit VII also possesses a signal sequence, but it is not cleaved after the import into mitochondria.

The reason for the detection of the 55-kD polypeptide by Poynton and collaborators and not by Mihara and Blobel (1980) or Lewin et al. (1980) is unclear. Lewin et al. suggested that the observation of the 55-kD polypeptide is due to the presence of a small proportion of contaminating antibodies against a highly antigenic polypeptide of unknown origin.

Complex III or Cytochrome-b-c$_1$ Complex

The last oligomeric complex containing mitochondrially synthesized subunits, complex III of the respiratory chain, includes cytochromes b and c_1 and catalyzes the oxidation of the ubiquinone (coenzyme Q) by cytochrome c. The purification of this complex from yeast mitochondria has encountered greater difficulties than that of the other complexes, and it has not yet been studied in comparable detail. The complex is made of seven polypeptides, only one of which (apocytochrome b) is highly hydrophobic (Katan et al. 1976b) and is synthesized on mitochondrial ribosomes (Katan et al. 1976a; Lin and Beattie 1976; Lin et al. 1978; Kreike et al. 1979). Cytochrome c_1, present in ρ^- mutants (Claisse and Pajot 1974), is synthesized on cytoplasmic ribosomes. The same is true of the other subunits. The subunit V is synthesized as a larger precursor, probably cleaved after its import into mitochondria (Cote et al. 1979).

Import of Cytoplasmically Synthesized Mitochondrial Polypeptides

Because they do not import mRNA, mitochondria must continuously import a large number of polypeptides, synthesized on the cytoplasmic ribosomes, to be eventually localized in the matrix at the internal side of the inner mitochondrial membrane or across it. Earlier, Butow and collaborators presented cytological and biochemical evidence for a specific binding of cytoplasmic ribosomes to the outer membrane of mitochondria (Kellems and Butow 1972, 1974; Kellems et al. 1974; Butow et al. 1975). They observed that the cytoplasmic ribosomes were sitting on the outer mitochondrial membrane precisely at the points of close contact with the inner membrane but not where the two membranes are separated. Inhibition by puromycin indicated that the ribosomes were bound to the outer membrane by the nascent polypeptide chain they synthesize. From these observations, they proposed that the mechanism of import of cytoplasmically synthesized polypeptides into mitochondria is a vectorial translation of the nascent polypeptide chain across the two mitochondrial membranes at the point of close apposition. This model once enjoyed wide acceptance but now needs to be reexamined in the light of more recent data. The vectorial transfer model predicts that no mitochondrial polypeptides (either mature or as precursor form) should be found in the cytoplasm (because they are imported into the mitochondria as they are synthesized) and that the import should be strictly coupled with protein synthesis. Furthermore, by analogy with vectorial translation in other systems, which usually proceeds via the transient formation of an aminoterminal signal sequence removed before translation is even completed, no discrete larger precursor of the cytoplasmically made mitochondrial polypeptides should exist in the cell.

In the case of the F1 ATPase protruding into the matrix, immunological evidence, pulse chase experiments, and aminoterminal labeling showed that each of the three larger subunits is synthesized as an individual precursor, larger than the mature product and present in the postribosomal supernatant (Maccacchini et al. 1979b; Schatz 1979; Lewin et al. 1980). If these precursors are incubated in the presence of purified mitochondria, they are converted to mature subunits that are no longer susceptible to external proteases, which indicates that they have been taken up by the mitochondria. Under the same conditions, the mature polypeptides are not taken up by the isolated mitochondria. Therefore, processing and uptake are concomitant but independent of protein synthesis.

A similar mechanism has also been found for other cytoplasmically synthesized polypeptides with different final destinations in the mitochondria. Cote et al. (1979) showed that some cytoplasmically made subunits of the cytochrome b-c_1 complex are also made as larger precursors (Table 8). The exact location of these subunits in the complex is not precisely determined, but it is probable that they are partially embedded in the inner membrane or associated with its outer face. The same is true for the four cytoplasmically synthesized subunits of cytochrome oxidase (mentioned earlier in the text). A larger precursor is also found in the cytoplasm for the cytochrome-c peroxidase, a polypeptide synthesized on cytoplasmic ribosomes and then imported into the intermembrane space (Schatz 1979).

Import and processing are energy-dependent mechanisms. Under complete ATP depletion of the matrix (by blocking simultaneously the oxidative phosphorylation and the adenyl nucleotide carrier), the precursors of the three largest subunits of ATPase are neither processed nor imported into mitochondria, whereas the cytochrome-c peroxidase, located in the intermembrane space where ATP is freely available, is still imported (Nelson and Schatz 1979).

This mechanism of import raises a number of questions on the nature, the specificity, and the location of the proteolytic enzyme(s). How are they, themselves, imported into the mitochondria and what is the nature of the receptor sites at the mitochondrial surface? Further studies are needed to answer these questions.

CONCLUDING REMARKS

Mitochondria are complex organelles whose formation necessitates the intricate cooperation of nuclear and mitochondrial genes and their products. The mitochondrial genome has now been characterized in great detail, both molecularly and genetically. If the amount of genetic information it contributes to the cell is limited, this information is, however, essential for the function of mitochondria, i.e., the production of energy to the cell through respiration and oxidative phosphorylation. It is probable that all of the mitochondrial genes directly involved in these functions are already known (three genes for cytochrome oxidase, two for ATPase, and one for apocytochrome b). Probably the most striking aspect of the mitochondrial genome, then, is this apparent discrepancy between the limited number of these genes and the existence of a complete genetic system to express this information, with a specific protein synthetic machinery, itself encoded in part by the mitochondrial genome. Contrary to the classic endosymbiotic theory for the origin of mitochondria, the universality of these functions among mitochondrial genomes of distant evolutionary organisms (e.g., yeast and human) argues against the gradual integration of mitochondrial genes into the nucleus (at least within the evolutionary scale of presently living organisms) and suggests that these genes must, for some reason, reside in the mitochondrial genome, hence, the existence of a complete genetic system.

One of these reasons may be sought in the specific features of the mitochondrial genome itself, which differs at almost every level of the organization and expression from both nuclear genomes and prokaryotic genomes. One of the most conspicuous differences is probably that mitochondria use a genetic code slightly different from the "universal" code. If the use of the UGA codon to specify tryptophan seems a universal feature to mitochondria of different organisms, a few additional departures from the universal code are also found in each case but seem specific to the organism (or at least to groups of related species). Possibly, the limited number of mitochondrial genes offers possibilities for evolution of the mitochondrial genetic code that are nonexistent for more complex genomes, and mitochondria may therefore represent the most highly evolved genetic system.

The mosaic structure of some mitochondrial genes classifies them as typical eukaryotic genes. But both the evolutionary conservation of the exon sequences and the fact that some introns may be optional within the same gene of the same species tends to suggest the late insertion of these introns into preexisting mitochondrial genes. That long reading frames are found in many mitochondrial introns raises a fundamental question about the origin of these DNA sequences. The fact that some of them are probably used, in continuity with the preceding exon(s), to encode hybrid polypeptides necessary for the splicing of the corresponding intron sequence reveals an original mode of organization of some mitochondrial genes with a nested set of coding sequences. Perhaps, the fact that transcription and translation occur in the same compartment in mitochondria accounts for some of these peculiarities.

The specificity of the mitochondrial genome is also apparent by the mode of inheritance followed by mitochondrial genes. In contrast to the very precise rules for the inheritance of nuclear genes at mitosis and meiosis, the rules for transmission, recombination, and segregation of mitochondrial genes are relatively lax, as if the multiplicity of the mitochondrial genomes in the cell compensated for the imprecision given to each individual genome.

Obviously, a number of important questions remain concerning the yeast mitochondrial genome (e.g., the significance of the intergenic sequences which, in total, represent more than half of all mtDNA sequences, the mechanism of ρ^- formation, the differentiation of mitochondria in response to oxygen pressure or catabolite repression, the nuclear genes that accomodate mitochondria). However, because of the unique property of *S. cerevisiae*, which has a mitochondrial genome entirely dispensable to the cell, yeast mitochondria represent a eukaryotic genetic system that is the most directly amenable to experimental analysis. This may be particularly important, for example, in unraveling the details of the protein synthetic machinery or in examining specific mechanisms of genetic regulation that are of general significance.

Note Added In Proof

A meeting on mitochondrial genes was recently held at Cold Spring Harbor Laboratory after this review was completed. The reader can refer to Slonimski et al. (1982).

ACKNOWLEDGMENTS

I am grateful to H. Blanc, G. Church, W. Gilbert, R. Morimoto, D. Morisato, P. Slonimski, and D. Wirth for support, discussions, and encouragement during the long period of this redaction. I gratefully acknowledge support from the Centre National de la Recherche Scientifique and from the National Science Foundation under their exchange program.

REFERENCES

Accoceberry, B. and A. Stahl. 1971. Chromatographic differences between the cytoplasmic and the mitochondrial tRNA of *Saccharomyces cerevisiae. Biochem. Biophys. Res. Commun.* **42:** 1235.

Alexander, N.J., P.S. Perlman, D.K. Hanson, and H.R. Mahler. 1980. Mosaic organization of a mitochondrial gene: Evidence from double mutants in the cytochrome *b* region of *Saccharomyces cerevisiae. Cell* **20:** 199.

Allen, N.E. and A.M. McQuillan. 1969. Target analysis of mitochondrial genetic units in yeasts. *J. Bacteriol.* **97:** 1142.

Arnberg, A.C., G.J.B. Van Ommen, L.A. Grivell, E.F.J. Van Bruggen, and P. Borst. 1980. Some yeast mitochondrial RNAs are circular. *Cell* **19:** 313.

Atchinson, B.A., K.B. Choo, R.J. Devenish, and A.W. Linnane. 1979. Biogenesis of mitochondria. 53. Physical map of genetic loci in the 21S ribosomal RNA region of mitochondrial DNA in *Saccharomyces cerevisiae. Mol. Gen. Genet.* **174:** 307.

Aufderheide, K.J. 1975. Cytoplasmic inheritance in *Saccharomyces cerevisiae:* Comparison of first zygotic budsite to mitochondrial inheritance patterns. *Mol. Gen. Genet.* **140:** 231.

Avner, P.R. and D.E. Griffiths. 1970. Oligomycin resistant mutants in yeast. *FEBS Lett.* **10:** 202.

———. 1973a. Studies on energy-linked reactions: Isolation and characterization of oligomycin resistant mutants of *Saccharomyces cerevisiae. Eur. J. Biochem.* **32:** 301.

———. 1973b. Studies on energy linked reactions: Genetic analysis of oligomycin resistant mutants of *Saccharomyces cerevisiae. Eur. J. Biochem.* **32:** 312.

Avner, P.R., D. Coen, B. Dujon and P.P. Slonimski. 1973. Mitochondrial genetics. IV. Allelism and mapping studies of oligomycin resistant mutants in *Saccharomyces cerevisiae. Mol. Gen. Genet.* **125:** 9.

Backhaus, B., R.J. Schweyen, F. Kaudewitz, and B. Dujon. 1978. On the formation of *rho⁻* petites in yeast. III. Effects of temperature on transmission and recombination of mitochondrial markers and on *rho⁻* cell formation in temperature sensitive mutants of *Saccharomyces cerevisiae. Mol. Gen. Genet.* **161:** 153.

Bacila, M., B.L. Horecker, and A.O.M. Stoppani, eds. 1978. *Biochemistry and genetics of yeast: Pure and applied aspects.* Academic Press, New York.

Baldacci, G., M. De Zamaroczy, and G. Bernardi. 1980. Excision sites in the GC clusters of the mitochondrial genome of yeast. *FEBS Lett.* **114**: 234.

Baldacci, G., E. Cundari, S. Francisci, and C. Palleschi. 1978. Variability of mitochondrial alanyl tRNAs isoaccepting species in *Saccharomyces cerevisiae. Bull. Mol. Biol. Med.* **3**: 243.

Baldacci, G., C. Falcone, L. Frontali, G. Macino, and C. Palleschi. 1975. tRNA and aminoacyl tRNA synthetases from wild type and petite yeast mitochondria. In *Molecular biology of nucleocytoplasmic relationships* (ed. S. Puiseux-Dao), p. 41. Elsevier, Amsterdam.

———. 1976. Isoaccepting tRNA$_{ser}$ in mitochondria from *Saccharomyces cerevisiae*: Mitochondrially coded and cytoplasmic species. In *Genetics and biogenesis of chloroplasts and mitochondria* (ed. T. Bücher et al.), p. 759. Elsevier/North-Holland, Amsterdam.

———. 1977. Differences in mitochondrial isoaccepting tRNAs from *Saccharomyces cerevisiae* as a function of growth conditions. In *Mitochondria 1977: Genetics and biogenesis of mitochondria* (ed. W. Bandlow et al.), p. 571. de Gruyter, Berlin.

Bandlow, W. 1976. On the formation of *rho⁻* petites in yeast. VI. Expression of a mitochondrial conditional mutation controlling petite formation in *Saccharomyces cerevisiae*. In *Genetics, biogenesis, and bioenergetics of mitochondria* (ed. W. Bandlow et al.), p. 179. de Gruyter, Berlin.

———. 1977. Derepression of mitochondrial enzymes in yeast wild type and a mitochondrial conditional mutant *tsm8*, retarded in oxidative adaptation. In *Mitochondria 1977: Genetics and biogenesis of mitochondria* (ed. W. Bandlow et al.), p. 531. de Gruyter, Berlin.

Bandlow, W. and R.J. Schweyen. 1975. On the mechanism of petite genesis in yeast. IV. Biochemical characterization of a conditional cytoplasmic mutant producing petites at restrictive temperature. *Biochem. Biophys. Res. Commun.* **67**: 1078.

Bandlow, W., R.J. Schewyn, K. Wolf, and F. Kaudewitz, eds. 1976. *Molecular biology of nucleocytoplasmic relationships*. de Gruyter, Berlin.

———. 1977. *Mitochondria 1977: Genetics and biogenesis of mitochondria*. de Gruyter, Berlin.

Barclay, B.J. and J.G. Little. 1978. Genetic damage during thymidylate starvation in *Saccharomyces cerevisiae. Mol. Gen. Genet.* **160**: 33.

Barrell, B.G., A.T. Bankier, and J. Drouin. 1979. A different genetic code in human mitochondria. *Nature* **282**: 189.

Bastos, R.N. and H.R. Mahler. 1974. Molecular mechanisms of mitochondrial genetic activity. Effects of ethidium bromide on the DNA and energetics of isolated mitochondria. *J. Biol. Chem.* **249**: 6617.

Baur, E. 1909. Das Wesen und die Erblichkeitsverhältnisse der "Varietates albomarginatae hort" von Pelargonium zonale. *Z. Vererbungsl.* **1**: 330.

Bech-Hansen, N.T. and G.H. Rank. 1972. Ethidium bromide resistance and petite induction in *S. cerevisiae. Can. J. Genet. Cytol.* **14**: 681.

Berlani, R.E., C. Pentella, G. Macino, and A. Tzagoloff. 1980. Assembly of the mitochondrial membrane system: Isolation of mitochondrial transfer ribonucleic acid mutants and characterization of transfer ribonucleic acid genes of *Saccharomyces cerevisiae. J. Bacteriol.* **141**: 1086.

Bernardi, G. 1976. Organization and evolution of the mitochondrial genome of yeast. *J. Mol. Evol.* **9**: 25.

Bernardi, G. and S.N. Timasheff. 1970. Optical rotatory dispersion and circular dichroism properties of yeast mitochondrial DNAs. *J. Mol. Biol.* **48**: 43.

Bernardi, G., G. Piperno, and G. Fonty. 1972. The mitochondrial genome of wild-type yeast cells. I. Preparation and heterogeneity of mitochondrial DNA. *J. Mol. Biol.* **65**: 173.

Bernardi, G., M. Faures, G. Piperno, and P.P. Slonimski. 1970. Mitochondrial DNAs from respiratory-sufficient and cytoplasmic respiratory-deficient mutant yeast. *J. Mol. Biol.* **48**: 23.

Bernardi, G., F. Carnevali, A. Nicolaieff, G. Piperno, and G. Tecce. 1968. Separation and characterization of a satellite DNA from a yeast cytoplasmic "petite" mutant. *J. Mol. Biol.* **37**: 493.

Bernardi, G., G. Baldacci, G. Faugeron-Fonty, C. Gaillard, R. Goursot, A. Huyard, M. Man-gin, R. Marotta, and M. De Zamaroczy. 1980. The petite mutation: Excision sequences, replication origins and suppressivity. In *The organization and expression of the mitochondrial genome* (ed. C. Saccone and A.M. Kroon), p. 21. Elsevier/North-Holland, Amsterdam.

Birchmeier, W., C.E. Kohler, and G. Schatz. 1976. Interaction of integral and peripheral membrane proteins: Affinity labelling of yeast cytochrome oxidase by modified cytochrome *c*. *Proc. Natl. Acad. Sci.* **73:** 4334.

Birky, C.W., Jr. 1973. On the origin of mitochondrial mutants: Evidence for intracellular selection of mitochondria in the origin of antibiotic-resistant cells in yeast. *Genetics* **74:** 421.

————. 1975a. Zygote heterogeneity and uniparental inheritance of mitochondrial genes in yeast. *Mol. Gen. Genet.* **141:** 41.

————. 1975b. Effects of glucose repression on the transmission and recombination of mitochondrial genes in yeast (*Saccharomyces cerevisiae*). *Genetics* **80:** 695.

————. 1978. Transmission genetics of mitochondria and chloroplasts. *Annu. Rev. Genet.* **12:** 471.

Birky, C.W., Jr., P.S. Perlman, and T.J. Byers, eds. 1975. *Genetics and biogenesis of mitochondria and chloroplasts.* Ohio University Press, Columbus.

Birky, C.W., Jr., C.A. Demko, P.S. Perlman, and R. Strausberg. 1978a. Uniparental inheritance of mitochondrial genes in yeast: Dependance on input bias of mitochondrial DNA and preliminary investigations of the mechanisms. *Genetics* **89:** 615.

Birky, C.W., Jr., R.L. Strausberg, P.S. Perlman, and J.L. Forster. 1978b. Vegetative segregation of mitochondria in yeast: Estimating parameters using a random model. *Mol. Gen. Genet.* **158:** 251.

Blamire, J., D.R. Cryer, D.B. Finkelstein, and J. Marmur. 1972. Sedimentation properties of yeast nuclear and mitochondrial DNA. *J. Mol. Biol.* **67:** 11.

Blamire, J., C.A. Michels, J.M. Walsh, and D.L. Friedenberg. 1976. Mitochondrial DNA in yeast recombination and subsequent modification following mating between a grande and a suppressive petite. *Mol. Gen. Genet.* **143:** 253.

Blanc, H. 1979. "Etude du phénomène de suppressivité et de la transformation chez *Saccharomyces cerevisiae*." Ph.D. thesis, Institut National Agronomique, Paris.

Blanc, H. and B. Dujon. 1980. Replicator regions of the yeast mitochondrial DNA responsible for suppressiveness. *Proc. Natl. Acad. Sci.* **77:** 3942.

————. 1981. Replicator regions of the yeast mitochondrial DNA active *in vivo* and in yeast transformants. In *Mitochondrial genes* (ed. P.P. Slonimski et al.). Cold Spring Harbor Laboratory, Cold Spring Harbor, New York. (In press.)

Bleeg, H.S., A. Leth-Bak, C. Christiansen, K.E. Smith, and A. Stenderup. 1972. Mitochondrial DNA and glucose repression in yeast. *Biochem. Biophys. Res. Commun.* **47:** 524.

Boardman, N.K., A.W. Linnane, and R.M. Smillie, eds. 1971. *Autonomy and biogenesis of mitochondria and chloroplasts.* Elsevier/North-Holland, Amsterdam.

Boker, E., F. Kaudewitz, K.V. Richmond, R.J. Schweyen, and D.Y. Thomas. 1976. Experimental variation of the number of mitochondrial genomes participating in mitochondrial genetics. In *Genetics, biogenesis and bioenergetics of mitochondria* (ed. W. Bandlow et al.), p. 99. de Gruyter, Berlin.

Bolotin, M., D. Coen, J. Deutsch, B. Dujon, P. Netter, E. Petrochilo, and P.P. Slonimski. 1971. La recombinaison des mitochondries chez *Saccharomyces cerevisiae*. *Bull. Inst. Pasteur* **69:** 215.

Bolotin-Fukuhara, M. 1979. Mitochondrial and nuclear mutations that affect the biogenesis of the mitochondrial ribosomes of yeast. I. Genetics. *Mol. Gen. Genet.* **177:** 39.

Bolotin-Fukuhara, M. and H. Fukuhara. 1976. Modified recombination and transmission of mitochondrial genetic markers in *rho* minus mutants of *Saccharomyces cerevisiae*. *Proc. Natl. Acad. Sci.* **73:** 4608.

Bolotin-Fukuhara, M., G. Faye, and H. Fukuhara. 1976. Localization of some mitochondrial mutations in relation to transfer and ribosomal RNA genes in *Saccharomyces cerevisiae*.

In *The genetic functions of mitochondrial DNA* (ed. C. Saccone and A.M. Kroon), p. 243. Elsevier/North-Holland, Amsterdam.

————. 1977. Temperature-sensitive respiratory-deficient mitochondrial mutations: Isolation and genetic mapping. *Mol. Gen. Genet.* **152**: 295.

Bonitz, S.G., G. Coruzzi, B.E. Thalenfeld, A. Tzagoloff, and G. Macino. 1980a. Assembly of the mitochondrial membrane system: Physical map of the *oxi3* locus of yeast mitochondrial DNA. *J. Biol. Chem.* **255**: 11922.

————. 1980b. Assembly of the mitochondrial membrane system: Structure and nucleotide sequence of the gene coding for subunit I of yeast cytochrome oxidase *J. Biol. Chem.* **255**: 11927.

Bonitz, S.G., R. Berlani, G. Coruzzi, M. Li, G. Macino, F.G. Nobrega, M.P. Nobrega, B.E. Thalenfeld, and A. Tzagoloff. 1980c. Codon recognition rules in yeast mitochondria. *Proc. Natl. Acad. Sci.* **77**: 3167.

Borst, P. 1972. Mitochondrial nucleic acids. *Annu. Rev. Biochem.* **41**: 333.

Borst, P. and R.A. Flavell. 1976. Properties of mitochondrial DNAs. In *Handbook of biochemistry and molecular biology: Nucleic acids* (ed. G.E. Fasman), vol. 2, p. 363. CRC Press, Cleveland.

Borst, P. and L.A. Grivell. 1978. The mitochondrial genome of yeast *Cell* **15**: 705.

Borst, P., C. Heyting, and J.P.M. Sanders. 1976a. The control of mitochondrial DNA synthesis in yeast petite mutants. In *Genetics and biogenesis of chloroplasts and mitochondria* (ed. T. Bücher et al.), p. 525. Elsevier/North-Holland, Amsterdam.

Borst, P., J.P.M. Sanders, and C. Heyting. 1976b. Genes on *Saccharomyces* DNA. In *Genetics, biogenesis and bioenergetics of mitochondria* (ed. W. Bandlow et al.), p. 85. de Gruyter, Berlin.

————. 1979. Biochemical methods to locate genes on the physical map of yeast mitochondrial DNA. *Methods Enzymol.* **56**: 182.

Borst, P., J.L. Bos, L.A. Grivell, G.S.P. Groot, C. Heyting, A.F.M. Moorman, J.P.M. Sanders, J.L. Talen, C.F. Van Kreijl, and G.J.B. Van Ommen. 1977. The physical map of yeast mitochondrial DNA: Anno 1977. In *Mitochondria 1977: Genetics and biogenesis of mitochondria* (ed. W. Bandlow et al.), p. 213. de Gruyter, Berlin.

Bos, J.L., C. Heyting, G. Van der Horst, and P. Borst. 1980a. The organization of repeating units in mitochondrial DNA from yeast petite mutants. *Curr. Genet.* **1**: 233.

Bos, J.L., K.A. Osinga, G. Van der Horst, and P. Borst. 1979. Nucleotide sequence of the mitochondrial structural genes for cysteine tRNA and histidine tRNA of yeast. *Nucleic Acids Res.* **6**: 3255.

Bos, J.L., C. Heyting, P. Borst, A.C. Arnberg, and E.F.J. van Bruggen. 1978a. An insert in the single gene for the large ribosomal RNA in yeast mitochondrial DNA. *Nature* **275**: 336.

Bos, J.L., C.F. Van Kreijl, F.H. Ploegaert, J.N.M. Mol, and P. Borst. 1978b. A conserved and unique (AT) rich segment in yeast mitochondrial DNA. *Nucleic Acids Res.* **5**: 4563.

Bos, J.L., K.A. Osinga, G. Van der Horst, N.B. Hecht, H.F. Tabak, G.J.B. Van Ommen, and P. Borst. 1980b. Splice point sequence and transcripts of the intervening sequence in the mitochondrial 21S ribosomal RNA gene of yeast. *Cell* **20**: 207.

Bücher, T., W. Neupert, W. Sebald, and S. Werner, eds. 1976. *Genetics and biogenesis of chloroplasts and mitochondria.* Elsevier/North-Holland, Amsterdam.

Burger, G., B. Lang, W. Bandlow, R.J. Schweyen, B. Backhaus, and F. Kaudewitz. 1976. Antimycin resistance in *Saccharomyces cerevisiae*: A new mutation on the mitDNA conferring antimycin resistance on the mitochondrial respiratory chain. *Biochem. Biophys. Res. Commun.* **72**: 1201.

Burkl, G., W. Demmer, H. Holzner, and E. Schweizer. 1976. Temperature sensitive nuclear "petite" mutants of *Saccharomyces cerevisiae.* In *Genetics, biogenesis and bioenergetics of mitochondria* (ed. W. Bandlow et al.), p. 39. de Gruyter, Berlin.

Butow, R.A. and R.L. Strausberg. 1979. Biochemical genetics of mitochondrial biogenesis. *Trends Biochem. Sci.* **4:** 110.

Butow, R.A., M.J. Ferguson, and A. Cederbaum. 1973. Low-temperature induction of respiratory deficiency in yeast mutants. *Biochemistry* **12:** 158.

Butow, R.A., W.F. Bennett, D.B. Finkelstein, and R.E. Kellems. 1975. Nuclear cytoplasmic interactions in the biogenesis of mitochondria in yeast. In *Membrane biogenesis* (ed. A. Tzagoloff), p. 155. Plenum Press, New York.

Butow, R.A., I.C. Lopez, H.-P. Chang, and F. Farrelly. 1980. The specification of var 1 polypeptide by the var 1 determinant. In *The organization and expression of the mitochondrial genome* (ed. C. Saccone and A.M. Kroon), p. 195. Elsevier/North-Holland, Amsterdam.

Cabral, F. and G. Schatz. 1978. Identification of cytochrome *c* oxidase subunits in nuclear yeast mutants lacking the functional enzyme. *J. Biol. Chem.* **253:** 4396.

Cabral, F., M. Solioz, Y. Rudin, G. Schatz, L. Clavilier, and P.P. Slonimski. 1978. Identification of the structural gene for yeast cytochrome *c* oxidase subunit II on mitochondrial DNA. *J. Biol. Chem.* **253:** 297.

Callen, D.F. 1974a. Recombination and segregation of mitochondrial genes in *Saccharomyces cerevisiae. Mol. Gen. Genet.* **134:** 49.

―――. 1974b. The effect of mating type on the polarity of mitochondrial gene transmission in *Saccharomyces cerevisiae. Mol. Gen. Genet.* **128:** 321.

―――. 1974c. Segregation of mitochondrially inherited antibiotic resistance genes in zygote cell lineages of *Saccharomyces cerevisiae. Mol. Gen. Genet.* **134:** 65.

Canaday, J., G. Dirheimer, and R.P. Martin. 1980. Yeast mitochondrial methionine initiator tRNA: Characterization and nucleotide sequence. *Nucleic Acids Res.* **8:** 1445.

Carignani, G., G. Dujardin, and P.P. Slonimski. 1979. Petite deletion map of the mitochondrial *oxi3* region in *Saccharomyces cerevisiae. Mol. Gen. Genet.* **167:** 301.

Caron, F., C. Jacq, and J. Rouviere-Yaniv. 1979. Characterization of a histone like protein extracted from yeast mitochondria. *Proc. Natl. Acad. Sci.* **76:** 4265.

Carnevali, F. and L. Leoni. 1972. Intramolecular heterogeneity of yeast mitochondrial DNA. *Biochem. Biophys. Res. Commun.* **47:** 1322.

Carnevali, F., G. Morpurgo, and G. Tecce. 1969. Cytoplasmic DNA from petite colonies of *Saccharomyces cerevisiae.* A hypothesis on the nature of the mutation. *Science* **163:** 1331.

Carnevali, F., L. Leoni, G. Morpurgo, and G. Conti. 1971. Induction of cytoplasmic "petite" mutation by antibacterial antibiotics. *Mutat. Res.* **12:** 357.

Casey, J., P. Gordon, and M. Rabinowitz. 1974a. Characterization of mitochondrial DNA from grande and petite yeasts by renaturation and denaturation analysis and by tRNA hybridization; evidence for internal repetition or heterogeneity in mitochonrial DNA population. *Biochemistry* **13:** 1059.

Casey, J.W., M. Cohen, M. Rabinowitz, H. Fukuhara, and G.S. Getz. 1972. Hybridization of mitochondrial transfer RNAs with mitochondrial and nuclear DNA of grande (wild type) yeast. *J. Mol. Biol.* **63:** 431.

Casey, J.W., H.-J. Hsu, G.S. Getz, M. Rabinowitz, and H. Fukuhara. 1974b. Transfer RNA genes in mitochondrial DNA of grande (wild type) yeast. *J. Mol. Biol.* **88:** 735.

Casey, J.W., H.-J. Hsu, M. Rabinowitz, G.S. Getz, and H. Fukuhara. 1974c. Transfer RNA genes in the mitochondrial DNA of cytoplasmic petite mutants of *Saccharomyces cerevisiae. J. Mol. Biol.* **88:** 717.

Cerletti, N. and G. Schatz. 1979. Cytochrome *c* oxidase from baker's yeast: Photoaffinity of subunits exposed to the lipid bilayer. *J. Biol. Chem.* **254:** 7746.

Choo, K.-J., P. Nagley, H.B. Lukins, and A.W. Linnane. 1977. Biogenesis of mitochondria. 47. Refined physical map of the mitochondrial genome of *Saccharomyces cerevisiae* determined by analysis of an extended library of genetically and molecularly defined petite mutants. *Mol. Gen. Genet.* **153:** 279.

Christiansen, G. and C. Christiansen. 1976. Comparison of the fine structure of mitochondrial

DNA from *Saccharomyces cerevisiae* and *S. carlsbergensis*: Electron microscopy of partially denatured molecules. *Nucleic Acids Res.* **3**: 465.

Christiansen, C., G. Christiansen, and A. Leth-Bak. 1974. Heterogeneity of mitochondrial DNA from *Saccharomyces carlsbergensis*: Renaturation and sedimentation studies. *J. Mol. Biol.* **84**: 65.

————. 1975. Heterogeneity of mitochondrial DNA from *Saccharomyces carlsbergensis*: Denaturation mapping by electron microscopy. *Nucleic Acids Res.* **2**: 197.

Christiansen, C., A.L. Leth-Bak, and A. Stenderup. 1971. Repetitive DNA in yeasts. *Nat. New Biol.* **231**: 176.

Church, G.M. and W. Gilbert. 1980. Yeast mitochondrial intron products required in trans for RNA splicing. In *Mobilization and reassembly of genetic information* (ed. D.R. Joseph et al.), p. 379. Academic Press, New York.

Church, G.M., P.P. Slonimski, and W. Gilbert. 1979. Pleiotropic mutations within two yeast mitochondrial cytochrome genes block mRNA processing. *Cell* **18**: 1209.

Claisse, M.L. and P. Pajot. 1974. Presence of cytochrome c_1 in cytoplasmic "petite" mutants of *Saccharomyces cerevisiae*. *Eur. J. Biochem.* **49**: 49.

Claisse, M.L., A. Spyridakis, and P.P. Slonimski. 1977. Mutations at any one of three unlinked mitochondrial genetic loci box1, box4 and box6 modify the structure of cytochrome *b* polypeptide(s). In *Mitochondria 1977: Genetics and biogenesis of mitochondria* (ed. W. Bandlow et al.), p. 337. de Gruyter, Berin.

Claisse, M.L., P.P. Slonimski, J. Johnson, and H.R. Mahler. 1980. Mutations within an intron and its flanking sites: Patterns of novel polypeptides generated by mutants in one segment of the cobbox region of yeast mitochondrial DNA. *Mol. Gen. Genet.* **177**: 375.

Claisse, M.L., A. Spyridakis, M.L. Wambier-Kluppel, P. Pajot, and P.P. Slonimski. 1978. Mosaic organization and expression of the mitochondrial DNA region controlling cytochrome *c* reductase and oxidase. II. Analysis of proteins translated from the box region. In *Biochemistry and genetics of yeast: Pure and applied aspects* (ed. M. Bacila et al.), p. 369. Academic Press, New York.

Clark-Walker, G.D. and G.L. Gabor Miklos. 1974. Mitochondrial genetics, circular DNA and the mechanism of the petite mutation in yeast. *Genet. Res.* **24**: 43.

Clark-Walker, G.D. and A.W. Linnane. 1966. In vivo differentiation of yeast cytoplasmic and mitochondrial protein synthesis with antibiotics. *Biochem. Biophys. Res. Commun.* **25**: 8.

Clavilier, L. 1976. Mitochondrial genetics. XII. An oligomycin resistant mutant localized at a new mitochondrial locus in *Saccharomyces cerevisiae*. *Genetics* **83**: 227.

Cobon, G.S., D.J. Groot-Obbink, R.M. Hall, R. Maxwell, M. Murphy, J. Rytka, and A.W. Linnane. 1976. Mitochondrial genes determining cytochrome *b* (complex III) and cytochrome oxidase function. In *Genetics and biogenesis of chloroplasts and mitochondria* (ed. T. Bücher et al.), p. 453. Elsevier/North-Holland, Amsterdam.

Coen, D., P. Netter, E. Petrochilo, and P.P. Slonimski. 1970. Mitochondrial genetics. I. Methodology and phenomenology. *Symp. Soc. Exp. Biol.* **24**: 449.

Cohen, M. and M. Rabinowitz. 1972. Analysis of grande and petite yeast mitochondrial DNA by tRNA hybridization. *Biochem. Biophys. Acta* **281**: 192.

Cohen, M., J. Casey, M. Rabinowitz, and G.S. Getz. 1972. Hybridization of mitochondrial transfer RNA and mitochondrial DNA in petite mutants of yeast. *J. Mol. Biol.* **63**: 441.

Colletti, E., L. Frontali, C. Palleschi, M. Wesolowski, and H. Fukuhara. 1979. Two isoaccepting seryl tRNAs coded by separate mitochondrial genes in yeast. *Mol. Gen. Genet.* **175**: 1.

Colson, A.M. and P.P. Slonimski. 1979. Genetic localization of diuron and mucidin resistant mutants relative to a group of loci of the mitochondrial DNA controlling coenzyme QH_2 cytochrome *c* reductase in *Saccharomyces cerevisiae*. *Mol. Gen. Genet.* **167**: 287.

Colson, A.M., G. Michaelis, E. Pratje, and P.P. Slonimski. 1979. Allelism and relationships between diuron-resistant, antimycin-resistant and funiculosin-resistant loci of the mitochondrial map in *Saccharomyces cerevisiae*. *Mol. Gen. Genet.* **167**: 299.

Colson, A.M., L.T. Van, B. Convent, M. Briquet, and A. Goffeau. 1977. Mitochondrial heredity of resistance to 3-(3,4-dichlorophenyl)-1,1-dimethylurea, an inhibitor of cytochrome b oxidation, in *Saccharomyces cerevisiae*. *Eur. J. Biochem.* **74**: 521.

Correns, C. 1909. Vererbungsversuche mit blas (gelb) grunen und bluntblattrigen sipen bei Mirabilis, Urtica und Lunaria. *Z. Vererbungsl.* **1**: 291.

Corruzzi, G. and A. Tzagoloff. 1979. Assembly of the mitochondrial membrane system: DNA sequence of subunit 2 of yeast cytochrome oxidase. *J. Biol. Chem.* **254**: 9324.

Corruzzi, G., M.K. Trembath, and A. Tzagoloff. 1978. Assembly of the mitochondrial membrane system: Mutations in the *pho2* locus of the mitochondrial genome of *S. cerevisiae*. *Eur. J. Biochem.* **92**: 279.

Cosson, J. and A. Tzagoloff. 1979. Sequence homologies of (guanosine + cytidine)-rich regions of mitochondrial DNA of *Saccharomyces cerevisiae*. *J. Biol. Chem.* **254**: 42.

Cote, C., M. Solioz, and G. Schatz. 1979. Biogenesis of the cytochrome bc_1 complex of yeast mitochondria. A precursor form of the cytoplasmically made subunit V. *J. Biol. Chem.* **254**: 1437.

Cottrell, S.F., M. Rabinowitz, and G.S. Getz. 1973. Mitochondrial DNA synthesis in a temperature sensitive mutant of DNA replication of *Saccharomyces cerevisiae*. *Biochemistry* **12**: 4374.

Criddle, R.S., L.W. Wheelis, M.K. Trembath, and A.W. Linnane. 1976a. Molecular and genetic events accompanying petite induction and recovery of respiratory competence induced by ethidium bromide. *Mol. Gen. Genet.* **144**: 263.

Criddle, R.S., C. Arulanandan, T. Edwards, R. Johnston, S. Scharf, and R. Enns. 1976b. Investigation of the oligomycin binding protein in yeast mitochondrial ATPase. In *Genetics and biogenesis of chloroplasts and mitochondria* (ed. T. Bücher et al.), p. 151. Elsevier/North-Holland, Amsterdam.

Cryer, D.R., C.D. Goldthwaite, S. Zinker, K.B. Lam, E. Storm, R. Hirschberg, J. Blamire, D.B. Finkelstein, and J. Marmur. 1974. Studies on nuclear and mitochondrial DNA of *Saccharomyces cerevisiae*. *Cold Spring Harbor Symp. Quant. Biol.* **38**: 17.

Cummings, D.J., I.B. Dawid, P. Borst, B.M. Weisman, and G.F. Fox, eds. 1979. *ICN-UCLA Symp. Mol. Cell. Biol.*, vol. 15. Academic Press, New York.

Darlison, M.G. and W.L. Lancashire. 1980. Genetics of oxidative phosphorylations: Allelism studies of mitochondrial loci in the *pho1-oli2* region of the genome. *Mol. Gen. Genet.* **180**: 227.

Davey, P.J., J.M. Haslam, and A.W. Linnane. 1970. Biogenesis of mitochondria. 12. The effects of aminoglycoside antibiotics on the mitochondrial and cytoplasmic protein synthesizing systems of *S. cerevisiae*. *Arch. Biochem. Biophys.* **136**: 54.

Dawes, I.W. and B.L.A. Carter. 1974. Nitrosoguanidine mutagenesis during nuclear and mitochondrial gene replication. *Nature* **250**: 709.

De Ronde, A., A.P.G.M. Van Loon, L.A. Grivell, and J. Kohli. 1980. *In vitro* suppression of UGA codons in mitochondrial mRNA. *Nature* **287**: 361.

Deutsch, J., B. Dujon, P. Netter, E. Petrochilo, P.P. Slonimski, M. Bolotin-Fukuhara, and D. Coen. 1974. Mitochondrial genetics. VI. The petite mutation in *Saccharomyces cerevisiae*: Interrelations between the loss of the rho^+ factor and the loss of the drug resistance mitochondrial genetic markers. *Genetics* **76**: 195.

Devenish, R.J., R.M. Hall, A.W. Linnane, and H.B. Lukins. 1979. Biogenesis of mitochondria. 52. Deletions in petite strains occuring in the mitochondrial gene for the 21S ribosomal RNA, that affect the properties of mitochondrial recombination. *Mol. Gen. Genet.* **174**: 297.

Devenish, R.J., K.J. English, R.M. Hall, A.W. Linnane, and H.B. Lukins. 1978. Biogenesis of mitochondria. 49. Identification and mapping of a new mitochondrial locus (*tsr1*) which maps within the polar region of the yeast mitochondrial genome. *Mol. Gen. Genet.* **161**: 251.

De Zamaroczy, M., G. Baldacci, and G. Bernardi. 1979. Putative origins of replication in the mitochondrial genome of yeast. *FEBS Lett.* **108**: 429.

Diatewa, M. and A.J.C. Stahl. 1980. Purification and subunit structure of mitochondrial phenylalanyl-tRNA synthetase from yeast. *Biochem. Biophys. Res. Commun.* **94:** 189.

DiFranco, A., J.P.M. Sanders, C. Heyting, P. Borst, and P.P. Slonimski. 1976. Restriction enzyme analysis and physical mapping of mitochondrial DNA from petite mutants, carrying a genetic marker for oligomycin or paromomycin resistance. In *The genetic functions of mitochondrial DNA* (ed. C. Saccone and A.M. Kroon), p. 291. Elsevier/North-Holland, Amsterdam.

Dockter, M.E., A. Steinemann, and G. Schatz. 1978. Mapping of yeast cytochrome *c* oxidase by fluorescence resonance energy transfer. *J. Biol. Chem.* **253:** 311.

Douglas, M.G. and R.A. Butow. 1976. Variant forms of mitochondrial translation products in yeasts: Evidence for location of determinants on mitochondrial DNA. *Proc. Natl. Acad. Sci.* **73:** 1083.

Douglas, M.G., D. Finkelstein, and R.A. Butow. 1979a. Analysis of products of mitochondrial protein synthesis in yeast: Genetic and biochemical aspects. *Methods Enzymol.* **56:** 58.

Douglas, M.G., Y. Koh, M.E. Dockter, and G. Schatz. 1977. Aurovertin binds to the *beta* subunit of yeast mitochondrial ATPase. *J. Biol. Chem.* **252:** 8333.

Douglas, M.G., R.L. Strausberg, P.S. Perlman, and R.A. Butow. 1976. Genetic analysis of mitochondrial polymorphic proteins in yeast. In *Genetics and biogenesis of chloroplasts and mitochondria* (ed. T. Bucher et al.), p. 435. Elsevier/North Holland, Amsterdam.

Douglas, M.G., Y. Koh, E. Ebner, E. Agsteribbe, and G. Schatz. 1979b. A nuclear mutation conferring aurovertin resistance to yeast mitochondrial adenosine triphosphatase. *J. Biol. Chem.* **254:** 1335.

Dujardin, G. and B. Dujon. 1979. Mutants in yeast affecting ethidium-bromide induced *rho*⁻ formation and their effects on transmission and recombination of mitochondrial genes. *Mol. Gen. Genet.* **171:** 205.

Dujardin, G., B. Robert, and L. Clavilier. 1978. Effect of hydroxyurea treatment on transmission and recombination of mitochondrial genes in *Saccharomyces cerevisiae*: A new method to modify the input of mitochondrial genes in crosses. *Mol. Gen. Genet.* **160:** 101.

Dujardin, G., P. Pajot, O. Groudinsky, and P.P. Slonimski. 1980a. Long range control circuits within mitochondria and between nucleus and mitochondria. I. Methodology and phenomenology of suppressors. *Mol. Gen. Genet.* **179:** 469.

Dujardin, G., O. Groudinsky, A. Kruszewska, P. Pajot, and P.P. Slonimski. 1980b. Cytochrome *b* messenger RNA maturase encoded in an intron regulates the expression of the split gene. III. Genetic and phenotypic suppression of intron mutations. In *The organization and expression of the mitochondrial genome* (ed. C. Saccone and A.M. Kroon), p. 157. Elsevier/North-Holland, Amsterdam.

Dujon, B. 1974. Recombination of mitochondrial genes in yeast. In *Mechanisms in recombination* (ed. R.F. Grell), p. 307. Plenum Press, New York.

———. 1976. Transmission, recombination and segregation of mitochondrial genes in *Saccharomyces cerevisiae*. In *Genetics, biogenesis and bioenergetics of mitochondria* (ed. W. Bandlow et al.), p. 1. de Gruyter, Berlin.

———. 1979. Mutants in a mosaic gene reveal functions for introns. *Nature* **282:** 777.

———. 1980. Sequence of the intron and flanking exons of the mitochondrial 21S rRNA gene of yeast strains having different alleles at the *omega* and *rib1* loci. *Cell* **20:** 185.

Dujon, B. and H. Blanc. 1980. Yeast mitochondria minilysates and their use to screen a collection of hypersuppressive *rho*⁻ mutants. In *The organization and expression of the mitochondrial genome* (ed. C. Saccone and A.M. Kroon), p. 33. Elsevier/North-Holland, Amsterdam.

Dujon, B. and G. Michaelis. 1974. Extrakaryotic inheritance. *Prog. Bot.* **36:** 236.

Dujon, B. and F. Michel. 1976. Genetic and physical characterization of a segment of the mitochondrial DNA involved in the control of genetic recombination. In *The genetic functions of mitochondrial DNA* (ed. C. Saccone and A.M. Kroon), p. 175. North-Holland, Amsterdam.

Dujon, B. and P.P. Slonimski. 1976. Mechanisms and rules for transmission, recombination and segregation of mitochondrial genes in *Saccharomyces cerevisiae*. In *Genetics and biogenesis of chloroplasts and mitochondria* (ed. T. Bücher et al.), p. 393. Elsevier/North-Holland, Amsterdam.

Dujon, B., H. Baranowska, and G. Dujardin. 1977a. Control of the genetic recombination of mitochondrial genes: Preliminary approaches. In *Mitochondria 1977: Genetics and biogenesis of mitochondria* (ed. W. Bandlow et al.), p. 53. de Gruyter, Berlin.

Dujon, B., A.M. Colson, and P.P. Slonimski. 1977b. The mitochondrial genetic map of *Saccharomyces cervisiae*: A literature compilation towards a unique map. In *Mitochondria 1977: Genetics and biogenesis of mitochondria* (ed. W. Bandlow et al.), p. 579. de Gruyter, Berlin.

Dujon, B., P.P. Slonimski, and L. Weill. 1974. Mitochondrial genetics. IX. A model for recombination and segregation of mitochondrial genomes in *Saccharomyces cerevisiae*. *Genetics* **78**:415.

Dujon, B., M. Bolotin-Fukuhara, D. Coen, J. Deutsch, P. Netter, P.P. Slonimski, and L. Weill. 1976. Mitochondrial genetics. XI. Mutations at the mitochondrial locus *omega* affecting the recombination of mitochondrial genes in *Saccharomyces cerevisiae*. *Mol. Gen. Genet.* **143**:131.

Dujon, B., A. Kruszewska, P.P. Slonimski, M. Bolotin-Fukuhara, D. Coen, J. Deutsch, P. Netter, and L. Weill. 1975. Mitochondrial genetics. X. Effects of UV irradiation on transmission and recombination of mitochondrial genes in *Saccharomyces cerevisiae*. *Mol. Gen. Genet.* **137**:29.

Ebner, E. and K.L. Maier. 1977. A protein inhibitor of mitochondrial adenosine triphosphatase (F1) from *Saccharomyces cerevisiae*. *J. Biol. Chem.* **252**:671.

Ebner, E. and G. Schatz. 1973. Mitochondrial assembly in respiration deficient mutants of *Saccharomyces cerevisiae*. III. A nuclear mutant lacking mitochondrial adenosine triphosphatase. *J. Biol. Chem.* **248**:5379.

Ebner, E., T.L. Mason, and G. Schatz. 1973a. Mitochondrial assembly in respiration-deficient mutants of *Saccharomyces cerevisiae*. II. Effect of nuclear and extrachromosomal mutations on the formation of cytochrome *c* oxidase. *J. Biol. Chem.* **248**:5369.

Ebner, E., L. Mennucci, and G. Schatz. 1973b. Mitochondrial assembly in respiration deficient mutants of *Saccharomyces cerevisiae*. I. Effect of nuclear mutations on mitochondrial protein synthesis. *J. Biol. Chem.* **248**:5360.

Eccleshall, T.R., R.B. Needleman, E.M. Storm, B. Buchterer, and J. Marmur. 1978. A temperature sensitive yeast mitochondrial mutant with altered cytochrome *c* oxidase subunit. *Nature* **273**:67.

Ehrlich, S.D., J.P. Thiery, and G. Bernardi. 1972. The mitochondrial genome of wild type yeast cells. III. The pyrimidine tracts of mitochondrial DNA. *J. Mol. Biol.* **65**:207.

Enns, R. and R.S. Criddle. 1977. Investigation of the structural arrangement of the protein subunits of mitochondrial ATPase. *Arch. Biophys. Biochem.* **183**:742.

Ephrussi, B. and S. Grandchamp. 1965. Etudes sur la suppressivité des mutants à déficience respiratoire de la levure. I. Existence au niveau cellulaire de divers degrés de suppressivité. *Heredity* **20**:1.

Ephrussi, B. and P.P. Slonimski. 1955. Yeast mitochondria: Subcellular units involved in the synthesis of respiratory enzymes in yeast. *Nature* **176**:1207.

Ephrussi, B., H. Hottinguer, and Y. Chimenes. 1949a. Action de l'acriflavine sur les levures. I. La mutation "petite colonie." *Ann. Inst. Pasteur* **76**:351.

Ephrussi, B., H. Hottinguer, J. Tavlitzki. 1949b. Action de l'acriflavine sur les levures. II. Etude genetique du mutant "petite colonie." *Ann. Inst. Pasteur* **76**:419.

Everett, T.D., E. Finzi, and D.S. Beattie. 1980. Stimulation of yeast mitochondrial protein synthesis by postpolysomal supernatants from yeast, rat liver and *E. Coli*. *Arch. Biophys. Biochem.* **200**:467.

Eytan, G.D. and G. Schatz. 1975. Cytochrome *c* oxidase from baker's yeast. V. Arrangement of the subunits in the isolated and membrane bound enzyme. *J. Biol. Chem.* **250**: 767.

Faugeron-Fonty, G., F. Culard, G. Baldacci, R. Goursot, A. Prunell, and G. Bernardi. 1979. The mitochondrial genome of wild-type yeast cells. VIII. The spontaneous cytoplasmic "petite" mutation. *J. Mol. Biol.* **134**: 493.

Fauman, M.A. and M. Rabinowitz. 1974. DNA-DNA hybridization studies of mitochondrial DNA of ethidium bromide induced petite mutants of yeast. *Eur. J. Biochem.* **42**: 67.

Fauman, M.A., M. Rabinowitz, and H.H. Swift. 1973. Comparison of the mitochondrial ribonucleic acid from a wild-type grande and a cytoplasmic petite yeast by ribonucleic acid-deoxyribonucleic acid hybridization. *Biochemistry* **12**: 124.

Faye, G. and F. Sor. 1977. Analysis of mitochondrial ribosomal proteins of *Saccharomyces cerevisiae* by two dimensional polyacrylamide gel electrophoresis. *Mol. Gen. Genet.* **155**: 27.

Faye, G., M. Bolotin-Fukuhara, and H. Fukuhara. 1976a. Mitochondrial mutations that affect mitochondrial transfer ribonucleic acid in *Saccharomyces cerevisiae*. In *Genetics and biogenesis of chloroplasts and mitochondria* (ed. T. Bücher et al.), p. 547. North-Holland, Amsterdam.

Faye, G., C. Kujawa, and H. Fukuhara. 1974. Physical and genetic organization of petite and grande yeast mitochondrial DNA. IV. In vivo transcription products of mitochondrial DNA and localization of 23S ribosomal RNA in petite mutants of *Saccharomyces cerevisiae*. *J. Mol. Biol.* **88**: 185.

Faye, G., N. Dennebouy, C. Kujawa, and C. Jacq. 1979a. Inserted sequence in the mitochondrial 23S ribosomal RNA gene of the yeast *Saccharomyces cerevisiae*. *Mol. Gen. Genet.* **168**: 101.

Faye, G., C. Kujawa, H. Fukuhara, and M. Rabinowitz. 1976b. Mapping of the mitochondrial 16S ribosomal RNA gene and its expression in the cytoplasmic petite mutants of *Saccharomyces cerevisiae*. *Biochem. Biophys. Res. Commun.* **68**: 476.

Faye, G., F. Sor, A. Glatigny, F. Lederer, and E. Lesquoy. 1979b. Comparison of amino-acid compositions of mitochondrial and cytoplasmic ribosomal proteins of *Saccharomyces cerevisiae*. *Mol. Gen. Genet.* **171**: 335.

Faye, G., C. Kujawa, B. Dujon, M. Bolotin-Fukuhara, K. Wolf, H. Fukuhara, and P.P. Slonimski. 1975. Localization of the gene coding for the mitochondrial 16S ribosomal RNA using *rho⁻* mutants of *Saccharomyces cerevisiae*. *J. Mol. Biol.* **99**: 203.

Faye, G., H. Fukuhara, C. Grandchamp, J. Lazowska, F. Michel, J. Casey, G.S. Stetz, J. Locker, M. Rabinowitz, M. Bolotin-Fukuhara, D. Coen, J. Deutsch, B. Dujon, P. Netter, and P.P. Slonimski. 1973. Mitochondrial nucleic acids in the petite colony mutants: Deletions and repetitions of genes. *Biochimie* **55**: 779.

Finkelstein, D.D. and R.A. Butow. 1979. Analysis of products of mitochondrial protein synthesis in yeast: Genetic and biochemical aspects. *Methods Enzymol.* **56**: 58.

Flury, U., H.R. Mahler, and F. Feldman. 1974. A novel respiration deficient mutant of *Saccharomyces cerevisiae*. I. Preliminary characterization of phenotype and mitochondrial inheritance. *J. Biol. Chem.* **249**: 6130.

Fonty, G., R. Goursot, D. Wilkie, and G. Bernardi. 1978. The mitochondrial genome of wild type yeast cells. VII. Recombination in crosses. *J. Mol. Biol.* **119**: 213.

Forster, J.L. and R.A. Kleese. 1975a. The segregation of mitochondrial genes in yeast. I. Analysis of zygote pedigrees of petite × grande crosses. *Mol. Gen. Genet.* **139**: 329.

―――.1975b. The segregation of mitochondrial genes in yeast. II. Analysis of zygote pedigrees of drug resistant × drug sensitive crosses. *Mol. Gen. Genet.* **139**: 341.

Foury, F. and A. Tzagoloff. 1976a. Localization on mitochondrial DNA of mutations leading to a loss of rutamycin-sensitive adenosine triphosphatase. *Eur. J. Biochem.* **68**: 113.

―――. 1976b. Assembly of the mitochondrial membrane system. XIX. Genetic characterization of mit⁻ mutants with deficiencies in cytochrome oxidase and coenzyme QH_2-cytochrome *c* reductase. *Mol. Gen. Genet.* **149**: 43.

————. 1978. Assembly of the mitochondrial membrane system. Genetic complementation of mit⁻ mutations in mitochondrial DNA of *Saccharomyces cerevisiae. J. Biol. Chem.* **253**: 3792.

Fox, T.D. 1979a. Five TGA stop codons occur within the translated sequence of the yeast mitochondrial gene for cytochrome *c* oxidase subunit II. *Proc. Natl. Acad. Sci.* **76**: 6534.

————. 1979b. Genetic and physical analysis of the mitochondrial gene for subunit II of yeast cytochrome *c* oxidase. *J. Mol. Biol.* **130**: 63.

Fukuhara, H. 1969. Relative proportions of mitochondrial and nuclear DNA in yeast under various conditions of growth. *Eur. J. Biochem.* **11**: 135.

Fukuhara, H. and M. Wesolowski. 1977. Preferential loss of a specific region of mitochondrial DNA by *rho⁻* mutation. In *Mitochondria 1977: Genetics and biogenesis of mitochondria* (ed. W. Bandlow et al.), p. 123. de Gruyter, Berlin.

Fukuhara, H., M. Faures, and C. Genin. 1969. Comparison of RNAs transcribed in vivo from mitochondrial DNA of cytoplasmic and chromosomal respiratory deficient mutants. *Mol. Gen. Genet.* **104**: 264.

Fukuhara, H., G. Faye, F. Michel, J. Lazowska, J. Deutsch, M. Bolotin-Fukuhara, and P.P. Slonimski 1974. Physical and genetic organization of petite and grande yeast mitochondrial DNA. I. Studies by RNA-DNA hybridization. *Mol. Gen. Genet.* **130**: 215.

Fukunaga, M. and K.L. Yielding. 1978. Propidium: Induction of petites and recovery from ethidium mutagenesis in *Saccharomyces cerevisiae. Biochem. Biophys. Res. Commun.* **84**: 501.

Gaillard, C., F. Strauss, and G. Bernardi. 1980. Excision sequences in the mitochondrial genome of yeast. *Nature* **283**: 218.

Gillham, N.W. 1977. *Organelle heredity.* Raven Press, New York.

Goldring, E.S, L.I. Grossman, and J. Marmur. 1971. Petite mutation in yeast. II. Isolation of mutants containing mitochondrial deoxyribonucleic acid of reduced size. *J. Bacteriol.* **107**: 377.

Goldring, E.S., L.I. Grossman, D. Krupnick, D.R. Cryer, and J. Marmur. 1970. The petite mutation in yeast. I. Loss of mitochondrial DNA during induction of petites with ethidium bromide. *J. Mol. Biol.* **52**: 323.

Goldthwaite, C.D., D.R. Cryer, and J. Marmur. 1974. Effect of carbon source on the replication and transmission of yeast mitochondrial genomes. *Mol. Gen. Genet.* **133**: 87.

Gordon, P. and M. Rabinowitz. 1973. Evidence for deletion and changed sequence in the mitochondrial deoxyribonucleic acid of a spontaneously generated petite mutant of *Saccharomyces cerevisiae. Biochemistry* **12**: 116.

Gordon, P., J. Casey, and M. Rabinowitz. 1974. Characterization of mitochondrial deoxyribonucleic acid from a series of petite yeast strains by deoxyribonucleic acid-deoxyribonucleic acid hybridization. *Biochemistry* **13**: 1067.

Goursot, R., M. De Zamaroczy, G. Baldacci, and G. Bernardi. 1980. Supersuppressive "petite" mutants of yeast. *Curr. Genet.* **1**: 173.

Griffiths, D.E. and R.L. Houghton. 1974. Studies on energy-linked reactions: Modified mitochondrial ATPase of oligomycin resistant mutants of *Saccharomyces cerevisiae. J. Biochem.* **46**: 157.

Griffiths, D.E., R.L. Houghton, and W.E. Lancashire. 1974. Mitochondrial genes and ATP-synthetase. In *The biogenesis of mitochondria: Transcriptional, translational and genetic aspects* (ed. A.M. Kroon and C. Saccone), p. 215. Academic Press, New York.

Griffiths, D.E., R.L. Houghton, W.E. Lancashire, and P.A. Meadows. 1975. Studies on energy linked reactions: Isolation and properties of mitochondrial venturicidin resistant mutants of *Saccharomyces cerevisiae. Eur. J. Biochem.* **51**: 393.

Grimes, G.W., H.R. Mahler, and P.S. Perlman. 1974. Nuclear gene dosage effects on mitochondrial mass and DNA. *J. Cell Biol.* **61**: 565.

Grivell, L.A. and A.F.M. Moorman. 1977. A structural analysis of the *oxi3* region of yeast mitDNA. In *Mitochondria 1977: Genetics and biogenesis of mitochondria* (ed. W. Bandlow et al.), p. 371. de Gruyter, Berlin.

Grivell, L.A., P. Netter, P. Borst, and P.P. Slonimski. 1973. Mitochondrial antibiotic resistance in yeast: Ribosomal mutants resistant to chloramphenicol, erythromycin and spiramycin. *Biochim. Biophys. Acta* **312**: 358.

Grivell, L.A., A.C. Arnberg, L.A.M. Hensgens, E. Roosendaal, G.J.B. Van Ommen, and E.F.J. Van Bruggen. 1980. Split genes on yeast mitochondrial DNA: Organization and expression. In *The organization and expression of the mitochondrial genome* (ed. C. Saccone and A.M. Kroon), p. 37. Elsevier/North-Holland, Amsterdam.

Grivell, L.A., A.C. Arnberg, P.H. Boer, P. Borst, J.L. Bos, E.F.J. Van Bruggen, G.S.P. Groot, N.B. Hecht, L.A.M. Hensgens, G.J.B. Van Ommen, and H.F. Tabak. 1979. Transcripts of yeast mitochondrial DNA and their processing. *ICN-UCLA Symp. Mol. Cell. Biol.* **15**: 305.

Groot, G.S.P. 1974. The biosynthesis of mitochondrial ribosomes in *Saccharomyces cerevisiae*. In *The biogenesis of mitochondria: Transcriptional, translational and genetic aspects* (ed. A.M. Kroon and C. Saccone), p. 443. Academic Press, New York.

Groot, G.S.P. and R.O. Poynton. 1975. Oxygen control of cytochrome c oxidase synthesis in isolated mitochondria from *Saccharomyces cerevisiae*. *Nature* **255**: 238.

Groot, G.S.P., T. Mason, and N. Van Harten-Loosbroek. 1979. Var1 is associated with the small ribosomal subunit of mitochondrial ribosomes in yeast. *Mol. Gen. Genet.* **174**: 339.

Groot, G.S.P., N. Van Harten-Loosbroek, and J. Kreike. 1978. Electrophoretic behavior of yeast mitochondrial translation products. *Biochem. Biophys. Acta* **517**: 457.

Groot, G.S.P., R.A. Flavell, G.J.B. Van Ommen, and L.A. Grivell. 1974. Yeast mitochondrial RNA does not contain poly(A). *Nature* **252**: 167.

Groot, G.S.P., L.A. Grivell, N. Van Harten-Loosbroek, J. Kreike, A.F.M. Moorman, and G.J.B. Van Ommen. 1977. The role of the mitochondrial genetic system in the biogenesis of the mitochondrial inner membrane. In *Structure and function of energy transducing membrane* (ed. K. Van Dam and B.F. Van Gelder), p. 177. Elsevier/North Holland Biomedical Press, Amsterdam.

Groot-Obbink, D.J., T.W. Spithill, R.J. Maxwell, and A.W. Linnane. 1977. Biogenesis of mitochondria. 48. Mikamycin resistance in *Saccharomyces cerevisiae*. A mitochondrial mutation conferring resistance to an antimycin A like contaminant in mikamycin. *Mol. Gen. Genet.* **151**: 127.

Groot-Obbink, D.J., R.M. Hall, A.W. Linnane, H.B. Lukins, B.C. Monk, T.W. Spithill, and M.K. Trembath. 1976. Mitochondrial genes involved in the determination of mitochondrial membrane proteins. In *The genetic functions of mitochondrial DNA* (ed. C. Saccone and A.M. Kroon), p. 163. Elsevier/North-Holland, Amsterdam.

Grossman, L.I., E.S. Goldring, and J. Marmur. 1969. Preferential synthesis of yeast mitochondrial DNA in the absence of protein synthesis. *J. Mol. Biol.* **46**: 367.

Grossman, L.I., D.R. Cryer, E.S. Goldring, and J. Marmur. 1971. The petite mutation in yeast. III. Nearest-neighbor analysis of mitochondrial DNA from normal and mutant cells. *J. Mol. Biol.* **62**: 565.

Guerineau, M. and C. Paoletti. 1975. Rearrangement of mitochondrial DNA molecules during the differentiation of mitochondria in yeast. II. Labeling studies on the precursor product relationship. *Biochimie* **57**: 931.

Guerineau, M., P.R. Avner, and P.P. Slonimski. 1974. Yeast episome: Oligomycin resistance associated with a small covalently closed non-mitochondrial circular DNA. *Biochem. Biophys. Res. Commun.* **61**: 462.

Guerineau, M., C. Granchamp, and P. Slonimski. 1975. Rearrangement of mitochondrial DNA molecules during the differentiation of mitochondria in yeast. I. Electron microscopic studies of size and shape. *Biochimie* **75**: 917.

Gunge, N. 1975. Genetic analysis of unequal transmission of the mitochondrial markers in *Saccharomyces cerevisiae*. *Mol. Gen. Genet.* **139**: 189.

————. 1976. Effects of elevation of strain ploidy on transmission and recombination of mito-chondrial drug resistance genes in *Saccharomyces cerevisiae*. *Mol. Gen. Genet.* **146:** 5.

Haid, A., G. Grosch, C. Schmelzer, R.J. Schweyen, and F. Kaudewitz. 1980. Expression of the split gene *cob* in yeast mtDNA. Mutational arrest in the pathway of transcript splicing. *Curr. Genet.* **1:** 155.

Haid, A., R.J. Schweyen, H. Bechmann, F. Kaudewitz, M. Solioz, and G. Schatz. 1979. The mitochondrial *cob* region in yeast codes for apocytochrome *b* and is mosaic. *Eur. J. Bio-chem.* **94:** 451.

Halbreich, A. and M. Rabinowitz. 1971. Isolation of *Saccharomyces cerevisiae* mitochondrial formyltetrahydrofolic acid:methionyl tRNA transformylase and the hybridization of mito-chondrial fMet-tRNA with mitochondrial DNA. *Proc. Natl. Acad. Sci.* **68:** 294.

Halbreich, A., P. Pajot, M. Foucher, C. Grandchamp, and P.P. Slonimski. 1980. A pathway of cytochrome *b* mRNA processing in yeast mitochondria: Specific splicing steps and an intron-derived circular RNA. *Cell* **19:** 321.

Hall, R.M., P. Nagley, and A.W. Linnane. 1976a. Biogenesis of mitochondria. XLII. Genetic analysis of the control of cellular mitochondrial DNA levels in *Saccharomyces cerevisiae*. *Mol. Gen. Genet.* **145:** 169.

Hall, R.M., M.K. Trembath, A.W. Linnane, L. Wheelis, and R.S. Criddle. 1976b. Factors affecting petite induction and the recovery of respiratory competence in yeast cells exposed to ethidium bromide. *Mol. Gen. Genet.* **144:** 253.

Hall, R.M., J.S. Mattick, P. Nagley, G.S. Cobon, F.W. Eastwood, and A.W. Linnane. 1977. The action of structural analogues of ethidium bromide on the mitochondrial genome of yeast. *Mol. Biol. Rep.* **3:** 443.

Handwerker, A., R.J. Schweyen, K. Wolf, and F. Kaudewitz. 1973. Evidence for an extrakary-otic mutation affecting the maintenance of the *rho* factor in yeast. *J. Bacteriol.* **113:** 1307.

Hanson, D.K., D.H. Miller, H. Mahler, N.J. Alexander, and P.S. Perlman. 1979. Regulatory interaction between mitochondrial genes. II. Detailed characterization of novel mutants mapping within one cluster in the *cob2* region. *J. Biol. Chem.* **254:** 2480.

Hendler, F.J., G. Padmanaban, J. Patzer, R. Ryan, and M. Rabinowitz. 1975. Yeast mitochon-drial RNA contains a short polyadenylic acid segment. *Nature* **258:** 357.

Hendler, F.J., A. Halbreich, S. Jakovcic, J. Patzer, S. Merten, and M. Rabinowitz. 1976. Characterization and translation of yeast mitochondrial RNA. In *Genetics and biogenesis of chloroplasts and mitochondria* (ed. T. Bücher et al.), p. 679. Elsevier/North-Holland, Amsterdam.

Hensgens, L.A.M., L.A. Grivell, P. Borst, and J.L. Bos. 1979. Nucleotide sequence of the mitochondrial structural gene for subunit 9 of yeast ATPase complex. *Proc. Natl. Acad. Sci.* **76:** 1663.

Heude, M., H. Fukuhara, and E. Moustacchi. 1979. Spontaneous and induced *rho* mutants of *Saccharomyces cerevisiae*: Patterns of loss of mitochondrial genetic markers. *J. Bacteriol.* **139:** 460.

Heyting, C. and H.H. Menke. 1979. Fine structure of the 21S ribosomal RNA region in yeast mitochondrial DNA. III. Physical location of mitochondrial genetic markers and the molec-ular nature of *omega*. *Mol. Gen. Genet.* **168:** 279.

Heyting, C. and J.P.M. Sanders. 1976. The physical mapping of some genetic markers in the 21S ribosomal region of the mitochondrial DNA of yeast. In *The genetic functions of mitochondrial DNA* (ed. C. Saccone and A.M. Kroon), p. 273. North-Holland, Amsterdam.

Heyting, C., J.L. Talen, P.J. Weijers, and P. Borst. 1979a. Fine structure of the 21S ribosomal RNA region in yeast mitochondrial DNA. II. The organisation of sequences in petite mitochondrial DNAs carrying genetic markers from the 21S region. *Mol. Gen. Genet.* **168:** 251.

Heyting, C., F.C.P.W. Meijlink, M.P. Verbeet, J.P.M. Sanders, J.L. Bos, and P. Borst. 1979b. Fine structure of the 21S ribosomal RNA region in yeast mitochondrial DNA. I. Construction of the physical map and localization of the cistron for the 21S mitochondrial ribosomal RNA. *Mol. Gen. Genet.* **168**:231.

Hixon, S.C., A.D. Burnham, and R.L. Irons. 1979. Reversal or protection by light of the ethidium bromide induced petite mutation in yeast. *Mol. Gen. Genet.* **169**:63.

Hollenberg, C.P. and P. Borst. 1971. Conditions that prevent rho⁻ induction by ethidium bromide. *Biochem. Biophys. Res. Commun.* **45**:1250.

Hollenberg, C.P., P. Borst, and E.F.J. Van Bruggen. 1970. Mitochondrial DNA. V. A 25μm closed circular duplex DNA molecule in wild-type yeast mitochondria. Structure and genetic complexity. *Biochim. Biophys. Acta* **209**:1.

———.1972a. Mitochondrial DNA from cytoplasmic petite mutants of yeasts. *Biochim. Biophys. Acta* **277**:35.

Hollenberg, C.P., P. Borst, R.A. Flavell, C.F. Van Kreijl, E.F.J. Van Bruggen, and A.C. Arnberg. 1972b. The unusual properties of mtDNA from a "low density" petite mutant of yeast. *Biochim. Biophys. Acta* **277**:44.

Howell, N., M.K. Trembath, A.W. Linnane, and H.B. Lukins. 1973. Biogenesis of mitochondria 30: An analysis of polarity of mitochondrial gene recombination and transmission. *Mol. Gen. Genet.* **122**:37.

Iwashima, A. and M. Rabinowitz. 1969. Partial purification of mitochondrial and supernatant DNA polymerase from *Saccharomyces cerevisiae*. *Biochim. Biophys. Acta* **178**:283.

Jacq, C., J. Lazowska, and P.P. Slonimski. 1980a. Sur un nouveau mećanisme de la régulation de l'expression génétique *C. R. Acad. Sci. Paris* **290**:89.

———. 1980b. Cytochrome *b* messenger RNA maturase encoded in an intron regulates the expression of the split gene. I. Physical location and base sequence of intron mutations. In *The organization and expression of the mitochondrial genome* (ed. C. Saccone and A.M. Kroon), p. 139. Elsevier/North-Holland, Amsterdam.

Jacq, C., C. Kujawa, C. Grandchamp, and P. Netter. 1977. Physical characterization of the difference between yeast mitDNA alleles *omega⁺* and *omega⁻*. In *Mitochondria 1977: Genetics and biogenesis of mitochondria* (ed. W. Bandlow et al.), p. 255. de Gruyter, Berlin.

James, A.P., B.F. Johnson, E.R. Inhaber, and N.T. Gridgeman. 1975. A kinetic analysis of spontaneous rho⁻ mutations in yeast. *Mutat. Res.* **30**:199.

Johnston, L.H. 1979. Nuclear mutations in *Saccharomyces cerevisiae* which increase the spontaneous mutation frequency in mitochondrial DNA. *Mol. Gen. Genet.* **170**:327.

Juliani, M.H., S. Hixon, and E. Moustacchi. 1976. Mitochondrial genetic damage induced in yeast by a photoreactivated furocoumarin in combination with ethidium bromide or ultraviolet. *Mol. Gen. Genet.* **145**:249.

Kaldma, J.A. 1975a. Study of mitochondrial recombination in yeast. I. Analysis of isomitochondrial crosses and the effect of cell mating type determining locus on recombination. *Genetika* (USSR), No. 9 **3**:111.

———. 1975b. Studies on mitochondrial recombination in yeast. II. Crosses of isochromosomal and anisomitochondrial strains. *Genetika* (USSR), No. 9 **8**:88.

———. 1975c. A new type of mitochondrial mutations in yeast: Temperature dependent antibiotic resistant mitochondrial mutations. *Genetika* (USSR), No. 9 **5**:151.

Katan, M.B., N.V. Harten-Loosbroek, and G.S.P. Groot. 1976a. The cytochrome b-c₁ complex of yeast mitochondria. Site of translation of the polypeptides in vivo. *Eur. J. Biochem.* **70**:409.

Katan, M.B., L. Pool, and G.S.P. Groot. 1976b. The cytochrome b-c₁ complex of yeast mitochondria. Isolation and partial characterization of the cytochrome b-c₁ complex and cytochrome *b*. *Eur. J. Biochem.* **65**:95.

Kellems, R.E. and R.A. Butow. 1972. Cytoplasmic type 80S ribosomes associated with yeast mitochondria. I. Evidence for ribosome binding sites on yeast mitochondria. *J. Biol. Chem.* **247**:8043.

———. 1974. Cytoplasmic type 80S ribosomes associated with yeast mitochondria. III.

Changes in the amount of bound ribosomes in response to changes in metabolic state. *J. Biol. Chem.* **249**: 3304.

Kellems, R.E., V.F. Allison, and R.A. Butow. 1974. Cytoplasmic type 80S ribosomes associated with yeast mitochondria. II. Evidence for the association of cytoplasmic ribosomes with the outer mitochondrial membrane in situ. *J. Biol. Chem.* **249**: 3297.

Keyhani, E. 1979. Identification of the structural gene for yeast cytochrome *c* oxidase subunit I on mitochondrial DNA. *Biochem. Biophys. Res. Commun.* **89**: 1212.

Kleese, R.A., R.C. Grotbeck, and J.R. Snyder. 1972a. Recombination among three mitochondrial genes in yeast (*Saccharomyces cerevisiae*). *J. Bacteriol.* **112**: 1023.

―――. 1972b. Two cytoplasmically inherited chloramphenicol resistance loci in yeast (*S. cerevisiae*). *Can. J. Genet. Cytol.* **14**: 713.

Klootwijk, J., I. Klein, and L.A. Grivell. 1975. Minimal post-transcriptional modification of yeast mitochondrial ribosomal RNA. *J. Mol. Biol.* **97**: 337.

Knight, J.E. 1980. New antibiotic resistance loci in the ribosomal region of yeast mitochondrial DNA. *Genetics* **91**: 69.

Kochko, A. 1979. "Expression d'un gene morcelé: Etude du mécanisme par l'analyse *cis-trans* lors de la complémentation in vivo au sein du gene mitochondrial codant pour l'apocytochrome *b* chez *Saccharomyces cerevisiae*." Ph.D. thesis, Université P. and M. Curie, Paris.

Kochko, A., A.M. Colson, and P.P. Slonimski. 1979. Expression en *cis*, lors de la complémentation, des exons du gene mosaique mitochondrial controlant le cytochrome *b* chez *Saccharomyces cerevisiae*. *Arch. Int. Physiol. Biochem.* **87**: 619.

Kotylak, Z. and P.P. Slonimski. 1976. Joint control of cytochrome *a* and *b* by a unique mitochondrial DNA region comprising four genetic loci. In *The genetic functions of mitochondrial DNA* (ed. C. Saccone and A.M. Kroon), p. 143. Elsevier/North-Holland, Amsterdam.

―――. 1977a. Mitochondrial mutants isolated by a new screening method based upon the use of the nuclear mutation *op1*. In *Mitochondria 1977: Genetics and biogeneis of mitochondria* (ed. W. Bandlow et al.), p. 83. de Gruyter, Berlin.

―――. 1977b. Fine structure genetic map of the mitochondrial DNA region controlling coenzyme QH_2-cytochrome *c* reductase. In *Mitochondria 1977: Genetics and biogenesis of mitochondria* (ed. W. Bandlow et al.), p. 161. de Gruyter, Berlin.

Kovac, L. 1974. Biochemical mutants: An approach to the mitochondrial energy coupling. *Biochim. Biophys. Acta.* **346**: 101.

Kovac, L. and K. Weisskova. 1968. Oxidative phosphorylation in yeast. III. ATPase activity of the mitochondrial fraction from a cytoplasmic respiratory deficient mutant. *Biochim. Biophys. Acta* **153**: 55.

Kovacova, V., J. Irmlerova, and L. Kovac. 1968. Oxidative phosphorylation in yeast. IV. Combination of a nuclear mutation affecting oxidative phosphorylation with cytoplasmic mutation to respiratory deficiency. *Biochim. Biophys. Acta* **162**: 157.

Kreike, J., H. Bechmann, F.J. Van Hemert, R.J. Schweyen, P.H. Boer, F. Kaudewitz, and G.S.P. Groot. 1979. The identification of apocytochrome *b* as a mitochondrial gene product and immunological evidence for altered apocytochrome *b* in yeast strains having mutations in the *cob* region of mitochondrial DNA. *Eur. J. Biochem.* **101**: 607.

Kroon, A.M. and C. Saccone, eds. 1974. *The biogenesis of mitochondria: Transcriptional, translational and genetic aspects.* Academic Press, New York.

―――. 1980. *The organization and expression of the mitochondrial genome.* Elsevier/North-Holland, Amsterdam.

Kuenzi, M.T. and R. Roth. 1974. Timing of mtDNA synthesis during meiosis in *Saccharomyces cerevisiae*. *Exp. Cell Res.* **85**: 377.

Lamb, A.J., G.D. Clark-Walker, and A.W. Linnane. 1968. The biogenesis of mitochondria. IV. The differentiation of mitochondrial and cytoplasmic protein synthesizing systems in vitro by antibiotics. *Biochim. Biophys. Acta* **161**: 415.

Lamouroux, P. 1979. "Complémentation *in vivo*: Moyen d' étude de l'expression du gene morcelé mitochondrial codant pour l'apocytochrome *b* chez *Saccharomyces cerevisiae*." Ph.D. thesis, Universite de Paris Sud, Orsay.

Lamouroux, A., P. Pajot, A. Kochko, A. Halbreich, and P.P. Slonimski. 1980. Cytochrome *b* messenger RNA maturase encoded in an intron regulates the expression of the split gene. II. *Trans* and *cis* acting mechanisms of mRNA splicing. In *The organization and expression of the mitochondrial genome* (ed. A.M. Kroon and C. Saccone), p. 153. Elsevier/North-Holland, Amsterdam.

Lancashire, W.E., 1976. Mitochondrial mutations conferring heat or cold sensitivity in *Saccharomyces cerevisiae*. In *Genetics and biogenesis of chloroplasts and mitochondria* (ed. T. Bücher et al.), p. 481. Elsevier/North-Holland, Amsterdam.

Lancashire, W.E. and D.E. Griffiths. 1975a. Studies on energy linked reactions: Isolation, characterization and genetic analysis of trialkyl tin resistant mutants of *Saccharomyces cerevisiae*. *Eur. J. Biochem.* **51**:377.

————. 1975b. Studies on energy linked reactions: Genetic analysis of venturicidin resistant mutants. *Eur. J. Biochem.* **51**:403.

Lancashire, W.E. and J.R. Mattoon. 1979. Genetics of oxidative phosphorylation: Mitochondrial loci determining ossamycin, venturicidin and oligomycin resistance in yeast. *Mol. Gen. Genet.* **176**:255.

Lazowska, J. and P.P. Slonimski. 1976. Electron microscopy analysis of circular repetitive mitochodrial DNA molecules from genetically characterized *rho⁻* mutants of *Saccharomyces cerevisiae*. *Mol. Gen. Genet.* **146**:61.

————. 1977. Site specific recombination in "petite colony" mutants of *Saccharomyces cerevisiae*. I. Electron microscopic analysis of the organization of recombinant DNA resulting from end to end joining of the two mitochondrial segments. *Mol. Gen. Genet.* **156**:163.

Lazowska, J., C. Jacq, and P.P. Slonimski. 1980. Sequence of introns and flanking exons in the wild type and *box3* mutants of the mitochondrial cytochrome *b* gene reveals an interlaced splicing protein coded by an intron. *Cell* **22**:333.

Lazowska, J., C. Jacq, S. Cebrat, and P.P. Slonimski. 1976. Fine structure map constructed by electron microscopy and restriction endonucleases of the mitochondrial DNA segment conferring erythromycin resistance. In *The genetic functions of mitochondrial DNA* (ed. C. Saccone and A.M. Kroon), p. 325. Elsevier/North-Holland, Amsterdam.

Lazowska, J., F. Michel, G. Faye, H. Fukuhara, and P.P. Slonimski. 1974. Physical and genetic organization of petite and grande yeast mitochondrial DNA. II. DNA-DNA hybridization studies and buoyant densities determinations. *J. Mol. Biol.* **85**:393.

Lee, C.P., G. Schatz, and L. Ernster, eds. 1979. *Membrane biogenesis.* Addison-Wesley, New York.

Lee, E.-H. and B.F. Johnson. 1977. Volume-related mitochondrial DNA synthesis in zygotes and vegetative cells of *Saccharomyces cerevisiae*. *J. Bacteriol.* **129**:1066.

Leff, J., and T.R. Eccleshall. 1978. Replication of bromodeoxyuridylate-substituted mitochondrial DNA in yeast. *J. Bacteriol.* **135**:436.

Leth-Bak, A., C. Christiansen, and A. Stenderup. 1969. Unusual physical properties of mitochondrial DNA in yeast. *Nature* **224**:270.

Levens, D., A. Lustig, and M. Rabinowitz. 1981a. Purification of mitochondrial RNA polymerase from *S. cerevisiae*. *J. Biol. Chem.* **256**:1474.

Levens, D., R. Morimoto, M. Rabinowitz. 1981b. Mitochondrial transcription complex from *S. cerevisiae*. *J. Biol. Chem.* **256**:1466.

Levens, D., B. Ticho, E. Acherman, and M. Rabinowitz. 1981c. Transcriptional initiation and 5' termini of yeast mitochondrial RNA. *J. Biol. Chem.* (in press).

Lewin, A., R. Morimoto, M. Rabinowitz, and H. Fukuhara. 1978. Restriction enzyme analysis of mitochondrial DNAs of petite mutants of yeast: Classification of petites and deletion mapping of mitochondrial genes. *Mol. Gen. Genet.* **163**:257.

Lewin, A., I. Gregor, T.L. Mason, N. Nelson, and G. Schatz. 1980. Cytoplasmically made subunits of yeast mitochondrial F1-ATPase and cytochrome *c* oxidase are synthesized as individual precursors, not as polyproteins. *Proc. Natl. Acad. Sci.* **77:** 3998.

Lewin, A., R. Morimoto, S. Merten, N. Martin, P. Berg, T. Christianson, D. Levens, and M. Rabinowitz. 1977. Physical mapping of mitochondrial genes and transcripts in *Saccharomyces cerevisiae*. In *Mitochondria 1977: Genetics and biogenesis of mitochondria* (ed. W. Bandlow et al.), p. 271. de Gruyter, Berlin.

Li, M. and A. Tzagoloff. 1979. Assembly of the mitochondrial membrane system: Sequences of yeast mitochondrial valine and an unusual threonine tRNA gene. *Cell* **18:** 47.

Lin, L.F. and D.S. Beattie. 1976. Purification and biogenesis of cytochrome *b* in baker's yeast. In *Genetics and biogenesis of chloroplasts and mitochondria* (ed. T. Bücher et al.), p. 281. Elsevier/North-Holland, Amsterdam.

Lin, L.F.H., L. Clejan, and D.S. Beattie. 1978. The synthesis of cytochrome *b* on mitochondrial ribosomes in baker's yeast. *Eur. J. Biochem.* **87:** 171.

Lindergreen, C.C., S. Nagai, and H. Nagai. 1958. Induction of respiratory deficiency in yeast by manganese, copper, cobalt and nickel. *Nature* **182:** 446.

Linnane, A.W. and P.B. Crowfoot. 1975. Biogenesis of yeast mitochondrial membranes. In *Membrane biogenesis* (ed. A. Tzagoloff), p. 99. Plenum Press, New York.

Linnane, A.W. and P. Nagley. 1978a. Mitochondrial genetics in perspective: The derivation of a genetic and physical map of the yeast mitochondrial genome. *Plasmid* **1:** 324.

Linnane, A.W., J.M. Haslam, H.B. Lukins, and P. Nagley. 1972. The biogenesis of mitochondria in microorganisms *Annu. Rev. Microbiol.* **26:** 163.

Linnane, A.W., A.J. Lamb, C. Christodoulou, and H.B. Lukins. 1968a. The biogenesis of mitochondria. VI. Biochemical basis of the resistance of *Saccharomyces cerevisiae* toward antibiotics which specifically inhibit mitochondrial protein synthesis. *Proc. Natl. Acad. Sci.* **59:** 1288.

Linnane, A.W., G.W. Saunders, E.B. Gingold, and H.B. Lukins. 1968b. The biogenesis of mitochondria. V. Cytoplasic inheritance of erythromycin resistance in *Saccharomyces cerevisiae*. *Proc. Natl. Acad. Sci.* **59:** 903.

Linnane, A.W., H.B. Lukins, P.L. Molloy, P. Nagley, J. Rytka, K.S. Sriprakash, and M.K. Trembath. 1976. Biogenesis of mitochondria: Molecular mapping of the mitochondrial genome of yeast. *Proc. Natl. Acad. Sci.* **73:** 2082.

Locker, J. and M. Rabinowitz. 1976. Electron microscopic analysis of mitochondrial DNA sequences from petite and grande yeast. In *The genetic functions of mitochondrial DNA* (ed. C. Saccone and A.M. Kroon), p. 313. North-Holland, Amsterdam.

Locker, J., A. Lewin, and M. Rabinowitz. 1979. The structure and organization of mitochondrial DNA from petite yeast. *Plasmid* **2:** 155.

Locker, J., M. Rabinowitz, and G.S. Getz. 1974a. Tandem inverted repeats in mitochondrial DNA of petite mutants of *Saccharomyces cerevisiae*. *Proc. Natl. Acad. Sci.* **71:** 1366.

———. 1974b. Electron microscopic and renaturation kinetic analysis of mitochondrial DNA of cytoplasmic petite mutants of *Saccharomyces cerevisiae*. *J. Mol. Biol.* **88:** 489.

Lukins, H.B., J.R. Tate, G.W. Saunders, and A.W. Linnane. 1973. The biogenesis of mitochondria 26. Mitochondrial recombination: The segregation of parental and recombinant mitochondrial genotypes during vegetative division of yeast. *Mol. Gen. Genet.* **120:** 17.

Maccacchini, M.L., Y. Rudin, and G. Schatz. 1979a. Transport of proteins across the mitochondrial outer membrane: A precursor form of the cytoplasmically made intermembrane enzyme cytochrome *c* peroxydase. *J. Biol. Chem.* **254:** 7468.

Maccacchini, M.L., Y. Rudin, G. Blobel, and G. Schatz. 1979b. Import of proteins into mitochondria: Precursor forms of the extramitochondrially made F1-ATPase subunits in yeast. *Proc. Natl. Acad. Sci.* **76:** 343.

Macino, G. and A. Tzagoloff. 1979a. Assembly of the mitochondrial membrane system: The

DNA sequence of a mitochondrial ATPase gene in *Saccharomyces cerevisiae. J. Biol. Chem.* **254:** 4617.

———. 1979b. Assembly of the mitochondrial membrane system: Two separate genes coding for threonyl-tRNA in the mitochondrial DNA of *Saccharomyces cerevisiae. Mol. Gen. Genet.* **169:** 183.

———. 1980. Assembly of the mitochondrial membrane system: Sequence analysis of a yeast mitochondrial ATPase gene containing the *oli-2* and *oli-4* loci. *Cell* **20:** 507.

Macino, G., G. Coruzzi, F. Nobrega, M. Li, and A. Tzagoloff. 1979. Use of the UGA terminator as a tryptophan codon in yeast mitochondria. *Proc. Natl. Acad. Sci.* **76:** 3784.

Mahler, H.R. 1973a. Structural requirements for mitochondrial mutagenesis. *J. Supramol. Struct.* **1:** 449.

———. 1973b. Biogenetic autonomy of mitochondria. *CRC Crit. Rev. Biochem.* **1:** 381.

———. 1976. Mitochondrial assembly: Attempts at resolution of complex functions. In *Mitochondria, bioenergetics, biogenesis and membrane structure* (ed. L. Packer and A. Gomez-Puyou), p. 213. Academic Press, New York.

Mahler, H.R. and R.N. Bastos. 1974a. Coupling between mitochondrial mutation and energy transduction. *Proc. Natl. Acad. Sci.* **71:** 2241.

———. 1974b. A novel reaction of mitochondrial DNA with ethidium bromide. *FEBS Lett.* **39:** 27.

Mahler, H.R. and K. Dawidowicz. 1973. Autonomy of mitochondria in *S. cerevisiae* in their production of messenger RNA. *Proc. Natl. Acad. Sci.* **70:** 111.

Mahler, H.R. and P.S. Perlman. 1972a. Effects of mutagenic treatment by ethidium bromide on cellular and mitochondrial phenotype. *Arch. Biochem. Biophys.* **148:** 115.

———. 1972b. Mitochondrial membranes and mutagenesis by ethidium bromide. *J. Supramol.*

Mahler, H.R., R.N. Bastos, U. Flury, C.C. Lin, and S.H. Phan. 1975. Mitochondrial biogenesis in fungi. In *Membrane biogenesis* (ed. A. Tzagoloff), p. 66. Plenum Press, New York.

Mahler, H.R., T. Bilinski, D. Miller, D. Hanson, P.S. Perlman, and C.A. Demko. 1976. Respiration deficient mutants with intact mitochondrial genomes: Casting a wider net. In *Genetics and biogenesis of chloroplasts and mitochondria* (ed. T. Bücher et al.), p. 857. Elsevier/North-Holland, Amsterdam.

Mahler, H.R., D. Hanson, D. Miller, T. Bilinski, D.M. Ellis, N.J. Alexander, and P.S. Perlman. 1977. Structural and regulatory mutations affecting mitochondrial gene products. In *Mitochondria 1977: Genetics and biogenesis of mitochondria* (ed. W. Bandlow et al.), p. 345. de Gruyter, Berlin.

Marcovich, H. 1951. Action de l'acriflavine sur les levures. VIII. Détermination du composant actif et étude de l'euflavine. *Ann. Inst. Pasteur* **81:** 452.

Marmiroli, N., C. Donnini, F.M. Restivo, F. Tassi, and P.P. Puglisi. 1980a. Analysis of *rho* mutability in *Saccharomyces cerevisiae*. II. Role of the mitochondrial protein synthesis. *Mol. Gen. Genet.* **177:** 589.

Marmiroli, N., F.M. Restivo, C. Donnini, L. Bianchi, and P.P. Puglisi. 1980b. Analysis of *rho* mutability in *Saccharomyces cerevisiae*. I. Effects of *mmc* and *pet*-ts alleles. *Mol. Gen. Genet.* **177:** 581.

Martin, N. and M. Rabinowitz. 1978. Mitochondrial tRNAs in yeast: Identification of isoaccepting tRNAs. *Biochemistry* **17:** 1628.

Martin, N., M. Rabinowitz, and H. Fukuhara. 1976. Isoaccepting mitochondrial glutamyl-tRNA species transcribed from different regions of the mitochondrial genome of *Saccharomyces cerevisiae. J. Mol. Biol.* **101:** 285.

———. 1977. Yeast mitochondrial DNA specifies tRNA for 19 amino acids. Deletion mapping of the tRNA genes. *Biochemistry* **16:** 4672.

Martin, N.C., D.L. Miller, J.E. Donelson, C. Sigurdson, J.L. Hartley, P.S. Moynihan, and H.D. Pham. 1979. Identification and sequencing of yeast mitochondrial tRNA genes in mitochondrial DNA-pBR322 recombinants. *ICN-UCLA Symp. Mol. Cell. Biol.* **15:** 357.

Martin, R.P., J.M. Schneller, A.J.C. Stahl, and G. Dirheimer. 1976. *Biochem. Biophys. Res. Commun.* **70:** 997.

———. 1977. Study of yeast mitochondrial tRNAs by two dimensional polyacrylamide gel electrophoresis: Characterization of isoaccepting species and search for imported cytoplasmic tRNAs. *Nucleic Acids Res.* **4:** 3497.

———. 1979. Import of nuclear deoxyribonucleic acid coded lysine-accepting transfer ribonucleic acid (anticodon CUU) into yeast mitochondria. *Biochemistry* **18:** 4600.

Martin, R.P., A.P. Sibler, R. Bordonne, J. Canaday, and G. Dirheimer. 1980. Nucleotide sequence and gene localization of yeast mitochondrial initiator tRNA^{fmet} and UGA decoding tRNA^{trp}. In *The organization and expression of the mitochondrial genome* (ed. C. Saccone and A.M. Kroon), p. 311. Elsevier/North-Holland, Amsterdam.

Martin, R.P., A.P. Sibler, J.M. Schneller, G. Keith, A.J.C. Stahl, and G. Dirheimer. 1978. Primary structure of yeast mitochondrial DNA coded phenylalanine tRNA. *Nucleic Acids Res.* **5:** 4579.

Marzuki, S., R.M. Hall, and A.W. Linnane. 1974. Induction of respiratory incompetent mutants by unsaturated fatty acid depletion in *Saccharomyces cerevisiae. Biochem. Biophys. Res. Commun.* **57:** 372.

Mason, T.L. and G. Schatz. 1973. Cytochrome *c* oxidase from baker's yeast. II. Site of translation of the protein components. *J. Biol. Chem.* **248:** 1355.

Mason, T., P. Boerner, and C. Biron. 1976. The use of double mutant strains containing both heat and cold sensitive mutations in studies of mitochondrial biogenesis. In *Genetics and biogenesis of chloroplasts and mitochondria* (ed. T. Bücher et al.), p. 239. Elsevier/North-Holland, Amsterdam.

Mason, T., M. Breitbart, and J. Meyers. 1979. Temperature sensitive mutations of *Saccharomyces cerevisiae* with defects in mitochondrial functions. *Methods Enzymol.* **56:** 131.

Mason, T.L., R.O. Poynton, D.C. Wharton, and G. Schatz. 1973. Cytochrome *c* oxidase from baker's yeast. I. Isolation and properties. *J. Biol. Chem.* **248:** 1346.

Mathews, S., R.J. Schweyen, and F. Kaudewitz. 1977. Preferential loss or retention of mitochondrial genes in *rho*⁻ clones. In *Mitochondria 1977: Genetics and biogenesis of mitochondria* (ed. W. Bandlow et al.), p. 133. de Gruyter, Berlin.

Mattick, J.S. and R.M. Hall. 1977. Replicative deoxyribonucleic acid synthesis in isolated mitochondria from *Saccharomyces cerevisiae. J. Bacteriol.* **130:** 973.

Mattick, J.S. and P. Nagley. 1977. Comparative studies of the effects of acridines and other petite inducing drugs on the mitochondrial genome of *Saccharomyces cerevisiae. Mol. Gen. Genet.* **152:** 267.

Mayer, V.W. and M.J. Legator. 1970. Induction by *N*-methyl-*N'*Nitro-*N*-nitrosoguanidine and UV light of petite mutants in aerobically and anaerobically cultivated *Saccharomyces cerevisiae. Mutat. Res.* **9:** 193.

Mehrotra, B.D. and H.R. Mahler. 1968. Characterization of some unusual DNA's from the mitochondria from certain "petite" strains of *Saccharomyces cerevisiae. Arch. Biochem. Biophys.* **128:** 685.

Merten, S., R.M. Synenki, J. Locker, T. Christianson, and M. Rabinowitz. 1980. Processing of precursors of 21S ribosomal RNA from yeast mitochondria *Proc. Natl. Acad. Sci.* **77:** 1417.

Michaelis, G. 1976. Cytoplasmic inheritance of antimycin A resistance in *Saccharomyces cerevisiae. Mol. Gen. Genet.* **146:** 133.

Michaelis, G. and E. Pratje. 1977. Mapping of two mitochondrial antimycin A resistance loci in *Saccharomyces cerevisiae. Mol. Gen. Genet.* **156:** 79.

Michaelis, G. and M. Somlo. 1976. Genetic analysis of mitochondrial biogenesis and function in *Saccharomyces cerevisiae. J. Bioenerg.* **8:** 93.

Michaelis, G., E. Petrochilo, and P.P. Slonimski. 1973. Mitochondrial genetics. III. Recombined molecules of mitochondrial DNA obtained from crosses between cytoplasmic petite mutants of *Saccharomyces cerevisiae*: Physical and genetic characterization. *Mol. Gen. Genet.* **123:** 51.

Michaelis, G., S. Douglass, M.J. Tsai, and R.S. Criddle. 1971. Mitochondrial DNA and suppressiveness of petite mutants in *Saccharomyces cerevisiae*. *Biochem. Genet.* **5**: 487.

Michaelis, G., E. Pratje, B. Dujon, and L. Weill. 1976a. Recombination of yeast mitochondrial DNA segments conferring resistance either to paromomycin or to chloramphenicol. In *Genetics, biogenesis and bioenergetics of mitochondria* (ed. W. Bandlow et al.), p. 49. de Gruyter, Berlin.

Michaelis, G., F. Michel, J. Lazowska, and P.P. Slonimski. 1976b. Recombined molecules of mitochondrial DNA obtained from crosses between cytoplasmic petite mutants of *Saccharomyces cerevisiae*: The stoichiometry of parental DNA repeats within the recombined molecule. *Mol. Gen. Genet.* **149**: 125.

Michel, F. 1974. Hysteresis and partial irreversibility of denaturation of DNA as a means of investigating the topology of base distribution constraints: Application to a yeast *rho⁻* (petite) mitochondrial DNA. *J. Mol. Biol.* **89**: 305.

———. 1981. "Etudes expérimentales et théoriques des transitions thermiques de l'DNA et leur application à l'étude génétique de l'DNA mitochondrial de levure." Ph.D. thesis, University of Paris.

Michel, F., C. Grandchamp, and B. Dujon. 1979. Genetic and physical characterization of a segment of yeast mitochondrial DNA involved in the control of genetic recombination. *Biochimie* **61**: 985.

Michel, F., J. Lazowska, G. Faye, H. Fukuhara, and P.P. Slonimski. 1974. Physical and genetic organization of petite and grande yeast mitochondrial DNA. III. High resolution melting and reassociation studies. *J. Mol. Biol.* **85**: 411.

Mihara, K. and G. Blobel. 1980. The four cytoplasmically made subunits of yeast mitochondrial cytochrome *c* oxidase are synthesized individually and not as a polyprotein. *Proc. Natl. Acad. Sci.* **77**: 4160.

Miller, P.L., ed. 1970. *Symp. Soc. Exp. Biol.* vol. **24**.

Miller, D.L., N.C. Martin, H.D. Phan, and J.E. Donelson. 1979. Sequence analysis of two yeast mitochondrial DNA fragments containing the genes for $tRNA_{UCR}^{ser}$ and $tRNA_{UUY}^{phe}$. *J. Biol. Chem.* **254**: 11735.

Mol, J.N.M., P. Borst, F.G. Grosveld, and J.H. Spencer. 1974. The size of the repeating unit of the repetitive mitochondrial DNA from a "low density" petite mutant of yeast. *Biochim. Biophys. Acta* **374**: 115.

Molloy, P.L., A.W. Linnane, and H.B. Lukins. 1975. Biogenesis of mitochondria: Analysis of deletion of mitochondrial antibiotic resistance markers in petite mutants of *Saccharomyces cerevisiae*. *J. Bacteriol.* **122**: 7.

———. 1976. Relative retention of mitochondrial markers in petite mutants: Mitochondrially determined differences between *rho⁺* strains. *Genet. Res.* **26**: 319.

Molloy, P.L., N. Howell, D.T. Plummer, A.W. Linnane, H.B. Lukins. 1973. Mitochondrial mutants of the yeast *Saccharomyces cerevisiae* showing resistance in vitro to chloramphenicol inhibition of mitochondrial protein synthesis. *Biochem. Biophys. Res. Commun.* **52**: 9.

Monnerot, M., R.J. Schweyen, and H. Fukuhara. 1977. Mapping of mutation tsm-8 with respect to transfer RNA genes on the mitochondrial DNA of *Saccharomyces cerevisiae*. *Mol. Gen. Genet.* **152**: 307.

Moorman, A.F.M., G.J.B. Van Ommen, and L. Grivell. 1978. Transcription in yeast mitochondria: Isolation and physical mapping of messenger RNAs for subunits of cytochrome *c* oxidase and ATPase. *Mol. Gen. Genet.* **160**: 13.

Moorman, A.F.M., F.N. Verkley, F.A.M. Asselbergs, and L.A. Grivell. 1977. Yeast mitochondrial messenger RNA is not correctly translated in heterologous cell free systems. In *Mitochondria 1977: Genetics and biogenesis of mitochondria* (ed. W. Bandlow et al.), p. 385. de Gruyter, Berlin.

Morimoto, R. and M. Rabinowitz. 1979a. Physical mapping of the *Xba*I, *Hinc*II, *Bgl*II, *Xho*I, *Sst*I and *Pvu*II restriction endonuclease cleavage fragments of mitochondrial DNA of *S. cerevisiae*. *Mol. Gen. Genet.* **170**: 11.

————. 1979b. Physical mapping of the yeast mitochondrial genome: Derivation of the fine structure and gene map of strain D273-10B and comparison with a strain (MH41-7B) differing in genome size. *Mol. Gen. Genet.* **170**: 25.

Morimoto, R., A. Lewin, and M. Rabinowitz. 1979a. Physical mapping and characterization of the mitochondrial DNA and RNA sequences from mit⁻ mutants defective in cytochrome oxidase peptide 1 (*oxi3*). *Mol. Gen. Genet.* **170**: 1.

Morimoto, R., A. Lewin, S. Merten, and M. Rabinowitz. 1976. Restriction endonuclease mapping and analysis of grande and mutant yeast mitochondrial DNA. In *Genetics and biogenesis of chloroplasts and mitochondria* (ed. T. Bücher et al.), p. 519. Elsevier/North-Holland, Amsterdam.

Morimoto, R., J. Locker, R.M. Synenki, and M. Rabinowitz. 1979b. Transcription, processing, and mapping of mitochondrial RNA from grande and petite yeast. *J. Biol. Chem.* **254**: 12641.

Morimoto, R., A. Lewin, H.-J. Hsu, M. Rabinowitz, and H. Fukuhara. 1975. Restriction endonuclease analysis of mitochondrial DNA from grande and genetically characterized cytoplasmic petite clones of *Saccharomyces cerevisiae*. *Proc. Natl. Acad. Sci.* **72**: 3868.

Morimoto, R., S. Merten, A. Lewin, N.C. Martin, and M. Rabinowitz. 1978. Physical mapping of genes on yeast mitochondrial DNA: Localization of antibiotic resistance loci and rRNA and tRNA genes. *Mol. Gen. Genet.* **163**: 241.

Mounolou, J.C., H. Jakob, and P.P. Slonimski. 1966. Mitochondrial DNA from yeast "petite" mutants: Specific changes of buoyant density corresponding to different cytoplasmic mutations. *Biochem. Biophys. Res. Commun.* **24**: 218.

————. 1968. Mutations of mitochondrial DNA. In *Biochemical aspects of the biogenesis of mitochondria* (ed. E.C. Slater et al.), p. 473. Adriatica Editrice, Bari.

Moustacchi, E. 1971. Evidence for nucleus independent steps in control of repair of mitochondrial damage. I. UV induction of the cytoplasmic "petite" mutation in UV sensitive nuclear mutants of *S. cerevisiae*. *Mutat. Res.* **114**: 50.

Murphy, M., S.J. Gukowski, S. Marzuki, H.B. Lukins, and A.W. Linnane. 1978. Mitochondrial oligomycin-resistance mutations affecting the proteolipid subunit of the mitochondrial adenosine triphosphatase. *Biochem. Biophys. Res. Commun.* **85**: 1283.

Nagai, S. 1969. High frequency production of respiratory mutants in yeast under nutritional deficiencies. *Mutat. Res.* **8**: 557.

Nagley, P. and A.W. Linnane. 1970. Mitochondrial DNA deficient petite mutants of yeast. *Biochem. Biophys. Res. Commun.* **39**: 989.

————. 1972a. Cellular regulation of mitochondrial DNA synthesis in *Saccharomyces cerevisiae*. *Cell Differ.* **1**: 143.

————. 1972b. Biogenesis of mitochondria. XXI. Studies on the nature of the mitochondrial genome in yeast: The degenerative effects of ethidium bromide on mitochondrial genetic information in a respiratory competent strain. *J. Mol. Biol.* **66**: 181.

Nagley, P. and J.S. Mattick. 1977. Mitochondrial DNA replication in petite mutants of yeast: Resistance to inhibition by ethidium bromide, berenil and euflavine. *Mol. Gen. Genet.* **152**: 277.

Nagley, P., K.S. Sriprakash, and A.W. Linnane. 1977. Structure, synthesis and genetics of yeast mitochondrial DNA. *Adv. Microbiol. Physiol.* **16**: 157.

Nagley, P., E.B. Gingold, H.B. Lukins, and A.W. Linnane. 1973. Biogenesis of mitochondria. XXV. Studies on the mitochondrial genomes of petite mutants of yeast using ethidium bromide as a probe. *J. Mol. Biol.* **78**: 335.

Nagley, P., J.S. Mattick, R.M. Hall, and A.W. Linnane. 1975. Biogenesis of mitochondria. 43. A comparative study of petite induction and inhibition of mitochondrial DNA replication in yeast by ethidium bromide and berenil. *Mol. Gen. Genet.* **141**: 291.

Nagley, P., K.S. Sriprakash, J. Rytka, K.B. Choo, M.K. Trembath, H.B. Lukins, and A.W. Linnane. 1976. Physical mapping of genetic markers in the yeast mitochondrial genome. In

The genetic functions of mitochondrial DNA (ed. C. Saccone and A.M. Kroon), p. 231. Elsevier/North-Holland, Amsterdam.

Nelson, N. and G. Schatz. 1979. Energy dependent processing of cytoplasmically made precursors to mitochondrial proteins. *Proc. Natl. Acad. Sci.* **76**: 4365.

Netter, P., E. Petrochilo, P.P. Slonimski, M. Bolotin-Fukuhara, D. Coen, J. Deutsch, and B. Dujon. 1974. Mitochondrial genetics. VII. Allelism and mapping studies of ribosomal mutants resistant to chloramphenicol, erythromycin and spiramycin in *Saccharomyces cerevisiae*. *Genetics* **78**: 1063.

Newlon, C.S. and W.L. Fangman. 1975. Mitochondrial DNA synthesis in cell cycle mutants of *Saccharomyces cerevisiae*. *Cell* **5**: 423.

Newlon, C.S., R.D. Ludescher, and S.K. Walter. 1979. Production of petites by cell cycle mutants of *Saccharomyces cerevisiae* defective in DNA synthesis. *Mol. Gen. Genet.* **169**: 189.

Nobrega, F.G. and A. Tzagoloff. 1980a. Assembly of the mitochondrial membrane system. Structure and location of the mitochondrial glutamic tRNA gene in *Saccharomyces cerevisiae*. *FEBS Lett.* **113**: 52.

———. 1980b. Assembly of the mitochondrial membrane system. Complete restriction map of the cytochrome *b* region of mitochondrial DNA in *Saccharomyces cerevisiae* D273-10B. *J. Biol. Chem.* **255**: 9821.

———. 1980c. Assembly of the mitochondrial membrane system. DNA sequence and organization of the cytochrome *b* gene in *Saccharomyces cerevisiae* D273-10B. *J. Biol. Chem.* **255**: 9828.

Nordstrom, K. 1967. Induction of the petite mutation in *Saccharomyces cerevisiae* by *N*-methyl *N'*-nitro-*N*-nitrosoguanidine. *J. Gen. Microbiol.* **48**: 277.

O'Brien, T.W. and D.E. Matthews. 1976. Mitochondrial ribosomes. In *Handbook of genetics* (ed. R.C. King), vol. 5, p. 535. Plenum Press, New York.

Ogur, M.S., S. Minckler, G. Lindegreen, and C. Lindegreen. 1952. The nucleic acids in a polyploid series of *Saccharomyces*. *Arch. Biochem. Biophys.* **40**: 175.

Ohashi, A. and G. Schatz. 1980. Stimulation of in vitro mitochondrial protein synthesis by yeast cytoplasmic extracts is caused by guanyl nucleotides *J. Biol. Chem.* **255**: 7740.

Ono, B.I., G. Fink, and G. Schatz. 1975. Mitochondrial assembly in respiration deficient mutants of *Saccharomyces cerevisiae*. *J. Biol. Chem.* **250**: 775.

Osinga, K.A., R.F. Evers, J.C. Van der Laan, and H.F. Tabak. 1981. A putative precursor for the small ribosomal RNA from mitochondria of *Saccharomyces cerevisiae*. *Nucleic Acids Res.* **9**: 1351.

Packer, L. and A. Gomez-Payou, eds. 1976. *Mitochondria, bioenergetics, biogenesis and membrane structure*. Academic Press, New York.

Padmanaban, G., F. Hendler, J. Patzer, R. Ryan, and M. Rabinowitz. 1975. Translation of RNA that contains polyadenylate from yeast mitochondria in an *Escherichia coli* ribosomal system. *Proc. Natl. Acad. Sci.* **72**: 4293.

Pajot, P., M.L. Wambier-Kluppel, and P.P. Slonimski. 1977. Cytochrome *c* reductase and cytochrome oxidase formation in mutants and revertants in the "box" region of the mitochondrial DNA. In *Mitochondria 1977: Genetics and biogenesis of mitochondria* (ed. W. Bandlow et al.), p. 173. de Gruyter, Berlin.

Pajot, P., M.L. Wambier-Kluppel, Z. Kotylak, and P.P. Slonimski. 1976. Regulation of cytochrome oxidase formation by mutations in a mitochondrial gene for cytochrome *b*. In *Genetics and biogenesis of chloroplasts and mitochondria* (ed. T. Bücher et al.), p. 443. Elsevier/North-Holland, Amsterdam.

Paoletti, C., H. Couder, and M. Guerineau. 1972. A yeast mitochondrial deoxyribonuclease stimulated by ethidium bromide. *Biochem. Biophys. Res. Commun.* **48**: 950.

Perlman, P.S. 1975. Cytoplasmic petite mutants in yeast: A model for the study of reiterated genetic sequences. In *Genetics and biogenesis of mitochondria and chloroplasts* (ed. C.W. Birky, Jr. et al.), p. 136. Ohio University Press, Columbus.

————. 1976. Genetic analysis of petite mutants of *Saccharomyces cerevisiae*: Transmissional types. *Genetics* **82**: 645.

Perlman, P.S. and C.W. Birky, Jr. 1974. Mitochondrial genetics in baker's yeast: A molecular mechanism for recombinational polarity and suppressiveness. *Proc. Natl. Acad. Sci.* **71**: 4612.

Perlman, P.S. and H.R. Mahler. 1971a. A premutational state induced in yeast by ethidium bromide. *Biochem. Biophys. Res. Commun.* **44**: 261.

————. 1971b. Molecular consequences of ethidium bromide mutagenesis. *Nat. New Biol.* **231**: 12.

————. 1973. Induction of respiratory deficient mutants in *Saccharomyces cerevisiae* by berenil. II. Characteristics of the process. *Mol. Gen. Genet.* **121**: 295.

Perlman, P.S., N.J. Alexander, D.K. Hanson, and H.R. Mahler. 1980. Mosaic genes in yeast mitochondria. In *Gene structure and expression* (ed. D.H. Dean et al.), p. 211. Ohio University Press, Columbus.

Perlman, P.S., M.G. Douglas, R.L. Strausberg, and R.A. Butow. 1977. Localization of genes for variant forms of mitochondrial proteins on mitochondrial DNA of *Saccharomyces cerevisiae. J. Mol. Biol.* **115**: 675.

Petes, T.D. and W.L. Fangman. 1973. Preferential synthesis of yeast mitochondrial DNA in alpha factor arrested cells. *Biochem. Biophys. Res. Commun.* **55**: 603.

Piñon, R., Y. Salts, and G. Simchen. 1974. Nuclear and mitochondrial DNA synthesis during yeast sporulation. *Exp. Cell Res.* **83**: 231.

Pinto, M., M. Guerineau, and C. Paoletti. 1975. Ethidium bromide mutagenesis in yeast: Protection by anaerobiosis *Mutat. Res.* **30**: 219.

Piperno, G., G. Fonty, and G. Bernardi. 1972. The mitochondrial genome of wild-type yeast cells. II. Investigations on the compositional heterogeneity of mitochondrial DNA. *J. Mol. Biol.* **65**: 191.

Plischke, M.E., R.C. von Borstel, R.D. Mortimer, and W.E. Cohn. 1976. Genetic markers and associated gene products in *Saccharomyces cerevisiae*. In *Handbook of biochemistry and molecular biology*, 3rd edition; (ed. G.D. Fasman), vol. 2, p. 767. Chemical Rubber, Cleveland.

Poynton, R.O. and G.S.P. Groot. 1975. Biosynthesis of polypeptides of cytochrome *c* oxidase by isolated mitochondria. *Proc. Natl. Acad. Sci.* **72**: 172.

Poynton, R.O. and J. Kavanagh. 1976. Regulation of mitochondrial protein synthesis by cytoplasmic proteins. *Proc. Natl. Acad. Sci.* **73**: 3947.

Poynton, R.O. and E. McKemmie. 1976. The assembly of cytochrome *c* oxidase from *Saccharomyces cerevisiae*. In *Genetics and biogenesis of chloroplasts and mitochondria* (ed. T. Bücher et al.), p. 207. Elsevier/North-Holland, Amsterdam.

————. 1979a. A polypeptide precursor to all four cytoplasmically translated subunits of cytochrome *c* oxidase from *Saccharomyces cerevisiae. J. Biol. Chem.* **254**: 6763.

————. 1979b. Post-translational processing and transport of the polyprotein precursor to subunit IV and VII of yeast cytochrome *c* oxidase. *J. Biol. Chem.* **254**: 6772.

Poynton, R.O. and G. Schatz. 1975a. Cytochrome *c* oxidase from baker's yeast. III. Physical characterization of isolated subunits and chemical evidence for two classes of polypeptides. *J. Biol. Chem.* **250**: 752.

————. 1975b. Cytochrome *c* oxidase from baker's yeast. IV. Immunological evidence for the participation of a mitochondrially synthesized subunit in enzymatic activity. *J. Biol. Chem.* **250**: 762.

Pratje, E. and G. Michaelis. 1976. Two mitochondrial antimycin A resistance loci in *Saccharomyces cerevisiae*. In *Genetics and biogenesis of chloroplasts and mitochondria* (ed. T. Bücher et al.), p. 467. Elsevier/North-Holland, Amsterdam.

————. 1977. Allelism studies of mitochondrial mutants resistant to antimycin A or funiculosin in *Saccharomyces cerevisiae. Mol. Gen. Genet.* **152**: 167.

Prunell, A. and G. Bernardi. 1974. The mitochondrial genome of wild type yeast cells. IV. Genes and spacers. *J. Mol. Biol.* **86**: 825.

———. 1977. The mitochondrial genome of wild-type yeast cells. VI. Genome organization. *J. Mol. Biol.* **110:** 53.

Prunell, A., H. Kopecka, F. Strauss, and G. Bernardi. 1977. The mitochondrial genome of wild type yeast cells. V. Genome evolution. *J. Mol. Biol.* **110:** 17.

Puiseux-Dao, S., ed. 1975. *Molecular biology of nuclear cytoplasmic relationships.* Elsevier/North-Holland, Amsterdam.

Putrament, A., H. Baranowska, and W. Prazmo. 1973. Induction by manganese of mitochondrial antibiotic resistance mutations in yeast. *Mol. Gen. Genet.* **126:** 357.

Putrament, A., H. Baranowska, A. Ejchart, and W. Jachymczk. 1977. Manganese mutagenesis in yeast. VI. Mn^{++} uptake, mtDNA replication and E^R induction. Comparison with other divalent cations. *Mol. Gen. Genet.* **151:** 69.

Putrament, A., H. Baranowska, A. Ejchart, and W. Prazmo. 1975a. Manganese mutagenesis in yeast. III. A practical application of manganese for the induction of mitochondrial antibiotic resistant mutations. *J. Gen. Microbiol.* **62:** 265.

———. 1975b. Manganese mutagenesis in yeast. IV. The effects of magnesium, protein synthesis inhibitors and hydroxyurea on Ant^R induction in mitochondrial DNA. *Mol. Gen. Genet.* **140:** 339.

Putrament, A., R. Polakowska, H. Baranowska, and A. Ejchart. 1976. On homozygotization of mitochondrial mutations in *Saccharomyces cerevisiae.* In *Genetics and biogenesis of chloroplasts and mitochondria* (ed. T. Bücher et al.), p. 415. Elsevier/North-Holland, Amsterdam.

Rabinowitz, M., G.S. Getz, J. Casey, H. Swift. 1969. Synthesis of mitochondrial and nuclear DNA in anaerobically grown yeast during the development of mitochondrial function in response to oxygen. *J. Mol. Biol.* **41:** 381.

Rabinowitz, M., J. Casey, P. Gordon, J. Locker, H.-J. Hsu, and G.S. Getz. 1974. Characterization of yeast grande and petite mitochondrial DNA by hybridization and physical techniques. In *The biogenesis of mitochondria: Transcriptional, translational and genetic aspects.* (ed. A.M. Kroon and C. Saccone), p. 89. Academic Press, New York.

Rabinowitz, M., S. Jakovcic, N. Martin, F. Hendler, A. Halbreich, A. Lewin, and R. Morimoto. 1976. Transcription and organization of yeast mitochondrial DNA. In *The genetic functions of mitochondrial DNA* (ed. C. Saccone and A.M. Kroon), p. 219. Elsevier/North-Holland, Amsterdam.

Rank, G.H. 1970a. Genetic evidence for "Darwinian" selection at the molecular level. I. The effect of the suppressive factor on cytoplasmically inherited erythromycin resistance in *Saccharomyces cerevisiae. Can. J. Genet. Cytol.* **12:** 129.

———. 1970b. Genetic evidence for "Darwinian" selection at the molecular level. II. Genetic analysis of cytoplasmically-inherited high and low suppressivity in *Saccharomyces cerevisiae. Can J. Genet. Cytol.* **12:** 340.

———. 1973. Recombination in 3-factor crosses of cytoplasmically inherited antibiotic resistance mitochondrial markers in *S. cerevisiae. Heredity* **30:** 265.

Rank, G.H. and N.T. Bech-Hansen. 1972a. Somatic segregation, recombination, asymmetrical distribution and complementation tests of cytoplasmically inherited antibiotic resistance mitochondrial markers in *S. cerevisiae. Genetics* **72:** 1.

———. 1972b. Genetic evidence for "Darwinian"selection at the molecular level. III. The effect of the suppressive factor on nuclearly and cytoplasmically inherited chloramphenicol resistance in *S. cerevisiae. Canad. J. Microbiol.* **18:** 1.

———. 1973. Single nuclear gene inherited cross resistance and collateral sensitivity to 17 inhibitors of mitochondrial function in *S. cerevisiae. Mol. Gen. Genet.* **126:** 93.

Reijnders, L. and P. Borst. 1972. The number of 4S RNA genes on yeast mitochondrial DNA. *Biochim. Biophys. Acta* **47:** 126.

Reijnders, L., P. Sloof, and P. Borst. 1973. The molecular weights of the mitochondrial ribosomal RNAs of *Saccharomyces carlsbergensis. Eur. J. Biochem.* **35:** 266.

Reijnders, L., C.M. Kleisen, L.A.Grivell, and P. Borst. 1972. Hybridization studies with yeast mitochondrial RNAs. *Biochim. Biophys. Acta* **272:** 396.

Roberts, H., W.M. Choo, M. Murphy, S. Marzuki, H.B. Lukins, and A.W. Linnane. 1979. mit⁻ mutations in the *oli2* region of mitochondrial DNA affecting the 20,000 dalton subunit of the mitochondrial ATPase in *Saccharomyces cerevisiae. FEBS Lett.* **108**: 501.

Ross, E., E. Ebner, R.O. Poynton, T.L. Mason, B. Ono, and G. Schatz. 1974. The biosynthesis of mitochondrial cytochromes. In *The biogenesis of mitochondria: Transcriptional, translational and genetic aspects* (ed. A.M. Kroon and C. Saccone), p. 477. Academic Press, New York.

Rubin, B.Y. and J. Blamire. 1977. Regulation of yeast mitochondrial DNA synthesis. I. Analysis of a mutant conditionally deficient in mitochondrial DNA metabolism. *Mol. Gen. Genet.* **156**: 41.

———. 1979. Regulation of yeast mitochondrial DNA synthesis. II. The kinetics and effect of growth media on the expression of the tpi mutation. *Mol. Gen. Genet.* **169**: 41.

Rubin, M.S. and A. Tzagoloff. 1973a. Assembly of the mitochondrial membrane system. IX. Purification, characterization and subunit structure of yeast and beef cytochrome oxidase. *J. Biol. Chem.* **248**: 4269.

———. 1973b. Assembly of the mitochondrial membrane system. X. Mitochondrial synthesis of three of the subunit proteins of yeast cytochrome oxidase. *J. Biol. Chem.* **248**: 4275.

Ryrie, I.J. and A. Gallagher. 1979. The yeast mitochondrial ATPase complex: Subunit composition and evidence for a latent protease contaminant. *Biochim. Biophys. Acta* **545**: 1.

Rytka, J., K.J. English, R.M. Hall, A.W. Linnane, and H.B. Lukins. 1976. The isolation and simultaneous physical mapping of mitochondrial mutations affecting respiratory complexes. In *Genetics and biogenesis of chloroplasts and mitochondria* (ed. T. Bücher et al.), p. 427. Elsevier/North-Holland, Amsterdam.

Saccone, C. and A.M. Kroon. 1976. *The genetic functions of mitochondrial DNA.* Elsevier/North-Holland, Amsterdam.

Sager, R. 1972. *Cytoplasmic genes and organelles.* Academic Press, New York.

Sanders, J.P.M. and P. Borst. 1977. The organization of genes in yeast mitochondrial DNA. IV. Analysis of (dA.dT) clusters in yeast mitochondrial DNA by poly(U) Sephadex chromatography. *Mol. Gen. Genet.* **157**: 263.

Sanders, J.P.M., C. Heyting and P. Borst. 1975. The organization of genes in yeast mitochondrial DNA. I. The genes for large and small ribosomal RNA are far apart. *Biochem. Biophys. Res. Commun.* **65**: 699.

Sanders, J.P.M., C. Heyting, A. Difranco, P. Borst, and P.P. Slonimski. 1976. The organization of genes in yeast mitochondrial DNA. In *The genetic functions of mitochondrial DNA* (ed. C. Saccone and A.M. Kroon), p. 259. Elsevier/North-Holland, Amsterdam.

Sanders, J.P.M., C. Heyting, M.P. Verbeet, F.C.P.W. Meijlink, and P. Borst. 1977a. The organization of genes in yeast mitochondrial DNA. III. Comparison of the physical maps of the mitochondrial DNAs from three wild-type *Saccharomyces* strains. *Mol. Gen. Genet.* **157**: 239.

Sanders, J.P.M., M.P. Verbeet, F.C.P.W. Meijlink, C. Heyting, and P. Borst. 1977b. The construction of the physical maps of three different *Saccharomyces* mitochondrial DNAs. *Mol. Gen. Genet.* **157**: 271.

Saunders, G.W., E.B. Gingold, M.K. Trembath, H.B. Lukins, and A.W. Linnane. 1970. Mitochondrial genetics in yeast: Segregation of a cytoplasmic determinant in crosses and its loss or retention in the petite. In *Autonomy and biogenesis of mitochondria and chloroplasts* (ed. A.W. Linnane et al.), p. 185. North-Holland, Amsterdam.

Schatz, G. 1968. Impaired binding of the mitochondrial adenosine triphosphatase in the cytoplasmic "petite" mutant of *Saccharomyces cerevisiae. J. Biol. Chem.* **243**: 2192.

———. 1976. The biogenesis of mitochondria: A review. In *Genetics, biogenesis and bioenergetics of mitochondria* (ed. W. Bandlow et al.), p. 163. de Gruyter, Berlin.

———. 1979. How mitochondria import proteins from the cytoplasm. *FEBS Lett.* **103**: 203.

Schatz, G. and T.L. Mason. 1974. The biogenesis of mitochondrial proteins. *Annu. Rev. Biochem.* **43**: 51.

Schatz, G.S., E. Haslbrunner, and H. Tuppy. 1964. Deoxyribonucleic acid associated with yeast mitochondria. *Biochem. Biophys. Res. Commun.* **15**: 127

Schneller, J.M., B. Accoceberry, and A.J.C. Stahl. 1975a. Fractionation of yeast mitochondrial tRNA^{tyr} and tRNA^{leu}. *FEBS Lett.* **53**: 44.

Schneller, J.M., C. Schneller, and A.J.C. Stahl. 1976a. Immunological study of yeast mitochondrial phenylalanyl-tRNA synthetase. In *Genetics and biogenesis of chloroplasts and mitochondria* (ed. T. Bücher et al.), p. 775. Elsevier/North-Holland, Amsterdam.

Schneller, J.M., A. Stahl, and H. Fukuhara. 1975b. Coding origin of isoaccepting tRNA in yeast mitochondria. *Biochimie* **57**: 1051.

Schneller, J.M., G. Faye, C. Kujawa, and A.J.C. Stahl. 1975c. Number of genes and base composition of mitochondrial tRNA from *Saccharomyces cerevisiae*. *Nucleic Acids Res.* **2**: 831.

Schneller, J.M., R. Martin, A. Stahl, and G. Dirheimer. 1975d. Studies of odd bases in yeast mitochondrial tRNA: Absence of the fluorescent "Y" base in mitochondrial DNA coded tRNA^{phe}, absence of 4-thiouridine. *Biochem. Biophys. Res. Commun.* **64**: 1046.

Schneller, J.M., C. Schneller, R. Martin, and A.J.C. Stahl. 1976b. Nuclear origin of specific yeast mitochondrial amino-acyl tRNA synthetases. *Nucleic Acids Res.* **3**: 1151.

Schweizer, E., W. Demmer, U. Holzner, and H.W. Tahedl. 1977. Control of mitochondrial inactivation of temperature sensitive *Saccharomyces cerevisiae* nuclear petite mutants. In *Mitochondria 1977: Genetics and biogenesis of mitochondria* (ed. W. Bandlow et al.), p. 91. de Gruyter, Berlin.

Schweyen, R.J. and F. Kaudewitz. 1976. On the formation of *rho⁻* petites in yeast. I. Multifactorial mitochondrial crosses *(rho⁺ × rho⁺)* involving a mutation conferring temperature sensitivity of rho factor stability. *Mol. Gen. Genet.* **149**: 311.

Schweyen, R.J., B. Weiss-Brummer, B. Backhaus, and F. Kaudewitz. 1977. The genetic map of the mitochondrial genome including the fine structure of *cob* and *oxi* clusters. In *Mitochondria 1977: Genetics and biogenesis of mitochondria* (ed. W. Bandlow et al.), p. 139. de Gruyter, Berlin.

———. 1978. The genetic map of the mitochondrial genome in yeast: Map positions of drug^R and mit⁻ markers as revealed from population analyses of *rho⁻* clones in *Saccharomyces cerevisiae*. *Mol. Gen. Genet.* **159**: 151.

Schweyen, R.J., U. Steyrer, F. Kaudewitz, B. Dujon, and P. Slonimski. 1976. Mapping of mitochondrial genes in *Saccharomyces cerevisiae*: Population and pedigree analysis of retention or loss of four genetic markers in *rho⁻* cells. *Mol. Gen. Genet.* **146**: 117.

Scragg, A.H. 1976. The isolation and properties of a DNA-directed RNA polymerase from yeast mitochondria. *Biochim. Biophys. Acta* **442**: 331.

Sebald, W. 1977. Biogenesis of mitochondrial ATPase. *Biochim. Biophys. Acta* **463**: 1.

Sebald, W. and E. Wachter. 1979. Amino-acid sequence of the putative protonophore of the energy transducing ATPase complex. In *29th Mosbacher Colloquium on Energy Conservation in Biological Membranes* (ed. G. Schaffer and M. Klingerberg), p. 228. Springer Verlag, Berlin.

Sebald, W., T. Graff, and H.B. Lukins. 1979a. The dicyclohexylcarbodiimide-binding protein of the mitochondrial ATPase complex from *Neurospora crassa* and *Saccharomyces cerevisiae*. *Eur. J. Biochem.* **93**: 587.

Sebald, W., E. Wachter, and A. Tzagoloff. 1979b. Identification of amino acid substitutions in the dicyclohexylcarbodiimide binding subunit of the mitochondrial ATPase complex from oligomycin resistant mutants of *Saccharomyces cerevisiae*. *Eur. J. Biochem.* **100**: 599.

Sels, A. and H. Jakob. 1967. Acceleration of cytochrome oxidase synthesis specific to zygotes from crosses between complementary respiratory deficient mutants of *S. cerevisiae*. *Biochim. Biophys. Res. Commun.* **28**: 453.

Sena, E.P., J. Welch, and S. Fogel. 1976. Nuclear and mitochondrial DNA replication during zygote formation and maturation in yeast. *Science* **194**: 433.

Sena, E.P., J.W. Welch, H. Halvorson, and S. Fogel. 1975. Nuclear and mitochondrial DNA replication during mitosis in *Saccharomyces cerevisiae*. *J. Bacteriol.* **123**: 497.

Senior, A.E. 1977. The structure of mitochondrial ATPase. *Biochem. Biophys. Acta* **301:** 249.

Sevarino, K.A. and R.O. Poynton. 1980. Mitochondrial membrane biogenesis: Identification of a precursor to yeast cytochrome *c* oxidase subunit II, an integral polypeptide. *Proc. Natl. Acad. Sci.* **77:** 142.

Shannon, C., R. Enns, L. Wheelis, K. Burchiel, and R.S. Criddle. 1973. Alterations in mitochondrial adenosine triphosphatase activity resulting from mutation of mitochondrial DNA. *J. Biol. Chem.* **248:** 3004.

Shepherd, M.H., B.A. Atchinson, and P. Nagley, 1978. Circular mitochondrial DNA molecules from petite mutants of *Saccharomyces cerevisiae*: Resolution by polyacrylamide gel electrophoresis. *Mol. Biol. Rep.* **4:** 101.

Sherman, F. 1963. Respiration deficient mutants of yeasts. I. Genetics. *Genetics* **48:** 375.

Sherman F. and P.P. Slonimski. 1964. Respiration deficient mutants of yeast. II. Biochemistry. *Biochim. Biophys. Acta* **90:** 1.

Sherman, F. and C.W. Lawrence. 1974. *Saccharomyces.* In *Handbook of genetics* (ed. R.C. King), vol. 1, p. 359. Plenum Press, New York.

Sibler, A.P., R.P. Martin, and G. Dirheimer. 1979. The nucleotide sequence of yeast mitochondrial histidine tRNA. *FEBS Lett.* **107:** 182.

Sierra, M.F. and A. Tzagoloff. 1973. Assembly of the mitochondrial membrane system: Purification of a mitochondrial product of the ATPase. *Proc. Natl. Acad. Sci.* **70:** 3155.

Singh, A., T.L. Mason, and R.A. Zimmerman. 1978. A cold sensitive mutation of *Saccharomyces cerevisiae* affecting assembly of the mitochondrial 50S ribosomal subunit. *Mol. Gen. Genet.* **161:** 143.

Slonimski, P. 1953. *La formation des enzymes respiratoires chez la levure.* Mason, Paris.

––––––. 1968. Biochemical studies of "mitochondria" in cytoplasmic mutants. In *Biochemical aspects of the biogenesis of mitochondria* (ed. E.C. Slater et al.), p. 475. Adriatica, Bari.

Slonimiski, P.P. and B. Ephrussi. 1949. Action de l'acriflavine sur les leuvres. V. Le système des cytochromes des mutants "petite colonie." *Ann. Inst. Pasteur* **77:** 47.

Slonimski, P.P. and J. Lazowska. 1977. Transposable segments of mitochondrial DNA: A unitary hypothesis for the mechanism of mutation, recombination, sequence reiteration and suppressiveness of yeast "petite colony" mutants. In *Mitochondria 1977: Genetics and biogenesis of mitochondria* (ed. W. Bandlow et al.), p. 39. de Gruyter, Berlin.

Slonimski, P. and A. Tzagoloff. 1976. Localization in yeast mitochondrial DNA of mutations expressed in a deficiency of cytochrome oxidase and/or coenzyme Q-H$_2$ cytochrome *c* reductase. *Eur. J. Biochem.* **61:** 27.

Slonimski, P.P., P. Borst, and G. Attardi. 1982. *Mitochondrial genes.* Cold Spring Harbor Laboratory, Cold Spring Harbor, New York. (In press.)

Slonimski, P.P., G. Perrodin, and J.H. Croft. 1968. Ethidium bromide induced mutation of yeast mitochondria: Complete transformation of cells into respiratory deficient nonchromosomal "petites". *Biochem. Biophys. Res. Commun.* **30:** 232.

Slonimski, P.P., M.L. Claisse, M. Foucher, C. Jacq, A. Kochko, A. Lamouroux, P. Pajot, G. Perrodin, A. Spyridakis, and M.L. Wambier-Kluppel. 1978a. Mosaic organization and expression of the mitochondrial DNA region controlling cytochrome *c* reductase and oxidase. III. A model of structure and function. In *Biochemistry and genetics of yeast: Pure and applied aspects* (ed. M. Bacila' et al.), p. 391. Academic Press, New York.

Slonimski, P.P., P. Pajot, C. Jacq, M. Foucher, G. Perrodin, A. Kochko, and A. Lamouroux. 1978b. Mosaic organization and expression of the mitochondrial DNA region controlling cytochrome *c* reductase and oxidase. I. Genetic, physical and complementation maps of the *box* region. In *Biochemistry and genetics of yeast: Pure and applied aspects* (ed. M. Bacila et al.), p. 339. Academic Press.

Solioz, M. and G. Schatz. 1979. Mutations in putative intervening sequences of the mitochondrial cytochrome *b* gene of yeast produce abnormal cytochrome *b* polypeptides. *J. Biol. Chem.* **254:** 9331.

Somlo, M. 1977. Effect of oligomycin on coupling in isolated mitochondria from oligomycin-resistant mutants of *Saccharomyces cerevisiae* carrying an oligomycin resistant ATP phosphohydrolase. *Arch. Biochem. Biophys.* **182**: 518.

Somlo, M. and J. Cosson. 1976. Mitochondrially encoded oligomycin-resistant mutants of *S. cerevisiae*: Structural integration of ATPase and phenotype. In *Genetics and biogenesis of chloroplasts and mitochondria* (ed. T. Bücher et al.), p. 143. Elsevier/North-Holland, Amsterdam.

Somlo, M. and M. Krupa. 1979. The mitochondrially encoded "phantom ATPase" of *S. cerevisiae*. II. A heterogeneous mitochondrial ATPase population in situ. *Mol. Gen. Genet.* **167**: 329.

Somlo, M., L. Clavilier, and M. Krupa. 1977. A phantom ATPase in a mitochondrially encoded mutant of *S. cerevisiae*. *Mol. Gen. Genet.* **156**: 289.

Somlo, M., P.R. Avner, J. Cosson, B. Dujon, and M. Krupa. 1974. Oligomycin sensitivity of ATPase studied as a function of mitochondrial biogenesis, using mitochondrially determined oligomycin resistant mutants of *Saccharomyces cerevisiae*. *Eur. J. Biochem.* **42**: 439.

Somlo, M., J. Cosson, L. Clavilier, M. Krupa, and I. Laporte. 1981. Identity problems concerning the membrane subunits of the mitochondrial ATPase of *Saccharomyces cerevisiae*. *Eur. J. Biochem.* (in press).

Sor, F. and G. Faye. 1979. Mitochondrial and nuclear mutations that affect the biogenesis of the mitochondrial ribosomes of yeast. II. Biochemistry. *Mol. Gen. Genet.* **177**: 47.

Spithill, T.W., P. Nagley, and A.W. Linnane. 1979. Biogenesis of mitochondria 51: Biochemical characterization of a mitochondrial mutation in *Saccharomyces cerevisiae* affecting the mitochondrial ribosome by conferring resistance to aminoglycoside antibiotics *Mol. Gen. Genet.* **173**: 159.

Spithill, T.W., K.J. English, P. Nagley, and A.W. Linnane. 1978. Altered mitochondrial ribosomes in a cold sensitive mutant of *Saccharomyces cerevisiae*. *Mol. Biol. Rep.* **4**: 83.

Spyridakis, A. and M.L. Claisse. 1978. Yeast cytochrome *b*: Structural modifications in yeast mitochondrial mutants. In *Plant mitochondria* (ed. G. Ducet and C. Lance), p. 11. Elsevier/North-Holland Biomedical Press, Amsterdam.

Sriprakash, K.S. and G.D. Clark-Walker. 1980. The size of yeast mitochondrial ribosomal RNAs. *Biochem. Biophys. Res. Commun.* **93**: 186.

Sriprakash, K.S., K.B. Choo, P. Nagley, and A.W. Linnane. 1976a. Physical mapping of mitochondrial rRNA genes. *Biochem. Biophys. Res. Commun.* **69**: 85.

Sriprakash, K.S., P.L. Molloy, P. Nagley, H.B. Lukins, and A.W. Linnane. 1976b. Biogenesis of mitochondria. XLI. Physical mapping of mitochondrial genetic markers in yeast. *J. Mol. Biol.* **104**: 485.

Storm, E.M. and J. Marmur. 1975. A temperature sensitive mitochondrial mutation of *Saccharomyces cerevisiae*. *Biochem. Biophys. Res. Commun.* **64**: 752.

Strausberg, R.L. and C.W. Birky, Jr. 1979. Recombination of yeast mitochondrial DNA does not require mitochondrial protein synthesis. *Curr. Genet.* **1**: 21.

Strausberg, R.L. and R.A. Butow. 1977. Expression of petite mitochondrial DNA in vivo: Zygotic gene rescue. *Proc. Natl. Acad. Sci.* **74**: 2715.

———. 1981. Gene conversion at the var1 locus on yeast mitochondrial DNA. *Proc. Natl. Acad. Sci.* **78**: 494.

Strausberg, R.L. and P.J. Perlman. 1978. The effect of zygotic bud position on the transmission of mitochondrial genes in *Saccharomyces cerevisiae*. *Mol. Gen. Genet.* **163**: 131.

Strausberg, R.L., R.D. Vincent, P.S. Perlman, and R.A. Butow. 1978. Asymmetric gene conversion at inserted segments on yeast mitochondrial DNA. *Nature* **276**: 577.

Stuart, K.D. 1970. Cytoplasmic inheritance of oligomycin and rutamycin resistance in yeast. *Biochem. Biophys. Res. Commun.* **39**: 1045.

Subik, J. 1975. Mucidin resistant antimycin A sensitive mitochondrial mutant of *Saccharomyces cerevisiae*. *FEBS Lett.* **59**: 273.

Subik, J. and A. Goffeau. 1980. Cytochrome *b* deficiency in a mitochondrial *muc1muc2* recombinant of *Saccharomyces cerevisiae*. *Mol. Gen. Genet.* **178**: 603.

Subik, J. and G. Takacsova. 1978. Genetic determination of ubiquinol-cytochrome c reductase. Mitochondrial locus *muc3* specifying resistance of *Saccharomyces cerevisiae* to mucidin. *Mol. Gen. Genet.* **161**: 99.

Subik, J., J. Kolarov, and L. Kovac. 1972. Obligatory requirement of intramitochondrial ATP for normal functioning of the eucaryotic cell. *Biochem. Biophys. Res. Commun.* **49**: 192.

Subik, J., V. Kovakova, and G. Takacsova. 1977. Mucidin resistance in yeast. Isolation, characterization and genetic analysis of nuclear and mitochondrial mucidin-resistant mutants of *Saccharomyces cerevisiae*. *Eur. J. Biochem.* **73**: 275.

Subik, J., G. Takacsova, and L. Kovac. 1978. Intramitochondrial ATP and cell functions. I. Growing yeast cells depleted of intramitochondrial ATP are losing mitochondrial genes. *Mol. Gen. Genet.* **166**: 103.

Suda, K. and A. Uchida. 1974. The linkage relationship of the cytoplasmic drug-resistance factors in *Saccharomyces cerevisiae*. *Mol. Gen. Genet.* **128**: 331.

Tabak, H.F., N.B. Hecht, H. Menke, and C.P. Hollenberg. 1979. The gene for the small ribosomal RNA on yeast mitochondrial DNA: Physical map, direction of transcription and absence of an intervening sequence. *Curr. Genet.* **1**: 33.

Takacsova, G., J. Subik, and Z. Kotylak. 1980. Localization of mucidin-resistant locus *muc3* on mitochondrial DNA with respect to ubiquinol cytochrome c reductase deficient *box* loci. Locus *muc3* is allelic to *box2*. *Mol. Gen. Genet.* **179**: 141.

Terpstra, P. and R.A. Butow. 1979. The role of var1 in the assembly of yeast mitochondrial ribosomes. *J. Biol. Chem.* **254**: 12662.

Terpstra, P., E. Zanders, and R.A. Butow. 1979. The association of var1 with the 38S mitochondrial ribosomal subunit in yeast. *J. Biol. Chem.* **254**: 12653.

Tewari, K.K., J. Jayaraman, and H.R. Mahler. 1965. Separation and characterization of mitochondrial DNA from yeast. *Biochem. Biophys. Res. Comm.* **21**: 141.

Thalenfeld, B.E. and A. Tzagoloff. 1980. Assembly of the mitochondrial membrane system: Sequence of the *oxi2* gene of yeast mitochondrial DNA. *J. Biol.. Chem.* **255**: 6173.

Thomas, D.Y. and D. Wilkie. 1968a. Recombination of mitochondrial drug resistance factors in *Saccharomyces cerevisiae*. *Biochem. Biophys. Res. Commun.* **30**: 368.

―――. 1968b. Inhibition of mitochondrial synthesis in yeast by erythromycin: cytoplasmic and nuclear factors controlling resistance. *Genet. Res.* **11**: 33.

Todd, R.D., T.A. Griesenbeck, and M.G. Douglas. 1980. The yeast mitochondrial adenosine triphosphatase complex: Subunit stoichiometry and physical characterization. *J. Biol. Chem.* **255**: 5461.

Todd, R.D., P.C. McAda, and M.G. Douglas. 1979. A nuclear mutation altering the assembly of the energy transducing membrane of yeast. *J. Biol. Chem.* **254**: 11134.

Trembath, M.K., G. Macino, and A. Tzagoloff. 1977. The mapping of mutations in tRNA and cytochrome oxidase genes located in the cap-par segment of the mitochondrial genome of *S. cerevisiae*. *Mol. Gen. Genet.* **158**: 35.

Trembath, M.K., C.L. Bunn, H.B. Lukins, and A.W. Linnane. 1973. Biogenesis of mitochondria. 27. Genetic and biochemical characterization of cytoplasmic and nuclear mutations to spiramycin resistance in *Saccharomyces cerevisiae*. *Mol. Gen. Genet.* **121**: 35.

Trembath, M.K., B.C. Monk, G.M. Kellerman, and A.W. Linnane. 1975a. Biogenesis of mitochondria 36: The genetic and biochemical analysis of a mitochondrially determined cold sensitive oligomycin resistant mutant of *Saccharomyces cerevisiae* with affected mitochondrial ATPase assembly. *Mol. Gen. Genet.* **141**: 9.

―――. 1975b. Biogenesis of mitochondria 40: Phenotypic suppression of a mitochondrial mutation by a nuclear gene in *Saccharomyces cerevisiae*. *Mol. Gen. Genet.* **140**: 333.

Trembath, M.K., P.L. Molloy, K.S. Sriprakash, G.J. Cutting, A.W. Linnane, and H.B. Lukins. 1976. Biogenesis of mitochondria 44: Comparative studies and mapping of mitochondrial oligomycin resistance mutations in yeast based on gene recombination and petite deletion analysis. *Mol. Gen. Genet.* **145**: 43.

Tsai, M.J., G. Michaelis, and R.S. Criddle. 1971. DNA dependent RNA polymerase from yeast mitochondria. *Proc. Natl. Acad. Sci.* **68**: 473.

Tzagoloff, A. 1969. Assembly of the mitochondrial membrane system. II. Synthesis of the mitochondrial adenosine triphosphatase F1. *J. Biol. Chem.* **244**: 5027.

———. 1970. Assembly of the mitochondrial membrane system. III. Function and synthesis of the oligomycin sensitivity conferring protein of yeast mitochondria. *J. Biol. Chem.* **245**: 1545.

———. 1975. *Membrane biogenesis.* Plenum Press, New York.

Tzagoloff, A. and A. Akai. 1972. Assembly of the mitochondrial membrane system. VIII. Properties of the products of mitochondrial protein synthesis in yeast. *J. Biol. Chem.* **247**: 6517.

Tzagoloff, A. and P. Meagher. 1971. Assembly of the mitochondrial membrane system. V. Properties of a dispersed preparation of the rutamycin sensitive adenosine triphosphatase of yeast mitochondria. *J. Biol. Chem.* **246**: 7338.

———. 1972. Assembly of the mitochondrial membrane system. VI. Mitochondrial synthesis of subunit proteins of the rutamycin-sensitive adenosine triphosphatase. *J. Biol. Chem.* **247**: 594.

Tzagoloff, A., A. Akai, and F. Foury. 1976a. Assembly of the mitochondrial membrane system. XVI. Modified form of the ATPase proteolipid in oligomycin-resistant mutants of *Saccharomyces cerevisiae. FEBS Lett.* **65**: 391.

Tzagoloff, A., A. Akai, and R.B. Needleman. 1975a. Properties of cytoplasmic mutants of *Saccharomyces cerevisiae* with specific lesions in cytochrome oxidase. *Proc. Natl. Acad. Sci.* **72**: 2054.

———. 1975b. Assembly of the mitochondrial membrane system: Isolation of nuclear and cytoplasmic mutants of *Saccharomyces cerevisiae* with specific defects in mitochondrial functions. *J. Bacteriol.* **122**: 826.

———. 1975c. Assembly of the mitochondrial membrane system: Characterization of nuclear mutants of *Saccharomyces cerevisiae* with defects in mitochondrial ATPase and respiratory enzymes. *J. Biol. Chem.* **250**: 8228.

Tzagoloff, A., A. Akai, and M.F. Sierra 1972. Assembly of the mitochondrial membrane system. VII. Synthesis and integration of F1 subunits into the rutamycin sensitive adenosine triphosphatase. *J. Biol. Chem.* **247**: 6511.

Tzagoloff, A., F. Foury, and A. Akai. 1976b. Resolution of the mitochondrial genome. In *The genetic functions of mitochondrial DNA* (ed. C. Saccone and A.M. Kroon), p. 155. Elsevier/North-Holland, Amsterdam.

———. 1976c. Genetic determination of mitochondrial cytochrome *b*. In *Genetics and biogenesis of chloroplasts and mitochondria* (ed. T. Bücher et al.), p. 419. Elsevier/North-Holland, Amsterdam.

———. 1976d. Assembly of the mitochondrial membrane system. XVIII. Genetic loci on mitochondrial DNA involved in cytochrome *b* biosynthesis. *Mol. Gen. Genet.* **149**: 33.

Tzagoloff, A., F. Foury, and G. Macino. 1978. Genetic loci and complementation groups of the mitochondrial genome. In *Biochemistry and genetics of yeast: Pure and applied aspects* (ed. M. Bacila et al.), p. 477. Academic Press, New York.

Tzagoloff, A., G. Macino, and W. Sebald. 1979a. Mitochondrial genes and translation products *Annu. Rev. Biochem.* **48**: 419.

Tzagoloff, A., M.S. Rubin, and M.F. Sierra. 1973. Biosynthesis of mitochondrial enzymes. *Biochim. Biophys. Acta.* **301**: 71.

Tzagoloff, A., G. Macino, N.P. Nobrega, and M. Li. 1979b. Organization of mitochondrial DNA in yeast. *ICN-UCLA Symp. Mol. Cell. Biol.* **15**: 339.

Uchida, A. and K. Suda. 1973. Ethidium bromide induced loss and retention of cytoplasmic drug resistance factors in yeast. *Mutat. Res.* **19**: 57.

———. 1976. Pattern of somatic segregation of the cytoplasmic drug resistance factors in yeast. *Mol. Gen. Genet.* **145**: 159.

————. 1978. Distribution of mitochondrially inherited drug-resistance genes to tetrads from young zygotes in yeast. *Mol. Gen. Genet.* **165**: 191.

Van Kreijl, C.F. and J.L. Bos. 1977. The repeating nucleotide sequence in the repetitive mitochondrial DNA from a "low density" petite mutant of yeast. *Nucleic Acids Res.* **4**: 2369.

Van Ommen, G.J.B., G.S.P. Groot, and P. Borst. 1977. Fine structure physical mapping of 4S RNA genes on mitochondrial DNA of *Saccharomyces cerevisiae*. *Mol. Gen. Genet.* **154**: 255.

Van Ommen, G.J.B., G.S.P. Groot, and L.A. Grivell. 1979. Transcription maps of mtDNAs of two strains of *Saccharomyces*: Transcription of strain-specific insertions; complex RNA maturation and splicing. *Cell* **18**: 511.

Van Ommen, G.J.B., P.H. Boer, G.S.P. Groot, M.De Haan, E. Roosendaal, L. Grivell, A. Haid, and R.J. Schweyen. 1980. Mutations affecting RNA splicing and the interaction of gene expression of the yeast mitochondrial loci *cob* and *oxi3*. *Cell* **20**: 173.

Vidova, M. and L. Kovac. 1972. Nalidixic acid prevents the induction of yeast cytoplasmic respiration deficient mutants by intercalating drugs. *FEBS Lett.* **22**: 347.

Vincent, R.D., P.S. Perlman, R.L. Strausberg, and R.A. Butow. 1980. Physical mapping of genetic determinants on yeast mitochondrial DNA affecting the apparent size of the var1 polypeptide. *Curr. Genet.* **2**: 27.

von Borstel, R.C. 1969. *Microb. Genet. Bull.* (Suppl.) **31**.

Wachter, E., W. Sebald, and A. Tzagoloff. 1977. Altered amino-acid sequence of the DCCD binding protein in the *oli1* resistant mutant D273-10B/A21 of *Saccharomyces cerevisiae*. In *Mitochondria 1977: Genetics and biogenesis of mitochondria* (ed. W. Bandlow et al.), p. 441. de Gruyter, Berlin.

Wakabayashi, K. 1974a. On the recombination of mitochondrial genes. *Proc. Jpn. Acad.* **50**: 396.

————. 1974b. Studies on the mitochondrial genes. I. The recombination of mitochondrial drug resistances. *J. Antibiot.* **27**: 373.

————. 1978. Segments of mitochondrial DNA of yeast carrying antibiotic resistances. *Mol. Gen. Genet.* **159**: 229.

Wakabayashi, K. and N. Gunge. 1970. Extrachromosomal inheritance of oligomycin resistance in yeast. *FEBS Lett.* **6**: 302.

Waxman, M.F., N. Eaton, and D. Wilkie. 1973. Effect of antibiotics on the transmission of mitochondrial factors in *Saccharomyces cerevisiae*. *Mol. Gen. Genet.* **127**: 277.

Weislogel, P.O. and R.A. Butow. 1970. Low temperature and chloramphenicol induction of respiratory deficiency in a cold sensitive mutant of *S. cerevisiae*. *Proc. Natl. Acad. Sci.* **67**: 52.

Weiss-Brummer, B., R. Guba, A. Haid, and R.J. Schweyen. 1979. Fine structure of *oxi1*, the mitochondrial gene coding for subunit II of yeast cytochrome *c* reductase. *Curr. Genet.* **1**: 75.

Wesolowski, M. and H. Fukuhara. 1979. The genetic map of transfer RNA genes of yeast mitochondria: Correction and extension. *Mol. Gen. Genet.* **170**: 261.

Weth, G. and G. Michaelis. 1974. The size of mitochondrial DNA from a cytoplasmic petite mutant of *Saccharomyces cerevisiae*. *Mol. Gen. Genet.* **135**: 269.

Whittaker, P.A. and M. Wright. 1972. Prevention by cycloheximide of petite mutation in yeast. *Biochem. Biophys. Res. Commun.* **48**: 1455.

Wilkie, D. 1963. The induction by monochromatic UV light of respiratory deficient mutants in aerobic and anaerobic cultures of yeast. *J. Mol. Biol.* **7**: 527.

Wilkie, D. and D.Y. Thomas. 1973. Mitochondrial genetic analysis by zygote cell lineages in *Saccharomyces cerevisiae*. *Genetics* **73**: 367.

Williamson, D.H. 1970. The effect of environmental and genetic factors on the replication of mitochondrial DNA in yeast. *Symp. Soc. Exp. Biol.* **24**: 247.

Williamson, D.H. and D.J. Fennell. 1974. Apparent dispersive replication of yeast mitochondrial DNA as revealed by density labelling experiments. *Mol. Gen. Genet.* **131**: 193.

Williamson, D.H. and E. Moustacchi. 1971. The synthesis of mitochondrial DNA during the cell cycle in the yeast *Saccharomyces cerevisiae*. *Biochem. Biophys. Res. Commun.* **42**: 195.

Williamson, D.H., N.G. Maroudas, and D. Wilkie. 1971a. Induction of the cytoplasmic petite mutation in *Saccharomyces cerevisiae* by the antibacterial antibiotics erythromycin and chloramphenicol. *Mol. Gen. Genet.* **111**: 209.

Williamson, D.H., E. Moustacchi, and D. Fennell. 1971b. A procedure for rapidly extracting and estimating the nuclear and cytoplasmic DNA components of yeast cells. *Biochim. Biophys. Acta* **238**: 369.

Williamson, D.H., L.H. Johnston, K.M.V. Richmond, and J.C. Game. 1977. Mitochondrial DNA and the heritable unit of the yeast mitochondrial genome: A review. In *Mitochondria 1977: Genetics and biogenesis of mitochondria* (ed. W. Bandlow et al.), p. 1. de Gruyter, Berlin.

Wintersberger, E. 1972. Isolation of a distinct rifamycin-resistant RNA polymerase from mitochondria of yeast, *Neurospora* and liver. *Biochem. Biophys. Res. Commun.* **48**: 1287.

Wintersberger, U. and H. Blutsch. 1976. DNA dependent DNA polymerase from yeast mitochondria: Dependence of enzyme activity on conditions of cell growth and properties of the highly purified polymerase. *Eur. J. Biochem.* **68**: 199.

Wintersberger, U. and E. Wintersberger. 1970. Studies on DNA polymerase from yeast 2. Partial purification and characterization of mitochondrial DNA polymerase from wild type and respiratory deficient yeast cells. *Eur. J. Biochem.* **13**: 20.

Wintersberger, U., J. Hirsch, A.M. Fink. 1974. Studies on nuclear and mitochondrial DNA replication in a temperature sensitive mutant of *Saccharomyces cerevisiae*. *Mol. Gen. Genet.* **131**: 291.

Wolf, K. and F. Kaudewitz 1976. Effect of caffeine on the rho^- induction with ethidium bromide in *Saccharomyces cerevisiae*. *Mol. Gen. Genet.* **146**: 89.

Wolf, K., B. Dujon, and P.P. Slonimski. 1973. Mitochondrial genetics. V. Multifactorial mitochondrial crosses involving a mutation conferring paromomycin-resistance in *Saccharomyces cerevisiae*. *Mol. Gen. Genet.* **125**: 53.

Woodrow, G. and G. Schatz. 1979. The role of oxygen in the biosynthesis of cytochrome *c* oxidase of yeast mitochondria. *J. Biol. Chem.* **254**: 6088.

Appendices

APPENDIX I
Genetic Nomenclature

Compiled by Fred Sherman
Department of Radiation Biology and Biophysics
University of Rochester School of Medicine and Dentistry
Rochester, New York 14642

A recommendation for the nomenclature used in yeast genetics was published as the November 1969 supplement to *Microbial Genetics*. This recommendation has been followed except for a few minor changes and the designation of the mating-type genes. The following are the updated and expanded rules of nomenclature:

1. Gene symbols are designated by three italicized letters, which should be consistent with the proposal of Demerec et al. (1966) whenever applicable, e.g., *ARG*. The genetic locus is identified by a number immediately following the gene symbol, e.g., *ARG2*. Alleles are designated by a number separated from the locus number by a hyphen, e.g., *arg2-6*. Although locus numbers must be consistent with the original assignment, allele numbers may be particular to each laboratory.
2. Gene clusters, complementation groups within a gene, or domains within a gene having different properties can be designated by capital letters following the locus number, e.g., *his4A, his4B*, etc.
3. Dominant and recessive genes are denoted by uppercase and lowercase letters, respectively, e.g., *SUP6* and *arg2*.
4. When there is no confusion, wild-type genes are designated simply as +; the + may follow the locus number to designate a specific wild-type gene, e.g., *sup6+* and *ARG2+*. Some publications prefer to use a superscript +, e.g., *ARG2$^+$*.
5. Although superscripts should be avoided, it is sometimes expedient to distinguish genes conferring resistance and sensitivity by superscript R and S, respectively. For example, the genes controlling resistance to canavanine sulfate (*can1*) and copper sulfate (*CUP1*) and their sensitive alleles could be denoted, respectively, as *can^{R}1, CUP^{R}1, CAN^{S}1*, and *cup^{S}1*.

6. Phenotypic designations sometimes are denoted by similar symbols in Roman type and by superscript + and −. For example, the independence and requirement for arginine can be denoted by Arg⁺ and Arg⁻, respectively.
7. Wild-type and mutant alleles of the mating-type and related loci do not follow the standard rules. The two wild-type alleles at the mating-type locus are designated *MATa* and *MATα*. The two complementation groups of the *MATα* locus are denoted *MATα1* and *MATα2*. Mutations of the *MAT* genes are denoted, e.g., *mata-1*, *matα1-1*, etc. The wild-type homothallic genes at the *HMR* and *HML* loci are denoted *HMRa*, *HMRα* *HMLa*, and *HMLα*. Mutations at these loci are denoted, e.g., *hmra-1*, *hmlα-1*, etc.
8. A detailed nomenclature for suppressors is described by Sherman (1982).
9. Where necessary, non-Mendelian genotypes can be distinguished from chromosomal genotypes by enclosure in brackets. Whenever applicable, it is advisable to employ the above rules for designating non-Mendelian genes and to avoid the use of Greek letters. However, it is less confusing to either retain the original symbols ρ^+, ρ^-, ψ^+, and ψ^- or to use their transliteration, *rho+*, *rho−*, *psi+*, and *psi−*, respectively. Detailed designations for mitochondrial mutants (Dujon, this volume) and killer strains (Wickner, this volume) are presented.

Note that in choosing designations for nuclear genes, care should be taken to avoid confusion with designations for non-Mendelian genes. The following is a guide to the use of genetic symbols:

ARG2, a locus
ARG2⁺, wild-type allele
arg2, a locus or any recessive mutation at the *ARG2* locus
arg2-9, a specific allele or mutation
Arg⁺, a strain not requiring arginine
Arg⁻, a strain requiring arginine

REFERENCES

Demerec, M., E.A. Adelberg, A.J. Clark, and P.E. Hartman. 1966. A proposal for a uniform nomenclature in bacterial genetics. *Genetics* **54**:61.
Sherman, F. 1982. Suppression in the yeast *Saccharomyces cerevisiae*. In *The molecular biology of the yeast* Saccharomyces. II. *Metabolism and gene expression* (ed. J. Strathern et al.), Cold Spring Harbor Laboratory, Cold Spring Harbor, New York. (In press.)

APPENDIX II

Genetic Map of
Saccharomyces cerevisiae

Compiled by Robert K. Mortimer and David Schild
Department of Biophysics and Medical Physics and
Donner Laboratory, University of California
Berkeley, California 94720

The map presented in this Appendix was prepared from a recent compilation of data from the literature and personal communications of a large number of investigators. Map distances were calculated by the method of maximum likelihood of Snow (96).[1] This approach is based on the earlier mapping formulas derived by Barratt et al. (5). Although the map contained in this Appendix is drawn to scale, accurate determinations of map distances should be obtained by referring to the original data published in *Microbiological Reviews* 44:519 (R.K. Mortimer and D. Schild 1980). Recent evidence (S. Klapholz and R.E. Esposito, pers. comm.) indicates that the genes assigned to chromosome XVII are located on chromosome XIV. Recent reviews of methods of genetic mapping in yeast are contained in this volume (Mortimer and Schild) and in a paper by Mortimer and Hawthorne (71). The gene products of many of the genes on the map are summarized in Appendix III of this volume and in the review by Plischke et al. (79).

---

[1]The assistance of R. Snow throughout the analysis of the data is greatly appreciated.

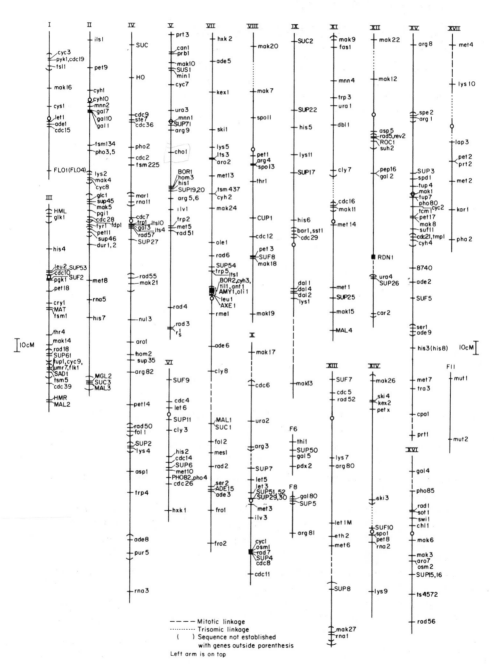

I
┬ cyc 3
├ pyk1,cdc19
├ tsl1

┼ mak16

┼ cys1
├ let1
├ ade1
├ cdc15

┼ FLO1(FLO4)

III
┼ HML
├ glk1

┼ his 4

├ leu2 SUP53
├ cdc10
├ pgk1 SUF2
├ pet18

├ cry1
├ MAT
├ tsm1
├ thr4
├ mak14
├ rad18
├ SUP61
├ tup1,cyc9,
├ umr7,flk1
├ SAD1
├ tsm5
├ cdc39
├ HMR
└ MAL2

II
┬ ils1

┼ pet9

┼ cyh1
├ cyh10
├ mnn2
├ ste7 gal7
├ gal10
├ gal1

┼ tsm134
├ pho3,5

┼ lys2
├ mak4
├ cyc8
├ glc1
├ sup45
├ mak5
├ pgi1
├ cdc28
├ trp1 ltsl0
├ gal3 lts4
├ pet11 rad57
├ sup46
├ dur1,2

┼ met8

┼ rna5

┼ his7

┼ aro1

┼ hom2
├ sup35

┼ arg82

┼ pet14

┼ rad50
├ fol1

┼ SUP2

┼ lys4

┼ asp1

┼ trp4

┼ ade8

┼ pur5

┼ MGL2
├ SUC3
├ MAL3

IV
┬ SUC

┼ HO

┼ cdc9
├ ste7
├ cdc36

┼ pho2

┼ cdc2
├ tsm225

┼ mar1
├ rna11

┼ cdc7
├ trp1
├ gal3
├ rad57
├ SUP27

┼ rad55
├ mak21

┼ nul3

┼ aro1

┼ hom2
├ sup35

V
┬ prt 3
├ can1
├ prb1
├ mak10
├ SUS1
├ min1
├ cyc7

┼ ura3
├ mnn1
├ SUP71
├ arg9

┼ chol

┼ BOR1
├ hom3
├ his1
├ SUP19,20
├ arg5,6

┼ ilv1

┼ trp2
├ met5
├ rad51

┼ ade6

┼ cly 8

VI
┬ SUF9

┼ cdc4
├ let6

┼ SUP11

┼ cly 3

┼ his2
├ cdc14
├ SUP6
├ met10
├ PHO82,pho4
├ cdc26

┼ hxk1

VII
┬ hxk 2

┼ ade5

┼ kex1

┼ ura3

┼ ski1

┼ lys5
├ lts3
├ aro2

┼ met13

┼ tsm437
├ cyh2

┼ mak24

┼ ole1

┼ rad6

┼ SUP54
├ trp5 ltsl
├ BOR2,cyh3,
├ fil1,ant1
├ AMY1,oli1
├ leu1
├ AXE1

┼ rme1

┼ MAL1
├ SUC1

┼ fol2

┼ mes1

┼ rad2

┼ ser2
├ ADE15
├ ade3

┼ fro1

┼ fro2

VIII
┬ mak20

┼ mak 7

┼ spoII

┼ pet1
├ arg4
├ spo13

┼ thr1

┼ CUP1

┼ pet 3
├ SUF8
├ mak18

X
┬ mak17

┼ cdc6

┼ ura2

┼ arg3

┼ SUP7

┼ let5
├ let3
├ SUP51,52
├ SUP29,30
├ met3
├ ilv 3

┼ cyc1
├ osm1
├ rad7
├ SUP4
├ cdc8

┼ cdc11

IX
┬ SUC2

┼ SUP22

┼ his5

┼ lys11

┼ SUP17

┼ his6
├ bar1,sst1
├ cdc29

┼ cdc12

┼ dal 1
├ dal4
├ dal2
├ lys1

┼ mak19

┼ makl3

F6
┼ thi1

┼ SUP50
├ gal 5

┼ pdx 2

F8
┼ gal80
├ SUP5

┼ arg 81

XI
┬ mak9
├ fas1

┼ mnn4

┼ trp 3
├ ura 1

┼ dbl1

┼ cly 7

┼ cdc16
├ makll

┼ met 14

┼ met 1

┼ SUP25

┼ mak15

┼ MAL 4

XIII
┬ SUF 7

┼ cdc 5
├ rad 52

┼ lys 7
├ arg80

┼ let1M

┼ eth 2

┼ met 6

┼ SUP8

┼ mak27
├ rna1

XII
┬ mak22

┼ mak12

┼ asp5
├ rad5,rev2
├ ROC1
├ suh2

┼ pep16
├ gal 2

┼ RDN1

┼ ura4
├ SUP26

┼ car 2

XIV
┬ mak26

┼ ski 4
├ kex2
├ pet x

┼ ski3

┼ SUF10
├ spo1
├ pet8
├ rna2

┼ lys9

XV
┬ arg 8

┼ spe 2
├ arg 1

┼ SUP 3
├ spd1
├ tup 4
├ mak1
├ tup7
├ pho80
├ tcm1 cyc2
├ pet17
├ mak 8
├ suf11
├ cdc21,tmp1
├ cyh4

┼ 8740

┼ ade 2

┼ SUF 5

┼ ser1
├ ade 9

┼ his3(his8)

XVI
┬ gal 4

┼ pho85

┼ rad 1
├ sot1
├ swi1
├ chl1

┼ mak6

┼ mak3
├ aro7
├ osm2
├ SUP15,16

┼ ts4572

┼ rad56

XVII
┬ met4

┼ lys 10

┼ lap 3

┼ pet 2
├ prt 2

┼ met 2

┼ kar 1

┼ pha 2

F11
┼ mut 1

┼ tra 3

┼ cpa1

┼ prt1

┼ mut 2

met 7

├ 10cM

┤ 10cM

– – – Mitotic linkage
. . . . Trisomic linkage
( ) Sequence not established
with genes outside parenthesis
Left arm is on top

List of Mapped Genes

GENE	MAP POSITION	REFERENCE
ade1	1R	69, 85
ade2	15R	69
ade3	7R	44, 69, 102
ade5	7L	69, 74, 110
ade6	7R	35, 80, 84a, 90
ade8	4R	2, 45, 70
ade9	15R	15, 69
ADE15	7R	44
AMY1	7L	62
ant1	7L	9
arg1	15L	10, 37
arg3	10L	Hilger, F., pers. comm.
arg4	8R	35, 69, 108, 109
arg5,6	5R	21, 57, 68, 69, 75
arg8	15L	37
arg9	5R	69
arg80	13R	Hilger, F., pers. comm.
arg81	F8	Hilger, F., pers. comm.
arg82	4R	Hilger, F., pers. comm.
aro1	4R	69, 70, 111
aro2	7L	69, 74, 90, 110
aro7	16R	36, 55, 77, 109
asp1	4R	45, 46
asp5	12R	18, 41, 69, 78
AXE1	7L	Sora, S., pers. comm.
bar1	9L	Sprague, G., and I. Herskowitz, pers. comm.
BOR1	5R	75
BOR2	7L	75
can1	5L	70, 89, 109
car2	12R	Hilger, F., pers. comm.
cdc2	4L	33, 49, 70, 111
cdc4	6L	16, 70
cdc5	13L	16, 70
cdc6	10L	48, Hilger, F., pers. comm.
cdc7	4L	70, 77
cdc8	10R	51, 70
cdc9	4L	33, 70
cdc10	3L	14, 70
cdc11	10R	51, 70
cdc12	8R	16, 109
cdc14	6R	16, 70
cdc15	1R	70
cdc16	11L	33, 111

GENE	MAP POSITION	REFERENCE
cdc19	1L	48
cdc21	15R	23, 109
cdc26	6R	48
cdc28	2R	12, 33, 109
cdc29	9L	33
cdc36	4L	Shuster, J., pers. comm.
cdc39	3R	Shuster, J., pers. comm.
chl1	16L	58
cho1	5R	3, 57
cly3	6R	70
cly7	11L	70
cly8	7R	70, 80
cpa1	15R	Hilger, F., pers. comm.
cry1	3R	27, 67, 93
CUP1	8R	35, 69
cyc1	10R	51, 70
cyc2	15R	85
cyc3	1L	85
cyc7	5L	88, 89
cyc8	2R	85
cyc9	3R	85
cyh1	2L	69
cyh2	7L	69, 74, 80, 90
cyh3	7L	69
cyh4	15R	69
cyh10	2R	91
cys1	1L	30, 85, 11
dal1	9R	11, 52
dal2	9R	11, 52
dal4	9R	11, 52
dbl1	11L	4, 22
dur1,2	2R	12
eth2	13R	65, 66
fas1	11L	8, 13
fdp1	2R	105
flk1	3R	Stark, H., pers. comm.
FLO1	1R	38, 85, 98
FLO4		see FLO1
fol1	4R	Game, J., and J. Little, pers. comm.
fol2	7R	Game, J., J. Little, and B. Rockmill, pers. comm.
fro1	7R	101, 102
fro2	7R	101, 102
gal1	2R	6, 18, 57, 69

GENE	MAP POSITION	REFERENCE		GENE	MAP POSITION	REFERENCE
gal2	12R	18, 69, 78				pers. comm.
gal3	4R	18, 42		*lys5*	7L	69, 74, 90, 110
gal4	16L	Haber, J., pers. comm.		*lys7*	13R	16, 32, 69, 70
				lys9	14R	69, 70, 110
gal5	F6	19		*lys10*	17L	70
gal7	2R	6, 18, 57, 69		*lys11*	9L	57, 69, 76
gal10	2R	6, 18, 57, 69		*mak1*	15C	109
gal80	F8	19, 20		*mak3*	16R	77, 109
glc1	2R	Pringle, J., pers. comm.		*mak4*	2R	109
				mak5	2R	109
glk1	3L	Maitra, P., and Z. Lobo, pers. comm.		*mak6*	16R	77, 109
				mak7	8L	107, 108, 109
his1	5R	21, 57, 69, 75, 87		*mak8*	15R	109
his2	6R	17, 36, 69, 70		*mak9*	11L	111
his3	15R	15, 69		*mak10*	5L	89, 109
his4	3L	14, 35, 102, 106		*mak11*	11L	111
his5	9L	57, 69, 76		*mak12*	12L	111
his6	9L	35, 70, 76		*mak13*	9R	108
his7	2R	69, 80		*mak14*	3R	111
his8		see *his3*		*mak15*	11R	111
HML	3L	32		*mak16*	1L	111
HMR	3R	32		*mak17*	10L	108
HO	4L	48, Kawasaki, G., pers. comm.		*mak18*	8R	108
				mak19	8R	108
hom2	4R	69, 70		*mak20*	8L	108
hom3	5R	21, 69, 75		*mak21*	4R	111
hxk1	6R	59, 60		*mak22*	12L	108
hxk2	7L	Maitra, P., pers. comm.		*mak24*	7L	108
				mak26	14L	108
ils1	2L	73		*mak27*	13R	111
ilv1	5R	39, 57, 69		*MAL1*	7R	44, 69, 102
ilv3	10R	36, 51, 69, 70		*MAL2*	3R	7, 35, 60
kar1	17R	Dutcher, S., pers. comm.		*MAL3*	2R	48, 64
				MAL4	11R	36, 69, 70, 111
kex1	7L	110		*mar1*	4L	49
kex2	14L	104, 110		*MAT*	3R	27, 35, 67, 69, 93
lap3	17R	Trumbly, R., pers. comm.		*mes1*	7R	Schild, D., pers. comm.
let1	1R	70		*met1*	11R	35, 69, 70
let1M	13R	72		*met2*	17R	70
let3	10L	72		*met3*	10R	51, 69, 77
let5	10L	72		*met4*	17L	70
let6	6L	72		*met5*	5R	57
leu1	7L	35, 60, 69, 90		*met6*	13R	65, 66
leu2	3L	14, 35, 57, 69		*met7*	15R	61
lts1	7L	91		*met8*	2R	12, 36, 69
lts3	7L	91		*met10*	6R	17, 48, 57, 60, 69
lts4	4R	91		*met13*	7L	69, 74, 90, 92, 110
lts10	4R	91		*met14*	11R	33, 40, 69, 111
lys1	9R	11, 35, 52, 57, 76		*MGL2*	2R	48, 64
lys2	2R	12, 35, 69, 80, 109		*min1*	5L	89
lys4	4R	Contopoulou, R.,		*mnn1*	5C	1

GENE	MAP POSITION	REFERENCE
mnn2	2R	4
mnn4	11L	4
mut1	F11	26
mut2	F11	26
nul3	4R	70
ole1	7L	84, 90
oli1	7L	86
osm1	10R	92
osm2	16R	92
pdx2	F6	35
pep16	12R	Jones, B., pers. comm.
pet1	8R	35, 69
pet2	17R	69, 70
pet3	8R	16, 70, 108, 109
pet8	14R	16, 69, 70, 110
pet9	2L	18, 69, 73
pet11	2R	12, 36, 69
pet14	4R	70, 111
pet17	15R	36, 70, 85, 109
pet18	3R	70, 109
petx	14L	110
pgi1	2R	64
pgk1	3R	50
pha2	17R	69, 70
pho2	4L	Toh-e, A., pers. comm.
pho3,5	2R	31, 94, 100, 103
pho4	6R	Toh-e, A., pers. comm.
pho80	15R	Bisson, L., pers. comm.
PHO82	6R	Toh-e, A., pers. comm.
pho85	16L	Toh-e, A., pers. comm.
prb1	5L	Jones, B., A. Mitchell, and G. Zubenko, pers. comm.
prt1	15R	70
prt2	17R	70
prt3	5L	70
pur5	4R	2
pyk1	1L	63, 85, 97
r⁷ₛ	5R	Eckardt, F., J. Games, and J. Little, pers. comm.
rad1	16L	58, 82, 83, 88
rad2	7R	Schild, D., pers. comm.

GENE	MAP POSITION	REFERENCE
rad3	5R	68, 95
rad4	5R	68, 95
rad5	12R	53
rad6	7L	24
rad7	10R	51
rad18	3R	70, 85
rad50	4R	Contopoulou, R., pers. comm
rad51	5R	68
rad52	13L	85
rad55	4R	70
rad56	16R	24
rad57	4R	24, 77
RDN1	12R	78
rev2		see rad5
rme1	7R	Sprague, G., and I. Herskowitz, pers. comm.
ROC1	12R	69
rna1	13R	70, 108, 111
rna2	14R	16, 70, 110
rna3	4R	70
rna5	2R	48, 70
rna11	4L	70, 111, Toh-e, A., pers. comm.
SAD1	3R	Kassir, Y., and I. Herskowitz, pers. comm.
ser1	15R	43, 61, 69
ser2	7R	44
ski1	7L	104
ski3	14L	104
ski4	14L	104
sot1	16L	82
spd1	15L	Dawes, I., pers. comm.
spe2	15L	10, 37
spo1	14R	Klapholz, S., and R. Easton-Esposito, pers. comm.
spo11	8L	Klapholz, S., and R. Easton-Esposito, pers. comm.
spo13	8R	Klapholz, S., and R. Easton-Esposito, pers. comm.
sst1	9L	Chan, R., pers. comm.
ste7	4L	Shuster, J., pers. comm.
SUC	4L	48
SUC1	7R	44, 69, 84a, 102

GENE	MAP POSITION	REFERENCE
SUC2	9L	57, 69, 76
SUC3	2R	48, 64
SUF2	3R	14
SUF5	15R	15
SUF7	13L	16
SUF8	8R	16
SUF9	6L	16
SUF10	14L	16
suf11	15R	Culbertson, M., pers. comm.
suh2	12R	70
SUP2	4R	70
SUP3	15L	36, 37
SUP4	10R	25, 51, 70
SUP5	F8	20
SUP6	6R	17, 36, 70
SUP7	10L	25, 36
SUP8	13R	70, 108
SUP11	6R	17, 36, 70
SUP15, 16	16R	36, 55, 77
SUP17	9L	76
SUP19, 20	5R	70, 77
SUP22	9L	76
SUP25	11R	36, 70
SUP26	12R	77
SUP27	4R	77
SUP29	10C	77
SUP30	10C	70
sup35	4R	69
sup45	2R	36, 94
sup46	2R	Ono, B., J. W. Stewart, and F. Sherman, pers. comm.
SUP50	F6	69, 70
SUP51	10C	70
SUP52	10L	56
SUP53	3L	81, Reed, C., and S. Liebman, pers. comm.
SUP54	7L	81, Reed, C., and S. Liebman, pers. comm.
SUP61	3R	7, 70
SUP71	5R	70
SUS1	5L	69
swi1	16L	29, Haber, J., and L. Rowe, pers. comm.
tcm1	15R	28
thi1	F6	20, 69, 70
thr1	8R	35, 69

GENE	MAP POSITION	REFERENCE
thr4	3R	50, 51, 54, 69, 99, 106
til1	7L	Stiles, J. I., and F. Sherman, pers. comm.
tmp1	15R	7a, 23, 109
tra3	15R	61
trp1	4R	18, 24, 34, 69, 70, 77, 111
trp2	5R	21, 57, 69, 75
trp3	11L	4, 8, 13, 69, 111
trp4	4R	45, 46, 70
trp5	7L	35, 69, 74, 75, 80, 84, 90, 92
ts4572	16R	Hilger, F., pers. comm.
tsl1 (tse)	1L	47, McCusker, J., and J. Haber, pers. comm.
tsm1	3R	Sprague, G., and I. Herskowitz, pers. comm.
tsm5	3R	Fink, G., pers. comm., McCusker J., and J. Haber, pers. comm.
tsm134	2R	31, 64, 70
tsm225	4L	Hilger, F., pers. comm.
tsm437	7L	70
tup1	3R	54, 106
tup4	15L	109
tup7	15R	Bisson, L., pers. comm.
tyr1	2R	12, 31, 69, 80, 85
umr7	3R	54
ura1	11L	4, 8, 13, 69, 111
ura2	10L	48, 70
ura3	5L	1, 3, 69, 70, 89
ura4	12R	70, 78
8740	15R	Hilger, F., pers. comm.

REFERENCES

1. Antalis, C., S. Fogel, and C.E. Ballou. 1973. Genetic control of yeast mannan structure: Mapping the first gene concerned with mannan biosynthesis. *J. Biol. Chem.* **248:** 4655.

2. Armitt, S. and R.A. Woods. 1970. Purine-excreting mutants of *Saccharomyces cerevisiae*. I. Isolation and genetic analysis. *Genet. Res.* **15:** 7.

3. Atkinson, K.D., B. Jensen, A.I. Kolat, E.M. Storm, S.A. Henry, and S. Fogel. 1980. Yeast mutants auxotrophic for ethanolamine or choline. *J. Bacteriol.* **141:** 558.

4. Ballou, D.L. 1975. Genetic control of yeast mannan structure: Mapping genes *mnn2* and *mnn4* in *Saccharomyces cerevisiae*. *J. Bacteriol.* **123:** 616.

5. Barratt, R.W. , D. Newmeyer, D.D. Perkins, and L. Garnjobst. 1954. Map construction in *Neurospora crassa*. *Adv. Genet.* **6:** 1.

6. Bassel, J. and R. Mortimer. 1971. Genetic order of the galactose structural genes in *Saccharomyces cerevisiae*. *J. Bacteriol.* **108:** 179.

7. Brandriss, M.C., J.W. Stewart, F. Sherman, and D. Botstein. 1976. Substitution of serine caused by a recessive lethal suppressor in yeast. *J. Mol. Biol.* **102:** 467.

7a. Brendel, M. and W.W. Fath. 1974. Isolation and characterization of mutants of *Saccharomyces cerevisiae* auxotrophic and conditionally auxotrophic for 5'-dTMP. *Z. Naturforsch.* **29:** 733.

8. Burkl, G., H. Castorph, and E. Schweizer. 1972. Mapping of a complex gene locus coding for part of the *Saccharomyces cerevisiae* fatty acid synthetase multienzyme complex. *Mol. Gen. Genet.* **119:** 315.

9. Cohen, J.D. and N.R. Eaton. 1979. Genetic analysis of multiple drug cross resistance in *Saccharomyces cerevisiae*: A nuclear-mitochondrial gene interaction. *Genetics* **91:** 19.

10. Cohn, M.S., C.W. Tabor, and H. Tabor. 1978. Isolation and characterization of *Saccharomyces cerevisiae* mutants deficient in *S*-adenosylmethionine decarboxylase, spermidine, and spermine. *J. Bacteriol.* **135:** 208.

11. Cooper, T.G., M. Gorski, and V. Turoscy. 1979. A cluster of three genes responsible for allantoin degradation in *Saccharomyces cerevisiae*. *Genetics* **92:** 383.

12. Cooper, T.G., C. Lam, and V. Turoscy. 1980. Structural analysis of the *dur* loci in *Saccharomyces cerevisiae*: Two domains of a single multifunctional gene. *Genetics* **94:** 555.

13. Culbertson, M.R. and S. Henry. 1973. Genetic analysis of hybrid strains trisomic for the chromosome containing a fatty acid synthetase gene complex (*fas1*) in yeast. *Genetics* **75:** 441.

14. Culbertson, M.R., L. Charnas, M.T. Johnson, and G.R. Fink. 1977. Frameshifts and frameshift suppressors in *Saccharomyces cerevisiae*. *Genetics* **86:** 745.

15. Culbertson, M.R., K.N. Underbrink, and G.R. Fink. 1981. Frameshift suppressors in *Saccharomyces cerevisiae*. II. Genetic properties of group II suppressors. *Genetics* **95:** 833.

16. Cummins, C.M., R.F. Gaber, M.R. Culbertson, R. Mann, and G.R. Fink. 1981. Frameshift suppression in *Saccharomyces cerevisiae*. II. Isolation and genetic properties of group III suppressors. *Genetics* **95:** 855.

17. di Caprio, L. and P.J. Hastings. 1976. Gene conversion and intragenic recombination at the *SUP6* locus and the surrounding region in *Saccharomyces cerevisiae*. *Genetics* **84:** 697.

18. Douglas, H.C. and D.C. Hawthorne. 1964. Enzymatic expression and genetic linkage of genes controlling galactose utilization in *Saccharomyces*. *Genetics* **49:** 837.

19. ———. 1966. Regulation of genes controlling synthesis of the galactose pathway enzymes in yeast. *Genetics* **54:** 911.

20. ———. 1972. Uninducible mutants in the *gali* locus of *Saccharomyces cerevisiae*. *J. Bacteriol.* **109:** 1139.

21. Fogel, S. and D.D. Hurst. 1967. Meiotic gene conversion in yeast tetrads and the theory of recombination. *Genetics* **57:** 455.

22. Friis, J. and P. Ottolenghi. 1970. The genetically determined binding of alcian blue by a minor fraction of yeast cell walls. *C.R. Trav. Lab. Carlsberg* **37**: 327.

23. Game, J.C. 1976. Yeast cell-cycle mutant *cdc21* is a temperature-sensitive thymidylate auxotroph. *Mol. Gen. Genet.* **146**: 313.

24. Game, J.C. and R.K. Mortimer. 1974. A genetic study of X-ray sensitive mutants in yeast. *Mutat. Res.* **24**: 281.

25. Gilmore, R. 1966. "Super-suppressors in *Saccharomyces cerevisiae*." Ph.D. thesis, University of California, Berkeley.

26. Gottlieb, D.J.C. and R.C. von Borstel. 1976. Mutators in *Saccharomyces cerevisiae, mut1-1, mut1-2*, and *mut2-1*. *Genetics* **83**: 655.

27. Grant, P.G., L. Sanchez, and A. Jimenez. 1974. Cryptopleurine resistance: Genetic locus for a 40S ribosomal component in *Saccharomyces cerevisiae*. *J. Bacteriol.* **120**: 1308.

28. Grant, P.G., D. Schindler, and J.E. Davies. 1976. Mapping of trichodermin resistance in *Saccharomyces cerevisiae*: A genetic locus for a component of the 60S ribosomal subunit. *Genetics* **83**: 667.

29. Haber, J.E. and B. Garvik. 1977. A new gene affecting the efficiency of mating type interconversions in homothallic strains of *Saccharomyces cerevisiae*. *Genetics* **87**: 33.

30. Halos, S. 1976. "Cysteine mutants in *Saccharomyces cerevisiae*." Ph.D. thesis, University of California, Berkeley.

31. Hansche, P.E., V. Beres, and P. Lange. 1978. Gene duplication in *Saccharomyces cerevisiae*. *Genetics* **88**: 673.

32. Harashima, S. and Y. Oshima. 1976. Mapping of the homothallic genes, HMα and HMa in *Saccharomyces* yeast. *Genetics* **84**: 437.

33. Hartwell, L.H., R.K. Mortimer, J. Culotti, and M. Culotti. 1973. Genetic control of the cell division cycle in yeast. V. Genetic analysis of *cdc* mutants. *Genetics* **74**: 267.

34. Hawthorne, D.C. 1955. The use of linear asci for chromosome mapping in *Saccharomyces*. *Genetics* **40**: 511.

35. Hawthorne, D.C. and R.K. Mortimer. 1960. Chromosome mapping in *Saccharomyces*: Centromere-linked genes. *Genetics* **45**: 1085.

36. ———. 1968. Genetic mapping of nonsense suppressors in yeast. *Genetics* **60**: 735.

37. Hilger, F and R.K. Mortimer. 1980. Genetic mapping of *arg1* and *arg8* in *Saccharomyces cerevisiae* by trisomic analysis combined with interallelic complementation. *J. Bacteriol.* **141**: 270.

38. Holmberg, S. and M.C. Kielland-Brandt. 1978. A ts mutant in yeast flocculation. *Carlsberg Res. Commun.* **43**: 37.

39. Hurst, D.D. and S. Fogel. 1964. Mitotic recombination and heteroallelic repair in *Saccharomyces cerevisiae*. *Genetics* **50**: 435.

40. Hwang, Y.L., G. Lindegren, and C.C. Lindegren. 1963. Mapping the eleventh centromere of *Saccharomyces*. *Can. J. Genet. Cytol.* **5**: 290.

41. Hwang, Y.L., G. Lindegren, and C.C. Lindegren. 1964. The twelfth chromosome of *Saccharomyces*. *Can. J. Genet. Cytol.* **6**: 373.

42. James, A.P. and B. Lee-Whiting. 1955. Radiation-induced genetic segregations in vegetative cells of diploid yeast. *Genetics* **40**: 826.

43. Johnston, J.R. 1962. "Studies in meiotic and mitotic recombination in *Saccharomyces cerevisiae*." Ph.D. thesis, University of California, Berkeley.

44. Jones, E.W. and K. Lam. 1973. Mutations affecting levels of tetrahydrofolate interconversion enzymes in *Saccharomyces cerevisiae*. II. Map positions on chromosome VII of *ade3-41* and *ADE15*. *Mol. Gen. Genet.* **123**: 209.

45. Jones, G.E. 1970. "L-Asparaginase degradation in *Saccharomyces cerevisiae*: Genetic control and biochemical properties of yeast L-asparaginase." Ph.D. thesis. University of California, Berkeley.

46. Jones, G.E. and R.K. Mortimer. 1970. L-asparaginase-deficient mutants of yeast. *Science* **167**: 181.

47. Kaback, D.B. and H.O. Halvorson. 1978. Ribosomal DNA magnification in *Saccharomyces cerevisiae*. *J. Bacteriol.* **134**: 237.

48. Kawasaki, G. 1979. "Karyotypic instability and carbon source effects in cell cycle mutants of *Saccharomyces cerevisiae*." Ph.D. thesis. University of Washington, Seattle.

49. Klar, A.J.S., S. Fogel, and K. MacLeod. 1979. *MARI*—A regulator of the *HMa* and *HMα* loci in *Saccharomyces cerevisiae*. *Genetics* **93**: 37.

50. Lam, K.-B. and J. Marmur. 1977. Isolation and characterization of *Saccharomyces cerevisiae* glycolytic pathway mutants. *J. Bacteriol.* **130**: 746.

51. Lawrence, C.W., F. Sherman, M. Jackson, and R.A. Gilmore. 1975. Mapping and gene conversion studies with the structural gene for iso-1-cytochrome *c* in yeast. *Genetics* **18**: 615.

52. Lawther, R.P., E. Riemer, B. Chojnacki, and T.G. Cooper. 1974. Clustering of the genes for allantoin degradation in *Saccharomyces cerevisiae*. *J. Bacteriol.* **119**: 461.

53. Lemontt, J.F. 1971. Mutants of yeast defective in ultraviolet light-induced mutation. *Genetics* **68**: 21.

54. Lemontt, J.F., D.R. Fugit, and V. Mackay. 1981. Pleiotropic mutations at the *tup1* locus that affect the expression of mating-type-dependent functions in *Saccharomyces cerevisiae*. *Genetics* **94**: 899.

55. Liebman, S.W., J.W. Stewart, and F. Sherman. 1975. Serine substitutions caused by an ochre suppressor in yeast. *J. Mol. Biol.* **94**: 595.

56. Liebman, S.W., J.W. Stewart, J.H. Parker, and F. Sherman. 1977. Leucine insertion caused by a yeast amber suppressor. *J. Mol. Biol.* **109**: 13.

57. Lindegren, C.C., G. Lindegren, E. Shult, Y.L. Hwang. 1962. Centromeres, sites of affinity and gene loci on the chromosomes of *Saccharomyces*. *Nature* **194**: 260.

58. Liras, P., J. McCusker, S. Mascioli, and J.E. Haber. 1978. Characterization of a mutation in yeast causing nonrandom chromosome loss during mitosis. *Genetics* **88**: 651.

59. Lobo, Z. 1976. "Sugar phosphorylating enzymes in yeast: Genetic and biochemical studies." Ph.D. thesis, University of Bombay, India.

60. Lobo, Z. and P.K. Maitra. 1977. Genetics of yeast hexokinase. *Genetics* **86**: 727.

61. Lowenstein, R. 1973. "Genetic studies in *Saccharomyces cerevisiae*." Ph.D. thesis, Cornell University, Ithaca, New York.

62. Lucchini, G., M.L. Carbone, M. Cocucci, and M.L. Sensi. 1979. Nuclear inheritance of resistance to antimycin A in *Saccharomyces cerevisiae*. *Mol. Gen. Genet.* **177**: 139.

63. Maitra, P.K. and Z. Lobo. 1977a. Pyruvate kinase mutants of *Saccharomyces cerevisiae*: Biochemical and genetic characterization. *Mol. Gen. Genet.* **152**: 193.

64. ———. 1977b. Genetic studies with a phosphoglucose isomerase mutant of *Saccharomyces cerevisiae*. *Mol. Gen. Genet.* **156**: 55.

65. Masselot, M. and H. de Robichon-Szulmajster. 1974. Methionine biosynthesis in *Saccharomyces cerevisiae*: Mutations at the regulatory locus *ETH2*. I. Genetic data. *Mol. Gen. Genet.* **129**: 339.

66. ———. 1975. Methionine biosynthesis in *Saccharomyces cerevisiae*. I. Genetical analysis of auxotrophic mutants. *Mol. Gen. Genet.* **139**: 121.

67. Meade, J.H., M.I. Riley, and T.R. Manney. 1977. Expression of cryptopleurine resistance in *Saccharomyces cerevisiae*. *J. Bacteriol.* **129**: 1428.

68. Morrison, D.P. 1978. "Repair parameters in mutator mutants of *Saccharomyces cerevisiae*." Ph.D. thesis, University of Alberta, Edmonton.

69. Mortimer, R.K. and D.C. Hawthorne. 1966. Genetic mapping in *Saccharomyces*. *Genetics* **53**: 165.

70. ———. 1973. Genetic mapping in *Saccharomyces*. IV. Mapping of temperature-sensitive genes and use of disomic strains in localizing genes. *Genetics* **74**: 33.

71. ———. 1975. Genetic mapping in yeast. *Methods Cell Biol.* **11**: 221.

72. Munz, P. 1972. Clone size distribution among haploid yeast segregants carrying a UV-induced recessive lethal mutation. *Arch. Genet.* **45**: 173.

73. McLaughlin, C.S. and L.H. Hartwell. 1969. A mutant of yeast with a defective methionyl-tRNA synthetase. *Genetics* **61**: 557.
74. Nakai, S. and R.K. Mortimer. 1969. Studies of the genetic mechanism of radiation-induced mitotic segregation in yeast. *Mol. Gen. Genet.* **103**: 329.
75. Nass, G. and K. Poralla. 1976. Genetics of borreliden resistant mutants of *Saccharomyces cerevisiae* and properties of their threonyl-tRNA-synthetase. *Mol. Gen. Genet.* **147**: 39.
76. Ono, B., J.W. Stewart, and F. Sherman. 1979a. Yeast UAA suppressors effective in ψ^+ strains: Serine-inserting suppressors. *J. Mol. Biol.* **128**: 81.
77. ———. 1979b. Yeast UAA suppressors effective in ψ^+ strains: Leucine-inserting suppressors. *J. Mol. Biol.* **132**: 507.
78. Petes, T.D. 1979. Meiotic mapping of yeast ribosomal DNA on chromosome XII. *J. Bacteriol.* **138**: 185.
79. Plischke, M.E., R.C. von Borstel, R.K. Mortimer, and W.E. Cohn. 1975. Genetic markers and associated gene products in *Saccharomyces cerevisiae*. In *Handbook of biochemistry and molecular biology*, vol. II, p. 767. CRC Press, Cleveland.
80. Plotkin, D.J. 1978. "Commitment to meiotic recombination: A temporal analysis." Ph.D. thesis, University of Chicago, Illinois.
81. Reed, C.R. and S.W. Liebman. 1979. New amber suppressors in *Saccharomyces cerevisiae*. *Genetics* **91**: 102.
82. Remer, S., A. Sherman, E. Kraig, and J.E. Haber. 1980. A suppressor of deoxythymidine monophosphate uptake mutants in *Saccharomyces cerevisiae*. *J. Bacteriol.* **138**: 638.
83. Resnick, M.A. 1969. Genetic control of radiation sensitivity in *Saccharomyces cerevisiae*. *Genetics* **62**: 519.
84. Resnick, M.A. and R.K. Mortimer. 1966. Unsaturated fatty acid mutants of *Saccharomyces cerevisiae*. *J. Bacteriol.* **92**: 597.
84a. Roman, H. 1956. A system selective for mutations affecting the synthesis of adenine in yeast. *C.R. Trav. Lab. Carlsberg.* **26**: 299.
85. Rothstein, R.J. and F. Sherman. 1981. Genes affecting the expression of cytochrome-*c* in yeast: Genetic mapping and genetic interactions. *Genetics* **94**: 871.
86. Saunders, G.W., G.H. Rank, B. Kustermann-Kuhn, and C.P. Hollenberg. 1979. Inheritance of multiple drug resistance in *Saccharomyces cerevisiae*: Linkage to *leu1* and analyses of 2 μm DNA in partial revertants. *Mol. Gen. Genet.* **175**: 45.
87. Savage, E. 1979. "A comparative analysis of recombination at the *his1* locus among five related diploid strains of *Saccharomyces cerevisiae*." Ph.D. thesis, University of Alberta, Edmonton.
88. Sherman, F. and C. Helms. 1978. A chromosomal translocation causing overproduction of iso-2-cytochrome *c* in yeast. *Genetics* **88**: 689.
89. Sherman, F., J.W. Stewart, C. Helms, and J.A. Downie. 1978. Chromosome mapping of the *CYC7* gene determining yeast iso-2-cytochrome *c*: Structural and regulatory regions. *Proc. Natl. Acad. Sci.* **75**: 1437.
90. Shult, E.E., G. Lindegren, and C.C. Lindegren. 1967. Hybrid specific linkage relations in *Saccharomyces*. *Can. J. Genet. Cytol.* **9**: 723.
91. Singh, A. and T.R. Manney. 1974. Genetic analysis of mutations affecting growth of *Saccharomyces cerevisiae* at low temperature. *Genetics* **77**: 651.
92. Singh, A. and F. Sherman. 1978. Deletions of the iso-1-cytochrome *c* and adjacent genes of yeast: Discovery of the *OSM1* gene controlling osmotic sensitivity. *Genetics* **89**: 653.
93. Skogerson, L., C. McLaughlin, and E. Wakatama. 1973. Modification of ribosomes in cryptopleurine-resistant mutants of yeast. *J. Bacteriol.* **116**: 818.
94. Smirnov, M., N. Krasnopevtseva, S. Inge-Vechtomov, and A. Yanoolaitis. 1974. Isolation of mutants for acid phosphatase in the yeast *Saccharomyces cerevisiae*. *Issl. Genet.* **5**: 59.
95. Snow, R. 1967. Mutants of yeast sensitive to ultraviolet light. *J. Bacteriol.* **94**: 571.

96. ——. 1979. Maximum likelihood estimation of linkage and interference from tetrad data. *Genetics* **92:** 231.

97. Sprague, G.F. 1977. Isolation and characterization of a *Saccharomyces cerevisiae* mutant deficient in pyruvate kinase activity. *J. Bacteriol.* **130:** 232.

98. Stewart, G.G. and I. Russell. 1977. The identification, characterization, and mapping of a gene for flocculation in *Saccharomyces* sp. *Can. J. Microbiol.* **23:** 441.

99. Takano, I. and Y. Oshima. 1970. Mutational nature of an allele specific conversion of the mating type by the homothallic gene HO in *Saccharomyces*. *Genetics* **65:** 421.

100. Ter-Avanesyan, M.D., S. Inge-Vechtomov, and M. Petrashan. 1974. Genetical and biochemical studies of acid phosphatases in yeast *Saccharomyces cerevisiae*. *Genetika* **10:** 101.

101. Thornton, R.J. 1978a. Investigations on the genetics of foaming in wine yeasts. *Eur. J. Appl. Microbiol. Biotechnol.* **5:** 103.

102. ——. 1978b. The mapping of two dominant genes for foaming in wine yeasts. *FEMS Microbiol. Lett.* **4:** 207.

103. Toh-e, A., Y. Ueda, S. Kakimoto, and Y. Oshima. 1973. Isolation and characterization of acid phosphatase mutants in *Saccharomyces cerevisiae*. *J. Bacteriol.* **113:** 727.

104. Toh-e, A., P. Guerry, and R. Wickner. 1978. Chromosomal super-killer mutants of *Saccharomyces cerevisiae*. *J. Bacteriol.* **136:** 1002.

105. van de Poll, K. and D. Schamhart. 1977. Characterization of a regulatory mutant of fructose 1,6-bisphosphatase in *Saccharomyces carlsbergensis*. *Mol. Gen. Genet.* **154:** 61.

106. Wickner, R.B. 1974. Mutants of *Saccharomyces cerevisiae* that incorporate deoxythymidine-5-monophosphate into deoxyribonucleic acid *in vivo*. *J. Bacteriol.* **117:** 252.

107. ——. 1978. Twenty-six chromosomal genes needed to maintain the killer double-stranded RNA plasmid in *Saccharomyces cerevisiae*. *Genetics* **88:** 419.

108. ——. 1979. Mapping chromosomal genes of *Saccharomyces cerevisiae* using an improved genetic mapping method. *Genetics* **92:** 803.

109. Wickner, R.B. and M.L. Leibowitz. 1976a. Chromosomal genes essential for replication of a double-stranded RNA plasmid of *Saccharomyces cerevisiae*: The killer character of yeast. *J. Mol. Biol.* **105:** 427.

110. ——. 1976b. Two chromosomal genes required for killing expression in killer strains of *Saccharomyces cerevisiae*. *Genetics* **82:** 429.

111. ——. 1979. *mak* Mutants of yeast: Mapping and characterization. *J. Bacteriol.* **140:** 154.

APPENDIX III

Genes of *Saccharomyces cerevisiae*

Compiled by James R. Broach
Department of Microbiology,
State University of New York
Stony Brook, New York 11794

This Appendix provides information on identified and characterized yeast genes. The core of this information is presented in part A, which is an alphabetical listing of all gene designations, providing for each designation any former or alternate designations, the phenotype of a strain containing a mutation in that gene, the gene product (if known), the map position (if known), and an appropriate literature reference. In addition, the information in part A has been extracted and indexed in alternative formats in parts B and C.

Part B is an alphabetical listing of all gene products given in part A, providing each with the corresponding gene designation. Thus, one can readily find, for example, that the gene encoding phosphoglycerate mutase is designated *GPM1*. To determine the phenotype of strains lacking phosphoglycerate mutase, one can then refer to part A under *GPM1*.

In Part C, the genes listed in part A are grouped by common function. For convenience, a list of categories by which the genes are grouped is provided on the first page of part C. The genes involved in each category follow. The listings are not exclusive; a particular gene may appear in several categories.

There are four sources of information provided in this Appendix. The authors of the various chapters have been of prime importance. Thus, much of the information has been extracted directly from individual chapters. Our second source was responses from a mass mailing to the yeast community, conducted in the Spring of 1980, in which we requested information on identification and/or characterization of yeast genes. We are grateful to all of those who took the time to respond to our request and we apologize to those we failed to reach in the mailing. Our third source was the compilation by R.K. Mortimer and D. Schild of mapped yeast genes, which appears in Appendix II of this monograph. Finally, to fill in the loose ends and to check the accuracy of our information, we have relied heavily on the gene list compiled by M.E. Plischke, R.C. von Borstel, R.K. Mortimer, and W.E. Cohn, which appeared in the 1976 edition of the *CRC Handbook of Biochemistry and Molecular Biology*. We would like to thank these authors for allowing us to use this information.

Information in part A of this listing is presented in the following manner.

Gene designation. The genes are identified by their wild-type designation. Thus, an uppercase or lowercase designation indicates the dominance characteristic of the wild-type allele (in relation to the more common mutant alleles of the gene).

Former or alternate designations. Any former or alternate designation for the listed gene is given in this column. A complete description of the gene is provided only in the listing for the preferred designation; the listing of a former designation merely refers one to the current designation. The absence of a cross reference in this column is *not* assurance that the gene is not allelic with some other gene described in this list, merely that such allelism has not been established.

Phenotype. The phenotype of strains containing a mutation in the designated gene is listed in this column. Generally, only phenotypes associated with the more common alleles of the gene are listed, although in some cases, additional allele-specific phenotypes are noted. In all cases, though, the phenotype described is that of a strain carrying a mutation in the listed gene, rather than that of a strain carrying the wild-type allele. This is true even if the phenotype of the wild-type allele is the characteristic normally associated with the gene. The phenotype listed for *MATa* is an example of this policy.

Gene Product. The product (if known) of the wild-type allele of the gene is given in this column. This would be more accurately described as "probable gene product" since, in many cases, a gene product is listed even though the evidence for such an assignment is suggestive but not absolutely compelling. Thus, one should examine the literature to evaluate the evidence oneself for this assignment before relying on it.

Reference. Whenever possible, the reference for a gene directs one to a chapter in this monograph in which the gene is discussed. Further information and actual literature references can be obtained from that chapter. In addition, an index listing pages on which individual genes are cited in this Monograph is provided following the standard index at the end of each volume. In the absence of a reference to this Monograph, a literature or personal communication reference is provided. Our goal in selecting references was to provide a starting point for obtaining additional information on a gene rather than an exhaustive compilation of the literature on that gene.

A. Yeast Genes Listed Alphabetically by Gene Designation

Gene Designation	Former/Alternate Designation	Phenotype	Gene Product	Map Position	References
AAS1	—	sensitive to amino acid analogs	—	—	52
AAS2	—	sensitive to amino acid analogs	—	—	52
AAS3	—	sensitive to amino acid analogs	—	—	52
ABI1	—	growth inhibited by excess α-amino-butyric acid	—	—	73
ABI2	—	growth inhibited by excess α-amino-butyric acid		—	73
ACC1	—	saturated fatty acid requirement	acetyl-CoA carboxylase	—	43
ACC2	—	saturated fatty acid requirement	apoacetyl-CoA carboxylase ligase	—	43
ACO1	*GLU1*	glutamate auxotrophy on glucose; unable to grow on nonfermentable carbon sources even in the presence of glutamate	aconitate hydratase	—	27
ACT1	—	lethal	actin	6L	104
ADC1	—	allyl-alcohol-resistant; antimycin-A-sensitive on glucose	alcohol dehydrogenase I	15L	27
ADE1	—	adenine auxotrophy; red	phosphoribosylaminoimidazole suc-cinocarboxyamide synthetase	1R	52
ADE2	—	adenine auxotrophy; red	phosphoribosylaminoimidazole carboxylase	15R	52
ADE3	—	adenine auxotrophy	methylenetetrahydrofolate dehy-drogenase; methylenetetrahydro-folate cyclohydrolase; formyltetra-hydrofolate synthetase	7R	52

Gene Designation	Former/Alternate Designation	Phenotype	Gene Product	Map Position	References
ADE4	—	adenine auxotrophy	amidophosphoribosyl transferase	—	52
ADE5	—	adenine auxotrophy	phosphoribosylglycinamide synthetase	7L	52
ADE6	—	adenine auxotrophy	phosphoribosyl-formylglycinamide synthetase	7R	52
ADE7	—	adenine auxotrophy	phosphoribosylaminoimidazole synthetase	—	52
ADE8	—	adenine auxotrophy	phosphoribosylglycinamide formyl transferase	4R	52
ADE9	—	adenine auxotrophy	—	15R	52
ADE10	—	adenine auxotrophy	—	—	52
ADE11	—	adenine auxotrophy	—	—	52
ADE12	—	adenine auxotrophy	adenylosuccinate synthetase	—	52
ADE13	—	adenine auxotrophy	adenylosuccinate lyase	—	52
ADM1	—	allyl-alcohol-resistant	mitochondrial alcohol dehydrogenase	—	27
ADR1	—	depending on the allele, either uninducible or partially constitutive alcohol dehydrogenase synthesis from *ADR2*	regulator of ADHII synthesis	—	27
ADR2	—	allyl-alcohol-resistant on glycerol	alcohol dehydrogenase II	—	27
adr3	—	constitutive alcohol dehydrogenase synthesis from *ADR2*	operator for *ADR2*	—	27
ADR4	—	partially constitutive alcohol dehydrogenase synthesis from *ADR2*	regulator of ADHII synthesis	—	27
AMT	see *MEP2*				
amy1	—	antimycin-resistant	—	7L	65
ANP1	—	osmotic-sensitive; sensitive to ANP, a hydrolytic product of chloramphenicol	—	5L	70
ANT1	see *FCY2*				
APP1		antibiotic-resistant	—	7L	15

ARG1	ARGGI	arginine auxotrophy	argininosuccinate synthetase	15L	52
ARG2	ARGA	arginine auxotrophy	acetyl CoA : glutamate-*N*-acetyl transferase (acetylglutamate synthetase)	—	52
ARG3	ARGF	arginine auxotrophy	ornithine carbamoyltransferase	10L	52
ARG4	ARGH	arginine auxotrophy	argininosuccinate lyase	8R	52
ARG5	ARGC	arginine auxotrophy	*N*-acetyl-γ-glutamyl-phosphate reductase	5R	52
ARG6	ARGB	arginine auxotrophy	*N*-acetylglutamate kinase	5R	52
ARG7	ARGE	arginine auxotrophy	acetylornithine-glutamate acetyl-transferase	—	52
ARG8	ARGD	arginine auxotrophy	acetylornithine aminotransferase	15L	52
ARG9	—	arginine auxotrophy	—	5R	52
ARG10	ARGII	arginine auxotrophy	—	—	52
ARG80	ARGRI	constitutive synthesis of several arginine biosynthetic enzymes	regulator of arginine biosynthetic enzymes	—	52
ARG81	ARGRII	same as *arg80*	regulator of arginine biosynthetic enzymes	—	52
ARG82	ARGRIII	same as *arg80*	regulator of arginine biosynthetic enzymes	—	52
arg84	argBC-O^c	constitutive synthesis of *N*-acetylgluta-mate kinase and *N*-acetyl-γ-glutamyl-phosphate reductase	operator/promoter for the *ARG5*/*ARG6* cluster	5R	48

ARGA	see ARG2
ARGB	see ARG6
argBC-O^c	see arg84
ARGC	see ARG5
ARGD	see ARG8
ARGE	see ARG7
ARGF	see ARG3

Gene Designation	Former/Alternate Designation	Phenotype	Gene Product	Map Position	References
ARGGI	see *ARG1*				
ARGGII	see *ARG10*				
ARO1A	—	phenylalanine + tyrosine + tryptophan + *p*-aminobenzoic acid auxotrophy	3-enol pyruvoylshikimate-5-phosphate synthase	4R	52
ARO1B	—	same as *aro1A*	shikimate kinase	4R	52
ARO1C	—	same as *aro1A*	3-dehydroquinate synthase	4R	52
ARO1D	—	same as *aro1A*	shikimate dehydrogenase	4R	52
ARO1E	—	same as *aro1A*	3-dehydroquinate dehydratase	4R	52
ARO2	—	phenylalanine + tyrosine + tryptophan + *p*-aminobenzoic acid auxotrophy	chorismate synthase	7L	52
ARO3	—	phenylalanine-sensitive; phenylalanine + tyrosine + tryptophan + *p*-aminobenzoic acid auxotrophy in *aro4* strains	deoxy-D-arabino-D-heptulosonate phosphate synthase (phenylalanine-sensitive)	—	52
ARO4	—	tyrosine-sensitive; phenylalanine + tyrosine + tryptophan + *p*-aminobenzoic acid auxotrophy in *aro3* strains	deoxy-D-arabino-D-heptulosonate phosphate synthase (tyrosine-sensitive)	—	52
ARO5	—	phenylalanine + tyrosine + tryptophan + *p*-aminobenzoic acid auxotrophy	—	—	52
ARO7	—	phenylalanine + tyrosine auxotrophy	chorismate mutase	16R	52
ASN1	*ASNA*	partial auxotrophy for asparagine; total auxotrophy in an *asn2* strain	asparagine synthetase A	—	52
ASN2	*ASNB*	asparagine auxotrophy in an *asn1* strain	asparagine synthetase B	—	52
ASN3	*ASNRS*	asparagine auxotrophy	asparaginyl tRNA synthetase	—	52
ASNA	see *ASN1*				
ASNB	see *ASN2*				
ASNRS	see *ASN3*				
ASP1	—	aspartate auxotrophy in *asp5* strains	asparaginase	4R	52
ASP5	—	asparatate auxotrophy	aspartate aminotransferase	12R	52

ASU1	—	prevents suppression by class-I ocher suppressors	—	—	99
ASU2	—	same as asu1	—	—	99
ASU3	—	same as asu1	—	—	99
ASU4	—	same as asu1	—	—	99
ASU5	—	same as asu1	—	—	99
ASU6	—	same as asu1	—	—	99
ASU7	—	same as asu1	—	—	99
ASU8	—	same as asu1	—	—	99
ASU9	—	prevents suppression by the omnipotent suppressors sup35 and sup45	—	—	60
AUR1	—	aurovertin-resistant growth on nonfermentable carbon source	ATPase, subunit 2	7L	20
AXE1	see SST1	axenomycin-resistant	—	—	108
BAR1	see HOM3		—	—	
BIN1	—	deficient for inhibitor of proteinase B	—	—	128
BIN2	—	deficient for inhibitor of proteinase B	—	—	128
BIN3	—	deficient for inhibitor of proteinase B	—	—	128
bor1	—	borrelidin-sensitive	threonyl-tRNA synthetase	7L	79
bor2	—	borrelidin-sensitive	—	—	79
bor3	—		—	—	
CAF1	—	caffeine-resistant	—	—	6
CAN1	—	resistant to canavanine	arginine permease	5R	18
cap1	—	resistant to chloramphenicol	—	—	63
CAR1	—	unable to use arginine as sole nitrogen source	arginase	—	17
CAR2	—	same as car1	ornithine-oxoacid aminotransferase (ornithine transaminase)	12R	17
CAR3	—	unable to use arginine or proline as sole nitrogen source	1-pyrroline dehydrogenase	—	17

Gene Designation	Former/Alternate Designation	Phenotype	Gene Product	Map Position	References
car80	$cargA^+O^h$	constitutive synthesis of arginase	operator/promoter region for CAR1	—	17
car81	$cargB^+O^h$	constitutive synthesis of ornithine-oxoacid aminotransferase (ornithine trans-aminase)	operator/promoter region for CAR2	12R	17
CAR82	CARGR	constitutive synthesis of arginase and ornithine transaminase	negative regulator of CAR1/CAR2	—	17
$cargA^+O^h$	see car80				
$cargB^+O^h$	see car81				
CARGR	see CAR82				
CAT1		unable to derepress gluconeogenic or glyoxylate shunt enzymes	—	—	27
CAT2		same as cat1	—	—	27
CAT80		glucose-repression-insensitive synthesis of invertase and α-glucosidase	—	—	27
CCR1	—	uninducible synthesis of gluconeogenic, glyoxylate cycle, and maltose utilization enzymes; unable to grow on nonfermentable carbon sources	—	—	27
CCR2	—	same as ccr1	—	—	27
CCR3	—	same as ccr1	—	5R	27
ccr80	—	derepressed gluconeogenic enzyme synthesis	—	—	27
CDC1	—	temperature-sensitive cell-cycle arrest; terminal phenotype: unbudded, uninucleate, G_1 cells with no spindle-pole-body satellite	—	—	87
CDC2	—	temperature-sensitive cell-cycle arrest; terminal phenotype: singly budded, uninucleate, G_1 or G_2 cells with complete, short spindles in nuclei situated at the mother-bud neck; sporulation-deficient	—	4L	24,41,87

CDC3	—	temperature-sensitive cell-cycle arrest; terminal phenotype: multiply budded, multinucleate, growing cells	—	—	87
CDC4	—	temperature-sensitive cell-cycle arrest; terminal phenotype: multiply budded, growing uninucleate, G$_1$ cells whose spindle pole bodies are duplicated but not separated; sporulation-deficient	—	6L	24,87
CDC5	—	temperature-sensitive cell-cycle arrest; terminal phenotype: singly budded, growing cells containing two nuclei, each with a single spindle pole body; sporulation-deficient	—	13L	24,87
CDC6	—	temperature-sensitive cell-cycle arrest; terminal phenotype: same as cdc2	—	10L	24,87
CDC7	—	temperature-sensitive cell-cycle arrest; terminal phenotype: same as cdc2; sporulation-deficient	—	4L	24,87
CDC8	—	temperature-sensitive cell-cycle arrest; terminal phenotype: same as cdc2; UV-sensitive; sporulation-deficient	—	10R	24,41,87
CDC9	—	temperature-sensitive cell-cycle arrest; terminal phenotype: same as cdc2; UV- and MMS-sensitive; sporulation-deficient; recombination-deficient	DNA ligase	4L	24,41,87
CDC10	—	temperature-sensitive cell-cycle arrest; terminal phenotype: same as cdc3	—	3L	87
CDC11	—	temperature-sensitive cell-cycle arrest; terminal phenotype: same as cdc3	—	10R	87

Gene Designation	Former/Alternate Designation	Phenotype	Gene Product	Map Position	References
CDC12	—	temperature-sensitive cell-cycle arrest; terminal phenotype: same as cdc3	—	8R	87
CDC13	—	temperature-sensitive cell-cycle arrest; terminal phenotype: same as cdc2, sporulation-deficient	—	—	24,87
CDC14	—	temperature-sensitive cell-cycle arrest; terminal phenotype: singly budded, growing, uninucleate, G_2 cells with complete, long spindles in nuclei extended through much of the length of the mother and bud; sporulation-deficient	—	6R	24,87
CDC15	—	temperature-sensitive cell-cycle arrest; terminal phenotype: same as cdc14	—	1R	87
CDC16	—	temperature-sensitive cell-cycle arrest; terminal phenotype: same as cdc2; sporulation-deficient	—	11L	24,87
CDC17	—	temperature-sensitive cell-cycle arrest; terminal phenotype: same as cdc2; sporulation-deficient	—	—	24,87
CDC18	—	temperature-sensitive cell-cycle arrest; terminal phenotype: same as cdc5	—	—	87
CDC19	see PYK1				
CDC20	—	temperature-sensitive cell-cycle arrest; terminal phenotype: same as cdc2; sporulation-deficient	—	—	24,87
CDC21	see TMP1				
CDC22	—	temperature-sensitive cell-cycle arrest; terminal phenotype: unbudded, uninucleate, G_1 cells with spindle-pole-body satellites	—	—	87

Gene		Function/Phenotype	Map position	References
CDC23	—	temperature-sensitive cell-cycle arrest; terminal phenotype: same as *cdc2*; sporulation-deficient	—	24,87
CDC24	—	temperature-sensitive cell-cycle arrest; terminal phenotype: unbudded cells that are unable to form chitin rings but continue growth, DNA synthesis, and nuclear division under restrictive conditions	—	87,97
CDC25	—	temperature-sensitive cell-cycle arrest; terminal phenotype: same as *cdc1*	—	87
CDC26	—	temperature-sensitive cell-cycle arrest; terminal phenotype: same as *cdc2*	6R	87
CDC27	—	temperature-sensitive cell-cycle arrest; terminal phenotype: same as *cdc14*	—	87
CDC28	—	temperature-sensitive cell-cycle arrest; terminal phenotype: same as *cdc22*; sporulation-deficient	2R	24,87
CDC29	—	temperature-sensitive cell-cycle arrest; terminal phenotype: same as *cdc1*	9L	87
CDC30	—	temperature-sensitive cell-cycle arrest; terminal phenotype: same as *cdc14*	—	87
CDC31	—	temperature-sensitive cell-cycle arrest; terminal phenotype: singly budded, growing, uninucleate, G_2 cells whose spindle pole bodies are not duplicated; X-ray-sensitive	—	41,87
CDC32	—	temperature-sensitive cell-cycle arrest; terminal phenotype: same as *cdc22*	—	87
CDC33	—	temperature-sensitive cell-cycle arrest; terminal phenotype: same as *cdc1*	—	87

Gene Designation	Former/Alternate Designation	Phenotype	Gene Product	Map Position	References
CDC34	—	temperature-sensitive cell-cycle arrest; terminal phenotype: same as cdc4	—	—	87
CDC35	—	temperature-sensitive cell-cycle arrest; terminal phenotype: same as cdc1	—	—	87
CDC36	—	temperature-sensitive cell-cycle arrest; terminal phenotype: same as cdc22	—	4L	87
CDC37	—	temperature-sensitive cell-cycle arrest; terminal phenotype: same as cdc22	—	—	87
CDC38	—	temperature-sensitive cell-cycle arrest; terminal phenotype: same as cdc22	—	—	87
CDC39	—	temperature-sensitive cell-cycle arrest; terminal phenotype: same as cdc22	—	3R	87
CDC40	—	temperature-sensitive cell-cycle arrest; terminal phenotype: same as cdc2; sporulation-deficient	—	—	24,87
CDC41	—	temperature-sensitive cell-cycle arrest; terminal phenotype: unbudded, uninucleate, G₁ cells	—	—	87
CDC42	—	temperature-sensitive cell-cycle arrest; terminal phenotype: same as cdc24	—	—	87
CDC43	—	temperature-sensitive cell-cycle arrest; terminal phenotype: same as cdc24	—	7L	87
CDC44	—	temperature-sensitive cell-cycle arrest; terminal phenotype: same as cdc2	—	—	87
CDC45	—	temperature-sensitive cell-cycle arrest; terminal phenotype: same as cdc2	—	—	87
CDC46	—	temperature-sensitive cell-cycle arrest; terminal phenotype: same as cdc2	—	—	87

Gene	Synonym	Phenotype	Gene product	Map position	Reference
CDC47	—	temperature-sensitive cell-cycle arrest; terminal phenotype: same as *cdc2*	—	—	87
CDC48	—	temperature-sensitive cell-cycle arrest; terminal phenotype: same as *cdc2*	—	—	87
CDC49	—	temperature-sensitive cell-cycle arrest; terminal phenotype: uninucleate cells with small buds	—	—	87
CDC50	—	temperature-sensitive cell-cycle arrest; terminal phenotype: same as *cdc49*	—	—	87
CGR1	—	glucose-resistant synthesis of catalase T	—	—	94
CGR2	—	glucose-resistant synthesis of catalase T	—	—	94
CGR4	—	glucose-resistant synthesis of catalase T	—	—	94
CHL1	—	spontaneous mitotic chromosome loss; increased mitotic recombination; reduced meiotic recombination	—	16L	77
CHO1	—	choline or ethanolamine requirement; defective in phosphatidylserine synthesis	—	5R	43
CHO2	see *INO4*				
CHO3	—	ethanolamine auxotrophy in a *chol* background	—	—	3
CHR1	—	chromate-resistant	sulfate permease	—	18,52
CIT1	*GLU3*	glutamate auxotrophy on glucose; unable to grow on nonfermentable carbon sources even in the presence of glutamate	citrate synthase	—	27
CLY3	—	cell lysis at 36°C	—	6R	76
CLY7	—	cell lysis at 36°C	—	11L	76
CLY8	—	cell lysis at 36°C	—	7R	76
CLY9	—	cell lysis at 36°C	—	3R	76
CMT	see *SIR3*				

Gene Designation	Former/Alternate Designation	Phenotype	Gene Product	Map Position	References
CON1	—	commitment to meiotic intragenic recombination absent; homozygote is sporulation-deficient	—	—	24
CON2	—	same as *con1*	—	—	24
CON3	—	same as *con1*	—	—	24
COR1	—	increased meiotic gene conversion; increased spontaneous mitotic recombination; increased spontaneous mutation rate	—	—	26,41
COR2	—	same as *cor1*	—	—	26,41
COR3	—	same as *cor1*	—	—	26,41
COR4	—	same as *cor1*	—	—	26,41
CPA1	—	arginine auxotrophy in *ura2C* strains or in excess uracil	carbamoyl-phosphate synthetase	15R	52
CPA2	*DUT1*	same as *cpa1*	carbamoyl-phosphate synthetase	—	52
CPS1	—	deficient in blocked dipeptide utilization in *prc1* background	carboxypeptidase S	—	124
CPU1	see *URA2C*				
CRY1	see *RME1*				
CSP1	—	cryptopleurine-resistant	—	3R	107
cup1	—	copper resistance	—	8R	75
CYC1	—	chlorolactate-resistant	iso-1-cytochrome *c*	10R	103
CYC2	—	chlorolactate-resistant; iso-1- and iso-2-cytochromes *c*-deficient; apocytochrome *c* present	—	15L	98,100
CYC3	—	same as *cyc2*	—	1L	98,100
CYC4	—	slight deficiency in iso-1- and iso-2-cytochromes *c*; deficient in prophyrin biosynthesis	—	—	98,125

Gene	Synonym	Product	Phenotype	Map position	Reference
CYC5	—		slight deficiency of iso-1- and iso-2-cytochromes c	—	98
CYC6	—		slight deficiency of iso-1- and iso-2-cytochromes c	—	98
CYC7	CYP3	iso-2-cytochrome c	deficient in iso-2-cytochrome c	15L	21,102
CYC8	—		overproduction of iso-2-cytochrome c; able to grow on lactate in a cyc1 background	2R	92
CYC9	UMR7, FLK1		same as cyc8	3R	92
CYH1	—		cycloheximide-resistant (1 ppm)	2L	120
CYH2	—	ribosomal protein L29	cycloheximide-resistant (6 ppm)	7L	29,120
CYH3	—		cycloheximide-resistant (1 ppm)	7L	120
cyh4	—		cycloheximide-resistant (0.5 ppm)	15R	120
CYH5	—		cycloheximide-resistant (1 ppm)	—	120
CYH6	—		cycloheximide-resistant (0.5 ppm)	—	120
CYH7	—		cycloheximide-resistant (1 ppm)	—	120
CYH8	—		cycloheximide-resistant (20 ppm)	—	120
CYH10	—		cycloheximide-resistant	2	105
cyp1	—		overproduction of iso-2-cytochrome c; able to grow on lactate in a cyc1 background	—	14
cyp2	—		same as cyp1	—	14
cyp3	see CYC7				
CYP4	—		same as cyp1	—	14
CYP5	—		same as cyp1	—	14
CYS1	—		cysteine auxotrophy	1L	52
CYS2	—		cysteine auxotrophy	—	52
CYT	see PET19				
CYT-P	see FCY2				
DAL1	—	allantoinase	unable to use allantoin as sole nitrogen source	9R	17

Gene Designation	Former/Alternate Designation	Phenotype	Gene Product	Map Position	References
DAL2	—	unable to use allantoin or allantoate as sole nitrogen source	allantoicase	9R	17
DAL4	—			9R	17,18
DAL5	see *UEP1*	unable to use allantoin as nitrogen source	allantoin permease	9R	17,18
DAL80	—	constitutive expression of allantoin degradative enzymes	—	11R	17
DAL81	—	unable to induce allantoin degradative enzymes	—	9R	17
DBL1	—	deficient in binding alcian blue dye	—	11L	5,30
DDS1	—	defective in DNA synthesis	—	—	50
DDS2	—	defective in DNA synthesis	—	—	50
DDS3	—	defective in DNA synthesis	—	—	50
DDS4	—	defective in DNA synthesis	—	—	50
DDS5	—	defective in DNA synthesis	—	—	50
DGR1	—	deoxyglucose-resistant sucrose fermentation	—	—	36
DIP1	—	cold-sensitive; defective in rRNA synthesis	—	—	115
DMT1	—	increased frequency of chromosome III rearrangement or deletion	—	—	72
DSF6	—	maltase-positive maltose nonfermenter; reduced maltose uptake	—	—	18
DSF7	—	same as *dsf6*	—	—	18
DSF17	—	same as *dsf6*	—	—	18
DSF21	—	same as *dsf6*	—	—	18
dsm1	—	suppresses the reduced premeiotic DNA synthesis and reduced meiotic recombination obtained in *spo8* strains; does not suppress sporulation-deficiency of *spo8* strains; has no phenotype in Spo$^+$ cells	—	7R	24

Gene	Synonym	Phenotype	Enzyme/Function	Map	Ref.
DUR1	—	unable to use urea as sole nitrogen source	urea carboxylase (hydrolyzing)	2R	17
DUR2	—	unable to use urea as sole nitrogen source	allophanate hydrolase	2R	17
DUR3	—	unable to use urea as sole nitrogen source	urea permease, active	8L	17,18
DUR4	—	unable to use urea as sole nitrogen source	urea permease, facilitative	—	17,18
DUT1	see CPS1				
DUT2	—	deficient in blocked dipeptide utilization	—	—	126
EB12	—	altered ethidium bromide and spontaneous induction of petites	—	—	22
EBI8	—	altered ethidium bromide induction of petites	—	—	22
EBI11	—	altered ethidium bromide and spontaneous induction of petites	—	—	22
EBI20	—	same as ebi11	—	—	22
EBI101	—	altered ethidium bromide induction of petites	—	—	22
ERG1	—	ergosterol requirement; accumulates squalene	squalene epoxidase	—	43
ERG2	POL2	nystatin-resistant; no auxotrophy	$\Delta 8 \to \Delta 7$ isomerase (in sterol biosynthesis)	—	43
ERG3	POL3, NYS3	nystatin-resistant; no auxotrophy	5-dehydrogenase (in sterol biosynthesis)	—	43
ERG4	—	nystatin-resistant	24-methylene sterol reductase	—	43
ERG5	POL5	nystatin-resistant; no auxotrophy	22-dehydrogenase (in sterol biosynthesis)	—	43
ERG6	POL1, NYS1	nystatin-resistant; no auxotrophy	24-methyl transferase (in sterol biosynthesis)	—	43
ERG7	—	temperature-sensitive growth; ergosterol requirement at intermediate temperatures	2,3-oxidosqualene cyclase (in sterol biosynthesis)	—	43
ERG8	—	same as erg7; defective in conversion of mevalonic acid to mevalonic acid pyrophosphate	—	—	43

669

Gene Designation	Former/Alternate Designation	Phenotype	Gene Product	Map Position	References
ERG9	—	same as *erg7*; defective in conversion of farnesyl pyrophosphate to squalene	—	—	43
ERG10	—	same as *erg9*	—	—	43
ERG11	*SG1*	same as *erg9*	—	—	43
ERG12	—	same as *erg8*	—	—	43
ETH2	see *SAM2*				
ETH10	see *SAM1*				
FAS1	—	saturated fatty acid requirement	fatty acid synthetase	11L	43
FAS2	—	saturated fatty acid requirement	fatty acid synthetase	—	43
FCY1	—	5-fluorocytosine resistant	cytosine deaminase	—	52
FCY2	*APP1*, *CYT-P*	resistant to 5-fluorocytosine and 4-aminopyrazole pyrimidine; decreased uptake of cytosine, adenine, and hypoxanthine	cytosine permease	—	18,52
FDP1	—	unable to ferment glucose, fructose, sucrose, or mannose	—	2R	27
FDR1	—	FdUMP resistance in *tup7* strains	—	—	11
FLK1	*CYC9*, *UMR7*	flocculent	—	3R	96
flo1	—	flocculent	—	1R	110
flo2	—	flocculent	—	—	61
FLO3	—	flocculent	—	—	61
flo4	—	flocculent	—	—	61
FOL1	*TMP3*	either thymidylate + adenine + histidine + methionine or folinic acid auxotrophy	serine hydroxymethyltransferase	4R	62
FOL2	—	same as *fol1*	—	7R	32
FRO1	—	frothing	—	7R	112,113
FRO2	—	frothing	—	7R	112,113
FSU1	—	suppressor of *FLO4*-induced flocculation	—	—	46
FSU2	—	suppressor of *FLO4*-induced flocculation	—	—	46

Gene		Phenotype	Gene product/function	Map position	Reference
FUI2	see *URD1*				
FUR1		5-fluorouracil- and 5-fluorocytosine-resistant	uracil phosphoribosyltransferase	—	52
FUR4	*URA-P*	5-fluorouracil- and 5-fluorouridine-resistant	uracil permease	—	18,52
GAL1	—	unable to ferment galactose	galactokinase	2R	82
GAL2	—	unable to ferment galactose	galactose permease	12R	82
GAL3	—	unable to ferment galactose	—	4R	82
GAL4	*gal81*	unable to ferment galactose; *GAL81* alleles: constitutive synthesis of galactose catabolic enzymes	transcriptional activator of *GAL1*, *GAL2*, *GAL7*, *GAL10*, and *MEL1*	16L	82
GAL5	see *PGM2*				
GAL7	—	unable to ferment galactose	galactose-1-phosphate uridylyltransferase	2R	82
GAL10	—	unable to ferment galactose	UDP-glucose-4-epimerase	2R	82
GAL80	—	constitutive synthesis of galactose catabolic enzymes; *GAL80*^{s} alleles: dominant noninducibility of galactose catabolic enzymes	regulator of *GAL4* activity	2L	82
gal81	see *GAL4*				
gal82	—	glucose-insensitive synthesis of galactose catabolic enzymes	—	—	68
gal83	—	same as *gal82*	—	5R	68
GAM1	—	γ- and X-ray-sensitive; increased spontaneous mutation rate	—	—	41
GAM2	—	increased spontaneous mutation rate	—	—	41
GAM3	—	γ- and X-ray-sensitive; increased spontaneous and X-ray-induced mutation rate	—	—	41
GAM4	—	increased spontaneous mutation rate	—	—	41
GAM5	—	same as *gam3*	—	—	41

Gene Designation	Former/Alternate Designation	Phenotype	Gene Product	Map Position	References
GAP1	—	D-histidine-resistant	general amino acid permease	—	18
GDC1	—	cannot use glycine to fulfill auxotrophy of *ser1* strains	glycine decarboxylase	—	52
GDH1	*GDHA, URE1*	glutamate auxotrophy; reduced growth on ammonia, serine, or allantoin as sole nitrogen source	glutamate dehydrogenase (NADP)	—	17,52
GDHA	see *GDH1*				
GDHCR	see *GLU80*				
GIA1	—	growth in agar		5R	70
GLC1	—	defective in glycogen and trehalose accumulation		2R	86
GLC2	—	defective in glycogen accumulation		—	86
GLC3	—	defective in glycogen accumulation		5	86
GLC4	—	same as *glc1*		7L	86
GLC5	—	same as *glc1*		—	86
GLC6	—	defective in glycogen accumulation		2R	86
GLC7	—	defective in glycogen accumulation		—	86
GLC8	—	defective in glycogen accumulation		—	86
GLK1	—	glucose nonfermenter in *hxk1 hxk2* background	glucokinase	3L	27
GLN1	—	glutamine auxotrophy	glutamine synthetase	16R	52
GLN3	—	glutamine auxotrophy	positive regulator of glutamine synthetase expression	—	52
GLR1	see *HXK2*				
GLU1	see *ACO1*				
GLU3	see *CIT1*				
GLU80	*URE2, GDHCR*	constitutive synthesis of enzymes subject to nitrogen repression	—	—	17

Gene	Synonym	Phenotype	Enzyme		Ref.
GPM1	*PGM*	glucose nonfermenter; able to grow on ethanol or glycerol	phosphoglycerate mutase	—	27
GRC1	—	glucosamine-resistant on lactate in *grc2* or *grc3* background		—	74
GRC2	—	glucosamine-resistant on lactate in *grc1* or *grc3* background		—	74
GRC3	—	glucosamine-resistant on lactate in *grc1* or *grc2* background		—	74
GUA1	—	guanine auxotrophy in excess adenine	guanosine monophosphate synthetase	—	52
GUT1	—	unable to grow on glycerol	glycerol kinase	—	27
GUT2	—	unable to grow on glycerol	glycerol-3-phosphate dehydrogenase	—	27
HEM1	*OLE3*	oleic acid plus methionine auxotrophy; nystatin resistance	5-aminolevulinate synthase	—	43
HEM2	*OLE4, HEM10*	same as *hem1*	5-aminolevulinate dehydratase (porphobilinogen synthase)	7R	43
HEM3	*OLE2, HEM11*	same as *hem1*	uroporphyrin-I synthase (porphobilinogen deaminase)	—	43
HEM4	*HEM13*	oleic acid plus ergosterol auxotrophy	coproporphyrinogenase (coproporphyrinogen III oxidase)	—	43
HEM5	*HEM14*	same as *hem4*	ferrochelatase	—	43
HEM6	—	same as *hem4*	coproporphyrinogen synthase (uroporphyrinogen decarboxylase)	—	43
HEM10	see *HEM2*				
HEM11	see *HEM3*				
HEM13	see *HEM4*				
HEM14	see *HEM5*				
HEX1	see *HXK2*				
HEX2	—	glucose-repression-insensitive synthesis of invertase and α-glucosidase		—	27

Gene Designation	Former/Alternate Designation	Phenotype	Gene Product	Map Position	References
HIP1	*HISP*	reduced histidine uptake; unable to supplement *his* auxotrophs with low levels of histidine	histidine permease	—	18
HIS1	—	histidine auxotrophy	ATP phosphoribosyltransferase	5R	52
HIS2	—	histidine auxotrophy	histidinol-phosphate phosphatase	6R	52
HIS3	—	histidine auxotrophy	imidazoleglycerolphosphate dehydratase	15R	52
HIS4A	—	histidine auxotrophy	phosphoribosyl-AMP cyclohydrolase	3L	52
HIS4B	—	histidine auxotrophy	phosphoribosyl-ATP pyrophosphohydrolase	3L	52
HIS4C	—	histidine auxotrophy	histidinol dehydrogenase	3L	52
HIS5	—	histidine auxotrophy	histidinol-phosphate aminotransferase	9L	52
HIS6	—	histidine auxotrophy	BBMII isomerase	9L	52
HIS7	—	histidine auxotrophy	glutamine amidotransferase (BBMIII to imidazoleglycerolphosphate)	2R	52
HISP	see *HIP1*				
HML	—	defective **a**- or α-mating-type cassette transposed to *MAT*	mating-type cassette, either **a** or α, transposable at high efficiency to *MAT* in homothallic (*HO*) strains	3L	44
HMR	—	defective **a**- or α-mating-type cassette transposed to *MAT*	mating-type cassette, either **a** or α, transposable at high efficiency to *MAT* in homothallic (*HO*) strains	3R	44
HO	—	defective in the efficient interconversion of mating type		4L	44
HOM2	—	either homoserine or threonine + methionine auxotrophy	aspartate-semialdehyde dehydrogenase	4R	52
HOM3	*bor1*	either homoserine or threonine + methionine auxotrophy; some alleles are borrelidin-sensitive	aspartokinase	5R	52

Gene	Synonyms	Phenotype	Gene product	Map position	Ref.
HOM6	—	same as hom2	homoserine dehydrogenase	—	52
HTA1	—	lethal in an hta2 background	histone H2A	—	55
HTA2	—	lethal in an hta1 background	histone H2A	—	55
HTB1	—	lethal in an htb2 background	histone H2B	—	93
HTB2	—	lethal in an htb1 background	histone H2B	—	93
HXK1	—	fructose-nonfermenter glucose-fermenter in hxk2 background; moderately 2-deoxyglucose-resistant in HXK2 background	hexokinase A or P1	6R	27
HXK2	GLR1, HEX1	fructose-nonfermenter glucose-fermenter in hxk1 background; moderately 2-deoxyglucose-resistant in HXK1 background; glucose-repression-insensitive synthesis of maltase, galactokinase, α-galactosidase, NADH–cytochrome-c reductase, and cytochrome-c oxidase	hexokinase B or P2	7L	27
ICL1	—	unable to grow on ethanol	isocitrate lyase	—	27
ILS1	—	lethal (identified alleles are temperature-sensitive); rapid cessation of protein synthesis following shift to nonpermissive temperature	isoleucyl-tRNA synthetase	2L	71
ILV1	—	isoleucine auxotrophy	threonine deaminase	5R	52
ILV2	—	isoleucine + valine auxotrophy	acetohydroxyacid synthase	—	52
ILV3	—	isoleucine + valine auxotrophy	dihydroxyacid dehydratase	10R	52
ILV4	—	isoleucine + valine auxotrophy	acetohydroxyacid isomeroreductase	—	52
ILV5	—	isoleucine + valine auxotrophy	acetohydroxyacid isomeroreductase	—	52
INO1	—	inositol requirement	inositol-1-phosphate synthase	—	43
INO2	—	inositol requirement	—	—	43
INO3	—	inositol requirement	—	—	43
INO4	—	inositol requirement; deficient in lipid methylation	S-adenosylmethionine: phosphatidyl-N, N-dimethylethanolamine methyltransferase	—	43

Gene Designation	Former/Alternate Designation	Phenotype	Gene Product	Map Position	References
INO5	—	inositol requirement	—	—	43
INO6	—	inositol requirement	—	—	43
INO7	—	inositol requirement	—	—	43
INO8	—	inositol requirement	—	—	43
INO9	—	inositol requirement	—	—	43
INO10	—	inositol requirement	—	—	43
ISI1	—	growth inhibited by excess isoleucine	—	—	73
ISI2	—	growth inhibited by excess isoleucine	—	—	73
ISI6	—	growth inhibited by excess isoleucine	—	—	73
ISI7	—	growth inhibited by excess isoleucine	—	—	73
KAR1	—	defective in nuclear fusion following conjugation	—	17R	16
KEX1	—	defective in killer toxin production; resistant to killer toxin	—	7L	118
KEX2	—	same as kex1; α-specific sterile; sporulation-defective	—	14L	44,118
KGD1	—	unable to grow on glycerol, lactate, pyruvate, or acetate; able to grow on glucose or ethanol	α-ketoglutarate dehydrogenase	—	27
krb1	—	functional suppressor of mak7 and pet18	—	—	118
KRE1	—	resistant to killer toxin; reduced binding of toxin	—	—	118
KRE2	—	same as kre1	—	—	118
KRE3	—	resistant to killer toxin; normal binding of toxin	—	—	118
LAP1	—	reduced leucine aminopeptidase activity	—	—	114
LAP2	—	reduced leucine aminopeptidase activity	—	—	114
LAP3	—	reduced leucine aminopeptidase activity	—	17R	114

LAP4	—	reduced leucine aminopeptidase activity	—	—	114
LET1	—	lethal	—	1R	76
LET1M	—	lethal	—	13R	78
LET3	—	lethal	—	10L	78
LET5	—	lethal	—	10L	78
LET6	—	lethal	—	6L	78
LEU1	—	leucine auxotrophy	α-isopropylmalate dehydratase	7L	52
LEU2	—	leucine auxotrophy	β-isopropylmalate dehydrogenase	3L	52
LEU3	—	reduced levels of αIPM dehydratase and βIPM dehydrogenase	—	—	52
LEU4	—	leaky leucine auxotrophy; some alleles are trifluoroleucine-resistant	α-isopropylmalate synthase	—	52
LTS1	—	unable to grow at 4°C	—	7L	105
LTS2	—	unable to grow at 4°C	—	—	105
LTS3	—	unable to grow at 4°C	—	7L	105
LTS4	—	unable to grow at 4°C	—	4R	105
LTS5	—	unable to grow at 4°C	—	—	105
LTS10	—	unable to grow at 4°C	—	4R	105
LYP1	LYSP	thiosine-resistant	lysine permease	—	18
LYS1	—	lysine auxotrophy	saccharopine dehydrogenase	9R	52
LYS2	—	lysine auxotrophy; resistant to α-amino-adipic acid	2-aminoadipate reductase	2R	52
LYS3	—	lysine auxotrophy	homoaconitase	—	52
LYS4	—	lysine auxotrophy	homoaconitase	4R	52
LYS5	—	lysine auxotrophy	2-aminoadipate reductase	7L	52
LYS6	—	lysine + glutamate auxotrophy	—	—	52
LYS7	—	lysine auxotrophy	homoaconitate hydratase	13R	52
LYS8	—	lysine + glutamate auxotrophy	—	—	52
LYS9	—	lysine auxotrophy	saccharopine reductase	14R	52
LYS10	—	lysine auxotrophy	—	17L	52

Gene Designation	Former/Alternate Designation	Phenotype	Gene Product	Map Position	References
LYS11	—	lysine auxotrophy	—	9L	52
LYS12	—	lysine auxotrophy	2-hydroxy-3-carboxyadipate dehydrogenase (homoisocitrate dehydrogenase)	—	52
LYS13	—	lysine auxotrophy	saccharopine reductase	—	52
LYS14	—	lysine auxotrophy	saccharopine reductase	—	52
LYS15	—	lysine auxotrophy	—	—	52
LYS16	—	lysine + glutamate auxotrophy	—	—	52
MAK1	—	loss of the K_1 killer plasmid; temperature-sensitive growth	—	15C	118
MAK3	—	loss of the K_1 killer plasmid; decreased L DS RNA; loss of [NEX]	—	16R	118
MAK4	—	loss of the K_1 killer plasmid	—	2R	118
MAK5	—	loss of the K_1 killer plasmid	—	2R	118
MAK6	—	loss of the K_1 killer plasmid	—	16R	118
MAK7	—	loss of the K_1 killer plasmid	—	8L	118
MAK8	—	loss of the K_1 killer plasmid	—	15R	118
MAK9	—	loss of the K_1 killer plasmid	—	11L	118
MAK10	—	loss of the K_1 killer plasmid; loss of the [HOK] plasmid	—	5R	118
MAK11	—	loss of the K_1 killer plasmid	—	11L	118
MAK12	—	loss of the K_1 killer plasmid	—	12L	118
MAK13	—	loss of the K_1 killer plasmid; slow growth	—	9R	118
MAK14	—	loss of the K_1 killer plasmid	—	3R	118
MAK15	—	loss of the K_1 killer plasmid; slow growth	—	11R	118
MAK16	—	loss of the K_1 killer plasmid; temperature-sensitive growth	—	1L	118
MAK17	—	loss of the K_1 killer plasmid; slow growth	—	10L	118

Gene	Synonym	Phenotype	Function	Chromosome	Reference
MAKI8	—	loss of the K₁ killer plasmid		8R	118
MAKI9	—	loss of the K₁ killer plasmid		8R	118
MAK20	—	loss of the K₁ killer plasmid		8L	118
MAK21	—	loss of the K₁ killer plasmid		4R	118
MAK22	—	loss of the K₁ killer plasmid, slow growth		12L	118
MAK23	—	loss of the K₁ killer plasmid		—	118
MAK24	—	loss of the K₁ killer plasmid		7L	118
MAK25	—	loss of the K₁ killer plasmid		—	118
MAK26	—	loss of the K₁ killer plasmid		14L	118
MAK27	—	loss of the K₁ killer plasmid		13R	118
MAK28	—	loss of the K₁ killer plasmid		10L	118
MAL1	—	maltose nonfermenter in the absence of other *MAL* loci	regulator of α-glucosidase synthesis	7R	27
MAL2	—	same as *mal1*	regulator of α-glucosidase synthesis	3R	27
MAL3	—	same as *mal1*	regulator of α-glucosidase synthesis	2R	27
MAL4	—	same as *mal1*	regulator of α-glucosidase synthesis	11R	27
MAL5	—	same as *mal1*	regulator of α-glucosidase synthesis	—	27
MAL6	—	same as *mal1*	regulator of α-glucosidase synthesis	8R	27
MAR1	*SIR2*	constitutive expression of mating-type information resident at *HML* and *HMR*		4L	44
MAR2	*SIR3*	same as *mar1*		12R	44
*MAT***a**	—	*mat***a**/*MAT*α strains are sporulation-deficient and mate as α cells	coregulator, with *MAT*α2-gene product of *MAT*α1 and **a**/α-specific mating-type genes	3R	44
*MAT*α1	—	sterile; *MAT***a**/*mat*α1 strains show normal sporulation	positive regulator of α-specific mating-type genes	3R	44
*MAT*α2	—	sterile; *MAT***a**/*mat*α2 strains are sporulation-deficient	negative regulator of **a**-specific mating-type genes and, in conjunction with *MAT***a**-gene product, negative regulator of *MAT*α1 and positive regulator of **a**/α-specific mating-type genes	3R	44

Gene Designation	Former/Alternate Designation	Phenotype	Gene Product	Map Position	References
MCR1	—	suppression of mitochondrially encoded chloramphenicol resistance	—	—	117
MCR2	—	same as mcr1	—	—	117
MCR3	—	same as mcr1	—	—	117
MDH1	—	reduced growth on glycerol	mitochondrial malate dehydrogenase	—	27
MEI1	—	sporulation-deficient; fails to initiate pre-meiotic DNA synthesis	—	—	24
MEI2	—	same as mei1	—	—	24
MEI3	—	same as mei1	—	—	24
MEL1	—	melibiose nonfermenter	α-galactosidase	—	27
MEP1	—	resistant to methylamine inhibition during growth on glutamate as sole nitrogen source	methylamine and ammonia permease, high capacity	—	18
MEP2	AMT	same as mep1	methylamine and ammonia permease, low capacity	—	18
MES1	—	lethal (identified alleles are temperature-sensitive) rapid cessation of protein synthesis following shift to nonpermissive temperature	methionyl-tRNA synthetase	7R	71
METP	—	ethionine-resistant	methionine permease	—	18
MET1	—	methionine auxotrophy	—	11R	52
MET2	—	methionine auxotrophy	homoserine acetyltransferase	17R	52
MET3	—	methionine auxotrophy	ATP: sulfate adenylyl transferase	10R	52
MET4	—	methionine auxotrophy	—	17L	52
MET5	—	methionine auxotrophy	sulfite reductase	5R	52
MET6	—	methionine auxotrophy	homocysteine methyltransferase	13R	52
MET7	—	methionine auxotrophy	—	15R	52
MET8	—	methionine auxotrophy	—	2R	52
MET9	—	methionine auxotrophy	—	—	52

Gene		Phenotype	Enzyme	Chromosome	Reference
MET10	—	methionine auxotrophy	sulfite reductase	6R	52
MET11	—	methionine auxotrophy	—	—	52
MET12	—	methionine auxotrophy	—	7L	52
MET13	—	methionine auxotrophy	—	11R	52
MET14	—	methionine auxotrophy	APS kinase (ATP: adenylylsulfate-3'-phosphotransferase)	—	52
MET15	—	methionine auxotrophy	—	—	52
MET16	—	methionine auxotrophy	PAPS reductase	—	52
MET17	—	methionine auxotrophy	PAPS reductase; homocysteine synthase	—	52
MET18	—	methionine auxotrophy	sulfite reductase	—	52
MET19	—	methionine auxotrophy	sulfite reductase	—	52
MET20	—	methionine auxotrophy	sulfite reductase	—	52
MET22	—	methionine auxotrophy	PAPS reductase	—	52
MET25	—	methionine auxotrophy; deficient in homocystein synthase, cysteine synthase, and α-cystathionine synthase	—	—	52
MGL1	—	unable to ferment α-methylglucose	α-methylglucosidase	2R	27
MGL2	—	unable to ferment α-methylglucose	α-methylglucoside permease	—	27
MGL3	—	unable to ferment α-methylglucose	α-methylglucosidase	—	27
MGL4	—	unable to ferment α-methylglucose	α-methylglucoside permease	—	27
MIA1	—	no cellular phenotype; altered mobility of iso-accepting tRNA species on RPC-5 chromotography		—	10
mic1	—	increased spontaneous mitotic recombination; increased spontaneous mutation rate; UV-sensitive; sporulation-deficient		—	25,41
mic5	—	increased spontaneous mitotic recombination; increased spontaneous mutation rate; UV-sensitive		—	25,41
mic8	—	same as *mic1*		—	25,41

Gene Designation	Former/Alternate Designation	Phenotype	Gene Product	Map Position	References
mic9	—	increased spontaneous mitotic recombination; increased spontaneous mutation rate; MMS-sensitive	—	—	25,41
mic12	—	increased spontaneous mitotic recombination; increased mutation rate	—	—	25,41
mic15	—	increased spontaneous mitotic recombination; increased spontaneous mutation rate; UV- and MMS-sensitive; sporulation-deficient	—	—	25,41
mic19	—	same as *mic15*	—	—	25,41
mic23	—	increased spontaneous mitotic recombination; UV- and MMS-sensitive; sporulation-deficient	—	—	25,41
MIM1	—	increased spontaneous [mit⁻] formation	—	—	49
MIM2	—	increased spontaneous [mit⁻] formation	—	—	49
MIM3	—	increased spontaneous [mit⁻] formation	—	—	49
MIN1	—	growth inhibited by excess methionine	—	5L	73
MKT1	—	loss of the K₂ killer plasmid in [NEX] strains only	—	—	118
MMS1	—	MMS-sensitive; sporulation-deficient	—	—	41
MMS2	—	MMS-sensitive; increased chemical-induced mutagenesis	—	—	41
MMS3	—	UV- and MMS-sensitive; resistant to UV, X-ray, and chemical mutagenesis	—	—	41
MMS4	—	MMS-sensitive; sporulation-deficient	—	—	41
MMS5	—	MMS-sensitive; sporulation-deficient	—	—	41
MMS6	—	UV- and MMS-sensitive	—	—	41

		Phenotype			Ref.
MMS7	—	X-ray-, UV-, and MMS-sensitive; sporulation-deficient; resistant to chemical mutagenesis	—	—	41
MMS8	—	X-ray- and MMS-sensitive; sporulation-deficient; increased spontaneous recombination	—	—	41
MMS9	—	same as *mms8*	—	—	41
MMS10	—	UV- and MMS-sensitive; sporulation-deficient	—	—	41
MMS11	—	X-ray-, UV-, and MMS-sensitive	—	—	41
MMS12	—	same as *mms7*	—	—	41
MMS13	—	same as *mms10*; increased spontaneous recombination	—	—	41
MMS14	—	X-ray- and MMS-sensitive; sporulation-deficient	—	—	41
MMS15	—	X-ray-, UV-, and MMS-sensitive; sporulation-deficient;	—	—	41
MMS16	—	same as *mms14*	—	—	41
MMS17	—	X-ray-, UV-, and MMS-sensitive; sporulation-deficient	—	—	41
MMS18	—	same as *mms10*	—	—	41
MMS19	—	UV- and MMS-sensitive; sporulation-deficient; increased spontaneous mutation rate	—	—	41
MMS20	—	same as *mms14*; resistant to chemical mutagenesis	—	—	41
MMS21	—	same as *mms15*; increased spontaneous mutation rate; increased spontaneous recombination	—	—	41
MMS22	—	MMS-sensitive; sporulation-deficient	—	—	41

683

Gene Designation	Former/Alternate Designation	Phenotype	Gene Product	Map Position	References
MNN1	—	mannoprotein lacks terminal $\alpha1 \rightarrow \alpha3$-linked mannose units in the core, outer chain, and 0-linked oligosaccharides	$\alpha1 \rightarrow 3$-mannosyltransferase	5L	4
MNN2	—	mannoprotein lacks side chains in the outer chain	$\alpha1 \rightarrow 2$-mannosyltransferase-II	2R	4
MNN3	—	branching of mannoprotein outer chain and length of 0-linked oligosaccharides are reduced	—	—	4
mnn4	—	mannoprotein lacks phosphate that is normally present as mannosylphosphate units	—	11L	4
MNN5	—	mannoprotein has single $\alpha1 \rightarrow 2$-linked mannose units as side chains in the outer chain	—	—	4
MNN6	—	same as *mnn4*	—	—	4
MNN7	—	mannoprotein has truncated outer chain	—	—	4
MNN8	—	same as *mnn7*	—	—	4
MNN9	—	same as *mnn7*	—	—	4
MNN10	—	same as *mnn7*	—	—	4
MOD5	—	antisuppression of class-I UAA suppressors	tRNA-isopentenyl transferase	—	99
mpr1	—	suppression of mitochondrially encoded paromomycin resistance	—	—	117
MUT1	—	increased spontaneous mutation rate	—	F11	41
MUT2	—	MMS-sensitive; increased spontaneous mutation rate	—	F11	41

Gene		Phenotype	Map	Ref.
MUT3	—	UV- and MMS-sensitive; increased spontaneous mutation rate	—	41
MUT4	—	same as *mut3*	—	41
MUT5	see *RAD51*		—	41
mut6	—	increased spontaneous mutation rate	—	41
MUT7	—	increased spontaneous mutation rate	—	41
MUT8	—	increased spontaneous mutation rate	—	41
MUT9	—	X-ray, UV-, and MMS-sensitive; increased spontaneous mutation rate	—	41
MUT10	—	increased spontaneous mutation rate	—	41
NUL3	—	nonspecific sterile	4R	76
NYS3	see *ERG3*		7L	43
OLE1	—	oleic acid requirement; fatty acid desaturase	7L	43
OLE2	see *HEM3*			
OLE3	see *HEM1*			
OLE4	see *HEM2*			
OLI1	—	resistant to oligomycin, venturicidin	7L	83,88,95
OLI2	—	resistant to oligomycin, venturicidin	—	83
OP1	see *PET9*			
OPI1	—	secretes inositol; constitutive for inositol-1-phosphate synthase	—	43
OPI2	—	same as *opi1*	—	43
OPI3	—	secretes inositol; not constitutive for inositol-1-phosphate synthase	—	43
OPI4	—	same as *opi1*	—	43
OSM1	—	sensitive to low osmotic pressure	10R	106
OSM2	—	sensitive to low osmotic pressure	16R	106
PAI1	—	altered regulation of proteinase A; temperature-sensitive and pH-sensitive growth	—	7

Gene Designation	Former/Alternate Designation	Phenotype	Gene Product	Map Position	References
PDC1	—	glucose nonfermenter	pyruvate decarboxylase	—	27
PDX2	—	pyridoxine auxotrophy	—	F6	39
PEP1	—	deficient for carboxypeptidase Y	—	—	51
PEP2	—	deficient for proteinase B and carboxypeptidase Y	—	—	51
PEP3	—	deficient for proteinases A and B and carboxypeptidase Y; some alleles temperature-sensitive-lethal	—	12R	51
PEP4	PHO9	defective in processing precursors of vacuolar enzymes; lacks nonspecific alkaline phosphatase activity; sporulation deficient	—	16	42,51
PEP5	—	same as pep3	—	—	51
PEP6	—	same as pep3	—	—	51
PEP7	—	same as pep3	—	—	51
PEP8	—	deficient for proteinases A and B and carboxypeptidase Y	—	—	51
PEP9	—	same as pep1	—	—	51
PEP10	—	same as pep1	—	—	51
PEP11	—	same as pep8	—	—	51
PEP12	—	same as pep3	—	15L	51
PEP13	—	same as pep8	—	—	51
PEP14	—	same as pep3	—	—	51
PEP15	—	same as pep8	—	—	51
PEP16	—	same as pep8	—	12R	51
PEP17	—	very low in proteinases A and B and carboxypeptidase Y	—	—	51

Gene		Phenotype		Map	References
PET1	—	will not grow on nonfermentable substrate; lacks cytochromes a, b, and c_1	—	8R	9,101
PET2	—	same as pet1	—	17R	8,9,101
PET3	—	will not grow on nonfermentable substrate; lacks cytochromes a and b	—	8R	8,9,101
PET4	—	respires but will not grow on nonfermentable substrate; levels of cytochromes a and b are substantially reduced	—	—	9,101
PET5	—	will not grow on nonfermentable substrate; lacks cytochrome a	—	—	101
PET6	—	same as pet1	—	—	9,101
PET7	—	same as pet1	—	—	8,101
PET8	—	will not grow on nonfermentable substrate; lacks cytochrome a and has reduced amounts of cytochromes b and c	—	14R	8,66
PET9	op1	will not grow on nonfermentable substrate; has normal amounts of all cytochromes but abnormal oxidative phosphorylation; petite lethal	—	2L	8,9,23,56
PET10	—	will not grow on nonfermentable substrate; lacks cytochromes b and c_1	—	—	9,89
PET11	—	will not grow on nonfermentable substrate	—	2R	75
PET12	—	same as pet1	—	—	9,66
PET13	—	same as pet8	—	—	9,66
PET14	—	same as pet11	—	4R	76
PET15	—	same as pet11	—	—	8
PET16	—	same as pet11	—	—	8
PET17	—	same as pet11	—	15R	40

Gene Designation	Former/Alternate Designation	Phenotype	Gene Product	Map Position	References
PET18	—	will not grow on nonfermentable substrate; loss of the K_1 killer plasmid; loss of [HOK] plasmid; temperature-sensitive growth	—	3R	76,118
PET19	CYT	will not grow on nonfermentable substrate; lacks cytochromes a, b, and c and accumulates coproporphyrin at 35°C	—	—	33
PET21	—	shows reduced growth on glycerol but has all cytochromes	—	—	57
PET22	—	will not grow on nonfermentable substrate; lacks cytochrome c_1 and has only trace amounts of cytochrome a	—	—	57
PET23	—	will not grow on glycerol but has all cytochromes	—	—	57
PET24	—	same as pet1	—	—	57
PET25	—	will not grow on nonfermentable substrate; lacks cytochromes a and b	—	—	57
PET26	—	same as pet5	—	—	57
PET29	—	same as pet23	—	—	57
PET30	—	will not grow on nonfermentable substrate; lacks cytochrome b	—	—	57
PET31	—	same as pet23	—	—	57
PET32	—	same as pet23	—	—	57
PET33	—	same as pet23	—	—	57
PET34	—	shows only traces of growth on glycerol but has all cytochromes	—	—	57

Gene	Synonym	Phenotype	Gene product	Chromosome	Reference
PET35	—	same as pet1		—	57
PET36	—	same as pet1		—	57
PET37	—	will not grow on nonfermentable substrate; has only trace amounts of cytochromes a, b, and c_1		—	57
PET38	—	same as pet1		—	57
PET39	—	same as pet5		—	57
PET40	—	same as pet5		—	57
PET41	—	same as pet1		—	57
PET42	—	same as pet5		—	57
PET43	—	same as pet23		—	57
PET44	—	will not grow on nonfermentable substrate; has only trace amounts of cytochrome c_1		—	57
PET45	—	same as pet1		—	57
PET46	—	will not grow on nonfermentable substrate; lacks cytochrome c_1		—	57
PET51	—	same as pet46		—	57
PFK1	—	glucose resistant in pyk1 background	phosphofructokinase	2R	27
PGI1	—	glucose nonfermenter, fructose fermenter	phosphoglucose isomerase	3R	27
PGK1	—	glucose nonfermenter, able to grow on ethanol or glycerol	phosphoglycerate kinase	3R	27
PGM	see GPM1				
PGM2	GAL5	galactose nonfermenter	phosphoglucomutase, isozyme 2	F6	27
PHA2	—	phenylalanine auxotrophy	prephenate dehydratase	17R	52
PHO1	PHOA	lacks both acid and alkaline phosphatases	positive factor required for PHO5 and PHO8 expression	—	82
PHO2	PHOB	lacks repressible acid phosphatase; unable to transport inorganic phosphate	positive factor required for PHO5 and PHO84 expression	4L	82
PHO3	PHOC	lacks constitutive acid phosphatase	acid phosphatase	2R	82

689

Gene Designation	Former/Alternate Designation	Phenotype	Gene Product	Map Position	References
PHO4	PHOD, pho82	unable to derepress inducible acid and alkaline phosphatases; unable to transport inorganic phosphate; PHO82 alleles: constitutive expression of inducible acid and alkaline phosphatases	activator of PHO5, PHO8, and PHO84 transcription	6R	82
PHO5	PHOE	lacks inducible acid phosphatase	acid phosphatase (inducible)	2R	82
PHO6	PHOF	lacks constitutive acid phosphatase	positive factor required for PHO3 expression	—	82
PHO7	PHOG	lacks constitutive acid phosphatase	positive factor required for PHO3 expression	—	82
PHO8	PHOH	lacks nonspecific alkaline phosphatase	alkaline phosphatase (nonspecific)	4R	82
PHO9	see PEP4				
PHO80	PHOR, TUP7	constitutive synthesis of inducible acid and alkaline phosphatases and phosphate permeases; incorporates exogenous dTMP into DNA in vivo; hypersensitive to BrdU and FUdR	repressor of PHO4 function	15R	11,82
PHO81	PHOS	unable to derepress inducible acid and alkaline phosphatases; certain alleles give dominant constitutive expression of the inducible phosphatases	modulator of PHO80/PHO85 activity	—	82
pho82	see PHO4				
pho83	phoP	cis-dominant constitutive expression of PHO5	promoter/operator region of PHO5	2R	82
PHO84	PHOT	unable to transport inorganic phosphate; constitutive synthesis of inducible acid and alkaline phosphatases	inorganic phosphate transport system	—	82
PHO85	PHOU	constitutive synthesis of inducible acid and alkaline phosphatases	repressor of PHO4 function	16L	82

Gene	Description		Ref
PHOA	see *PHO1*		—
PHOB	see *PHO2*		—
PHOC	see *PHO3*		—
PHOD	see *PHO4*		—
PHOE	see *PHO5*		—
PHOF	see *PHO6*		—
PHOG	see *PHO7*		—
PHOH	see *PHO8*		—
PHOI	see *PHO9*		—
phoO	see *pho82*		—
phoP	see *pho83*		—
PHOR	see *PHO80*		—
PHOS	see *PHO81*		—
PHOT	see *PHO84*		—
PHOU	see *PHO85*		—
PLI1	smooth colonies; increased lactic acid utilization		31
PLI2	same as *PLI1*		31
PLI3	same as *PLI1*		31
PLI4	same as *PLI1*		31
PLI5	same as *PLI1*		31
PLI6	same as *PLI1*		31
PLI7	same as *PLI1*		31
PLI8	same as *PLI1*		31
PLI9	same as *PLI1*		31
PLI10	same as *PLI1*		31
PLI11	same as *PLI1*		31
PLI12	same as *PLI1*		31
PMI1	mannose nonfermenter; mannose-sensitive on glycerol	phosphomannose isomerase	27

Gene Designation	Former/Alternate Designation	Phenotype	Gene Product	Map Position	References
POL2	see ERG2				
POL3	see ERG3				
POL5	see ERG5				
PRA2	—	low in proteinase-A activity	—	—	53
PRB1	—	reduced sporulation; reduced protein turnover	proteinase B	5L	122,127
PRB2	—	low in proteinase-B activity	—	—	127
PRB3	—	low in proteinase-B activity	—	—	127
PRB4	—	low in proteinase-B activity	—	—	127
PRC1	—	low in proteinase C	proteinase. C (carboxypeptidase Y)	—	123
PRO1	—	proline auxotrophy	glutamyl kinase or glutamyl-phosphate reductase (glutamate-semialdehyde dehydrogenase)	—	52
PRO2	—	proline auxotrophy	glutamyl kinase or glutamyl-phosphate reductase (glutamate-semialdehyde dehydrogenase)	—	52
PRO3	—	proline auxotrophy	pyrroline-5-carboxylate reductase	—	52
PRT1	—	defective in initiation of polypeptide chains (identified alleles are temperature-sensitive)	—	15R	38
PRT2	—	defective in elongation of polypeptide chain (identified alleles are temperature-sensitive)	—	17R	38
PRT3	—	same as prt2	—	5L	38
PSO1	—	X-ray-, UV-, nitrogen-mustard-, and 8-methoxypsoralen + UV-sensitive; sporulation-deficient; reduced spontaneous and induced mutagenesis	—	—	41
PSO2	—	nitrogen-mustard- and 8-methoxypsoralen + UV-sensitive; reduced induced mutation rate	—	—	41

Gene	Synonym	Gene product	Phenotype	Map	Reference
PSO3	—		same as pso2; sporulation-deficient; resistant to induced mutagenesis	—	41
PT-R1	see THP1				
PUR2	—		excretes purines; resistant to 8-azaadenine and 8-azaguanine	—	62
PUR3	—		same as pur2	—	62
PUR4	—		same as pur2	—	62
PUR5	—		same as pur2	4R	62
PUT1	—	proline oxidase	unable to use proline as sole nitrogen source	—	17
PUT2	—	pyrroline-5-carboxylate dehydrogenase	same as put1	—	17
PUT3	—		constitutive synthesis of proline catabolic enzymes	—	17
PUT4	—	proline permease	unable to use proline as sole nitrogen source	—	58
PYK1	CDC19	pyruvate kinase	glucose nonfermenter, able to grow on ethanol or lactate; glucose sensitive; some alleles show temperature-sensitive cell cycle arrest with terminal phenotypes similar to cdc1 strains	1L	27,87
RAD1	—		X-ray-, UV-, MMS-, and nitrogen-mustard-sensitive; increased spontaneous and UV-induced mutation rates; increased UV-induced mitotic recombination	16L	41
RAD2	—		UV-, MMS-, and nitrogen-mustard-sensitive; increased UV- and chemical-induced mutation rate; increased UV-induced mitotic recombination	7R	41

Gene Designation	Former/Alternate Designation	Phenotype	Gene Product	Map Position	References
RAD3	—	UV- and MMS-sensitive; increased spontaneous and induced mutation rates; increased spontaneous and induced mitotic recombination	—	5R	41
RAD4	—	UV- and MMS-sensitive; increased UV- and chemically induced mutation rates; increased UV-induced mitotic recombination	—	5R	41
RAD5	REV2	UV-sensitive; increased spontaneous mutation rate; decreased induced mutation rate; increased UV- and X-ray-induced mitotic recombination	—	12R	41
RAD6	—	X-ray-, UV-, MMS-, and nitrogen-mustard-sensitive; increased spontaneous mutation rate; decreased induced mutation rate; increased spontaneous and induced mitotic recombination; sporulation-deficient	—	7L	41
RAD7	—	UV-sensitive; increased UV-induced mutation rate	—	10R	41
RAD8	—	X-ray-, UV- and MMS-sensitive; resistant to UV and X-ray mutagenesis	—	—	41
RAD9	—	X-ray-, UV-, MMS-, and nitrogen-mustard-sensitive; resistant to UV and chemical mutagenesis	—	—	41
RAD10	—	X-ray-, UV-, and MMS-sensitive; increased induced mutation rate	—	—	41
RAD11	—	X-ray-, UV-, and MMS-sensitive	—	—	41
RAD12	—	same as rad10	—	—	41

RAD13	—	UV- and MMS-sensitive; resistant to UV mutagenesis	—	41
RAD14	—	UV- and MMS-sensitive; increased induced mutation rate	—	41
RAD15	—	X-ray- and UV-sensitive; resistant to UV-, X-ray- and chemical mutagenesis	—	41
RAD16	—	same as rad14	—	41
RAD17	—	X-ray-, UV-, and MMS-sensitive; increased UV-induced mutation rate	—	41
RAD18	—	X-ray, UV, MMS-, and nitrogen-mustard-sensitive; increased spontaneous mutation rate; resistant to induced mutagenesis; increased spontaneous and induced mitotic recombination	3R	41
RAD19	—	UV- and MMS-sensitive; sporulation-deficient	—	41
RAD20	—	same as rad19	—	41
RAD21	—	same as rad19	—	41
RAD22	—	UV- and MMS-sensitive	5L	41
RAD23	—	UV-sensitive; increased spontaneous mutation rate	—	41,70
RAD50	—	X-ray-, UV-, and MMS-sensitive; sporulation-deficient; resistant to UV- and chemical-induced mutagenesis; increased spontaneous mitotic recombination; decreased induced mitotic recombination	4R	24,25,41
RAD51	MUT5	X-ray-, UV-, and MMS-sensitive; sporulation-deficient; increased spontaneous mutation rate; decreased UV and chemical mutagenesis; decreased induced mitotic recombination	5R	24,25,41

Gene Designation	Former/Alternate Designation	Phenotype	Gene Product	Map Position	References
RAD52	REC2	X-ray-, UV-, and MMS-sensitive; sporulation-deficient; increased spontaneous mutation rate; decreased X-ray and chemical mutagenesis; reduced spontaneous and induced mitotic recombination	—	13L	24,25,41
RAD53	—	X-ray-, UV-, and MMS-sensitive; sporulation-deficient; decreased X-ray and chemical mutagenesis; decreased induced mitotic recombination	—	—	24,25,41
RAD54	—	same as rad51	—	—	24,25,41
RAD55	—	X-ray-, UV-, and MMS-sensitive; sporulation-deficient; increased spontaneous mutation rate; decreased UV and chemical mutagenesis; increased X-ray and chemically induced mitotic recombination	—	4R	24,25,41
RAD56	—	X-ray-, UV-, and MMS-sensitive; increased spontaneous mutation rate; decreased UV and chemical mutagenesis; decreased induced mitotic recombination	—	16R	24,25,41
RAD57	—	X-ray-, UV-, and MMS-sensitive; sporulation-deficient; increased spontaneous mutation rate; decreased induced mutagenesis; decreased induced mitotic recombination	—	4R	24,25,41
RDN1	—	—	rRNA	12R	84

Gene	Synonym	Phenotype	Product	Reference
REC1	—	reduced UV- and X-ray-induced mitotic recombination	—	25,41
REC2	see RAD52			
REC3	—	reduced spontaneous and UV- and X-ray-induced mitotic recombination; sporulation-deficient	—	25,41
REC5	—	X-ray-sensitive; reduced UV- and X-ray-induced mitotic recombination	—	25,41
REC6	—	same as rec1	—	25,41
REC7	—	X-ray- and UV-sensitive; reduced UV- and X-ray-induced mitotic recombination	—	25,41
REM1	—	increased spontaneous mitotic recombination; increased spontaneous mutation rate; sporulation-deficient	—	25,41
REV1	—	X-ray- and UV-sensitive; reduced UV, X-ray, and chemical mutagenesis; increased induced mitotic recombination	—	25,41
REV2	see RAD5			
REV3	—	UV- and X-ray-sensitive; reduced spontaneous and induced mutation rate; increased mitotic recombination	—	25,41
REX1	—	cannot express resistance to killer toxin	—	118
RIB1	—	riboflavin auxotrophy	—	81
RIB2	—	riboflavin auxotrophy	—	81
RIB3	—	riboflavin auxotrophy	—	81
RIB4	—	riboflavin auxotrophy	riboflavin synthase	81
RIB5	—	riboflavin auxotrophy	—	81
RIB7	—	riboflavin auxotrophy	—	81
RME1	CSP1	allows sporulation of diploids homozygous at MAT	—	44

Gene Designation	Former/Alternate Designation	Phenotype	Gene Product	Map Position	References
RNA1	—	reduced transport of poly(A)-containing RNA to the cytoplasm; accumulation of unspliced precursors of tRNAs (identified alleles are temperature-sensitive)	—	13R	116
RNA2	—	defective in biosynthesis of rRNA; defective in processing of spliced mRNAs (identified alleles are temperature-sensitive)	—	14R	91,116
RNA3	—	defective in biosynthesis of rRNA (identified alleles are temperature-sensitive)	—	4R	116
RNA4	—	same as rna3	—	—	116
RNA5	—	same as rna3	—	2R	116
RNA6	—	same as rna3	—	—	116
RNA7	—	same as rna3	—	—	116
RNA8	—	same as rna3	—	—	116
RNA9	—	same as rna3	—	—	116
RNA10	—	same as rna3	—	—	116
RNA11	—	same as rna3	—	4R	116
roc1	—	Roccal-resistant	—	12R	75
RPO2	RPOB1	reduced polymerase-II activity in vitro; slightly cordycepin-resistant	RNA polymerase II, B220 subunit	—	121
RPOB1	see RPO2				
RRP1	—	temperature-sensitive lethal; defective in processing the 27S rRNA precursor	—	—	2
sad	—	allows sporulation of MATα/MATα diploids due to chromosome-III rearrangement yielding a hybrid MAT/HMR cassette	—	—	44

Gene	Synonym	Gene product	Phenotype	Map	Ref.
SAL1	—		increased efficiency of suppression by non-sense suppressors	—	99
SAL2	—		same as *sal1*	—	99
SAL3	—		same as *sal1*; cold-sensitive growth	—	99
SAL4	—		same as *sal1*	—	99
SAL5	—		same as *sal1*	—	99
SAM1	*ETH10*	*S*-adenosylmethionine synthetase	ethionine-resistant; constitutive synthesis of several methionine biosynthetic enzymes	—	52
SAM2	*ETH2*	*S*-adenosylmethionine synthetase	same as *sam1*	13R	52
SAMP3	see *SAP3*				
SAP3	*SAMP3*	*S*-adenosylmethionine permease	*S*-adenosylethionine-resistant	—	18
SCA	—		permits sporulation of *MAT*a/*MAT*a and *MAT*α/*MAT*α homozygotes	—	24,44
SDH1	—	succinate dehydrogenase	unable to grow on nonfermentable carbon sources	—	27
SEC1	—		secretion-defective; temperature-sensitive for growth; defective bud growth at the nonpermissive temperature; accumulates secretory proteins and membrane-bound cytoplasmic vesicles at the nonpermissive temperature	—	97
SEC2	—		same as *sec1*	—	97
SEC3	—		same as *sec1*	—	97
SEC4	—		same as *sec1*	—	97
SEC5	—		same as *sec1*	—	97
SEC6	—		same as *sec1*	—	97
SEC7	—		secretion-defective; temperature-sensitive for growth; defective bud growth at the nonpermissive temperature; accumulates secretory proteins and elaborated Golgi apparatus at the nonpermissive temperature	—	97

Gene Designation	Former/Alternate Designation	Phenotype	Gene Product	Map Position	References
SEC8	—	same as sec1	—	—	97
SEC9	—	same as sec1	—	—	97
SEC10	—	same as sec1	—	—	97
SEC11	—	temperature-sensitive for growth; secretion-defective at the nonpermissive temperature; no accumulation of secretory proteins at the nonpermissive temperature	—	—	97
SEC12	—	temperature-sensitive for growth; secretion-defective at the nonpermissive temperature; defective bud growth; accumulation of secretory proteins and elaborated endoplasmic reticulum at the nonpermissive temperature	—	—	97
SEC13	—	same as sec12	—	—	97
SEC14	—	same as sec7	—	—	97
SEC15	—	same as sec1	—	—	97
SEC16	—	same as sec12	—	—	97
SEC17	—	same as sec12	—	—	97
SEC18	—	same as sec12	—	—	97
SEC19	—	same as sec11	—	—	97
SEC20	—	same as sec12	—	—	97
SEC21	—	same as sec12	—	—	97
SEC22	—	same as sec12	—	—	97
SEC23	—	same as sec12	—	—	97
SEL	—	selenate-resistant	sulfate permease	—	18,52
SER1	—	serine or glycine auxotrophy in glucose media	phosphoserine aminotransferase	15R	52
SER2	—		phosphoserine phosphatase	7R	52
SGI	see ERG11	same as ser1			

Gene	Synonyms	Description	Enzyme	Location	Ref
SIR1	—	constitutive expression of mating-type information resident at *HML* and *HMR*		—	44
SIR2	*MAR1*	same as *sir1*		4L	44
SIR3	*CMT, STE8, MAR2*	same as *sir1*		—	44
SIR4	*STE9*	same as *sir1*		—	44
SKI1	—	superkiller; suppresses *mak* mutations		7L	118
SKI2	—	superkiller; suppresses *mak* mutations		—	118
SKI3	—	superkiller; suppresses *mak* mutations		14L	118
SKI4	—	superkiller; suppresses *mak* mutations		14L	118
SNM1	—	UV- and nitrogen-mustard-sensitive		—	41
SNM2	—	UV- and nitrogen-mustard-sensitive		—	41
SNM3	—	UV- and nitrogen-mustard-sensitive		—	41
SNM6	—	UV- and nitrogen-mustard-sensitive		—	41
SNM9	—	UV- and nitrogen-mustard-sensitive		—	41
SNM103	—	UV- and nitrogen-mustard-sensitive		—	41
SNM106	—	UV- and nitrogen-mustard-sensitive		—	41
SOT1	—	suppresses thymidylate uptake of *tup* strains		16L	90
SPD1	—	nitrogen-repression-resistant sporulation		15L	24
SPE1	—	reduced growth rate in the absence of polyamines; reduced level of ornithine decarboxylase		—	52
SPE2	—	spermidine or spermine auxotrophy; cannot sporulate or maintain killer plasmid in absence of spermine or spermidine	*S*-adenosylmethionine decarboxylase	15L	52
SPE3	—	same as *spe2*	spermidine synthase (putrecine aminopropyl transferase)	—	52
SPE4	—	suppressor of *spe10*	spermine synthase (spermidine aminopropyl transferase)	—	52

Gene Designation	Former/Alternate Designation	Phenotype	Gene Product	Map Position	References
SPE10	—	polyamine auxotrophy on defined medium; sporulation-deficient in absence of polyamines; killer plasmid loss in absence of polyamines; reduced levels of ornithine decarboxylase	—	—	52
SPE40	—	suppressor of spe10	—	—	52
SPO1	—	sporulation-defective; cells arrest at first spindle-pole-body duplication	—	14R	24
SPO2	—	sporulation-defective; cells arrest at tetra-nucleate stage	—	—	24
SPO3	—	sporulation-defective; cells arrest with un-enclosed nuclei and anucleate aneuploid spores	—	—	24
SPO4	—	sporulation-defective; same as spo3	—	—	24
SPO5	—	sporulation-defective; same as spo3	—	—	24
SPO7	—	sporulation-defective; no premeiotic DNA synthesis occurs; cells arrest prior to first spindle-pole-body duplication	—	—	24
SPO8	—	sporulation-defective; premeiotic DNA synthesis reduced; cells arrest prior to first spindle-pole-body duplication	—	—	24
SPO9	—	sporulation-defective; same as spo7	—	—	24
SPO10	—	sporulation-defective; cells arrest prior to first spindle-pole-body duplication and accumulate aggregates of synaptonemal complexes	—	—	24
SPO11	—	sporulation-defective; intragenic and intergenic recombination absent	—	8L	24

Gene	Description		Location	Ref.
SPO12	reductional division seldom occurs during meiosis, resulting in two-spored asci with diploid or near diploid genomes	—	—	24
SPO13	same as spo12	—	8R	24
srn1	suppressor of rna2, rna6, rna8	—	—	116
srs1	suppression of trimethoprim sensitivity of rad6 and rad18 strains	—	—	59
srs2	suppression of trimethoprim and UV-sensitivity of rad6 and rad18 strains	—	—	59
SRP1	suppressor of rna2	—	—	116
SST1	supersensitive to α-factor (in MATa background)	BAR1	9L	111
SST2	supersensitive to α-factor in MATa background and to a-factor in MATα background	—	—	111
STE2	a-specific sterile; α-factor-resistant	—	—	44
STE3	α-specific sterile	—	—	44
STE4	nonspecific sterile; α-factor-resistant	—	—	44
STE5	nonspecific sterile; α-factor-resistant	—	—	44
STE6	a-specific sterile	—	—	44
STE7	nonspecific sterile; α-factor resistant	—	—	44
STE8		see SIR3		
STE9		see SIR4		
STE11	nonspecific sterile; α-factor-resistant	—	—	44
STE12	nonspecific sterile; α-factor-resistant	—	—	44
STE13	α-specific sterile	—	—	44
stk1	inefficient interconversion of mating type in homothallic strains	—	3R	44
SUC		see SUC8		

Gene Designation	Former/Alternate Designation	Phenotype	Gene Product	Map Position	References
SUC1	—	sucrose nonfermenter in strains having no other SUC loci	β-fructofuranosidase (invertase)	7R	1,13,27
SUC2	—	same as suc1	β-fructofuranosidase (invertase)	9L	13,27
SUC3	—	same as suc1	β-fructofuranosidase (invertase)	2R	13,27,35
SUC4	—	same as suc1	β-fructofuranosidase (invertase)	—	13,27
SUC5	—	same as suc1	β-fructofuranosidase (invertase)	—	13,27
SUC6	—	same as suc1	β-fructofuranosidase (invertase)	—	13,27
SUC7	—	same as suc1	β-fructofuranosidase (invertase)	—	13,27
SUC8	SUC	same as suc1	β-fructofuranosidase (invertase)	4L	27
suf1	—	frameshift suppressor (group IIA)	$tRNA_3^{Gly}$	15R	99
suf2	—	frameshift suppressor (group III)	$tRNA^{Pro}$	3L	19,99
suf3	—	frameshift suppressor (group IIB)		4R	99
suf4	—	frameshift suppressor (group IIA)	—	—	99
suf5	—	frameshift suppressor (group IIB)	$tRNA_3^{Gly}$	15R	99
suf6	—	frameshift suppressor (group IIA)	$tRNA_1^{Gly}$	17R	99
suf7	—	frameshift suppressor	$tRNA_3^{Gly}$	13L	99
suf8	—	frameshift suppressor	—	8R	99
suf9	—	frameshift suppressor	—	6L	99
suf10	—	frameshift suppressor (group III)	$tRNA^{Pro}$	14L	19,99
SUF11	see SUP35				
SUF12	—	frameshift suppressor	—	15R	99
SUF13	—	omnipotent frameshift suppressor		15R	19
SUF14	—	omnipotent frameshift suppressor	—	17L	19
suf15	—	frameshift suppressor (group II)	—	7R	19
suf16	—	frameshift suppressor (group IIC)	—	3R	19
suf17	—	frameshift suppressor (group IIA)	—	15L	19
suf18	—	frameshift suppressor (group IIC)	—	6R	19
suf19	—	frameshift suppressor (group IIC)	—	5L	19
suf20	—	frameshift suppressor (group IIC)	—	6R	19
suf21	—	frameshift suppressor (group IIC)	—	16R	19

Gene	Synonym	Description	tRNA		Map	Ref
suf22	—	frameshift suppressor (group II)		—	13L	19
suf23	—	frameshift suppressor (group II)		—	10R	19
suf24	—	frameshift suppressor (group II)		—	4R	19
suf25	—	frameshift suppressor (group II)		—	—	19
suh1	—	allele-specific suppressor of his2-1		—	—	99
suh2	—	allele-specific suppressor of his2-1		—	12R	99
sup2	—	nonsense suppressor (ocher, amber types identified)	$tRNA^{Tyr}$	—	4R	34,99
sup3	—	nonsense suppressor (ocher type identified)	$tRNA^{Tyr}$		15L	34,99
sup4	—	same as sup2	$tRNA^{Tyr}$		10R	34,99
sup5	—	same as sup2	$tRNA^{Tyr}$		F8	34,99
sup6	—	same as sup2	$tRNA^{Tyr}$		6R	34,99
sup7	—	same as sup2	$tRNA^{Tyr}$		10L	34,99
sup8	—	same as sup2	$tRNA^{Tyr}$		13R	34,99
sup11	—	nonsense suppressor (ocher, amber types identified)	$tRNA^{Tyr}$		6R	34,99
sup15	see sup16					
sup16	suQ5, sup15	nonsense suppressor (ocher type identified)	$tRNA^{Ser}_{UCA}$		16R	34,99
sup17	—	same as sup16	$tRNA^{Ser}_{UCA}$		9L	34,99
sup19	sup20	same as sup16	$tRNA^{Ser}_{UCA}$		5R	34,99
sup20	see sup19					
sup22	—	nonsense suppressor, serine-inserting (ocher, amber types identified)		—	9L	34,99
sup26	—	nonsense suppressor (ocher type identified)	$tRNA^{Leu}$		12R	34,99
sup27	—	same as sup26	$tRNA^{Leu}$		4R	34,99
sup28	—	same as sup26	$tRNA^{Leu}$		—	34,99
sup29	—	same as sup26	$tRNA^{Leu}$		10L	34,99
sup30	see sup29					
sup32	—	same as sup26	$tRNA^{Leu}$		—	34,99
sup33	—	same as sup26	$tRNA^{Leu}$		—	34,99

Gene Designation	Former/Alternate Designation	Phenotype	Gene Product	Map Position	References
SUP35	SUP2, SUPP, SUF12	omnipotent (frameshift and nonsense) suppressor	—	4R	99
SUP45	SUP1, SUPQ	omnipotent suppressor	—	2R	99
sup46		omnipotent suppressor		2R	99
sup51	see sup52				
sup52		nonsense suppressor (amber type identified)	$tRNA^{Leu}$	10I	34,99
sup53		same as sup52	$tRNA^{Leu}$	3L	34,99
sup54		same as sup52	$tRNA^{Leu}$	7L	34,99
sup55		same as sup52	$tRNA^{Leu}$	—	34,99
sup56		same as sup52	$tRNA^{Leu}$	1R	34,99
sup57		same as sup52	$tRNA^{Leu}$	—	34,99
sup61	sup-RL1	same as sup2	$tRNA^{Ser}_{UCG}$	3R	34,99
sup71		nonsense suppressor (UGA type, class II, identified)	—	5R	99
sup72		same as sup71	—	2R	99
sup73		same as sup71	—	10L	99
sup74		same as sup71	—	10L	99
sup75		same as sup71	—	2L	99
sup76		same as sup71	—	7R	99
sup77		same as sup71	—	7R	99
sup78		same as sup71	—	13R	99
sup79		same as sup71	—	13L	99
sup80		same as sup71	—	4R	99
sup85		nonsense suppressor (UGA type, class I, identified)	—	5R	99
sup86		same as sup85	—	12R	99
sup87		same as sup85	—	2R	99
sup88		same as sup85	—	4R	99
sup90		same as sup85	—	9L	99

Gene	Synonym	Phenotype	Gene product	Map	Ref.
SUP-RL1	see *sup61*				
suQ5	see *sup16*				
sas1	—	suppressor of *ser1*	—	5L	52
SW11	—	defective in homothallic mating-type inter-conversion	—	16L	44
TCM1	—	tricodermin-resistant	ribosomal protein L3	15R	28
TH11	—	thiamine auxotrophy	—	F6	39
THP1	*PT-R1*	pyrithiamine-resistant	thiamine permease	—	18
THR1	—	threonine auxotrophy	homoserine kinase	8R	52
THR4	—	threonine auxotrophy	threonine synthetase	3R	52
TIL1	—	thiaisoleucine resistant	—	7L	100
TMP1	*CDC21*	thymidylate auxotrophy; some alleles show temperature-sensitive cell cycle arrest with terminal phenotypes similar to *cdc2*; sporulation-deficient	thymidylate synthetase	15R	12,87
TMP2	—	either thymidylate + adenine + histidine + methionine or 5-formyltetrahydrofolate auxotrophy	dihydrofolate reductase	—	52
TMP3	see *FOL1*				
TP11	—	glucose nonfermenter; able to grow on ethanol or glycerol	triosephosphate isomerase	—	27
TRA3	—	triazole-alanine-resistant; elevated synthesis of amino acid biosynthetic enzymes; temperature-sensitive cell-cycle arrest; terminal phenotype: unbudded; uninucleate G_1 cells		15R	52,87
TRM1	—	no cellular phenotype; tRNA-deficient in m_2^2G	—	—	85
TRM2	—	no cellular phenotype; tRNA-deficient in ribothymine	—	—	47

Gene Designation	Former/Alternate Designation	Phenotype	Gene Product	Map Position	References
TRP1	—	tryptophan auxotrophy	phosphoribosylanthranilate isomerase	4R	52
TRP2	—	tryptophan auxotrophy	anthranilate synthase	5R	52
TRP3	—	tryptophan auxotrophy	anthranilate synthase; indole-3-glycerolphosphate synthase	11L	52
TRP4	—	tryptophan auxotrophy	anthranilate phosphoribosyltransferase	4R	52
TRP5	—	tryptophan auxotrophy	tryptophan synthase	7L	52
TSL1	—	temperature-sensitive lethal	—	1L	54
TSM1	—	temperature-sensitive lethal	—	3R	109
TSM5	—	temperature-sensitive lethal	—	3R	69
TSM134	—	temperature-sensitive lethal	—	2R	37,67,76
TSM225	—	temperature-sensitive lethal	—	4L	45
TSM437	—	temperature-sensitive lethal	—	7L	76
TUB2	—	benomyl-resistant; some alleles are conditional-lethal	β-tubulin	6L	80
TUP1	—	incorporates exogenous dTMP into DNA in vivo; some alleles may exhibit pleiotropic growth defects	—	3R	119
TUP2	—	incorporates exogenous dTMP into DNA in vivo	—	—	119
TUP3	—	same as tup2	—	—	119
TUP4	—	same as tup2	—	15L	119
TUP5	—	same as tup2	—	—	11
TUP7	see PHO80				
TYR1	—	tyrosine auxotrophy	prephenate dehydrogenase	2R	52
UEP1	DAL5	unable to use allantoate as nitrogen source; reduced uptake of ureidosuccinate	ureidosuccinate permease	—	18

Gene	Synonyms	Phenotype	Gene product	Chromosome	Reference
UMR1	—	reduced UV-induced mutagenesis		—	41
UMR2	—	reduced UV-induced mutagenesis; sporulation-deficient		—	41
UMR3	—	same as *umr2*		—	41
UMR4	—	same as *umr2*		—	41
UMR5	—	same as *umr1*		—	41
UMR6	—	same as *umr2*		—	41
UMR7	*FLK1, CYC9*	same as *umr2*		3R	41
UPF1	—	enhancement of suppression by group-II frameshift suppressors		—	99
UPF2	—	same as *upf1*		—	99
URA1	—	uracil auxotrophy	dihydroorotate dehydrogenase	11L	52
URA2B	—	uracil auxotrophy	asparate carbamoyltransferase	10L	52
URA2C	*CPU1*	uracil auxotrophy in *cpa1* or *cpa2* strains or in excess arginine	carbamoyl-phosphate synthetase	10L	52
URA3	—	uracil auxotrophy	orotidine-5'-phosphate decarboxylase	5L	52
URA4	—	uracil auxotrophy	dihydroorotase	12R	52
URA5	—	uracil auxotrophy	orotate phosphoribosyltransferase	—	52
URAP	see *FUR4*				
URD1	*URID-P FUI1*	5-fluorouridine-resistant	uridine permease	—	18,52
URD2	—	5-fluorouridine-resistant	uridine kinase	—	52
URD3	—	5-fluorouridine-resistant	uridine nucleosidase	—	52
URE1	see *GDH1*				
URE2	see *GLU80*				
URID-P	see *URD1*				

B. Alphabetical Listing of Gene Products

Gene Product	Gene Designation
acetohydroxyacid isomeroreductase	*ILV4, ILV5*
acetohydroxyacid synthase	*ILV2*
acetyl-CoA carboxylase	*ACC1*
acetyl-CoA: glutamate-*N*-acetyl transferase	*ARG2*
N-acetyl-γ-glutamyl-phosphate reductase	*ARG5*
N-acetylglutamate kinase	*ARG6*
N-acetylglutamate synthase	*ARG2*
acetylornithine aminotransferase	*ARG8*
acetylornithine-glutamate acetyl-transferase	*ARG7*
aconitate hydratase	*ACO1*
actin	*ACT1*
activator	see "regulator, positive"
S-adenosylmethionine decarboxylase	*SPE2*
S-adenosylmethionine permease	*SAP3*
S-adenosylmethionine: phosphatidyl-*N*, *N*-dimethyl-ethanolamine methyltransferase	*INO4*
S-adenosylmethionine synthetase	*SAM1, SAM2*
adenylosuccinate lyase	*ADE13*
adenylosuccinate synthetase	*ADE12, ADE13*
alcohol dehydrogenase I	*ADC1*
alcohol dehydrogenase II	*ADR2*
alcohol dehydrogenase, mitochondrial	*ADM1*
allantoicase	*DAL2*
allantoin permease	*DAL4*
allantoinase	*DAL1*
allophanate hydrolase	*DUR2*
amidophosphoribosyl transferase	*ADE4*
amino acid permease	*GAP1*
2-aminoadipate reductase	*LYS2, LYS5*
5-aminolevulinate dehydratase	*HEM2*
5-aminolevulinate synthase	*HEM1*
ammonia permease	*MEP1, MEP2*
anthranilate phosphoribosyltransferase	*TRP4*
anthranilate synthase	*TRP2, TRP3*
apoacetyl-CoA carboxylase ligase	*ACC2*
APS kinase	*MET14*
arginase	*CAR1*
arginine permease	*CAN1*
argininosuccinate lyase	*ARG4*
argininosuccinate synthetase	*ARG1, ARG10*
asparaginase	*ASP1*
asparagine synthetase A	*ASN1*
asparagine synthetase B	*ASN2*
asparaginyl tRNA synthetase	*ASN3*
aspartate aminotransferase	*ASP5*
aspartate carbamoyltransferase	*URA2B*
aspartate-semialdehyde dehydrogenase	*HOM2*

Gene Product	Gene Designation
aspartokinase	*HOM3*
ATP: adenylylsulfate-3′-phosphotransferase	*MET14*
ATP phosphoribosyltransferase	*HIS1*
ATP: sulfate adenylyl transferase	*MET3*
ATP sulfurylase	*MET3*
ATPase, subunit 2	*AUR1*
BBMII isomerase	*HIS6*
carbamoyl-phosphate synthetase (arginine-sensitive)	*CPA1, CPA2*
carbamoyl-phosphate synthetase (uracil-sensitive)	*URA2C*
carboxypeptidase S	*CPS1*
carboxypeptidase Y	*PRC1*
chorismate mutase	*ARO7*
chorismate synthase	*ARO2*
citrate synthase	*CIT1*
coproporphyrinogen-III oxidase	*HEM4*
coproporphyrinogen synthase	*HEM6*
coproporphyrinogenase	*HEM4*
cytochrome *c*, isozyme 1	*CYC1*
cytochrome *c*, isozyme 2	*CYC7*
cytosine deaminase	*FCY1*
cytosine permease	*FCY2*
3-dehydroquinate dehydratase	*ARO1E*
3-dehydroquinate synthase	*ARO1C*
deoxy-D-arabino-D-heptulosonate phosphate synthase (phenylalanine-sensitive)	*ARO3*
deoxy-D-arabino-D-heptulosonate phosphate synthase (tyrosine-sensitive)	*ARO4*
dihydrofolate reductase	*TMP2*
dihydroorotase	*URA4*
dihydroorotate dehydrogenase	*URA1*
dihydroxyacid dehydratase	*ILV3*
N-dimethylethanolamine methyltransferase	*INO4*
DNA ligase	*CDC9*
3-enol pyruvoylshikimate-5-phosphate synthase	*ARO1A*
fatty acid desaturase	*OLE1*
fatty acid synthetase	*FAS1, FAS2*
ferrochelatase	*HEM5*
formyltetrahydrofolate synthetase	*ADE3*
β-fructofuranosidase	*SUC1–SUC8*
galactokinase	*GAL1*
galactose permease	*GAL2*
galactose-1-phosphate uridylyl transferase	*GAL7*
α-galactosidase	*MEL1*
glucokinase	*GLK1*
glutamate dehydrogenase (NADP)	*GDH1*
glutamate-semialdehyde dehydrogenase	*PRO1* or *PRO2*
glutamine amidotransferase (BBMIII to imidazolegly-cerolphosphate)	*HIS7*

Gene Product	Gene Designation
glutamine synthetase	*GLN1*
glutamyl kinase	*PRO1* or *PRO2*
glutamyl-phosphate reductase	*PRO1* or *PRO2*
glycerol kinase	*GUT1*
glycerol-3-phosphate dehydrogenase	*GUT2*
glycine decarboxylase	*GDC1*
guanosine monophosphate synthetase	*GUA1*
hexokinase A or P1	*HXK1*
hexokinase B or P2	*HXK2*
histidine permease	*HIP1*
histidinol dehydrogenase	*HIS4C*
histidinol-phosphate aminotransferase	*HIS5*
histidinol-phosphate phosphatase	*HIS2*
histone H2A	*HTA1, HTA2*
histone H2B	*HTB1, HTB2*
homoaconitase	*LYS3, LYS4*
homoaconitate hydratase	*LYS7*
homocysteine methyltransferase	*MET6*
homocysteine synthase	*MET17*
homoisocitrate dehydrogenase	*LYS12*
homoserine acetyltransferase	*MET2*
homoserine dehydrogenase	*HOM6*
homoserine kinase	*THR1*
2-hydroxy-3-carboxyadipate dehydrogenase	*LYS12*
imidazoleglycerolphosphate dehydratase	*HIS3*
indole-3-glycerolphosphate synthase	*TRP3*
inositol-1-phosphate synthase	*INO1*
invertase	*SUC1–SUC8*
isocitrate lyase	*ICL1*
isoleucyl-tRNA synthetase	*ILS1*
isomerase, $\Delta 8 \rightarrow \Delta 7$	*ERG2*
α-isopropylmalate dehydratase	*LEU1*
β-isopropylmalate dehydrogenase	*LEU2*
α-isopropylmalate synthase	*LEU4*
α-ketoglutarate dehydrogenase	*KGD1*
lysine permease	*LYP1*
malate dehydrogenase, mitochondrial	*MDH1*
$\alpha 1 \rightarrow 3$-mannosyltransferase	*MNN1*
$\alpha 1 \rightarrow 2$-mannosyltransferase II	*MMN2*
mating-type cassette	*HML, HMR*
methionine permease	*METP*
methionyl-tRNA synthetase	*MES1*
methylamine permease	*MEP1, MEP2*
24-methylene sterol reductase	*ERG4*
α-methylglucosidase	*MGL1, MGL3*
α-methylglucoside permease	*MGL2, MGL4*
methylenetetrahydrofolate cyclohydrolase	*ADE3*
methylenetetrahydrofolate dehydrogenase	*ADE3*

Gene Product	Gene Designation
24-methylsteroltransferase	*ERG6*
operator/promoter for *ADR2*	*adr3*
operator/promoter for *ARG5/ARG6*	*arg84*
operator/promoter for *CAR1*	*car80*
operator/promoter for *CAR2*	*car81*
operator/promoter for *PHO5*	*pho83*
ornithine carbamoyltransferase	*ARG3*
ornithine oxoacid aminotransferase	*CAR2*
ornithine transaminase	*CAR2*
orotate phosphoribosyltransferase	*URA5*
orotidine-5′-phosphate decarboxylase	*URA3*
2,3-oxidosqualene cyclase	*ERG7*
PAPS reductase	*MET16, MET17,*
	MET22
phosphatase, acid, constitutive	*PHO3*
phosphatase, acid, inducible	*PHO5*
phosphatase, alkaline	*PHO8*
phosphate permease	*PHO84*
phosphofructokinase	*PFK1*
phosphoglucomutase, isozyme 2	*PGM2*
phosphoglucose isomerase	*PGI1*
phosphoglycerate kinase	*PGK1*
phosphoglycerate mutase	*GPM1*
phosphomannose isomerase	*PMI1*
phosphoribosyl-AMP cyclohydrolase	*HIS4A*
phosphoribosyl-ATP pyrophosphohy-	
drolase	*HIS4B*
phosphoribosylaminoimidazole car-	
boxylase	*ADE2*
phosphoribosylaminoimidazole suc-	
cinocarboxamide synthetase	*ADE1*
phosphoribosylaminoimidazole syn-	
thetase	*ADE7*
phosphoribosylanthranilate iso-	
merase	*TRP1*
phosphoribosyl-formylglycinamide	
synthetase	*ADE6*
phosphoribosylglycinamide formyl transferase	*ADE8*
phosphoribosylglycinamide synthetase	*ADE5*
phosphoserine aminotransferase	*SER1*
phosphoserine phosphatase	*SER2*
porphobilinogen deaminase	*HEM3*
porphobilinogen synthase	*HEM2*
prephenate dehydratase	*PHA2*
prephenate dehydrogenase	*TYR1*
proline oxidase	*PUT1*
proline permease	*PUT4*
proteinase B	*PRB1*

Gene Product	Gene Designation
proteinase C	*PRC1*
putrecine aminopropyl transferase	*SPE3*
pyrroline-5-carboxylate dehydrogenase	*PUT2*
pyrroline-5-carboxylate reductase	*PRO3*
1-pyrroline dehydrogenase	*CAR3*
pyruvate decarboxylase	*PDC1*
pyruvate kinase	*PYK1*
regulator, negative, of arginine biosynthetic enzymes	*ARG80–ARG82*
regulator, negative, of *CAR1/CAR2*	*CAR82*
regulator, negative, of *DAL1/DAL2*	*DAL80*
regulator, negative, of *GAL4* activity	*GAL80*
regulator, negative, of *PHO4* activity	*PHO80, PHO85*
regulator, negative, of *PHO80/PHO85* activity	*PHO81*
regulator of *ADR2* activity	*ADR4*
regulator of *MATα1* and **a**/α-mating-type genes	*MATa1*
regulator of *MATα1*, **a**-specific-, and **a**/α-specific mating-type genes	*MATα2*
regulator, positive, of *ADR2*	*ADR1*
regulator, positive, of α-specific-mating-type genes	*MATα1*
regulator, positive, of constitutive acid phosphatase	*PHO6–PHO7*
regulator, positive, of *DAL1/DAL2*	*DAL81*
regulator, positive, of *GAL1, GAL2, GAL7, GAL10,* and *MEL1*	*GAL4*
regulator, positive, of *GLN1*	*GLN3*
regulator, positive, of inducible acid and alkaline phosphatases	*PHO1*
regulator, positive, of inducible acid phosphatase and phosphate transport	*PHO2*
regulator, positive, of maltase	*MAL1–MAL6*
regulator, positive, of *PHO5, PHO8,* and *PHO84*	*PHO4*
repressor	see "regulator, negative"
riboflavin synthase	*RIB5*
ribosomal protein L3	*TCM1*
ribosomal protein L29	*CYH2*
RNA polymerase II; B220 subunit	*RPO2*
rRNA	*RDN1*
saccharopine dehydrogenase	*LYS1*
saccharopine reductase	*LYS9, LYS13, LYS14*
serine acetyltransferase	*CYS1, CYS2*
serine hydroxymethyltransferase	*FOL1*
shikimate dehydrogenase	*ARO1D*
shikimate kinase	*ARO1B*
spermidine aminopropyl transferase	*SPE4*

Gene Product	Gene Designation
spermidine synthase	*SPE3*
spermine synthase	*SPE4*
squalene epoxidase	*ERG1*
sterol Δ8 → Δ7 isomerase	*ERG2*
sterol 5-dehydrogenase	*ERG3*
sterol 22-dehydrogenase	*ERG5*
sterol 24-methyltransferase	*ERG6*
succinate dehydrogenase	*SDH1*
sulfate permease	*CHR1, SEL1*
sulfite reductase	*MET5, MET10,*
	MET18–MET20
thiamine permease	*THP1*
threonine deaminase	*ILV1*
threonine synthetase	*THR4*
threonyl-tRNA synthetase	*bor3*
thymidylate synthetase	*TMP1*
triosephosphate isomerase	*TPI1*
tRNA$_1^{Gly}$	*suf5*
tRNA$_3^{Gly}$	*suf1, suf4, suf6*
tRNA^{Leu}	*sup26–sup29,*
	sup32, sup33,
	sup52–sup57
tRNA^{Pro}	*suf2, suf10*
tRNA$_{UCA}^{Ser}$	*sup16, sup17,*
	sup19
tRNA$_{UCG}^{Ser}$	*sup61*
tRNA^{Tyr}	*sup2–sup8,*
	sup11
tRNA-isopentenyl transferase	*MOD5*
tRNA synthetase, asparaginyl	*ASN3*
tRNA synthetase, isoleucyl	*ILS1*
tRNA synthetase, methionyl	*MES1*
tRNA synthetase, threonyl	*bor5*
tryptophan synthase	*TRP5*
β-tubulin	*TUB2*
UDP-glucose 4-epimerase	*GAL10*
uracil permease	*FUR4*
uracil phosphoribosyltransferase	*FUR1*
urea carboxylase (hydrolyzing)	*DUR1*
urea permease, active	*DUR3*
urea permease, facilitative	*DUR4*
ureidosuccinate permease	*UEP1*
uridine kinase	*URD2*
uridine nucleosidase	*URD3*
uridine permease	*URD1*
uroporphyrin-I synthase	*HEM3*
uroproporphyrinogen decarboxylase	*HEM6*

C. Genes Grouped by Common Function

1. **Cell Cycle**
2. **Mating/Conjugation**
3. **Sporulation**
4. **Recombination**
5. **Mutagenesis/Repair**
6. **Killer Factor**
7. **Mitochondrial Functions**
8. **Carbon Metabolism**
   a. Glycolysis/Citric Acid Cycle
   b. Disaccharide Utilization
   c. Galactose
   d. Glycogen
   e. Glucose Repression
9. **Nitrogen Metabolism**
   a. Allantoin/Urea Utilization
   b. Amino Acid Catabolism
   c. Regulation of Nitrogen Metabolism
10. **Amino Acid Metabolism**
    a. General Control
    b. Arginine
    c. Aromatic Amino Acids
    d. Aspartate/Asparagine
    e. Cysteine
    f. Glutamate/Glutamine
    g. Histidine
    h. Homoserine
    i. Isoleucine/Valine
    j. Leucine
    k. Lysine
    l. Methionine
    m. Phenylalanine
    n. Proline
    o. Serine/Glycine
    p. Threonine
    q. Tryptophan
    r. Tyrosine
11. **Phosphorous Metabolism**
12. **Lipid Metabolism**
13. **Mannan Biosynthesis**
14. **Colony Morphology**
15. **Secretion**
16. **Transport/Permeases**
17. **Protein Synthesis**
    a. Components of Translation
    b. Resistance to Inhibitors
    c. Suppressors/Antisuppressors
18. **Nucleic Acid Metabolism**
    a. Uracil
    b. Thymidylate
    c. Purines
    d. RNA Biosynthesis
    e. DNA Replication
19. **Vitamin Metabolism**
20. **Proteases/Peptidases**
21. **Resistance (Selectable Markers)**

1. CELL CYCLE

CDC1-CDC18	*HTA2*	*PYK1*	*TRA3*
CDC20	*HTB1*	*STE2-STE7*	
CDC22-CDC50	*HTB2*	*STE11-STE13*	
HTA1	*KAR1*	*TMP1*	

2. MATING/CONJUGATION

CDC4	*HMR*	*MATα1*	*SST1*
CDC28	*HO*	*MATα2*	*SST2*
CDC33	*KAR1*	*NUL3*	*STE2-STE7*
CDC34	*KEX2*	*RME*	*STE11-STE13*
CDC36-CDC39	*MAR1*	*sad*	*stk1*
DMT1	*MAR2*	*SCA*	*SWI1*
HML	*MATa*	*SIR1-SIR4*	

3. SPORULATION

CDC2	*CON1-CON3*	*MMS13-MMS22*	*RME*
CDC4-CDC9	*dsm1*	*PEP4*	*SCA*
CDC13	*KEX1*	*PRB1*	*SPD1*
CDC14	*MEI1-MEI3*	*PRC1*	*SPE2*
CDC16	*mic8*	*PSO1*	*SPE4*
CDC17	*mic15*	*PSO3*	*SPE10*
CDC20	*mic19*	*RAD6*	*SPE40*
CDC23	*MMS1*	*RAD19-RAD22*	*SPO1-SPO5*
CDC28	*MMS4*	*RAD50-RAD57*	*SPO7-SPO13*
CDC36-CDC40	*MMS5*	*REC3*	*TMP1*
CHL1	*MMS7-MMS10*	*REM1*	*UMR2-UMR4*

4. RECOMBINATION

CDC9	*mic5*	*MMS13*	*REM1*
CHL1	*mic8*	*MMS21*	*REV1*
CON1-CON3	*mic9*	*RAD1-RAD6*	*REV3*
COR1-COR4	*mic15*	*RAD18*	*SPO1-SPO5*
DMT1	*mic19*	*RAD50-RAD57*	*SPO7-SPO13*
dsm1	*mic23*	*REC1*	
MEI1-MEI3	*MMS8*	*REC3*	
mic1	*MMS9*	*REC5-REC7*	

5. MUTAGENESIS/REPAIR

CDC2	mic8	MUT6–MUT10	SNM6
CDC8	mic9	PSO1–PSO3	SNM9
CDC9	mic12	RAD1–RAD23	SNM103
CDC31	mic15	RAD50–RAD57	SNM106
COR1–COR4	mic19	REC5	srs1
DMT1	mic23	REC7	srs2
GAM1–GAM5	MIM1–MIM3	REV1	UMR1–UMR7
mic1	MMS1–MMS22	REV3	
mic5	MUT1–MUT4	SNM1–SNM3	

6. KILLER FACTOR

KEX1	MAK1	REX1	SPE10
KEX2	MAK3–MAK28	SKI1–SKI4	SPE40
krb1	MKT1	SPE2	
KRE1–KRE3	PET18	SPE4	

7. MITOCHONDRIAL FUNCTIONS

ADM1	cyp1	EBI101	MIM1–MIM3
amy1	cyp2	GRC1–GRC3	mpr1
ANT1	CYP4	HEM1–HEM6	PET1–PET19
AUR1	CYP5	HXK2	PET21–PET26
CGR1	EBI2	ICL1	PET29–PET46
CGR2	EBI8	krb1	PET51
CGR4	EBI11	MCR1–MCR3	PLI1–PLI12
CYC1–CYC9	EBI20	MDH1	SDH1

8. CARBON METABOLISM
a. Glycolysis/Citric Acid Cycle

ACO1	CCR80	HXK2	PGM2
ADC1	CIT1	ICL1	PMI1
ADM1	GLK1	KGD1	PYK1
ADR1–ADR4	GPM1	MDH1	SDH1
CAT1	GUT1	PDC1	TPI1
CAT2	GUT2	PFK1	
CAT80	HEX2	PGI1	
CCR1–CCR3	HXK1	PGK1	

b. Disaccharide Utilization

DGR1	DSF17	MAL1–MAL6	SUC1–SUC8
DSF6	DSF21	MGL1–MGL4	
DSF7	FDP1		

c. Galactose

GAL1–GAL4	*GAL10*	*gal82*	*MEL1*
GAL7	*GAL80*	*gal83*	*PGM2*

d. Glycogen

GLC1–GLC8

e. Glucose Repression

CAT1	*CCR80*	*DGR1*	*HXK2*
CAT2	*CGR1*	*FDP1*	
CAT80	*CGR2*	*GRC1–GRC3*	
CCR1–CCR3	*CGR4*	*HEX2*	

9. NITROGEN METABOLISM

a. Allantoin/Urea Utilization

DAL1	*DAL4*	*DAL81*	*DUR1–DUR4*
DAL2	*DAL80*		

b. Amino Acid Catabolism

CAR1–CAR3	*car80–CAR82*	*PUT1–PUT4*

c. Regulation of Nitrogen Metabolism

GDH1	*GLU80*	*MEP1*	*MEP2*

10. AMINO ACID METABOLISM

a. General Control

AAS1–AAS3	*GAP1*	*TRA3*

b. Arginine

ARG1–ARG10	*ARG84*	*CAR1–CAR3*	*CPA1*
ARG80–ARG82	*CAN1*	*car80–CAR82*	*CPA2*

c. Aromatic Amino Acids

ARO1A–ARO1E	*ARO7*	*TRP1–TRP5*	*TYR1*
ARO2–ARO5	*PHA2*		

d. Aspartate/Asparagine

ASN1–ASN3	*ASP1*	*ASP5*

e. Cysteine

CYS1   *CYS2*

f. Glutamate/Glutamine

ACO1   *GDH1*   *GLN3*   *GLU80*
CIT1   *GLN1*

g. Histidine

FOL1   *HIP1*   *suh1*   *TMP2*
FOL2   *HIS1–HIS7*   *suh2*   *TRA3*

h. Homoserine

HOM2   *HOM3*   *HOM6*

i. Isoleucine/Valine

ABI1   *ILV1–ILV5*   *ISI6*   *TIL1*
ABI2   *ISI1*   *ISI7*
ILS1   *ISI2*

j. Leucine

LEU1–LEU4

k. Lysine

LYP1   *LYS1–LYS16*

l. Methionine

CHR1   *HOM6*   *METP*   *SEL*
FOL1   *MES1*   *MIN1*   *TMP2*
FOL2   *MET1–MET20*   *SAM1*
HOM2   *MET22*   *SAM2*
HOM3   *MET25*   *SAP3*

m. Phenylalanine

ARO1A–ARO1E   *ARO2–ARO5*   *ARO7*   *PHA2*

n. Proline

CAR3   *PRO1–PRO3*   *PUT1–PUT4*

o. Serine/Glycine

GDC1 *SER1* *SER2* *sus1*

p. Threonine

bor3 *HOM3* *THR1* *THR4*
HOM2 *HOM6*

q. Tryptophan

ARO1A–ARO1E *ARO2–ARO5* *TRP1–TRP5*

r. Tyrosine

ARO1A–ARO1E *ARO2–ARO5* *ARO7* *TYR1*

11. PHOSPHOROUS METABOLISM

PEP4 *PHO80* *PHO81* *pho83–PHO85*
PHO1–PHO8

12. LIPID METABOLISM

ACC1 *CHO3* *FAS2* *OLE1*
ACC2 *ERG1–ERG12* *HEM1–HEM6* *OPI1–OPI4*
CHO1 *FAS1* *INO1–INO10*

13. MANNAN BIOSYNTHESIS

MMN1–MMN10

14. COLONY MORPHOLOGY

DBL1 *flo1–flo4* *FSU2* *PLI1–PLI12*
FLK1 *FSU1* *GIA1*

15. SECRETION

KEX1 *PEP4* *PRC1* *SUC1–SUC8*
KEX2 *PHO5* *SEC1–SEC23*
MEL1 *PRB1*

16. TRANSPORT/PERMEASES

CAN1	DUR4	METP	SEL
CHR1	FCY2	MGL2	THP1
DAL4	FUR4	MGL4	TUP1–TUP5
DSF6	GAP1	PHO4	UEP1
DSF7	HIP1	PHO80	URD1
DSF17	LYP1	PHO84	
DSF21	MEP1	PUT4	
DUR3	MEP2	SAP3	

17. PROTEIN SYNTHESIS

a. Components of Translation

ASN3	suf1–SUF11	sup19	sup61
bor3	suf15–suf25	sup22	sup71–sup80
ILS1	sup2–sup8	sup26–sup29	sup85–sup90
MES1	sup11	sup32	
PRT1–PRT3	sup16	sup33	
RDN1	sup17	sup52–sup57	

b. Resistance to Inhibitors

CRY1	CYH1–CYH8	CYH10	TCM1

c. Suppressors/Antisuppressors

ASU1–ASU9	sup2–sup8	sup32	sup71–sup80
MOD5	sup11	sup33	sup85–sup90
SAL1–SAL5	sup16	SUP35	sus1
suf1–SUF11	sup17	SUP45	UPF1
SUF13–suf25	sup19	sup46	UPF2
suh1	sup22	sup52–sup57	
suh2	sup26–sup29	sup61	

18. NUCLEIC ACID METABOLISM

a. Uracil

FCY1	FUR1	UEP1	URD1–URD3
FCY2	FUR4	URA1–URA5	

b. Thymidylate

FDR1	FOL2	SOT1	TMP2
FOL1	PHO80	TMP1	TUP1–TUP5

c. Purines

ADE1–ADE13	*FOL1*	*GUA1*	*TMP2*
FCY2	*FOL2*	*PUR2–PUR5*	

d. RNA Biosynthesis

DIP1	*srn1*	*sup17*	*sup61*
MIA1	*SRP1*	*sup19*	*sup71–sup80*
MOD5	*suf1–suf11*	*sup22*	*sup85–sup90*
RDN1	*suf15–suf25*	*sup26–sup29*	*TRM1*
RNA1–RNA11	*sup2–sup8*	*sup32*	*TRM2*
RPO2	*sup11*	*sup33*	
RRP1	*sup16*	*sup52–sup57*	

e. DNA Replication

CDC2	*CDC28*	*DDS1–DDS5*	*HTB2*
CDC4	*CDC34*	*HTA1*	*PHO80*
CDC6–CDC9	*CDC36–CDC40*	*HTA2*	*TMP1*
CDC22	*CDC44–CDC48*	*HTB1*	*TUP1–TUP5*

19. VITAMIN METABOLISM

FOL1	*OPI1–OPI4*	*RIB7*	*THP1*
FOL2	*PDX2*	*THI1*	
INO1–INO10	*RIB1–RIB5*		

20. PROTEASES/PEPTIDASES

BIN1–BIN3	*LAP1–LAP4*	*PRA2*	*PRC1*
CPS1	*PAI1*	*PRB1–PRB4*	
DUT2	*PEP1–PEP17*		

21. RESISTANCE (SELECTABLE MARKERS)

ADC1	*CYH1–CYH8*	*LEU4*	*SEL*
ADM1	*CYH10*	*LYP1*	*STE2*
ADR2	*DGR1*	*LYS2*	*STE4*
amy1	*ERG2–ERG6*	*MEP1*	*STE5*
ANT1	*FCY1*	*MEP2*	*STE7*
AUR1	*FCY2*	*METP*	*STE11*
AXE1	*FDR1*	*OLI1*	*STE12*
CAF1	*FUR1*	*OLI2*	*TCM1*
CAN1	*FUR4*	*PUR2–PUR5*	*THP1*
cap1	*GAP1*	*roc1*	*TIL1*
CHR1	*GRC1–GRC3*	*RPO2*	*TRA3*
CRY1	*HEM1–HEM3*	*SAM1*	*TUB2*
cup1	*HXK1*	*SAM2*	*URD1–URD3*
CYC1–CYC3	*HXK2*	*SAP3*	

REFERENCES

1. Abrams, B.B., R.A. Hackel, J. Mizunaga, and J.O. Lampen. 1978. *J. Bacteriol.* **135:** 809.
2. Andrew, C., A.K. Hopper, and B.D. Hall. 1976. *Mol. Gen. Genet.* **144:** 29.
3. Atkinson, K. (pers. comm.)
4. Ballou, C.E. 1982. In *The molecular biology of the yeast* Saccharomyces. II. *Metabolism and gene expression* (ed. J. Strathern et al.). Cold Spring Harbor Laboratory, Cold Spring Harbor, New York. (In press.)
5. Ballou, D.L. 1975. *J. Bacteriol.* **123:** 616.
6. Bard, M. 1980. *J. Bacteriol.* **141:** 999.
7. Beck, I., G.R. Fink, and D. Wolf. 1980. *J. Biol. Chem.* **255:** 4821.
8. Beck, J.C., J.R. Mattoon, D.C. Hawthorne, and F. Sherman. 1968. *Proc. Natl. Acad. Sci.* **60:** 186.
9. Beck J.C., J.H. Parker, W.X. Balcavage, and J.R. Mattoon. 1971. In *Autonomy and biogenesis of mitochondria and chloroplasts* (ed. N.K. Boardman et al.), p. 194. North Holland, New York.
10. Bell, J.R., R.Y.C. Lo, and S.K. Quay. 1979. *Mol. Gen. Genet.* **153:** 145.
11. Bisson, L. and J. Thorner (pers. comm.)
12. Bisson, L. and J. Thorner. 1977. *J. Bacteriol.* **132:** 44.
13. Carlson, M., B. Osmond, and D. Botstein (pers. comm.)
14. Clavilier, L., G. Pere, and P.P. Slominski. 1969. *Mol. Gen. Genet.* **104:** 195.
15. Cohen, J.D. and N.R. Eaton. 1979. *Genetics* **91:** 19.
16. Conde, J. and G.R. Fink. 1976. *Proc. Natl. Acad. Sci.* **73:** 3651.
17. Cooper, T. 1982. Nitrogen metabolism. In *The molecular biology of the yeast* Saccharomyces. II. *Metabolism and gene expression* (ed. J. Strathern et al.). Cold Spring Harbor Laboratory, Cold Spring Harbor, New York. (In press.)
18. Cooper, T. 1982. Transport in *Saccharomyces cerevisiae.* In *The molecular biology of the yeast* Saccharomyces. II. *Metabolism and gene expression* (ed. J. Strathern et al.). Cold Spring Harbor Laboratory, Cold Spring Harbor, New York. (In press.)
19. Culbertson, M. (pers. comm.)
20. Douglas, M., Y. Koh, E. Ebner, E. Agsteribbe, and G. Schatz. 1979. *J. Biol. Chem.* **250:** 8228.
21. Downie, J.A., J.W. Stewart, N. Brockman, A.M. Schweingruber, and F. Sherman. 1977. *J. Mol. Biol.* **113:** 369.
22. Dujardin, G. and B. Dujon. 1979. *Mol. Gen. Genet.* **171:** 205.
23. Dujon (this volume)
24. Esposito and Klapholz (this volume)
25. Esposito and Wagstaff (this volume)
26. Fogel et al. (this volume)
27. Fraenkel, D.G. 1982. In *The molecular biology of the yeast* Saccharomyces. II. *Metabolism and gene expression* (ed. J. Strathern et al.). Cold Spring Harbor Laboratory, Cold Spring Harbor, New York. (In press.)
28. Fried, H.M. and J.R. Warner. 1981. *Proc. Nat. Acad. Sci.* **78:** 238.
29. Fried, H.M. and J.R. Warner (pers. comm.)
30. Friis, J. and P. Ottolenjhs. 1970. *C.R. Trav. Lab. Carlsberg* **37:** 327.
31. Galzey, P. and C. Bizeau. 1966. *Ann. Technol. Agric.* **15:** 289.
32. Game, J.C. (pers. comm.)
33. Gunge, N., T. Sugimura, and M. Iwasaki. 1967. *Genetics* **57:** 213.
34. Guthrie, C. and J. Abelson. 1982. In *The Molecular Biology of the yeast* Saccharomyces. II. *Metabolism and gene expression* (ed. J. Strathern et al.). Cold Spring Harbor Laboratory, Cold Spring Harbor, New York. (In press.)
35. Hackel, R.A. 1975. *Mol. Gen. Genet.* **140:** 361.
36. Hackel, R.A. and N.H. Khan. 1978. *Mol. Gen. Genet.* **164:** 295.

37. Hansche, P.E., V. Beres, and P. Lange. 1978. *Genetics* **88**:673.
38. Hartwell, L. and C. McLaughlin. 1968. *J. Bacteriol.* **96**:1664.
39. Hawthorne, D. and R. Mortimer. 1960. *Genetics* **45**:1085.
40. Hawthorne, D.C. and R.K. Mortimer. 1968. *Genetics* **60**:735.
41. Haynes and Kunz (this volume).
42. Hemmings, B.A., G.S. Zubenko, A. Hasilik, and E.W. Jones. 1981. *Proc. Natl. Acad. Sci.* **78**:435.
43. Henry, S. 1982. In *The molecular biology of the yeast* Saccharomyces. II. *Metabolism and gene expression* (ed. J. Strathern et al.). Cold Spring Harbor Laboratory, Cold Spring Harbor, New York. (In press.)
44. Herskowitz and Oshima (this volume)
45. Hilger, F. (pers. comm.)
46. Holmberg, S. 1978. *Carlsberg Res. Commun.* **43**:401.
47. Hopper, A. (pers. comm.)
48. Jacobs, P., J.C. Jauniaux, and M. Grenson. 1980. *J. Mol. Piol.* **139**:691.
49. Johnston, L. 1979. *Mol. Gen. Genet.* **170**:327.
50. Johnston, L.H. and J.C. Game. 1978. *Mol. Gen. Genet.* **161**:205.
51. Jones, E. 1977. *Genetics* **85**:23.
52. Jones, E. and G. Fink. 1982. In *The molecular biology of the yeast* Saccharomyces. II. *Metabolism and gene expression* (ed. J. Strathern et al.). Cold Spring Harbor Laboratory, Cold Spring Harbor, New York. (In press.)
53. Jones, E.W., G.S. Zubenko, R.R. Parker, B.A. Hemmings, and A. Hasilik. 1981. *Alfred Benzon Symp.* **16**:182.
54. Kaback, D.B. and H.O. Halvorson. 1978. *J. Bacteriol.* **134**:237.
55. Kolodrubetz, D., J. Choe, M. Rykowski, and M. Grunstein (pers. comm.)
56. Kovac, L., T.M. Lachowicz, and P.L. Slominski. 1967. *Science* **158**:1564.
57. Lachowicz, T.M., Z. Kotylak, J. Kolodynski, and Z. Sniegocka. 1969. *Arch. Immunol. Ther. Exp.* **17**:72.
58. Lasko, P. and M. Brandriss. 1981. *J. Bacteriol.* (in press).
59. Lawrence, C. 1979. *J. Bacteriol.* **139**:866.
60. Leibman, S. (pers. comm.)
61. Lewis, C.W., J.R. Johnston, and P.A. Martin. 1976. *J. Inst. Brew.* **82**:158.
62. Little, J.G. and R. Haynes. 1979. *Mol. Gen. Genet.* **168**:141.
63. Linnane, A.W., A.J. Lamb, C. Christodoulou, and H.B. Lukins. 1968. *Proc. Natl. Acad. Sci.* **59**:1288.
64. Lomax, C.A. and R.A. Woods. 1973. *Mol. Gen. Genet.* **120**:139.
65. Lucchini, G., M.L. Carbone, M. Cocucci, and M.L. Sensi. 1979. *Mol. Gen. Genet.* **177**:139.
66. Mackler, B., H.C. Douglas, S. Will, D.C. Hawthorne, and H.R. Mahler. 1965. *Biochemistry* **4**:2016.
67. Maitra, P.K. and Z. Lobo. 1977. *Mol. Gen. Genet.* **156**:55.
68. Matsumoto, K. and Y. Oshima. 1981. *J. Mol. Cell. Biol.* **1**:83.
69. McCuster, J. and J. Haber (pers. comm.)
70. McKnight, G. and F. Sherman (pers. comm.)
71. McLaughlin, C. and L. Hartwell. 1969. *Genetics* **61**:557.
72. Melnick, L.M. and J. Blamire. 1978. *Mol. Gen. Genet.* **160**:157.
73. Meuris, P. 1969. *Genetics* **63**:569.
74. Michels, C.A. and S.D. Mishra (pers. comm.)
75. Mortimer, R.K. and D.C. Hawthorne. 1966. *Genetics* **53**:165.
76. Mortimer, R. and D. Hawthorne. 1973. *Genetics* **74**:35.
77. Mortimer and Schild (this volume)
78. Munz, P. 1972. *Arch. für Genetik* **45**:173.
79. Nass, G. and K. Poralla. 1976. *Mol. Gen. Genet.* **147**:39.

80. Neff, N. and D. Botstein (pers. comm.)
81. Oltmanns, O. and A. Bacher. 1972. *J. Bacteriol.* **110:** 818.
82. Oshima, Y. 1982. In *The molecular biology of the yeast* Saccharomyces. II. *Metabolism and gene expression* (ed. J. Strathern et al.). Cold Spring Harbor Laboratory, Cold Spring Harbor, New York. (In press.)
83. Parker, J.H., I.R. Trimble, and J.R. Mattoon. 1968. *Biochem. Biophys. Res. Commun.* **33:** 590.
84. Petes, T.D. 1979. *J. Bacteriol.* **138:** 185.
85. Phillips, J.H. and K. Kjellin-Straby. 1967. *J. Mol. Biol.* **26:** 509.
86. Pringle, J. (pers. comm.)
87. Pringle and Hartwell (this volume)
88. Rank, G.H. and N.T. Bech-Hansen. 1973. *Mol. Gen. Genet.* **126:** 93.
89. Reilly, C. and F. Sherman. 1965. *Biochim. Biophys. Acta* **95:** 640.
90. Remer, S., A. Sherman, E. Kraig, and J.E. Haber. 1980. *J. Bacteriol.* **138:** 638.
91. Roshbash, M., P.K. Harris, J.L. Wolford, and J.L. Teen. 1981. *Cell* **24:** 679.
92. Rothstein, R.J. and F. Sherman. 1980. *Genetics* **94:** 871.
93. Rykowski. M., J.W. Wallis, J. Choe, and M. Grunstein. 1981. *Cell* **25:** 477.
94. Rytka, J., A. Sledziewski, J. Lukaszkiewicz, and T. Bilinski. 1978. *Mol. Gen. Genet.* **160:** 51.
95. Saunders, G.W., G.H. Rank, B. Kustermann-Kuhn, and C. Hellenberg. 1979. *Mol. Gen. Genet.* **175:** 45.
96. Schamhart, D.H.J., A.M.A. Ten Berge, and K.W. van de Poll. 1975. *J. Bacteriol.* **121:** 747.
97. Schekman and Novick. 1982. In *The molecular biology of the yeast* Saccharomyces. II. *Metabolism and gene expression* (ed. J. Strathern et al.), Cold Spring Harbor Laboratory, Cold Spring Harbor, New York. (In press.)
98. Sherman, F. 1964. *Genetics* **49:** 39.
99. Sherman, F. 1982. In *The molecular biology of the yeast* Saccharomyces. II. *Metabolism and gene expression* (ed. J. Strathern et al.). Cold Spring Harbor Laboratory, Cold Spring Harbor, New York. (In press.)
100. Sherman, F. (pers. comm.)
101. Sherman, F. and P.L. Slonimski. 1964. *Biochim. Biophys. Acta.* **90:** 1.
102. Sherman, F., J.W. Stewart, C. Helms, and J.A. Downie. 1978. *Proc. Natl. Acad. Sci.* **75:** 1437.
103. Sherman, F., J.W. Stewart, E. Margoliash, J. Parker, and W. Campbell. 1966. *Proc. Natl. Acad. Sci.* **55:** 1458.
104. Shortle, D. and D. Botstein (pers. comm.)
105. Singh, A. and T. Manney. 1974. *Genetics* **77:** 651.
106. Singh, A. and F. Sherman. 1978. *Genetics* **89:** 653.
107. Skorgerson, L., C. McLaughlin, and E. Wakatama. 1973. *J. Bacteriol.* **116:** 818.
108. Sora, S. (pers. comm.)
109. Sprague, G. and I. Herskowitz (pers. comm.)
110. Stewart, G. and I. Russell. 1977. *Can. J. Microbiol.* **23:** 441.
111. Thorner (this volume)
112. Thornton, R.J. 1978a. *Eur. J. Appl. Microbiol. Biotechnol.* **5:** 103.
113. Thornton, R.J. 1978b. *FEMS Microbiol. Lett.* **4:** 207.
114. Trumbly, R. 1980. Ph.D. thesis, University of California, Davis.
115. Ursic, D. and J. Davies. 1979. *Mol. Gen. Genet.* **175:** 313.
116. Warner, J.R. 1982. In *The molecular biology of the yeast* Saccharomyces. II. *Metabolism and gene expression* (ed. J. Strathern et al.). Cold Spring Harbor Laboratory, Cold Spring Harbor, New York. (In press.)
117. Waxman, M.F., J.A. Knight, and P.S. Perlman. 1979. *Mol. Gen. Genet.* **167:** 243.
118. Wickner (this volume)
119. Wickner, R.B. 1974. *J. Bacteriol.* **117:** 252.

120. Wilkie, D. and B.K. Lee. 1965. *Genet. Res.* **6**:130.
121. Winsor, B., F. Lacroute, A. Ruet, and A. Sentenac. 1979. *Mol. Gen. Genet.* **173**:145.
122. Wolf, D. and C. Ehmann. 1979. *Eur. J. Biochem.* **98**:375.
123. Wolf, D. and G.R. Fink. 1975. *J. Bacteriol.* **123**:1150.
124. Wolf, D., C. Ehmann, and I. Beck. 1979. In *Biological functions of proteinases* (ed. H. Holzer and H. Tschesche), p. 55. Springer-Verlag, Berlin.
125. Woods, R.A., H.K. Sanders, M. Briquet, F. Foury, B.E. Drysdele, and J.R. Mattoon. 1975. *J. Biol. Chem.* **250**:4090.
126. Zubenko, G. and E. Jones (pers. comm.)
127. Zubenko, G.S., A.P. Mitchell, and E.W. Jones. 1979. *Proc. Natl. Acad. Sci.* **76**:2395.
128. Zubenko, G.S., R. Parker, and E. Jones (pers. comm.)

Subject Index

and mitochondrial DNA replication, 47, 590

and mitotic recombination, 345–351, 357–359

resistance to X-rays during, 378–379

S phase, 98, 100
  and DNA replication, 38, 44–45
  and histone transcription, 37–38, 123
  length of, 35–36, 38, 129
  in meiosis vs. mitosis, 228, 270
  transition into, 36–37

and spindle formation, 67–68

and sporulation competency, 218

stage-specific processes of, 98–121

synchronization of, 49, 105, 130–131, 358–359

temporal map of, 99

timing of, and bud emergence, 36

2μ plasmid replication during, 46, 459, 462, 465

Cell-division-cycle mutants (*cdc* mutants)
  and agglutination, 169
  and bud emergence, 67–68, 70, 113
  and chromosome loss, 19–20
  and conjugation, 104, 107, 131–133, 135, 172
  and cytokinesis, 70–71, 215, 256
  definition and characterization of, 6, 101–102, 105, 112–114, 117, 122
  and DNA replication, 36, 38–40, 230
    of autonomously replicating sequences, 42
    of mitochondrial DNA, 47, 104–107, 124, 590
    of 2μ plasmid DNA, 45–46
  execution points of, 102–103
  and functional sequence mapping of the cell cycle, 114–119
  and G$_1$ arrest, 36–37, 39
  growth and, 122
  and hyper-recombinant phenotype, 365
  and initiation of cell cycle, 125–126, 132–133
  and karyogamy, 104, 107–108, 133, 135
  and killer, maintenance of, 425
  list of, 104–111
  and meiosis, 74, 131–132, 215, 223–224, 258–260
  and mitotic recombination, induction of, 359
  in nuclear division, 39–40, 64
  number of, 103–112
  and ρ^- mutants, 534
  and sensitivity to mutagens, 388
  sequencing of, 37
  and S phase, 36–37, 39–40
  in spindle function, 37, 67–68
  and spindle pole body, 67–68, 112–113, 131, 170
  and sporulation, 79, 131, 230, 256, 269–272

synthesis of gene products of, 120–121, 124, 130

temporal mapping of, 102–103

Cell lineage. *See* Pedigree analysis

Cell membrane. *See* Plasma membrane

Cell permeability, 170, 471–472

Cell size
  control of, 125, 127, 133
  and start of cell-division cycle, 126–129, 132–135

Cell wall
  and α-factor receptor, 157
  and bud emergence, 70–71
  and bud scar, 70
  and cell size, 127
  composition of, 153
  and conjugation, 71, 145, 147, 167–171
  and cytokinesis, 70
  and killer toxin receptor, 437–438
  removal of, 130, 160, 168, 472
  and shmoo formation, 157
  and sporulation, 131
  synthesis of, during growth, 122–123

Centromere, 24, 29, 66, 131

Centromere linkage, 13–16

Chiasma, 12

Chiasma interference, 13–14, 16, 317

Chitin, synthesis of
  after α-factor treatment, 167–168
  and bud scar, 70
  cdc mutants affecting, 112–113
  in cell-division cycle, 98, 100, 103, 118–119, 122–123
  during conjugation, 171
  and cytokinesis, 70, 101
  and shmoo formation, 157
  and spindle pole body, 134
  and spore walls, 78

Chloramphenicol
  and mitochondrial protein synthesis, 431, 507, 512, 559, 599
  mutants resistant to, 507, 555, 560, 562–563, 572
  and ρ^- mutations, 590–591

Chondriolites, 479

Chondriome, 474–477, 480, 484, 486

Chromatid interference, 12

Chromatin
  complexes of, 230–231
  condensation of, 60–62, 80
  DNase I sensitivity of, 31
  and spindle, 64, 66
  structure and organization of, 29–31
  transcription of (extent), 31
  and 2μ plasmid, 452

Chromosomal imbalance, 17

Chromosome(s)
  behavior of, during sporulation, 221–224, 230–231

Gene Index